HANDBOOK OF TABLEAU METHODS

HANDBOOK OF TABLEAU METHODS

Edited by

MARCELLO D'AGOSTINO

Università di Ferrara, Ferrara, Italy

DOV M. GABBAY

King's College, London, United Kingdom

REINER HÄHNLE

Universität Karlsruhe, Germany

and

JOACHIM POSEGGA

Deutsche Telekom AG, Research Centre, Darmstadt, Germany

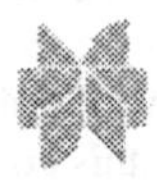

KLUWER ACADEMIC PUBLISHERS

DORDRECHT / BOSTON / LONDON

A C.I.P. Catalogue record for this book is available from the Library of Congress.

ISBN 978-90-481-5184-4

Published by Kluwer Academic Publishers,
P.O. Box, 3300 AA Dordrecht, The Netherlands.

Sold and distributed in North, Central and South America
by Kluwer Academic Publishers,
101 Philip Drive, Norwell, MA 02061, U.S.A.

In all other countries, sold and distributed
by Kluwer Academic Publishers,
P.O. Box 322, 3300 AH Dordrecht, The Netherlands.

Printed on acid-free paper

CONTENTS

PREFACE

Recent years have been blessed with an abundance of logical systems, arising from a multitude of applications. A logic can be characterised in many different ways. Traditionally, a logic is presented via the following three components:

1. an intuitive non-formal motivation, perhaps tie it in to some application area
2. a semantical interpretation
3. a proof theoretical formulation.

There are several types of proof theoretical methodologies, Hilbert style, Gentzen style, goal directed style, labelled deductive system style, and so on. The tableau methodology, invented in the 1950s by Beth and Hintikka and later perfected by Smullyan and Fitting, is today one of the most popular, since it appears to bring together the proof-theoretical and the semantical approaches to the presentation of a logical system and is also very intuitive. In many universities it is the style first taught to students.

Recently interest in tableaux has become more widespread and a community crystallised around the subject. An annual tableaux conference is being held and proceedings are published. The present volume is a *Handbook of Tableaux* presenting to the community a wide coverage of tableaux systems for a variety of logics. It is written by active members of the community and brings the reader up to frontline research. It will be of interest to any formal logician from any area.

The first chapter by Melvin Fitting contains a general introduction to the subject which can help the reader in finding a route through the following chapters, but can also be read on its own as a 'crash course' on tableaux, concentrating on the key ideas and the historical background. In Chapter 2, focusing on classical propositional logic, Marcello D'Agostino explores and compares the main types of tableau methods which appear in the literature, paying special attention to variants and 'improvements' of the original method. Chapter 3, by Reinhold Letz, investigates the impact of tableaux on the challenging problems of classical quantification theory and in Chapter 4 Bernhard Beckert looks at various methodologies for equality reasoning. The treatment of non-classical tableaux is started in Chapter 5, by Lincoln Wallen and Arild Waaler, which deals with the fundamental topic of intuitionistic logic. An area in which the tableau methodology is proving particularly useful is that of modal logics. These, together with their temporal 'neighbours' are studied in Chapter 6 by Rajeev Goré. Chapter 7, by Marcello

D'Agostino, Dov Gabbay and Krysia Broda, discusses the new family of 'substructural logics' which include the well known relevance logic and linear logic. Another frontier of today's logical research where a tableau-style approach seems especially well-suited is that of non-monotonic logics which are treated in Chapter 8 by Nicola Olivetti. Chapter 9, by Reiner Hähnle, offers a treatment of many-valued logics, a traditional topic which has now re-gained popularity owing to its important computer science applications. Finally, in Chapter 10, Joachim Posegga and Peter Schmitt, present an intriguing 'minimalist' approach to the implementation of classical tableaux. The volume concludes with an extensive annotated bibliography compiled by Graham Wrightson.

Some of the chapters overlap considerably in their contents. This was a deliberate choice, motivated primarily by the need for making each chapter self-contained. However, we believe that different expositions of the same notions and ideas can also contribute to a deeper understanding of the subject.

ACKNOWLEDGEMENTS

We would like to thank all our friends and colleagues who helped with the Handbook project. Among them are the various chapter second readers: Roy Dyckhoff; Luis Fariñas del Cerro; Jaakko Hintikka; Wilfrid Hodges; Donald Loveland; Daniele Mundici; and Paliath Narendran.

We would also like to thank Mrs Jane Spurr for her usual efficient and dedicated production and administration work.

Finally thanks are due to our Kluwer Editor, Mrs A. Kuipers for her patience and understanding while waiting for the completion of the book.

The Editors
M. D'Agostino, D. M. Gabbay, R. Hähnle and J. Posegga
Ferrara, London, Karlsruhe and Darmstadt

MELVIN FITTING

INTRODUCTION

1 GENERAL INTRODUCTION

1.1 What Is A Tableau?

This chapter is intended to be a prolog setting the stage for the acts that follow—a bit of background, a bit of history, a bit of general commentary. And the thing to begin with is the introduction of the main character. What is a tableau?

It will make the introductions easier if we first deal with a minor nuisance. Suppose we know what a tableau is—what do we call several of them: 'tableaus' or 'tableaux?' History and the dictionary are on the side of 'tableaux'. On the other hand, language evolves and tends to simplify; there is a clear drift toward 'tableaus'. In this chapter we will use 'tableaus', with the non-judgemental understanding that either is acceptable. This brings our trivial aside to a cloeaux.

Now, what is a tableau? In its everyday meaning it is simply a picture or a scene, but of course we have something more technical in mind. A tableau method is a formal proof procedure, existing in many varieties and for several logics, but always with certain characteristics. First, it is a refutation procedure: to show a formula X is valid we begin with some syntactical expression intended to assert it is not. How this is done is a detail, and varies from system to system. Next, the expression asserting the invalidity of X is broken down syntactically, generally splitting things into several cases. This part of a tableau procedure—the *tableau expansion stage*—can be thought of as a generalization of disjunctive normal form expansion. Generally, but not always, it involves moves from formulas to subformulas. Finally there are rules for *closing* cases: impossibility conditions based on syntax. If each case closes, the tableau itself is said to be *closed*. A closed tableau beginning with an expression asserting that X is not valid is a tableau proof of X.

There is a second, more semantical, way of thinking about the tableau method, one that, perhaps unfortunately, has played a lesser role thus far: it is a search procedure for models meeting certain conditions. Each branch of a tableau can be considered to be a partial description of a model. Several fundamental theorems of model theory have proofs that can be extracted from results about the tableau method. Smullyan developed this approach in [1968], and it was carried further by Bell and Machover in [Bell and Machover, 1977]. In automated theorem-proving, tableaus can be used, and sometimes are used, to generate counter-examples. The connection between the two roles for tableaus—as a proof procedure and as a model search procedure—is simple. If we use tableaus to search for a model in which X is false, and we produce a closed tableau, no such model exists, so X must be valid.

M. D'Agostino et al. (eds.), Handbook of Tableau Methods, 1–43.

This is a bare outline of the tableau method. To make it concrete we need syntactical machinery for asserting invalidity, and syntactical machinery allowing a case analysis. We also need syntactical machinery for closing cases. All this is logic dependent. We will give examples of several kinds as the chapter progresses, but in order to have something specific before us now, we briefly present a tableau system for classical logic.

1.2 Classical Propositional Tableaus as an Example

In their current incarnation, tableau systems for classical logic are generally based on the presentation of Raymond Smullyan in [Smullyan, 1968]. We follow this in our sketch of a *signed tableau system* for classical propositional logic. The chapter by D'Agostino continues the discussion of propositional logic via tableaus. (Throughout the rest of this handbook, *unsigned* tableaus are generally used for classical logic, but signs play a significant role when other logics are involved, and classical logic provides the simplest context in which to introduce them.)

First, we need syntactical machinery for asserting the invalidity of a formula, and for doing a case analysis. For this purpose two *signs* are introduced: T and F, where these are simply two new symbols, not part of the language of formulas. *Signed formulas* are expressions of the form $F\,X$ and $T\,X$, where X is a formula. The intuitive meaning of $F\,X$ is that X is *false* (in some model); similarly $T\,X$ intuitively asserts that X is *true*. Then $F\,X$ is the syntactical device for (informally) asserting the invalidity of X: a tableau proof of X begins with $F\,X$.

Next we need machinery—rules—for breaking signed formulas down and doing a case division. To keep things simple for the time being, let us assume that $\neg$ and $\supset$ are the only connectives. This will be extended as needed. The treatment of negation is straightforward: from $T\,\neg X$ we get $F\,X$ and from $F\,\neg X$ we get $T\,X$. These rules can be conveniently presented as follows.

Negation $\qquad \dfrac{T\,\neg X}{F\,X} \qquad\qquad \dfrac{F\,\neg X}{T\,X}$

The rules for implication are somewhat more complex. From truth tables we know that if $X \supset Y$ is *false*, X must be *true* and Y must be *false*. Likewise, if $X \supset Y$ is *true*, either X is *false* or Y is *true*; this involves a split into two cases. Corresponding syntactic rules are as follows.

Implication $\qquad \dfrac{T\,X \supset Y}{F\,X \mid T\,Y} \qquad\qquad \dfrac{F\,X \supset Y}{\begin{array}{c} T\,X \\ F\,Y \end{array}}$

The standard way of displaying tableaus is as downward branching trees with signed formulas as node labels—indeed, the tableau method is often referred to as the *tree method*. Think of a tree as representing the disjunction of its branches, and a branch as representing the conjunction of the signed formulas on it. Since a node may be common to several branches, a formula labeling it, in effect, occurs as a constituent of several conjunctions, while being written only once. This amounts to a kind of structure sharing.

When using a tree display, a tableau expansion is thought of temporally, and one talks about the *stages* of constructing a tableau, meaning the stages of growing a tree. The rules given above are thought of as branch-lengthening rules. Thus, a branch containing $T\,\neg X$ can be lengthened by adding a new node to its end, with $F\,X$ as label. Likewise a branch containing $F\,X \supset Y$ can be lengthened with two new nodes, labelled $T\,X$ and $F\,Y$ (take the node with $F\,Y$ as the child of the one labelled $T\,X$). A branch containing $T\,X \supset Y$ can be split—its leaf is given a new left and a new right child, with one labelled $F\,X$, the other $T\,Y$. This is how the schematic rules above are applied to trees.

An important point to note: the tableau rules are non-deterministic. They say what can be done, not what must be done. At each stage we choose a signed formula occurrence on a branch and apply a rule to it. Since the order of choice is arbitrary, there can be many tableaus for a single signed formula. Sometimes a prescribed order of rule application is imposed, but this is not generally considered to be basic to a tableau system.

Here is the final stage of a tableau expansion beginning with (that is, *for*) the signed formula $F\,(X \supset Y) \supset ((X \supset \neg Y) \supset \neg X)$.

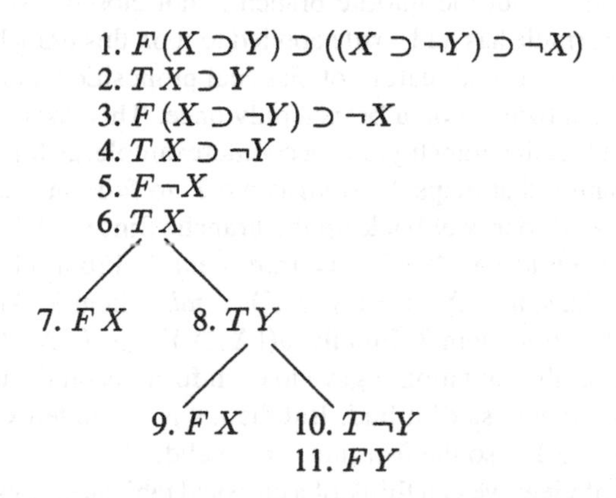

1. $F\,(X \supset Y) \supset ((X \supset \neg Y) \supset \neg X)$
2. $T\,X \supset Y$
3. $F\,(X \supset \neg Y) \supset \neg X$
4. $T\,X \supset \neg Y$
5. $F\,\neg X$
6. $T\,X$
7. $F\,X$
8. $T\,Y$
9. $F\,X$
10. $T\,\neg Y$
11. $F\,Y$

In this we have added numbers for reference purposes. Items 2 and 3 are from 1 by $F \supset$; 4 and 5 are from 3 by $F \supset$; 6 is from 5 by $F\neg$; 7 and 8 are from 2 by $T \supset$; 9 and 10 are from 4 by $T \supset$; 11 is from 10 by $T\neg$.

Finally, the conditions for closing off a case—declaring a branch closed—are simple. A branch is closed if it contains $T\,A$ and $F\,A$ for some formula A. If each branch is closed, the tableau is closed. A closed tableau for $F\,X$ is a tableau proof of X. The tableau displayed above is closed, so the formula $(X \supset Y) \supset ((X \supset$

$\neg Y) \supset \neg X)$ has a tableau proof.

It may happen that no tableau proof is forthcoming, and we can think of the tableau construction as providing us with counterexamples. Consider the following attempt to prove $(X \supset Y) \supset ((\neg X \supset \neg Y) \supset Y)$.

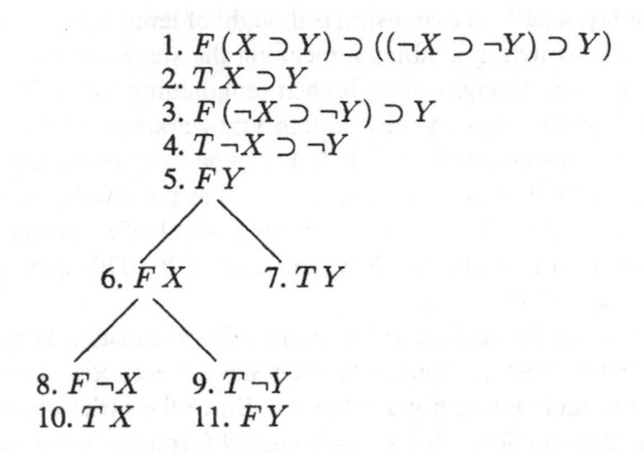

1. $F\,(X \supset Y) \supset ((\neg X \supset \neg Y) \supset Y)$
2. $T\,X \supset Y$
3. $F\,(\neg X \supset \neg Y) \supset Y$
4. $T\,\neg X \supset \neg Y$
5. $F\,Y$
6. $F\,X$
7. $T\,Y$
8. $F\,\neg X$
10. $T\,X$
9. $T\,\neg Y$
11. $F\,Y$

Items 2 and 3 are from 1 by $F \supset$, as are 4 and 5 from 3. Items 6 and 7 are from 2 by $T \supset$, as are 8 and 9 from 4. Finally 10 is from 8 by $F\neg$, and 11 is from 9 by $T\neg$. The leftmost branch is closed because of 6 and 10. Likewise the rightmost branch is closed because of 5 and 7. But the middle branch is not closed. Notice that every non-atomic signed formula has had a rule applied to it on this branch—there is nothing left to do. (This is a special feature of classical propositional logic: it is sufficient to apply a rule to a formula on a branch only once. This does not apply generally to all logics.) In fact the branch yields a counterexample, as follows. Let v be a propositional valuation that maps X to *false* and Y to *false* in accordance with 6 and 11. Now, we work our way back up the branch. Since $v(Y) = \textit{false}$, $v(\neg Y) = \textit{true}$, item 9. Then $v(\neg X \supset \neg Y) = \textit{true}$, item 4. From this and the fact that $v(Y) = \textit{false}$ we have $v((\neg X \supset \neg Y) \supset Y) = \textit{false}$, item 3. Also since $v(X) = \textit{false}$, $v(X \supset Y) = \textit{true}$, item 2. Finally, $v((X \supset Y) \supset ((\neg X \supset \neg Y) \supset Y)) = \textit{false}$, item 1. Notice, the valuation v gave to each formula on the unclosed branch the truth value the branch 'said' it had. But then, v is a counterexample to $(X \supset Y) \supset ((\neg X \supset \neg Y) \supset Y)$, so the formula is not valid.

From a different point of view, we can think of a classical tableau simply as a set of sets of signed formulas: a tableau is the set of its branches, and a branch is the set of signed formulas that occur on it. Semantically, we think of the outer set as the disjunction of its members, and these members, the inner sets, as conjunctions of the signed formulas they contain. Considered this way, a tableau is a generalization of disjunctive normal form (a generalization because formulas more complex than literals can occur). Now, the tableau construction process can be thought of as a variation on the process for converting a formula into disjunctive normal form.

In set terms, instead of lengthening branches, we expand sets. For example, suppose $C = \{Z_1, Z_2, \ldots, Z_n\}$ is a set of signed formulas, and one of the members is $T \neg X$. Then the set $\{Z_1, Z_2, \ldots, Z_n, F X\}$ is said to *follow from* C. Similarly if the set contains $F \neg X$. If C contains $F X \supset Y$ then the set $\{Z_1, Z_2, \ldots, Z_n, T X, F Y\}$ follows from C. Finally, if C contains $T X \supset Y$ then the pair of sets $\{Z_1, Z_2, \ldots, Z_n, F X\}, \{Z_1, Z_2, \ldots, Z_n, T Y\}$ follows from C. Next, if D_1 and D_2 are tableaus (each represented as a set of sets of signed formulas) D_2 follows from D_1 if it is like D_1 but with one of its members C replaced with the set or sets that follow from it. Taken this way, a *tableau expansion* for a signed formula S is a sequence of tableaus, beginning with $\{\{S\}\}$, each tableau after the first following from its predecessor.

1.3 Abstract Data Types vs Implementations

Computer Science has made us familiar with the distinction between an abstract data type and an implementation of it. The notion of a *list*, with appropriate operations on it, is an abstract data type. It can be concretely implemented using an array, or using a linked structure, or in other ways as well. Which is better, which is worse? It depends on the intended application. But this abstract/concrete distinction is actually an old one. Before there were electronic computers, there were human computors, generally using mechanical devices like slates, paper, and slide rules. Before there were algorithms designed for computers, there were algorithms designed for humans, using the devices at hand. What we now call a good implementation of an abstract data type was once called *good notation*. Think, for instance, of the distinction between Roman numerals and Arabic notation. They both implement the same data type—non-negative integers—but one is more efficient for algorithmic purposes than the other.

Tableaus were described abstractly above: sets of sets of signed formulas. One of the things that helped make them popular was a good concrete implementation, a good notation: the tree display of a tableau. It is space-saving, since there is structure sharing, that is, formulas common to several branches are written only once. It is time-saving, since formulas are not copied over and over as the tree grows. Instead the state of a tree at any given moment represents a member of a tableau expansion sequence. Trees present a display that people find relatively easy to grasp, at least if it is not too big. And most importantly, there are by-hand algorithms that are wonderfully suited for use with tableaus as trees. In creating a tree tableau one has the same sense of calculation that one has when adding or multiplying using place-value notation.

Still, from an abstract point of view the tree display of signed formulas is not a tableau but an implementation of a tableau. It is not the only one possible; Manna and Waldinger use quite a different notation for their version of tableaus [1990; 1993]. The connection method can be thought of as based on tableaus, with trees replaced by more general graphs; see [Bibel *et al.*, 1987] for details. Indeed, we will see as this Chapter develops, that the tree display itself underwent considerable evolution before reaching its current form. Also, for some logics, the straightforward tree version may not be best possible. For instance, with certain logics formulas get removed as well as added to branches, and with others they can come and go several times. For these, something more elaborate than just a growing tree is appropriate. Finally, what is good for hand calculation may not be at all useful for machine implementation. This point will be taken up in Hähnle's Chapter and elsewhere. But perhaps the basic point of this digression is a simple one: trees implement tableaus. They do so quite well for many purposes—so well that they are often thought of as *being* tableaus. This is too restrictive, especially today when computers are being used to explore a wide variety of logics. Be willing to experiment.

1.4 What Good Is a Tableau System?

There are many kinds of proof procedures for many kinds of logics. What advantages does a tableau system have? Let us begin with what might be called the 'practical' ones.

The classical propositional system presented in Section 1.2 can be used to 'calculate' in a way that Hilbert systems, say, can not. Each signed formula that is added comes directly from other signed formulas that are present. There is a choice of which signed formula to work with next, but there are always a finite number of choices—a bounded non-determinism, if you will. By contrast, since modus ponens is generally a Hilbert system rule, to prove Y we must find an X for which both X and $X \supset Y$ are provable. There are infinitely many possibilities for X—an unbounded non-determinism. Also note that the X we need may be considerably more complicated than the Y we are after, unlike with classical tableaus, where the formulas added are always simpler than the formulas they come from. Further, it can be shown that if the rules are applied in a classical tableau argument in a *fair* way, all proof attempts will succeed if any of them do. Thus any choices we make affect efficiency, not success. This means tableaus are well-suited for the discovery of proofs, either by people or by machines.

Once quantifiers are added, as in Section 2.6, complications to this simple picture arise. There are infinitely many ways of applying some of the quantifier rules to a signed formula. If we systematically apply all rules in a fair way, it is still possible to show that a proof will be found, if a proof exists. But now, if a proof does not exist, the tableau expansion process will never terminate. Thus we get a semi-decision procedure—but after all, this is best possible. There is a problem for automation of tableaus that stems from the quantifier rules: systematically trying closed term after closed term to instantiate a universal quantifier is a terribly inefficient method. Fortunately there is a way around it, using so-called *free variable tableaus* and unification. This will be discussed in later chapters.

Suitability for proof discovery is something that applies equally well to resolution, and to several other techniques that have been worked out over the years. A peculiar advantage that tableaus have is that it seems to be easier to develop tableau systems for new logics than it is to develop other automatable proof procedures.

This may be because tableaus tend to relate closely to the semantic ideas underlying a logic—or maybe the reasons lie elsewhere. What we are describing is an empirical observation, not a mathematical truth. Of course, once a tableau proof procedure has been created for a logic, it may be possible to use it to develop an automatable proof procedure of a different sort. Something like this is at the heart of Maslov's method [Mints, 1991]. But even so, tableaus provide a good starting point.

On a pedagogical level, tableaus can be used to provide quite appealing proofs of metatheoretical results about a logic. Take the issue of proving completeness as an example. One common way of showing a proof procedure is complete is to make use of maximal consistent sets. Such an approach is quite general, and can be applied to tableau proof procedures as readily as to Hilbert systems—see the proof of the Model Existence Theorem in [Fitting, 1996]. But there is another approach to proving completeness that is much more intuitive. Start constructing a tableau expansion for $F\,X$. Apply rules fairly: systematically apply each applicable rule. If no closed tableau is produced, it can be shown that the resulting tableau contains enough information to construct a countermodel to X. We saw a propositional example of this kind in Section 1.2. This is a nice feature indeed.

Other basic theorems about logic can also be given equally perspicuous tableau proofs. Smullyan, in [1968], gives proofs of the compactness theorem, various interpolation theorems, and the Model Existence Theorem, all using tableaus in an essential way. Bell and Machover carried this even further [1977]. Of course, we are speaking of classical logic, but similar arguments often carry over to other logics that have tableau proof procedures.

Finally, tableaus are well-suited for computer implementation. Their history in this respect is somewhat curious. We will have more to say on this topic as we discuss the history of tableaus, which we do throughout the rest of this Chapter. (For another presentation of tableau history, see Anellis [1990; 1991].)

2 CLASSICAL HISTORY

Tableau history essentially begins with Gentzen. For classical logic, ignoring issues of machine implementation, it culminates with Smullyan. Here we discuss this portion of the development of our subject. In order to keep clutter down, we confine things to classical propositional logic (and occasionally intuitionistic propositional logic). This is sufficient to illustrate differences between systems and to follow their evolution. We will generally re-prove the same formula that we did in Section 1.2, to allow easy comparison of the various systems.

2.1 Gentzen

In his short career Gentzen made several fundamental contributions to logic, see [Szabo, 1969]. The one that concerns us here is his 1935 introduction of the *se-*

quent calculus in [Gentzen, 1935]. Before this, Hilbert-style, or axiomatic, proof procedures were the norm. In a sense, a Hilbert system characterizes a logic as a whole—it is difficult to separate out the role of individual connectives since several of them may appear in each axiom. What Gentzen contributed was a formulation of both classical and intuitionistic logics with a clear separation between structural rules (essentially characterizing deduction in the abstract) and specific rules for each connective and quantifier. Further, each connective and quantifier has exactly two kinds of rules; roughly, for its introduction and for its elimination. We say roughly because this terminology is more appropriate for the natural deduction systems that Gentzen also introduced in [1935], but the essential idea is basically the same.

We'll sketch Gentzen's system, and make some comments on it. This should be familiar ground to most logicians. But we also note that, as with most things, the true beginnings of our subject are fuzzy. Gentzen's ideas grew out of earlier work of Paul Hertz [1929]. Even the famous *cut* rule is a special case of Hertz's *syllogism* rule. See [Szabo, 1969] for further discussion of this.

The Classical Sequent Calculus

First a new construct is introduced, the *sequent*. A sequent is an expression of the form:

$$X_1, \ldots, X_n \to Y_1, \ldots, Y_k$$

where $X_1, \ldots, X_n, Y_1, \ldots, Y_k$ are formulas. The arrow, $\to$, is a new symbol. It is understood that either (or both) of n and k may be 0. Informally, think of the sequent above as asserting: the disjunction of Y_1, ..., Y_k follows from the conjunction of $X_1, \ldots, X_n$.

The system has axioms, and rules of derivation. Axioms, or *initial sequents* as Gentzen called them, are sequents of the form $A \to A$, where A is a formula. Next, Gentzen has seven *structural rules*. We give six of them here; the seventh, *cut* will be discussed in a section of its own. In these and later rules, Γ, Δ, Θ, and Λ are sequences of formulas, possibly empty.

Thinning $\dfrac{\Gamma \to \Theta}{X, \Gamma \to \Theta}$ $\qquad$ $\dfrac{\Gamma \to \Theta}{\Gamma \to \Theta, X}$

Contraction $\dfrac{X, X, \Gamma \to \Theta}{X, \Gamma \to \Theta}$ $\qquad$ $\dfrac{\Gamma \to \Theta, X, X}{\Gamma \to \Theta, X}$

Interchange $\dfrac{\Delta, Y, X, \Gamma \to \Theta}{\Delta, X, Y, \Gamma \to \Theta}$ $\qquad$ $\dfrac{\Gamma \to \Theta, Y, X, \Lambda}{\Gamma \to \Theta, X, Y, \Lambda}$

Next we give Gentzen's rules for the connectives $\neg$, $\wedge$, and $\supset$. The rules for $\vee$ are dual to those for $\wedge$ and are omitted.

Negation $\qquad \dfrac{X,\Gamma \to \Theta}{\Gamma \to \Theta,\neg X} \qquad \dfrac{\Gamma \to \Theta, X}{\neg X,\Gamma \to \Theta}$

Conjunction $\qquad \dfrac{\Gamma \to \Theta, X \quad \Gamma \to \Theta, Y}{\Gamma \to \Theta, X \wedge Y} \qquad \dfrac{X,\Gamma \to \Theta}{X \wedge Y,\Gamma \to \Theta} \qquad \dfrac{Y,\Gamma \to \Theta}{X \wedge Y,\Gamma \to \Theta}$

Implication $\qquad \dfrac{X,\Gamma \to \Theta, Y}{\Gamma \to \Theta, X \supset Y} \qquad \dfrac{\Gamma \to \Theta, X \quad Y,\Delta \to \Lambda}{X \supset Y,\Gamma,\Delta \to \Theta,\Lambda}$

Proofs are displayed in tree form, root at bottom. Each leaf must be labelled with an axiom; each non-leaf must be labelled with a sequent that follows from the labels of its children by one of the rules of derivation. A proof of the sequent $\to X$ is considered to be a proof of the formula X. Here is an example, a proof of $(X \supset Y) \supset ((X \supset \neg Y) \supset \neg X)$, with explanations added.

$$\dfrac{X \to X \qquad \dfrac{Y \to Y}{\neg Y, Y \to}\ \textit{Negation}}{X \supset \neg Y, X, Y \to}\ \textit{Implication}$$

$$\dfrac{X \supset \neg Y, X, Y \to}{X \supset \neg Y, Y, X \to}\ \textit{Interchange}$$

$$\dfrac{X \supset \neg Y, Y, X \to}{Y, X \supset \neg Y, X \to}\ \textit{Interchange}$$

$$\dfrac{Y, X \supset \neg Y, X \to}{Y, X, X \supset \neg Y \to}\ \textit{Interchange}$$

$$\dfrac{X \to X \qquad Y, X, X \supset \neg Y \to}{X \supset Y, X, X, X \supset \neg Y \to}\ \textit{Implication}$$

$$\dfrac{X \supset Y, X, X, X \supset \neg Y \to}{X, X \supset Y, X, X \supset \neg Y \to}\ \textit{Interchange}$$

$$\dfrac{X, X \supset Y, X, X \supset \neg Y \to}{X, X, X \supset Y, X \supset \neg Y \to}\ \textit{Interchange}$$

$$\dfrac{X, X, X \supset Y, X \supset \neg Y \to}{X, X \supset Y, X \supset \neg Y \to}\ \textit{Contraction}$$

$$\dfrac{X, X \supset Y, X \supset \neg Y \to}{X, X \supset \neg Y, X \supset Y \to}\ \textit{Interchange}$$

$$\dfrac{X, X \supset \neg Y, X \supset Y \to}{X \supset \neg Y, X \supset Y \to \neg X}\ \textit{Negation}$$

$$\dfrac{X \supset \neg Y, X \supset Y \to \neg X}{X \supset Y \to (X \supset \neg Y) \supset \neg X}\ \textit{Implication}$$

$$\dfrac{X \supset Y \to (X \supset \neg Y) \supset \neg X}{\to (X \supset Y) \supset ((X \supset \neg Y) \supset \neg X)}\ \textit{Implication}$$

in the sequent calculus are displayed beginning with axioms, ending with the sequent to be proved. Although it was probably of minor importance to Gentzen, others soon realized that by turning the rules upside-down, a proof discovery system resulted. Given any sequent, there are a limited number of sequents from which it could be derived. Try deriving them—this reduces the problem to a simpler one, since premises of rules involve subformulas of their conclusions. Thus the discovery of a Gentzen-style proof is a much more mechanical thing than it is with Hilbert systems. It is not hard to extract a formal algorithm for decidability of both classical and intuitionistic propositional logics.

When using the rules backward, it is useful to think 'negatively' instead of 'positively'.' That is, suppose we want to show a sequent, say $X_1, X_2 \to Y_1, Y_2, A \wedge B$ is provable. Well, suppose it is not. By one of the Conjunction rules, either

$X_1, X_2 \to Y_1, Y_2, A$ or $X_1, X_2 \to Y_1, Y_2, B$ is not provable. Continue working backward in this way, until a contradiction (an axiom is not provable) is reached.

This backward way of thinking makes it easy to see in what way the sequent calculus relates to later tableau systems. Recall, we think of a sequent as informally saying the disjunction of the right side follows from the conjunction of the left. Then, if we did not have $X_1, X_2 \to Y_1, Y_2, A \wedge B$, everything on the left 'holds' in some model, and nothing on the right does. That is, denying the sequent informally amounts to assuming the satisfiability of $\{T\,X_1, T\,X_2, F\,Y_1, F\,Y_2, F\,A \wedge B\}$. From this, using a Conjunction rule backwards, we have the satisfiability of one of $\{T\,X_1, T\,X_2, F\,Y_1, F\,Y_2, F\,A\}$ or $\{T\,X_1, T\,X_2, F\,Y_1, F\,Y_2, F\,B\}$. At this point we can represent things using a set of sets of formulas,

$$\Big\{\{T\,X_1, T\,X_2, F\,Y_1, F\,Y_2, F\,A\}, \{T\,X_1, T\,X_2, F\,Y_1, F\,Y_2, F\,B\}\Big\}$$

where the outer set is thought of disjunctively and the inner sets conjunctively. This leads back to our set version of tableaus, in Section 1.1.

Cut and the Structural Rules

We gave six structural rules above. The combined effect of Contraction and Interchange is that we can think of the *sequences* of formulas on either side of a sequent arrow as *sets* of formulas. This, combined with Thinning, allows us to use an apparently more general axiom schema: $\Gamma \to \Delta$, where Γ and Δ have a formula in common. These days, it is not uncommon to find sequent calculi formulated with sets instead of sequences, or with Gentzen's axiom scheme modified, or some combination of these. Kleene, in [Kleene, 1950], gives three different versions, G1, G2, and G3, differing primarily on structural details.

The rules for conjunction and for implication are not like each other. There are three conjunction rules but only two implication rules. This can be remedied, if desired. It can be shown that an equivalent system results if the two rules for introducing a conjunction on the left of an arrow are replaced by the following single rule:

$$\frac{X, Y, \Gamma \to \Theta}{X \wedge Y, \Gamma \to \Theta}$$

Proof of the equivalence of the two formulations uses the structural rules in an essential way.

Girard realized that the structural rules are not minor, but central. They are essential for proving the equivalence of the two versions of the conjunction rules, as we just saw. By dropping Thinning and Contraction (and making other changes as well) Girard devised *Linear Logic* [Girard, 1986; Troelstra, 1992]. Other so-called substructural logics, such as *Relevance Logic* [Dunn, 1986], arise in similar ways. Note that, without the structural rules, there are two different ways of introducing conjunction (and disjunction). Since these are no longer equivalent, substructural logics, in fact, have two notions of conjunction and disjunction.

We have left the Cut rule for last, since its role is both important and unique. The Cut rule is a kind of transitivity condition; in the following, the formula X is cut away.

Cut

$$\frac{\Gamma \rightarrow \Theta, X \quad X, \Delta \rightarrow \Lambda}{\Gamma, \Delta \rightarrow \Theta, \Lambda}$$

All the other Gentzen rules have a special, remarkable property: the subformula property. Each formula appearing above the line of a rule is a subformula of some formula appearing below the line. It is this on which decidability results in the propositional case rest. It is this that makes the construction of proofs seem mechanical. The Cut rule violates the subformula property: X appears above the line, and disappears below. If Cut is allowed, the system is in many ways less appealing.

Why, then, have a Cut rule at all? In showing Gentzen's formulation is at least as strong as a Hilbert axiom system, we must do two things: we must show each Hilbert axiom is Gentzen provable, and we must show each Hilbert rule of inference preserves Gentzen provability. All this is straightforward, except for modus ponens. However, if the Cut rule is available, it is easy to show that modus ponens preserves Gentzen provability. Consequently, it is enough to show Gentzen's systems with and without the Cut rule are equivalent—a result usually referred to as 'Cut eliminability'.

One can show Cut is eliminable by showing that Gentzen systems with and without cut are both sound and complete. Gentzen did not proceed this way, essentially because completeness proofs for first-order logic are non-constructive, and constructivity was a key part of Gentzen's motivation. Instead, Gentzen gave what today we would describe as an algorithm for removing Cuts from a proof, together with a termination argument. Cut elimination has become the centerpoint of proof theory.

Allowing Cuts in an automated proof system is, in a sense, allowing the use of Lemmas. Cut elimination says they are not necessary. On the other hand, an analysis of Gentzen's proof of Cut elimination shows that, when removing the use of Lemmas, proof length can grow exponentially. Clearly this is an important issue.

Intuitionistic Logic

In [Gentzen, 1935] Gentzen showed something of the versatility inherent in the sequent calculus by giving an intuitively plausible system for intuitionistic logic, as well as one for classical logic. (Wallen's chapter contains a full treatment of theorem-proving in intuitionistic systems.) Intuitionistic logic is meant to be constructive—in particular, a proof of $X \vee Y$ should be either a proof of X or a proof of Y. This is different than in classical logic where one can have a proof of $X \vee \neg X$ without having either a proof or a disproof of X. Now, recall that the right-hand side of a sequent is interpreted as a disjunction. Then, intuitionistically, we should be able to say *which* member of the disjunction is a consequence of the left-hand side. This, of course, is all quite informal, but it led to Gentzen's dramatic modification of the sequent calculus rules: allow at most one formula to appear on the right of an arrow. Gentzen showed this gave a system that was equivalent to an axiomatic formulation of intuitionistic logic, by making use of his Cut elimination theorem. Nothing else was possible, since there was no known semantics for intuitionistic logic at that time.

Gentzen's Immediate Heirs

Gentzen's introduction of the sequent calculus was enormously influential, and similar formulations were soon introduced (after the void of World War II) for other kinds of logics. We briefly sketch some of the early developments.

Beginning in 1957, Ohnishi and Matsumoto gave calculi for several modal logics [1957; 1959; 1964] and [Ohnishi, 1961; Matsumoto, 1960]. We describe their system for $S4$ as a representative example. (See Goré's Chapter for an extended discussion of the role of tableaus in modal theorem-proving.) We take $\Box$ as primitive, and for a sequence Γ of formulas, we write $\Box\Gamma$ for the result of prefixing each formula in Γ with $\Box$. Now we add to Gentzen's rules the following.

S4 $\dfrac{X, \Gamma \to \Theta}{\Box X, \Gamma \to \Theta}$ $\qquad\qquad$ $\dfrac{\Gamma \to X}{\Box\Gamma \to \Box X}$

Note that the second of the rules for introducing $\Box$ allows only a single formula X on the right, analogous to Gentzen's rules for intuitionistic logic. This should come as no surprise, since there are close connections between $S4$ and intuitionistic logic.

Since semantical methods were not much used in modal logic at the time, the equivalence of these systems with Hilbert style ones was via a translation procedure, making essential use of Cut elimination. The sequent formulations were, in turn, used to obtain decision procedures for the logics.

At about the same time, Kanger also gave sequent style formulations for some modal logics [1957]. His system for $S5$ is of special interest because it introduced a new piece of machinery: propositional formulas were indexed with positive integers. These integers can be thought of as corresponding to the possible worlds of Kripke models, though this is not how Kanger thought of them. If X is indexed with n, it is written as X^n. The Gentzen rules are modified so that in a conjunction rule, for instance, a conjunction receives the same index as its conjuncts (which must be the same). Then the following two rules are added.

S5

$$\frac{X^m, \Box X^n, \Gamma \to \Theta}{\Box X^n, \Gamma \to \Theta} \qquad \frac{\Gamma \to \Theta, X^n}{\Gamma \to \Theta, \Box X^m}$$

Where $m \neq n$.

Where no formula with index n occurs within the scope of $\Box$ in Γ or Θ.

This is an early forerunner of the now widespread practice of adding extra machinery to sequent and tableau systems. We will see more examples later on. The Ohnishi and Matsumoto systems, and the Kanger systems, can be found in some detail in [Feys, 1965].

In 1967 Rousseau [1967] treated many-valued logics using Gentzen methods. (The chapter by Hähnle discusses current tableau theorem-provers for many-valued logics.) The basic ideas are relatively simple. A classical sequent, $X_1, \ldots, X_n \to Y_1, \ldots, Y_k$ is considered satisfiable if, under some valuation, either some X_i is *false* or some Y_i is *true*. Consider the left-hand side as the falses's and the right-hand side as the true's. Then a sequent is satisfiable if one of the false's is *false* or one of the true's is *true*. Rousseau extended this to an m-valued logic—say the truth values are $0, 1, \ldots, m-1$. Now a sequent is an expression:

$$\Gamma_0 \mid \Gamma_1 \mid \ldots \mid \Gamma_{m-1}$$

where each Γ_i is a sequence of formulas. The sequent is considered satisfiable if some member of Γ_i has truth value i. This reading of a sequent suggests appropriate rules. For instance, Gentzen's axiom schema, $A \to A$, turns into the schema $A \mid A \mid \ldots \mid A$. Rousseau gave a method of producing connective rules, and showed soundness and completeness. For many-valued logics, the more interesting issue is that of quantification, which we do not touch on here, though it was discussed by Rousseau.

2.2 *Beth*

Gentzen's motivation was proof-theoretic. He was more-or-less explicitly analyzing proofs, and his work has become the foundation of modern proof theory. There is no attempt at a completeness or soundness argument in his paper—only constructive proofs of equivalence with other formalisms. Beth, on the other hand, was motivated by semantic concerns [Beth, 1955; Beth, 1956]. In 1955 he introduced the terminology, 'semantic tableau', and thought of one as a systematic attempt to find a counter-example. To quote from [Beth, 1955]:

> "If such a counter-example is found, then we have a negative answer to our problem. And if it turns out that no suitable counter-example can be found, then we have an affirmative answer. In this case, however, we must be sure that no suitable counter-example whatsoever is available; therefore, we ought not to look for a counter-example in a

> haphazard manner, but we must rather try to construct one in a systematic way. Now there is indeed a systematic method for constructing a counter-example, if available; it consists in drawing up a *semantic tableau*."

Beth arranged his counter-example search in the form of a table with two columns, one labelled 'Valid,' the other, 'Invalid.' Perhaps 'True' and 'False' (in some model) would be more accurate. To determine whether Y is a consequence of $X_1, \ldots, X_n$, begin by placing $X_1, \ldots, X_n$ in the *Valid* column, and Y in the *Invalid* one. This corresponds to beginning with the conjecture that Y is *not* a consequence of $X_1, \ldots, X_n$, rather like using the sequent calculus backward. Next, systematically break down the formulas in each column. For example, if $A \wedge B$ appears in the *Valid* column, add both A and B to it. Similarly if $A \vee B$ appears in the *Invalid* column, add both A and B to it. If $\neg A$ occurs in a column, add A to the other one. Things get a little awkward with disjunctive cases however. If $A \vee B$ occurs in the *Valid* column we should be able to add one of A or B—the problem is which. Beth's solution was to split the *Valid* column in two, thus displaying both possibilities. Of course a corresponding split has to be made in the *Invalid* column. Since further splitting might occur, it is necessary to label the various columns, to keep the *Valid* and the *Invalid* columns that belong together properly associated. In practice this can be hard to follow if there are many cases, but the principle is certainly clear.

If a semantic tableau is constructed as outlined above, there are basically only two possible outcomes. It may happen that a formula appears in both the *Valid* and the *Invalid* columns, which indicates an impossibility. If the tableau system really does embody a thorough, systematic analysis, such an impossibility tells us there are no models in which $X_1, \ldots, X_n$ are true but Y is not; that is, there are no counter-examples, and so Y must be a consequence of $X_1, \ldots, X_n$. The other possibility is that no such contradiction ever appears. In this case, Beth observed, the tableau itself supplies all the necessary information to produce a counter-example, and so Y is not a consequence of $X_1, \ldots, X_n$.

The description above is correct in the propositional case, and Beth's method supplies a decision procedure. In the first-order case things are more complex since if there are no counterexamples, a tableau construction may never terminate. If this happens we still generate the information to construct a model, but only in the limit. A rigorous treatment of this point requires some care, and we gloss over it.

Here is an example of a Beth semantic tableau—a proof once again of the familiar tautology $(X \supset Y) \supset ((X \supset \neg Y) \supset \neg X)$.

Valid	Invalid
	(1) $(X \supset Y) \supset ((X \supset \neg Y) \supset \neg X)$
(2) $X \supset Y$	(3) $(X \supset \neg Y) \supset \neg X$
(4) $X \supset \neg Y$	(5) $\neg X$
(6) X	
(i): (8) Y	(i)
$(i)(iii)$: (10) $\neg Y$	$(i)(iii)$: (11) Y
$(i)(iv)$	$(i)(iv)$: (9) X
(ii)	(ii): (7) X

In the tableau above, we begin with 1 in the *Invalid* column, which gives 2 in the *Valid*, and 3 in the *Invalid* ones. From 3 we conclude 4 is *Valid* and 5 is *Invalid*. Likewise 5 produces 6. Now things become more complicated. If $X \supset Y$ is true in a model, either X is not true there, or Y is. Then formula 2 causes a split into two cases, labelled i and ii, one placing X in the *Invalid* column, formula 7, the other placing Y in the *Valid* column, formula 8. Likewise formula 4 causes another split, creating subcases iii and iv, with formula 9 on the *Invalid* side and formula 10 on the *Valid* one. Formula 10 in turn yields formula 11. Now we see we have arrived at a contradictory tableau. Case ii is impossible because of formulas 6 and 7. Case iii is impossible because of 8 and 11, and case iv is impossible because of 6 and 9; this means case i is impossible. Since each subcase is impossible, or *closed* as Beth called it, the tableau itself is closed; there are no counter-examples; $(X \supset \neg Y) \supset \neg X$ *is* a consequence of $X \supset Y$.

Tableaus and Consequences

Beth recognized that tableaus make it possible to give proofs of results about classical logic that are intuitively satisfying. In his book [Beth, 1959] Beth used tableaus to show a *subformula principle* saying that if a formula has a proof, it has one in which only subformulas of it occur. He derived a version of Herbrand's Theorem, and Gentzen's Extended Hauptsatz. He explicitly discussed the relationship between tableaus and the sequent calculus. He even gave a tableau-based proof of Gentzen's Cut Elimination Theorem (a proof with a fundamental flaw, as it happens). And he considered the relationship between tableaus and natural deduction.

Among Beth's fundamental consequences is his famous Definability Theorem of 1953 [1953; 1953a], relating implicit and explicit definability in first-order logic. Statements of it can be found in many places, in particular in [Chang and Keisler, 1990; Smullyan, 1968; Fitting, 1996]; details are beside the point here. It is usual to derive Beth's Theorem from the Craig Interpolation Theorem and Beth takes this route in his book (this was not his original proof however). In turn, Beth uses tableaus to prove the Craig Lemma (not Craig's original proof either). The reason this is mentioned here is to illustrate Beth's almost physical sense of the tableau mechanism. The following is a quote from the beginning of his proof of Craig's Lemma.

> "Let us suppose that the semantic tableau for a certain sequent (f) is closed. We consider the tableau as a system of communicating vessels. The left and right columns are considered as tubes which are connected at the bottom of the apparatus. The formulas U create a downward pressure in the left tubes and likewise the formulas V create a downward pressure in the right tubes; these various pressures result in a state of equilibrium.
>
> This picture suggests the construction, for each of the formulas U and V, of a formula U^0 or V^0 which sums up the total contribution of U or V to the balance of pressures."

The proof continues somewhat more technically. These days tableaus are often used to prove versions of Craig's Lemma, but never in quite as picturesque a fashion.

Intuitionistic Logic Again

Gentzen's approach to intuitionistic logic could not be based on semantics, since none was available at that time. An algebraic/topological semantics was developed soon after [Tarski, 1938; Rasiowa, 1951; Rasiowa, 1954], but this was not particularly satisfactory as an explication of intuitionistic ideas to a classical mathematician. In 1956 Beth provided a much more intuitively appealing semantics [Beth, 1956; Beth, 1959], known today as *Beth models.* (These have been largely superseded by an alternative semantics due to Kripke.) What concerns us here is that, at the same time, Beth introduced a tableau system for intuitionistic logic. He presented this in the form of a sequent calculus and, unlike in Gentzen's system, several formulas could appear on the right of an arrow, at least sometimes. He explicitly noted it could be used as written, as a sequent calculus, or upsidedown, as a tableau system.

Beth's intuitionistic tableau calculus introduced a new element: there were two kinds of branching, conjunctive and disjunctive. When a branch splits disjunctively, closure of one of the new branches is enough to close the original one. With conjunctive branching, on the other hand, both branches must close for the original branch to be closed. Conjunctive branching is the only kind that occurs in classical tableaus. Intuitionistic propositional logic has a higher degree of computational complexity than classical propositional logic, and this can be traced to the possibility of disjunctive branching.

Beth gave a constructive proof of the equivalence of his system with that of Gentzen. More interestingly, he proved the soundness and completeness of his intuitionistic tableau system with respect to his semantics. Since he was concerned with intuitionism as a philosophy of mathematics, he explicitly considered which points of his completeness proof would be problematic for an intuitionist. He found this centered on the *Tree Theorem*, essentially König's Lemma, though he does not call it that.

We do not give Beth's intuitionistic tableau system here. It is easier to present the basic ideas using signed formulas, and we do so later.

2.3 Hintikka

It has been noted that scientific advances come when the times are ready, and often occur to several people simultaneously. The work of Beth and Hintikka is such an event, with Hintikka's primary paper, [Hintikka, 1955], appearing in 1955, the same year as Beth's. (See also [Hintikka, 1953].) Like Beth, and unlike Gentzen, Hintikka was motivated by semantic concerns: the idea behind a proof of X is that it is a systematic attempt to construct a model in which $\neg X$ is true; if the attempt fails, X has been established as valid. Or, as Hintikka puts it:

> "...we interpreted all proofs of logical truth in a seemingly negative way, *viz., as proofs of impossibility of counter-examples*."

As expected, a proof attempt proceeds by breaking formulas down into constituent parts.

> "...the typical situation is one in which we are confronted by a complex formula (or sentence) the truth or falsity of which we are trying to establish by inquiring into its components. Here the rules of truth operate from the complex to the simple: they serve to tell us what, under the supposition that a given complex formula or sentence is true, can be said about the truth-values of its components."

Model Sets

The essentially new element in Hintikka's treatment was the *model set*. It makes possible a considerable simplification in the proof of completeness for tableaus by abstracting properties of satisfiability of formula sets out of details of the tableau construction process. And it suggests the possibility of extensions to modal logics, which Hintikka himself later developed. As usual, we illustrate the ideas via classical propositional logic. First, though, we mention a pecularity of Hintikka's treatment: he assumed all negations occur at the atomic level. Any non-atomic occurrence of a negation symbol was taken to be eliminable, via the usual negation normal form rules. Likewise, implication was not taken as primitive.

Now, suppose we have a classical propositional model, $\mathcal{M}$. Associate with it the set μ of propositional formulas that are true in it. We can say things like: $X \wedge Y \in \mu$ if and only if $X \in \mu$ and $Y \in \mu$. This if-and-only-if assertion can be divided into two implications; Hintikka's insight was to see that one of these implications determines the other. To be more precise, consider the following two sets of conditions.

(C.0)(a) If $A \in \mu$, then not $\neg A \in \mu$, where A is atomic.

(C.1)(a) If $X \wedge Y \in \mu$, then $X \in \mu$ and $Y \in \mu$.

(C.2)(a) If $X \vee Y \in \mu$, then $X \in \mu$ or $Y \in \mu$.

(C.0)(b) If not $\neg A \in \mu$, then $A \in \mu$, where A is atomic.

(C.1)(b) If $X \in \mu$ and $Y \in \mu$, then $X \wedge Y \in \mu$.

(C.2)(b) If $X \in \mu$ or $Y \in \mu$, then $X \vee Y \in \mu$.

If μ is a set meeting all the conditions above, both (a) and (b), there is clearly a model $\mathcal{M}$ whose true formulas are exactly those of μ. Now, Hintikka proved that if μ is known to meet only the (a) conditions, this is still enough—any such set can be extended to one meeting the (b) conditions as well, and hence corresponds to some model. This led Hintikka to call sets meeting the (a) conditions *model sets*. Today they are sometimes called *downward saturated sets* or even *Hintikka sets*. Using this terminology, we have the following.

Hintikka's Lemma Every downward saturated set is satisfiable.

In Section 1.2 we saw that an unclosed tableau branch can be used to generate a model provided that on it all possible rules have been applied. In a natural sense, the set of signed formulas on such a branch is a version of Hintikka's notion of model set, and our creation of a model amounts to a special case of Hintikka's proof of his Lemma. Of course as we stated it, Hintikka's Lemma is for propositional logic. He actually proved a version for first-order logic; it extends to admit equality, and to various non-classical logics as well.

The Hintikka Approach

Suppose we wish to establish the validity of some formula, say that of our old friend $(X \supset Y) \supset ((X \supset \neg Y) \supset \neg X)$. Hintikka's idea is simple: show $\neg[(X \supset Y) \supset ((X \supset \neg Y) \supset \neg X)]$ is not satisfiable, and do this by showing it belongs to no model (downward saturated) set. Do this by supposing otherwise, and deriving a contradiction. Recall, Hintikka assumed formulas are in negation normal form, so if we begin with a downward saturated set μ with $\neg[(X \supset Y) \supset ((X \supset \neg Y) \supset \neg X)] \in \mu$, we are really assuming the following.

1. $(\neg X \vee Y) \wedge ((\neg X \vee \neg Y) \wedge X) \in \mu$

From 1, by (C.1)(a) we have:

2. $\neg X \vee Y \in \mu$

3. $(\neg X \vee \neg Y) \wedge X \in \mu$

Then from 3 we get:

4. $\neg X \vee \neg Y \in \mu$

5. $X \in \mu$

Now, by 2, either $\neg X \in \mu$ or $Y \in \mu$. If the first of these held, we would have $X \notin \mu$, by (C.0)(a), contradicting 5. Consequently we have

6. $Y \in \mu$

By 4, either $\neg X \in \mu$ or $\neg Y \in \mu$. But, using (C.0)(a), the first of these possibilities contradicts 5, and the second contradicts 6. Thus we have arrived at a contradiction—no such μ can exist.

2.4 Lis

Beth and Hintikka each had all the pertinent parts of tableaus as we know them, but their systems were not 'user-friendly'. Beth proposed a graphical representation for tableaus, see Section 2.2, but his two-column tables, with two-column subtables (and subsubtables, and so on) are not handy in practice. Hintikka, in effect, used a tree structure but with sets of formulas at nodes, requiring much recopying. Notational simplification was the essential next step in the development of tableaus and, just as with the preceding stage, it was taken independently by two people: Zbigniew Lis and Raymond Smullyan. Lis published his paper [Lis, 1960] in 1960, but in Polish (with Russian and English summaries), in *Studia Logica*. At that time there was a great gulf fixed between the East and the West in Europe, and Lis's ideas did not become generally known. They were subsequently rediscovered and extended by Smullyan, culminating in his 1968 book [Smullyan, 1968]. The work of Lis himself only came to general attention in the last few years.

Lis, following Beth, divided formulas into two categories, Beth's 'valid' and 'invalid'. But Lis did so not by separating them into columns, but keeping them together and distinguishing them by 'signs'. Lis used arithmetical notation, + or −, for Beth's two categories. (He also used a formal numeration system to record which formulas followed from which—we ignore this aspect of his system.) He then stated the following rules (we only give the propositional ones).

(i) If $\pm\neg X$, then $\mp X$.

(ii) Conjunctive Rules

a) If $+(X \wedge Y)$, then $+X, +Y$.
b) If $-(X \vee Y)$, then $-X, -Y$.
c) If $-(X \supset Y)$, then $+X, -Y$.

(iii) Disjunctive Rules

a) If $\dfrac{-(X \wedge Y)}{-X \mid -Y}$ then

b) If $\dfrac{+(X \vee Y)}{+X \mid +Y}$ then

c) If $\dfrac{+(X \supset Y)}{-X \mid +Y}$ then

These rules are intended to be used in the same way the 'T' and 'F' signed rules were in Section 1.2, though his display of trees was rather like Beth's tables. Lis also gave rules for quantifiers, for equality, and even for definite descriptions.

In addition to the system of semantic tableaus using signs, Lis also presented what he called a *natural deduction* system—what we would call an *unsigned* tableau system. For this, drop all occurrences of the $+$ sign, and replace occurrences of the $-$ sign with occurrences of negation, $\neg$. (This makes half of rule (i) redundant.)

2.5 *Smullyan*

It is through Smullyan's 1968 book *First-Order Logic* [1968] that tableaus became widely known. They also appeared in the 1967 textbook [Jeffrey, 1967], which was directed at beginning logic students. Smullyan's book was preceded by [Smullyan, 1963; Smullyan, 1965; Smullyan, 1966] in which the still unknown contributions of Lis were rediscovered, deepened, and extended. Smullyan called his version 'analytic tableau,' meaning by this that the subformula principle is a central feature. Smullyan used tableaus as the basis of a general treatment of classical logic, including an analysis of the variety of completeness proofs. Drawing together ideas from several sources, and adding new ones of his own, quite an elegant treatment resulted.

Unifying Notation

Like Lis, Smullyan introduced both signed and unsigned tableau systems. Where Lis used $+$ and $-$ as signs, Smullyan used T and F, but the essential idea is the same. But Smullyan, instead of treating these as parallel, similar systems, abstracted their common features. He noted that signed formulas act either *conjunctively* or *disjunctively* (in the propositional case—quantification adds two more categories). He grouped the conjunctive cases together as *type A* formulas, and the disjunctive ones as *type B*, using α for a generic type A formula and β as generic type B. For each of these, two *components* were defined: α_1 and α_2 for type A; β_1 and β_2 for type B. Smullyan's tables for both the signed and the unsigned versions are as follows.

α	α_1	α_2
$T(X \wedge Y)$	$T X$	$T Y$
$F(X \vee Y)$	$F X$	$F Y$
$F(X \supset Y)$	$T X$	$F Y$
$T \neg X$	$F X$	$F X$
$F \neg X$	$T X$	$T X$
$(X \wedge Y)$	X	Y
$\neg(X \vee Y)$	$\neg X$	$\neg Y$
$\neg(X \supset Y)$	X	$\neg Y$
$\neg\neg X$	X	X

β	β_1	β_2
$F(X \wedge Y)$	$F X$	$F Y$
$T(X \vee Y)$	$T X$	$T Y$
$T(X \supset Y)$	$F X$	$T Y$
$\neg(X \wedge Y)$	$\neg X$	$\neg Y$
$(X \vee Y)$	X	Y
$(X \supset Y)$	$\neg X$	Y

The idea is, in any interpretation an α is true if and only if both α_1 and α_2 are true; and a β is true if and only if at least one of β_1 or β_2 is true. (A signed formula $T X$ is true in an interpretation if X has the value *true*; likewise $F X$ is true if X has the value *false*.)

Today negation is often left out of the conjunctive/disjunctive classification, not because it is mathematically inappropriate, but because it introduces redundancy if one is attempting to automate semantic tableaus. This was not a concern of Smullyan's. In [Fitting, 1996] these tables are extended to include all other binary connectives except for equivalence and exclusive-or (the dual to equivalence), which follow a different pattern. But if one is willing to weaken the subformula principle somewhat, even these can be included, using the following definitions (in which we use $\not\equiv$ for exclusive-or).

α	α_1	α_2
$T(X \equiv Y)$	$T(X \supset Y)$	$T(Y \supset X)$
$F(X \not\equiv Y)$	$T(X \supset Y)$	$T(Y \supset X)$
$(X \equiv Y)$	$(X \supset Y)$	$(Y \supset X)$
$\neg(X \not\equiv Y)$	$(X \supset Y)$	$(Y \supset X)$

β	β_1	β_2
$F(X \equiv Y)$	$F(X \supset Y)$	$F(Y \supset X)$
$T(X \not\equiv Y)$	$F(X \supset Y)$	$F(Y \supset X)$
$\neg(X \equiv Y)$	$\neg(X \supset Y)$	$\neg(Y \supset X)$
$(X \not\equiv Y)$	$\neg(X \supset Y)$	$\neg(Y \supset X)$

The effects of uniform notation are quite lovely: all the classical propositional tableau rules for extending branches reduce to the following pair.

$$\frac{\alpha}{\begin{array}{c}\alpha_1\\ \alpha_2\end{array}} \qquad\qquad \frac{\beta}{\beta_1 \mid \beta_2}$$

Smullyan took this abstract approach considerably further, eventually doing away with formulas altogether. In [Smullyan, 1970] the essence of the tableau approach to classical logic was distilled, and in [Smullyan, 1973] this was extended, to intuitionistic and modal logic.

The Role of Signs

From the beginning of the subject the connection between tableaus and the sequent calculus was clear. Loosely, a tableau proof is a sequent proof backwards. In the sequent calculus we show a sequent is valid; in a tableau system we show a formula (or a set of formulas) is unsatisfiable. So, in order to make the sequent/tableau relationship clear, we need a suitable translation between sequents and (finite) sets of formulas. In one direction things are simple: map the sequent $X_1, \ldots, X_n \rightarrow Y_1, \ldots, Y_k$ to the set $\{X_1, \ldots, X_n, \neg Y_n, \ldots, \neg Y_k\}$. Then the sequent is valid if and only if the corresponding set is unsatisfiable. But a problem arises in going the other way, from a set to a sequent. Given the set $\{X, \neg Y\}$, say, it could have come from the sequent $X \rightarrow Y$ or from $X, \neg Y \rightarrow$, and for more complex sets the number of possibilities can be much greater. Thus we have a many-one mapping. While one can work with this, it makes things unnecessarily complicated. Signed formulas deal with this problem nicely.

Following Smullyan [1968], if $S = \{T\,X_1, \ldots, T\,X_n, F\,Y_1, \ldots, F\,Y_k\}$ is a set of signed formulas, let $|S|$ be the sequent $X_1, \ldots, X_n \rightarrow Y_1, \ldots, Y_k$. If we think of the strings of formulas on the left and the right of the sequent arrow as *sets* rather than as *sequences*, thus ignoring the structural rules, this defines a one-one translation between sequents and sets of signed formulas. What is more, using uniform notation, the sequent calculus rules can be presented in the following abstract form (again, omitting the structural rules).

Axioms $\qquad |S, T\,X, F\,X|$

Inference Rules $\qquad \dfrac{|S, \alpha_1, \alpha_2|}{|S, \alpha|} \qquad \dfrac{|S, \beta_1| \quad |S, \beta_2|}{|S, \beta|}$

This more abstract approach makes it possible, for instance, to give a uniform proof of cut elimination, one that applies to both tableaus and the sequent calculus, rather than proving it for one and deriving it for the other as a consequence. From this point of view, tableau and sequent proofs *are* the same thing, which is what everyone suspected all along.

Cut and Analytic Cut

Just as with the sequent calculus, one can introduce a Cut rule for tableaus, and show it can be eliminated from proofs. The sequent calculus formulation of Cut was given in Section 2.1; for tableaus it has a much simpler appearance. Here are signed and unsigned versions.

Cut $$\frac{}{T\,X \mid F\,X} \qquad \frac{}{X \mid \neg X}$$

That is, at any time during a signed tableau construction, a branch may be split, with $T\,X$ added to one fork, and $F\,X$ added to the other, for *any* formula X, and similarly for the unsigned version. Clearly this violates the subformula principle. But Smullyan also considered what he called *Analytic Cut*, which is simply the Cut rule as given above, but with the restriction that X must be a subformula of some formula already appearing on the branch. Like unrestricted Cut, Analytic Cut also shortens proofs, and it clearly does not violate the subformula principle. It lends itself well to automation, and has played some role in this area.

2.6 *The Complications Quantifiers Add*

In our survey of various proof systems above, we ignored quantifiers. The reason is simple: all the systems treat quantifiers more-or-less the same way, so differences between systems can be illustrated sufficiently well at the propositional level. Now it is time to say a little about them. (A full treatment of first-order tableaus can be found in the chapter by Letz, Chapter 3.)

Quantifier rules for classical logic are deceptively simple. If $(\forall x)\varphi(x)$ is true in some model, then $\varphi(c)$ is also true for any closed term c. On the other hand, if $(\forall x)\varphi(x)$ is false in a model, there is some member of the domain of the model for which $\varphi(x)$ does not hold. Then, if d is a new constant symbol, we can interpret it to designate some member of the domain for which $\varphi(x)$ fails, and so $\varphi(d)$ will be false in the model. Note that the model in which $\varphi(d)$ is false is not the original model, since we had to re-interpret d, but since d was chosen to be a constant symbol that was new to the proof, this has no effect on how formulas already appearing are interpreted.

Now we give quantifier tableau rules, using Smullyan's uniform notation. In stating things we use the informal convention that, if $\varphi(x)$ is a formula and c is a constant symbol, $\varphi(c)$ is like $\varphi(x)$ but with occurrences of c substituted for all free occurrences of the variable x. Quantified formulas are classified into type C, universal, and type D, existential, with γ as generic type C, and δ as generic type D.

γ	$\gamma(c)$
$T\,(\forall x)\varphi(x)$	$T\,\varphi(c)$
$F\,(\exists x)\varphi(x)$	$F\,\varphi(c)$
$(\forall x)\varphi(x)$	$\varphi(c)$
$\neg(\exists x)\varphi(x)$	$\neg\varphi(c)$

δ	$\delta(c)$
$T\,(\exists x)\varphi(x)$	$T\,\varphi(c)$
$F\,(\forall x)\varphi(x)$	$F\,\varphi(c)$
$(\exists x)\varphi(x)$	$\varphi(c)$
$\neg(\forall x)\varphi(x)$	$\neg\varphi(c)$

Now, the quantifier rules are these.

$$\frac{\gamma}{\gamma(c)}$$

Where c is any constant symbol whatever.

$$\frac{\delta}{\delta(c)}$$

Where c is a constant symbol that is new to the branch.

Here is an example of a classical first-order proof, of $(\forall x)(\forall y)R(x,y) \supset (\forall z)R(z,z)$. In it, 2 and 3 are from 1 by $F\supset$; 4 is from 3 by $F\forall$; 5 is from 2 by $T\forall$; and 6 is from 5 by $T\forall$. Notice that when the $F\forall$ rule was applied, the constant symbol c had not yet been used.

1. $F\,(\forall x)(\forall y)R(x,y) \supset (\forall z)R(z,z)$
2. $T\,(\forall x)(\forall y)R(x,y)$
3. $F\,(\forall z)R(z,z)$
4. $F\,R(c,c)$
5. $T\,(\forall y)R(c,y)$
6. $T\,R(c,c)$

For a sequent calculus formulation things are reversed from the tableau version. Existential quantifiers, instead of introducing new constant symbols, remove constant symbols. Using the notation of Section 2.5, here are sequent rules for classical quantifiers.

$$\frac{|S,\gamma(c)|}{|S,\gamma|}$$

Where c is any constant symbol.

$$\frac{|S,\delta(c)|}{|S,\delta|}$$

Where c is a constant symbol that does not occur in $\{S,\delta\}$.

As we said, the quantifier rules are deceptively simple. Since there are (we assume) infinitely many constant symbols available, if the γ-rule can be applied at all, it can be applied in infinitely many different ways. This means a tableau can never be completed in a finite number of steps. Essentially, this is the source of the undecidability of classical first-order logic.

3 MODERN HISTORY

After reaching a stable form in the classical case, the next stage in the development of tableaus was the extension to various non-classical logics. In this section we sketch a few such systems, say how they came about, and present the intuitions behind them. The particular systems chosen illustrate the variety of extra machinery that has been developed for and added to tableau systems: reinterpreting signs, generalizing signs, modifying closure rules, allowing trees to change in ways other than simple growth, adjoining 'side' information, and using pairs of coupled trees.

3.1 Intuitionistic Logic

Sequent calculi for intuitionistic logic were around from the beginning—Gentzen and Beth both developed them—so it is not surprising that a tableau version would be forthcoming (see Wallen's chapter). The first explicitly presented as such seems to be in the 1969 book of Fitting [1969]. In this the signed tableau system of Smullyan was adapted, with the signs given a new informal interpretation. In the resulting tableau system, proof trees were allowed to shrink as well as grow.

For both Lis and Smullyan, signs primarily were a device to keep track of left and right sides of sequents, without explicitly using sequent notation. Signs also had an intuitive interpretation that was satisfying: $T\,X$ and $F\,X$ can be thought of as asserting that X is true or false in a model. But now, think of $T\,X$ as informally meaning that X is *intuitionistically* true, that is, X has been given a proof that an intuitionist would accept. Likewise think of $F\,X$ as asserting the opposite: X has not been given an intuitionistically acceptable proof. (This is quite different from assuming X is intuitionistically refutable, by the way.) Some tableau rules are immediately suggested. For instance, intuitionists read disjunction constructively: to prove $X \vee Y$ one should either prove X or prove Y (see [Heyting, 1956]). Then if we have $T\,X \vee Y$ in a tableau, informally $X \vee Y$ has been intuitionistically proved, hence this is the case for one of X or Y, so the tableau branch splits to $T\,X$ and $T\,Y$. Likewise if we have $F\,X \vee Y$, we do not have an intuitionistically acceptable proof of $X \vee Y$, so we can have neither a proof of X nor of Y, and so we can add both $F\,X$ and $F\,Y$ to the branch. That is, intuitionistic rules for disjunction look like classical ones! The same is the case for conjunction. But things begin to get interesting with implication. We quote Heyting [1956].

> "The *implication* $\mathfrak{p} \rightarrow \mathfrak{q}$ can be asserted, if and only if we possess a construction $\mathfrak{r}$, which, joined to any construction proving $\mathfrak{p}$ (supposing that the latter be effected), would automatically effect a construction proving $\mathfrak{q}$. In other words, a proof of $\mathfrak{p}$, together with $\mathfrak{r}$, would form a proof of $\mathfrak{q}$."

If $T\,X \supset Y$ occurs in a tableau, informally we have a proof of $X \supset Y$, and so we have a way of converting proofs of X into proofs of Y. Then, in our present state of knowledge, either we are not able to prove X, or we are, in which case we can provide a proof of Y as well. That is, the tableau branch splits to $F\,X$ and $T\,Y$, just as it does classically.

Now suppose $F\,X \supset Y$ occurs in a tableau. Then intuitively, we do not have a mechanism for converting proofs of X into proofs of Y. This does not say anything at all about whether we are able to prove X. What it says is that someday, not necessarily now, we may discover a proof of X without being able to convert it to a proof of Y. That is, *someday* we could have both $T\,X$ and $F\,Y$. We are talking about a possible future state of our mathematical lives. Now, as we move into the future, what do we carry with us? If we *have not* proved some formula Z, this is not necessarily a permanent state of things—tomorrow we may discover a proof. But if we *have* proved Z, tomorrow this will still be so—a proof remains a proof. Thus, when passing from a state to a possible future state, signed formulas of the form $T\,Z$ should remain with us; signed formulas of the form $F\,Z$ need not. This suggests the following tableau rule: if a branch contains $F\,X \supset Y$, add both $T\,X$ and $F\,Y$, but first delete all signed formulas on the branch that have an F sign. (Negation has a similar analysis.)

We must be a little careful with this notion of formula deletion, though. The tree representation for tableaus that we have been using marks the presence of a node by using a formula as a label. If we simply delete formulas, information about node existence and tableau structure could be lost. What we do instead, when using this representation of tableaus, is leave deleted formulas in place, but check them off, placing a $\surd$ in front of them. (Of course, when using the set of sets representation for tableaus from Section 1.2, things are simpler: just replace one set by another, since there is no structure sharing.) There is still one more problem, though. It may happen that a formula should be deleted on one branch, but not on another, and using the tree representation for tableaus, its presence on both branches might be embodied in a single formula occurrence. In this case, check it off where it occurs, and add a fresh, unchecked occurrence to the end of the branch on which it should not be deleted.

To state the intuitionistic rules formally we use notation from [Fitting, 1969], if S is a set of signed formulas, let S_T be the set of T-signed members of S. We write $S, F\,Z$ to indicate a tableau branch containing the signed formula $F\,Z$, with S being the set of remaining formulas on the branch. Now, the propositional intuitionistic rules are these.

Conjunction $\qquad \dfrac{S, T\,X \wedge Y}{S, T\,X, T\,Y} \qquad \dfrac{S, F\,X \wedge Y}{S, F\,X \mid S, F\,Y}$

Disjunction $\qquad \dfrac{S, T\,X \vee Y}{S, T\,X \mid S, T\,Y} \qquad \dfrac{S, F\,X \vee Y}{S, F\,X, F\,Y}$

Implication $\qquad \dfrac{S, T\,X \supset Y}{S, F\,X \mid S, T\,Y} \qquad \dfrac{S, F\,X \supset Y}{S_T, T\,X, F\,Y}$

Negation $\dfrac{S, T\neg X}{S, F X}$ $\dfrac{S, F\neg X}{S_T, T X}$

Then unlike with classical tableaus, as far as usable formulas are concerned, intuitionistic tableaus can shrink as well as grow. If a branch contains, say, both $F\, X \supset Y$ and $F\, A \supset B$, using an implication rule on one will destroy the other. It is possible to make a bad choice at this point and miss an available proof. For completeness sakc, both possibilities must be explored. This is the analog of Beth's *disjunctive* branching, see Section 2.2. As a matter of fact, the tableau rules above correspond to Beth's rules in the same way that the classical tableau rules of Lis and Smullyan correspond to Beth's classical rules.

A tableau branch is closed if it contains $T\, X$ and $F\, X$, for some formula X, *where neither is a deleted signed formula.* Of course the intuitive idea is somewhat different than in the classical case now: the contradiction is that an intuitionist has both verified and failed to verify X. Nonetheless, a contradiction is still a contradiction.

We conclude with two examples, a non-theorem and a theorem. The non-theorem is $(\neg X \supset X) \supset X$. A tableau proof attempt begins with $F\,(\neg X \supset X) \supset X$. An application of the $F \supset$ rule causes this very F-signed formula to be deleted, and produces 2 and 3 below. Use of $T \supset$ on 2 gives 4 and 5. The right branch is closed, but closure of the left branch is impossible since an application of $F\neg$ to 4 causes deletion of 3.

√ 1. $F\,(\neg X \supset X) \supset X$
2. $T\,\neg X \supset X$
3. $F\,X$

4. $F\,\neg X$ 5. $T\,X$

The formula $(\neg X \supset X) \supset \neg\neg X$, on the other hand, is a theorem. A proof of it begins with $F\,(\neg X \supset X) \supset \neg\neg X$, then continues as follows.

√ 1. $F\,(\neg X \supset X) \supset \neg\neg X$
2. $T\,\neg X \supset X$
√ 3. $F\,\neg\neg X$
4. $T\,\neg X$

5. $F\,\neg X$ 6. $T\,X$
7. $F\,X$

An application of the $F \supset$ rule deletes 1 and adds 2 and 3. Then $F\neg$ applied to 3 deletes it and adds 4. The $T \supset$ rule applied to 2 adds 5 and 6, and finally, the $T\neg$ rule applied to 4 adds 7. The tableau is closed because of 4 and 5, and 6 and 7, none of which are checked off.

At the cost of a small increase in the number of signs, Miglioli, Moscato, and Ornaghi have created a tableau system for intuitionistic logic that is more efficient than the one presented above [Miglioli *et al.*, 1988; Miglioli *et al.*, 1993; Miglioli *et al.*, 1994]. In addition to the signs T and F, one more sign, F_c, is introduced. In terms of Kripke models, we can think of $T\,X$ as true at a possible world if X is true there, in the sense customary with intuitionistic semantics. Likewise we can think of $F\,X$ as true at a possible world if X is *not* true there. But now, think of $F_c\,X$ as true at a world if $\neg X$ is true there. This requires additional tableau rules, but reduces duplications inherent in the tableau system without the additional sign. Without going into details, this should suggest some of the flexibility made possible by the use of signed formulas, a topic to be continued in the next section.

3.2 *Many-Valued Logic*

The signs of a signed tableau system can be reinterpreted, as in intuitionistic logic, and they can be extended, as happened in many-valued logic (see Olivetti's chapter). Finitely-valued Łukasiewicz logics were given a tableau treatment by Suchon [1974]. Surma considered a more general situation, [1984], and this was further developed by Carnielli in [1987; 1991]. (The paper [Carnielli, 1987] contains an error in the quantifier rules which is corrected in [Carnielli, 1991].) In these papers, the two signs of Lis and Smullyan were extended to a larger number, with one sign for each truth value of the logic. Essentially, this is the tableau version of the many-valued sequent calculus of [Rosseau, 1967], discussed in Section 2.1. To show a formula X is a theorem, one must construct a closed tableau for $V\,X$, where V is a sign corresponding to the truth value v, for each non-designated value v.

While this is a natural idea, in practice it means several tableaus may need to be constructed for a single validity proof. Also, the rules themselves tend to be complicated. Suppose, for instance, we consider Kleene's strong three-valued logic, a well known logic. We can take as the three truth values $\{\mathit{false}, \bot, \mathit{true}\}$, where $\bot$ is intended to represent 'unknown.' Disjunction is easily characterized: order the truth values by: $\mathit{false} < \bot < \mathit{true}$; then disjunction-of is simply maximum-of. Suppose we introduce F, U, and T as signs corresponding to the three truth values. Then one of the Carnielli rules for disjunction is the following.

$$\frac{U\,X \vee Y}{\begin{array}{c|c|c} F\,X & U\,X & U\,X \\ U\,Y & U\,Y & F\,Y \end{array}}$$

More recently, Hähnle showed that many-valued logics could often be treated more efficiently by tableaus if *sets* of truth values were used as signs, rather than single truth values, [Hähnle, 1990; Hähnle, 1991; Hähnle, 1992]. Think of $S\,X$, where S is a set, as asserting that the truth value of X (in some model) is a member of S. Then to show X is valid one needs a single closed tableau, beginning with $\overline{D}\,X$, where $\overline{D}$ is the set of non-designated truth values. The following is a typical example of a Hähnle-style rule, again for the strong Kleene logic.

$$\frac{\{F,U\}\,X \vee Y}{\{F,U\}\,X \mid \{F,U\}\,Y}$$

As this example suggests, once sets are used as signs, a generalized uniform notation becomes possible. In many cases, this works quite well, [Hähnle, 1991]. We have not discussed quantification, which is a central issue in many-valued logics.

3.3 Modal Logic

What is now called *relational semantics* for modal logic was fully developed by the mid-sixties, drawing on work of Kanger [1957], Hintikka [1961; 1962], and Kripke [1959; 1963a; 1963b; 1965]. This led to a renewed interest in modal logic itself, and to the development of tableau systems for various such logics (covered by Goré in his chapter). See [Fitting, 1993] for a general overview of the subject. Kripke himself gave Beth-style tableaus for the modal logics he treated. In fact, his completeness proofs for axiom systems proceeded by showing equivalence to Beth tableau systems (using cut elimination) then proving completeness for these by a systematic tableau style construction. This was complex and hard to follow, and soon Henkin-style completeness arguments became standard. A tree-style system for S4 appeared in [Fitting, 1969], similar to the intuitionistic system of that book. Systems for several modal logics, based on somewhat different principles, appeared in [Fitting, 1972], and for temporal logics in [Rescher and Urquhart, 1971]. But the most extensive development was in Fitting's 1983 book [Fitting, 1983] which, among other things, gave tableau systems for dozens of normal and non-normal modal systems. We sketch a few to give an idea of the style of treatment, and the intuition behind the tableaus.

Destructive Tableau Systems

Fitting extended Smullyan's uniform notation to the modal case. Here is a signed-formula version. The idea is: a ν formula is true at a possible world if and only if the corresponding ν_0 is true at every accessible world; a π formula is true at a possible world if the corresponding π_0 is true at some accessible world.

ν	ν_0
$T\,\Box X$	$T\,X$
$F\,\Diamond X$	$F\,X$

π	π_0
$F\,\Diamond X$	$F\,X$
$T\,\Box X$	$T\,X$

Next, if S is a set of signed formulas, a set $S^{\#}$ is defined. The idea is, if the members of S are true at a possible world, and we move to a 'generic' accessible world, the members of $S^\#$ should be true there. The definition of $S^\#$ differs from modal logic to modal logic. We give the version for K, the smallest normal modal logic.

$$S^\# = \{\nu_0 \mid \nu \in S\}$$

The rules for K are exactly as in the classical Smullyan system, together with the following 'destructive' rule.

$$\frac{S, \pi}{S^\#, \pi_0}$$

Unlike the other rules (but exactly like the intuitionistic rules in Section 3.1), this one modifies a whole branch. If S, π is the set of signed formulas on a branch, the whole branch can be replaced with $S^\#, \pi_0$. Since this removes formulas, and modifies others, it is an information-loosing rule, hence the description 'destructive'. We continue to use the device of checking off deleted formulas in trees.

Here is a simple example of a proof in this system, of $\Box(X \supset Y) \supset (\Box X \supset \Box Y)$. It begins as follows.

$$\begin{array}{lll}
\surd & 1. & F\,\Box(X \supset Y) \supset (\Box X \supset \Box Y) \\
\surd & 2. & T\,\Box(X \supset Y) \\
\surd & 3. & F\,\Box X \supset \Box Y \\
\surd & 4. & T\,\Box X \\
\surd & 5. & F\,\Box Y \\
 & 6. & T\,X \supset Y \\
 & 7. & T\,X \\
 & 8. & F\,Y
\end{array}$$

Here 2 and 3 are from 1, and 4 and 5 are from 3 by $F \supset$. Now take 5 as π, and 1 through 4 as S, and apply the modal rule. Formula 1 is simply deleted; 2 is deleted but 6 is added; 3 is deleted; 4 is deleted but 7 is added (at this point, S has been replaced by $S^\#$); and finally 5 is deleted but 8 is added (this is π_0). Now an application of $T \supset$ to 6 produces a closed tableau.

The underlying intuition is direct. All rules, except the modal one, are seen as exploring truth at a single world. The modal rule corresponds to a move from a world to an alternative one. A soundness argument can easily be based on this. Completeness can be proved using either a systematic tableau construction or a maximal consistent set approach. As is the case with both classical and intuitionistic tableaus, interpolation theorems and related results can be derived from the tableau formulation.

Several other normal modal logics can be treated by modifying the definition of $S^\#$, or by adding rules, or both. For instance, the logic $K4$ (adding transitivity to the model conditions) just requires a change in a definition, to the following.

$$S^{\#} = \{\nu, \nu_0 \mid \nu \in S\}$$

Generally speaking, modal logics that have tableau systems of this kind can not have a semantics whose models involve *symmetry* of the accessibility relation. Interestingly enough, though, such logics can often be given tableau systems in this style if a cut rule is allowed, and in fact a *semi-analytic* version is enough. Semi-analyticity extends and weakens the notion of analytic cut, but is still not as broad as the unrestricted version. See Fitting, [1983], for more details.

Various regular but non-normal logics can be dealt with by restricting rule applicability (see [Fitting, 1983; Fitting, 1993] for a definition of regularity). For instance, if we use the system for K above, but restrict the modal rule to those cases in which $S^{\#}$ is non-empty, we get a tableau system for the smallest regular logic C. It is even possible to treat such quasi-regular logics as $S2$ and $S3$ by similar techniques.

Making Accessibility Explicit

In the various tableau systems described in the previous section, possible worlds were implicit, not explicit. Other approaches have brought possible worlds visibly into the picture. Hughes and Cresswell [1968], for instance, have a system of *diagrams* which are tableau-like, and involve boxes representing possible worlds, with arrows representing accessibility. In 1972 Fitting [1972] gave tableau systems using *prefixes* in which the idea was to designate possible worlds in such a way that syntactical rules determined accessibility. In a straightforward way, prefixes correspond to the Hughes and Cresswell boxes. The notion of prefixes is at its simplest for $S5$, where we can take as prefixes just natural numbers. The resulting system can be seen as a direct descendant of the sequent system of Kanger, discussed in Section 2.1.

A *prefixed signed formula* for $S5$ is just $n\ Z$, where n is a non-negative integer, and Z is a signed formula. The α- and β- rules from Sction 2.5 are modified in a direct way: conclude $n\,\alpha_1$ and $n\,\alpha_2$ from $n\,\alpha$, and similarly for β. The following modal rules are used (similarity to quantifier rules is intentional). This time there is no notion of formula deletion.

$$\frac{n\,\nu}{k\,\nu_0}$$

Where k is any non-negative integer.

$$\frac{n\,\pi}{k\,\pi_0}$$

Where k is any non-negative integer that is new to the branch.

Prefixes should be thought of as names for possible worlds. The system for $S5$ is particularly simple because the logic is characterized by models in which every world is accessible from every other. For other modal logics *sequences* of integers are used, where the intuition is: *extension-of* corresponds to *accessible-from*. This builds on an abstract of Fitch concerning modal natural deduction systems. See [Fitting, 1972; Fitting, 1983] for details.

There are many ways explicit reference to possible worlds can be incorporated into tableaus. The most obvious is to simply record accessibility information directly, in a side table. Other techniques have also been used, motivated by automation concerns. By such methods very general theorem proving mechanisms can be created. There is a drawback however. One of the nice features of tableaus is the extra information they can provide about the logic, most notably, proofs of interpolation theorems. Such proofs are not available once explicit possible worlds appear. On the other hand, decision procedures can often be easier to describe and program, using the additional machinery.

3.4 Relevance Logic

Fault has been found with tautologies like $(P \wedge \neg P) \supset Q$, since the antecedent and the consequent are not related to each other. A sense of dissatisfaction with such things led to the creation of the family of *relevance logics*; see [Dunn, 1986] for a survey of the subject. While the semantics for relevance logic is generally quite complicated, that for so-called *first-degree entailment* is rather simple—this is the fragment in which one considers only formulas of the form $A \supset B$ where A and B do not contain implications. The reason this is of interest here is that the method of 'coupled trees' developed for it shows yet another way of working with tableaus.

Smullyan presented a system of 'linear reasoning' for first-order classical logic in [Smullyan, 1968]. The system was tableau based, and was motivated by Craig's original proof of his Interpolation Theorem. Something very much like this, but for propositional classical logic only, appeared in a pedagogically nice form in Jeffrey [Jeffrey, 1967], under the name of *coupled trees*. It is Jeffrey's version with which we begin.

Suppose we want to give a classical propositional logic proof of $X \supset Y$, but instead of constructing a closed tableau for $F\,X \supset Y$, we do the following. Completely construct *two* tableaus, $\mathcal{T}_1$ for $T\,X$, and $\mathcal{T}_2$ for $T\,Y$. The open branches of $\mathcal{T}_1$ represent all the ways in which X could be true, and similarly for $\mathcal{T}_2$. Let us say a branch θ_1 *covers* a branch θ_2 if every signed atomic formula on θ_2 also occurs on θ_1. Suppose every open branch of $\mathcal{T}_1$ covers some open branch of $\mathcal{T}_2$—intuitively, each of the ways X could be true must also be a way in which Y is true. Then we have argued for the validity of $X \supset Y$. In effect, we are thinking of the proof of $X \supset Y$ as beginning with X, breaking it down by constructing $\mathcal{T}_1$ for $T\,X$, making the transition from $\mathcal{T}_1$ to $\mathcal{T}_2$ using the covering condition, then building up to Y using the tableau $\mathcal{T}_2$ for $T\,Y$ backwards. Such a proof technique is complete for implications.

To illustrate the technique, here is a coupled tableau proof of $(P \supset Q) \supset (\neg Q \supset \neg P)$.

$$\begin{array}{ccc} & 1.\, T\, P \supset Q & \\ & \swarrow \quad \searrow & \\ 2.\, F\, P & & 3.\, T\, Q \end{array} \qquad \begin{array}{ccc} & 4.\, T \neg Q \supset \neg P & \\ & \swarrow \quad \searrow & \\ 5.\, F \neg Q & & 6.\, T \neg P \\ 7.\, T\, Q & & 8.\, F\, P \end{array}$$

We do not describe the construction of the two tableaus, which is elementary. But note that the left branch of tableau one covers the right branch of tableau two, via formulas 2 and 8, and the right branch of tableau one covers the left branch of tableau two, via formulas 3 and 7. Thus we have a correctly constructed coupled tableau argument.

There are some problems with this intuitively simple idea, however. The formula $P \supset (Q \vee \neg Q)$ is classically valid, though not provàble by the technique described above. To get around this Jeffrey added a device that amounts to allowing applications of the cut rule in the tableau for the antecedant of an implication. Also, closed branches are ignored and, while this may seem natural at first encounter, even closed branches contain information. It is the restriction of the covering condition to only open branches that allows a coupled tree argument for $(P \wedge \neg P) \supset Q$, which is the standard example of a tautology not acceptable to relevance logicians.

Dunn's proposal in [1976] is to simplify Jeffrey's system to the extreme. To show $X \supset Y$, completely construct tableaus $\mathcal{T}_1$ for $T\,X$ (not using the Cut rule) and $\mathcal{T}_2$ for $T\,Y$, and see if *every* branch of $\mathcal{T}_1$, closed or not, covers a branch of $\mathcal{T}_2$. Dunn showed this gave a sound and complete proof procedure for first-degree entailment.

What sense can be made semantically of using tableau branches even if closed? Suppose, instead of working in the ideal setting of classical logic, we work in something more like the real world. We may have information that tells us a proposition P is *true*, or *false*. But equally well, we may have no information about P at all, neither *true* nor *false*, or we may have contradictory information, both *true* and *false*. In effect we are using a four-valued logic whose truth values are all subsets of $\{\mathit{false}, \mathit{true}\}$. This is a logic that was urged as natural for computer science in Belnap [1977]. Dunn showed that a simple semantics for first-degree entailment could be given using this four-valued logic: $X \supset Y$ is a valid first-degree entailment if and only if, under every valuation v in the four-valued logic, if $v(X)$ is at least true (has *true* as a member) then $v(Y)$ is also at least true.

The relationship between the four-valued logic and tableaus is simple: if a branch θ of a tableau has had all applicable tableau rules applied to it we can think of θ as determining a four-valued valuation as follows. Map P to $\{\mathit{true}\}$ if $T\,P$ is on θ but $F\,P$ is not; map P to $\{\mathit{false}\}$ if $F\,P$ is on θ but $T\,P$ is not; map P to $\emptyset$ if neither $T\,P$ nor $F\,P$ is on θ; and map P to $\{\mathit{false}, \mathit{true}\}$ if both $T\,P$ and $F\,P$ are on θ. Indeed, a similar 'ambiguation' can be developed starting with many-valued logics other than the classical, two-valued one. The technique is fairly general.

There are other approaches to relevance logic that make use of tableaus, each with additional features of interest. Hähnle [1992] gives a formulation of first-degree entailment using many-valued tableaus with sets of truth values as signs; see Section 3.2. Also, Schröder [1992] gives a tableau system in which additional bookkeeping machinery is introduced to check that each occurrence of a propositional variable was actually used to close a branch. Relevance logic itself is part of the more general subject of *substructural logic*, covered by D'Agostino *et al.*.

4 POST-MODERN HISTORY

In our subject, as in every other, post-modernism begins before modernism ends, in fact, before it starts. By the *post-modern period* for tableaus we mean the period of their automation (involving issues discussed by Hähnle in his Chapter). Indeed, Beth had machine theorem-proving very much in mind [Beth, 1958], though this did not have a lasting influence. It is curious that resolution and tableaus in their current form appeared within a few years of each other [Lis, 1960; Robinson, 1965; Smullyan, 1968]. Robinson, who invented resolution, was primarily interested in automation, Smullyan and Lis were not interested in automation at all. Perhaps this accounts for much of the emphasis on resolution in the automated theorem-proving community. But another determinant was more technical—nobody seems to have connected tableau methods and unification for a long time, without which only toy examples are possible. Still, all along there has been a subcurrent of interest in the uses of tableaus for automated theorem-proving—today it has become a major stream. We will sketch the swelling of this interest—it is complex, with basic ideas occurring independently several times. We will not bring our history up to today because present developments are many and are continuing to appear at a rapid rate. We lay the historical foundations for today's activity. We also confine the discussion to 'pure' tableau issues—we do not consider the increasingly fruitful relationships between tableaus and other theorem-proving mechanisms like connection graphs or resolution.

4.1 The Beginnings

Even though resolution has historically dominated automated deduction, among the first implemented theorem provers are some based on tableau ideas. In 1957–58, Dag Prawitz, Håkan Prawitz, and Neri Voghera developed a tableau-based system that was implemented on a Facit EDB [Prawitz *et al.*, 1960]. At approximately the same time, the summer of 1958, Hao Wang proposed a family of theorem provers based on the sequent calculus, which he then implemented on an IBM 704 [Wang, 1960]. The first of Wang's programs, for classical propositional logic, proved all the approximately 220 propositional theorems of Russell and Whitehead's *Principia Mathematica* in 3 minutes! This was quite a remarkable achievement for 1958.

The γ, or universal quantifier, rule is a major source of difficulties for first-order tableau implementations. It allows us to pass from γ to $\gamma(t)$ for *any* t, but how do we know which t will be a useful choice? Eventually, unification solved this problem, as we will see in the next section, but unification was not available in the 1950's. Prawitz and his colleagues worked with a formalization having constant symbols but no function symbols, which simplified the structure of terms (representation of strings, terms, and formulas was non-trivial on these early machines.) Then the γ rule was implemented to simply instantiate to $\gamma(c)$ for every constant symbol c that had been introduced by δ rule applications. In general, the set of such constant symbols grows without bound, which is the source of the undecidability of first-order logic. But also, as the authors note, this method creates many useless instantiations, and introduces an exponential growth factor into tableau construction, thus limiting the theorems provable by their implementation to rather simple examples. Wang discussed essentially the same idea, but did not actually carry out an implementation.

Instead of a full first-order system, Wang implemented a calculus for a decidable fragment—the AE formulas (with equality). A formula X is an AE formula if X is in prenex form, and all universal quantifiers precede any existential quantifiers. Since the rules for putting a formula into prenex form are not hard to mechanize, we can extend the definition to include those formulas that convert to AE form. The AE class is a decidable logic, and a complete tableau procedure for it is remarkably simple. Suppose, for instance, that we have a typical AE formula, $(\forall x_1)(\forall x_2)(\exists y_1)(\exists y_2)\varphi(x_1, x_2, y_1, y_2)$, and we attempt a tableau proof. We have two δ-rule applications to begin with, each introducing a new constant symbol, say c_1 and c_2. Thus the tableau begins as follows.

1. $F\,(\forall x_1)(\forall x_2)(\exists y_1)(\exists y_2)\varphi(x_1, x_2, y_1, y_2)$
2. $F\,(\forall x_2)(\exists y_1)(\exists y_2)\varphi(c_1, x_2, y_1, y_2)$
3. $F\,(\exists y_1)(\exists y_2)\varphi(c_1, c_2, y_1, y_2)$

Except for the γ case, classical tableau rules need only be applied to formulas once. Making use of this fact, we now have only γ-rules to apply. Suppose we apply them in all possible ways, *using only the constant symbols* c_1 *and* c_2 (four applications in all). After this, we only use propositional rules. It is rather easy to show that if this does not produce a proof, no proof is possible. More generally, without compromising completeness, γ-rule applications in proofs of AE formulas can be limited to constant symbols that were previously introduced by δ-rule applications, all of which must come first. It follows that a boundable number of γ-rule applications is always enough. (If there are no initial universal quantifiers, add a dummy one, then apply the procedure just outlined.)

Wang implemented the AE system just described. Of the 158 first-order propositions with equality in *Principia Mathematica* his program proved 139 of them. Subsequent modifications made possible proofs of all the theorems of *9 to *13 of *Principia* in about four minutes. Although Wang only worked with a decidable portion of first-order logic,

> "A rather surprising discovery, which tends to indicate our general ignorance of the extensive range of decidable subdomains, is the absence of any theorem of the predicate calculus in *Principia* which does not fall within the simple decidable subdomain of the AE predicate calculus."

It is not clear what this says about the work of Russell and Whitehead, but it is a curious discovery.

One interesting experiment that Wang undertook was to have the computer generate propositional formulas at random, test them for theoremhood, and print out those that passed an *ad hoc* test for being 'nontrivial'. In a way, the experiment was a failure, because 14,000 propositions were formed and tested in one hour, and 1000 were retained as nontrivial. The mass of data was simply too great. It is interesting just how hard it is to say what is interesting, and why.

This work seems to have had few direct successors. Possibly the introduction of the Davis-Putnam method, and then resolution, drew research attention elsewhere. Popplestone, in [1967], implemented a Beth tableau style theorem prover and specifically noted its relationship with Wang's version. The universal quantifier rule was still seen as a central problem, and heuristics were introduced to deal with it. In 1978 Mogilevskii and Ostroukhov [Mogilevskii and Ostroukhov, 1978] implemented (in ALGOL) a Smullyan-style theorem prover, but only for propositional classical logic, though they mention variations for $S4$ and for intuitionistic logic.

4.2 *Dummy Variables and Unification*

It is universally recognized that the γ or universal quantifier rule is the most problematic for a first-order tableau implementation. The rule allows passage from γ to $\gamma(t)$ for *any* term t. Without guidance on what term or terms to choose, automation is essentially hopeless. Systematically trying everything would, of course, yield a complete theorem-prover, but one that is hopelessly inefficient. Wang avoided the problem by confining his theorem-prover to a subsystem of full first-order logic. In the meantime Robinson, anticipated by Herbrand, introduced *unification* into automated theorem-proving [Robinson, 1965] and this, or its weaker cousin *matching*, serves very nicely as an appropriate tool for dealing with the universal quantifier problem. The idea, simply expressed, is to modify the γ-rule so that it reads: from γ pass to $\gamma(x)$, where x is a new free variable (a dummy, to use terminology from [Prawitz, 1960]). Then we use unification to discover what is a good choice for x—a good choice being something that will aid in tableau closure. For instance, if a branch contains $T\,P(t)$ and $F\,P(u)$, a substitution that unifies t and u will close the branch. What is wanted is a substitution that will simultaneously close all branches.

Unification was introduced independently into tableau theorem proving by several people, beginning with the 1974 paper of Cohen, Trilling, and Wegner [Cohen *et al.*, 1974]. While their paper was primarily devoted to presenting the virtues of ALGOL-68, it in fact gave a first-order theorem-prover based on Beth tableaus. It was written using a systematically-try-everything approach to the γ-rule, but then the introduction of Skolem functions and unification were specifically considered. This paper was followed by Bowen in [1980; 1982] and Broda [1980] (neither of which seems to be aware of [Cohen *et al.*, 1974]), both motivated by logic programming issues. Bowen used a sequent calculus, though he noted relationships with work of Beth and Smullyan; Broda used semantic tableaus directly. Wrightson in [1984b], Reeves in [1985], and Fitting in [1986] also explicitly brought unification into the picture, while Oppacher and Suen in their HARP theorem-prover of 1988 [Oppacher and Suen, 1988] use matching, only moving to unification when 'necessitated by the presence of complex terms'.

The technique of using dummy variables to deal with universal quantifiers was described in too simple a way above. If we restrict things so that the rule, passing from γ to $\gamma(x)$, can be applied to a given formula only once, an incomplete theorem-prover results. No general upper limit on the number of applications can be set (or else first-order logic would be decidable). On the other hand, if we place no restrictions on which unifiers we can accept, an unsound system can result. The problem is this. Applications of the δ-rule require introduction of new constants. If we use free variables in γ-rules and delay determination of their ultimate values, we don't know what is new and what is not when δ-rule applications come up. Wrightson [1984b] and Reeves [1985] deal with this difficulty essentially by imposing constraints on the unification process. Neither Bowen nor Broda discuss the issue explicitly, though they may have had a similar device in mind.

Matching with constraints gets quite complicated when function symbols are present. Of course the problem of what to do with the δ-rule can be easily avoided by Skolemizing away quantifier occurrences that would lead to δ-rule applications before the proof actually starts. If this is done, a rather simple sound and complete proof procedure combining tableaus and unification results. Such an approach was discussed by Reeves in [1985]. An implementation of tableaus involving initial Skolemization, written in LISP, was presented by Fitting in [1986].

4.3 *Run-Time Skolemization*

We noted that the combination of unification and tableaus leads to problems with the δ-rule, and that Skolemization provides one possible way out. Unfortunately, classical logic (more generally, many-valued logic with a finite number of truth values) is virtually the only first-order logic in which one can Skolemize a formula ahead of time. Tableau systems have been developed for a wide variety of logics, and this problem with the δ-rule could limit their usefulness. Fortunately there is a modification that works for many non-classical logics. It is commonly known as *run-time Skolemization*, a name whose significance will soon become apparent.

Suppose we are using the version of the γ-rule described in the previous section, passing from γ to $\gamma(x)$, where x is a new free variable. This kind of tableau system is sometimes referred to as a *free-variable tableau system*. In applying the δ-rule, passing from δ to $\delta(t)$, we need to be able to guarantee that the term t will be new to the branch, no matter how free variables are instantiated. A simple way of ensuring this is to take for t the expression $f(x_1, \ldots, x_n)$, where f is a new function symbol and $x_1, \ldots, x_n$ are all the free variables that occur on the branch. Clearly, we can think of the introduction of this term as part of a Skolemization process that goes on simultaneously with the tableau construction. In effect, Skolem functions come with an implicit 'time stamp' and this makes the technique suitable for many modal and similar logics.

Run-time Skolemization for tableaus seems to have first appeared in 1987 in Schmitt's THOT system [Schmitt, 1987], though it probably occurred to others around the same time. It is not necessary for classical logic, but its use does eliminate a preprocessing step, so it was included in the Prolog implementation of [Fitting, 1996]. For non-classical logics, however, Skolemization ahead of time is generally impossible, so the run-time version used by Fitting in 1988 for modal logics [Fitting, 1988] was essential.

Soon after Fitting's book [1996] appeared, Hähnle and Schmitt noted that the version of run-time Skolemization used was unnecessarily inefficient. In [Hähnle and Schmitt, 1993] they showed it is enough to take as a rule: from δ pass to $\delta(f(x_1, \ldots, x_n))$ where $x_1, \ldots, x_n$ are all the free variables *that occur in* δ. This work was extended in [Beckert, Hähnle and Schmitt, 1993]. Subsequently Shankar [Shankar, 1992] used a proof-theoretic analysis to show that a more complicated restriction on free variables—but still simpler than using all those that occur on a branch—suffices for first-order intuitionistic logic. Shankar also observed that his way of analyzing a tableau system will yield similar results for modal and other logics as well.

4.4 Where Now

If a logic has a tableau system at all, it probably will be the basis for the first theorem-prover to be implemented for it, though other kinds of theorem-provers may follow in time. For a given logic, tableaus may or may not turn out to be the best possible, most efficient approach to automated theorem-proving. Nevertheless, tableaus will continue to have a central role because they are relatively easy to develop, and in turn can be used to help create theorem-provers based on other methodologies. Many people have noted connections with resolution; in fact Maslov's method readily converts tableau systems to resolution-style systems [Mints, 1991]. There is a clear relationship with the connection method explored, among other places, in [Wrightson, 1984b; Wrightson, 1984a]. Wallen's 1990 book on non-classical theorem-proving [Wallen, 1990] exploits a relationship between tableaus and the matrix method. Examples continue to appear in the literature.

The development of theorem-provers that are not tableau-based, but are derived from them, is a topic of current research. The field is developing rapidly, so anything more specific we say about it will be out of date by the time this appears in print. I don't know what comes after post-modernism generally, but for tableau theorem-proving, maybe we have reached it.

5 CONCLUSIONS

We have given a general overview of how tableaus began and developed. Their history is much like their appearance—branching and re-branching. Ideas occurred independently more than once; researchers influenced each other directly and indirectly. Details are less important than the general picture of tableau history, beginning with Gentzen, growing to encompass semantical ideas with Beth and Hintikka, becoming an elegant tool with Lis and especially Smullyan, extending to many logics, developing relationships with other proof techniques, and suggesting exciting automation possibilities. Some of these topics will be explored further in subsequent chapters.

With all this, there is much we have not discussed. Relationships with logic programming were not mentioned though connections are many ([Fitting, 1994] will serve as a representative example). Equality is a central topic in logic, but we said little about it (see Beckert's chapter). The system of Lis [1960] includes a complete set of rules for equality, but it was little known at the time. Jeffrey's book [1967] presents essentially the same rules. Reeves discusses the topic in [1987] and there is a theoretical treatment in Fitting's book [1996]. Much has occurred since then—we cannot discuss it adequately here. Higher order logic has not been mentioned, though tableau systems for it exist. Toledo, [Toledo, 1975], investigates tableau systems for arithmetic that use the ω-rule, allowing infinite branching. Andrews, [1986], develops tableaus for type theory. Smith, [1993], presents two tableau systems for monadic higher-order logic. Van Heijenoort did research on tableau systems for higher order logic, but this has not yet been published.

The invention of tableau systems will continue, simply because they are easier to think of than other formulations. There is something inherently natural about them, whether they grow out of proof theory as with Gentzen, or out of semantics as with Beth and Hintikka. The increasing interest in non-classical theorem-proving has brought tableaus to a position of prominence, because they exist for many, many logics. The creation of logics, the development of tableau systems for them, all are very active areas of research. May the history of tableaus need rewriting in another generation.

ACKNOWLEDGEMENTS

I want to thank Perry Smith and Reiner Hähnle for their suggestions, which considerably improved this chapter.

CUNY, New York

REFERENCES

[Andrews, 1986] P. B. Andrews. *An Introduction to Mathematical Logic and Type Theory: To Truth Through Proof.* Academic Press, Orlando, Florida, 1986.

[Anellis, 1990] I. Anellis. From semantic tableaux to Smullyan trees: the history of the falsifiability tree method. *Modern Logic 1*, 1, 36–69, 1990.

[Anellis, 1991] I. Anellis. Erratum, from semantic tableaux to Smullyan trees: the history of the falsifiability tree method. *Modern Logic 2*, 2 (Dec. 1991), 219, 1991.

[Beckert, Hähnle and Schmitt, 1993] B. Beckert, R. Hähnle and P. H. and Schmitt. The *even more* liberalized δ-rule in free variable semantic tableaux. In *Proceedings of the third Kurt Gödel Colloquium KGC'93, Brno, Czech Republic* (aug 1993), G. Gottlob, A. Leitsch, and D. Mundici, Eds., Springer LNCS 713, pp. 108–119, 1993.

[Bell and Machover, 1977] J. L. Bell and M. Machover. *A Course in Mathematical Logic.* North-Holland, Amsterdam, 1977.

[Belnap, 1977] N. D. Belnap Jr. A useful four-valued logic. In *Modern Uses of Multiple-Valued Logic*, J. M. Dunn and G. Epstein, Eds. D. Reidel, Dordrecht and Boston, pp. 8–37, 1977.

[Beth, 1953a] E. W. Beth. On Padoa's method in the theory of definition. *Indag. Math. 15*, 330–339, 1953.

[Beth, 1953] E. W. Beth. Some consequences of the theorem of Löwenheim-Skolem-Gödel-Malcev. *Indag. Math. 15*, 1953.

[Beth, 1955] E. W. Beth. Semantic entailment and formal derivability. *Mededelingen der Kon. Ned. Akad. v. Wet. 18*, 13, 1955. new series.

[Beth, 1956] E. W. Beth. Semantic construction of intuitionistic logic. *Mededelingen der Kon. Ned. Akad. v. Wet. 19*, 11, 1956. new series.

[Beth, 1958] E. W. Beth. On machines which prove theorems. *Simon Stevin Wissen-Natur-Kundig Tijdschrift 32*, 49–60, 1958. Reprinted in [Siekmann and Wrighson, 1983] vol. 1, pp 79 – 90.

[Beth, 1959] E. W. Beth. *The Foundations of Mathematics.* North-Holland, Amsterdam, 1959. Revised Edition 1964.

[Bibel *et al.*, 1987] W. Bibel, F. Kurfess, K. Aspetsberger, P. Hintenaus and J. Schumann. Parallel inference machines. In *Future Parallel Computers*, P. Treleaven and M. Vanneschi, Eds. Springer, Berlin, pp. 185–226, 1987.

[Bowen, 1982] K. A. Bowen. Programming with full first-order logic. In *Machine Intelligence*, Hayes, Michie, and Pao, Eds., vol. 10, pp. 421–440, 1982.

[Bowen, 1980] K. A. Bowen. Programming with full first order logic. Tech. Rep. 6-80, Syracuse University, Syracuse, NY, Nov. 1980.

[Broda, 1980] K. Broda. The relation between semantic tableaux and resolution theorem provers. Tech. Rep. DOC 80/20, Imperial College of Science and Technology, London, Oct. 1980.

[Carnielli, 1987] W. A. Carnielli. Systematization of finite many-valued logics through the method of tableaux. *Journal of Symbolic Logic 52*, 2, 473–493, 1987.

[Carnielli, 1991] W. A. Carnielli. On sequents and tableaux for many-valued logics. *Journal of Non-Classical Logic 8*, 1, 59–76, 1991.

[Chang and Keisler, 1990] C. C. Chang and H. J. Keisler. *Model Theory*, third ed. North-Holland Publishing Company, 1990.

[Cohen *et al.*, 1974] J. Cohen, L. Trilling and P. Wegner. A nucleus of a theorem-prover described in ALGOL-68. *International Journal of Computer and Information Sciences 3*, 1, 1–31, 1974.

[Dunn, 1976] J. M. Dunn. Intuitive semantics for first-degree entailments and 'coupled trees'. *Philosophical Studies 29*, 149–168, 1976.

[Dunn, 1986] J. M. Dunn. Relevance logic and entailment. In *Handbook of Philosophical Logic*, D. Gabbay and F. Guenthner, Eds., vol. 3. Kluwer, Dordrecht, 1986, ch. III.3, pp. 117–224.

[Feys, 1965] R. Feys. *Modal Logics*. No. IV in Collection de Logique Mathématique, Sèrie B. E. Nauwelaerts (Louvain), Gauthier-Villars (Paris), Joseph Dopp, editor. 1965.

[Fitting, 1969] M. C. Fitting. *Intuitionistic Logic Model Theory and Forcing*. North-Holland Publishing Co., Amsterdam, 1969.

[Fitting, 1972] M. C. Fitting. Tableau methods of proof for modal logics. *Notre Dame Journal of Formal Logic 13*, 237–247, 1972.

[Fitting, 1983] M. C. Fitting *Proof Methods for Modal and Intuitionistic Logics*. D. Reidel Publishing Co., Dordrecht, 1983.

[Fitting, 1986] M. C. Fitting A tableau based automated theorem prover for classical logic. Tech. rep., Herbert H. Lehman College, Bronx, NY 10468, 1986.

[Fitting, 1988] M. C. Fitting First-order modal tableaux. *Journal of Automated Reasoning 4*, 191–213, 1988.

[Fitting, 1996] M. C. Fitting *First-Order Logic and Automated Theorem Proving*. Springer-Verlag, 1996. (First edition, 1990.)

[Fitting, 1993] M. C. Fitting Basic modal logic. In *Handbook of Logic in Artificial Intelligence and Logic Programming*, D. M. Gabbay, C. J. Hogger, and J. A. Robinson, Eds., vol. 1. Oxford University Press, Oxford, pp. 368–448, 1993.

[Fitting, 1994] M. C. Fitting Tableaux for logic programming. *Journal of Automated Reasoning 13*, 175–188, 1994.

[Gentzen, 1935] G. Gentzen. Untersuchungen über das logische Schliessen. *Mathematische Zeitschrift 39*, 176–210, 405–431, 1935. English translation, 'Investigations into logical deduction', in [Szabo, 1969].

[Girard, 1986] J. Y. Girard. Linear logic. *Theoretical Computer Science 45*, 159–192, 1986.

[Hähnle, 1990] R. Hähnle. Towards an efficient tableau proof procedure for multiple-valued logics. In *Proceedings Workshop on Computer Science Logic, Heidelberg*, vol. 533 of *LNCS*, Springer-Verlag, pp. 248–260, 1990.

[Hähnle, 1991] R. Hähnle. Uniform notation of tableaux rules for multiple-valued logics. In *Proceedings International Symposium on Multiple-Valued Logic, Victoria*, IEEE Press, pp. 238–245, 1991.

[Hähnle, 1992] R. Hähnle. *Automated Theorem Proving in Multiple-Valued Logics*, vol. 10 of *International Series of Monographs on Computer Science*. Oxford University Press, 1993.

[Hähnle and Schmitt, 1993] R. Hähnle and P. H. Schmitt. The liberalized δ-rule in free variable semantic tableaux. *Journal of Automated Reasoning, to appear*, 1993.

[Herz, 1929] P. Hertz. Über Axiomensysteme für beliebige Satzsysteme. *Mathematische Annalen 101*, 457–514, 1929.

[Heyting, 1956] A. Heyting. *Intuitionism, an Introduction*. North-Holland, Amsterdam, 1956. Revised Edition 1966.

[Hintikka, 1953] J. Hintikka. A new approach to sentential logics. *Soc. Scient. Fennica, Comm. Phys.-Math. 17*, 2, 1953.

[Hintikka, 1955] J. Hintikka. Form and content in quantification theory. *Acta Philosophica Fennica – Two Papers on Symbolic Logic 8*, 8–55, 1955.

[Hintikka, 1961] J. Hintikka. Modality and quantification. *Theoria 27*, 110–128, 1961.

[Hintikka, 1962] J. Hintikka. *Knowledge and Belief*. Cornell University Press, 1962.

[Hughes and Cresswell, 1968] G. E. Hughes and Cresswell. *An Introduction to Modal Logic*. Methuen and Co., London, 1968.

[Jeffrey, 1967] R. C. Jeffrey. *Formal Logic: Its Scope and Limits*. McGraw-Hill, New York, 1967.

[Kanger, 1957] S. G. Kanger. Provability in logic (Acta Universitatis Stockholmiensis, Stockholm Studies in Philosophy, 1). Almqvist and Wiksell, Stockholm, 1957.

[Kleene, 1950] S. C. Kleene. *Introduction to Metamathematics*. D. Van Nostrand, North-Holland, P. Noordhoff, 1950.

[Kripke, 1959] S. Kripke. A completeness theorem in modal logic. *Journal of Symbolic Logic 24*, 1–14, 1959.

[Kripke, 1963a] S. Kripke. Semantical analysis of modal logic I, normal propositional calculi. *Zeitschrift für mathematische Logik und Grundlagen der Mathematik 9*, 67–96, 1963.

[Kripke, 1963b] S. Kripke. Semantical considerations on modal logics. *Acta Philosophica Fennica, Modal and Many-valued Logics*, 83–94, 1963.

[Kripke, 1965] S. Kripke. Semantical analysis of modal logic II, non-normal modal propositional calculus. In *The Theory of Models*, J. W. Addison, L. Henkin, and A. Tarski, Eds. North-Holland, Amsterdam, pp. 206–220, 1965.

[Lis, 1960] Z. Lis. Wynikanie semantyczne a wynikanie formalne (logical consequence, semantic and formal). *Studia Logica 10*, 39–60, 1960. Polish, with Russian and English summaries.

[Manna and Waldinger, 1990] Z. Manna and R. Waldinger. *The Logical Basis for Computer Programming*. Addison-Wesley, 1990. 2 vols.

[Manna and Waldinger, 1993] Z. Manna and R. Waldinger. *The Deductive Foundations of Computer Programming*. Addison-Wesley, 1993.

[Matsumoto, 1960] K. Matsumoto. Decision procedure for modal sentential calculus S3. *Osaka Mathematical Journal 12*, 167–175, 1960.

[Miglioli *et al.*, 1988] P. Miglioli, U. Moscato and M. Ornaghi. An improved refutation system for intuitionistic predicate logic. Rapporto interno 37/88, Dipartimento di Scienze dell'Informazione, Universitá degli Studi di Milano, 1988.

[Miglioli *et al.*, 1993] P. Miglioli, U. Moscato and M. Ornaghi. How to avoid duplications in refutation systems for intuitionistic logic and Kuroda logic. Rapporto interno 99/93, Dipartimento di Scienze dell'Informazione, Universitá degli Studi di Milano, 1993.

[Miglioli *et al.*, 1994] P. Miglioli, U. Moscato and M. Ornaghi. An improved refutation system for intuitionistic predicate logic. *Journal of Automated Reasoning 13*, 3, 361–373, 1994.

[Mints, 1991] G. Mints. Proof theory in the USSR 1925 - 1969. *Journal of Symbolic Logic 56*, 2 , 385–424, 1991.

[Mogilevskii and Ostroukhov, 1978] G. L. Mogilevskii and D. A. Ostroukhov. A mechanical propositional calculus using Smullyan's analytic tables. *Cybernetics 14*, 526–529, 1978. Translation from *Kibernetika*, 4, 43–46, 1978.

[Ohnishi, 1961] M. Ohnishi. Gentzen decision procedures for Lewis's systems S2 and S3. *Osaka Mathematical Journal 13*, 125–137, 1961.

[Ohnishi and Matsumoto, 1957] M. Ohnishi and K. Matsumoto. Gentzen method in modal calculi I. *Osaka Mathematical Journal 9*, 113–130, 1957.

[Ohnishi and Matsumoto, 1959] M. Ohnishi and K. Matsumoto. Gentzen method in modal calculi II. *Osaka Mathematical Journal 11*, 115–120, 1959.

[Ohnishi and Matsumoto, 1964] M. Ohnishi and K. Matsumoto. A system for strict implication. *Annals of the Japan Assoc. for Philosophy of Science 2*, 183–188, 1964.

[Oppacher and Suen, 1988] F. Oppacher and E. Suen. HARP: A tableau-based theorem prover. *Journal of Automated Reasoning 4*, 69–100, 1988.

[Popplestone, 1967] R. J. Popplestone. Beth-tree methods in automatic theorem-proving. In *Machine Intelligence*, N. L. Collins and D. Michie, Eds., vol. 1. American Elsevier, New York, pp. 31–46, 1967.

[Prawitz, 1960] D. Prawitz. An improved proof procedure. *Theoria 26* (1960). Reprinted in [Siekmann and Wrighson, 1983] vol. 1, pp 162 – 199.

[Prawitz *et al.*, 1960] D. Prawitz, H. Prawitz and N. Voghera. A mechanical proof procedure and its realization in an electronic computer. *Journal of the ACM 7*, 102–128, 1960.

[Rasiowa, 1951] H. Rasiowa. Algebraic treatment of the functional calculi of Heyting and Lewis. *Fundamenta Mathematica 38*, 1951.

[Rasiowa, 1954] H. Rasiowa. Algebraic models of axiomatic theories. *Fundamenta Mathematica 41*, 1954.

[Reeves, 1985] S. V. Reeves. *Theorem-proving by Semantic Tableaux*. PhD thesis, University of Birmingham, 1985.

[Reeves, 1987] S. V. Reeves. Adding equality to semantic tableaux. *Journal of Automated Reasoning 3*, 225–246, 1987.

[Rescher and Urquhart, 1971] N. Rescher and A. Urquhart. *Temporal Logic*. Springer-Verlag, 1971.

[Robinson, 1965] J. A. Robinson. A machine-oriented logic based on the resolution principle. *Journal of the ACM 12*, 23–41, 1965.

[Rosseau, 1967] G. Rousseau. Sequents in many valued logic I. *Fundamenta Mathematica 60*, 23–33, 1967.

[Schmitt, 1987] P. H. Schmitt. The THOT theorem prover. Tech. Rep. TR–87.09.007, IBM Heidelbert Scientific Center, 1987.

[Schröder, 1992] Schröder, J. Körner's criterion of relevance and analytic tableaux. *Journal of Philosophical Logic 21*, 2, 183–192, 1992.

[Shankar, 1992] N. Shankar. Proof search in the intuitionistic sequent calculus. In *Automated Deduction — CADE-11*, D. Kapur, Ed., no. 607 in Lecture Notes in Artificial Intelligence. Springer-Verlag, Berlin, pp. 522–536, 1992.

[Siekmann and Wrighson, 1983] J. Siekmann and G. Wrightson, Eds. *Automation of Reasoning*. Springer-Verlag, Berlin, 1983. 2 vols.

[Smith, 1993] P. Smith. Higher-Order Logic, Model Theory, Recursion Theory, and Proof Theory. Unpublished Manuscript, 1993.

[Smullyan, 1963] R. M. Smullyan. A unifying principle in quantification theory. *Proceedings of the National Academy of Sciences 49*, 6, 828–832, 1963.

[Smullyan, 1965] R. M. Smullyan. Analytic natural deduction. *Journal of Symbolic Logic 30*, 123–139, 1965.

[Smullyan, 1966] R. M. Smullyan. Trees and nest structures. *Journal of Symbolic Logic 31*, 303–321, 1966.

[Smullyan, 1968] R. M. Smullyan. *First-Order Logic*. Springer-Verlag, 1968. Revised edition, Dover Press, NY, 1994.

[Smullyan, 1970] R. M. Smullyan. Abstract quantification theory. In *Intuitionism and Proof Theory, Proceedings of the Summer Conference at Buffalo N. Y. 1968*, A. Kino, J. Myhill, and R. E. Vesley, Eds. North-Holland, Amsterdam, pp. 79–91, 1970.

[Smullyan, 1973] R. M. Smullyan. A generalization of intuitionistic and modal logics. In *Truth, Syntax and Modality, Proceedings of the Temple University Conference on Alternative Semantics*, H. Leblanc, Ed. North-Holland, Amsterdam, pp. 274–293, 1973.

[Suchoón, 1974] W. Suchoń. La méthode de Smullyan de construire le calcul n-valent des propositions de Łukasiewicz avec implication et négation. *Reports on Mathematical Logic, Universities of Cracow and Katowice 2*, 37–42, 1974.

[Surma, 1984] S. J. Surma. An algorithm for axiomatizing every finite logic. In *Computer Science and Multiple-Valued Logics*, D. C. Rine, Ed. North-Holland, Amsterdam, 1977, pp. 143–149. Revised edition, 1984.

[Szabo, 1969] M. E. Szabo., Ed. *The Collected Papers of Gerhard Gentzen*. North-Holland, Amsterdam, 1969.

[Tarski, 1938] A. Tarski. Der Aussagenkalkül und die Topologie. *Fundamenta Mathematica 31*, 103–34, 1938. Reprinted as 'Sentential calculus and topology' in [Tarski, 1956].

[Tarski, 1956] A. Tarski. *Logic, Semantics, Metamathematics*. Oxford, 1956. J. H. Woodger translator.

[Toledo, 1975] S. Toledo. *Tableau Systems for First Order Number Theory and Certain Higher Order Theories*, vol. 447 of *Lecture Notes in Mathematics*. Springer-Verlag, Berlin, 1975.

[Troelstra, 1992] A. S. Troelstra. *Lectures on Linear Logic*. No. 29 in CSLI Lecture Notes. CSLI, 1992.

[Wallen, 1990] L. A. Wallen. *Automated Deduction in Nonclassical Logics*. The MIT Press, 1990.

[Wang, 1960] H. Wang. Toward mechanical mathematics. *IBM Journal for Research and Development 4* (1960), 2–22. Reprinted in *A Survey of Mathematical Logic*, Hao Wang, North-Holland, pp 224 – 268, 1960; and in [Siekmann and Wrighson, 1983], vol 1, pp 244 – 264.

[Wrightson, 1984a] G. Wrightson. Non-classical theorem proving using links and unification in semantic tableaux. Tech. Rep. CSD-ANZARP-84-003, Victoria University, Wellington, NZ, 1984.

[Wrightson, 1984b] G. Wrightson. Semantic tableaux, unification and links. Tech. Rep. CSD-ANZARP-84-001, Victoria University, Wellington, New Zealand, 1984.

MARCELLO D'AGOSTINO

TABLEAU METHODS FOR CLASSICAL PROPOSITIONAL LOGIC

1 INTRODUCTION

Traditionally, a mathematical problem was considered 'closed' when an algorithm was found to solve it 'in principle'. In this sense the deducibility problem of classical propositional logic was already 'closed' in the early 1920's, when Wittgenstein and Post independently devised the well-known decision procedure based on the truth-tables. As usually happens this *positive* result ended up killing any theoretical interest in the subject. In contrast, first order logic was spared by a *negative* result. Its undecidability, established by Church and Turing in the 1930's, made it an 'intrinsically' interesting subject forever. If no decision procedure can be found, that is all decision procedures have to be partial, the problem area is in no danger of saturation: it is always possible to find 'better' methods (at least for certain purposes).

This point of view reflected a situation in which logical investigations were largely motivated by philosophical and mathematical (as opposed to computational) problems. Algorithms were mainly regarded as ideal objects churning out their solutions in a 'platonic' world unaffected by realistic limitations of time and space. The emergence and rapid growth of computing science over the last few decades has brought about a dramatic change of attitude. Algorithms are devised not to be 'contemplated' but to be *implemented*, and an algorithm which solves a problem 'in principle' might require unrealistic resources in terms of space and time. For instance, the truth-table method requires checking 2^n rows for a formula containing n distinct propositional variables. Therefore it quickly becomes infeasible even for relatively small values of n. So, in some sense, it is *not* a decision procedure for full propositional logic, but only for formulas containing a 'small' number of variables.

This quest for feasibility has deeply affected the subject of computability theory and, as a result, has led to a refinement of many traditional logical problems. Not only is the question of *decidability* refined by considering the efficiency of decision procedures, but also the question of *completeness* is refined by considering the *length of proofs* in a given system. Suppose we have proved that a system **S** is complete for classical propositional logic. It may well be, however, that for certain classes of tautologies the length of their *shortest* **S**-proof grows beyond any feasible limit, even for relatively 'short' tautologies. Hence, in some sense, the proof-system **S** is *not* complete.

Such refinements of old 'closed' problems have revitalised a subject which seemed to be saturated and deserve only a brief mention as a stepping stone to the

M. D'Agostino et al. (eds.), Handbook of Tableau Methods, 45–123.

'real thing'. Open questions of theoretical computer science such as $\mathcal{P} =?\mathcal{NP}$ and $\mathcal{NP} =?co\text{-}\mathcal{NP}$ (see [Garey and Johnson, 1979] and [Stockmeyer, 1987]), have been shown to be equivalent to quantitative problems about decidability and proof-length in propositional logic. Moreover, the widespread conjecture that $\mathcal{P} \neq \mathcal{NP}$, although unproven, is presently providing the surrogate of a negative result. If true, it would imply that there is no *feasible* (i.e. polynomial time, see Section 2.3 below) decision procedure for classical propositional logic. Since it is believed to be true by most researchers, as a matter of fact it is playing the same stimulating role as the undecidability result has played with regard to first order logic.

Such quantitative questions about propositional logic are related to a plethora of problems which are of the utmost practical and theoretical importance. One example is the simplification of Boolean formulae. As Alasdair Urquhart put it [Urquhart, 1992] 'Here the general problem takes the form: how many logical gates do we need to represent a given Boolean function? This is surely as simple and central a logical problem that one could hope to find; yet, in spite of Quine's early contributions, the whole area has been simply abandoned by most logicians, and is apparently thought to be fit only for engineers[...] our lack of understanding of the simplification problem retards our progress in the area of the complexity of proofs.' In fact, the development of the NP-completeness theory has brought to light a web of connections between propositional logic and a variety of hard computational problems from different important areas of theoretical and technological research. From this point of view, the design of improved proof systems and proof procedures for propositional logic is *objectively* a central theme in discrete mathematics, quite independent of its impact on automated theorem proving and before any first-order issue is raised.

This central role played by logic (and by propositional logic in particular) over the last few decades has resulted in a greater awareness of the computational aspects of logical systems and a closer (or at least a fresh) attention to their proof-theoretical presentations. Traditionally two proof systems were considered equivalent if they proved exactly the same theorems. The other extreme of this view has been vividly put forward by Gabbay:

> "[...] a logical system $\mathbf{L}$ is not just the traditional consequence relation $\vdash$, but a pair $(\vdash, \mathbf{S}_\vdash)$, where $\vdash$ is a mathematically defined consequence relation (i.e. a set of pairs (Δ, Q) such that $\Delta \vdash Q$) and $\mathbf{S}_\vdash$ is an algorithmic system for generating all those pairs. Thus, according to this definition, classical logic $\vdash$ perceived as a set of tautologies together with a Gentzen system $\mathbf{S}_\vdash$ is not the same as classical logic together with the two-valued truth-table decision procedure $\mathbf{T}_\vdash$ for it. In our conceptual framework, $(\vdash, \mathbf{S}_\vdash)$ is *not the same logic* as $(\vdash, \mathbf{T}_\vdash)$.[1]"

[1] [Gabbay, 1996, Section 1.1].

Gabbay's proposal seems even more suggestive when considerations of computational complexity enter the picture. Different proof-theoretical algorithms for generating the same consequence relation may have different complexities. Even more interesting: algorithmic characterizations which appear to be 'close' to each other from the proof-theoretical point of view may show dramatic differences as far as their complexity is concerned.

In this chapter we shall adopt this point of view as a conceptual grid to explore the landscape of propositional tableau methods.

2 PRELIMINARIES

2.1 Partial Valuations and Semi-valuations

Let us call a *partial valuation* any partial function $V : \mathbf{F} \mapsto \{1, 0\}$ where 1 and 0 stand as usual for the truth-values (*true* and *false* respectively) and $\mathbf{F}$ stands for the sets of all formulae. It is convenient for our purposes to represent partial functions as total functions with values in $\{1, 0, *\}$ where $*$ stands for the 'undefined' value. By a *total* valuation we shall mean a partial valuation which for no formula A yields the value $*$. For every formula A we say that *A is true under V* if $V(A) = 1$, *false under V* if $V(A) = 0$ and *undefined under V* if $V(A) = *$. We say that a *sequent* $\Gamma \vdash \Delta$ *is true under* V if $V(A) = 0$ for some $A \in \Gamma$ or $V(A) = 1$ for some $A \in \Delta$. We say that it is *false under* V if $V(A) = 1$ for all $A \in \Gamma$ and $V(A) = 0$ for all $A \in \Delta$.

A *Boolean valuation*, is regarded from this point of view as a special case of a partial valuation of $\mathbf{F}$, namely one which is total and is faithful to the usual truth-table rules, i.e. for all formulae A and B:

1. $V(\neg A) = 1$ iff $V(A) = 0$
2. $V(A \vee B) = 1$ iff $V(A) = 1$ or $V(B) = 1$
3. $V(A \wedge B) = 1$ iff $V(A) = 1$ and $V(B) = 1$
4. $V(A \rightarrow B) = 1$ iff $V(A) = 0$ or $V(B) = 1$

The consequence relation associated with classical propositional logic can be expressed in terms of *sequents*, i.e. expressions of the form

(1) $A_1, \ldots, A_n \vdash B_1, \ldots, B_m$

(where the A_i's and the B_i's are formulae) with the same informal meaning as the formula

$$A_1 \wedge \ldots \wedge A_n \rightarrow B_1 \vee \ldots \vee B_m.$$

The sequence to the left of the turnstyle is called 'the antecedent' and the sequence to the right is called 'the succedent'. Let us say that a valuation V *falsifies* a sequent $\Gamma \vdash \Delta$ if V makes true all the formulae in Γ and false all the formulae in

Δ. In the 'no-countermodel' approach to validity we start from a sequent $\Gamma \vdash \Delta$ intended as a valuation problem and try to find a countermodel to it — at the propositional level a Boolean valuation which falsifies it. In fact, there is no need for *total* valuations, but it is sufficient to construct some partial valuation satisfying certain closure conditions. Such partial valuations are known in the literature under different names and shapes. Following Prawitz [1974] we shall call them 'semivaluations':

DEFINITION 1. A (Boolean) *semivaluation* is a partial valuation V satisfying the following conditions:

1. if $V(\neg A) = 1$, then $V(A) = 0$;
2. if $V(\neg A) = 0$, then $V(A) = 1$.
3. if $V(A \vee B) = 1$, then $V(A) = 1$ or $V(B) = 1$;
4. if $V(A \vee B) = 0$, then $V(A) = 0$ and $V(B) = 0$;
5. if $V(A \wedge B) = 1$, then $V(A) = 1$ and $V(B) = 1$;
6. if $V(A \wedge B) = 0$, then $V(A) = 0$ or $V(B) = 0$;
7. if $V(A \to B) = 1$, then $V(A) = 0$ or $V(B) = 1$;
8. if $V(A \to B) = 0$, then $V(A) = 1$ and $V(B) = 0$;

The crucial property of semivaluations is that they can be readily extended to Boolean valuations as stated by the following lemma whose proof is left to the reader.

LEMMA 2. *Every semivaluation can be extended to a Boolean valuation.*

Each stage of our attempt to construct a semivaluation which falsifies a given sequent $\Gamma \vdash \Delta$ can, therefore, be described as a partial valuation. We start from the partial valuation which assigns 1 to all the formulae in Γ and 0 to all the formulae in Δ, and try to refine it step by step, taking care that the classical rules of truth are not infringed. If we eventually reach a partial valuation which is a semivaluation, by virtue of Lemma 2 we have successfully described a *countermodel* to the original sequent. Otherwise we have to ensure that *no* refinement of the initial partial valuation will ever lead to a semivaluation.

The search space, in this process, is a set of partial valuations which are naturally ordered by the approximation relationship (in Scott's sense [1970]):

$$V \sqsubseteq V' \text{ if and only if } V(A) \preceq V'(A) \text{ for all formulae } A$$

where $\preceq$ is the usual partial ordering over the truth-values $\{1, 0, *\}$, namely

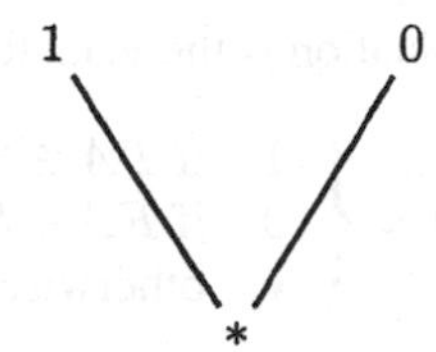

The set of all these partial valuations, together with the approximation relationship defined above, forms a lattice with a bottom element consisting of the 'empty' valuation ($V(A) = *$ for all formulae A). It is convenient to 'top' this lattice by adding an 'overdefined' element $\top$. This 'fictitious' element of the lattice does not correspond to any real partial valuation and is used to provide a least upper bound for pairs of partial valuations which have no common refinement[2]. Hence we can regard the equation

$$V \sqcup V' = \top$$

as meaning intuitively that V and V' are *inconsistent*.

2.2 Expansion Systems

We assume a 0-order language defined as usual. We shall denote by X, Y, Z (possibly with subscripts) arbitrary *signed formulae* (s-formulae for short), i.e. expressions of the form TA or FA where A is a formula. The *conjugate* of an s-formula is the result of changing its sign (so TA is the conjugate of FA and viceversa). Sets of *signed* formulae will be denoted by S, U, V (possibly with subscripts). We shall use the upper case greek letters $\Gamma, \Delta, \ldots$ for sets of *unsigned* formulae. We shall often write S, X for $S \cup \{X\}$ and S, U for $S \cup U$. Given a formula A, the set of its *subformulae* is defined in the usual way. We shall call *subformulae of an s-formula* $\mathsf{s}A$ ($\mathsf{s} = T, F$) all the formulae of the form TB or FB where B is a subformula of A. For instance TA, TB, FA, FB will all be subformulae of $TA \vee B$.

DEFINITION 3. We say that an s-formula X is *satisfied* by a Boolean valuation V if $X = TA$ and $V(A) = 1$ or $X = FA$ and $V(A) = 0$. A *set* S of s-formulae is *satisfiable* if there is a Boolean valuation V which satisfies all its elements.

A set of s-formulae S is *explicitly inconsistent* if S contains both TA and FA for some formula A.

Sets of s-formulae correspond to the partial valuations of the previous section in the obvious way (we shall omit the adjective 'partial' from now on): given a set

[2] Namely partial valuations V and V' such that for some A in their common domain of definition $V(A) = 1$ and $V'(A) = 0$.

S of s-formulae its associated valuation is the valuation V_S defined as follows:

$$V_S(A) = \begin{cases} 1 & \text{if } TA \in S \\ 0 & \text{if } FA \in S \\ * & \text{otherwise} \end{cases}$$

(An explicitly inconsistent set of s-formulae is associated with the inconsistent valuation denoted by $\top$.) Conversely, given a partial valuation V its associated set of s-formulae will be the set S_V containing TA for every formula A such that $V(A) = 1$, and FA for every formula A such that $V(A) = 0$ (and nothing else).

The sets of s-formulae corresponding to semivaluations are known in the literature as *Hintikka sets* or *downward saturated sets.* The sets of s-formulae corresponding to Boolean valuations are often called *truth sets* or *saturated sets.*

The translation of Lemma 2 in terms of s-formulae is known as the (propositional) Hintikka lemma.

LEMMA 4 (Propositional Hintikka lemma). *Every propositional Hintikka set is satisfiable.*

In other words, every Hintikka set can be extended to a truth set.

We shall now define the notion of *expansion system* which generalizes the tableau method.

DEFINITION 5.

1. An *expansion rule* R *of type* $\langle n \rangle$, with $n \geq 1$, is a computable relation between sets of s-formulae and n-tuples of sets of s-formulae satisfying the following condition:

   $$R(S_0, (S_1, \ldots, S_n)) \Longrightarrow \text{for every truth-set } S, \text{ if } S_0 \subseteq S, \text{ then } S_i \subseteq S \text{ for some } S_i.$$

   If $n = 1$ we say that the rule is of *linear type*, otherwise we say that the rule is of *branching type.*

2. An *expansion system* $\mathbf{S}$ is a finite set of expansion rules.

3. We say that a set of s-formulae S' is an *expansion* of S under a rule R if S' belongs to the image of S under R.

4. Let R be an expansion rule of type $\langle n \rangle$. A set of s-formulae S is *saturated under* R or R*-saturated* if for every n-tuple $S_1, \ldots, S_n$ such that $R(S, (S_1, \ldots, S_n)$, we have that $S_i \subseteq S$ for at least one S_i.

5. Let $\mathbf{S}$ be an expansion system. A set of s-formulae S is $\mathbf{S}$*-saturated* if it is R-saturated for every rule R of $\mathbf{S}$.

The rules of an expansion system are intended as rules which help us in our search for a countermodel to a given sequent $\Gamma \vdash A$. When we apply these rules systematically, we construct a *tree* whose nodes are sets of s-formulae. The notion of **S**-*tree for* a given set of s-formulae S is defined inductively as follows:

DEFINITION 6.

1. S is an **S**-tree for S;
2. if $\mathcal{T}$ is an **S**-tree for S, and R is a rule of **S** such that $R(S, (S_1, \ldots, S_n))$, then the following tree is also an **S**-tree for S:

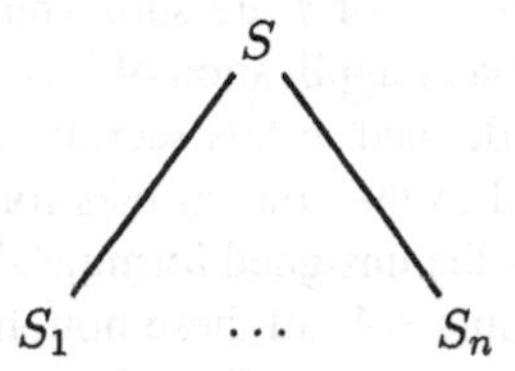

3. nothing else is an **S**-tree for S.

A branch ϕ of an **S**-tree $\mathcal{T}$ is *closed* if the set $\bigcup_{i=1}^{n} S_i$, where $S_1, \ldots, S_n$ are the sets of s-formulae associated with its nodes, is an explicitly inconsistent set. Otherwise, we say that ϕ is *open*. An **S**-*tree* $\mathcal{T}$ is *closed* if every branch of $\mathcal{T}$ is closed and *open* otherwise.

The rules of the expansion systems we shall consider in the sequel will have the following general form:

> *If a valuation V satisfies $\sigma(X_1), \ldots, \sigma(X_n)$, where $X_1, \ldots, X_n$ are s-formulae and σ is a uniform substitution, then V satisfies also all the formulae of one of the sets $\sigma(S_1), \ldots, \sigma(S_n)$, where $S_1, \ldots, S_n$ are* finite *sets of s-formulae depending on $X_1, \ldots, X_n$ and $\sigma(S_i)$ denotes the result of applying the substitution σ to every formula of S_i*

This formulation suggests that, whenever S is finite we can make **S**-trees 'slimmer' by representing an **S**-tree for S as a tree whose nodes are labelled with s-formulae rather than sets of s-formulae. First we generate the one-branch tree whose nodes are labelled with the s-formulae in S (taken in an arbitrary sequence). Then, at each application of an **S**-rule of type n to a branch ϕ we split ϕ into n distinct branches each of which extends ϕ by means of a sequence of s-formulae as prescribed by the rule.

So if we deal with such 'slim' **S**-trees, a typical **S**-rule can be represented as follows:

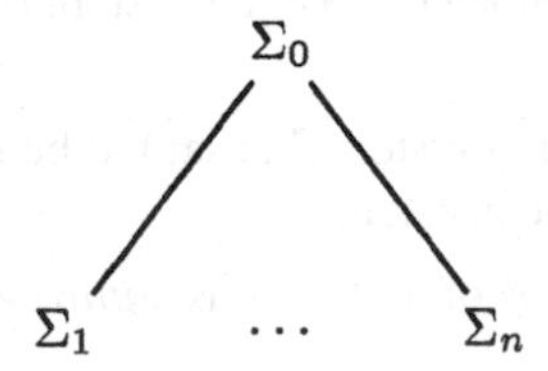

where $\Sigma_0, \Sigma_1, \ldots, \Sigma_n$ are vertical arrays of *schematic* s-formulae (i.e. schemas of s-formulae which can be instantiated by s-formulae). The array Σ_0 characterizes the branches ϕ to which the rule can be applied, namely those branches such that S_ϕ contains at least one instance of every schematic s-formula in Σ_0. The arrays $\Sigma_1, \ldots, \Sigma_n$ characterize, in a similar way, the extensions of ϕ resulting from an application of the rule.

Given a rule R of an expansion system **S**, we shall say that an application of R to a branch ϕ is *analytic* when it has the *subformula property*, i.e. if all the new s-formulae appended to the end of ϕ are subformulae of s-formulae occurring in ϕ. A *rule* R is *analytic* if every application of it is analytic.

In fact, all the notions defined in this section, as well as in the previous one, are by no means restricted to the context of s-formulae. If we interpret TA as meaning the same thing as the unsigned formula A and FA as meaning the same thing as the unsigned formula $\neg A$, all these notions can be reformulated in terms of unsigned formulae with minor modifications (which we leave to the reader). The only change which might have some significance from the technical viewpoint concerns the notion of *analytic* application of a rule. We shall say that an unsigned formula A is a *weak subformula* of an unsigned formula B if A is either a subformula of B or the negation of a subformula of B. If R is an expansion rule for *unsigned* formulae, we shall say that an application of R to a branch ϕ is *analytic* when it has the *weak subformula property*, that is when all the new unsigned formulae appended to the end of ϕ are weak subformulae of formulae occurring in ϕ. However, we shall often neglect this distinction and speak of subformulae and subformula property also when we mean weak subformulae and weak subformula property, provided this creates no confusion. Accordingly, we shall speak of 'formulae' to mean either s-formulae or unsigned formulae depending on the context.

DEFINITION 7. An expansion system **S** is said to be *confluent* if it has the following property:

Conf If Γ is an unsatisfiable set of formulae, every **S**-tree for Γ can be expanded into a closed one.

Confluence is a crucial property for an expansion system. It ensures that we never take 'the wrong path' in our attempt to construct a closed tableau. Whatever move we may make, it will not prevent us from succeeding if there is a closed tableau for the input set of formulae. In other words, the *order* in which the expansion rules are applied does not affect the final result.

Let us say that a set of formulae Γ is **S**-inconsistent if there exists a closed **S**-tree for Γ.

DEFINITION 8. An expansion system **S** is said to be *complete* if every unsatisfiable set of formulae Γ is **S**-inconsistent.

We now say that a *branch* ϕ of an **S**-tree is *saturated* if the set of all formulae occurring in ϕ is **S**-saturated. An **S**-*tree* $\mathcal{T}$ is said to be *completed* if every branch of $\mathcal{T}$ is either closed or saturated.

The following proposition establishes a sufficient condition for both confluence and completeness of an expansion system:

PROPOSITION 9. *An expansion system* **S** *is complete* and *confluent if it satisfies the following conditions:*

1. *for every set of formulae* Γ *there exists a completed* **S**-*tree for* Γ;
2. *if* ϕ *is an open* **S**-*saturated branch of an* **S**-*tree, the set* Γ_ϕ *of the formulae occurring in* ϕ *is a Hintikka set.*

Proof. Assume **S** satisfies conditions 1 and 2 above. Suppose a set Γ of formulae is **S**-consistent, that is every **S**-tree for Γ is open. By hypothesis there exists a completed **S**-tree, say $\mathcal{T}$, for Γ which must also be open. Hence this completed **S**-tree has an open **S**-saturated branch ϕ. Again by hypothesis, the set Γ_ϕ of the formulae occurring in ϕ is a Hintikka set, and therefore Γ_ϕ is satisfiable. Since $\Gamma \subseteq \Gamma_\phi$, it follows that Γ is also satisfiable. So **S** is complete. Now, let $\mathcal{T}$ be a tableau for an unsatisfiable set Γ of formulae and let ϕ be an open branch of $\mathcal{T}$. Let us denote by Γ_ϕ the set of all formulae occurring in ϕ. Since $\Gamma \subseteq \Gamma_\phi$, Γ_ϕ is also unsatisfiable. By 1 above, there is a completed **S**-tree for Γ_ϕ which must be closed. ■

2.3 Background on Computational Complexity

The subject of computational complexity can be seen as a refinement of the traditional theory of computability. The refinement, which is motivated by practical considerations and above all by the rapid development of computer science, consists of replacing the fundamental question, 'Is the problem P computationally solvable?' with the question, 'Is P solvable within bounded resources (time and space)?'. Workers in computational complexity agree in identifying the class of 'practically solvable' or 'feasible' problems with the class $\mathcal{P}$ of the problems that can be solved by a Turing machine within polynomial time, i.e. time bounded above by a polynomial in the length of the input.

Most computational problems can be viewed as language-recognition problems i.e. problems which ask whether or not a word over a given alphabet is a member of some distinguished set of words. For instance, the problem of deciding whether a formula of the propositional calculus is a tautology can be identified with the set TAUT of all the words over the alphabet of propositional calculus which express tautologies, and an algorithm which solves the problem is one which decides, given a word over the alphabet, whether or not it belongs to TAUT. So the class $\mathcal{P}$ can be described as the class of the languages which can be recognized in polynomial time by a Turing machine.

The rationale of this identification of feasible problems with sets in $\mathcal{P}$ is that, as the length of the input grows, exponential time algorithms require resources which quickly precipitate beyond any practical constraint. Needless to say, an exponential time algorithm may be preferable in practice to a polynomial time algorithm with running time, say, n^{1000}. However, the notion of polynomial time computability is theoretically useful because it is particularly robust: it is invariant under any reasonable choice of models of computation. In fact, there is an analog of the Church-Turing thesis in the field of computational complexity, namely the thesis that a Turing machine can simulate any 'reasonable' model of computation with at most a polynomial increase in time and space. Moreover polynomial time computability is invariant under any reasonable choice of 'encoding scheme' for the problem under consideration. Finally, 'natural problems', i.e. problems which arise in practice and are not specifically constructed in order to defy the power of our computational devices, seem to show a tendency to be either intractable or solvable in time bounded by a polynomial of reasonably low degree.

The analog of the class $\mathcal{P}$, when non-deterministic models of computation are considered, for example non-deterministic Turing machines,[3] is the class $\mathcal{NP}$ of the problems which are 'solved' in polynomial time by some non-deterministic algorithm. The class $\mathcal{NP}$ can be viewed as the class of all languages L such that, for every word $w \in L$, there is a 'short' proof of its membership in L, where 'short' means that the length of the proof is bounded above by some polynomial function of the length of w. (See [Garey and Johnson, 1979] and [Stockmeyer, 1987] for definitions in terms of non-deterministic Turing machines.) The central role played by propositional logic in theoretical computer science is related to the following well-known results [Cook, 1971; Cook and Reckhow, 1974]:

1. There is a deterministic polynomial time algorithm for the tautology problem if and only if $\mathcal{P} = \mathcal{NP}$.

2. There is a non-deterministic polynomial time algorithm for the tautology problem if and only if $\mathcal{NP}$ is closed under complementation.

As far as the first result is concerned, the theory of $\mathcal{NP}$-completeness[4] is providing growing evidence for the conjecture that $\mathcal{P} \neq \mathcal{NP}$, which would imply that no *proof procedure* can be uniformly feasible for the whole class of tautologies (it can of course be feasible for a number of infinite subclasses of this class).

The second result involves the notion of a *proof system* rather than the notion of a proof procedure. The following definitions are due to Cook and Reckhow [Cook and Reckhow, 1974] (Σ^* denotes the set of all finite strings or 'words' over the alphabet Σ):

DEFINITION 10. If $L \subset \Sigma^*$, a *proof system* for L is a function $f : \Sigma_1^* \mapsto L$ for

[3] See [Garey and Johnson, 1979] and [Stockmeyer, 1987].

[4] We refer the reader to [Garey and Johnson, 1979].

some alphabet Σ_1, where $f \in \mathcal{L}$ (the class of functions computable in polynomial time).

The condition that $f \in \mathcal{L}$ is intended to ensure that there is a feasible way, when given a string over Σ_1, of checking whether it represents a proof and what it is a proof *of.* So, for example, a proof system **S** is associated with a function f such that $f(x) = A$ if x is a string of symbols which represents a legitimate proof of A in **S**. If x does not represent a proof in **S**, then $f(x)$ is taken to denote some fixed tautology in L.

DEFINITION 11. A proof system f is *polynomially bounded* if there is a polynomial $p(n)$ such that for all $y \in L$, there is an $x \in \Sigma_1^*$ such that $y = f(x)$ and $|x| \leq p(|y|)$, where $|z|$ is the length of the string z.

This definition captures the idea of a proof system in which, for every element of L, there *exists* a 'short' proof of its membership in L. If a proof system is polynomially bounded, this does not imply (unless $\mathcal{P} = \mathcal{NP}$) that there is a proof procedure based on it (namely a deterministic version) which is polynomially bounded. On the other hand if a proof system is *not* polynomially bounded, *a fortiori* there is no polynomially bounded proof procedure based on it.

The question of whether a proof system is polynomially bounded or not is one concerning its *absolute* complexity. Most conventional proof systems for propositional logic have been shown *not* to be polynomially bounded by exhibiting for each system some infinite class of 'hard examples' which have no polynomial size proofs. One consequence of these results (for an overview see [Urquhart, 1995]), is that we should not expect a complete proof system to be feasible and should be prepared either to give up completeness and restrict our language in order to attain feasibility (this is the line chosen for some of the resolution-based applications), or to appeal to suitable *heuristics*, namely *fallible* 'strategies' to guide our proofs. In fact the results mentioned above imply that heuristics alone is not sufficient if we want to be able to obtain proofs expressible in some conventional system. So we should be prepared to use heuristics *and* give up completeness. However the importance of the complexity analysis of proof systems is by no means restricted to the $\mathcal{P}$ versus $\mathcal{NP}$ question. Nor should we conclude that all conventional systems are to be regarded as equivalent and that the only difference is caused by the heuristics that we use. On the contrary, besides the questions concerning the absolute complexity of proof systems, there are many interesting ones concerning their *relative* complexity which are computationally significant even when the systems have been proved intractable. As far as automated deduction is concerned, such questions of relative complexity may be relevant, before any heuristic considerations, to the choice of an appropriate formal system to start with.

In the study of the relative complexity of proof systems, we have to provide a *quantitative* version of some basic qualitative notions. Let **S** be a proof system for propositional logic. We write

$$\Gamma \overset{n}{\vdash}_{\mathbf{S}} A$$

to mean that there is a proof π of A from Γ in the system **S** such that $|\pi| \leq n$ (where $|\pi|$ denotes as usual the *length* of π intended as a string of symbols over the alphabet of **S**).

Suppose that, given two systems **S** and **S**′, there is a function g such that for all Γ, A:

$$(2) \quad \Gamma \overset{n}{\vdash}_{\mathbf{S}'} A \Longrightarrow \Gamma \overset{g(n)}{\vdash}_{\mathbf{S}} A$$

we are interested in the rate of growth of g for particular systems **S** and **S**′. Positive results about the above relation are usually obtained by means of *simulation procedures*:

DEFINITION 12. If $f_1 : \Sigma_1^* \mapsto L$ and $f_2 : \Sigma_2^* \mapsto L$ are proof systems for L, a *simulation* of f_1 in f_2 is a computable function $h : \Sigma_1^* \mapsto \Sigma_2^*$ such that $f_2(h(x)) = f_1(x)$ for all $x \in \Sigma_1^*$.

Negative results consist of *lower bounds* for the function g.

An important special case of the relation in (2) occurs when $g(n)$ is a polynomial in n. This can be shown by exhibiting a simulation function (as in definition 12) h such that for some polynomial $p(n)$, $|h(x)| \leq p(|x|)$ for all x. In this case **S** is said to *polynomially simulate*,[5] or shortly *p-simulate*, **S**′. The simulation function h is then a mapping from proofs in **S**′ to proofs in **S** which preserves feasibility: if **S**′ is a polynomially bounded system for L, so is **S** (where L can be any infinite subset of TAUT). The p-simulation relation is a quasi-ordering and its symmetric closure is an equivalence relation. We can therefore order proof systems and put them into equivalence classes with respect to their relative complexity. Systems belonging to the same equivalence class can be considered as having 'essentially' (i.e. up to a polynomial) the same complexity. On the other hand if **S** p-simulates **S**′, but there is no p-simulation in the reverse direction, we can say that, as far as the length of proofs is concerned, **S** is essentially more efficient than **S**′ — **S** is polynomially bounded for every $L \subset$ TAUT for which **S**′ is polynomially bounded but the opposite is not true; therefore **S** has a larger 'practical' scope than **S**′.

The study of the relative complexity of proof systems was started by Cook and Reckhow [1974; 1979]. Later on, some open questions were settled and new ones have been raised (see [Urquhart, 1995] for a survey and a map of the p-simulation relation between a number of conventional proof systems). Results expressed in terms of polynomial simulation, in the Cook and Reckhow tradition, refer to the relative length of *minimal* proofs in different systems and do not say much about the relative difficulty of *proof-search*. In other words, 'easy' proofs may be rather hard to find! A crucial consideration is therefore the size of the search space in

[5] Our definition of p-simulation is the same as the one used in [Buss, 1987] and is slightly different from the original one given, for instance, in [Cook and Reckhow, 1979]. However it is easy to see that the two definitions serve exactly the same purposes as far as the study of the relative complexity of proof systems is concerned.

which such' easy' proofs are to be found. In some cases we can define systematic procedures to explore the search space efficiently, so that a 'speed-up' in proof-length can be translated into a similar speed-up in proof-search. In other cases, we may not know of any such systematic procedure and the *existence* of shorter proofs may not help us developing more efficient proof procedures. So the significance of these results for automated deduction must be considered case by case.

3 SMULLYAN'S TABLEAUX

3.1 Tableau Expansion Rules

The tableau rules have been introduced in Fitting's chapter. For the reader's convenience we list them all in Table 1 both in their signed and unsigned versions. Let us consider the signed rules first. We can distinguish between *linear* rules,

Signed rules

$$\frac{TA\land B}{\begin{array}{l}TA\\ TB\end{array}} \qquad \frac{FA\land B}{FA \mid FB} \qquad \frac{TA\lor B}{TA \mid TB} \qquad \frac{FA\lor B}{\begin{array}{l}FA\\ FB\end{array}}$$

$$\frac{TA\to B}{FA \mid TB} \qquad \frac{FA\to B}{\begin{array}{l}TA\\ FB\end{array}} \qquad \frac{T\neg A}{FA} \qquad \frac{F\neg A}{TA}$$

Unsigned rules

$$\frac{A\land B}{\begin{array}{l}A\\ B\end{array}} \qquad \frac{\neg(A\land B)}{\neg A \mid \neg B} \qquad \frac{A\lor B}{A \mid B} \qquad \frac{\neg(A\lor B)}{\begin{array}{l}\neg A\\ \neg B\end{array}}$$

$$\frac{A\to B}{\neg A \mid B} \qquad \frac{\neg(A\to B)}{\begin{array}{l}A\\ \neg B\end{array}} \qquad \frac{\neg\neg A}{A}$$

Table 1. Tableau rules for signed and unsigned formulae

which apply to formulae of type α (in Smullyan's unifying notation, see Fitting's introduction), and *branching* rules which apply to formulae of type β. Some of the linear rules have two conclusions, which are asserted 'conjunctively', i.e. they are both true at the same time if the premise is true. We could, of course, describe these rules as pairs of single-conclusion rules. For instance the rules for eliminating true conjunctions is equivalent to the following pair of single-conclusion rules:

$$\frac{TA \wedge B}{TA} \qquad\qquad \frac{TA \wedge B}{TB}$$

In Smullyan's unifying notation, this amounts to decomposing the two-conclusion rule

$$\frac{\alpha}{\begin{array}{c}\alpha_1\\ \alpha_2\end{array}}$$

into the two single-conclusion rules:

$$\frac{\alpha}{\alpha_1} \qquad \frac{\alpha}{\alpha_2}$$

So, we can describe the tableau method as an example of expansion system, with eight rules of type $\langle 1 \rangle$ — those for eliminating the α-formulae — and three rules of type $\langle 2 \rangle$ — those for eliminating the β-formulae. All the rules are *analytic* in the sense explained above: every (signed) formula which occurs as a conclusion is a (signed) subformula of the (signed) formula which occurs as premiss of the rule.

3.2 Confluence and Completeness

THEOREM 13. *A set* Γ *of formulae is unsatisfiable if and only if there is a closed tableau for* Γ.

Proof. The if-direction of this theorem (*soundness* of tableaux) depends on the fact that the tableau expansion rules preserve the following property: any valuation satisfying all the formulae in a set S must also satisfy all the formulae contained in at least one branch of any tableau for S. Hence if a tableau for S is closed, no valuation can satisfy all the formulae in S.

The only-if direction (*completeness* of tableaux) follows from Proposition 9 above. To see why, observe that a branch ϕ of a tableau is saturated only if it satisfies the following two conditions:

1. For every α-formula in ϕ, both α_1 and α_2 also belong to ϕ.

2. For every β-formula in ϕ, either β_1 or β_2 also belongs to ϕ.

So, the set of formulae in an open saturated branch is a Hintikka set (see Chapter 1). We now show that for *every* set Γ of formulae we can construct a completed tableau for it. This will imply, by Proposition 9 that the tableau method is both confluent and complete.

Let us then say that a node n is *fulfilled* if (1) it is an atomic s-formula or (2) it is of type α and both α_1 and α_2 occur in all the branches passing through n or (3) it is of type β and for every branch ϕ passing through the node either β_1 or β_2 occurs in ϕ. Clearly a tableau is completed if and only if every node in it is fulfilled.

A simple procedure for constructing a completed tableau is the following one (from [Smullyan, 1968, [pp. 33–34]). Let Γ be a set of s-formulae arranged in a denumerable sequence $X_1, X_2, \ldots$. Start the tree with X_1. This node constitutes the *level 1*. Then fulfil[6] the origin and append X_2 to every open branch. Call all the nodes so obtained *nodes of level 2*. At the i-th step fulfil all the nodes of level $i-1$ and append X_i to the end of each open branch. So every node gets fulfilled after a finite number of steps. The procedure either terminates with a closed tableau or runs forever. (Notice that if Γ is finite, the procedure always terminates). In the latter case we 'obtain' an *infinite* tree which is a completed tableau for Γ. ■

3.3 Tableaux as a Decision Procedure

The *Decision Problem for Propositional Logic* is the set of all questions 'is B a consequence of Γ?' (where B is a formula and Γ a finite set of formulae) or, more in general, 'is the sequent $\Gamma \vdash \Delta$ true?' A *decision procedure* is a systematic method, or *algorithm*, to answer all these questions. We have seen that in classical logic this problem is reduced to the *satisfiability* problem, i.e. to the set of questions 'is Γ satisfiable?' for any finite set of formulae Γ

Notice that when the original set Γ is finite, the procedure described in the previous section is indeed a decision procedure. It always terminates in a finite number of steps either with a closed tableau, if Γ is unsatisfiable, or with an open completed tableau, if Γ is satisfiable. If Γ is allowed to be infinite, it is a *semi-decision* procedure, i.e. it is guaranteed to terminate only if S is unsatisfiable, while if S is satisfiable, it runs forever.

3.4 Tableaux and the Sequent Calculus

Tableaux have been introduced in Chapter 1 which also mentions their close relation to Gentzen's sequent calculus. Here we elaborate on this topic and show how the tableau rules correspond to a semantic interpretation of (a suitable variant of) Gentzen's sequent rules.

Invertible Sequent Calculi

Gentzen introduced the sequent calculi **LK** and **LJ** as well as the natural deduction calculi **NK** and **NJ** in his famous 1935 paper [Gentzen, 1935]. Apparently he considered the sequent calculi as technically more convenient for metalogical investi-

[6] By 'fulfilling a node' we mean applying the appropriate tableau rule so that the node becomes fulfilled.

gation.[7] In particular he thought that they were 'especially suited to the purpose' of proving the *Hauptsatz* [his 'Fundamental Theorem'] and that their form was 'largely determined by considerations connected with [this purpose]'.[8] He called these calculi 'logistic' because, unlike the natural deduction calculi, they do not involve the introduction and subsequent discharge of assumptions, but deal with formulae which are 'true *in themselves*, i.e. whose truth is no longer *conditional* on the truth of certain assumption formulae'.[9] Such 'unconditional' formulae are *sequents* (see p. 47 above), i.e. expressions of the form

(3) $A_1, \ldots, A_n \vdash B_1, \ldots, B_m$

In the case of intuitionistic logic the succedent (the expression to the right of the turnstile) may contain at most one formula. In this chapter we shall consider the classical system and focus on the *propositional* rules.

Although Gentzen considered the antecedent and the succedent as *sequences*, it is often more convenient to use *sets*, which eliminates the need for 'structural' rules to deal with permutations and repetitions of formulae.[10] Table 2 shows the rules of Gentzen's **LK** once sequences have been replaced by sets (we use Γ, Δ, etc. for sets of formulae and write Γ, A as an abbreviation of $\Gamma \cup \{A\}$). A proof of a sequent $\Gamma \vdash \Delta$ consists of a tree of sequents built up in accordance with the rules and on which all the leaves are axioms. Gentzen's celebrated *Hauptsatz* says that the cut rule can be eliminated from proofs. This obviously implies that the cut-free fragment is complete. Furthermore, one can discard the last structural rule left — the thinning rule — and do without structural rules altogether without affecting completeness, provided that the axioms are allowed to have the more general form

$$\Gamma, A \vdash \Delta, A.$$

This well-known variant corresponds to Kleene's system **G4** [Kleene, 1967, chapter VI].

Gentzen's rules have become a paradigm both in proof-theory and in its applications. This is not without reason. First, like the natural deduction calculi, they provide a precise analysis of the logical operators by specifying how each operator can be introduced in the antecedent or in the succedent of a sequent.[11] 'Second, their form ensures the validity of the *Hauptsatz*: each proof can be transformed

[7] [Gentzen, 1935], p. 69.

[8] [Gentzen, 1935, p. 89].

[9] [Gentzen, 1935], p. 82.

[10] This reformulation is adequate for classical and intuitionistic logic, but not if one wants to use sequents for some other logic, like relevance or linear logic, in which the number of occurrences of formulae counts. For such logics the antecedent and the succedent are usually represented as *multisets*; see [Thistlewaite *et al.*, 1988] and [Avron, 1988].

[11] Whereas in the natural deduction calculi there are, for each operator, an introduction and an elimination rule, in the sequent calculi there are only introduction rules and the eliminations take the form of introductions in the antecedent. Gentzen seemed to consider the difference between the two formulations as a purely technical aspect (see [Sundholm, 1983]). He also suggested that the rules of the

Axioms

$$A \vdash A$$

Structural rules

$$\frac{\Gamma \vdash \Delta}{\Gamma, \Theta \vdash \Delta, \Lambda}\text{[Thinning]} \qquad \frac{\Gamma, A \vdash \Delta \quad \Gamma \vdash \Delta, A}{\Gamma \vdash \Delta}\text{[Cut]}$$

Operational rules

$$\frac{\Gamma, A \vdash \Delta \quad \Gamma, B \vdash \Delta}{\Gamma, A \vee B \vdash \Delta}\text{[I-}\vee\text{left]} \qquad \frac{\Gamma \vdash \Delta, A \quad \Gamma \vdash \Delta, B}{\Gamma \vdash \Delta, A \wedge B}\text{[I-}\wedge\text{right]}$$

$$\frac{\Gamma, A, B \vdash \Delta}{\Gamma, A \wedge B \vdash \Delta}\text{[I-}\wedge\text{left]} \qquad \frac{\Gamma \vdash \Delta, A, B}{\Gamma \vdash \Delta, A \vee B}\text{[I-}\vee\text{right]}$$

$$\frac{\Gamma \vdash \Delta, A \quad \Gamma, B \vdash \Delta}{\Gamma, A \rightarrow B \vdash \Delta}\text{[I-}\rightarrow\text{left]} \qquad \frac{\Gamma, A \vdash \Delta, B}{\Gamma \vdash \Delta, A \rightarrow B}\text{[I-}\rightarrow\text{right]}$$

$$\frac{\Gamma \vdash \Delta, A}{\Gamma, \neg A \vdash \Delta}\text{[I-}\neg\text{left]} \qquad \frac{\Gamma, A \vdash \Delta}{\Gamma \vdash \Delta, \neg A}\text{[I-}\neg\text{right]}$$

Table 2. Sequent rules for classical propositional logic

into one which is cut-free, and cut-free proofs enjoy the *subformula property*: every sequent in the proof tree contains only subformulae of the formulae in the sequent to be proved.

From a conceptual viewpoint this property represents the notion of a purely *analytic* or 'direct' argument[12]: 'no concepts enter into the proof other than those contained in its final result, and their use was therefore essential to the achievement of that result' [Gentzen, 1935, p. 69], so that 'the final result is, as it were, gradually built up from its constituent elements' [Gentzen, 1935, p.88]. Third, a cut-free system, like Kleene's **G4**, seems to be particularly suited to a 'backward' search

natural deduction calculus could be seen as *definitions* of the operators themselves. In fact he argued that the introduction rules alone are sufficient for this purpose and that the elimination rules are 'no more, in the final analysis, than consequences of these definitions' [Gentzen, 1935, p. 80]. He also observed that this 'harmony' is exhibited by the intuitionistic calculus but breaks down in the classical case. For a thorough discussion of this subtle meaning-theoretical issue the reader is referred to the writings of Michael Dummett and Dag Prawitz, in particular [Dummett, 1978] and [Prawitz, 1978].

[12] On this point see [Statman, 1977].

for proofs: instead of going from the axioms to the endsequent, one can start from the endsequent and use the rules in the reverse direction, going from the conclusion to suitable premises from which it can be derived.

This method, which is clearly reminiscent of Pappus' 'theoretical analysis',[13] works only in virtue of an important property of the rules of **G4** which is described in Lemma 6 of [Kleene, 1967], namely their *invertibility*.

DEFINITION 14. A rule is *invertible* if the provability of the sequent below the line in each application of the rule implies the provability of all the sequents above the line.

As was early recognized by the pioneers of Automated Deduction[14], if a logical calculus has to be employed for this kind of 'backward' proof-search it is important that its rules be invertible: this allows us to stop as soon as we reach a sequent that we can recognize as unprovable (for instance one containing only atomic formulae and in which the antecedent and the succedent are disjoint) and conclude that the initial sequent is also unprovable. We should notice that the absence of the thinning rule is crucial in this context. In fact, it is easy to see that *the thinning rule is not invertible*: the provability of its conclusion does not imply, in general, the provability of the premise.

As far as classical logic is concerned, a system like **G4** admits of an interesting semantic interpretation. Let us say that a sequent $\Gamma \vdash \Delta$ is *valid* if every situation (i.e. a Boolean valuation) which makes all the formulae in Γ true, also makes true at least one formula in Δ. Otherwise if some situation makes all the formulae in Γ true and all the formulae in Δ false, we say that the sequent is *falsifiable* and that the situation provides a *countermodel* to the sequent.

According to this semantic viewpoint we prove that a sequent is valid by ruling out all possible falsifying situations. So a sequent $\Gamma \vdash \Delta$ represents a *valuation problem*: find a Boolean valuation which falsifies it. The *soundness* of the rules ensures that a valuation which falsifies the conclusion must also falsify at least one of the premises. Thus, if applying the rules backwards we reach an axiom in every branch, we are allowed to conclude that no falsifying valuation is possible (since no valuation can falsify an axiom) and that the endsequent is therefore valid. On

[13]The so-called 'method of analysis' was largely used in the mathematical practice of the ancient Greeks, its fullest description can be found in Pappus (3rd century A.D.), who writes:

> Now analysis is a method of taking that which is sought as though it were admitted and passing from it through its consequences in order, to something which is admitted as a result of synthesis; for in analysis we suppose that which is sought be already done, and we inquire what it is from which this comes about, and again what is the antecedent cause of the latter, and so on until, by retracing our steps, we light upon something already known or ranking as a first principle; and such a method we call analysis, as being a solution backwards ([Thomas, 1941] pp. 596–599.).

This is the so-called *directional* sense of analysis. The idea of an 'analytic method', however, is often associated with another sense of 'analysis' which is related to the 'purity' of the concepts employed to obtain a result and, in the framework of proof-theory, to the *subformula principle* of proofs.

[14]See for instance [Matulis, 1962].

the other hand, the *invertibility* of the rules allows us to stop as soon as we reach a falsifiable sequent and claim that any falsifying valuation provides a countermodel to the endsequent. Again, if the thinning rule were allowed, we would not be be able, in general, to retransmit falsifiability back to the endsequent. So, if employed in bockward search procedures, the thinning rule may result in the loss of crucial semantic information.

The construction of an analytic tableau for a sequent $\Gamma \vdash \Delta$ closely corresponds to the systematic search for a countermodel outlined above except that Smullyan's presentation uses trees of formulae instead of trees of sequents. Their correspondence with the rules of invertible sequent calculi is straightforward and is illustrated in [Smullyan, 1968] and [Smullyan, 1968b] and [Sundholm, 1983].

3.5 *Tableaux and Natural Deduction*

Is Natural Deduction Really 'Natural'?

Gentzen's natural deduction was intended by his author as a deductive system which 'comes as close as possible to actual reasoning'[15], and so it is still regarded by most of the logic community, as shown by the prominent role it plays in most undergraduate courses (see the Appendix for a listing of the natural deduction rules). On the other hand, there is considerable accordance in regarding natural deduction as unsuitable for automated deduction, because it does not appear to share the 'mechanical' character of other proof systems such as tableaux and resolution. As has been convincingly argued by Wilfried Sieg, this widespread opinion is to a large extent the result of a prejudice and (normal) natural deduction *does admit*, at least in principle, of proof-search strategies comparable to those adopted in more popular approaches to automated deduction (see [Sieg, 1993]).

Our main argument in favour of tableau methods versus natural deduction methods is of a different nature and puts in question the received view that the latter are more 'natural' than the former. In fact, natural deduction rules do *not* capture the *classical meaning* of the logical operators. This was already remarked in [Prawitz, 1965], where the author pointed out that natural deduction rules are nothing but a reading of Heyting's explanations of the *constructive meaning* of the logical operators [Heyting, 1956].[16] In particular the rules for the conditional are based

[15][Gentzen, 1935], p. 68.

[16]In the classical system of natural deduction, as argued in [Prawitz, 1971], '*classical* deductive operations are then analysed as consisting of the constructive ones plus a principle of indirect proof for atomic sentences'(p. 244), and 'one may doubt whether this is the proper way of analysing classical inferences'(p. 244–245). Prawitz suggested that a good candidate for this role could be the classical sequent calculus, since its rules are 'closer to the classical meaning of the logical constants'(p. 245). More recently this criticism of natural deduction has been taken up and extended in [Girard *et al.*, 1989] and [Cellucci, 1992].

on an interpretation of this operator which has nothing to do with the classical interpretation based on the truth-tables.[17]

This mismatch between the natural deduction rules and the classical meaning of the logical operators is responsible for the fact that rather simple classical tautologies become quite hard to prove within the natural deduction framework, in the sense that their simplest proofs must involve 'tricks' which are far from being natural.

Let us consider, for instance, one of *de Morgan's laws*

$$\neg(P \wedge Q) \rightarrow (\neg P \vee \neg Q)$$

and try to prove it in the classical natural deduction system with the *classical reductio ad absurdum* rule. It is not difficult to check that the simplest proof one can obtain is the following:

$$\dfrac{\dfrac{\neg(P \wedge Q)^1 \qquad \dfrac{\dfrac{\dfrac{\neg(\neg P \vee \neg Q)^2 \quad \dfrac{\neg P^3}{\neg P \vee \neg Q}\,3}{\mathbf{F}}\,2,3}{P}\,2 \qquad \dfrac{\dfrac{\neg(\neg P \vee \neg Q)^2 \quad \dfrac{\neg Q^4}{\neg P \vee \neg Q}\,4}{\mathbf{F}}\,2,4}{Q}\,2}{P \wedge Q}\,2}{\dfrac{\mathbf{F}}{\neg P \vee \neg Q}\,1}\,2,1}{\neg(P \wedge Q) \rightarrow (\neg P \vee \neg Q)}\,\emptyset$$

where the numerals on the right of the inference lines mean that the conclusion depends on the assumptions marked with the same numerals.

Notice that we obtain a rather unnatural deduction whatever variant of 'natural' deduction we may decide to use, although probably the least bad of all is obtained by means of the variant which *explicitly* incorporates the principle of bivalence in the form of the 'classical dilemma' rule:

[17] This does not apply, however, to the multi-conclusion system proposed by [Cellucci, 1992] which is, in fact, a mixture of natural deduction and sequent calculus.

$$
\dfrac{\dfrac{\dfrac{\dfrac{\dfrac{\neg(P\wedge Q)^1 \quad \dfrac{P^2 \quad Q^3}{P\wedge Q}\,{\scriptstyle 2,3}}{\mathbf{F}}\,{\scriptstyle 1,2,3}}{\neg P}\,{\scriptstyle 1,3}}{\neg P\vee\neg Q}\,{\scriptstyle 1,3} \qquad \dfrac{\neg Q^3}{\neg P\vee\neg Q}\,{\scriptstyle 3}}{\neg P\vee\neg Q}\,{\scriptstyle 1}}{\neg(P\wedge Q)\rightarrow(\neg P\vee\neg Q)}\,{\scriptstyle\emptyset}
$$

Another example of a tautology whose natural deduction proofs are extremely contrived is Peirce's law: $((P\rightarrow Q)\rightarrow P)\rightarrow P$. The reader can verify that any attempt to prove this classical tautology within the natural deduction framework leads to logical atrocities. Here is a typical proof:

$$
\dfrac{\dfrac{\dfrac{\dfrac{(P\rightarrow Q)\rightarrow P^1 \qquad \dfrac{\dfrac{\dfrac{\neg P^2 \quad P^3}{\mathbf{F}}\,{\scriptstyle 2,3}}{Q}\,{\scriptstyle 2,3}}{P\rightarrow Q}\,{\scriptstyle 2}}{P}\,{\scriptstyle 1,2} \qquad \neg P^2}{\mathbf{F}}\,{\scriptstyle 1,2}}{P}\,{\scriptstyle 1}}{((P\rightarrow Q)\rightarrow P)\rightarrow P}\,{\scriptstyle\emptyset}
$$

It should be emphasized that the main problem here does not lie in the complexity of proofs (i.e. their length with respect to the length of the tautology to be proved): in fact, if we look at the class of tautologies which generalizes the de Morgan law proved above, namely $T_i = \neg(P_1\wedge\ldots\wedge P_i)\rightarrow\neg P_1\vee\ldots\vee\neg P_i$, it is easy to verify that the length of their shortest proofs is linear in the length of the tautology under consideration, both in the system with the classical dilemma and in the system with classical reductio. It is rather their *contrived character* which is disturbing, and this depends on the fact that the natural deduction rules do not capture the *classical* meaning of the logical operators. Since these rules are closely related to the *constructive* meaning of the operators, we would not expect such logical atrocities when we use them to prove *intuitionistic* tautologies. (Notice that both the de Morgan law we have considered above and Peirce's law are *not* intuitionistically valid.) Consider, for instance, the other (intuitionistically valid) de Morgan law:

$$\neg P\vee\neg Q\rightarrow\neg(P\wedge Q).$$

One can easily obtain a 'really natural' intuitionistic proof of this tautology as follows:

$$\dfrac{\dfrac{\neg P \vee \neg Q^1 \qquad \dfrac{\dfrac{\neg P^3 \qquad \dfrac{P \wedge Q^2}{P}\,2}{\mathbf{F}}\,2,3}{\neg(P \wedge Q)}\,3 \qquad \dfrac{\dfrac{\neg Q^5 \qquad \dfrac{P \wedge Q^4}{Q}\,4}{\mathbf{F}}\,4,5}{\neg(P \wedge Q)}\,5}{\neg(P \wedge Q)}\,1}{\neg P \vee \neg Q \to \neg(P \wedge Q)}\,\emptyset$$

These examples illustrate our claim that *'natural' deduction is really natural only for intuitionistic tautologies* (and not necessarily for all of them).[18] The difficulties of natural deduction with classical tautologies depend on the fact that a sentence and its negation are not treated symmetrically, whereas this is exactly what one would expect from the classical meaning of the logical operators.

The tableau rules, on the contrary, treat a sentence and its negation symmetrically. For instance, they allow for a very simple and natural proof of the non-intuitionistic de Morgan law:

$$\begin{array}{c}
\neg(\neg(P \wedge Q) \to (\neg P \vee \neg Q)) \\
\neg(P \wedge Q) \\
\neg(\neg P \vee \neg Q) \\
\neg\neg P \\
\neg\neg Q \\
P \\
Q \\
/ \quad \backslash \\
\neg P \qquad \neg Q \\
\times \qquad \times
\end{array}$$

We stress once again that it is not the length of proofs which is at issue here. Although the tableau proofs for the class of tautologies which generalizes this de Morgan law are, strictly speaking, shorter than the corresponding natural deduction proofs, the difference in length is negligible, since in both cases one obtains proofs whose length is linear in the length of the given tautology. What is at issue is that the tableau proofs are more natural than those based on the 'natural' deduction rules, as a result of the symmetrical treatment of sentences with respect to their

[18] On this point see [Girard *et al.*, 1989] and [Cellucci, 1992].

negations. As we shall see, however, the symmetry of the tableau rules is by no means sufficient to guarantee, in general, that such rules *always* lead to proofs which are natural from the classical point of view.

As for the relationship between natural deduction proofs and tableau refutations, the previous discussion strongly suggests that no straightforward step-by-step simulation procedure should be available, owing mainly to the very different treatment of negation in these two systems. However, one can indicate a route which leads from a natural deduction proof to a corresponding tableau refutation via the equivalence of both to a suitable proof in the sequent calculus. A detailed discussion of this correspondence can be found in [Scott, 1981] and [Sundholm, 1983], which makes a crucial use of the cut rule and of its admissibility in the sequent calculus without cut. Indeed one would expect tableau refutations and *normal* natural deduction proofs to correspond to each other in a more direct way, without having to appeal to the power of the cut rule (and of its admissibility). Despite some optmistic remarks contained in Beth's seminal paper on tableaux ([Beth, 1955]), the problem is one of the trickiest in the area of proof transformation. The weak link is the correspondence between the cut-free sequent calculus and normal natural deduction which is far from being straightforward. A significant step towards bridging the gap has been made by Sieg's *intercalation calculus*, a variant of the sequent calculus obtained by reversing the natural deduction rules, which can be used to search for a natural deduction proof in a mechanical way by means of a procedure very similar to the tableau saturation procedure (see [Sieg, 1993] for the details).

3.6 Tableaux and Resolution

By a *literal* l we mean, as usual, a formula which is either a propositional variable (*positive* literal) or the negation of a propositional variable (*negative* literal). A clause is a disjunction of literals, e.g. $l_1 \vee \cdots \vee l_k$. Given the commutativity and idempotence of disjunction, a clause can be regarded as a *set* of literals and denoted simply by listing its elements; for example, the clause $l_1 \vee l_2 \vee l_3$ can be written as $l_1 l_2 l_3$. We shall also write AB to denote the concatenation of clauses A and B. The *complement* of a literal l, denoted by l' is equal to $\neg P$ if $l = P$ and to P if $l = \neg P$.

At the propositional level, the *resolution rule* amounts to the following basic inference:

$$\frac{\begin{array}{c}Al\\ Bl'\end{array}}{AB}$$

which is nothing but a simple form of the familiar cut rule. (To see this, simply think of a clause A as a sequent in which the antecedent is the list of the negative literals of A and the succedent is the list of its positive literals; then the resolution rule is exactly the cut rule.) Its first formulations date back to the origin of modern

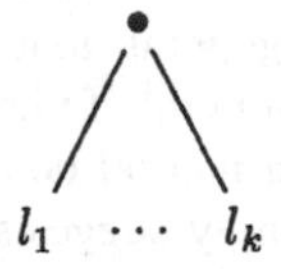

Figure 1. Tableau rule for a clause $l_1 \vee \cdots \vee l_k$

logic[19]. However, the resolution rule has been widely known under this name since the publication of Robinson's epoch-making paper ([Robinson, 1965]). A *resolution refutation* of a set of clauses Γ consists in a derivation of the *empty clause*, denoted by Λ, starting from the clauses in Γ and using the resolution rule as the only rule of inference. Such derivations can take different formats, namely that of *sequences* or *trees* of clauses. The more basic version is the one which yields refutations in tree format. This way of representing resolution refutations is clearly the least concise, but perhaps the more perspicuous. In the sequel we shall follow the exposition in [Urquhart, 1995].

A *tree resolution* refutation for a set of clauses Γ is a binary tree where each leaf is labelled with a clause of Γ, each interior node with immediate successors labelled Al and Bl' (where l is any literal, A and B are clauses)is labelled with AB, and the root is labelled with Λ (the emtpy clause). We shall define the size of a tree resolution refutation as the number of leaves in the tree. This is clearly a good measure since the total number of nodes in the tree is given by $2N - 1$, where N is the number of leaves.

Smullyan's tableaux can also be used very perspicuously to generate refutations of sets of clauses. However, we shall see that, without appropriate enhancements, they cannot p-simulate resolution refutations even in tree format. In order to refute sets of clauses we can use a version of Smullyan's tableaux with only one decomposition rule which is an adaptation of the standard β-decomposition (see Figure 1). Given a set Γ of clauses, a tableau for Γ is a tree in which the interior nodes are associated with clauses in Γ as follows: if a node n is associated with a given clause A, then its children are labelled with the literals in A. Notice that the node associated with a clause A is *not* labelled with A. Indeed, the clause associated with a node n is the disjunction of the literals labelling the children of n. The origin of the tree is labelled with the whole set Γ of input clauses. In Figure 2 we show a tableau for the set of clauses $\{P \vee Q, \neg P \vee R, \neg P \vee \neg R, \neg Q \vee S, \neg Q \vee \neg S\}$

[19]See for instance [Schroeder-Heister, 1997], which finds its anticipation in some posthumous papers by Gottlob Frege.

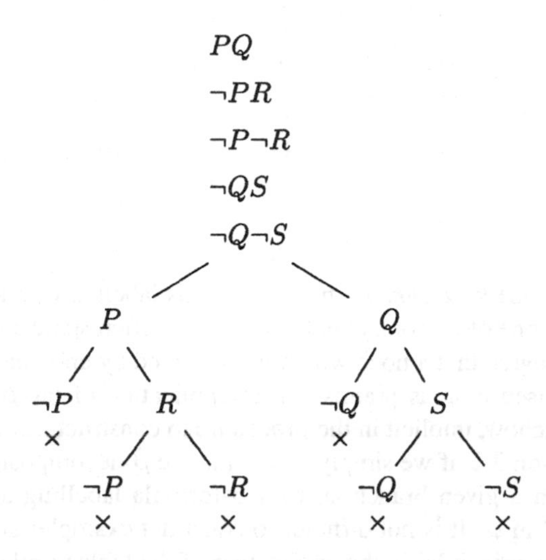

Figure 2. A tableau for the set of clauses $\{P \vee Q, \neg P \vee R, \neg P \vee \neg R, \neg Q \vee S, \neg Q \vee \neg S\}$

3.7 Proof-search with Smullyan's Tableaux

Regularity

Suppose that in the construction of a tableau $\mathcal{T}$ we come across a branch ϕ containing both a node k labelled with a formula β and and a node m labelled with β_i (with $i = 1$ or $i = 2$). Among the many useless moves that one can make when applying the tableau rules mechanically, one of the worst consists in applying the β-decomposition rule to the formula β in ϕ, because one of the two branches generated by the rule application will contain another copy of β_i. So, eventually, the sub-tableau generated by the formulae in ϕ will have the following form:

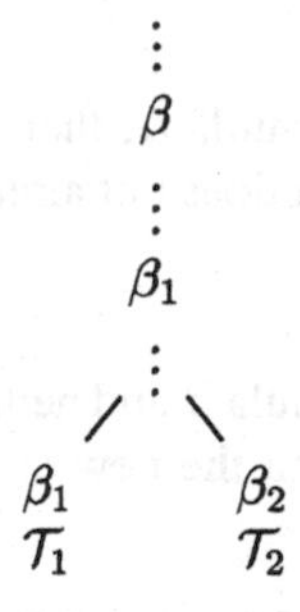

Now, if this sub-tableau is a completed tableau for Γ_ϕ (i.e. the set of formulae in the original ϕ), so is the following more concise one:

$$\begin{array}{c} \vdots \\ \beta \\ \vdots \\ \beta_1 \\ \vdots \\ \mathcal{T}_1 \end{array}$$

Let us say that a node k *subsumes* a node m if m is labelled with a formula β and k is labelled with one of its components β_i. We can often spare a large number of branchings if we agree that a node which is subsumed by another node in a given branch ϕ is *not* used in ϕ as premiss of an application of the β-decomposition rule. This is, somehow, implicit in the procedure to construct a completed tableau described in Section 3.2, if we simply agree that the β-decomposition rule should not be applied, in a given branch ϕ, to a β-formula labelling a node which is already 'fulfilled' in ϕ. It is not difficult to construct examples showing that this obvious 'economy principle' in the application of the tableau rules may, on some occasions, shorten proofs (or in general completed tableaux) by an exponential factor.

The restriction on the structure of a tableau that we have just described is usually incorporated into a more general restriction called *regularity*. A tableau is said to be *regular* if no formula occurs more than once in each branch. Indeed, it seems that not only would regularity stop us from applying the β-decomposition rule to a node which is 'subsumed' by another node, but it would also stop us from applying the α-decomposition rule, interpreted as a two-conclusion rule, whenever just one of the α_i is already in the branch. This would clearly affect completeness, for instance it would make it impossibile to generate a closed tableau for the inconsistent formula $(P \wedge Q) \wedge (P \wedge \neg Q)$. However, there are several obvious ways of imposing the regularity condition without affecting completeness. One way consists in interpreting the α-decomposition rule as a *pair* of one-conclusion rules, as suggested in Section 3.1. Another way consists in wording the tableau rules so that regularity is satisfied by definition. For instance, the α rule can be reformulated as follows:

> if a branch ϕ contains a formula α, then ϕ can be expanded by appending to it every α_i which does not already occur in it.

And the β rule as follows:

> if a branch ϕ contains a formula β and neither β_1 nor β_2 already occur in it, then ϕ can be split into the new expanded branches ϕ, β_1 and ϕ, β_2.

Again, the systematic procedure of Section 3.2 can be construed as implicitly incorporating the regularity restriction, if by 'fulfilling a node' we mean applying to it the relevant rule in the way just described. It is not difficult to turn the informal argument given above into a proof showing that closed tableaux of minimal size are regular, i.e.

FACT 15. If Γ is an unsatisfiable set of formulae, every minimal size closed tableau for Γ is regular.

As far as the propositional rules are concerned, the regularity restriction implies the basic condition called *strictness*, namely that no formula is used more than once as a premiss of a rule application. Things are different in the first order case for which we refer the reader to Letz's chapter in this *Handbook*.

Priority Strategies

The simplest priority strategy was presented by Smullyan in his well-known book on tableaux [Smullyan, 1968]. It consists in giving priority to α-formulae over β-formulae in the expansion of a tableau. The reason is simple: α-decompositions do not create new branches, so if we execute them, whenever possible, before the β-decompositions, the resulting trees will be much 'slimmer' and avoid a great deal of duplication. To see this it is sufficient to observe the examples in Figures 3 and 4.

Another priority strategy inspired by the same principle, namely that of avoiding branching whenever possible, consists in giving priority to α-decompositions and to β-decompositions applied to β's such that either β_1' or β_2' (the complements of β_1 and β_2) also occurs in the same branch. Obviously one of the two branches generated by such β-decompositions will close immediately.

Notice that this simple strategy— that we might call, following [Vellino, 1989], *clash priority strategy*— is equivalent to allowing the following 'derived' rules with two premisses:

$$\frac{\begin{array}{c}\beta\\ \beta_1'\end{array}}{\beta_2} \qquad \frac{\begin{array}{c}\beta\\ \beta_2'\end{array}}{\beta_1} .$$

Again, it is easy to observe that the clash priority strategy may enable us to shorten considerably the resulting tableaux.

Restriction Strategies

Let us focus on clausal tableaux. A plausible strategy to reduce the size of refutations seems that of imposing that each clause decomposition (except the first one) closes a branch in the tableau. This amounts to requiring that each clause decomposed by an application of the tableau rule at a given node k labelled with a literal contains a literal which 'clashes', i.e. is the complement of, the literal labelling an ancestor of k. A tableau constructed in accordance with this restriction is called *path connected* (it is also called *ancestor clash restricted* in [Vellino, 1989]).

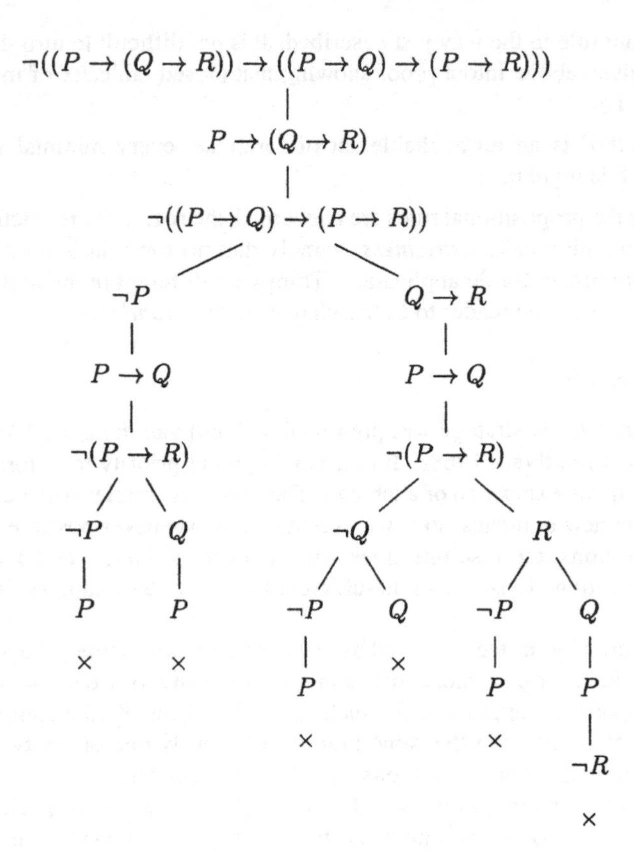

Figure 3. This tableau is constructed by decomposing the formulae in the order in which they appear in the tree, without giving priority to the α-decompositions

The example in Figure 5 shows that the path connection strategy may lead on some occasions to more concise refutations. This is clearly the case when the input set of clauses S is not minimally inconsistent. The path connection strategy does not affect completeness, but it does affect confluence. To see why, suppose we tried to construct the path connected tableau of Figure 5 starting with the decomposition of R, S. If the path connection restriction were applied, the construction of the tableau would terminate without generating a closed tableau. So the *order* in which the rules are applied is crucial.

One might be tempted to think that, in order to restore confluence, it is sufficient to remove 'redundant' clauses (such as the clause RS in our previous example) from the input set by observing that, if a clause C in the input set S contains a 'pure literal', i.e. a literal l which is not complemented in any other

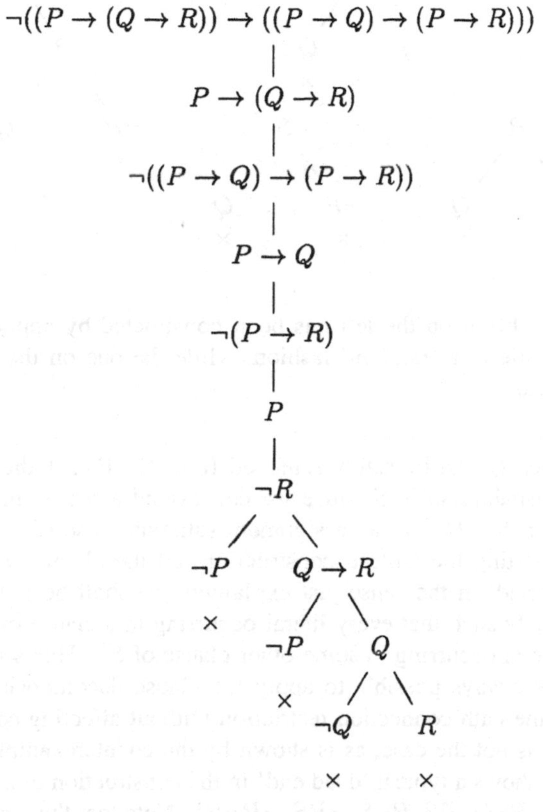

$$\neg((P \to (Q \to R)) \to ((P \to Q) \to (P \to R)))$$

$$P \to (Q \to R)$$

$$\neg((P \to Q) \to (P \to R))$$

$$P \to Q$$

$$\neg(P \to R)$$

$$P$$

$$\neg R$$

$$\neg P \quad\quad Q \to R$$

$$\times \quad\quad \neg P \quad\quad Q$$

$$\times \quad\quad \neg Q \quad\quad R$$

$$\times \quad\quad \times$$

Figure 4. This tableau is constructed giving priority to the α-decompositions over the β-decompositions

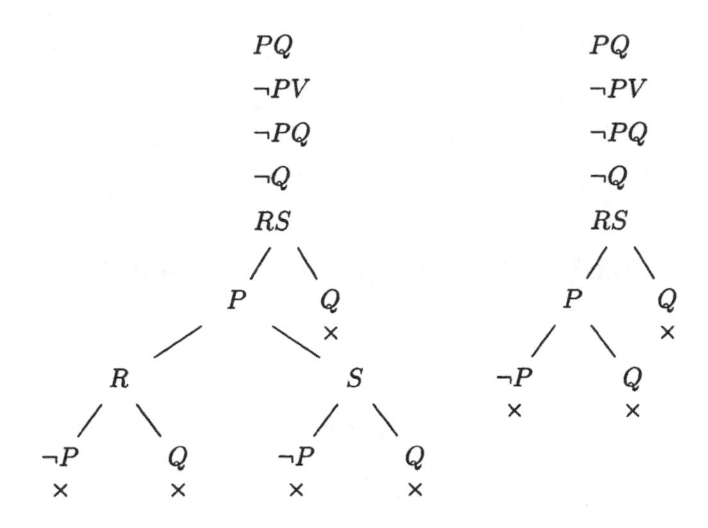

Figure 5. The tableau on the left has been constructed by applying the clause decomposition rule in a 'random' fashion, while the one on the right is a path connected tableau

clause of S, then C can be safely removed from S. For, if the set of clauses $S \setminus \{C\}$ is satisfiable, so is S, since we can extend any assignment satisfying all the clauses in $S \setminus \{C\}$ to an assignment satisfying also C. Thus, if we remove, before starting the tableau construction, all the clauses which are obviously 'unconnected' in the sense just explained, we shall be left with a set S' of clauses, namely such that every literal occurring in a clause of S' is complemented by a literal occurring in some other clause of S'. This seems to suggest that it might be always possible to apply the clause decomposition rule in accordance with the path connection restriction without affecting confluence. Unfortunately this is not the case, as is shown by the counterexample illustrated in Figure 6 which shows a typical 'dead end' in the construction of a tableau for the set $\Gamma = \{\neg PQ, P\neg Q, RS, R\neg S, \neg RS, \neg R\neg S\}$. Note that the input set does not contain 'pure literals' and yet the two open branches ending with $\neg Q$ and $\neg P$ cannot be further expanded without violating the path connection restriction. In this example we simply hit a satisfible subset of the input set which is not connected with the unsatisfiable subset and, yet, contains no pure literal.

The failure of confluence in path connected tableaux implies that their completeness cannot be proved by means of a systematic proof procedure leading to branch saturation and, therefore, alternative procedures, based on some other way of exploring the search space, are needed. A detailed discussion of such alternative proof procedures can be found in Letz's chapter on First Order Tableaux.

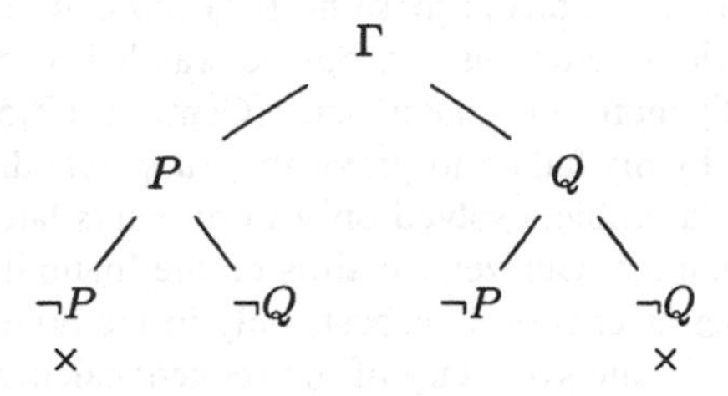

Figure 6. A 'dead end' for the path connection strategy

The path connection restriction is closely related on the one hand to Bibel's *connection method* [Bibel, 1982] and on the other to Loveland's *model elimination* [Loveland, 1978]. This relation is also discussed in Letz's chapter.

A stronger restriction requires that every application of the clause decomposition rule (except the first one) generates at least a literal that clashes with its *immediate* predecessor. In other words, if the rule is applied to a clause C in a branch ϕ whose last node is labelled with a literal, say l, then C must contain the literal l'. A tableau constructed in accordance with this restriction is called *tightly connected* (or *parent clash restricted* in [Vellino, 1989]). The tight connection strategy is also complete, although it obviously inherits non-confluence from the path connection strategy.

It must be observed that neither strategy is optimal: on some occasions both generate tableau which are not of minimal size. A class of examples is presented in [Vellino, 1989, pp.24–25]. Another class of examples showing that minimal connected tableaux can be exponentially larger than unrestricted tableaux is presented in [Letz *et al.*, 1994].

3.8 Anomalies and Limitations

The Redundancy of Cut-free Proofs

Gentzen said that the essential property of a cut-free proof is that 'it is not roundabout' [Gentzen, 1935, p.69]. By this he meant that: 'the final result is, as it were, gradually built up from its constituent elements. The proof represented by the derivation is not roundabout in that it contains only concepts which recur in the final result' [Gentzen, 1935, p.88]. The importance of Gentzen's achievement can hardly be overestimated: by obeying the subformula principle, his sequent calculus provided a partial realization of a time-honoured ideal, that of a purely analytical method of deduction, which had played such an important role in the history of logic and philosophy. However, it must also be remarked that, in designing the sequent rules, Gentzen was admittedly more concerned with simplifying the proof of his main metalogical result (the *Hauptsatz*) than with optimizing the formal representation of proofs. For this purpose he believed that the natural deduction calculus would provide a better answer, but he was led to discard the latter in favour of a 'technically more convenient' (cf. [Gentzen, 1935]) solution, namely the sequent calculus, by his failure to prove the *Hauptsatz* directly for the natural deduction system (a problem solved only thirty years later by Dag Prawitz). As discussed in Section 3.5, Gentzen's claims on the 'naturalness' of the natural deduction calculus can be endorsed, at best, only in the domain of intuitionistic logic. For classical logic, the symmetry of the sequent calculus and its 'mechanical' flavour seemed to provide a suitable approach to the automation of deduction. Today we take Gentzen's *Hauptsatz* for granted and are much more sensitive to the choice of carefully designed rules of inference, allowing for a terse presentation of classical deductions which aims to be at the same time perspicuous and (reasonably) efficient. From this point of view, as we shall argue in the sequel, Gentzen's rules are far from being ideal.

Let us consider, as a simple example, a cut-free proof of the sequent:

$$A \vee B, A \vee \neg B, \neg A \vee C, \neg A \vee \neg C \vdash \emptyset$$

expressing the fact that the antecedent is inconsistent. A minimal proof is illustrated in Figure 7 (we write the proof upside-down according to the interpretation of the sequent rules as reduction rules in the search for a counterexample; by $\Gamma \vdash$, we mean $\Gamma \vdash \emptyset$ and consider as axioms all the sequents of the form $\Gamma, A, \neg A \vdash$). Such a proof is, in some sense, redundant when it is interpreted as a (failed) systematic search for a countermodel to the endsequent (i.e. a model of the antecedent): the subtree $\mathcal{T}_1$ encodes the information that *there are no countermodels which make A true*, but this information cannot be used in other parts of the tree and, in fact, $\mathcal{T}_2$ still tries (in its left subtree) to construct a countermodel which makes A true, only to show again that such a countermodel is impossible.

Notice that (i) the proof in the example is minimal; (ii) the redundancy does not depend on our representation of the proof as *a tree*: the reader can easily check that all the sequents which label the nodes are different from each other and, as a result, the proof would have the same size if represented as a sequence or as a directed acyclic graph. The only way to obtain a non-redundant proof in the form of a sequence or a directed acyclic graph of sequents would be by using the thinning rule as in Figure 8.

In this case the proof obtained by employing thinning is not much shorter because of the simplicity of the example considered. Yet, it illustrates the use of thinning in *direct* proofs in order to eliminate redundancies. However, for the reasons discussed in Section 3.4, the thinning rule is *not* suitable for backward proof search.

The situation is perhaps clearer if we represent the proof in the form of a closed tableau *à la* Smullyan (see Figure 9). It is easy to see that such a tableau shows the same redundancy as the sequent proof given before.

This intrinsic redundancy of the cut-free analysis is responsible in many cases

$$A \vee B, A \vee \neg B, \neg A \vee C, \neg A \vee \neg C \vdash$$

$$A, A \vee \neg B, \neg A \vee C, \neg A \vee \neg C \vdash \qquad B, A \vee \neg B, \neg A \vee C, \neg A \vee \neg C \vdash$$

$$\mathcal{T}_1 \qquad \mathcal{T}_2$$

Where $\mathcal{T}_1 =$

$$A, A \vee \neg B, \neg A \vee C, \neg A \vee \neg C \vdash$$

$$A, A \vee \neg B, \neg A, \neg A \vee \neg C \vdash \qquad A, A \vee \neg B, C, \neg A \vee \neg C \vdash$$

$$A, A \vee \neg B, C, \neg A \vdash \qquad A, A \vee \neg B, C, \neg C \vdash$$

and $\mathcal{T}_2 =$

$$B, A \vee \neg B, \neg A \vee C, \neg A \vee \neg C \vdash$$

$$B, A, \neg A \vee C, \neg A \vee \neg C \vdash \qquad B, \neg B, \neg A \vee C, \neg A \vee \neg C \vdash$$

$$B, A, \neg A, \neg A \vee \neg C \vdash \qquad B, A, C, \neg A \vee \neg C \vdash$$

$$B, A, C, \neg A \vdash \qquad B, A, C, \neg C \vdash$$

Figure 7. A minimal cut-free proof of the sequent which occurs at the root of the top tree

(1)	$A, C, \neg A \vdash$	Axiom
(2)	$A, C, \neg C \vdash$	Axiom
(3)	$A, C, \neg A \vee \neg C \vdash$	From (1) and (2)
(4)	$A, \neg A, \neg A \vee \neg C \vdash$	Axiom
(5)	$A, \neg A \vee C, \neg A \vee \neg C \vdash$	From (4) and (3)
(6)	$B, A, \neg A \vee C, \neg A \vee \neg C \vdash$	From (5) by thinning
(7)	$B, \neg B, \neg A \vee C, \neg A \vee \neg C \vdash$	Axiom
(8)	$B, A \vee \neg B, \neg A \vee C, \neg A \vee \neg C \vdash$	From (6) and (7)
(9)	$A, A \vee \neg B, \neg A \vee C, \neg A \vee \neg C \vdash$	From (5) by thinning
(10)	$A \vee B, A \vee \neg B, \neg A \vee C, \neg A \vee \neg C \vdash$	From (8) and (9)

Figure 8. A proof in linear format making use of the thinning rule

for explosive growth in the size of the search tree. Moreover, it is *essential*: it does not depend on any particular proof-search procedure (it affects *minimal* proofs) but only on the use of the cut-free rules. In the rest of this chapter this point will be examined in detail.

The Culprit

Can we think of a more economical way of organizing our search for a countermodel? of avoiding the basic redundancy of the cut-free analysis? We must first identify the culprit. Our example contains a typical pattern of cut-free refutations which can be described as follows: the subtree $\mathcal{T}_1$ searches for possible countermodels which make A true. If the search is successful, the original sequent is not valid and the problem is solved. Otherwise there is *no* countermodel which makes A true (i.e. if we restrict ourselves to classical bivalent models, every countermodel, if any, must make A false). In both cases it is pointless, while building up $\mathcal{T}_2$, to try to construct (as we do if our search is governed by Gentzen's rules) countermodels which make A true, because this kind of countermodel is already sought in $\mathcal{T}_1$.

In general, we may have to reiterate this redundant pattern an arbitrary number of times, depending on the composition of our input set of formulae. For instance, if a branch contains n disjunctions $A \vee B_1, \ldots, A \vee B_n$ which are all to be analysed in order to obtain a closed subtableau, it is often the case that the *shortest* tableau has to contain higly redundant configurations like the one shown in figure 10: where the subtree $\mathcal{T}^*$ has to be repeated n times. Each copy of $\mathcal{T}^*$ may, in turn, contain a similar pattern. It is not difficult to see how this may rapidly lead to a combinatorial explosion which is by no means related to any 'intrinsic difficulty' of the problem considered but only to the redundant behaviour of the tableau rules. Let us discuss in some more detail the relation between the inefficiency of these rules and their failure to simulate 'analytic cut' inferences.

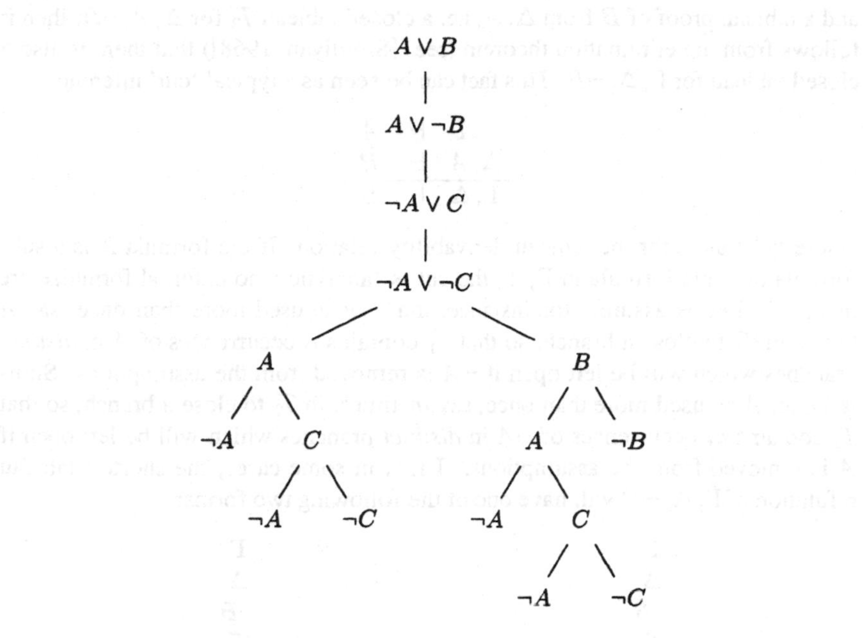

Figure 9. A closed tableau for $\{A \vee B, A \vee \neg B, \neg A \vee C, \neg A \vee \neg C\}$

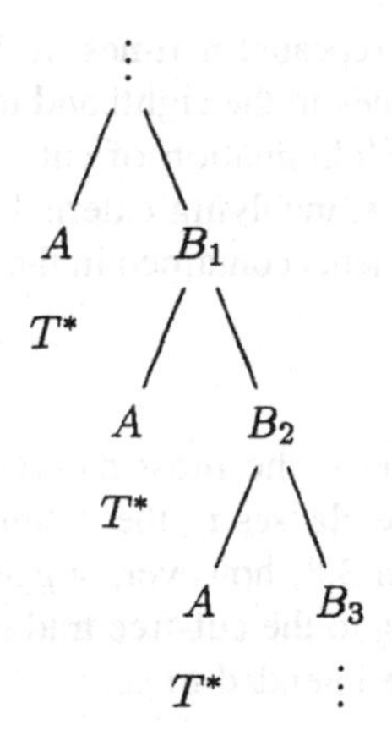

Figure 10. Redundancy of tableau refutations

Suppose there is a tableau proof of A from Γ, i.e. a closed tableau $\mathcal{T}_1$ for $\Gamma, \neg A$ and a tableau proof of B from Δ, A, i.e. a closed tableau $\mathcal{T}_2$ for $\Delta, A, \neg B$; then it follows from the elimination theorem (see [Smullyan, 1968]) that there is also a closed tableau for $\Gamma, \Delta, \neg B$. This fact can be seen as a typical 'cut' inference:

$$\frac{\begin{array}{rcl}\Gamma & \vdash & A \\ \Delta, A & \vdash & B\end{array}}{\begin{array}{rcl}\Gamma, \Delta & \vdash & B\end{array}}$$

where '$\vdash$' stands for the tableau derivability relation. If the formula A is a subformula of some formula in Γ, Δ, the cut is 'analytic': no external formulae are involved. Let us assume, for instance, that $\neg A$ is used more than once, say n times, in $\mathcal{T}_1$ to close a branch, so that $\mathcal{T}_1$ contains n occurrences of A in *distinct* branches which will be left open if $\neg A$ is removed from the assumptions. Similarly, let A be used more than once, say m times, in $\mathcal{T}_2$ to close a branch, so that $\mathcal{T}_2$ contains m occurrences of $\neg A$ in *distinct* branches which will be left open if A is removed from the assumptions. Then, in some cases, the shortest tableau refutation of $\Gamma, \Delta, \neg B$ will have one of the following two forms:

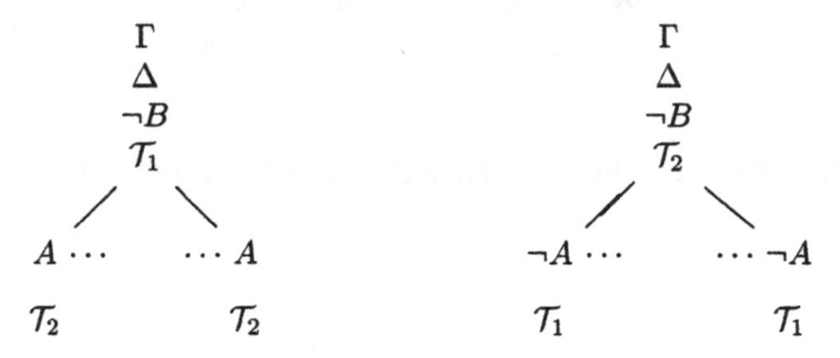

where the subrefutation $\mathcal{T}_2$ is repeated n times in the lefthand tree and the subrefutation $\mathcal{T}_1$ is repeated m times in the righthand tree. When A is a subformula of some formula in Γ, Δ, the 'elimination of cuts' from the tableau proof does not remove 'impure' inferences, involving external formulae, but 'pure' analytic inferences, involving only elements contained in the data.

Searching for a Countermodel

In a sense, the tableau method is the most natural expansion system, since its rules correspond exactly to the clauses in the definition of a semivaluation. The examples discussed in Section 3.8, however, suggests that, in another, analytic tableaux constructed according to the cut-free tradition are not well-suited to the nature of the problem they are intended to solve. In this section we shall render this claim more precise.

As we have seen in Section 2.2, an expansion system can be seen as a (non-deterministic) algorithm to search through the space of all possible semivaluations for one which satisfies a given set of formulae. We have observed, however, that the search space has a natural structure of its own. It is therefore reasonable to require that the rules we adopt in our systematic search reflect this structure. This can be made precise as follows: given an **S**-tree $\mathcal{T}$ we can associate with each node n of $\mathcal{T}$, the set Δ_n of the s-formulae occurring in the path from the root to n or, equivalently, the partial valuation v_n which assigns 1 to all the formulae A such that $T(A)$ occurs in the path to n, and 0 to all the formulae A such that $F(A)$ occurs in the path to n (and leaves all the other formulae undefined). Let $\preceq_{\mathcal{T}}$ the partial ordering defined by $\mathcal{T}$ on the set of its nodes (i.e. for all nodes n_1, n_2, $n_1 \preceq_{\mathcal{T}} n_2$ if an only if n_1 is a predecessor of n_2), and $\sqsubseteq$ the partial ordering or partial valuations defined above. Clearly

(4) $\quad n_1 \preceq_{\mathcal{T}} n_2 \Longrightarrow v_{n_1} \sqsubseteq v_{n_1}.$

It would be desirable to require that the converse also holds, namely

(5) $\quad n_1 \not\preceq_{\mathcal{T}} n_2 \Longrightarrow v_{n_1} \not\sqsubseteq v_{n_1}.$

So that for every pair of nodes n_1, n_2 belonging to *different* branches, the associated partial valuations are incomparable, and therefore

$$n_1 \preceq_{\mathcal{T}} n_2 \Longleftrightarrow v_{n_1} \sqsubseteq v_{n_2}.$$

In other words the relations between the nodes in an **S**-tree correspond to the relations between the associated partial valuations. If a tree does not satisfy (5), i.e. $v_{n_1} \sqsubseteq v_{n_2}$ for some pair of nodes n_1, n_2 belonging to *different* branches, then it is obviously redundant for the reasons discussed in section 3.8: if v_{n_1} can be extended to a semivaluation, we have found a countermodel to the original sequent and the problem is solved. Otherwise, if no extension of v_{n_1} is a semivaluation, the same applies to v_{n_2}.

The importance of avoiding this kind of redundancy is both conceptual and practical. A redundant tree does *not* reflect the structure of the semantic space of partial valuations which it is supposed to explore and this very fact has disastrous *computational* consequences: redundant systems are ill-designed, from an algorithmic point of view, in that, in some cases, they force us to repeat over and over again what is essentially the same computational process.

It is easy to see that the non-redundancy condition *is not satisfied by the tableau method* (and in general by cut-free Gentzen systems). We can therefore say that, in some sense, *such systems are not natural for classical logic.*[20]

Are Tableaux an Improvement on Truth-tables?

The intrisic redundancy of tableau refutations has some embarassing consequences from the computational viewpoint. These emerge in the most striking fashion

[20]This suggestion may be contrasted with Prawitz's suggestion, advanced in [Prawitz, 1974], that 'Gentzen's calculus of sequents may be understood as *the* natural system for generating logical truths'.

when we compare the complexity of tableau proofs with that of truth-tables. The truth-table method, introduced by Wittgenstein in his celebrated *Tractatus Logico-Philosophicus*, provides a decision procedure for propositional logic which is immediately implementable on a machine. However this time-honoured method is usually mentioned only to be immediately dismissed because of its incurable inefficiency. Beth himself, who was (in the 1950's) one of the inventors of tableaux, also stressed that they 'may be considered in the first place as a more convenient presentation of the familiar truth-table analysis'.[21] Beth's opinion is echoed in the widespread yet unsubstantiated claim that the truth-table method is clearly and generally less efficient than the tableau method and relates to it as an exhaustive search relates to 'smarter' procedures which makes use of shortcuts. Here are some typical quotations on this topic:

> Is there a better way of testing whether a proposition A is a tautology than computing its truth table (which requires computing at least 2^n entries where n is the number of proposition symbols occurring in A)? One possibility is to work backwards, trying to find a truth assignment which makes the proposition false. In this way, one may detect failure much earlier. This is the essence of Gentzen [cut-free] systems[22] ...

> The truth-table test is straightforward but needlessly laborious when statement letters are numerous, for the number of cases to be searched doubles with each additional letter (so that, e.g., with 10 letters there are over 1,000 cases). The truth-tree [i.e tableau] test [...] is equally straightforward but saves labor by searching whole blocks of cases at once.[23]

This appraisal, however, may turn out to be rather unfair. In fact, the situation is not nearly as clear-cut as it appears. In the next section we shall show that, contrary to expectations (and Beth's opinion), there are examples for which the *complete* truth-tables perform incomparably better than the *standard* tableau method. For these examples it is, of course, true — to pursue the above quotation — that with 10 propositional letters there are over 1,000 cases (exactly 1,024) to check. However it turns out that a *minimal* closed tableau must contain more than 3,000,000 branches. With 20 letters there are slightly more than one million cases in the truth-table analysis (which is still feasible on today's computers) but over 2×10^{18} branches in a minimal closed tableau.

Technically speaking, these examples serve the purpose of filling a gap in the classification of conventional proof systems in terms of the p-simulation relation

---

[21] [Beth, 1958, page 82].
[22] [Gallier, 1986, pages 44–45].
[23] [Jeffrey, 1967, page 18].

ship: *truth-tables and Smullyan's analytic tableaux are incomparable proof systems.*[24]

The key observation is that the complexity of tableau proofs depends essentially on the *length* (i.e. total number of symbols) of the formula to be decided, whereas the complexity of truth-tables depends only on the number of *distinct propositional variables* which occur in it. If an expression is 'fat',[25] i.e. its length is large compared to the number of distinct variables in it, the number of branches generated by its tableau analysis may be large compared to the number of rows in its truth-table.

An extreme example is represented by the sequence of 'truly fat' expressions in conjunctive normal form, defined as follows: given a sequence of k atomic variables $P_1, \ldots, P_k$, consider all the possible clauses containing as members, for each $i = 1, 2, \ldots, k$, either P_i or $\neg P_i$ and no other member. There are 2^k of such clauses. Let $H_{P_1,\ldots,P_k}$ denote the set containing these 2^k clauses. The expression $\bigwedge H_{P_1,\ldots,P_k}$ is unsatisfiable. For instance, $\bigwedge H_{P_1,P_2}$ is the following expression in CNF:

$$(P_1 \vee P_2) \wedge (P_1 \vee \neg P_2) \wedge (\neg P_1 \vee P_2) \wedge (\neg P_1 \vee \neg P_2)$$

Notice that in this case the truth-table procedure contains as many rows as clauses in the expressions, namely 2^k. In other words, this class of expressions is not 'hard' for the truth-table method. However we shall prove that it is hard for the tableau method.

Recall that, in order to deal with clauses, we redefine the tableau $\vee$-elimination rule as follows:

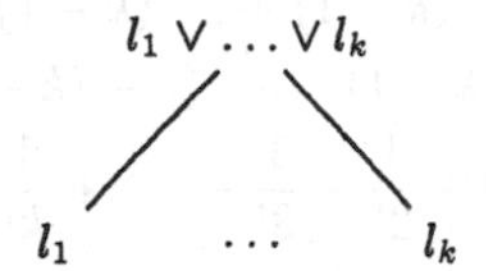

Since the formula $\bigwedge H_{P_1,\ldots,P_k}$ is in CNF, we can assume that the first part of the tableau consists of a sequence of applications of the $\wedge$-elimination rule until the conjunction is decomposed into a sequence of clauses. For our purposes we can consider these clauses as 'assumption' formulas and forget about the $\wedge$-elimination steps.

PROPOSITION 16. *Every closed tableau for the set $H_{P_1,\ldots,P_k}$ contains more than $k!$ distinct branches.*

Proof. Let T be a minimal closed tableau for $H_{P_1,\ldots,P_k}$. We assume that all the nodes labelled with assumption clauses precede the other nodes. So the tableau

[24]As Alasdair Urquhart has pointed out, this appears to be the only known example of such an incomparability among conventional proof-systems.

[25]This rather fancy terminology is used in [Dunham and Wang, 1976].

starts with a sequence of 2^k nodes, each labelled with an assumption. Let n_0 be the last node of this sequence. Every $n \geq n_0$ has exactly k children $m_1, \ldots, m_k$ (resulting from the decomposition of an assumption clause) such that m_i ($i = 1, \ldots, k$) is labelled either with P_i or with $\neg P_i$. So for every node n and every $j = 1, \ldots, k$, either P_j or $\neg P_j$ is among the children of n. Let $Q_1, \ldots, Q_k$ be an arbitrary permutation of $P_1, \ldots, P_k$. There must be a path $n_1, \ldots, n_k$ such that the labelling formulas form a sequence $\pm Q_1, \ldots, \pm Q_k$, where $\pm Q_i$ is either Q_i or $\neg Q_i$. Such a path is obviously open, since $Q_i \neq Q_j$ for $i \neq j$. Moreover, distinct permutations are associated with distinct paths.

Hence T contains at least as many distinct paths as there are permutations of $P_1, \ldots, P_k$, namely at least $k!$ distinct paths. ■

Notice that $k!$ grows faster than any polynomial function of 2^k, and there are only 2^k rows in the truth-table analysis[26] of $H_{P_1,\ldots,P_k}$. This means, in practice, that for this class of examples even the most naive program implementing the old truth-table method will run incomparably faster than *any* program implementing some proof-search procedure (no matter how smart) based on the tableau rules.

This implies that analytic tableaux cannot p-simulate the truth-table method. Notice also that for $H_{P_1,\ldots,P_k}$ there are resolution refutations in which the number of steps is linear in the number of clauses.

The previous argument can be easily adapted to give an exact lower bound on the size of a tableau. Let $C(k)$ be the number of *interior* (i.e. non-leaf) nodes in a *minimal* closed tableau for $H_{P_1,\ldots,P_k}$. Then, a simple analysis shows that $C(k)$ is determined exactly by the following equation:

$$\begin{aligned} C(k) &= k + k \cdot (k-1) + k \cdot (k-1)(k-2) + \ldots + \\ &\quad +k \cdot (k-1) \cdot \ldots \cdot (k-(k-1)) \\ &= k! \cdot (1 + \frac{1}{2!} + \frac{1}{3!} + \cdots + \frac{1}{k!}) \end{aligned}$$

One immediate consequence of this fact is that analytic tableaux do not provide, at least in their standard form given them by Smullyan, a uniform improvement on the truth-table method as far as computational complexity is concerned. Indeed, there is no "feasible" simulation in either direction (since truth-tables cannot p-simulate tableaux). Even worse, there are natural and well-defined subproblems of the tautology problem which are tractable for the truth-tables and intractable for *any* procedure based on analytic tableaux. Namely: the class of 'truly fat' tautologies, in which the number of variables is of order $\log n$, where n is the length of the tautology, are decidable in polynomial time by the truth-tables, but cannnot be decided in polynomial time—as shown by our examples—if we use

[26]To be precise, the truth-table analysis involves $O(k \cdot n \cdot 2^k)$ steps, where n is the total number of occurrences of atomic letters and operators in the CNF expression, which is $O(k \cdot 2^k)$, so that the overall complexity is $O(k^2 \cdot 2^{2k})$.

Smullyan's analytic tableau rules (no matter how 'smart' we are in applying these rules).

In general, given a decidable logic L which admits of characterization by means of m-valued truth-tables, the complexity of the semantic decision procedure for L is essentially $O(k \cdot n \cdot m^k)$ where n is the length of the input formula and k is the number of distinct variables in it. This is a natural upper bound on the decision problem for L which should be at least equalized by any proof procedure. Analytic tableaux, and their equivalent cut-free sequent calculus in tree form, do not meet this requirement. As we shall see, this is not just an oddity due to a careful choice of artificial examples but the result of a *fundamental* inadequacy of the tableau rules which affects, at different degrees, every tableau refutation: the 'fatter' the formulas (namely the smaller the number of variables relative to the overall length of the formulas), the more likely for the truth-tables to beat their more quoted rivals.

The cause of this behaviour is not difficult to detect if we look at the way in which the two methods enumerate the possible cases to be checked in order to test a formula for validity. While all the cases enumerated in the truth-table analysis are *mutually exclusive*, this is not, in general, true of the tableau analysis. When applying the typical tableau branching rules:

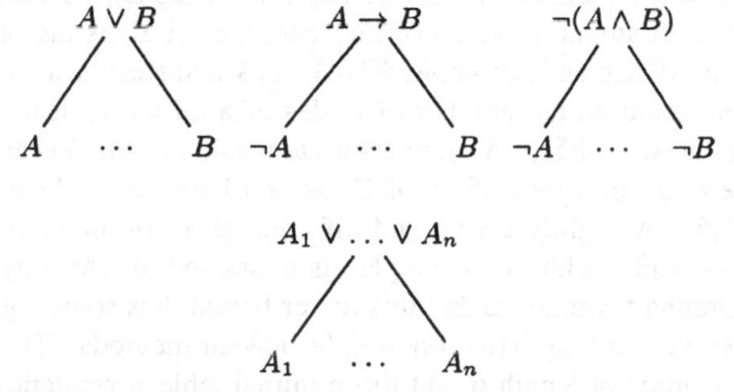

for expressions in clausal form, it is obvious that the branches do not represent mutually inconsistent 'possible models'.[27] As a result, when expanding the tree, we may well (and often do) end up considering more cases than is necessary. For instance, after we apply the rule for analysing $A \vee B$, when we expand the tableau below B we may have to consider possible models in which A is true. But all these possibile models are already enumerated below A, so that the enumeration process is redundant.

More Hard Examples for Smullyan's Tableaux

Another class of examples which are hard both for tableaux and for the truth-tables is described in [Cook and Reckhow, 1974]. These examples consist of sets

[27]By a 'possible model' we simply mean a set of formulas which *may* be embedded into a Hintikka set, namely the set of formulas which occur in an open path.

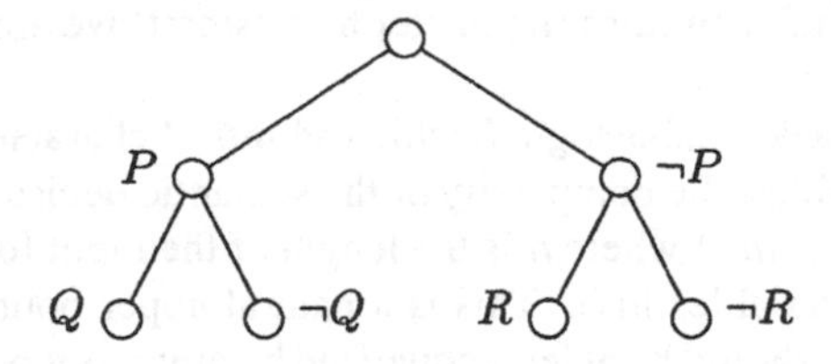

Figure 11. $\Sigma(\mathcal{T}) = \{PQ, P\neg Q, \neg PR, \neg P\neg R\}$

of clauses associated with binary trees as follows. Consider a binary tree $\mathcal{T}$ where each node, except the root, is labelled with a distinct literal, and sibling nodes are labelled with complementary literals. We stipulate that, for each pair of sibling nodes, the one on the left is labelled with a positive literal and the one on the right is labelled with the corresponding negative literal. Now, we can associate with each branch ϕ the clause containing exactly the literals in ϕ. In this way, the whole tree $\mathcal{T}$ can be taken as representing the set $\Sigma(\mathcal{T})$ of the clauses associated with its branches. A simple example is shown in Figure 11. Let Σ_n be the set of clauses associated with a complete binary tree of depth n. Thus Σ_n contains 2^n clauses and $2^n - 1$ distinct atomic letters. For instance, the set Σ_2 is the one illustrated in Figure 11. In [Cook and Reckhow, 1974] Cook and Reckhow report (without proof) a lower bound on the number of nodes of a closed tableau Σ_n which is exponential in the size of Σ_n. A more accurate version of this lower bound, with a proof, is the result of a joint effort of Cook and Urquhart and can be found in [Urquhart, 1995]. A slightly earlier and different proof of an exponential lower bound for Cook and Reckhow's examples is contained in [Murray and Rosenthal, 1994]. Urquhart comments that this lower bound 'has some significance for automated theorem proving based on simple tableau methods. The set Σ_6 contains only 64 clauses of length 6, but the minimal tableau refutation for Σ_6 has 10,650,056,950,806 interior nodes. This shows that any pratical implementation of the tableau method must incorporate routines to eliminate repetition in tableau construction'.[28]

On the other hand the above examples have very small resolution refutations, even in tree format. To see this, just observe that we can label the nodes of a complete binary tree of depth n with suitable clauses to obtain a tree resolution refutation of Σ_n. For this purpose it is sufficient to label each node a with the clause consisting of the literals occurring in the path from the root to a (and therefore the root with the empty clause). It is easy to check that the resulting tree provides the required refutation and its size is linearly related to the size of Σ_n. This is sufficient to prove the following:

[28] [Urquhart, 1995], p. 435.

THEOREM 17. *Smullyan's tableaux cannot p-simulate tree resolution.*

On the contrary, it is not difficult to show that tree resolution *can* p-simulate Smullyan's tableaux for clauses (for a simple argument see [Urquhart, 1995]).

Are the Tableau Rules Really 'Classical'?

Beth's aim, in his 1955 paper, where he introduced his version of semantic tableaux [Beth, 1955], was to point out the "close connections between semantics and derivability theory" so as to avoid the difficulties which were usually encountered in the proof of the completeness theorem [pp. 318,313]. He stressed the close correspondence between the formal rules in a Gentzen-type system and the semantic-oriented rules of the tableau method [pp. 318,322-23] so that, adopting his approach, proofs of completeness become trivial and 'such celebrated and profound results as Herbrand's Theorem, the Theorem of Löwenheim-Skolem-Gödel, Gentzen's Subformula Theorem and Extended Hauptsatz, and Bernay's Consistency Theorem are [...] within (relatively) easy reach" [p. 318]. Morever, he claimed that the Gentzen-type formal proofs obtained in a 'purely mechanical manner', by transformation of semantic tableaux, turned out to be, contrary to expectations, not clumsy or cumbersome but "remarkably concise', and could "even be proved to be, in a sense, the shortest ones which are possible" [p. 323]. In this way he thought he had reached a formal method which was 'in complete harmony with the standpoint of [classical] semantics' [p. 317].

We have already shown that Beth was wrong in claiming that tableaux constitute a uniform improvement on the truth-table method. How about this other claim on the close correspondence between the tableau rules and classical semantics?

The classical notion of truth is governed by two basic principles: the principle of *Non-contradiction* (no proposition can be true and false at the same time) and the principle of *Bivalence* (every proposition is either true or false, and there are no other possibilities). While the former principle is clearly embodied in the rule for closing a branch, there is no rule in the tableau method (and in cut-free Gentzen systems) which corresponds to the Principle of Bivalence. While enumerating all the possible cases, the tableau rules allow for the possibility of a proposition's being something else other than true or false. Suppose our semantics is 3-valued, with the truth-value 'undefined' ($*$) along with 'true' (1) and 'false' (0). Suppose also that the truth-tables are extended in a way which preserves the classical rules of truth, namely:

A	B	$A \vee B$	$A \wedge B$	$A \rightarrow B$	$\neg A$
1	1	1	1	1	0
*	1	1	*	1	*
0	1	1	0	1	1
1	*	1	*	*	
*	*	*	*	*	
0	*	*	0	1	
1	0	1	0	0	
*	0	*	0	*	
0	0	0	0	1	

Then the tableau rules are still *sound* for this semantics, and in fact for *any* many-valued semantics which preserves the classical equivalences. A closed tableau for $\neg A$ shows that A cannot be false. A closed tableau for $\Gamma, \neg A$ shows that A cannot be false if all the formulas in Γ are true. So, if we define logical consequence as follows

$$\Gamma \vdash A \text{ iff } A \text{ is non-false when all the formulas in } \Gamma \text{ are true}$$

the logic defined by this many-valued semantics collapses into classical logic.[29] This last move amounts to restricting the possible models to the classical bivalent ones. So bivalence is re-introduced *at the end* of the analysis in order to exclude non-classical models, but nowhere is it used in the search for these models.

There are historical reasons for this 'elimination of bivalence' from the semantics of classical proof-theory. Gentzen introduced the sequent calculi **LK** and **LJ**, as well as the natural deduction calculi **NK** and **NJ**, in his famous 1935 paper [Gentzen, 1935]. Apparently he considered the sequent calculi as technically more convenient for metalogical investigation [Gentzen, 1935, p. 69]. In particular he thought that they were 'especially suited to the purpose' of proving the *Hauptsatz*, and that their form was 'largely determined by considerations connected with [this purpose]'[p. 89].

Gentzen's rules have become a paradigm both in proof-theory and in its applications. This is not without reason. First, like the natural deduction calculi, they provide a precise analysis of the logical operators by specifying how each operator can be introduced in the antecedent or in the succedent of a sequent. Second, their form ensures the validity of the *Hauptsatz*: each proof can be transformed into one which is cut-free, and cut-free proofs enjoy the *subformula property*: every sequent in the proof tree contains only subformulae of the formulae in the sequent to be proved.

However, Gentzen's rules are not the only possible way of analysing the classical operators and not necessarily the best. Their form was influenced by con-

[29]This kind of semantics validates all the rules of the sequent calculus except the cut rule, and has been used by Schütte, and more recently by Girard, for proof-theoretical investigations into the sequent calculus. See the interesting discussion in [Girard, 1987].

siderations which were partly philosophical, partly technical. In the first place Gentzen wanted to set up a formal system which 'comes as close as possible to actual reasoning' [Gentzen, 1935, p. 68]. In this context he introduced the natural deduction calculi in which the inferences are analysed essentially in a constructive way and classical logic is obtained by adding the law of excluded middle in a purely external manner. Then he recognized that the special position occupied by this law would have prevented him from proving the *Hauptsatz* in the case of classical logic. So he introduced the sequent calculi as a technical device in order to enunciate and prove the *Hauptsatz* in a convenient form [p. 69] both for intuitionistic and classical logic. These calculi still have a strong deduction-theoretic flavour and Gentzen did not show any sign of considering the relationship between the classical calculus **LK** and the semantic notion of entailment.

What is, then, the semantic counterpart of the cut-elimination theorem? If we think of Gentzen's sequent rules in semantic terms, and adopt the usual intepretation, namely define $\Gamma \vdash \Delta$ as valid if and only if, for every model M, at least one formula in Δ is true in M whenever all formulas in Γ are true in M, it is easy to see that the sequent rules can be read upside-down as rules for constructing a countermodel to the endsequent, exactly as the tableau rules. However, the rule which the *Hauptsatz* shows to be eliminable is the cut rule:

$$\frac{\Gamma, A \vdash \Delta \qquad \Gamma \vdash \Delta, A}{\Gamma \vdash \Delta}$$

If we read this rule upside-down, following the same semantic interpretation that we adopt for the operational rules, then what the cut rule says is:

In all models and for all propositions A, either A is true or A is false.

But this is the Principle of Bivalence, one of the two *fundamental* principles which characterize the classical notion of truth. In contrast, none of the rules of the cut-free fragment implies bivalence (as is shown by the three-valued semantics for this fragment). The elimination of cuts from proofs is, so to speak, the elimination of bivalence from the underlying semantics. But, as Smullyan once remarked: 'The real importance of cut-free proofs is not the elimination of cuts per se, but rather that such proofs obey the subformula principle' [Smullyan, 1968a, p. 560]. As we shall see the subformula principle, as well as any other desirable property of the cut-fre proofs, do not require such a drastic move.

4 EXTENSIONS OF SMULLYAN'S TABLEAUX

The refinements described in this section have in common that they *extend* the method of analytic tableaux (in Smullyan's formulation) by means of additional rules which are *redundant*, i.e. are not required for completeness, but may solve some of the anomalies of the original rules.

4.1 Tableaux with Merging

The computational redundancy discussed in Section 3.8 is well-known to anybody who has worked with Smullyan's tableaux in the area of automated deduction. It is usually avoided by augmenting the standard tableau rules with extra 'control' features which stop the expansion of redundant paths in the tree or license the generation of 'lemmas' to be used in closing redundant branches. One consequence of the separation result in Proposition 16 is that some correction of the tableau rules is not just a discretional 'optimization' step, but a *necessary condition* for a respectable tableau-like system.

One of the 'enhancements' of Smullyan's tableaux which is taken into consideration in the area of automated deduction is called *merging* (see [Vellino, 1989] for a discussion of this method in connection with clausal tableaux, and [Broda, 1992] for an extension to first order logic). This technique can be easily described in terms of our discussion of the redundancy of analytic tableaux. If for any two nodes n, m which lie on different branches we have

$$\Delta_n \subseteq \Delta_m$$

then one can stop pursuing the branch through m without loss of soundness.

More precisely, we can mark a branch of a tableau $\mathcal{T}$ as *checked* if the set of formulae occurring in it is a superset of the set of formulae occurring in another (unchecked) branch of $\mathcal{T}$. We say that a branch of $\mathcal{T}$ is *M-saturated* if it is either saturated or checked and that $\mathcal{T}$ is *M-completed* if every branch is *M-saturated.* Morever, we say that $\mathcal{T}$ is *M-closed* if every branch of $\mathcal{T}$ is closed or checked.

The soundness of Tableaux with Merging (TM for short) is a corollary of the following lemma:

LEMMA 18. *Every* M-closed *tableau can be expanded into a closed tableau.*

Proof. This lemma can be proved as follows. Let $\mathcal{T}$ be an M-closed tableau and let k be the ending node of a checked branch ϕ. Then there must be a node m such that the set of formulae occurring in the path to m is a subset of the set of formulae occurring in ϕ. We say that the checked branch ϕ is *justified* by m. Let $\mathcal{T}_m$ be the sub-tableau generated by m. Now, either $\mathcal{T}_m$ is closed or it is M-closed and at least one of its branches is checked. None of the checked branches of $\mathcal{T}_m$ can be justified by k or by a predecessor of k (by construction of $\mathcal{T}$). This implies that there must be a checked branch which is justified by a node n such that the sub-tableau $\mathcal{T}_n$ generated by n is closed. By appending T_n to n we obtain an M-completed tableau $\mathcal{T}'$ with a number of checked branches which is strictly less than the number of checked branches in $\mathcal{T}$. By repeating this operation a sufficient number of times, we eventually obtain a closed tableau. ■

Merging is clearly an *ad hoc* method especially devised to remedy the problems highlighted in Section 3.8. It is easy to see that if merging is employed, the hard examples of Section 3.8 can be solved by tableaux of polynomial size. This is sufficient to establish the following fact:

COROLLARY 19. *The standard tableau method cannot p-simulate the tableau method with merging.*

4.2 *Tableaux and Lemmas*

Another way of solving the anomalies of Smullyan's tableaux consists in using *lemmas* in construction of a tableau. Suppose a node k of a tableau $\mathcal{T}$ is the origin of a closed subtableau. Then, if k is labelled with a formula A, and Γ is the set of the formulae occurring in the path that leads to A, we can say that $\Gamma \vdash \neg A$. Hence, we can use $\neg A$ as a 'lemma' and append this formula to the end of all the branches which share the set of formulae Γ, namely all the branches passing through any sibling of k. This operation is called 'lemma generation'. Let us call *L-tableau* a tableau constructed by making use of 'lemmas'. Then, it is easy to see that:

PROPOSITION 20. *There exists a closed tableau for* Γ *if and only if there exists a closed L-tableau for* Γ.

Now, it is not difficult to realize that the operation that we have called 'lemma generation' can be soundly performed even if the sub-tableau generated by k has not yet been constructed. For, either it is possible to construct a closed sub-tableau below k, or every sub-tableau below k must be open. In the first case, the sub-tableau justfies the use of the lemma $\neg A$ as explained above. In the second case, the tableau will anyway remain open and the addition of the 'lemmas' in the other branches will not change this situation.

So, what is usually called 'lemma generation' is equivalent to replacing the branching rules of the standard tableau method with corresponding asymmetric ones. For instance, the rules for eliminating disjunction might have the following form:

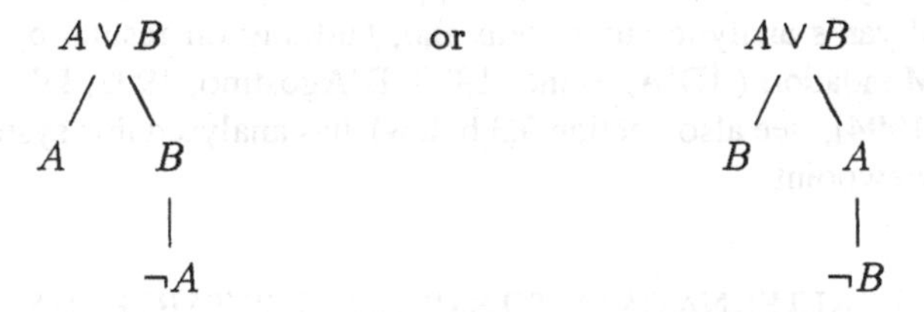

We shall call this extended method 'tableaux with lemma generation' (TLM for short).

Again, it is not difficult to see that all the hard examples of section 3.8 have polynomial size refutations if these asymmetric rules are employed.

COROLLARY 21. *The standard tableau method cannot p-simulate TLM.*

4.3 Tableaux with Analytic Cut

Since the computational anomaly of the tableau rules is related to their failure to simulate 'analytic cut' inferences, the most direct way of solving it seems to consist in adding an analytic cut rule to Smullyan's rules. This analytic cut rule has the following forms depending on whether we deal with signed or unsigned formulae:

$$\frac{}{TA \mid FA} \qquad\qquad \frac{}{A \mid \neg A}$$

where A is a subformula of some formula occurring above in the branch to which this rule is applied. The relation between this rule and the cut rule of the classical sequent calculus (with the cut formulae restricted to subformulae) should be clear from our discussion in Section 3.8: Suppose $\Gamma \vdash A, \Delta$, i.e. there is a closed tableau for $\Gamma \cup \{\neg A\} \cup \{\neg B \mid B \in \Delta\}$, and $\Gamma, A \vdash \Delta$, i.e. there is a closed tableau for $\Gamma \cup \{A\} \cup \{\neg B \mid B \in \Delta\}$. Then it is easy to see that the above branching rule allows us to construct a closed tableau for $\Gamma \vdash \{\neg B \mid B \in \Delta\}$, so showing that $\Gamma \vdash \Delta$.

Richard Jeffrey [Jeffrey, 1967] noticed the importance of this rule, that he called 'punt', in combination with the other rules of the tableau method. Smullyan [Smullyan, 1968a] once presented a sequent system with an analytic cut rule as the only proof rule, the others being replaced by suitable axioms, and stressed that such a system preserves the subformula property (which, in Gentzen's approach, provided the main motivation for the cut-free sequent calculus). Neither, however, noticed that the use of analytic cut can considerably shorten proofs. To show this, it is sufficient to observe that tableaux with the addition of the analytic cut rule (TAC for short) can linearly simulate both TM and TLM. It then follows from either Corollary 19 or Corollary 21 that:

COROLLARY 22. *Smullyan's tableaux cannot p-simulate TAC.*

More recently, Cellucci [Cellucci, to appear] has presented a tableau-like version of Smullyan's analytic cut system and, building on results by the present author and Mondadori ([D'Agostino, 1990; D'Agostino, 1992; D'Agostino and Mondadori, 1994], see also Section 5.3 below), has analysed this system from the complexity viewpoint.

5 ALTERNATIVES TO SMULLYAN'S TABLEAUX

In this section we shall discuss two tableau systems which are not classified as extensions of Smullyan's Tableaux (in the sense of the previous section), but as *alternative* systems, because their rules are different from Smullyan's rules and *no rule is redundant.* They still fall, however, within the category of expansion systems and, therefore, belong to the same family of proof systems as Smullyan's Tableaux.

5.1 The Davis–Putnam Procedure

The Davis-Putnam procedure was introduced in 1960 [Davis and Putnam, 1960] and later refined in [Davis *et al.*, 1992]. It was meant as an efficient theorem proving method[30] for (prenex normal form) first-order logic, but it was soon recognized that it combined an efficient test for truth-functional validity with a wasteful search through the Herbrand universe.[31] This situation was later remedied by the emergence of unification. However, at the propositional level, the procedure is still considered among the most efficient, and is clearly connected with the resolution method, so that Robinson's resolution [Robinson, 1965] can be viewed as a (non-deterministic) combination of the Davis-Putnam propositional module and unification, in a single inference rule.

The proof-system underlying the DPP consists of the following two expansion rules:

$$\frac{}{l \mid l'} \qquad \frac{\begin{array}{c}Al \\ l'\end{array}}{A}$$

where A is a (possibly empty) clause, l is a literal and l' denotes the complement of l. We call the first rule *splitting rule* and the second rule *unit resolution rule*. Clearly the splitting rule is nothing but a cut rule restricted to literals. We shall denote the empty clause by Λ. Then any application of the unit-resolution rule with an empty C, that is with l and l' as premisses, will yield Λ as conclusion.

The DPP is carried out in two stages. The first is a 'pre-processing' stage whose purpose is to arrange the input clauses and eliminate obvious redundancies. It consists in:

1. removing any repetitions from the input clauses;
2. removing any clause containing a 'pure' literal, i.e. a literal which is not complemented in any other clause;
3. deleting any clause that contains both a literal and its complement.

It is obvious that if S is a set of clauses and S' is the set that results from it after performing the three steps we have just described, S is satisfiable if and only if S' is satisfiable.

This pre-processing stage is not strictly necessary but can be useful to simplify the input. Let us now turn to describe the essential steps of the procedure. We shall refer to these steps as Step 1, Step 2, etc., but the order in which we list them is immaterial. The steps of the procedure are meant to give us instructions about how

[30] The version given in [Davis and Putnam, 1960] was not in fact a completely deterministic procedure: it involved the choice of which literal to eliminate at each step. Such choices may crucially affect the complexity of the resulting refutation.

[31] See [Davis, 1983].

the two expansion rules given above are to be applied. Some of these steps will involve *marking* some clauses as *fulfilled* in a given branch ϕ. Clauses fulfilled in ϕ cannot be used any more in an application of the unit-resolution rule in the same branch ϕ (but can be used in other branches). The actual mechanism of marking clauses as fulfilled does not concern us here, the only essential point is that it keeps track of the branch in which a given clause is marked as fulfilled. Clauses which are not so marked we shall call *unfulfilled*. A *DPP-tree for* S, where S is a set of clauses, is either a one-branch tree whose nodes are the clauses in S, or a tree which results from a given DPP-tree by performing one of the following steps:

Step 1. Let ϕ be a branch of the given tree; if an unfulfilled literal l belongs to ϕ, then

a. mark as fulfilled in ϕ all the clauses in ϕ of the form Cl (including l itself);

b. mark as fulfilled in ϕ all the clauses in ϕ of the form Cl' and apply the unit-resolution rule to each of them with l' as auxiliary premiss.

Step 2. Let ϕ be a branch such that some unfulfilled clauses in ϕ contain the literal l, but no clause in ϕ contains the complementary literal l'. Then mark as fulfilled in ϕ all the clauses in ϕ containing l.

Step 3. Let ϕ be a branch containing two clauses C_1 and C_2 such that C_1 subsumes C_2, that is every literal in C_1 is contained also in C_2, and C_2 is unfulfilled. Then mark C_2 as fulfilled in ϕ.

Step 4. Let ϕ be a branch such that some unfulfilled clauses in ϕ contain the literal L and some other unfulfilled clauses in ϕ contain the complementary literal l'. Then apply the splitting rule to l, that is append to ϕ both l and l' as immediate successors. We call l the *cut literal* of this application of the splitting rule.

Notice that part a of Step 1, if we except the marking of the literal l, is not strictly necessary since it is subsumed by Step 3. However, it is convenient to execute it in the context of Step 1, since it involves the same kind of pattern-matching which is required anyway to carry out part b. This allows us to reduce as much as possible the set of unfulfilled clauses in the given branch with no extra cost in terms of efficiency.

The DPP can be described as a procedure which generates a DPP-tree as follows:

PROCEDURE 23. INPUT: A set S of clauses and a linear ordering R of the literals occurring in S.

1. **Start** with the one-branch tree consisting of all the clauses in S (taken in any order);

2. Consider the tree $\mathcal{T}$ generated by the preceding step; **if** $\mathcal{T}$ is closed, **then stop**: S is unsatisfiable;

3. **else** pick up the leftmost branch ϕ of the DPP-tree generated by the preceding step;

4. **if** Step i (with $i = 1, 2, 3$) is applicable to ϕ, **then** apply Step i to ϕ and return to 2 above;

5. **if** none of the Steps 1–3 is applicable to ϕ, **then**

   (5.1) **if** Step 4 is applicable to ϕ, **then** apply Step 4 to ϕ, choosing as cut literal (among those satisfying the condition for applying Step 4) the one which comes first in the ordering R, and return to 2 above;

   (5.2) **else stop**: S is satisfiable.

The soundness and completeness of this procedure is discussed in detail in [Fitting, 1996]. We conclude by remarking that the DPP is still regarded as one of the most efficient decision procedures for classical propositional logic. Indeed, the ordering which determines which literals should be used as cut literals when applying the splitting rule is crucial for the performance of the DPP. A 'wrong' choice of this order may lead to combinatorial explosion even when a short proof is available and the procedure can find it given the 'right' order of literals. Of course deciding which order is 'right' or optimal is a non-trivial combinatorial problem which may turn out to be intractable.

5.2 The KE System

The discussion in Section 3.8 leaves us with the following problem:

PROBLEM 24. Is there a refutation system which, though being 'close' to the tableau method, is not affected by the anomalies of cut-free systems?

Observe that the non-redundancy condition (5) on p. 81 is obviously fulfilled by any tree-method satisfying the stronger condition that distinct branches define *mutually exclusive* situations, i.e. contain inconsistent sets of formulae. The simplest rule of the branching type which generates mutually inconsistent branches is a 0-premise rule, that we call PB, corresponding to the principle of bivalence:

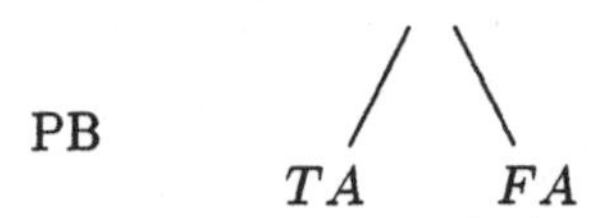

A version for unsigned formulae can be obtained in the obvious way, as a 0-premise branching rule with A and $\neg A$ as alternative conclusions.

We could, of course, simply 'throw in' this rule, leaving the tableau rules unchanged. The rule PB is nothing but a semantic reading of the cut rule as a principle of bivalence. Moreover, it is not difficult, by using the above rule in conjunction with the usual tableau rules, to construct short refutations of the 'hard examples' described in the previous section. Indeed, any proof system including a cut rule can polynomially simulate the truth-tables (see below, Section 6). Thus all the anomalies of Smullyan's tableaux would be solved simply by (re)-introducing cut as a primitive rule. However, such a 'solution' is not satisfactory. In the first place, it is clearly *ad hoc*. The tableau method is complete without the cut rule and it is not clear *when* such a rule should be used in a systematic refutation procedure. The 'mechanics' of the tableau method does not seem to accommodate the cut rule in any natural way. Secondly, the standard branching rules still do not satisfy our non-redundancy condition (5), which we identified as a *desideratum* for a well-designed refutation system.

The above considerations suggest that in a well-designed tableau method: (a) the cut rule (PB) should not be redundant and (b) it should be the *only* branching rule. Thus, a good solution to Problem 24 may consist in overturning the cut-free tradition: instead of eliminating cuts from proofs we assign the cut rule a central role and reformulate the elimination rules accordingly.

Given (a) and (b) above, Problem 24 reduces to the following:

PROBLEM 25. Are there simple elimination rules of linear type which combined with PB yield a refutation system for classical logic?

However, since a 0-premise rule like PB can introduce arbitrary formulae, we need also to solve:

PROBLEM 26. Can we restrict ourselves to *analytic* applications of PB, i.e. applications which do not violate the subformula property, without affecting completeness?

Furthermore, if Problem 26 has a positive solution, we shall be anyway left with a large choice of formulae as potential conclusions of an application of PB (let us call them *PB-formulae*). This may be a problem for the development of systematic refutation procedures. Thus we have to address also the following:

PROBLEM 27. Can we further restrict the set of potential PB-formulae so as to allow for simple systematic refutation procedures, like the standard procedure for the tableau method?

To solve Problem 25 we need only to notice that the following eleven facts hold true under any boolean valuation:

1. If $A \vee B$ is true and A is false, then B is true.
2. If $A \vee B$ is true and B is false, then A is true.
3. If $A \vee B$ is false then both A and B are false.
4. If $A \wedge B$ is false and A is true, then B is false.
5. If $A \wedge B$ is false and B is true, then A is false.
6. If $A \wedge B$ is true then both A and B are true.

7. If $A \to B$ is true and A is true, then B is true.

8. If $A \to B$ is true and B is false, then A is false.

9. If $A \to B$ is false, then A is true and B is false.

10. If $\neg A$ is true, then A is false.

11. If $\neg A$ is false, then A is true.

These facts can immediately be used to provide a set of expansion rules of the linear type which, with the addition of PB, constitute a complete set of rules for classical propositional logic.

These rules characterize the propositional fragment of the system **KE** (first proposed in [Mondadori, 1988; Mondadori, 1988a]) and are shown in Table 3. Notice that those with two signed formulae below the line represent *a pair* of expansion rules of the linear type, one for each signed formula. The rules involving the logical operators will be called *elimination rules* or *E-rules*.[32]

In contrast with the tableau rules for the same logical operators, the E-rules are all of the linear type and are *not* a complete set of rules for classical propositional logic. The reason is easy to see. The E-rules, intended as 'operational rules' which govern our use of the logical operators do not say anything about the bivalent structure of the intended models. If we add the rule PB as the only rule of the branching type, completeness is achieved. So PB is *not eliminable* in the system **KE**.[33] As pointed out above (p. 89), there is a close correspondence between the semantic rule PB and the cut rule of the sequent calculus.

We call an application of PB a *PB-inference* and the formulae which are the conclusions of the PB-inference *PB-formulae*. Finally, if TA and FA are the conclusions of a given PB-inference, we shall say that PB has been applied to the formula A.

REMARK 28. Notice that all the linear rules can be easily simulated by the tableau rules. This, of course, does not apply to PB.

DEFINITION 29. Let $S = \{X_1, \ldots, X_m\}$ be a set of signed formulae. Then $\mathcal{T}$ *is a* **KE**-*tree for* S if there exists a finite sequence $(\mathcal{T}_1, \mathcal{T}_2, \ldots, \mathcal{T}_n)$ such that (i) $\mathcal{T}_1$

[32]Quite independently, and with a different motivation, Cellucci [Cellucci, 1987] formulates the same set of rules (although he does not use signed formulae). Surprisingly, the two-premise rules in the above list were already discovered by Chrysippus who claimed them to be the fundamental rules of reasoning ('anapodeiktoi'), except that disjunction was interpreted by him in an exclusive sense. Chrysippus also maintained that his 'anapodeiktoi' formed a complete set of inference rules ('the indemonstrables are those of which the Stoics say that they need no proof to be maintained. [...] They envisage many indemonstrables but especially five, from which it seems all others can be deduced'. See [Blanché, 1970], pp.115–119 and [Bochensky, 1961], p.126).

[33]This does not mean, however, that the **KE**-rules are unsuitable for non-classical logics. In fact, the classical signs can be used in a non-standard interpretation in order to yield appropriate rules for a variety of non-classical logics. For many-valued logics see [Hähnle, 1992]. For substructural logics see [D'Agostino and Gabbay, 1994] and Chapter 7 of this *Handbook*.

Disjunction Rules

$$\frac{TA \vee B \quad FA}{TB}\,\mathrm{E}T\vee 1 \qquad \frac{TA \vee B \quad FB}{TA}\,\mathrm{E}T\vee 2 \qquad \frac{FA \vee B}{FA \quad FB}\,\mathrm{E}F\vee$$

Conjunction Rules

$$\frac{FA \wedge B \quad TA}{FB}\,\mathrm{E}F\wedge 1 \qquad \frac{FA \wedge B \quad TB}{FA}\,\mathrm{E}F\wedge 2 \qquad \frac{TA \wedge B}{TA \quad TB}\,\mathrm{E}T\wedge$$

Implication Rules

$$\frac{TA \to B \quad TA}{TB}\,\mathrm{E}T\to 1 \qquad \frac{TA \to B \quad FB}{FA}\,\mathrm{E}T\to 2 \qquad \frac{FA \to B}{TA \quad FB}\,\mathrm{E}F\to$$

Negation Rules

$$\frac{T\neg A}{FA}\,\mathrm{E}T\neg \qquad \frac{F\neg A}{TA}\,\mathrm{E}F\neg$$

Principle of Bivalence

$$\frac{}{TA \mid FA}\,\mathrm{PB}$$

Table 3. The **KE**-rules for signed formulae

is a one-branch tree consisting of the sequence of $X_1, \ldots, X_m$; (ii) $\mathcal{T}_n = \mathcal{T}$, and (iii) for each $i < n$, $\mathcal{T}_{i+1}$ results from $\mathcal{T}_i$ by an application of a rule of **KE**.

DEFINITION 30.

1. Given a **KE**-tree $\mathcal{T}$ of signed formulae, a branch ϕ of $\mathcal{T}$ is *closed* if for some atomic formula P, both TP and FP are in ϕ. Otherwise it is *open*.
2. A **KE**-tree $\mathcal{T}$ of signed formulae is *closed* if each branch of $\mathcal{T}$ is closed. Otherwise it is *open*.
3. A tree $\mathcal{T}$ is a **KE**-*refutation of* S if $\mathcal{T}$ is a closed **KE**-tree for S.
4. A tree $\mathcal{T}$ is a **KE**-*proof of* A *from a set* Γ *of formulae* if $\mathcal{T}$ is a **KE**-refutation of $\{TB | B \in \Gamma\} \cup \{FA\}$.

REMARK 31. It is easy to prove that if a branch ϕ of $\mathcal{T}$ contains both TA and FA for some non-atomic formula A, ϕ can be extended *by means of the E-rules only* to a branch ϕ' which is atomically closed in the sense of the previous definition. Hence, in what follows we shall consider a branch closed as soon as both TA and FA appear in it, for an arbitrary formula A.

We can, of course, give a version of **KE** which works with unsigned formulae. The rules are shown in Table 4. It is intended that all definitions be modified in the obvious way. We can see from the unsigned version that the two-premise rules correspond to familiar principles of inference: *modus ponens, modus tollens, disjunctive syllogism* and the dual of disjunctive syllogism. Thus **KE** can be seen as a system of classical natural deduction using elimination rules only. However, the classical operators are analysed as such and not as 'stretched' versions of the constructive ones (like in [Gentzen, 1935] and [Prawitz, 1965]).

In Figure 12 we give a **KE**-refutation (using unsigned formulae) and compare it with a minimal tableau for the same set of formulae; the reader can compare the different structure of the two refutations and the crucial use of (the unsigned version of) PB to eliminate the redundancy exhibited by the tableau refutation. Notice that the thicker subtree in the tableau refutation is clearly redundant.

It is convenient to use Smullyan's unifying notation in order to reduce the number of cases to be considered (for which see Fitting's Introduction). So the E-rules of **KE** can be 'packed' into the following three rules (where β_i^c, $i = 1, 2$ denotes the *conjugate* of β_i):

Rule A $\quad \dfrac{\alpha}{\begin{array}{c}\alpha_1 \\ \alpha_2\end{array}}$

Rule B1 $\quad \dfrac{\begin{array}{c}\beta \\ \beta_1^c\end{array}}{\beta_2}$ $\qquad$ Rule B2 $\quad \dfrac{\begin{array}{c}\beta \\ \beta_2^c\end{array}}{\beta_1}$

Disjunction Rules

$$\frac{\begin{array}{c}A \vee B\\ \neg A\end{array}}{B}\ \mathrm{E}\vee 1 \qquad \frac{\begin{array}{c}A \vee B\\ \neg B\end{array}}{A}\ \mathrm{E}\vee 2 \qquad \frac{\neg(A \vee B)}{\begin{array}{c}\neg A\\ \neg B\end{array}}\ \mathrm{E}\neg\vee$$

Conjunction Rules

$$\frac{\begin{array}{c}\neg(A \wedge B)\\ A\end{array}}{\neg B}\ \mathrm{E}\neg\wedge 1 \qquad \frac{\begin{array}{c}\neg(A \wedge B)\\ B\end{array}}{\neg A}\ \mathrm{E}\neg\wedge 2 \qquad \frac{A \wedge B}{\begin{array}{c}A\\ B\end{array}}\ \mathrm{E}\wedge$$

Implication Rules

$$\frac{\begin{array}{c}A \to B\\ A\end{array}}{B}\ \mathrm{E}{\to}1 \qquad \frac{\begin{array}{c}A \to B\\ \neg B\end{array}}{\neg A}\ \mathrm{E}{\to}2 \qquad \frac{\neg(A \to B)}{\begin{array}{c}A\\ \neg B\end{array}}\ \mathrm{E}\neg\to$$

Negation Rule

$$\frac{\neg\neg A}{A}\ \mathrm{E}\neg\neg$$

Principle of Bivalence

$$\frac{}{A \mid \neg A}\ \mathrm{PB}$$

Table 4. **KE** rules for unsigned formulae

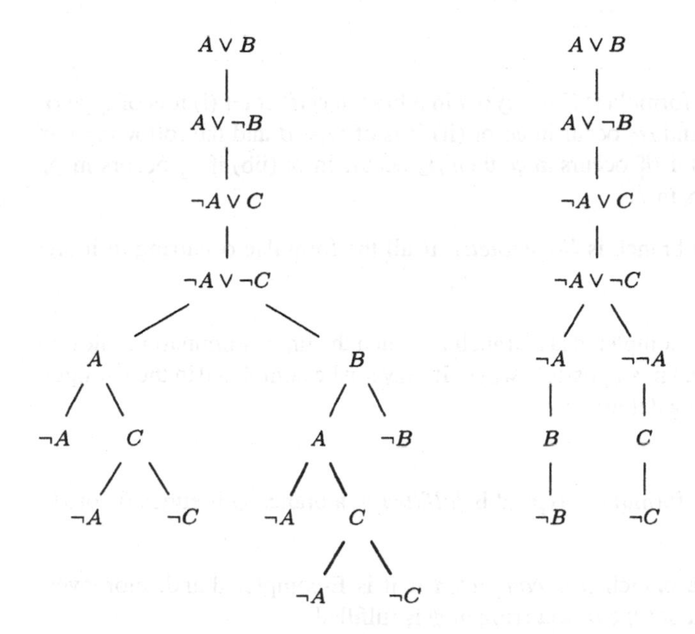

Figure 12. A minimal tableau (on the left) and a **KE**-refutation (on the right) of $\{A \vee B, A \vee \neg B, \neg A \vee C, \neg A \vee \neg C\}$. The thick subtree in the tableau refutation is redundant.

The Rule A can be seen as a pair of branch-expansion rules, one with conclusion α_1 and the other with conclusion α_2. In each application of the rules, the signed formulae α and β are called *major premisses*. In each application of rules B1 and B2 the signed formulae β_i^c, $i = 1, 2$, are called *minor premisses* (rule A has no minor premiss).

REMARK 32. The unifying notation can be easily adapted to unsigned formulae: simply delete all the signs 'T' and replace all the signs 'F' by '$\neg$'. The 'packed' version of the rules then suggests a more economical version of **KE** for unsigned formulae when β_i^c is taken to denote the *complement* of β_i defined as follows: the *complement* of an *unsigned formula* A, is equal to $\neg B$ if $A = B$ and to B if $A = \neg B$. In this version the rules E$\vee$1, E$\vee$2 and E$\rightarrow$2 become:

$$\frac{\begin{array}{c}A \vee B\\ A^c\end{array}}{B} \qquad \frac{\begin{array}{c}A \vee B\\ B^c\end{array}}{A} \qquad \frac{\begin{array}{c}A \rightarrow B\\ B^c\end{array}}{A^c}$$

This version is to be preferred for practical applications.

We now outline a simple refutation procedure for **KE** that we call *the canonical procedure*. First we define some related notions.

DEFINITION 33.

1. We say that a formula is *E-analysed* in a branch ϕ if either (i) it is of type α and both α_1 and α_2 occur in ϕ; or (ii) it is of type β and the following are satisfied: (iia) if β_1^c occurs in ϕ, then β_2 occurs in ϕ; (iib) if β_2^c occurs in ϕ, then β_1 occurs in ϕ.

2. We say that a branch is *E-completed* if all the formulae occurring in it are E-analysed.

A branch which is E-completed is a branch in which the linear elimination rules of **KE** have been applied in all possible ways. It may not be completed in the stronger sense of the following definition.

DEFINITION 34.

1. We say that a formula of type β is *fulfilled* in a branch ϕ if either β_1 or β_2 occurs in ϕ.

2. We say that a branch ϕ is *completed* if it is E-completed and, moreover, every formula of type β occurring in ϕ is fulfilled.

3. We say that a **KE**-*tree* is *completed* if all its branches are completed.

PROCEDURE 35. The *canonical procedure for* **KE** starts from the one-branch tree consisting of the initial formulae and applies the **KE**-rules until the resulting tree is either closed or completed. At each stage of the construction the following steps are performed:

1. select an open branch ϕ which is not yet completed (in the sense of Definition 34);

2. if ϕ is not E-completed, expand ϕ by means of the E-rules until it becomes E-completed;

3. if the resulting branch ϕ' is neither closed nor completed then

   (3.1) select a formula of type β which is not yet fulfilled (in the sense of Definition 34) in the branch;

   (3.2) apply PB with β_1 and β_1^c (or, equivalently, β_2 and β_2^c) as PB-formulae and go to step 1.

   otherwise, go to step 1.

PROPOSITION 36. *The canonical procedure is complete.*

Proof. It is easy to see that the canonical procedure eventually yields a **KE**-tree which is either closed or completed. The only crucial observation is that when one applies the rule PB as prescribed, with β_i and β_i^c as PB-formulae (with i equals 1 or 2 and β a non-fulfilled formula occurring in the branch), then once the two resulting branches are E-completed as a result of performing step 2, each of them will contain either β_1 or β_2. Therefore, eventually, each branch will be either closed or completed. Since the formulae in a completed branch form a Hintikka set, the completeness of **KE** follows immediately via Hintikka's lemma. ■

Of course, what we have called 'the canonical procedure' is not, strictly speaking, a completely deterministic algorithm. Some steps involve a choice and different strategies for making these choices will lead to different algorithms. However it describes a 'mechanical version' of **KE** which is sufficient for our purpose of comparing it with the tableau method. We can look at it as to a *restricted proof system*, where the power of the cut rule is severely limited to allow for easy proof-search.[34] The next three corollaries unfold the content of Proposition 36.

COROLLARY 37 (Analytic cut property). *If S is unsatisfiable, then there is a closed* **KE***-tree $\mathcal{T}'$ for S such that all the applications of PB preserve the subformula property.*

Proof. All the PB-formulae involved in the canonical procedure are subformulae of formulae previously occurring in the branch. ■

Let us call *analytic* the applications of PB which preserve the subformula property, and *analytic restriction of* **KE** the system obtained by restricting PB to analytic applications. The above corollary says that the analytic restriction of **KE** is complete. Since the elimination rules preserve the subformula property, the subformula principle follows immediately. A constructive proof of the subformula principle, which yields a procedure for transforming any **KE**-proof into an equivalent **KE**-proof which enjoys the subformula property, is given in [Mondadori, 1988a].

In fact the use of PB in the canonical procedure is even more restricted than it appears from the above corollary. Consider the following definition of *strongly analytic* application of PB.

DEFINITION 38. An application of PB in a branch ϕ of a **KE**-tree is *strongly analytic* if the PB-formulae of this application are β_i and β_i^c for some $i = 1, 2$ and some non-fulfilled formula of type β occurring in ϕ. A **KE**-*tree* is *strongly analytic* if it contains only strongly analytic applications of PB.

All the applications of PB in the canonical procedure are strongly analytic. Thus, it follows from the completeness of the canonical procedure that:

[34] The reader should be aware that this is not the *only* mechanical procedure and not necessarily *the best*.

COROLLARY 39. *If S is unsatisfiable, then there is a closed strongly analytic* **KE**-*tree for S.*

In other words, our argument establishes the completeness of the restricted system obtained by replacing the 'liberal' version of PB with one which allows only strongly analytic applications. In general, analytic applications of PB do not need to be strongly analytic. The PB-formulae may well be subformulae of some formula occurring above in the branch, and yet not satisfy the strongly analytic restriction. Therefore, Corollary 39 is stronger than Corollary 37. One can ask whether the strongly analytic restriction is as powerful, from the complexity viewpoint, as the analytic restriction. This problem is still open.

In fact, the canonical procedure imposes even more control on the applications of the cut rule PB by requiring that it is applied on a branch only when the linear elimination rules are no further applicable, i.e. when the branch is E-completed. This strategy avoids unnecessary branchings which may increase the size of proofs, much as, in the standard tableau method, the basic strategy consisting in applying the α-elimination rule before the β-elimination rule, allows for slimmer tableaux.[35] Such a stricter notion of an analytic **KE**-tree is captured by the following definition:

DEFINITION 40. We say that an application of PB in a branch ϕ is *canonical* if (i) it is strongly analytic, and (ii) ϕ is E-completed. We also say that a **KE**-*tree is canonical* if all the applications of PB in it are canonical.

Canonical **KE**-trees are exactly those which are generated by our canonical procedure described above. Hence:

COROLLARY 41. *If S is unsatisfiable, then there is a closed canonical* **KE**-*tree for S.*

The canonical procedure for **KE** is closely related to the Davis–Putnam procedure. It is not difficult to see that, if we extend our language to deal with 'generalized' (n-ary) disjunctions and conjunctions, the Davis–Putnam procedure (in the version of [Davis *et al.*, 1992] which is also the one exposed in [Chang and Lee, 1973, Section 4.6] and in Fitting's textbook [Fitting, 1996, Section 4.4]), can be represented as a special case of the canonical procedure for **KE**. So, from this point of view, the canonical procedure for **KE** provides a generalization of the Davis–Putnam procedure which does not require reduction in clausal form. To see that the DPP is a special case of the canonical procedure for **KE** observe that, if we restrict ourselves to formulae in clausal form, every branch in a canonical **KE**-tree performs what essentially is a *unit-resolution* refutation [Chang, 1970]. On the

[35] Since in our approach *all* the elimination rules are linear, it does not make any difference whether, in applying them, we give priority to formulae of type α or β. Of course, one may describe a similar strategy for the standard tableau method, by giving priority to α-expansions or β-expansions applied to formulae β such that either β_1^c or β_2^c occurs above in the branch. In this way one of the branches of the β-expansion closes immediately.

other hand, in this special case, PB corresponds to the splitting rule of the Davis–Putnam procedure (when represented in tree form, as in [Chang and Lee, 1973, Section 4.6] and in [Fitting, 1996, Section 4.4]). All we need is to generalize our language so as to include n-ary disjunctions, with arbitrary n, and modify the rule for $\vee$-elimination in the obvious way.

Notice that the system resulting from this generalized version of **KE** by disallowing the branching rule PB (so that all the rules are linear) includes unit resolution as a *special case*. This restricted version of **KE** can then be seen as an extension of unit resolution (it is therefore a complete system for Horn clauses, although it is not confined to formulae in clausal form). Although its scope largely exceeds that of unit resolution, the procedure is still *feasible* as shown by Proposition 45 below.

5.3 *The Relative Complexity of* **KE**

In this section we discuss the complexity of *analytic* **KE**-refutations. We first show that, unlike the tableau method, the analytic **KE**
indexanalytic restriction of **KE** system can p-simulate the truth-tables . Next we briefly discuss the complexity of proof-search in a hierarchy of subsystems arising by allowing only a fixed number of applications of the cut-rule (PB). Then we compare the analytic, strongly analytic and canonical restrictions of **KE** with other analytic proof systems (i.e. systems obeying the subformula principle). Some systems of deduction—such as tableau method and Gentzen's sequent calculus without cut—yield only analytic proofs. Others—such as Natural Deduction, Gentzen's sequent calculus with cut and **KE**— allow for a more general notion of proof which includes non-analytic proofs, although in all these cases the systems obtained by restricting the potentially non-analytic rules to analytic applications are still complete. Since we are interested, for theoretical and practical reasons, in analytic proofs, we shall pay special attention to simulation procedures which preserve the subformula property.

DEFINITION 42. The *length* of a proof π, denoted by $|\pi|$ is the total number of symbols occurring in π (intended as a string).

The λ-*complexity*, of π, denoted by $\lambda(\pi)$, is the *number of lines* in the proof π (each 'line' being a sequent, a formula, or any other expression associated with an inference step, depending on the system under consideration). Finally the ρ-*complexity* of π, denoted by $\rho(\pi)$ is the length (total number of symbols) of a line of maximal length occurring in π.

Our complexity measures are obviously connected by the relation

$$|\pi| \leq \lambda(\pi) \cdot \rho(\pi).$$

Now, observe that the λ-measure is sufficient to establish negative results about the p-simulation relation, but is not sufficient in general for positive results. It may, however, be adequate also for positive results whenever one can show that the ρ-measure (the length of lines) is not significantly increased by the simulation procedure under consideration. All the procedures that we shall consider in the sequel will be of this kind. So we shall forget about the ρ-measure and identify the complexity of a proof system with the λ-measure (this property is called 'polynomial transparency' in the general framework developed in [Letz, 1993].

As said before, we are interested in the complexity not of proofs in general but of *analytic proofs*. We shall then speak of the *analytic restriction* of a system, i.e. the rules are restricted to applications which preserve the subformula property.

Notice that the analytic restrictions of Gentzen's sequent calculus and natural deduction are strictly more powerful than the cut-free sequent calculus and normal natural deduction respectively. For instance, the analytic restriction of the sequent calculus allows cuts provided they are restricted to subformulae.

In what follows we shall consider the version of **KE** which uses *unsigned* formulae.

KE *and the Truth-tables*

The complexity of the truth-table procedure for a given formula A is sometimes measured by the number of rows in the complete truth-table for that formula, i.e. 2^k, where k is the number of distinct atomic formulae in A. In fact, a better way of measuring the complexity of the truth-tables takes into account also the *length* of the formula to be tested. In any case, it is important to notice that the complexity of the truth-table procedure is *not always* exponential in the length of the formula. It is so only when the number of distinct atoms approaches the length of the formula. On the contrary, the complexity of tableau proofs depends more crucially on the length of the formula. Therefore, there might be (and in fact there are, as pointed out in Section 3.8 above) examples which are 'easy' for the truth-tables and 'hard' for the tableau method. We described this situation as unnatural. We can turn these considerations into a positive criterion and require that a well-designed proof system should be able, at least, to p-simulate the truth-tables, i.e. the most basic semantical and computational characterization of classical propositional logic.

DEFINITION 43. Let us say that a proof system is *standard* if its complexity (i.e. minimal proof-length) is $O(n^c \cdot 2^k)$, where n is the length of the input formula, c is a fixed constant and k the number of distinct variables occurring in it.

The above definition requires that the complexity of the truth-tables be an upper bound on the complexity of an acceptable proof system.

PROPOSITION 44. *The analytic restriction of* **KE** *is a standard proof system. In fact, for every tautology A of length n and containing k distinct variables, there is a* **KE**-*refutation $\mathcal{T}$ of $\neg A$ with $\lambda(\mathcal{T}) = O(\lambda(\mathcal{T})) = O(n \cdot 2^k)$*

Proof. [Sketch] There is an easy **KE**-simulation of the truth-table procedure. First apply PB to all atomic letters on each branch. This generates a tree with

2^k branches. Then each branch can be closed by means of a **KE**-tree of linear size. (Hint: each truth-table rule can be simulated in a fixed number of steps.) Notice that the applications of PB required for the proof, though analytic, are not strongly analytic. ■

It is easy to show that the task of saturating a branch under the E-rules is computationally easy:

PROPOSITION 45. *Let ϕ be a branch containing m nodes, each of which is an occurence of a formula of degree d_m. The task of E-completing a branch ϕ (i.e. saturating it under the* **KE** *elimination rules) can be performed in polynomial time, more precisely in time $O(n^2)$ where $n = \sum_{m \in \phi} d_m$.*

Proof. The proposition trivially follows from the fact that the E-rules have the subformula property and there are no more that n distinct subformulae of the formulae in ϕ. It is not difficult to see that at most $O(n^2)$ pattern matchings need to be performed. ■

Therefore, the complexity of a tautology depends entirely on the number of PB-branchings which are required in order to complete the tree. Let us call **KE**(k) the system obtained from **KE** by allowing at most k *nested* analytic applications of the cut-rule (i.e. the maximum number of cut-formulae on each branch is k). We have just shown that **KE**(0) has a decision procedure which runs in time $O(n^2)$. The set for which **KE**(0) is complete includes the horn-clause fragment of propositional logic (however **KE**(0) is not restricted to clausal form logic). Moreover, we can show that:

PROPOSITION 46. *For every fixed k,* **KE**(k) *has a polynomial time decision procedure.*

It is obvious that the set of tautologies for which **KE**(k) is complete tends to TAUT, the set of all the tautologies, as k tends to infinity. The crucial point is that low-degree cut-bounded systems are powerful enough for a wide range of applications. Notice that in this approach, the source of the complexity of proving a tautology in the system is clearly identified, and quite large fragments of classical logic can be covered with a very limited number of applications of the branching rule PB (even with no applications of PB, most of the 'textbook examples' can be proved).

KE *Versus the Tableau Method*

First we notice that, given a tableau refutation $\mathcal{T}$ of Γ we can effectively construct a *strongly analytic* **KE**-refutation $\mathcal{T}'$ of Γ which is not essentially longer.

PROPOSITION 47. *If there is a tableau proof $\mathcal{T}$ of A from Γ, then there is a* strongly analytic **KE**-*proof $\mathcal{T}'$ of A from Γ such that $\lambda(\mathcal{T}') \leq 2\lambda(\mathcal{T})$.*

Proof. Observe that the elimination rules of **KE**, combined with PB, can easily simulate the branching rules of the tableau method. For instance in the case of the branching rule for eliminating disjunctions either of the following two simulations can be used (all the other cases are similar):

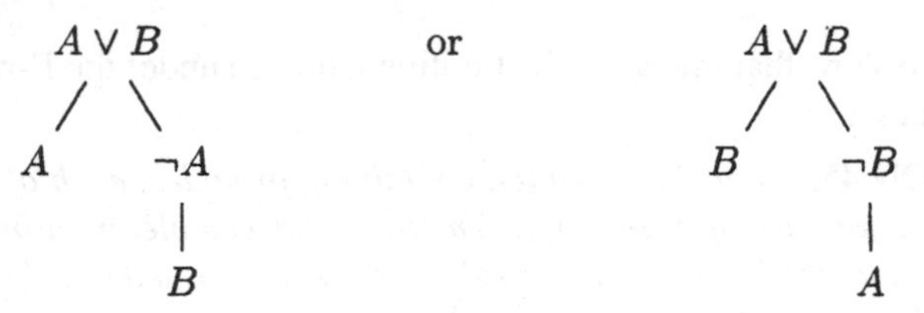

Notice that the applications of PB are strongly analytic. Such a simulation lengthens the original tableau by one node for each application of a branching rule. Since the linear rules of the tableau method are also rules of **KE**, it follows that there is a **KE**-refutation $\mathcal{T}'$ of Γ such that $\lambda(\mathcal{T}') \leq \lambda(\mathcal{T}) + k$, where k is the number of applications of branching rules in $\mathcal{T}$. Since k is obviously $\leq \lambda(\mathcal{T})$, then $\lambda(\mathcal{T}') \leq 2\lambda(\mathcal{T})$. ■

It also follows from Proposition 47 and Proposition 45 that:

PROPOSITION 48. *The canonical restriction of* **KE** *can p-simulate the tableau method.*

Notice that the **KE**-simulation contains more information than the simulated tableau. In the **KE**-simulation of the branching rules, one of the branches contains a formula which does not occur in the corresponding branch of the tableau. These additional formulae may allow for the closure of branches that, in the simulated tableau, remain open and are closed only after a redundant computational process. So, while all the tableau rules can be easily simulated by means of **KE**-rules, **KE** includes a rule, namely PB, which cannot be easily simulated by means of the tableau rules. Although it is well-known that the addition of this rule to the tableau rules does not increase the stock of inferences that can be shown valid (since PB is classically valid and the tableau method is classically complete), its absence, in some cases, is responsible for an explosive growth in the size of tableau proofs. In Section 3.8 we have already identified the source of this combinatorial explosion in the fact that the tableau rules fail to simulate even *analytic* cut inferences. When a rule like PB is available, simulating any 'cut' inference is relatively inexpensive in terms of proof size as is shown by the diagram below:

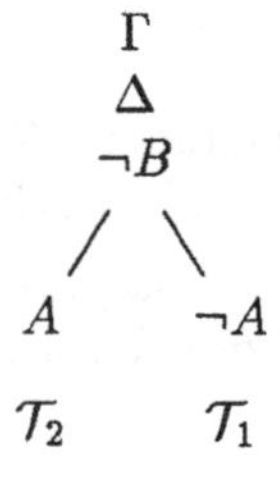

On the other hand if PB is not in the stock of rules, reproducing a cut inference (even an analytic one) may be much harder, as we have shown in Section 3.8, and require a great deal of duplication in the construction of a closed tableau. We have already discussed a class of hard examples in Section 3.8 which are easy not only for (the canonical restriction of) **KE** but also for the truth-table method. By Proposition 16 above, the tableau method cannot p-simulate the truth-tables. Therefore, in spite of their similarity, the tableau method cannot polinomially simulate **KE**. This fact already follows from Proposition 16, Propositions 44 and 47. Moreover, it is not difficult to see that the class of hard examples for the tableau method used in the proof of Proposition 16 is 'easy' also for the *canonical restriction* of **KE**. Figure 13 shows a canonical **KE**-refutation of the set of clauses H_{P_1,P_2,P_3}. It is apparent that, in general, the number of branches in the **KE**-tree for $H_{P_1,\ldots,P_k}$, constructed according to the same pattern, is exactly 2^{k-1} (which is the number of clauses in the expression divided by 2) and that the refutation trees have size $O(k \cdot 2^k)$.

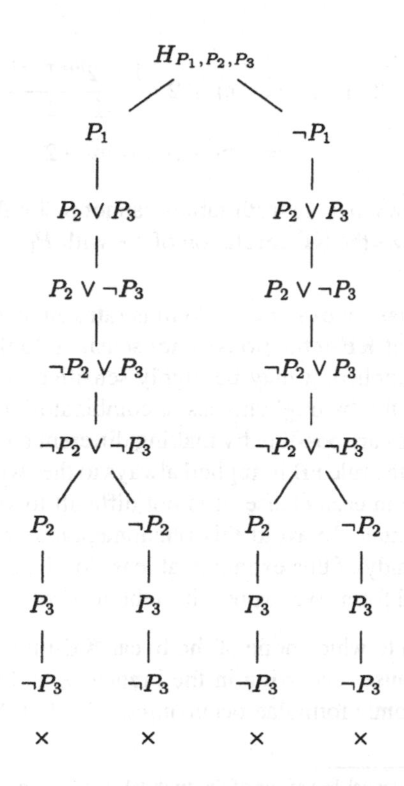

Figure 13. A canonical **KE**-refutation of H_{P_1,P_2,P_3}

This is sufficient to establish:

PROPOSITION 49. *The tableau method cannot p-simulate the canonical restriction of* **KE**.

A simple analysis shows that also the hard examples for Smullyan's tableaux used in the Section 3.8 to prove Theorem 17 are easy for the canonical restriction of **KE**. Indeed, there are short canonical **KE**-refutations of Σ_n which contain $2^n + 2^n n - 2$ nodes. Such refutations have the following form: start with Σ_n. This will be a set containing $m(= 2^n)$ clauses of which $m/2$ start with, say, P^1 and the remaining $m/2$ with its negation. Apply PB to $\neg P^1$. This creates a branching with $\neg P^1$ in one branch and $\neg\neg P^1$ in the other. Now, on the first branch, by means of $m/2$ applications of the rule E$\vee$1 we obtain a set of formulae which is of the same form as Σ_{n-1}. Similarly on the second branch we obtain another set of the same form as Σ_{n-1}. By reiterating the same procedure we eventually produce a closed tree for the original set Σ_n. It is easy to see that the number of nodes generated by the refutation can be calculated as follows (where m is the number of formulae in Σ_n, namely 2^n):

$$\begin{aligned}\lambda(\mathcal{T}) = m + \sum_{i=1}^{\log m-1} 2^i + m &= m + 2 \cdot \frac{1 - 2^{\log m - 1}}{1-2} + m \cdot (\log m - 1) \\ &= m + m \log m - 2.\end{aligned}$$

This result also shows that the truth-tables cannot p-simulate **KE** in non-trivial cases.[36] Figure 14 shows the **KE**-refutation of Σ_3 with $P_1, \ldots, P_7$ as atomic variables.

REMARK 50. This class of examples also illustrates an interesting phenomenon: while the complexity of **KE**-refutations is not sensitive to the order in which the elimination rules are applied, it may be highly sensitive to the choice of the PB formulae. If we make the 'wrong' choices, a combinatorial explosion may result when 'short' refutations are possible by making different choices. If, in Cook and Reckhow's examples, the rule PB is applied always to the 'wrong' atomic variable, namely to the last one in each clause, it is not difficult to see that the size of the tree becomes exponential. To avoid this phenomenon an obvious criterion suggests itself from the study of this example, at least for the case in which the input formulae are in clausal form. We express it in the form of a *heuristic principle*:

HP Let ϕ be a branch to which none of the linear **KE**-rules is applicable. Let S_ϕ be the set of clauses occurring in the branch ϕ, and let $P_1, \ldots, P_k$ be the list of all the atomic formulae occurring in S_ϕ. Let N_{P_l} be the number of

[36] We mean that the exponential behaviour of the truth-tables in this case does not depend *only* on the large number of variables but also on the logical structure of the expressions. So these examples are essentially different from the examples which are usually employed in textbooks to show that the truth-tables are intractable (a favourite one is the sequence of expressions $A \vee \neg A$ where A contains an increasing number of variables).

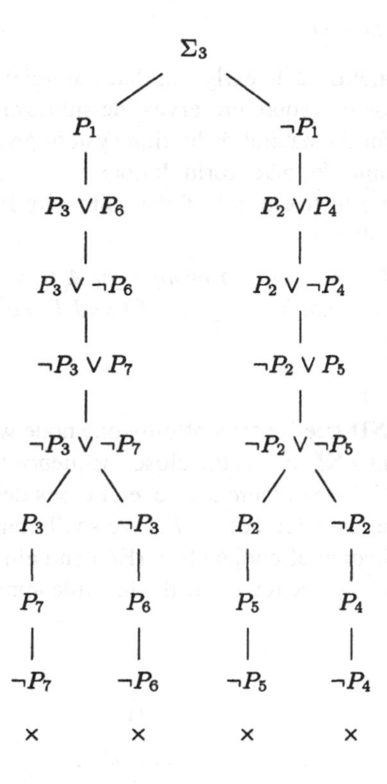

Figure 14. A **KE**-refutation of Σ_3

clauses C such that P_I or $\neg P_i$ occurs in C. Then apply PB to an atom P_i such that N_{P_i} is maximal.

A detailed study of proof-search in the **KE**-system will have to involve more sophisticated criteria for the choice of the PB-formulae. Here we just stress that the simulation of the tableau rules by means of the **KE**-rules is independent of the choice of the PB-formulae. On the other hand a good choice may sometimes be crucial for generating essentially shorter proofs than those generated by the tableau method (sometimes it does not matter at all: the examples of Proposition 16 which are 'hard' for the tableau method, are 'easy' for the canonical restriction of **KE** no matter how the PB-formulae are chosen). In any case, our discussion shows that analytic cuts are often essential for the existence of short refutations *with the subformula property*.[37]

[37] This can be taken as further evidence in support of Boolos' plea for not eliminating cut [Boolos, 1984]. In that paper he gives a natural example of a class of first order inference schemata which are

KE *Versus Natural Deduction*

It can also be shown that **KE** can linearly simulate natural deduction (in tree form). Moreover the simulation procedure preserves the subformula property. We shall sketch this procedure for the natural deduction system given in [Prawitz, 1965], the procedure being similar for other formulations.

We want to give an effective proof of the following Proposition (where **ND** stands for Natural Deduction):

PROPOSITION 51. *If there is an* **ND**-*proof* $\mathcal{T}$ *of* A *from* Γ, *then there is a* **KE**-*proof* $\mathcal{T}'$ *of* A *from* Γ *such that* $\lambda(\mathcal{T}') \leq 3\lambda(\mathcal{T})$ *and* $\mathcal{T}'$ *contains only formulae* A *such that* A *occurs in* $\mathcal{T}$.

Proof. By induction on $\lambda(\mathcal{T})$.
If $\lambda(\mathcal{T}) = 1$, then the **ND**-tree consists of only one node which is an assumption, say C. The corresponding **KE**-tree is the closed sequence $C, \neg C$.
If $\lambda(\mathcal{T}) = k$, with $k > 1$, then there are several cases depending on which rule has been applied in the last inference of $\mathcal{T}$. We shall consider only the cases in which the rule is elimination of conjunction (E$\wedge$) and elimination of disjunction (E$\vee$), and leave the others to the reader. If the last rule applied in $\mathcal{T}$ is E$\wedge$, then $\mathcal{T}$ has the form:

$$\mathcal{T} = \begin{array}{c} \Delta \\ \mathcal{T}_1 \\ A \wedge B \\ \hline A \end{array}$$

By inductive hypothesis there is a **KE**-refutation $\mathcal{T}_1'$ of $\Delta, \neg(A \wedge B)$ such that $\lambda(\mathcal{T}_1') \leq 3\lambda(\mathcal{T}_1)$. Then the following **KE**-tree:

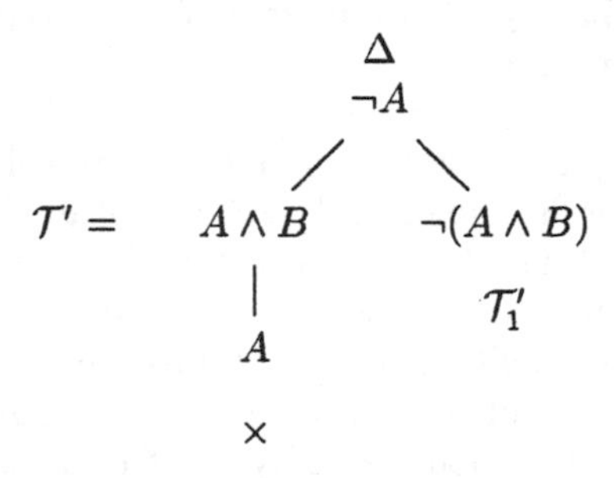

'hard' for the tableau method while admitting 'easy' (*non-analytic*) natural deduction proofs. Boolos' examples are a particularly clear illustration of the well-known fact that the elimination of cuts from proofs in a system in which cuts are eliminable can greatly increase the complexity of proofs. (For a related technical result see [Statman, 1978].) **KE** provides an elegant solution to Boolos' problem by making cut non-eliminable while preserving the subformula property of proofs. Our discussion also shows that eliminating *analytic* cuts can result in a combinatorial explosion.

is the required **KE**-proof and it is easy to verify that $\lambda(\mathcal{T}') \leq 3\lambda(\mathcal{T})$.

If the last rule applied in $\mathcal{T}$ is the rule of elimination of disjunction, then $\mathcal{T}$ has the form:

$$\mathcal{T} = \begin{array}{ccc} \Delta_1 & \Delta_2, [A] & \Delta_3, [B] \\ \mathcal{T}_1 & \mathcal{T}_2 & \mathcal{T}_3 \\ A \vee B & C & C \\ \hline & C & \end{array}$$

By inductive hypothesis there are **KE**-refutations $\mathcal{T}_1'$ of $\Delta_1, \neg(A \vee B)$, $\mathcal{T}_2'$ of $\Delta_2, A, \neg C$, and $\mathcal{T}_3'$ of $\Delta_3, B, \neg C$, such that $\lambda(\mathcal{T}_i') \leq 3\lambda(\mathcal{T}_i)$, $i = 1, 2, 3$. Then the following **KE**-tree:

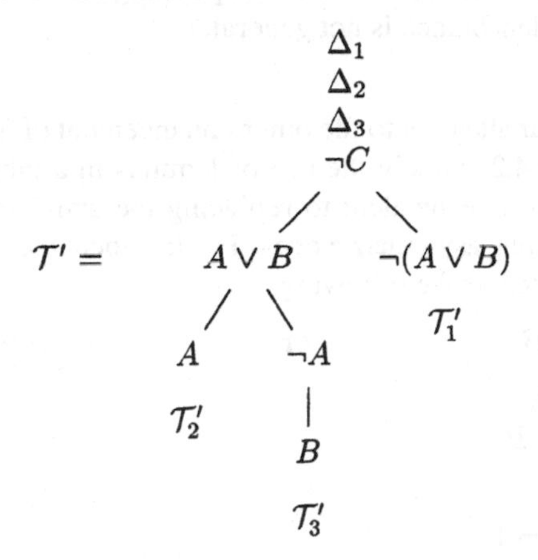

is the required proof and it is easy to verify that $\lambda(\mathcal{T}') \leq 3\lambda(\mathcal{T})$. ■

KE *and Other Refinements of Analytic Tableaux*

We have seen that the technique called 'merging' is an *ad hoc* method for solving some of the anomalies of Smullyan's tableaux, in particular what we have identified as the 'redundancy' of Smullyan's rules (see Section 3.8 above). In **KE**, the use of PB as the sole branching rule, instead of the standard branching rules of tableaux, *removes* this redundancy by making all the branches in the tree *mutually inconsistent*. As a result, the redundant branches which are stopped via merging are simply *not* generated in a **KE**-tree (see the example in Figure 15). The example shows that the applications of PB required to simulate merging are *strongly analytic*. These considerations establish the following fact:

PROPOSITION 52. *The strongly analytic restriction of* **KE** *linearly simulates the tableau method with merging.*

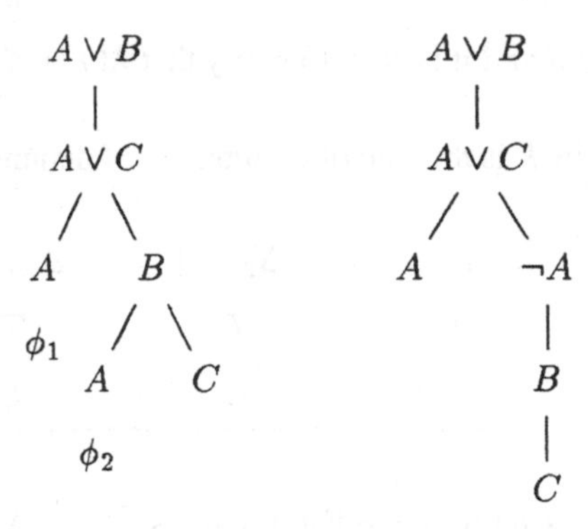

Figure 15. In the tableau in the left-hand side the branch ϕ_2 is not further expanded (it is 'merged' with branch ϕ_1). In the corresponding **KE**-tree shown in the right-hand side, the redundant branch is not generated.

Let us now turn our attention to the other enhancement of Smullyan's tableaux, discussed in Section 4.2, namely the use of lemmas in a tableau refutation. We have that this method is equivalent to replacing the standard tableau branching rules with corresponding asymmetric ones. For instance, the rules for eliminating disjunction are replaced by the following:

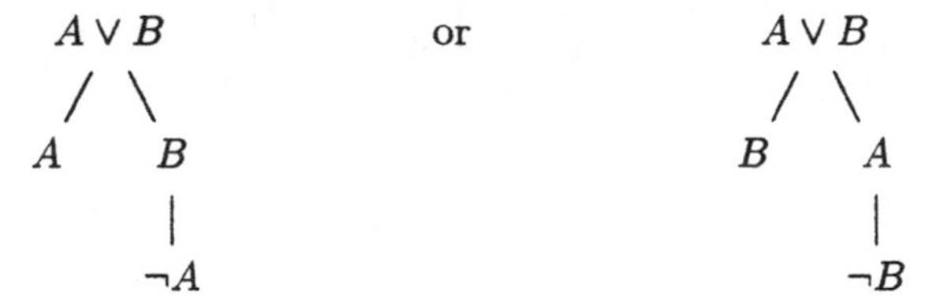

It is obvious that such asymmetric rules are equivalent to a combination of strongly analytic applications of PB and of the **KE**-elimination rules, as shown in the diagram used to show Proposition 47 above. Hence:

PROPOSITION 53. *The strongly analytic restriction of* **KE** *linearly simulates the tableau method with lemma generation.*

We remark that it appears misleading, however, to refer to tableaux with the asymmetric rules as to 'tableaux with lemma generation'. The use of this terminology conveys the idea that the additional formula appendend to one side of the branching rules is a 'lemma', namely that the sub-tableau below the other side is closed or may be closed. In contrast, tableau rules can be used also for enumerating all the models of a satisfiable set of formulae, and not only for refuting unsatisfiable sets. In the former case the sub-tableau in question may well be open, so that the appended formula is not a 'lemma'. The fact that we can append it without loss of soundness, does not depend on its being a lemma, but merely on the principle of bivalence. In any case, the appended formula provides additional information which can be used below to close every branch containing its complement. As a result, the enumeration of models is non-redundant, in the sense that it does not generate models which are subsumed one by the other, since all the models resulting from the enumeration are mutually exclusive.

Thus the strongly analytic restriction of **KE** is sufficient to simulate efficiently the best known enhancements of the tableau method, simply by making an appropriate use of the (analytic) cut rule. We stress once again that the **KE**-system cannot be identified with its strongly analytic restriction. In fact, less restricted notions of proof may be more powerful from the complexity viewpoint and still obey the subformula principle.

6 A MORE GENERAL VIEW

As seen in the previous section, the speed-up of **KE** over the tableau method can be traced to the fact that **KE**, unlike the tableau method, can easily simulate inference steps based on 'cuts', as shown on p. 80 and p. 109. The cut rule expresses the transitivity property of deduction and it is natural to require that a 'simple' procedure for implementing this property be *feasible*. Concatenation of proofs is feasible in the ordinary deductive practice and should be feasible in any realistic formal model of this practice. We are, therefore, led to formulate the following condition which has to be satisfied by any acceptable formal model of the notion of a classical proof:

> **Strong Transitivity Principle (STP):** *Proof-concatenation should be feasible and uniform. More precisely, let $|T|$ denote the size of the proof T (*ie *the number of steps in it). There must be a uniform, structural procedure, which, given a proof T_1 of A from Γ_1, B and a proof T_2 of B from Γ_2, yields a proof T_3 of A from Γ_1, Γ_2 such that $|T_3| = |T_1| + |T_2| + c$, for some fixed constant c depending on the system.*

In other word we require that the transitivity property be a valid derived rule in a particularly strict sense.

Cleraly **KE** satisfies our strong transitivity principle, the required procedure being the one described in the previous section. Some formal systems (like the tableau method and the cut-free sequent calculus) do not satisfy the STP, some others satisfy it in a non-obvious way. Notice that the existence of a uniform procedure for grafting proofs of subsidiary conclusions, or *lemmata*, in the proof of a theorem is just one condition of the STP. The other condition requires that such a procedure be computationally easy. In Natural Deduction, for instance, replacing *every occurrence* of an assumption A with its proof provides an obvious grafting method which, though very perspicuous, is highly inefficient, leading to much unwanted duplication in the resulting proof-tree. We also require the method to be also computationally *feasible*. So, the standard way of grafting proofs of subsidiary conclusions in natural deduction proofs, though providing a uniform method, does not satisfy the feasibility condition. The rules of natural deduction, however, permit us to bypass this difficulty and produce a method satisfying the whole of STP. Consider the rule of *Non-Constructive Dilemma* (NCD):

$$\frac{\begin{array}{c}\Gamma, [A] \\ \vdots \\ B\end{array} \qquad \begin{array}{c}\Delta, [\neg A] \\ \vdots \\ B\end{array}}{B}$$

This is a derived rule in Prawitz's style Natural Deduction which yields classical logic if added to the intuitionistically valid rules (see [Tennant, 1978, section 4.5]). By 'derived rule' here we mean that every application of NCD can be eliminated in a fixed number of steps (the expression 'derived rule' is often used in the literature in a much more liberal sense), as shown by the following construction:

$$\frac{A \vee \neg A \qquad \begin{array}{c}\Gamma, [A] \\ \vdots \\ B\end{array} \qquad \begin{array}{c}\Delta, [\neg A] \\ \vdots \\ B\end{array}}{B}$$

(Notice that $A \vee \neg A$ can always be proved in a fixed number of steps.)

NCD is a classical version of the cut rule. We can show Natural Deduction to satisfy the STP by means of the following construction:

$$\frac{\begin{array}{c}\Gamma, [A] \\ \vdots \\ B\end{array} \qquad \dfrac{\dfrac{\begin{array}{c}\Delta \\ \vdots \\ A\end{array} \qquad [\neg A]}{F}}{B}}{B}$$

Notice that the construction does not depend on the number of occurrences of the assumption A in the subproof of B from Γ, A.

In analogy with the STP we can formulate a condition requiring a proof system to simulate efficiently another form of cut which holds for classical systems and is closely related to the rule PB:

(C) Let π_1 be a proof of B from Γ, A and π_2 be a proof of B from $\Delta, \neg A$. Then there is a uniform method for constructing from π_1 and π_2 a proof π_3 of B from Γ, Δ such that $\lambda(\pi_3) \leq \lambda(\pi_1) + \lambda(\pi_2) + c$ for some constant c.

Similarly, the next condition requires that proof systems can efficiently simulate the *ex falso* inference scheme:

(XF) Let π_1 be a proof of A from Γ and π_2 be a proof of $\neg A$ from Δ. Then there is a uniform method for constructing from π_1 and π_2 a proof π_3 of B from Γ, Δ, for any B, such that $\lambda(\pi_3) \leq \lambda(\pi_1) + \lambda(\pi_2) + c$ for some constant c.

So, let us say that a proof system is a *classical cut system* if (i) it is sound and complete for classical logic, and (ii) it satisfies conditions (C) and (XF). It is easy to show that every classical cut system satisfies also the STP and, therefore, allows for an efficient implementation of the transitivity property of ordinary proofs. Notice that the definition of a classical cut system is very general and does not assume anything about the form of the inference rules. For instance, as shown above, classical natural deduction in *tree* form is a classical cut system.

The next Proposition shows that *every* classical cut system can simulate **KE** without a significant increase in proof complexity. The simulation procedure shows that the cut inferences are restricted to formulae which occur in the simulated proof.

PROPOSITION 54. *If* **S** *is a classical cut system, then* **S** *can linearly simulate* **KE***.*

To prove the proposition it is convenient to assume that our language includes a 0-ary operator F (Falsum) and that the proof systems include suitable rules to deal with it.[38] For **KE** this involves only adding the obvious rule which allows us to append F to any branch containing both A and $\neg A$ for some formula A, so that every closed branch in a **KE**-tree ends with a node labelled with F. The assumption is made only for convenience's sake and can be dropped without consequence. Moreover we shall make the obvious assumption that, for every system **S**, the complexity of a proof of A from A in **S** is equal to 1.

Let $\tau(\pi)$ denote the number of nodes *generated* by a **KE**-refutation π of Γ (i.e. the assumptions are not counted).[39] Let **S** be a classical cut system. If **S** is complete, then for every rule r of **KE** there is an **S**-proof π_r of the conclusion of r from its premises. Let $b_1 = Max_r(\lambda(\pi_r))$ and let b_2 and b_3 be the constants representing, respectively, the λ-cost of simulating classical cut in **S**–associated with condition (C) above—and the λ-cost of simulating the *ex falso* inference scheme in **S**–associated with condition (XF) above. As mentioned before, every classical cut system satisfies also condition STP and it is easy to verify that the constant associated with this condition, representing the λ-cost of simulating 'absolute' cut in **S**, is $\leq b_2 + b_3 + 1$. We set $c = b_1 + b_2 + b_3 + 1$.

The proposition is an immediate consequence of the following lemma:

LEMMA 55. *For every classical cut system* **S***, if there is a* **KE***-refutation* π *of* Γ*, then there is an* **S***-proof* π' *of* F *from* Γ *with* $\lambda(\pi') \leq c \cdot \tau(\pi)$*.*

---

[38] Systems which are not already defined over a language containing F can usually be redefined over such an extended language without difficulty.

[39] The reader should be aware that our τ-measure applies to **KE**-refutations and *not* to trees: the same tree can represent different refutations yielding different values of the τ-measure.

Proof. The proof is by induction on $\tau(\pi)$, where π is a **KE**-refutation of Γ. $\tau(\pi) = 1$. Then Γ is explicitly inconsistent, i.e. contains a pair of complementary formulae, say B and $\neg B$, and the only node generated by the refutation is F, which is obtained by means of an application of the **KE**-rule for F to B and $\neg B$. Since there is an **S**-proof of the **KE**-rule for F, we can obtain an **S**-proof π' of the particular application contained in π simply by performing the suitable substitutions and $\lambda(\pi') \leq b_1 < c$.

$\tau(\pi) > 1$. Case 1. The **KE**-refutation π has the form:

$$\begin{array}{c} \Gamma \\ C \\ \mathcal{T}_1 \end{array}$$

where C follows from premises in Γ by means of an E-rule. So there is a **KE**-refutation π_1 of Γ, C such that $\tau(\pi) = \tau(\pi_1) + 1$. By inductive hypothesis, there is an **S**-proof π_1' of F from Γ, C such that $\lambda(\pi_1') \leq c \cdot \tau(\pi_1)$. Moreover, there is an **S**-proof π_2 of C from the premises from which it is inferred in π such that $\lambda(\pi_2) \leq b_1$. So, from the hypothesis that **S** is a classical cut system, it follows that there is an **S**-proof π' of F from Γ such that

$$\begin{aligned} \lambda(\pi') &\leq c \cdot \tau(\pi_1) + b_1 + b_2 + b_3 + 1 \\ &\leq c \cdot \tau(\pi_1) + c \\ &\leq c \cdot (\tau(\pi_1) + 1) \\ &\leq c \cdot \tau(\pi) \end{aligned}$$

Case 2. π has the following form:

$$\begin{array}{ccc} & \Gamma & \\ / & & \backslash \\ C & & \neg C \\ \mathcal{T}_1 & & \mathcal{T}_2 \end{array}$$

So there are **KE**-refutations π_1 and π_2 of Γ, C and $\Gamma, \neg C$ such that $\tau(\pi) = \tau(\pi_1) + \tau(\pi_2) + 2$. Now, by inductive hypothesis there is an **S**-proof π_1' of F from Γ, C and an **S**-proof π_2' of F from $\Gamma, \neg C$ with $\lambda(\pi_i') \leq c \cdot \tau(\pi_i), i = 1, 2$. Since **S** is a classical cut system, it follows that there is an **S**-proof π' of F from Γ such that

$$\begin{aligned} \lambda(\pi') &\leq c \cdot \tau(\pi_1) + c \cdot \tau(\pi_2) + b_2 \\ &< c \cdot \tau(\pi_1) + c \cdot \tau(\pi_2) + c \\ &< c \cdot (\tau(\pi_1) + \tau(\pi_2) + 2) \end{aligned}$$

$$< \quad c \cdot \tau(\pi)$$

■

Notice that in the simulation used in the proof of the previous proposition the cut formula is always a formula which occurs in the **KE**-refutation. So, if the latter enjoys the subformula property, its simulation in the cut system will be such that the cut formulae are all subformulae of the conclusion or of the assumptions. This holds true even if the proof as a whole does not enjoy the subformula property (for instance when the cut system in question is a Hilbert-style system). In fact, the simulation given above provides a means for transforming every proof-procedure based on the **KE**-rules into a proof-procedure of comparable complexity based on the rules of the given cut system.

DEFINITION 56. We say that a proof system **S** is an *analytic cut system* if (i) **S** is the analytic restriction of a classical cut system, and (ii) **S** is complete for classical logic.

Then, the proof of Proposition 54 also shows that:

COROLLARY 57. *Every analytic cut system can linearly simulate the analytic restriction of* **KE**.

This implies that it is only the possibility of representing (analytic) *cuts* and not the form of the *operational rules* which is crucial from a complexity viewpoint.

It follows from the above corollary and Proposition 44 that:

COROLLARY 58. *Every analytic cut system is a standard proof system. In fact, for every tautology A of length n and containing k distinct variables there is a proof π, with $\lambda(\pi) = O(n \cdot 2^k)$.*

Moreover, Corollary 57, Proposition 47 and Proposition 49 imply that:

COROLLARY 59. *Every analytic cut system can linearly simulate the tableau method, but the tableau method cannot p-simulate any analytic cut system.*

Since (the classical version of) Prawitz's style natural deduction is a classical cut system, it follows that it can linearly simulate **KE**. Moreover, for all these systems, the simulation preserves the subformula property (i.e. it maps analytic proofs to analytic proofs). Therefore Proposition 54, together with Proposition 51, imply that Prawitz's style natural deduction and **KE** can linearly simulate each other with a procedure which preserves the subformula property. Corollary 59 implies that both these systems cannot be p-simulated by the tableau method, even if we restrict our attention to analytic proofs (the tableau method cannot p-simulate any analytic cut system). Finally Corollary 58 implies that, unlike the tableau method, the analytic restriction of both these systems have the same *upper bound* as the truth-table method.

ACKNOWLEDGEMENTS

I wish to thank Reinhold Letz for reading several versions of the present chapter and offering valuable comments. Thanks are also due to Krysia Broda, Dov Gabbay and Marco Mondadori for many helpful suggestions. Special thanks to Alasdair Urquhart and André Vellino for making available interesting papers and lecture notes.

Dipartimento di Scienze Umane, Università di Ferrara, Italy.

APPENDIX

NATURAL DEDUCTION AND VARIANTS

$$\frac{P \wedge Q}{P} \qquad \frac{P \wedge Q}{Q} \qquad \frac{P \quad Q}{P \wedge Q}$$

$$\frac{P}{P \vee Q} \qquad \frac{Q}{P \vee Q} \qquad \frac{P \vee Q \quad \begin{matrix} \Psi_1, [P] \\ \vdots \\ R \end{matrix} \quad \begin{matrix} \Psi_2, [Q] \\ \vdots \\ R \end{matrix}}{R}$$

$$\frac{\begin{matrix} \Psi, [P] \\ \vdots \\ Q \end{matrix}}{P \to Q} \qquad \frac{P \to Q \quad P}{Q}$$

$$\frac{\begin{matrix} \Psi, [P] \\ \vdots \\ \mathbf{F} \end{matrix}}{\neg P} \qquad \frac{P \quad \neg P}{\mathbf{F}} \qquad \frac{\mathbf{F}}{Q}$$

The notation $[P]$ means that the conclusion of the rule does not depend on the assumption P. The rules listed above are the intuitionistic rules. Classical logic is obtained by adding one of the following three rules:

$$\frac{}{P \vee \neg P} \qquad \frac{\neg\neg P}{P} \qquad \frac{\begin{matrix} [\neg P] \\ \vdots \\ \mathbf{F} \end{matrix}}{P}$$

The last of these rules is the *classical reductio ad absurdum*. Another variant of the classical calculus is described in [Tennant, 1978] where the author uses the following classical *rule of dilemma*:

$$\frac{\begin{array}{cc}[P] & [\neg P]\\ \vdots & \vdots \\ Q & Q\end{array}}{Q}$$

REFERENCES

[Avron, 1988] A. Avron. The semantics and proof theory of linear logic. *Theoretical Computer Science*, 57:161–184, 1988.

[Beth, 1955] E. W. Beth. Semantic entailment and formal derivability. *Mededelingen der Koninklijke Nederlandse Akademie van Wetenschappen*, 18:309–342, 1955.

[Beth, 1958] E. W. Beth. On machines which prove theorems. *Simon Stevin Wissen Natur-kundig Tijdschrift*, 32:49–60, 1958. Reprinted in [Siekmann and Wrightson, 1983], vol. 1, pages 79–90.

[Bibel, 1982] W. Bibel. *Automated Theorem Proving*. Vieweg, Braunschweig, 1982.

[Blanché, 1970] R. Blanché. *La logique et son histoire*. Armand Colin, Paris, 1970.

[Bochensky, 1961] I. M. Bochensky. *A History of Formal Logic*. University of Notre Dame, Notre Dame (Indiana), 1961.

[Boolos, 1984] G. Boolos. Don't eliminate cut. *Journal of Philosophical Logic*, 7:373–378, 1984.

[Broda, 1992] K. Broda. The application of semantic tableaux with unification to automated deduction. Ph.D thesis. Technical report, Department of Computing, Imperial College, 1992.

[Broda *et al.*, 1998] K. Broda, M. Finger and A. Russo. LDS-Natural deduction for substructural logics. Technical report DOC97-11. Imperial College, Department of Computing, 1997. Short version presented at WOLLIC 96, *Journal of the IGPL*, 4:486–489.

[Buss, 1987] S. R. Buss. Polynomial size proofs of the pigeon-hole principle. *The Journal of Symbolic Logic*, 52:916–927, 1987.

[Cellucci, to appear] C. Cellucci. Analytic cut trees. To appear in the *Journal of the IGPL*.

[Cellucci, 1987] C. Cellucci. Using full first order logic as a programming language. In *Rendiconti del Seminario Matematico Università e Politecnico di Torino. Fascicolo Speciale 1987*, pages 115–152, 1987. Proceedings of the conference on 'Logic and Computer Science: New Trends and Applications'.

[Cellucci, 1992] C. Cellucci. Existential instantiation and normalization in sequent natural deduction. *Annals of Pure and Applied Logic*, 58:111–148, 1992.

[Chang, 1970] C. L. Chang. The unit proof and the input proof in theorem proving. *Journal of the Association for Computing Machinery*, 17:698–707, 1970.

[Chang and Lee, 1973] C. .L Chang and R. C. T. Lee. *Symbolic Logic and Mechanical Theorem Proving*. Academic Press, Boston, 1973.

[Cook, 1971] S. A. Cook. The complexity of theorem proving procedures. In *Proceedings of the 3rd Annual Symposium on the Theory of Computing*, 1971.

[Cook and Reckhow, 1974] S. A. Cook and R. Reckhow. On the length of proofs in the propositional calculus. In *Proceedings of the 6th Annual Symposium on the Theory of Computing*, pages 135–148, 1974.

[Cook and Reckhow, 1979] S. A. Cook and R. Reckhow. The relative efficiency of propositional proof systems. *The Journal of Symbolic Logic*, 44:36–50, 1979.

[D'Agostino, 1990] M. D'Agostino. Investigations into the complexity of some propositional calculi. PRG Technical Monographs 88, Oxford University Computing Laboratory, 1990.

[D'Agostino, 1992] M. D'Agostino. Are tableaux an improvement on truth-tables? *Journal of Logic, Language and Information*, 1:235–252, 1992.

[D'Agostino and Gabbay, 1994] M. D'Agostino and D. Gabbay. A generalisation of analytic deduction via labelled deductive systems. Part 1: basic substructural logics. *Journal of Automated Reasoning*, 13:243–281, 1994.

[D'Agostino and Mondadori, 1994] M. D'Agostino and M. Mondadori. The taming of the cut. *Journal of Logic and Computation*, 4:285–319, 1994.

[Davis, 1983] M. Davis. The prehistory and early history of automated deduction. In *[Siekmann and Wrightson, 1983]*, pages 1–28. 1983.

[Davis *et al.*, 1992] M. Davis, G. Logemann, and D. Loveland. A machine program for theorem proving. *Communications of the Association for Computing Machinery*, 5:394–397, 1962. Reprinted in [Siekmann and Wrightson, 1983], pp. 267–270.

[Davis and Putnam, 1960] M. Davis and H. Putnam. A computing procedure for quantification theory. *Journal of the Association for Computing Machinery*, 7:201–215, 1960. Reprinted in [Siekmann and Wrightson, 1983], pp. 125–139.

[Dummett, 1978] M. Dummett. *Truth and other Enigmas*. Duckworth, London, 1978.

[Dunham and Wang, 1976] B. Dunham and H. Wang. Towards feasible solutions of the tautology problem. *Annals of Mathematical Logic*, 10:117–154, 1976.

[Fitting, 1996] M. Fitting. *First-Order Logic and Automated Theorem Proving*. Springer-Verlag, Berlin, 1996. First edition, 1990.

[Gabbay, 1996] D. M. Gabbay. Labelled Deductive Systems, Volume 1 - Foundations. Oxford University Press, 1996.

[Gallier, 1986] J. H. Gallier. *Logic for Computer Science*. Harper & Row, New York, 1986.

[Garey and Johnson, 1979] M. R. Garey and D. S. Johnson. *Computers and Intractability. A Guide to the theory of NP-Completeness*. W. H. Freeman & Co., San Francisco, 1979.

[Gentzen, 1935] G. Gentzen. Unstersuchungen über das logische Schliessen. *Math. Zeitschrift*, 39:176–210, 1935. English translation in [Szabo, 1969].

[Girard, 1987] J. Y. Girard. *Proof Theory and Logical Complexity*. Bibliopolis, Napoli, 1987.

[Girard *et al.*, 1989] J.-Y. Girard, Y. Lafont and P. Taylor. *Proofs and Types*. Cambridge Tracts in Theoretical Computer Science, Cambridge University Press, 1989.

[Hähnle, 1992] R. Hähnle. *Tableau-Based Theorem-Proving in Multiple-Valued Logics*. PhD thesis, Department of Computer Science, University of Karlsruhe, 1992.

[Heyting, 1956] A. Heyting. *Intuitionism*. North-Holland, Amsterdam, 1956.

[Jeffrey, 1967] R. C. Jeffrey. *Formal Logic: its Scope and Limits*. McGraw-Hill Book Company, New York, second edition, 1981. first edition 1967.

[Kleene, 1967] S. C. Kleene. *Mathematical Logic*. John Wiley & Sons, Inc., New York, 1967.

[Letz, 1993] R Letz.On the Polynomial Transparency of Resolution. In *Proceedings of the 13th International Joint Conference on Artificial Intelligence(IJCAI)*, Chambery, pp. 123–129, 1993.

[Letz *et al.*, 1994] R Letz, K. Mayr and C. Goller. Controlled integration of the cut rule into connection tableaux calculi. *Journal of Automated Reasoning*, 13:297–337, 1994.

[Loveland, 1978] D. W. Loveland. *Automated Theorem Proving: A Logical Basis*. North Holland, 1978.

[Matulis, 1962] V. A. Matulis. Two versions of classical computation of predicates without structural rules. *Soviet Mathematics*, 3:1770–1773, 1962.

[Mondadori, 1988] M. Mondadori. Classical analytical deduction. Annali dell'Università di Ferrara; Sez. III; Discussion paper 1, Università di Ferrara, 1988.

[Mondadori, 1988a] M. Mondadori. Classical analytical deduction, part II. Annali dell'Università di Ferrara; Sez. III; Discussion paper 5, Università di Ferrara, 1988.

[Murray and Rosenthal, 1994] N. V. Murray and E. Rosenthal. On the computational intractability of analytic tableau methods. *Bulletin of the IGPL*, 1994.

[Prawitz, 1965] D. Prawitz. *Natural Deduction. A Proof-Theoretical Study*. Almqvist & Wilksell, Uppsala, 1965.

[Prawitz, 1971] D. Prawitz. Ideas and results in proof theory. In *Proceedings of the II Scandinavian Logic Symposium*, pages 235–308, Amsterdam, 1971. North-Holland.

[Prawitz, 1974] D. Prawitz. Comments on Gentzen-type procedures and the classical notion of truth. In A. Dold and B. Eckman, editors, *ISILC Proof Theory Symposium. Lecture Notes in Mathematics, 500*, pages 290–319, Springer. Berlin, 1974.

[Prawitz, 1978] D. Prawitz. Proofs and the meaning and completeness of the logical constants. In J. Hintikka, I. Niinduoto, and E. Saarinen, editors, *Essays on Mathematical and Philosphical Logic*, pages 25–40. Reidel, Dordrecht, 1978.

[Robinson, 1965] J. A. Robinson. A machine-oriented logic based on the resolution principle. *Journal of the Association for Computing Machinery*, 12:23–41, 1965.

[Schroeder-Heister, 1997] P. Schroeder-Heister. Frege and the resolution calculus. *History and Philosophy of Logic*, 18:95–108, 1997.

[Scott, 1970] D. Scott. Outline of a mathematical theory of computation. PRG Technical Monograph 2, Oxford University Computing Laboratory, Programming Research Group, 1970. Revised and expanded version of a paper under the same title in the Proceedings of the Fourth Annual Princeton Conference on Information Sciences and System, 1970.

[Scott, 1981] D. Scott. Notes on the formalization of logic. Study aids monographs, n. 3, University of Oxford, Subfaculty of Philosophy, 1981. Compiled by Dana Scott with the aid of David Bostock, Graeme Forbes, Daniel Isaacson and Gören Sundholm.

[Sieg, 1993] W. Sieg. Mechanism and search. aspects of proof theory. Technical report, AILA preprint, 1993.

[Siekmann and Wrightson, 1983] J. Siekmann and G. Wrightson, editors. *Automation of Reasoning*. Springer-Verlag, New York, 1983.

[Smullyan, 1968] R. Smullyan. *First-Order Logic*. Springer, Berlin, 1968.

[Smullyan, 1968a] R. M. Smullyan. Analytic cut. *The Journal of Symbolic Logic*, 33:560–564, 1968.

[Smullyan, 1968b] R. M. Smullyan. Uniform Gentzen systems. *The Journal of Symbolic Logic*, 33:549–559, 1968.

[Statman, 1978] R. Statman. Bounds for proof-search and speed-up in the predicate calculus. *Annals of Mathematical Logic*, 15:225–287, 1978.

[Statman, 1977] R. Statman. Herbrand's theorem and Gentzen's notion of a direct proof. In J. Barwise, editor, *Handbook of Mathematical Logic*, pages 897–912. North-Holland, Amsterdam, 1977.

[Stockmeyer, 1987] L. Stockmeyer. Classifying the computational complexity of problems. *The Journal of Symbolic Logic*, 52:1–43, 1987.

[Sundholm, 1983] G. Sundholm. Systems of deduction. In D. Gabbay and F. Guenthner, editors, *Handbook of Philosophical Logic*, volume I, chapter I.2, pages 133–188. Reidel, Dordrecht, 1983.

[Szabo, 1969] M. Szabo, editor. *The Collected Papers of Gerhard Gentzen*. North-Holland, Amsterdam, 1969.

[Tennant, 1978] N. Tennant. *Natural Logic*. Edimburgh University Press, Edinburgh, 1978.

[Thistlewaite *et al.*, 1988] P. B. Thistlewaite, M. A. McRobbie and B. K. Meyer. *Automated Theorem Proving in Non Classical Logics*. Pitman, 1988.

[Thomas, 1941] I. Thomas, editor. *Greek Mathematics*, volume 2. William Heinemann and Harvard University Press, London and Cambridge, Mass., 1941.

[Urquhart, 1992] A. Urquhart. Complexity of proofs in classical propositional logic. In Y. Moschovakis, editor, *Logic from Computer Science*, pages 596–608. Springer-Verlag, 1992.

[Urquhart, 1995] A. Urquhart. The complexity of propositional proofs. *The Bulletin of Symbolic Logic*, 1:425–467, 1995.

[Vellino, 1989] A. Vellino. *The Complexity of Automated Reasoning*. PhD thesis, University of Toronto, 1989.

REINHOLD LETZ

FIRST-ORDER TABLEAU METHODS

INTRODUCTION

In this chapter tableau systems for classical first-order logic are presented. We introduce the tableau calculi due to Smullyan as well as tableaux with free variables and unification due to Fitting. Special emphasis is laid on the presentation of refined tableau systems for clause logic which consist of more condensed inference rules and hence are particularly suited for automated deduction.

The material is organized in six sections. The first section provides the general background on first-order logic with function symbols; the syntax and the classical model-theoretic semantics of first-order logic are introduced and the fundamental properties of variable substitutions are described. In Section 2 we study the most important normal forms and normal-form transformations of first-order logic, including Skolemization. Furthermore, the central rôle of Herbrand interpretations is emphasized. In the third section, we turn to the traditional tableau systems for first-order logic due to Smullyan. Using uniform notation, tableau calculi for first-order sentences are introduced. We prove Hintikka's Lemma and the completeness of first-order tableaux, by applying a systematic tableau procedure. We also consider some basic refinements of tableaux: strictness, regularity and the Herbrand restriction. The fourth section concentrates on the crucial weakness of traditional tableau systems with respect to proof search. It lies in the nature of the standard γ-rule, which enforces that instantiations have to be chosen too early. An approach to remedy this weakness is to permit free variables in a tableau which are treated as placeholders for terms, as so-called 'rigid' variables. The instantiation of rigid variables then is guided by unification. Free-variable tableaux require a generalization of the quantifier rules. Unfortunately, systematic procedures for free-variable tableaux cannot be deviced as easily as for sentence tableaux. Therefore, typically, tableau enumeration procedures are used instead. In Section 5, we concentrate on formulae in clausal form. Clause logic permits a more condensed representation of tableaux and hence a redefinition of the tableau rules. Following the tableau enumeration approach of free-variable tableaux, the crucial demand is to reduce the number of tableaux to be considered by the search procedure. The central such concept in clause logic is the notion of a connection, which can be employed to guide the tableau construction. Since connection tableaux are not confluent, systematic tableau procedures cannot be applied. Moreover, a simulation of general clausal tableaux is not possible. Therefore a fundamentally different completeness proof has to be given. We also compare the connection tableau framework with other calculi from the area of automated deduction. In Section 6 methods are mentioned which can produce significantly shorter tableau proofs. The techniques can be subdivided into three different classes. The mechanisms of the first type

M. D'Agostino et al. (eds.), Handbook of Tableau Methods, 125–196.

are centered around the (backward) cut rule. Second, so-called liberalizations of the δ rule are discussed, which may lead to even nonelementarily smaller tableau proofs. Finally, we consider a special condition under which free variables in a tableau can be treated as universally quantified on a branch.

1 CLASSICAL FIRST-ORDER LOGIC

The theory of first-order logic is a convenient and powerful formal abstraction from expressions and concepts occurring in natural language, and, most significantly, in mathematical discourse. In this section we present the syntax and semantics of first-order logic with function symbols. Furthermore, the central modification operation on first-order expressions is introduced, the replacement of variables, and some of its invariances are studied.

1.1 Syntax of First-order Logic

Propositional logic deals with sentences and their composition, hence the alphabet of a propositional language consists of only three types of symbols, propositional variables, logical symbols, and punctuation symbols. First-order logic does not stop at the sentence level, it can express the *internal* structure of sentences. In first-order logic, the logical structure and content of assertions of the following form can be studied that have no natural formalization in propositional logic.

EXAMPLE 1. If every person that is not rich has a rich father, then some rich person must have a rich grandfather.

In order to express such formulations, a first-order alphabet has to provide symbols for denoting objects, functions, and relations. Furthermore, it must be possible to make universal or existential assertions, hence we need quantifiers. Altogether, the alphabet or *signature* of a first-order language will be defined as consisting of six disjoint sets of symbols.

DEFINITION 2 (First-order signature). A *first-order signature* is a pair $\Sigma = \langle A, a \rangle$ consisting of a denumerably infinite alphabet A and a partial mapping a: $A \longrightarrow \mathbb{N}_0$, associating natural numbers with certain symbols in A, called their *arities*, such that A can be partitioned into the following six pairwise disjoint sets of symbols.

1. An infinite set $\mathcal{V}$ of *variables*, without arities.

2. An infinite set of *function symbols*, all with arities such that there are infinitely many function symbols of every arity. Nullary function symbols are called *constants*.

3. An infinite set of *predicate symbols*, all with arities such that there are infinitely many predicate symbols of every arity.

4. A set of *connectives* consisting of five distinct symbols $\neg$, $\wedge$, $\vee$, $\rightarrow$, and $\leftrightarrow$, the first one with arity 1 and all others binary. We call $\neg$ the *negation symbol*, $\wedge$ is the *conjunction symbol*, $\vee$ is the *disjunction symbol*, $\rightarrow$ is the *material implication symbol*, and $\leftrightarrow$ is the *material equivalence symbol*,

5. A set of *quantifiers* consisting of two distinct symbols $\forall$, called the *universal quantifier*, and $\exists$, called the *existential quantifier*, both with arity 2.

6. A set of *punctuation symbols* consisting of three distinct symbols without arities, which we denote with the symbols '(', ')', and ','.

NOTATION 3. Normally, variables and function symbols will be denoted with lower-case letters and predicate symbols with upper-case letters. Preferably, we use for variables letters from 'u' onwards; for constants the letters 'a', 'b', 'c', 'd', and 'e'; for function symbols with arity ≥ 1 the letters 'f', 'g' and 'h'; and for predicate symbols the letters 'P', 'Q' and 'R'; nullary predicate symbols shall occasionally be denoted with lower-case letters. Optionally, subscripts will be used. We do not distinguish between symbols and unary strings consisting of symbols, the context will clear up possible ambiguities. We will always talk *about* symbols of first-order languages and never give examples of concrete expressions *within* a specific object language.

Given a first-order signature Σ, the corresponding *first-order language* is defined inductively[1] as a set of specific strings over the alphabet of the signature. In our presentation of first-order languages we use prefix notation for the representation of terms and atomic formulae, and infix notation for the binary connectives. Let in the following $\Sigma = \langle A, a \rangle$ be a fixed first-order signature.

DEFINITION 4 (Term (inductive)).

1. Every variable in A is said to be a *term over* Σ.

2. If f is an n-ary function symbol in A with $n \geq 0$ and $t_1, \ldots, t_n$ are terms over Σ, then the concatenation $f(t_1, \ldots, t_n)$ is a *term over* Σ.

DEFINITION 5 (Atomic formula). If P is (the unary string consisting of) an n-ary predicate symbol in A with $n \geq 0$ and $t_1, \ldots, t_n$ are terms over Σ, then the concatenation $P(t_1, \ldots, t_n)$ is an *atomic formula*, or *atom, over* Σ.

NOTATION 6. Terms of the form $a()$ and atoms of the form $P()$ are abbreviated by writing just a and P, respectively.

DEFINITION 7 (Formula (inductive)).

1. Every atom over Σ is a *formula over* Σ.

---

[1] In inductive definitions we shall, conveniently, omit the explicit formulation of the necessity condition.

2. If Φ and Ψ are formulae over Σ and x is (the symbol string consisting of) a variable in A, then the following concatenations are also *formulae over* Σ:
$\neg\Phi$, called the *negation* of Φ,
$(\Phi \wedge \Psi)$, called the *conjunction* of Φ *and* Ψ,
$(\Phi \vee \Psi)$, called the *disjunction* of Φ *and* Ψ,
$(\Phi \rightarrow \Psi)$, called the *material implication* of Ψ *by* Φ,
$(\Phi \leftrightarrow \Psi)$, called the *material equivalence* of Φ *and* Ψ,
$\forall x\Phi$, called the *universal quantification* of Φ *in* x, and
$\exists x\Phi$, called the *existential quantification* of Φ *in* x.

DEFINITION 8 ((Well-formed) expression). All terms and formulae over Σ are called *(well-formed) expressions over* Σ.

DEFINITION 9 (First-order language). The set of all (well-formed) expressions over Σ is called the *first-order language over* Σ.

DEFINITION 10 (Complement). The *complement* of any negated formula $\neg\Phi$ is Φ and the *complement* of any unnegated formula Φ is its negation $\neg\Phi$; we denote the complement of a formula Φ with $\sim\Phi$.

DEFINITION 11 (Literal). Every atomic formula and every negation of an atomic formula is called a *literal.*

Recalling the assertion given in Example 1: 'if every person that is not rich has a rich father, then some rich person must have a rich grandfather', a possible (abstracted) first-order formalization would be the following formula.

EXAMPLE 12. $\forall x(\neg R(x) \rightarrow R(f(x))) \rightarrow \exists x(R(x) \wedge R(f(f(x))))$.

DEFINITION 13 (Subexpression). If an expression Φ is the concatenation of strings $W_1, \ldots, W_n$, in concordance with the Definitions 4 to 7, then any expression among these strings is called an *immediate subexpression* of Φ. The sequence obtained by deleting all elements from $W_1, \ldots, W_n$ that are not expressions is called the *immediate subexpression sequence* of Φ. Among the strings $W_1, \ldots, W_n$ there is a unique string W whose symbol is a connective, a quantifier, a function symbol, or a predicate symbol; W is called the *dominating symbol* of Φ. An expression Ψ is said to be a *subexpression* of an expression Φ if the pair $\langle\Psi, \Phi\rangle$ is in the transitive closure of the immediate subexpression relation. Analogously, the notions of (immediate) subterms and (immediate) subformulae are defined.

EXAMPLE 14. According to our conventions of denoting symbols and strings, a formula of the form $P(x, f(a, y), x)$ has the immediate subexpression sequence $x, f(a, y), x$; the immediate subexpressions x and $f(a, y)$; the subexpressions x, $f(a, y)$, a, and y; and, lastly, P as dominating symbol.

We have to provide a means for addressing different occurrences of symbols and subexpressions in an expression E. One could simply address occurrences by giving the first and last word positions in E. Although this way occurrences of symbols and subexpressions in an expression could be uniquely determined, this notation has the disadvantage that whenever expressions are modified, e.g., by concatenating them or by replacing an occurrence of a subexpression, then the addresses of the occurrences may change completely. We will use a notation which is more robust concerning concatenations of and replacements in expressions. This notation is motivated by a *symbol tree* representation of logical expressions, as displayed in Figure 1.

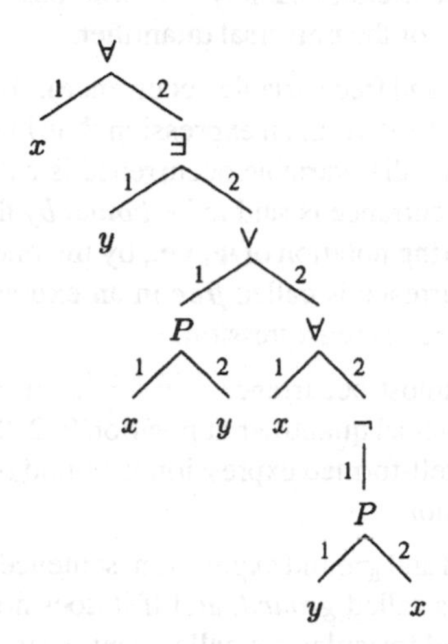

Figure 1. Symbol tree of the formula $\forall x \exists y (P(x, y) \vee \forall x \neg P(y, x))$

Each occurrence of a symbol or a subexpression in an expression can be uniquely determined by a *sequence of natural numbers* that encodes the edges to be followed in the symbol tree. Formally, tree positions can be defined as follows.

DEFINITION 15 (Position (inductive)). For any expression E,

1. if s is the dominating symbol of an expression E, then the *position* both of E and of the dominating occurrence of s *in* E is the empty sequence $\emptyset$.

2. if $E_1, \ldots, E_n$ is the immediate subexpression sequence of E and if $p_1, \ldots, p_n$ is the position of an occurrence of an expression or a symbol W in E_i, $1 \leq i \leq n$, then the *position* of that occurrence of W *in* E is the sequence $i, p_1, \ldots, p_n$.

An occurrence of a symbol or an expression W with position $p_1, \ldots, p_n$ in an expression E is denoted with ${}^{p_1,\ldots,p_n}W$.

For example, the occurrences of the variable x in the formula $\forall x \exists y (P(x, y) \vee \forall x \neg P(y, x))$ are ${}^{1}x$, ${}^{2,2,1,1}x$, ${}^{2,2,2,1}x$, and ${}^{2,2,2,2,1,2}x$. It is essential to associate variable occurrences in an expression with occurrences of quantifiers.

DEFINITION 16 (Scope of a quantifier occurrence). If ${}^{p_1,\ldots,p_m}Q$ is the occurrence of a quantifier in a formula Φ, then the occurrence of the respective quantification ${}^{p_1,\ldots,p_m}Qx\Psi$ is called the *scope* of ${}^{p_1,\ldots,p_m}Q$ *in* Φ; all occurrences ${}^{p_1,\ldots,p_m,p_{m+1},\ldots,p_n}W$, $m \leq n$, of symbols or expressions in Φ are said to be *in the scope* of ${}^{p_1,\ldots,p_m}Q$ *in* Φ.

Referring to the formula in Figure 1, the occurrence ${}^{2,2,1}P(x,y)$ is in the scope of only one quantifier occurrence, namely, ${}^{\emptyset}\forall$, whereas ${}^{2,2,2,2,1}P(y,x)$ is in the scope of both occurrences of the universal quantifier.

DEFINITION 17 (Bound and free variable occurrence). If an occurrence of a variable ${}^{p_1,\ldots,p_m,p_{m+1},\ldots,p_n}x$, $m < n$, in an expression Φ is in the scope of a quantifier occurrence ${}^{p_1,\ldots,p_m}Q$, then that variable occurrence is called a *bound occurrence* of x in Φ; the variable occurrence is said to be *bound by* the rightmost such quantifier occurrence in the string notation of Φ, i.e., by the one with the greatest index $m < n$. A variable occurrence is called *free* in an expression if is not bound by some quantifier occurrence in the expression.

Accordingly, the rightmost occurrence ${}^{2,2,2,2,1,2}x$ of x in the formula in Figure 1 is bound by the universal quantifier at position $2,2,2$. Note that every occurrence of a variable in a well-formed expression is bound by at most one quantifier occurrence in the expression.

DEFINITION 18 (Closed and ground expression, sentence). If an expression does not contain variables, it is called *ground*, and if it does not contain free variables, it is termed *closed*. Closed formulae are called *sentences*.

DEFINITION 19 (Closures of a formula). Let Φ be a formula and $\{x_1, \ldots, x_n\}$ the set of free variables of Φ, then the sentence $\forall x_1 \cdots \forall x_n \Phi$ is called a *universal closure* of Φ, and the sentence $\exists x_1 \cdots \exists x_n \Phi$ is called an *existential closure* of Φ.

NOTATION 20. In order to gain readability, we shall normally spare brackets. As usual, we permit to omit outermost brackets. Furthermore, for arbitrary binary connectives $\circ_1, \circ_2$, any formula of the shape $\Phi \circ_1 (\Psi \circ_2 \Xi)$ may be abbreviated by writing just $\Phi \circ_1 \Psi \circ_2 \Xi$ (right bracketing). Accordingly, if brackets are missing, the dominating infix connective is always the leftmost one.

1.2 Semantics of Classical First-order Logic

Now we are going to present the classical model-theoretic semantics of first-order logic due to [Tarski, 1936]. In contrast to propositional logic, where it is sufficient to work with Boolean valuations and where the atomic formulae can be treated as the basic meaningful units, the richer structure of the first-order language requires a finer analysis. In first-order logic the basic semantic components are the denotations of the terms, a collection of objects termed *universe*.

DEFINITION 21 (Universe). Any non-empty collection[2] of objects is called a *universe.*

The function symbols and the predicate symbols of the signature of a first-order language are then interpreted as functions and relations over such a universe.

NOTATION 22. For every universe $\mathcal{U}$, we denote with $\mathcal{U}_{\mathcal{F}}$ the collection of mappings $\bigcup_{n\in\mathbb{N}_0} \mathcal{U}^n \longrightarrow \mathcal{U}$, and with $\mathcal{U}_{\mathcal{P}}$ the collection of relations $\bigcup_{n\in\mathbb{N}_0} \mathfrak{P}(\mathcal{U}^n)$ with $\mathfrak{P}(\mathcal{U}^n)$ being the power set of $\mathcal{U}$. Note that any nullary mapping in $\mathcal{U}_{\mathcal{F}}$ is from the singleton set $\{\emptyset\}$ to $\mathcal{U}$, and hence, subsequently, will be identified with the single element in its image. Any nullary relation in $\mathcal{U}_{\mathcal{P}}$ is just an element of the two-element set $\{\emptyset, \{\emptyset\}\}$ $(= \{0,1\}$, according to the Zermelo-Fraenkel definition of natural numbers). We call the sets $\emptyset$ and $\{\emptyset\}$ *truth values*, and abbreviate them with $\perp$ and $\top$, respectively.

This way the mapping of atomic formulae to truth values as performed for the case of propositional logic is captured as a special case by the more general framework developed now. In the following, we denote with $\mathcal{L}$ a first-order language, with $\mathcal{V}$, $\mathcal{F}$, and $\mathcal{P}$ the sets of variables, function symbols, and predicate symbols in the signature of $\mathcal{L}$, respectively, and with $\mathcal{T}$ and $\mathcal{W}$ the sets of terms and formulae in $\mathcal{L}$, respectively.

DEFINITION 23 (First-order structure, interpretation). A *(first-order) structure* is a pair $\langle \mathcal{L}, \mathcal{U} \rangle$ consisting of a first-order language $\mathcal{L}$ and a universe $\mathcal{U}$. An *interpretation for* a first-order structure $\langle \mathcal{L}, \mathcal{U} \rangle$ is a mapping $\mathcal{I}\colon \mathcal{F} \cup \mathcal{P} \longrightarrow \mathcal{U}_{\mathcal{F}} \cup \mathcal{U}_{\mathcal{P}}$ such that

1. $\mathcal{I}$ maps every n-ary function symbol in $\mathcal{F}$ to an n-ary function in $\mathcal{U}_{\mathcal{F}}$.

2. $\mathcal{I}$ maps every n-ary predicate symbol in $\mathcal{P}$ to an n-ary relation in $\mathcal{U}_{\mathcal{P}}$.

Since formulae may contain free variables, the notion of variable assignments will be needed.

DEFINITION 24 (Variable assignment). A *variable assignment from* a first-order language $\mathcal{L}$ *to* a universe $\mathcal{U}$ is a mapping $\mathcal{A}\colon \mathcal{V} \longrightarrow \mathcal{U}$.

Once an interpretation and a variable assignment have been fixed, the meaning of any term and any formula in the language is uniquely determined.

DEFINITION 25 (Term assignment (inductive)). Let $\mathcal{I}$ be an interpretation for a structure $\langle \mathcal{L}, \mathcal{U} \rangle$, and let $\mathcal{A}$ be a variable assignment from $\mathcal{L}$ to $\mathcal{U}$. The *term assignment* of $\mathcal{I}$ *and* $\mathcal{A}$ is the mapping $\mathcal{I}^{\mathcal{A}}\colon \mathcal{T} \longrightarrow \mathcal{U}$ defined as follows.

1. For every variable x in $\mathcal{V}$: $\mathcal{I}^{\mathcal{A}}(x) = \mathcal{A}(x)$.

2. For every constant c in $\mathcal{F}$: $\mathcal{I}^{\mathcal{A}}(c) = \mathcal{I}(c)$.

[2]Whenever the term 'collection' will be used, no restriction is made with respect to the cardinality of an aggregation, whereas the term 'set' indicates that only denumerably many elements are contained.

3. If f is a function symbol of arity $n > 0$ and $t_1, \ldots, t_n$ are terms, then

$$\mathcal{I}^{\mathcal{A}}(f(t_1, \ldots, t_n)) = \mathcal{I}(f)(\mathcal{I}^{\mathcal{A}}(t_1), \ldots, \mathcal{I}^{\mathcal{A}}(t_n)).$$

Finally, we come to the assignment of truth values to formulae.

DEFINITION 26 (Formula assignment). (by simultaneous induction) Let $\mathcal{I}$ be an interpretation for a structure $\langle \mathcal{L}, \mathcal{U} \rangle$, and let $\mathcal{A}$ be a variable assignment from $\mathcal{L}$ to $\mathcal{U}$. The *formula assignment* of $\mathcal{I}$ *and* $\mathcal{A}$ is the mapping $\mathcal{I}^{\mathcal{A}}$: $\mathcal{W} \longrightarrow \{\top, \bot\}$ defined as follows. Let Φ and Ψ denote arbitrary formulae of $\mathcal{L}$.

1. For any nullary predicate symbol P in the signature of $\mathcal{L}$: $\mathcal{I}^{\mathcal{A}}(P) = \mathcal{I}(P)$.

2. If P is a predicate symbol of arity $n > 0$ and $t_1, \ldots, t_n$ are terms, then

$$\mathcal{I}^{\mathcal{A}}(P(t_1, \ldots, t_n)) = \begin{cases} \top & \text{if } \langle \mathcal{I}^{\mathcal{A}}(t_1), \ldots, \mathcal{I}^{\mathcal{A}}(t_n) \rangle \in \mathcal{I}(P) \\ \bot & \text{otherwise.} \end{cases}$$

3. $$\mathcal{I}^{\mathcal{A}}((\Phi \vee \Psi)) = \begin{cases} \top & \text{if } \mathcal{I}^{\mathcal{A}}(\Phi) = \top \text{ or } \mathcal{I}^{\mathcal{A}}(\Psi) = \top \\ \bot & \text{otherwise.} \end{cases}$$

4. $$\mathcal{I}^{\mathcal{A}}(\neg \Phi) = \begin{cases} \top & \text{if } \mathcal{I}^{\mathcal{A}}(\Phi) = \bot \\ \bot & \text{otherwise.} \end{cases}$$

5. $$\mathcal{I}^{\mathcal{A}}((\Phi \wedge \Psi)) = \mathcal{I}^{\mathcal{A}}(\neg(\neg \Phi \vee \neg \Psi)).$$

6. $$\mathcal{I}^{\mathcal{A}}((\Phi \rightarrow \Psi)) = \mathcal{I}^{\mathcal{A}}((\neg \Phi \vee \Psi)).$$

7. $$\mathcal{I}^{\mathcal{A}}((\Phi \leftrightarrow \Psi)) = \mathcal{I}^{\mathcal{A}}(((\Phi \rightarrow \Psi) \wedge (\Psi \rightarrow \Phi))).$$

8. A variable assignment is called an x-*variant* of a variable assignment if both assignments differ at most in the value of the variable x.

$$\mathcal{I}^{\mathcal{A}}(\forall x \Phi) = \begin{cases} \top & \text{if } \mathcal{I}^{\mathcal{A}'}(\Phi) = \top \text{ for all } x\text{-variants } \mathcal{A}' \text{ of } \mathcal{A} \\ \bot & \text{otherwise.} \end{cases}$$

9. $$\mathcal{I}^{\mathcal{A}}(\exists x \Phi) = \mathcal{I}^{\mathcal{A}}(\neg \forall x \neg \Phi).$$

If S is a set of first-order formulae, with $\mathcal{I}^{\mathcal{A}}(S) = \top$ we express the fact that $\mathcal{I}^{\mathcal{A}}(\Phi) = \top$, for all formulae $\Phi \in S$.

Particularly interesting is the case of interpretations for sentences. From the definition of formula assignments (items 8 and 9) it follows that, for any sentence and any interpretation $\mathcal{I}$, the respective formula assignments are all identical, and hence do not depend on the variable assignments. Consequently, for sentences, we shall speak of *the* formula assignment of an interpretation $\mathcal{I}$, and write it $\mathcal{I}$, too. Possible ambiguities between an interpretation and the corresponding formula assignment will be clarified by the context.

To comprehend the manner in which formula assignments give meaning to expressions, see Example 27. The example illustrates how formulae are interpreted in which an occurrence of a variable is in the scopes of different quantifier occurrences. Loosely speaking, Definition 26 guarantees that variable assignments obey 'dynamic binding' rules (in terms of programming languages), in the sense that a variable assignment to a variable x for an expression Φ is *overwritten* by a variable assignment to the same variable x in a subexpression of Φ.

EXAMPLE 27. Consider two sentences $\Phi = \forall x(\exists x P(x) \wedge Q(x))$ and $\Psi = \forall x \exists x(P(x) \wedge Q(x))$. Given a universe $\mathcal{U} = \{u_1, u_2\}$, and an interpretation $\mathcal{I}(P) = \mathcal{I}(Q) = \{u_1\}$, then $\mathcal{I}(\Phi) = \bot$ and $\mathcal{I}(\Psi) = \top$.

The central semantic notion is that of a model.

DEFINITION 28 (Model). Let $\mathcal{I}$ be an interpretation for a structure $\langle \mathcal{L}, \mathcal{U} \rangle$, A a collection of variable assignments from $\mathcal{L}$ to $\mathcal{U}$, and Φ a first-order formula. We say that $\mathcal{I}$ is an *A-model for* Φ if $\mathcal{I}^{\mathcal{A}}(\Phi) = \top$, for every variable assignment $\mathcal{A} \in A$; if $\mathcal{I}$ is an A-model for Φ and A is the collection of all variable assignments, then $\mathcal{I}$ is called a *model for* Φ. If $\mathcal{I}$ is an (A-)model for every formula in a set of first-order formulae S, then we also call $\mathcal{I}$ an *(A-)model* for S.

The notions of satisfiability and validity abstract from the consideration of specific models.

DEFINITION 29 (Satisfiability, validity). Let Γ be a (set of) formula(e) of a first-order language $\mathcal{L}$ (and A a collection of variable assignments). The set Γ is called *(A-)satisfiable* if there exists an (A-)model for Γ. We call Γ *valid* if every interpretation is a model for Γ.

DEFINITION 30 (Implication, equivalence). Let Γ and Δ be two (sets of) first-order formulae.

1. We say that Γ *implies* Δ, written $\Gamma \models \Delta$, if every model for Γ is a model for Δ; obviously, if $\Gamma = \emptyset$, then Δ is valid, and we simply write $\models \Delta$.

2. Γ *strongly implies* Δ if, for every universe $\mathcal{U}$ and every variable assignment $\mathcal{A}$ from $\mathcal{L}$ to $\mathcal{U}$: every $\{\mathcal{A}\}$-model for Γ is an $\{\mathcal{A}\}$-model for Δ.

If Γ and Δ (strongly) imply each other, they are called *(strongly) equivalent.*

Note that according to this definition any first-order formula is equivalent to any-one of its universal closures. Obviously, for (sets of) *sentences*, implication and strong implication coincide. Furthermore, the notion of material (object-level) implication and the strong (meta-level) implication concept of first-order formulae are related as follows.

THEOREM 31 (Implication Theorem). *Given two first-order formulae Φ and Ψ, Φ strongly implies Ψ if and only if the formula $\Phi \to \Psi$ is valid.*

Proof. For the 'if'-part, assume $\Phi \rightarrow \Psi$ be valid. Let $\mathcal{A}$ be any variable assignment and $\mathcal{I}$ an arbitrary $\{\mathcal{A}\}$-model for Φ. By Definition 26, $\mathcal{I}^{\mathcal{A}}(\Phi) = \bot$ or $\mathcal{I}^{\mathcal{A}}(\Psi) = \top$. By assumption, $\mathcal{I}^{\mathcal{A}}(\Phi) = \top$; hence $\mathcal{I}^{\mathcal{A}}(\Psi) = \top$, and $\mathcal{I}$ is an $\{\mathcal{A}\}$-model for Ψ. For the 'only-if'-part, suppose that Φ strongly implies Ψ. Let $\mathcal{A}$ be any variable assignment and $\mathcal{I}$ an arbitrary interpretation. Now, either $\mathcal{I}^{\mathcal{A}}(\Phi) = \bot$; then, by Definition 26, $\mathcal{I}^{\mathcal{A}}(\Phi \rightarrow \Psi) = \top$. Or, $\mathcal{I}^{\mathcal{A}}(\Phi) = \top$; in this case, by assumption, $\mathcal{I}(\Psi) = \top$; hence $\mathcal{I}^{\mathcal{A}}(\Phi \rightarrow \Psi) = \top$. Consequently, in either case $\mathcal{I}$ is an $\{\mathcal{A}\}$-model for $\Phi \rightarrow \Psi$. ■

It is obvious that strongly equivalent formulae can be substituted for each other in any context without changing the meaning of the context.

LEMMA 32 (Replacement Lemma). *Given two strongly equivalent formulae F and G and any formula Φ with F as subformula, if the formula Ψ can be obtained from Φ by replacing an occurrence of F in Φ with G, then Φ and Ψ are strongly equivalent.*

Another more subtle useful replacement property is the following.

LEMMA 33. *If $\models F \rightarrow G$, then $\models \forall x F \rightarrow \forall x G$.*

Proof. Assume $\models F \rightarrow G$. Let $\mathcal{I}$ be any interpretation and $\mathcal{A}$ any variable assignment with $\mathcal{I}^{\mathcal{A}}(\forall x F) = \top$. Then, for all x-variants $\mathcal{A}'$ of $\mathcal{A}$: $\mathcal{I}^{\mathcal{A}'}(F) = \top$ and, by assumption, $\mathcal{I}^{\mathcal{A}'}(G) = \top$. Consequently, $\mathcal{I}^{\mathcal{A}}(\forall x G) = \top$. ■

The subsequently listed basic strong equivalences between first-order formulae can also be demonstrated easily.

PROPOSITION 34. *Let F, G, and H be arbitrary first-order formulae. All formulae of the following structures are valid.*

(1) $\neg\neg F \leftrightarrow F$

(2) $\neg(F \wedge G) \leftrightarrow (\neg F \vee \neg G)$ *(De Morgan law for $\wedge$)*

(3) $\neg(F \vee G) \leftrightarrow (\neg F \wedge \neg G)$ *(De Morgan law for $\vee$)*

(4) $(F \vee (G \wedge H)) \leftrightarrow ((F \vee G) \wedge (F \vee H))$ *($\vee$-distributivity)*

(5) $(F \wedge (G \vee H)) \leftrightarrow ((F \wedge G) \vee (F \wedge H))$ *($\wedge$-distributivity)*

(6) $\neg\exists x F \leftrightarrow \forall x \neg F$ *($\exists\forall$-conversion)*

(7) $\neg\forall x F \leftrightarrow \exists x \neg F$ *($\forall\exists$-conversion)*

(8) $\forall x(F \wedge G) \leftrightarrow (\forall x F \wedge \forall x G)$ *($\forall\wedge$-distributivity)*

(9) $\exists x(F \vee G) \leftrightarrow (\exists x F \vee \exists x G)$ *($\exists\vee$-distributivity)*

We conclude this part with proving a technically useful property of variable assignments.

DEFINITION 35. Two variable assignments $\mathcal{A}$ and $\mathcal{B}$ are said to *overlap on* a set of variables V if for all $x \in V : \mathcal{A}(x) = \mathcal{B}(x)$.

NOTATION 36. For any mapping f, its modification by changing the value of x to u, i.e. $(f \setminus \langle x, f(x)\rangle) \cup \langle x, u\rangle$, will be denoted with f^x_u.

PROPOSITION 37. *Let Φ be a formula of a first-order language $\mathcal{L}$ with V being the set of free variables in Φ, and $\mathcal{U}$ a universe. Then, for any two variable assignments $\mathcal{A}$ and $\mathcal{B}$ from $\mathcal{L}$ to $\mathcal{U}$ that overlap on V:*

(1) $\mathcal{I}^{\mathcal{A}}(\Phi) = \mathcal{I}^{\mathcal{B}}(\Phi)$, and

(2) if $\Phi = \exists x\Psi$, then $\{u \in \mathcal{U} \mid \mathcal{I}^{\mathcal{A}^x_u}(\Psi) = \top\} = \{u \in \mathcal{U} \mid \mathcal{I}^{\mathcal{B}^x_u}(\Psi) = \top\}$.

Proof. (1) is obvious from Definition 26 of formula assignments. For (2), consider an arbitrary element $u \in \mathcal{U}$ with $\mathcal{I}^{\mathcal{A}^x_u}(\Psi) = \top$. Since $\mathcal{A}^x_u$ and $\mathcal{B}^x_u$ overlap on $V \cup \{x\}$, by (1), $\mathcal{I}^{\mathcal{B}^x_u}(\Psi) = \top$, which proves the set inclusion in one direction. The reverse direction holds by symmetry. ■

1.3 Variable Substitutions

The concept of variable substitutions, which we shall introduce next, is the basic modification operation performed on logical expressions. Let in the following denote $\mathcal{T}$ the set of terms and $\mathcal{V}$ the set of variables of a first-order language.

DEFINITION 38 ((Variable) substitution). A *(variable) substitution* is any mapping $\sigma : V \longrightarrow \mathcal{T}$ where V is a finite subset of $\mathcal{V}$ and $x \neq \sigma(x)$, for every x in the domain of σ. A substitution is called *ground* if no variables occur in the terms of its range.

DEFINITION 39 (Binding). Any element $\langle x, t\rangle$ of a substitution, abbreviated x/t, is called a *binding*. We say that a binding x/t is *proper* if the variable x does not occur in the term t.

Now, we consider the application of substitutions to logical expressions.

DEFINITION 40 (Instance). If Φ is any expression and σ is a substitution, then the *σ-instance of Φ*, written $\Phi\sigma$, is the expression obtained from Φ by simultaneously replacing every occurrence of each variable $x \in$ domain(σ) that is free in Φ by the term $\sigma(x)$. If Φ and Ψ are expressions and there is a substitution σ with $\Psi = \Phi\sigma$, then Ψ is called an *instance of* Φ. Similarly, if S is a (collection of) set(s) of formulae, then $S\sigma$ denotes the (collection of the) set(s) of σ-instances of its elements.

As a matter of fact, bound variable occurrences are not replaced. Furthermore, we are interested in substitutions which preserve the models of a formula. In order to preserve modelhood for arbitrary logical expressions, the following property is sufficient.

DEFINITION 41 (Free substitution). A substitution σ is said to be *free for* an expression Φ provided, for every free occurrence ${}^s x$ of a variable in Φ, all variable occurrences in ${}^s x\sigma$ are free in $\Phi\sigma$.

While no bound variable occurrence can vanish when a substitution is applied, for free substitutions, no additional bound variable occurrences are imported. This means that the following proposition holds.

PROPOSITION 42. *A substitution σ is free for an expression Φ if and only if any variable occurs bound at the same positions in Φ and in $\Phi\sigma$.*

Bringing in additional bound variables can lead to unsoundness, as shown with the following example.

EXAMPLE 43. Consider a formula Φ of the form $\exists x(P(x,y,z) \land \neg P(y,y,z))$ and the substitutions $\sigma_1 = \{y/z\}$ and $\sigma_2 = \{y/x\}$. While σ_1 is free for Φ and $\Phi \models \Phi\sigma_1 = \exists x(P(x,z,z) \land \neg P(z,z,z))$, σ_2 is not, and indeed $\Phi \not\models \Phi\sigma_2 = \exists x(P(x,x,z) \land \neg P(x,x,z))$.

The following fundamental result relates substitutions and interpretations.

NOTATION 44. If $\mathcal{I}$ is an interpretation, $\mathcal{A}$ a variable assignment, and $\sigma = \{x_1/t_1, \ldots, x_n/t_n\}$ a substitution, then we denote with $\mathcal{A}\sigma_{\mathcal{I}}$ the variable assignment $\mathcal{A}^{x_1}_{\mathcal{I}^{\mathcal{A}}(t_1)} \cdots {}^{x_n}_{\mathcal{I}^{\mathcal{A}}(t_n)}$ using Notation 36. If the underlying interpretation is clear from the context, we will sometimes omit the subscript and simply write $\mathcal{A}\sigma$.

LEMMA 45. *If σ is a substitution that is free for a first-order expression E, then, for any interpretation $\mathcal{I}$ and any variable assignment $\mathcal{A}$: $\mathcal{I}^{\mathcal{A}}(E\sigma) = \mathcal{I}^{\mathcal{A}\sigma}(E)$.*

Proof. The proof is by induction on the structural complexity of the expression E. First, for any term, the result is immediate from the Definition 25 of term assignments. The cases of quantifier-free formula are also straightforward from items (1) – (7) of the Definition 26 of formula assignments. We consider the case of a universal formula $\forall xF$ in more detail. $\mathcal{I}^{\mathcal{A}}(\forall xF\sigma) = \top$ if and only if (by item (8) of formula assignment) for all x-variants $\mathcal{A}'$ of $\mathcal{A}$: $\top = \mathcal{I}^{\mathcal{A}'}(F\sigma) =$ (by the induction hypothesis) $\mathcal{I}^{\mathcal{A}'\sigma}(F)$ iff (since σ is assumed to be free for F) for all x-variants $\mathcal{A}\sigma'$ of $\mathcal{A}\sigma$: $\mathcal{I}^{\mathcal{A}\sigma'}(F) = \top$ iff (by item (8) of formula assignment) $\mathcal{I}^{\mathcal{A}\sigma}(\forall xF) = \top$. The existential case is similar. ∎

Now we can state a very general soundness result for the application of substitutions to logical expressions. Let $\mathcal{V}$ denote the set of variables of the underlying first-order language.

PROPOSITION 46 (Substitution soundness). *Given a first-order formula Φ and a substitution $\sigma = \{x_1/t_1, \ldots, x_n/t_n\}$ that is free for Φ, let $V \subseteq \mathcal{V}$ be any set of variables containing $x_1, \ldots, x_n$ and A any collection of all variable assignments* *that overlap on $\mathcal{V} \setminus V$. If an interpretation $\mathcal{I}$ is an A-model for Φ, then $\mathcal{I}$ is an A-model for $\Phi\sigma$.*

Proof. Consider an arbitrary variable assignment $\mathcal{A} \in A$. Now, by assumption, A contains all variable assignments that overlap on $\mathcal{V} \setminus V$ where V is any superset of $\{x_1, \ldots, x_n\}$, therefore, $\mathcal{A}\sigma \in A$ and hence $\mathcal{I}^{\mathcal{A}}(\Phi\sigma) = \top$. Since σ is free for Φ, Lemma 45 can be applied which yields that $\mathcal{I}^{\mathcal{A}\sigma}(\Phi) = \mathcal{I}^{\mathcal{A}}(\Phi\sigma) = \top$. ■

As a special instance of this proposition we obtain the following corollary (simply set $V = \mathcal{V}$ and A will be the collection of all variable assignments).

COROLLARY 47. *For any formula Φ and any substitution σ which is free for $\Phi : \Phi \models \Phi\sigma$.*

DEFINITION 48 (Composition of substitutions). Assume σ and τ to be substitutions. Let σ' be the substitution obtained from the set $\{\langle x, t\tau\rangle \mid x/t \in \sigma\}$ by removing all pairs for which $x = t\tau$, and let τ' be that subset of τ which contains no binding x/t with $x \in \text{domain}(\sigma)$. The substitution $\sigma' \cup \tau'$, written $\sigma\tau$, is called the *composition* of σ and τ.

PROPOSITION 49. *Let σ, τ and θ be arbitrary substitutions and Φ any logical expression such that σ is free for Φ.*

1. $\sigma\emptyset = \emptyset\sigma = \sigma$, *for the empty substitution* $\emptyset$.
2. $(\Phi\sigma)\tau = \Phi(\sigma\tau)$.
3. $(\sigma\tau)\theta = \sigma(\tau\theta)$.

Proof. (1) is immediate. For (2) consider any free occurrence ${}^{s}x$ of a variable x in Φ. We distinguish three cases. First, $x \notin \text{domain}(\sigma)$ and $x \notin \text{domain}(\tau)$; then ${}^{s}(x\sigma)\tau = {}^{s}x = {}^{s}x(\sigma\tau)$. If, secondly, $x \notin \text{domain}(\sigma)$ but $x \in \text{domain}(\tau)$, then ${}^{s}(x\sigma)\tau = {}^{s}x\tau = {}^{s}x(\sigma\tau)$. Lastly, assume $x \in \text{domain}(\sigma)$; as σ was assumed free for Φ, no variable occurrence in ${}^{s}x\sigma$ is bound in $\Phi\sigma$, therefore ${}^{s}(x\sigma)\tau = {}^{s}x(\sigma\tau)$. Since ${}^{s}x$ was chosen arbitrary and only free variable occurrences were modified, we have the result for Φ. For (3) let x be any variable. The repeated application of (2) yields that $x((\sigma\tau)\theta) = (x(\sigma\tau))\theta = ((x\sigma)\tau)\theta = (x\sigma)(\tau\theta) = x(\sigma(\tau\theta))$. This means that the substitutions $(\sigma\tau)\theta$ and $\sigma(\tau\theta)$ map every variable to the same term, hence they are identical. ■

Summarizing these results, we have that $\emptyset$ acts as a left and right identity for composition; (2) expresses that under the given assumption substitution application and composition permute; and (3), the associativity of substitution composition, permits to omit parentheses when writing a composition of substitutions. As a consequence of (1) and (3), the set of substitutions with the composition operation forms a semi-group.

2 NORMAL FORMS AND NORMAL FORM TRANSFORMATIONS

A *logical problem* for a first-order language consists in the task of determining whether a relation holds between certain first-order expressions. For an *efficient solution* of a logical problem, it is very important to know whether it is possible to restrict attention to a proper sublanguage of the first-order language. This is because certain sublanguages permit the application of more efficient solution techniques than available for the full first-order format. For classical logic, this can be strongly exploited by using prenex and Skolem forms.

2.1 Prenex and Skolem Formulae

DEFINITION 50 (Prenex form). A first-order formulae Φ is said to be a *prenex formula* or in *prenex form* if it has the structure $Q_1x_1 \cdots Q_nx_nF$, $n \geq 0$, where the Q_i, $1 \leq i \leq n$, are quantifiers and F is quantifier-free. We call F the *matrix* of Φ.

PROPOSITION 51. *For every first-order formula Φ there is a formula in prenex form which is strongly equivalent to Φ.*

Proof. We give a constructive method to transform any formula Φ over the connectives $\neg$, $\wedge$, and $\vee$ into prenex form—by the definition of formula assignment, the connectives $\leftrightarrow$ and $\rightarrow$ can be eliminated before, without affecting strong equivalence. If Q is any quantifier, $\forall$ or $\exists$, with $\bar{Q}$ we denote the quantifier $\exists$ respectively $\forall$. Now, for any formula which is not in prenex form, one of the following two cases holds.

1. Φ has a subformula of the structure $\neg QxF$; then, by Proposition 34(6) and (7), and the Replacement Lemma (Lemma 32), the formula Ψ obtained from Φ by substituting all occurrences of $\neg QxF$ in Φ by $\bar{Q}x\neg F$ is strongly equivalent to Φ.
2. Φ has a subformula Ψ of the structure $(QxF \circ G)$ or $(G \circ QxF)$ where $\circ$ is $\wedge$ or $\vee$; let y be a variable not occurring in Φ, then, clearly Ψ and $\Psi' = Qy(F\{x/y\} \circ G)$ or $Qy(G \circ F\{x/y\})$, respectively, are strongly equivalent; therefore, by the Replacement Lemma, the formula obtained by replacing all occurrences of Ψ in Φ by Ψ' is strongly equivalent to Φ.

Consequently, in either case one can move quantifiers in front without affecting strong equivalence, and after finitely many iterations prenex form is achieved. ■

It is obvious that, except for the case of formulae containing $\leftrightarrow$, the time needed for carrying out this procedure is bounded by a polynomial in the size of the input, and the resulting prenex formula has less than double the size of the initial formula. The removal of $\leftrightarrow$, however, can lead to an exponential increase of the formula size (see [Reckhow, 1976]).

DEFINITION 52 (Skolem form). A first-order formula Φ is said to be a *Skolem formula* or in *Skolem form* if it has the form $\forall x_1 \cdots \forall x_n F$ and F is quantifier-free.

The possibility of transforming any first-order formula into Skolem form is fundamental for the field of automated deduction. This is because the removal of existential quantifiers facilitates a particularly efficient computational treatment of first-order formulae. Furthermore, the single step in which an existential quantifier is removed, occurs as a basic component of *any* calculus for full first-order logic.

DEFINITION 53 (Skolemization). Let S be a set of formulae containing a formula Φ with the structure $\forall y_1 \cdots \forall y_m \exists y F$, $m \geq 0$, and $x_1, \ldots, x_n$ the variables that are free in $\exists y F$. If f is an n-ary function symbol not occurring in any formula of S (we say that f is *new to* S), then the formula $\forall y_1 \cdots \forall y_m (F\{y/f(x_1, \ldots, x_n)\})$ is named a *Skolemization* of Φ *w.r.t.* S.

We have introduced a general form of Skolemization which is applicable to arbitrary, not necessarily closed, sets of first-order formulae in prenex form. This is necessary for the free-variable tableau systems developed in Section 4. When moving to a Skolemization of a formula, for any variable assignment $\mathcal{A}$, the collection of $\{\mathcal{A}\}$-models does not increase.

PROPOSITION 54. *Given a formula Φ of a first-order language $\mathcal{L}$, if Ψ is a Skolemization of Φ w.r.t. a set of formulae S, then Ψ strongly implies Φ.*

Proof. Let $\Phi = \forall y_1 \cdots \forall y_m \exists y F$. First, we show that, for *any* term t, the subformula $\exists y F$ of Φ is strongly implied by $F' = F\{y/t\}$. Let $\mathcal{A}$ be any variable assignment and $\mathcal{I}$ any $\{\mathcal{A}\}$-model for F'. By Lemma 45, $\mathcal{I}^{\mathcal{A}\{y/t\}}(F) = \top$. Since $\mathcal{A}\{y/t\}$ is an y-variant of $\mathcal{A}$, by the definition of formula assignments, $\mathcal{I}^{\mathcal{A}}(\exists y F) = \top$. Then, a repeated application of Lemma 33 yields that $\Psi = \forall y_1 \cdots \forall y_m F'$ strongly implies Φ. ■

When moving to a Skolemization of a formula, the collection of models may decrease, however. Consequently, for the transformation of formulae into Skolem form, equivalence must be sacrificed, but the preservation of A-satisfiability can be guaranteed, for any collection A of variable assignments.

PROPOSITION 55. *Let Ψ be a Skolemization of a formula Φ w.r.t. a set of formulae S and A any collection of variable assignments. If S is A-satisfiable, then $S \cup \{\Psi\}$ is A-satisfiable.*

Proof. By assumption, $\Phi \in S$ has the structure $\forall y_1 \cdots \forall y_m \exists y F$ $(m \geq 0)$; Ψ has the form $\forall y_1 \cdots \forall y_m F'$ where $F' = F\{y/f(x_1, \ldots, x_n)\}$; f is an n-ary function symbol new to S; and $x_1, \ldots, x_n$ are the free variables in $\exists y F$. Now let A be any collection of variable assignments that has an A-model $\mathcal{I}$ for S. $\mathcal{I}$ need not be an A-model for Ψ, but we show that with a modification of merely the meaning of the function symbol f an A-model for $S \cup \{\Psi\}$ can be specified. First, we define a total and disjoint partition P on the collection of variable assignments A by grouping together all elements in A that overlap on the variables

$x_1, \ldots, x_n$. By Proposition 37, for any two variable assignments $\mathcal{B}$ and $\mathcal{C}$ in any element of P: $\{u \in \mathcal{U} \mid \mathcal{I}^{\mathcal{B}^y_u}(F) = \top\} = \{u \in \mathcal{U} \mid \mathcal{I}^{\mathcal{C}^y_u}(F) = \top\}$, i.e., for any element of the partition P, the collection of objects "with the property" F is unique; we abbreviate with $\mathcal{U}_{u_1,\ldots,u_n}(F)$ the collection of objects determined by the variable assignments that map $x_1, \ldots, x_n$ to the objects $u_1, \ldots, u_n$, respectively. By the assumption of $\mathcal{I}$ being an A-model for Φ, none of these collections of objects is empty. In order to be able to identify elements in those possibly nondenumerable collections, which is necessary to define a mapping, we have to assume the existence of a well-ordering[3] $\prec$ on $\mathcal{U}$. For any collection $M \subseteq \mathcal{U}$, let μM denote the smallest element modulo $\prec$. Now we can define a total n-ary mapping f: $\mathcal{U}^n \longrightarrow \mathcal{U}$ by setting $\mathsf{f}(u_1, \ldots, u_n) = \mu\mathcal{U}_{u_1,\ldots,u_n}(F)$ and an interpretation $\mathcal{I}_\Psi = \mathcal{I}^f_\mathsf{f}$ (using Notation 36). Since the symbol f does not occur in any formula of S, $\mathcal{I}_\Psi$ is an A-model for S. To realize that $\mathcal{I}_\Psi$ is an A-model for Ψ, too, consider an arbitrary variable assignment $\mathcal{A} \in A$. Clearly, $\mathcal{I}^{\mathcal{A}}_\Psi(\exists y F) = \top$. Let $P_\mathcal{A}$ be the element of the partition P that contains $\mathcal{A}$. Define the variable assignment $\mathcal{A}' = \mathcal{A}^y_{\mathsf{f}(\mathcal{A}(x_1),\ldots,\mathcal{A}(x_n))}$. $\mathcal{I}^{\mathcal{A}'}_\Psi(F) = \top$ and hence $\mathcal{I}^{\mathcal{A}'}_\Psi(F\{y/f(x_1, \ldots, x_n)\}) = \top$. Now $\mathcal{A}$ and $\mathcal{A}'$ are identical except for the value of y, but y does not occur free in $F\{y/f(x_1, \ldots, x_n)\}$, therefore, $\mathcal{I}^{\mathcal{A}}_\Psi(F\{y/f(x_1, \ldots, x_n)\}) = \top$. ∎

THEOREM 56 (Skolemization Theorem). *Given a formula Φ of a first-order language $\mathcal{L}$, let Ψ be a Skolemization of Φ w.r.t. a set of formulae S and A any collection of variable assignments. S is A-satisfiable if and only if $S \cup \{\Psi\}$ is A-satisfiable.*

Proof. Immediate from the Propositions 54 and 55. ∎

Concerning the space and time complexity involved in a transformation into Skolem form, the following estimate can be formulated.

PROPOSITION 57. *Given a prenex formula Φ of a first-order language $\mathcal{L}$, if Ψ is a Skolem formula obtained from Φ via a sequence of Skolemizations, then* size(Ψ), *i.e., the length of the string Ψ, is smaller than* size(Φ)^2, *and the run time of the Skolemization procedure is polynomially bounded by the size of Φ.*

Proof. Every variable occurrence in Φ is bound by exactly one quantifier occurrence in Φ, and every variable occurrence in an inserted Skolem term is bound by a universal quantifier. This entails that, throughout the sequence of Skolemization steps, whenever a variable occurrence is replaced by a Skolem term, then no variable occurrence *within* an inserted Skolem term is substituted afterwards.

[3] A total ordering $\prec$ on a collection of objects S is a *well-ordering on* S if every non-empty subcollection M of objects from S has a smallest element modulo $\prec$. Note that supposing the existence of a well-ordering amounts to assuming the *axiom of choice* (for further equivalent formulations of the axiom of choice, consult, for example, [Krivine, 1971]).

Moreover, the arity of each inserted Skolem function is bounded by the number of free variables in Φ plus the number of variables in the quantifier prefix of Φ. Therefore, the output size is quadratically bounded by the input size. Since in the Skolemization operation merely variable replacements are performed, any deterministic execution of the Skolemization procedure can be done in polynomial time. ■

Prenexing and Skolemization only work for classical logic, but not for intuitionistic or other logics. In those cases, more sophisticated methods are needed to encode the nesting of the connectives and quantifiers. Some of those are considered in [Wallen, 1989; Ohlbach, 1991], and in Chapter 5 of this book (see also [Prawitz, 1960] and [Bibel, 1987]).

2.2 *Herbrand Interpretations*

The standard theorem proving procedures are based on the following obvious proposition.

PROPOSITION 58. *Given a set of sentences* Γ *and a sentence* F. $\Gamma \models F$ *if and only if* $\Gamma \cup \{\neg F\}$ *is unsatisfiable.*

Accordingly, the problem of determining whether a sentence is logically implied by a set of sentences can be reformulated as an unsatisfiability problem. Demonstrating the unsatisfiability of a set of formulae of a first-order language $\mathcal{L}$, however, means to prove, for any universe $\mathcal{U}$, that no interpretation for $\langle \mathcal{L}, \mathcal{U} \rangle$ is a model for the set of formulae. A further fundamental result for the efficient computational treatment of first-order logic is that, for formulae in Skolem form, it is sufficient to examine only the interpretations for one particular domain, the *Herbrand universe* of the set of formulae. Subsequently, let $\mathcal{L}$ denote a first-order language and $a_{\mathcal{L}}$ a fixed constant in the signature of $\mathcal{L}$.

DEFINITION 59 (Herbrand universe (inductive)). Let S be (a set of) formula(e) of $\mathcal{L}$. With S_C we denote the set of constants occurring in (formulae of) S. The *constant base* of S is S_C if S_C is non-empty, and the singleton set $\{a_{\mathcal{L}}\}$ if $S_C = \emptyset$. The *function base* S_F of S is the set of function symbols occurring in (formulae of) S with arities > 0. Then the *Herbrand universe* of S is the set of terms defined inductively as follows.

1. Every element of the constant base of S is in the *Herbrand universe* of S.
2. If $t_1, \ldots, t_n$ are in the Herbrand universe of S and f is an n-ary function symbol in the function base of S, then the term $f(t_1, \ldots, t_n)$ is in the *Herbrand universe* of S.

DEFINITION 60 (Herbrand interpretation). Given a (set of) formula(e) S of a first-order language $\mathcal{L}$ with Herbrand universe $\mathcal{U}$. A *Herbrand interpretation for* S is an interpretation $\mathcal{I}$ for the pair $\langle \mathcal{L}, \mathcal{U} \rangle$ meeting the following properties.

1. $\mathcal{I}$ maps every constant in S_C to itself.

2. $\mathcal{I}$ maps every function symbol f in S_F with arity $n > 0$ to the n-ary function that maps every n-tuple of terms $\langle t_1, \ldots, t_n \rangle \in \mathcal{U}^n$ to the term $f(t_1, \ldots, t_n)$.

PROPOSITION 61. *For any (set of) first-order formulae S in Skolem form, if S has a model, then it has a Herbrand model.*

Proof. Let $\mathcal{I}'$ be an interpretation with arbitrary universe $\mathcal{U}'$ which is a model for S, and let $\mathcal{U}$ denote the Herbrand universe of S. First, we define a mapping h: $\mathcal{U} \longrightarrow \mathcal{U}'$, as follows.

1. For every constant $c \in \mathcal{U}$: $h(c) = \mathcal{I}'(c)$.
2. For every term $f(t_1, \ldots, t_n) \in \mathcal{U}$: $h(f(t_1, \ldots, t_n)) = \mathcal{I}'(f)(h(t_1), \ldots, h(t_n))$.

Next, we define a Herbrand interpretation $\mathcal{I}$ for S.

3. For every n-ary predicate symbol P, $n \geq 0$, and any n-tuple of objects $\langle t_1, \ldots, t_n \rangle \in \mathcal{U}^n$: $\langle t_1, \ldots, t_n \rangle \in \mathcal{I}(P)$ if and only if $\langle h(t_1), \ldots, h(t_n) \rangle \in \mathcal{I}'(P)$.

Now let $\mathcal{A}$ be an arbitrary variable assignment $\mathcal{A}$ from $\mathcal{L}$ to $\mathcal{U}$. With $\mathcal{A}'$ we denote the functional composition of $\mathcal{A}$ and h. It can be verified easily by induction on the construction of formulae that $\mathfrak{I}'^{\mathcal{A}'}(S) = \top$ entails $\mathfrak{I}^{\mathcal{A}}(S) = \top$. The induction base is evident from the definition of $\mathcal{I}$, and the induction step follows from Definition 26. Consequently, $\mathcal{I}$ is a model for S. ■

For formulae in Skolem form, the Herbrand universe is always rich enough to be used as a *representative* for any other universe, and the question whether a model exists can always be solved by restricting attention to Herbrand interpretations. For formulae that are not in Skolem form, this does not work, as illustrated with the following simple example.

EXAMPLE 62. The formula $\exists x(P(x) \land \neg P(a))$ is satisfiable, but it has no Herbrand model.

The fact that Herbrand interpretations are sufficient for characterizing modelhood in the case of Skolem formulae can be used for proving the *Löwenheim-Skolem theorem.*

THEOREM 63 (Löwenheim-Skolem theorem). *Every satisfiable (set of) first-order formula(e) S has a model with a countable universe.*

Proof. Given any satisfiable (set of) first-order formula(e) S, let S' be a (set of) first-order formula(e) obtained from S by prenexing and Skolemization.[4] By

[4]If S is an infinite set of formulae and the Herbrand universe of S already contains almost all function symbols of a certain arity, then it may be necessary to move to an extended first-order language $\mathcal{L}'$ that contains enough function symbols of every arity.

Propositions 51 and 55, S' must be satisfiable, too. Then, by Proposition 61, there exists a Herbrand model $\mathcal{I}$ for S' with a countable Herbrand universe, since obviously every Herbrand universe is countable. By Propositions 51 and 54, $\mathcal{I}$ must be a model for S. ■

Working with Herbrand interpretations has the advantage that interpretations can be represented in a very elegant manner.

DEFINITION 64 (Herbrand base). Given a (set of) formula(e) S of a first-order language $\mathcal{L}$ with Herbrand universe $\mathcal{U}$. The *predicate base* S_P of S is the set of predicate symbols occurring in (formulae of) S. The *Herbrand base* of S, written $\mathcal{B}_S$, is the set of all atomic formulae $P(t_1, \ldots, t_n)$, $n \geq 0$, with $P \in S_P$ and $t_i \in \mathcal{U}$, for every $1 \leq i \leq n$.

NOTATION 65. Every Herbrand interpretation $\mathcal{H}$ of a (set of) formula(e) S can be uniquely represented by a subset H of the Herbrand base $\mathcal{B}_S$ of S by defining

$$\mathcal{H}(L) = \begin{cases} \top & \text{if } L \in H \\ \bot & \text{otherwise} \end{cases}$$

for any ground atom $L \in \mathcal{B}_S$. We shall exploit this fact and occasionally use subsets of the Herbrand base for denoting Herbrand interpretations.

2.3 *Formulae in Clausal Form*

After prenexing and Skolemizing a formula, it is a standard technique in automated deduction to transform the resulting formula into *clausal form*.

DEFINITION 66 (Clause). Any formula c of the form $\forall x_1 \cdots \forall x_n(L_1 \vee \cdots \vee L_m)$, with $m \geq 1$ and the L_i being literals, is a *clause*. Each literals L_i is said to be *contained in c*.

DEFINITION 67 (Clausal form (inductive)).

1. Any clause is *in clausal form*.
2. If F is in clausal form and c is a clause, then $c \wedge F$ is *in clausal form*.

PROPOSITION 68. *For any first-order formula Φ in Skolem form there exists a strongly equivalent formula Ψ in clausal form.*

Proof. Let F be the matrix of a first-order formula Φ in Skolem form. We perform the following four equivalence preserving operations. First, by items 7 and 6 of Definition 26 of formula assignment, successively, the connectives $\leftrightarrow$ and $\rightarrow$ are removed. Secondly, the negation signs are pushed immediately before atomic formulae, using recursively Proposition 34(1) and de Morgan's laws (2) and (3). Finally, apply $\vee$-distributivity from left to right until no conjunction is dominated by a disjunction. The resulting formula is in clausal form. ■

It can easily be verified that, even for matrices not containing $\leftrightarrow$, the given transformation may lead to an exponential increase of the formula size. There exists no equivalence preserving polynomial transformation of a matrix into clausal form, even if $\leftrightarrow$ does not occur in the matrix (see [Reckhow, 1976]). But there are polynomial transformations if logical equivalence is sacrificed. Those transformations (see [Eder, 1985; Boy de la Tour, 1990]) are satisfiability and unsatisfiability preserving, and the transformed formula logically implies the source formula, so that the typical logical problems—unsatisfiability detection or model finding if possible—can be solved by considering the transformed formula.

3 FIRST-ORDER SENTENCE TABLEAUX

The *tableau method* was introduced by Beth in [1955; 1959] and elaborated by Hintikka in [1955] and others, but the most influential standard format was given by Smullyan in [1968] (cf. Section 2 in the first chapter of this book). Therefore, the tableau calculus for closed first-order formulae, which is developed in this section, will be essentially Smullyan's.

3.1 Quantifier Elimination in Unifying Notation

The tableau method for propositional logic exploits the fact that all propositional formulae that are not literals can be partitioned into two syntactic types, a *conjunctive type*, called the *α-type*, or a *disjunctive type*, named the *β-type*. Accordingly, only two inference rules are needed, the α- and the β-rule. This uniformity extends to the first-order language, in that just two more syntactic types are needed to capture all first-order formulae, the *universal type*, called the *γ-type*, and the *existential type*, named the *δ-type*. Likewise, just two further inference rules will be needed, called the γ- and the δ-rule.

Altogether, this results in the following classification and decomposition schema for first-order *sentences*—arbitrary first-order formulae containing free variables will be treated in the next section. To any first-order sentence F of any *connective* type (α or β) a *sequence* of sentences different from F will be assigned, called the α- or *β-subformulae sequence* of F, respectively, as defined in Table 1, for all formulae over the connectives $\neg$, $\vee$, $\wedge$ and $\rightarrow$. Note that, by exploitation of the associativity of the connectives $\vee$ and $\wedge$, we permit subformulae sequences of more than two formulae. This straightforward generalization speeds up the decomposition of formulae.

While the subformula sequence and hence the decomposition of any formula of a connective type is always finite, this cannot be achieved in general. A first-order sentence F of any *quantification* type (γ or δ) has possibly infinitely many γ- or *δ-subformulae*, respectively, as defined in Table 2 where t ranges over the set of ground terms and c over the set of constants of a first-order language.

Conjunctive		Disjunctive	
α	α-subformulae sequence	β	β-subformulae sequence
$\neg\neg F$	F		
$F_1 \wedge \cdots \wedge F_n$	$F_1, \ldots, F_n$	$F_1 \vee \cdots \vee F_n$	$F_1, \ldots, F_n$
$\neg(F_1 \vee \cdots \vee F_n)$	$\neg F_1, \ldots, \neg F_n$	$\neg(F_1 \wedge \cdots \wedge F_n)$	$\neg F_1, \ldots, \neg F_n$
$\neg(F \to G)$	$F, \neg G$	$F \to G$	$\neg F, G$

Table 1. Connective types and α-, β-subformulae sequences

Universal		Existential	
γ	γ-subformulae	δ	δ-subformulae
$\forall xF$	$F\{x/t\}$	$\exists xF$	$F\{x/c\}$
$\neg\exists xF$	$\neg F\{x/t\}$	$\neg\forall xF$	$\neg F\{x/c\}$

Table 2. Quantification types and γ-, δ-subformulae

DEFINITION 69. Any α-, β-, γ-, or δ-subformula F' of a formula F is named an *immediate tableau subformula* of F. A formula F' is called a *tableau subformula* of F if the pair $\langle F', F\rangle$ is in the transitive closure of the immediate tableau subformula relation.

Obviously, the decomposition schema guarantees that all tableau subformulae of a sentence are sentences. Moreover, the decomposition rules have the following fundamental proof-theoretic property.

DEFINITION 70 (Formula complexity). The *formula complexity* of a formula F is the number of occurrences of formulae in F.

PROPOSITION 71. *Every tableau subformula of a formula F has a smaller formula complexity than F.*

This assures that there can be no infinite decomposition sequences.

NOTATION 72. We shall often use suggestive meta-symbols for naming formulae of a certain type. Thus, a formula of the α- or β-type will be denoted with 'α' or 'β', and the formulae in its subformula sequence with 'α_1',...,'α_n' or 'β_1',...,'β_n', respectively; for a formula of the γ- or δ-type and its subformula w.r.t. a term t, we will write 'γ' or 'δ' and '$\gamma(t)$' or '$\delta(t)$', respectively.

As in the propositional case, first-order tableaux are particular *formula trees*, i.e, ordered trees with the nodes labelled with formulae. We do not formally introduce trees, and we permit trees to be infinite. Trees will be viewed as *downwardly* growing from the root. Furthermore, the following abbreviations will be used.

DEFINITION 73. If a formula is the label of a node on a branch B of a formula tree T, we say that F *appears* or *is on* B and *in* T. With $B \oplus F_1 \mid \cdots \mid F_n$ we mean the result of attaching $n > 0$ new successor nodes $N_1, \ldots, N_n$, in this order, fanning out of the end of B and labelled with the formulae $F_1, \ldots, F_n$, respectively. Any such sequence $N_1, \ldots, N_n$ is termed a *family* in T. We shall often treat the branch B of a formula tree as the set of formulae appearing on B. All nodes above a node N on a branch are called its *ancestors*, the ancestor immediately above N is termed its *predecessor*. If two nodes in a tableau are labelled with complementary formulae, we shall also call the nodes *complementary*.

Based on the developed formula decomposition schema, first-order tableaux for *sentences* are defined inductively.

DEFINITION 74 (Sentence tableau (inductive)). Let S be any set of sentences of a first-order language $\mathcal{L}$.

- Every one-node formula tree labelled with a formula from S is a *sentence tableau for* S.
- If B is a branch of a sentence tableau T for S, then the formula trees obtained from the following five expansion rules are all *sentence tableaux for* S:

  (α) $B \oplus \alpha_i$, if α_i is an α-subformula of a formula α on B,[5]

  (β) $B \oplus \beta_1 \mid \cdots \mid \beta_n$, if $\beta_1, \ldots, \beta_n$ is the β-subformula sequence of a formula β on B,

  (γ) $B \oplus \gamma(t)$, if t is a ground term and a formula γ occurs on B,

  (δ) $B \oplus \delta(c)$, if c is a constant new to S and to the formulae in T, and a formula δ appears on B,

  (F) $B \oplus F$, for any formula $F \in S$.

If T is a sentence tableau for a singleton set $\{\Phi\}$, we also say that T is a sentence tableau for the *formula* Φ. When a decomposition rule is performed on a formula F at a node N, we say that F or N *is used*.

$$\frac{\alpha}{\alpha_i} \qquad \frac{\beta}{\beta_1 \mid \cdots \mid \beta_n} \qquad \frac{\gamma}{\gamma(t)} \qquad \frac{\delta}{\delta(c)}$$

(for any ground term t) (for any new constant c)

Figure 2. Inference rules of sentence tableaux for formulae

Obviously, if the input set contains just one formula, the *formula rule* denoted with (F) can be omitted. The inference rules of sentence tableaux for formulae are summarized in Figure 2.

[5]Note that this rule is slightly more flexible than the standard α-rule as presented in [Smullyan, 1968] according to which B has to be modified to $B \oplus \alpha_1 \oplus \cdots \oplus \alpha_n$ in a single inference step. The need for this will become clear at the end of the section when we introduce the regularity refinement.

DEFINITION 75 (Closed tableau). A branch B of a tableau is called *(atomically) closed* if an (atomic) formula F and its negation appear on B, otherwise the branch is termed *(atomically) open*. Similarly, a node N is called *(atomically) closed* if all branches through N are (atomically) closed, and *(atomically) open* otherwise. Finally, a tableau is termed *(atomically) closed* if its root node is (atomically) closed, otherwise the tableau is called *(atomically) open*.

$$(1)\ \neg\exists x(\forall y\forall z P(y,f(x,y,z)) \rightarrow (\forall y P(y,f(x,y,x)) \wedge \forall y\exists z P(g(y),z)))$$

$\gamma(1)$

$$(2)\ \neg(\forall y\forall z P(y,f(a,y,z)) \rightarrow (\forall y P(y,f(a,y,a)) \wedge \forall y\exists z P(g(y),z)))$$

$\alpha(2)$

$$(3)\ \forall y\forall z P(y,f(a,y,z))$$

$\alpha(2)$

$$(4)\ \neg(\forall y P(y,f(a,y,a)) \wedge \forall y\exists z P(g(y),z))$$

$\beta(4)$

$(5)\ \neg\forall y P(y,f(a,y,a))$	$(6)\ \neg\forall y\exists z P(g(y),z)$
$\delta(5)$	$\delta(6)$
$(7)\ \neg P(b,f(a,b,a))$	$(10)\ \neg\exists z P(g(b),z)$
$\gamma(3)$	$\gamma(10)$
$(8)\ \forall z P(b,f(a,b,z))$	$(11)\ \neg P(g(b),f(a,g(b),a))$
$\gamma(8)$	$\gamma(3)$
$(9)\ P(b,f(a,b,a))$	$(12)\ \forall z P(g(b),f(a,g(b),z))$
	$\gamma(12)$
	$(13)\ P(g(b),f(a,g(b),a))$

Figure 3. An atomically closed sentence tableau

In Figure 3, a larger sentence tableau for a first-order sentence is displayed that illustrates the application of each tableau rule. For every tableau expansion step, the respective type of tableau expansion rule and the used ancestor node are annotated at the connecting vertices. Note that all branches of the tableau are atomically closed. A closed sentence tableau for a set of sentences S represents a correct proof of the unsatisfiability of S. The correctness of the tableau approach as a proof method for first-order sentences is based on the fact that the decomposition rules are satisfiability preserving.

PROPOSITION 76. *Let S be any satisfiable set of first-order sentences.*

(1) If $\alpha \in S$, then $S \cup \{\alpha_i\}$ is satisfiable, for every α-subformula α_i of α.

(2) If $\beta \in S$, then $S \cup \{\beta_i\}$ is satisfiable, for some β-subformula β_i of β.

(3) If $\gamma \in S$, then $S \cup \{\gamma(t)\}$ is satisfiable, for any ground term t.

(4) If $\delta \in S$, then $S \cup \{\delta(c)\}$ is satisfiable, for any constant c that is new to S.

Proof. Items (1) and (2) are immediate from the definition of formula assignment; (3) is a consequence of the soundness of substitution application (Proposition 46); lastly, since δ is assumed as closed and c is new to S, $\delta(c)$ is a Skolemization of δ w.r.t. S, hence (4) follows from Proposition 55. ■

PROPOSITION 77 (Soundness of sentence tableaux). *If a set of sentences S is satisfiable, then every sentence tableau for S has an open branch.*

Proof. We use the following notation. A branch of a tableau for a set of formulae S is called *satisfiable* if $S \cup B$ is satisfiable where B is the set of formulae on the branch. Clearly, every satisfiable branch must be open. We prove, by induction on the number of tableau expansion steps, that every sentence tableau for a satisfiable set of sentences S has a satisfiable branch. The induction base is evident. For the induction step, consider any tableau T for S generated with $n+1$ expansion steps. Let T' be a tableau for S from which T can be obtained by a single expansion step. By the induction assumption, T' has a satisfiable branch B. Now, either T contains B, in which case T' has a satisfiable branch. Or B is expanded; in this case, Proposition 76 guarantees that one of the new branches in T' is satisfiable. ■

A fundamental proof-theoretic advantage of the tableau method over *synthetic* proof systems like axiomatic calculi [Hilbert and Ackermann, 1928] is the *analyticity* of the decomposition rules. The formulae in a tableaux are in the reflexive-transitive closure of the tableau subformula relation on the input set. For certain formula classes, this permits the generation of decision procedures based on tableaux, which will be discussed below.

3.2 Completeness of Sentence Tableaux

First-order logic differs from propositional logic in that there are no decision procedures for the logical status of a set of formulae, but merely *semi-decision* procedures. More precisely, there exist effective mechanical methods for verifying the unsatisfiability of sets of first-order formulae (or the logical validity of first-order formulae[6]), whereas, when subscribing to *Church's Thesis*, the satisfiability of sets of first-order formulae (or the non-validity of first-order formulae) cannot be effectively recognized.[7]

[6] This result was first demonstrated by Gödel in [1930].

[7] Thus settling the *undecidability* of first-order logic, which was proved by Church in [1936] and Turing in [1936].

The tableau calculus represents such an effective mechanical proof method. In this part, we will provide a completeness proof of sentence tableaux. An essential concept used in this proof is that of a *downward saturated* set of sentences.

DEFINITION 78 (Downward saturated set). Let S be a set of first-order sentences and $\mathcal{U}$ the Herbrand universe of S. The set S is called *downward saturated* provided:

1. if S contains an α, then it contains all its α-subformulae,
2. if S contains a β, then it contains at least one of its β-subformulae,
3. if S contains a γ, then it contains all $\gamma(t)$ with $t \in \mathcal{U}$,
4. if S contains a δ, then it contains a $\delta(c)$ with c being a constant in $\mathcal{U}$.

DEFINITION 79 (Hintikka set). By an *(atomic) Hintikka set* we mean a downward saturated set which does not contain an (atomic) formula and its negation.

LEMMA 80 (Hintikka's Lemma). *Every atomic Hintikka set (and hence every Hintikka set) is satisfiable.*

Proof. Let S be an atomic Hintikka set. We show that some Herbrand interpretation of S is a model for S. Let H denote the set of ground literals in S. First, since H does not contain an atomic formula and its negation, it defines a Herbrand interpretation $\mathcal{H}$, using Notation 65. We show, by induction on the formula complexity, that the formula assignment of $\mathcal{H}$ maps all formulae in S to $\top$. The induction base is evident. For the induction step, assume that $\mathcal{H}(F) = \top$, for all formulae F in S with formula complexity $< n$. Consider any non-literal formula $F \in S$ with formula complexity n. The formula complexity of every tableau subformula of F is $< n$.

1. If F is an α, then, by the definition of downward saturation, every α_i is in S. Since, by the induction assumption, $\mathcal{H}(\alpha_i) = \top$, $\mathcal{H}(F) = \top$.
2. If F is a β, then, again by the definition of downward saturation, some β_i is in S. By the induction assumption, $\mathcal{H}(\beta_i) = \top$, therefore, $\mathcal{H}(F) = \top$.
3. If F is a $\gamma = \forall x F'$, by the downward saturatedness of S and the induction assumption, $\mathcal{H}(\gamma(t)) = \top$, for any term t in the Herbrand universe $\mathcal{U}$ of S. Since $\mathcal{U}$ is the universe of $\mathcal{H}$ and since $\mathcal{H}$ (being a Herbrand interpretation) maps every term t to itself, for all variable assignments $\mathcal{A}$ to $\mathcal{U}$, $\mathcal{H}^{\mathcal{A}}(F') = \mathcal{H}(F'\{x/\mathcal{A}(x)\}) = \top$. Therefore, $\mathcal{H}(F) = \top$.
4. Finally, if F is a δ, by downward saturation and the induction assumption $\mathcal{H}(\delta(c)) = \top$, for some constant $c \in \mathcal{T}$, therefore $\mathcal{H}(F) = \top$. ■

After these preliminaries, we can come back to tableaux. The tableau calculus is indeterministic, i.e., many possible expansion steps are possible in a certain situation. We are now going to demonstrate that the tableau construction can be made completely deterministic and yet it can be guaranteed that the tableau will eventually close if the set of input formulae is unsatisfiable. Such tableaux are called *systematic tableaux*. In order to make the expansion deterministic, we have to determine,

1. from where the next formula has to be taken, and

2. for the case of the quantifier rules, to which closed term the respective variable has to be instantiated.

Furthermore, since systematic tableaux shall be introduced for the most general case in which the set of input formulae may be infinite, we have to provide means for making sure that any formula in the set will be taken into account in the tableau construction, if necessary.

For the node selection, we equip the nodes of tableaux with an additional number label, expressing whether the formula at the node can be used for a tableau expansion step or not. If a node carries a number label, then the formula at the node will be a possible candidate for a tableau expansion step, otherwise not.

DEFINITION 81 (Usable node). If a tableau T contains nodes with number labels, then from all the nodes labelled with the smallest number the leftmost one with minimal tableau depth is called *the usable node* of T; otherwise T has *no usable node*.

For the term selection needed in the quantifier rules, we employ a total ordering on the set of closed terms. Both selection functions together can be used to uniquely determine the next tableau expansion steps. Concerning the fairness problem in case the set of input formulae be infinite, we use an additional total ordering on the formulae.

DEFINITION 82 (Systematic tableau (sequence) (inductive)). Let π be a mapping from $\mathbb{N}_0$ onto the set of ground terms and $\prec$ a total ordering on the set of formulae of a first-order language $\mathcal{L}$, respectively, and S any set of closed first-order formulae. The *systematic tableau sequence* of S *w.r.t.* π *and* $\prec$ is the following sequence $\mathcal{T}$ of tableaux for S. Let Φ be the smallest formula in S modulo $\prec$.

- The one-node tableau T_0 with root formula Φ and number label 0 is the first element of $\mathcal{T}$.

- If T_n is the n-th element in $\mathcal{T}$ and has nodes with number labels, let N be the usable node of T_n with formula F and number k. Furthermore, if some formula in S is not on some branch passing through N, let G denote the smallest such formula modulo $\prec$. Now expand each open branch B passing through N to:

1. $B\,[\,\oplus G\,]\oplus\alpha_1\oplus\cdots\oplus\alpha_n$, if F is of type α with α-subformula sequence $\alpha_1,\ldots,\alpha_n$,
2. $B\,[\,\oplus G\,]\oplus\beta_1 \mid \cdots \mid \beta_n$, if F is of type β with β-subformula sequence $\beta_1,\ldots,\beta_n$,
3. $B\,[\,\oplus G\,]\oplus\gamma(\pi(k))$, if F is of type γ,
4. $B\,[\,\oplus G\,]\oplus\delta(c)$ if F is of type δ and c is the smallest constant modulo π not occurring in T.

Then give every newly attached node the number label 0 if its formula label is not a literal. Next, remove the number labels from all nodes that have become atomically closed through the expansion steps. Finally, if F is not of type γ, remove the number label from N; otherwise change the number k at N to $k+1$. The tableau resulting from the entire operation is the $n+1$-st element of the sequence $\mathcal{T}$.

- If T_n has no usable node, it is the last element of $\mathcal{T}$.

In Figure 4, a closed systematic tableau is shown with $\pi(0) = a$ and $\pi(1) = b$. The following structural property of sentence tableaux plays an important role. We formulate it generically, for any system of tableau inference rules.

DEFINITION 83 (Nondestructiveness). A tableau calculus C is called *nondestructive* if, whenever a tableau T can be deduced from a tableau T' according to the inference rules of C, then T' is an initial segment of T; otherwise C is called destructive.

Since obviously the calculus of sentence tableaux is nondestructive, one can form the (tree) union of all the tableaux in a systematic tableau sequence.

DEFINITION 84 (Saturated systematic tableau). Let $\mathcal{T}$ be a systematic tableau sequence for a set of first-order formulae S. With T^* we denote the smallest formula tree containing all tableaux in $\mathcal{T}$ as initial segments; T^* is called a *saturated systematic tableau* of S.

PROPOSITION 85. *For any (atomically) open branch B of a saturated systematic tableau, the set of formulae on B is a(n atomic) Hintikka set.*

Proof. Let B be any (atomically) open branch of a saturated systematic tableau. According to the definition of systematic tableau, it is guaranteed that the branch B satisfies the following condition: for any formula F on B,

1. if F is of type α, then all α-subformulae of F must be on B;
2. if F is of type β, then some β-subformula of F must be on B;
3. if F is of type γ, then all γ-subformulae of F must be on B;
4. if F is of type δ, then some δ-subformulae of F must be on B.

$$
\begin{array}{c}
(1)\ \exists y \exists z \forall x (P(x,y) \wedge (P(z,x) \rightarrow \neg P(y,y))) \\
\big|\ \delta(1) \\
(2)\ \exists z \forall x (P(x,a) \wedge (P(z,x) \rightarrow \neg P(a,a))) \\
\big|\ \delta(2) \\
(3)\ \forall x (P(x,a) \wedge (P(b,x) \rightarrow \neg P(a,a))) \\
\big|\ \gamma(3) \\
(4)\ (P(a,a) \wedge (P(b,a) \rightarrow \neg P(a,a))) \\
\big|\ \alpha(4) \\
(5)\ P(a,a) \\
\big|\ \alpha(4) \\
(6)\ (P(b,a) \rightarrow \neg P(a,a)) \\
\beta(6)
\end{array}
$$

$$
\begin{array}{cc}
(7)\ \neg P(b,a) & (8)\ \neg P(a,a) \\
\big|\ \gamma(3) & \\
(9)\ (P(b,a) \wedge (P(b,b) \rightarrow \neg P(a,a))) & \\
\big|\ \alpha(9) & \\
(10)\ P(b,a) & \\
\big|\ \alpha(9) & \\
(11)\ (P(b,b) \rightarrow \neg P(a,a)) &
\end{array}
$$

Figure 4. An atomically closed systematic tableau

So the set S of formulae on B is downward saturated. Since B is (atomically) open, no (atomic) formula and its negation are in S, hence S is a(n atomic) Hintikka set. ■

Now the refutational completeness of tableaux is straightforward.

THEOREM 86 (Sentence tableau completeness). *If S is an unsatisfiable set of first-order sentences, then there exists a finite atomically closed sentence tableau for S.*

Proof. Let T be a saturated systematic tableau for S. First, we show that T must be atomically closed. Assume, indirectly, that T contained an atomically open branch B. Then, by Proposition 85, there would exist an atomic Hintikka set for the set S' of formulae on B and, by Hintikka's Lemma, a model $\mathcal{I}$ for S'. Now, by the definition of saturated systematic tableau, $S \subseteq S'$, hence $\mathcal{I}$ would be a model for S, contradicting the unsatisfiability assumption. This proves that every branch of T must be atomically closed. In order to recognize the finiteness of T, note that the closedness of any branch of a systematic tableau entails that it cannot have a branch of infinite length. Since the branching rate of any tableau is finite, (by König's Lemma) T must be finite. ■

The generality of our systematic tableau procedure permits an easy proof of a further fundamental property of first-order logic.

THEOREM 87 (Compactness Theorem). *Any unsatisfiable set of first-order sentences has a finite unsatisfiable subset.*

Proof. Let S be any unsatisfiable set of first-order sentences. By Theorem 86, there exists a finite closed tableau T for S. Let S' be the set of formulae in S appearing in T. S' is finite and, by the soundness of tableaux, S' must be unsatisfiable. ■

Sentence tableaux can also be used to illustrate the basic Herbrand-type property of first-order logic that with any unsatisfiable set of prenex formulae one can associate unsatisfiable sets of ground formulae as follows.

DEFINITION 88. A tableau is called *quantifier preferring* if on any branch all applications of quantifier rules precede applications of the connective rules. Such a tableau begins with a single branch containing only quantifier rule applications up to a node N below which only connective rules are applied; the set of formulae on this branch up to the node N is called *the initial set* of the tableau, and the set of ground formulae in the initial set is termed *the initial ground set* of T.

It is evident that we can reorganize any tableau for a set of prenex formulae in such a way that it is quantifier preferring, without increasing its size or affecting its closedness. None of those properties is guaranteed to hold for sets containing formulae which are not in prenex form.

PROPOSITION 89. *If T is a closed quantifier preferring tableau for a set S of first-order formulae, then the initial ground set of T is unsatisfiable.*

Since, for any unsatisfiable set S of prenex formulae, a closed quantifier preferring tableau exists, we can associate with S the collection of the initial ground sets of all closed quantifier preferring sentence tableaux for S. The sets in this collection, in particular the ones with minimal complexity, play an important rôle as a complexity measure.

DEFINITION 90. The *Herbrand complexity* of an unsatisfiable set S of prenex formulae is the minimum of the complexities[8] of the initial ground sets of all closed quantifier preferring sentence tableaux for S.

Since the quantifier rules of tableaux are not specific to the tableau calculus, the Herbrand complexity can be used as a calculus-independent refutation complexity measure for unsatisfiable sets of formulae. This measure may also be extended to

[8]As the complexity of a set of formulae one may take the sum of the occurrences of symbols in the elements of the set.

formulae which are not in prenex form, by working with transformations of the formulae in prenex form (see also [Baaz and Leitsc, 1992]).

Next, we come to an important proof-theoretic virtue of sentence tableaux, which we introduce generically for any system of tableau rules.

DEFINITION 91 (Confluence). A tableau calculus C is called *proof confluent* or just *confluent* (for a class of formulae) if, for any unsatisfiable set S of formulae (from the class), from any tableau T for S constructed with the rules of C a closed tableau for S can be constructed with the rules of C.

Loosely speaking, a confluent (tableau) calculus does never run into dead ends.[9]

PROPOSITION 92. *Sentence tableaux are confluent for first-order sentences.*

Proof. Let T be any sentence tableau generated for an unsatisfiable set of sentences S. By the completeness of sentence tableaux, for any branch B in T, there exists a closed sentence tableau T_B for $S \cup B$. At the leaf of any branch B of T, simply repeat the construction of T_B. ■

In the subsequent sections, we shall introduce tableau calculi and procedures that are not confluent and for which no systematic procedures of the type presented in this section exist. Nonconfluence may have strong consequences on the termination behaviour and the functionality of a calculus, particularly, when one is interested in decision procedures or in model generation for (sublanguages of) first-order logic. As we shall see, the lack of confluence may also require completely different approaches towards proving completeness.

3.3 *Refinements of Tableaux*

The calculus of sentence tableaux permits the performance of certain inference steps that are redundant in the sense that they do not contribute to the closing of the tableau. In order to avoid such redundancies, one can restrict the tableau rules and/or impose conditions on the tableau structure. First, we discuss the notion of strictness which is a refinement of the tableau rules. Adapted to our framework, it reads as follows.

DEFINITION 93 (Strict tableau). A tableau construction is *strict* if

- every β- and δ-node is used only once on a branch and
- for any α-node, any occurrence of an α_i in α is used only once a branch.

The strictness condition is motivated by the definition of systematic tableaux, since obviously any systematic tableau construction is strict. Consequently, the

[9]Note that the notion of confluence used here slightly differs from its definition in the area of term rewriting (see, e.g., [Huet, 1980].

strictness condition is completeness-preserving. Strictness can also be implemented very efficiently by simply labelling nodes (or occurrences of tableau subformulae at nodes) as already used on a branch. But strictness does not perform optimal redundancy elimination, since it does not prevent that one and the same formula may appear twice on a branch. This is particularly detrimental if it happens in a β-rule application where new branches with new proof obligations are produced, which obviously is completely useless. A stronger tableau restriction concerning the connective rules is achieved with the following *structural* condition.

DEFINITION 94 (Regular tableau). A formula tree is called *regular* if, on no branch, a formula appears more than once.

The main reason why regularity has not been used in the traditional presentation of tableaux lies in the different definition of the α-rule here and there. We permit that only *one* α-subformula can be attached, whereas the traditional format requires to append *all* α-subformulae at once, one below the other. It is straightforward to realize that regularity is not compatible with the traditional definition of the α-rule. An obvious example is the unsatisfiable formula $(p \wedge q) \wedge (p \wedge \neg q)$, for which no closed regular tableau exists if the traditional α-rule is used. Since (w.r.t. the connective rules) the regularity condition is a more powerful mechanism of avoiding suboptimal proofs than the strictness condition, we have generalized the α-rule in order to achieve compatibility with regularity.[10] The following fact demonstrates that tableaux which are irregular can be safely ignored.

DEFINITION 95. The *size of a formula tree* is the sum of the symbol occurrences in the formulae at the nodes of the tree.

PROPOSITION 96. *Every closed sentence tableau of minimal size is regular.*

Proof. We show the contraposition, i.e., that every closed irregular sentence tableau T is not minimal in size. Let T be any closed irregular sentence tableau for a set S and N any node dominated by a node labelled with the same formula. Prune the branch by taking out the node N and its brothers (if existing) and attach the successors of N (if existing) to the predecessor of N. The branch remains closed, the resulting formula tree T' is smaller in size and (according to Definition 74) T' is a sentence tableau for S. ■

In order to integrate the δ-rule restriction of strictness, we call a tableau *strictly regular* if it is strict and regular. The regularity restriction can easily be integrated into systematic tableaux, by simply omitting the attachment of nodes $B \oplus F_1 \mid \cdots \mid F_n$ if one of the F_i already is on the branch B.

A further fundamental refinement of sentence tableau concerns the γ-rule.

[10] This is an interesting illustration of the fact that an unfortunate presentation of inference rules can block certain obvious pruning mechanisms.

DEFINITION 97 (Herbrand tableau). *Herbrand tableaux* are defined like sentence tableaux, but with the γ-rule replaced by the *Herbrand γ-rule*:

$$(\gamma_{\mathcal{H}}) \qquad \frac{\gamma}{\gamma(t)} \qquad \text{where } t \text{ is from the Herbrand universe of the branch.}$$

The Herbrand restriction on the γ-rule may significantly improve the termination behaviour of sentence tableaux, as illustrated with the formula F given in Example 98. The formula is satisfiable. But unfortunately, infinitely large sentence tableau can be constructed for F, as shown in Figure 98, since the γ-rule can be applied again and again using the formula (4). Any strict Herbrand tableau construction terminates, since the number of ground terms that can be selected for (4) is finite.

EXAMPLE 98. $F = \neg\forall x(\exists y P(x,y) \rightarrow \exists y P(y,x))$

Sentence tableau	Herbrand tableau
(1) $\neg\forall x(\exists y P(x,y) \rightarrow \exists y P(y,x))$	(1) $\neg\forall x(\exists y P(x,y) \rightarrow \exists y P(y,x))$
$\mid \delta(1)$	$\mid \delta(1)$
(2) $\neg(\exists y P(a,y) \rightarrow \exists y P(y,a))$	(2) $\neg(\exists y P(a,y) \rightarrow \exists y P(y,a))$
$\mid \alpha(2)$	$\mid \alpha(2)$
(3) $\exists y P(a,y)$	(3) $\exists y P(a,y)$
$\mid \alpha(2)$	$\mid \alpha(2)$
(4) $\neg\exists y P(y,a)$	(4) $\neg\exists y P(y,a)$
$\mid \delta(3)$	$\mid \delta(3)$
(5) $P(a,b)$	(5) $P(a,b)$
$\mid \gamma(4)$	$\mid \gamma_{\mathcal{H}}(4)$
(6) $\neg P(a,a)$	(6) $\neg P(a,a)$
$\mid \gamma(4)$	$\mid \gamma_{\mathcal{H}}(4)$
(7) $\neg P(b,a)$	(7) $\neg P(b,a)$
$\mid \gamma(4)$	all Herbrand terms selected for (4)
(8) $\neg P(c,a)$	
$\vdots$	

Figure 5. Sentence and Herband tableau for Example 98

The Herbrand restriction on tableaux is as reasonable as regularity, since it preserves minimal proof size.

PROPOSITION 99. *For every (atomically) closed sentence tableau T for a set S, there exists a(n atomically) closed Herbrand tableau T' for S with less or equal size than T.*

Proof. Without increasing the size, we can rearrange T in such a way that all formula rule applications are performed first. Now consider any γ-step in the tableau that is not Herbrand, attaching a formula $\gamma(t)$ to the leaf N of a branch B. Replace any occurrence of t below N with a constant from the Herbrand universe of B. Obviously, the modified formula tree is (atomically) closed and does not increase in size. It is straightforward to prove, e.g., by induction on the tableau inference steps, that the formula tree is a sentence tableau for S. Finitely many applications of this operation produce a(n atomically) closed Herbrand tableau T' for S equal or smaller in size than T. ■

PROPOSITION 100. *Strictly regular Herbrand tableaux are confluent for first-order sentences.*

Proof. Let T be any strictly regular Herbrand tableau for an unsatisfiable set of sentences S. By the completeness of sentence tableaux, there exists a closed sentence tableau T' for S. Simply repeat the construction of T', at any leaf of T. Now modify the resulting sentence tableau, as described in the proofs of Propositions 96 and 99. The procedure results in a closed strictly regular Herbrand tableaux T'' for S. Since the modification operation is performed from the leaves towards the root, it does not affect the inital tree T, hence T'' is as desired. ■

The Herbrand tableau rule also has an effect on the *systematic* tableau construction. Since the Herbrand universe may increase during branch expansion, the enumeration of ground terms must be organized differently.

DEFINITION 101 (Systematic Herbrand tableau). *Systematic Herbrand tableaux* are defined like systematic tableaux except that the γ-rule application is controlled differently. Whenever a γ at a node N is selected, for any atomically open branch B through N, select the smallest term t (modulo the ordering π) from the Herbrand universe of B that has not been selected at N on B; if all terms from the Herbrand universe of B have already been selected at N on B, γ cannot be used for expanding the current leaf of B.

In particular, this entails that, for different branches, different Herbrand terms may be selected in the systematic tableau construction. Imposing the Herbrand restriction on systematic tableaux preserves completeness, since Proposition 85 also holds for Herbrand tableaux.

PROPOSITION 102. *Every (atomically) open branch B of a saturated regular systematic Herbrand tableau is a(n atomic) Hintikka set, moreover, the set of atoms on B defines a Herbrand model for B.*

Proof. See the proof of Proposition 85. ■

Herbrand tableaux provide a higher functionality than sentence tableaux, since a larger class of first-order formulae can be decided.

DEFINITION 103 (Weak Skolem, datalogic form). A sentence Φ is said to be in *weak Skolem form* if Φ has no tableau subformula of type γ that has a tableau subformula of type δ. A sentence Φ is said to be in *(weak) datalogic form* if Φ is in (weak) Skolem form, respectively, and Φ has no function symbol of arity > 0.

The set of weak datalogic formulae is a generalization of the Bernays–Schönfinkel class [1928].

PROPOSITION 104. *Every strictly regular Herbrand tableau for any finite set S of weak datalogic formulae is finite.*

Proof. The formula structure and the tableau rules guarantee that only δ-formulae can appear on a branch which occur as subformulae in the elements of S. Since S is assumed as finite, this entails that the number of δ-formulae on any branch must be finite. Because of the strictness condition on the δ-rule, only finitely many new constants can occur on a branch. Since no functions symbols of arity > 0 occur in the elements of S, the Herbrand universe of any branch must be finite, and hence the set of formulae occurring on a branch. Regularity then assures that also the length of any branch must be finite. ■

Both properties demonstrate that Herband tableaux are decision procedures for the class.

PROPOSITION 105. *Given any finite set S of weak datalogic formulae, any regular systematic Herbrand tableau construction terminates,*

- *either with a closed tableau if S is unsatisfiable,*
- *or with an open branch which defines a Herbrand model for S.*

4 FREE-VARIABLE TABLEAUX

The tableau approach is traditionally useful as an elegant format for *presenting* proofs. With the increasing importance of automatic deduction, however, the question arises whether the tableau paradigm is also suited for proof' *search*. In principle, systematic tableau procedures could be used for this purpose. But systematic procedures, even regular Herbrand ones, are still too inefficient for a broad application. As an illustration, see the tableau displayed in Figure 3, which is not systematic. A systematic tableau would be much larger. The essential difference concerns the applications of the γ-rule. Consider, e.g., the γ-step from node (10) $\neg\exists z P(g(b), z)$ to node (11) $\neg P(g(b), f(a, g(b), a))$ in which the 'right' substitution $\{z/f(a, g(b), a))\}$ has been selected. Since a systematic procedure has to enumerate *all* (Herbrand) instances in a systematic and therefore 'blind' manner, it would normally perform the substitution $\{z/f(a, g(b), a))\}$ much later. The obvious weakness of the γ-rule is that it enforces to perform ground instantiations too early, at a time when it is not clear whether the substitution will contribute to the closing of a branch. The natural approach for overcoming this problem is to postpone the term selection completely by permitting free variables in a tableau and to determine the instances later when they can be used to immediately close a branch. The free variables are then treated in a *rigid* manner, i.e., they are not being considered universally quantified but as placeholders for arbitrary (ground) terms. This view of free variables dates back to work of Prawitz [1960], was applied by Bibel [1981] and Andrews [1981], and incorporated into tableaux, for example, by Fitting [1996] (see also [Reeves, 1987]). In this section, we will investigate this approach. Closure of a branch means producing two *complementary* formulae, i.e., a formula and its negation, on the branch. Since we can confine ourselves to atomic closures, the problem can be reduced to finding a substitution σ such that for two literals K and L on the branch: $K\sigma = {\sim}L\sigma$. So one has to integrate *unification* into the tableau calculus.

4.1 *Unification*

Unification is one of the most successful advances in automated deduction, because it permits to make instantiation optimal with respect to generality. Unification will be introduced here for arbitrary finite sets of quantifier-free expressions.

DEFINITION 106 (Unifier). For any finite set S of quantifier-free expressions and any substitution σ, if $|S\sigma| = 1$,[11] then σ is called a *unifier for* S. If a unifier exists for a set S, then S is called *unifiable*.

Subsequently, we will always assume that S denotes finite sets of quantifier-free expressions. The general notion of a unifier can be subclassified in certain useful ways.

DEFINITION 107 (Most general unifier). If σ and τ are substitutions and there is a substitution θ such that $\tau = \sigma\theta$, then we say that σ is *more general* than τ. A unifier for a set S is called a *most general unifier*, MGU for short, if σ is more general than any unifier for S.

Most general unifiers have the nice property that any unifier for two atoms can be generated from a most general unifier by further composition. This qualifies MGUs as a useful instantiation vehicle in many inference systems. The central unifier concept in automated deduction, however, is the following.

DEFINITION 108 (Minimal unifier). If a unifier σ for a set S has the property that for every unifier τ for S: $|\sigma| \leq |\tau|$, then we say that σ is a *minimal unifier for* S.

For a minimal unifier the number of substituted variables is minimal.

EXAMPLE 109. Given the set of terms $S = \{x, f(y)\}$, the two substitutions $\sigma = \{y/x, x/f(x)\}$ and $\tau = \{x/f(y)\}$ are both MGUs for S, but only τ is a minimal unifier.

[11] With $|M|$ we denote the cardinality of a set M.

In fact, every minimal unifier is a most general unifier, as will be shown in the Unification Theorem (Theorem 118) below. How can we a find a minimal unifier for a given set? For this purpose, the procedurally oriented concept of a *computed unifier* will be developed.

DEFINITION 110 (Disagreement set). Let S be a finite set of quantifier-free expressions. A *disagreement set* of S is any two-element set $\{E_1, E_2\}$ of expressions such that the dominating symbols of E_1 and E_2 are distinct and E_1 and E_2 occur at the same position as subexpressions in two of the expressions in S.

EXAMPLE 111. The set of terms $S = \{x, g(a, y, u), g(z, b, v)\}$ has the following disagreement sets: $\{a, z\}, \{y, b\}, \{u, v\}, \{x, g(a, y, u)\}, \{x, g(z, b, v)\}$.

Obviously, a set of expressions S has a disagreement set if and only if $|S| > 1$. The following facts immediately follow from the above definitions.

PROPOSITION 112. *If σ is a unifier for a set S and D is a disagreement set of S, then σ unifies D, each member of D is a term, and D contains a variable which does not occur in the other term of D.*

The last item of the proposition expresses that any binding formed from any disagreement set of a unifiable set must be a proper binding. Operationally, the examination whether a binding is proper is called the *occurs-check.* A particularly useful technical tool for proving the Unification Theorem below is the Decomposition Lemma.

LEMMA 113 (Decomposition lemma). *Let σ be a unifier for a set S with $|S| > 1$ and let $\{x, t\}$ be any disagreement set of S with $x \neq x\sigma$. If $\tau = \sigma \setminus \{x/x\sigma\}$, then $\sigma = \{x/t\}\tau$.*

Proof. First, since σ unifies any disagreement set of S, $x\sigma = t\sigma$. By Proposition 112, x does not occur in t, which gives us $t\sigma = t\tau$. Consequently, $x\sigma = t\tau$ and $x \neq t\tau$. Furthermore, $x \notin \text{domain}(\tau)$, and by the composition of substitutions, $\{x/t\}\tau = \{x/t\tau\} \cup \tau$. Putting all this together yields the chain $\{x/t\}\tau = \{x/t\tau\} \cup \tau = \{x/x\sigma\} \cup \tau = \sigma$. ■

Now we shall introduce a concept which captures the elementary operation performed when making a set of expressions equal by instantiation. It works by eliminating exactly one variable x from all expressions of the set and by replacing this variable with another term t from a disagreement set $\{x, t\}$ of S provided that x does not occur in t.

DEFINITION 114 (Variable elimination and introduction). If S is a finite set of expressions such that from the elements of one of its disagreement sets a proper binding x/t can be formed, then $S\{x/t\}$ is said to be *obtainable from S* by a *variable elimination w.r.t. x/t.*

PROPOSITION 115. *Let S be any finite set of quantifier-free expressions and let V_S be the set of variables occurring in S.*

1. *If S is unifiable, so are all sets obtainable from S by a variable elimination.*
2. *Only finitely many sets can be obtained from S by a variable elimination.*
3. *If S' has been obtained from S by a variable elimination w.r.t. a binding $\{x/t\}$ and $V_{S'}$ is the set of variables occurring in S', then $|S'| \leq |S|$ and $V_{S'} = V_S \setminus \{x\}$.*
4. *The transitive closure of the relation*

   $$\{\langle S', S\rangle \mid S' \textit{ can be obtained from } S \textit{ by a variable elimination step}\}$$

   is well-founded where S and S' are arbitrary finite sets of quantifier-free expressions, i.e., there are no infinite sequences of successive variable elimination steps.

Proof. For the proof of (1), let $S' = S\{x/t\}$ be obtained from S by a variable elimination w.r.t. to the binding x/t composed from a disagreement set of S, and suppose σ unifies S. Since σ unifies every disagreement set of S, it follows that $x\sigma = t\sigma$. Let $\tau = \sigma \setminus \{x/x\sigma\}$. By the Decomposition Lemma (Lemma 113), we have $\{x/t\}\tau = \sigma$. Therefore, $S(\{x/t\}\tau) = (S\{x/t\})\tau = S'\tau$. Hence τ unifies S'. For (2) note that since there are only finitely many disagreement sets of S and each of them is finite, only finitely many proper bindings are induced, and hence only finitely many sets can be obtained by a variable elimination. To recognize (3), let $S' = S\{x/t\}$ be any set obtained from S by a variable elimination. Then S' is the result of replacing any occurrence of x in S by the term t. Therefore, $|S'| \leq |S|$, and, since x/t is proper and t already occurs in S, we get $V_{S'} = V_S \setminus \{x\}$. Lastly, (4) is an immediate consequence of (3). ■

Now we are able to introduce the important notion of a computed unifier, which is defined by induction on the cardinality of the unifier.

DEFINITION 116 (Computed unifier (inductive)).

1. $\emptyset$ is a (the only) *computed unifier for* any singleton set of quantifier-free expressions.
2. If a substitution σ of cardinality n is a computed unifier for a finite set S' and S' can be obtained from S by a variable elimination w.r.t. a binding x/t, then the substitution $\sigma \cup \{x/t\sigma\} = \{x/t\}\sigma$ of cardinality $n+1$ is a *computed unifier for S*.

The definition of a computed unifier can be seen as a declarative specification of an algorithm for *really computing* a unifier for a given set of expressions, which we will present now using a procedural notation. The procedure is a generalization of the algorithm given by Robinson in [1965].[12]

[12]Historically, the first unification procedure was given by Herbrand in [1930].

DEFINITION 117 (Unification algorithm). Let S be any finite set of quantifier-free expressions. $\sigma_0 = \emptyset$, $S_0 = S$, and $k = 0$. Then go to 1.

1. If $|S_k| = 1$, output σ_k as a computed unifier for S. Otherwise select a disagreement set D_k of S_k and go to 2.

2. If D_k contains a proper binding, choose one, say x/t; then set $\sigma_{k+1} = \sigma_k\{x/t\}$, set $S_{k+1} = S_k\{x/t\}$, increment k by 1 and go to 1. Otherwise output 'not unifiable'.

Note that the unification algorithm is a nondeterministic procedure. This is because there may be several different choices for a disagreement set and for a binding. Evidently, the unification procedure can be directly read off from the definition of a computed unifier: it just successively performs variable elimination operations, until either there are no variable elimination steps possible, or the resulting set is a singleton set. Conversely, the notion of a computed unifier is an adequate declarative specification of the unification algorithm. It follows immediately from Proposition 115 (1) and (4) that each unifier output of the unification algorithm is indeed a computed unifier and that the procedure terminates, respectively.

We shall demonstrate now that the notions of a minimal and a computed unifier coincide, and that both of them are most general unifiers.

THEOREM 118 (Unification Theorem). *Let S be any unifiable finite set of quantifier-free expressions.*

1. *If σ is a minimal unifier for S, then σ is a computed unifier for S.*

2. *If σ is a computed unifier for S, then σ is a minimal unifier for S.*

3. *If σ is a computed unifier for S, then σ is an* MGU *for S.*

Proof. We will prove (1) to (3) by induction on the cardinalities of the respective unifiers. First, note that $\emptyset$ is the only minimal and computed unifier for any singleton set of quantifier-free expressions S and that $\emptyset$ is an MGU for S. Assume the result to hold for any set of expressions with minimal and computed unifiers of cardinalities $\leq n$. For the induction step, suppose S has only minimal or computed unifiers of cardinality $> n \geq 0$. Let σ be an arbitrary unifier for S and x/t any proper binding from a disagreement set of S with $x \neq x\sigma$ (which exists by Proposition 112). Let $S' = S\{x/t\}$ and set $\tau = \sigma \setminus \{x/x\sigma\}$, which is a unifier for S', by the Decomposition Lemma (Lemma 113). For the proof of (1), let σ be a minimal unifier for S. We first show that τ is minimal for S'. If θ' is any minimal unifier for S', then $\theta = \{x/t\}\theta'$ is a unifier for S and all variables in domain(θ') occur in S'. Therefore, the Decomposition Lemma can be applied yielding that $\theta' = \theta \setminus \{x/x\theta\}$. And from the chain $|\theta'| = |\theta| - 1 \geq |\sigma| - 1 = |\tau|$ it follows that τ is a minimal unifier for S'. Since $|\tau| \leq n$, by the induction assumption, τ is a computed unifier for S'. Hence, by definition, $\sigma = \{x/t\}\tau$ is a computed unifier for S. For (2) and (3), let σ be a computed unifier for S. Then, by definition, τ is a computed unifier for S'. Let θ be an arbitrary unifier for S. Since x is in some disagreement set of S, either $x \in \text{domain}(\theta)$ or there is a variable y and $y/x \in \theta$. Define

$$\eta = \begin{cases} \theta & \text{if } x \in \text{domain}(\theta) \\ \theta\{x/y\} & \text{otherwise.} \end{cases}$$

Since $x \in \text{domain}(\eta)$, the Decomposition Lemma yields that if $\eta' = \eta \setminus \{x/x\eta\}$, then $\{x/t\}\eta' = \eta$, and η' is a unifier for S'. The minimality of σ can be recognized as follows. By the induction assumption, τ is minimal for S'. Then consider the chain

$$|\theta| = |\eta| = |\eta'| + 1 \geq |\tau| + 1 = |\sigma|.$$

For (3), note that τ is an MGU for S', by the induction assumption, i.e., there is a substitution γ: $\eta' = \tau\gamma$. On the other hand, $\theta = \theta\{x/y\}\{y/x\}$, hence there is a substitution ν: $\theta = \eta\nu$. This gives us the chain

$$S\theta = S\eta\nu = S\{x/t\}\eta'\nu = S\{x/t\}\tau\gamma\nu = S\sigma\gamma\nu$$

demonstrating that σ is an MGU for S. This completes the proof of the Unification Theorem. ■

Concerning terminology, notions are treated differently in the literature (see [Lassez *et al.*, 1988] for a comparison). We have chosen a highly indeterministic presentation of the unification algorithm, it is even permitted to select between alternative disagreement sets. Furthermore, we have stressed the importance of minimal unifiers. Therefore our Unification Theorem is stronger than normally presented, it also states that *each* minimal unifier indeed can be computed by the unification algorithm.

Polynomial Unification

Unification is the central ingredient applied in all advanced proof systems for first-order logic. As a consequence, the complexity of unification is a lower bound for the complexity of each advanced calculus. While the cardinality of a most general unifier σ for a set of expressions S is always bounded by the number of variables in S, the range of the unifier may contain terms with a size exponential with respect to the size of the initial expressions. Of course, this would also entail that $S\sigma$ contains expressions with an exponential size. The following class of examples demonstrates this fact.

EXAMPLE 119. If P is an $(n+1)$-ary predicate symbol and f a binary function symbol, then, for every $n > 1$, define S_n as the set containing the atomic formulae

$$P(x_1, x_2, \ldots, x_n, x_n), \text{ and}$$
$$P(f(x_0, x_0), f(x_1, x_1), \ldots, f(x_{n-1}, x_{n-1}), x_n).$$

Obviously, any unifier for an S_n must contain a binding x_n/t such that the number of symbol occurrences in t is greater than 2^n. As a consequence, we have the problem of exponential space and, therefore, also of exponential time, when working with such structures. Different solutions have been proposed for doing unification polynomially. In [Paterson and Wegman, 1978], a linear unification algorithm is presented. Furthermore, a number of 'almost' linear algorithms have been developed, for example, in [Huet, 1976] and [Martelli and Montanari, 1976; Martelli and Montanari, 1982]. Similar to the early approach in [Herbrand, 1930], all of the mentioned efficient algorithms reduce the unification problem to the problem of solving a set of equations. However, all of those procedures—particularly the one in [Paterson and Wegman, 1978]—need sophisticated and expensive additional data structures, which render them not optimal for small or average sized expressions. Therefore, Corbin and Bidoit rehabilitated Robinson's exponential algorithm by improving it with little additional data structure up to a quadratic worst-case complexity [Corbin and Bidoit, 1983; Letz, 1993]. This algorithm turns out to be very efficient in practice.

We cannot treat polynomial unification in detail here, but we give the essential two ideas contained in any of the mentioned polynomial unification algorithms.

1. The representation of expressions has to be generalized from strings or trees to (directed acyclic) graphs. This way, the space complexity can be reduced from exponential to linear, as shown with the directed acyclic graph representing the term $x_n\sigma$ in the example above:

$$\underbrace{f \rightrightarrows f \rightrightarrows \ldots f \rightrightarrows}_{n-\text{times}} x_0$$

2. In order to reduce the time complexity, which may still be exponential even if graphs are used, since, in the worst case, there are exponentially many paths through such a graph, the following will work. One must remember

    - which pairs of expressions had already been unified in the graph (e.g. during the unification of $x_n\sigma$ with itself at the last argument positions of the atoms),
    - and in occurs-checks: for which expressions the occurrence of the respective variable was already checked (e.g. during the check whether x_n occurs in $f(x_{n-1}, x_{n-1})\sigma$ at the n-th argument positions of the atoms),

    and one must not repeat those operations. How sophisticated this is organized determines whether the worst-case complexity can be reduced to linear or just to quadratic time.

4.2 *Generalized Quantifier Rules*

Using the unification concept, the γ-rule of sentence tableaux can be modified in such a way that instantiations of γ-formulae are delayed until a branch can be immediately closed. Two further modification have to be performed. On the one hand, since now free variables occur in the tableau, one has to generalize the δ-rule to full Skolemization in order to preserve soundness.

EXAMPLE 120. Consider the satisfiable formula $\exists y(\neg P(x,y) \wedge P(x,x))$. An application of the δ-rule of sentence tableaux would result in an unsatisfiable formula $\neg P(x,a) \wedge P(x,x)$.

One the other hand, substitutions have to be applied to the formulae in a tableau. With $T\sigma$ we denote the result of applying a substitution σ to the formulae in a tableau T. Before defining tableau with unification, we introduce a tableau system in which arbitrary substitutions can be applied. This system will serve as a very general reference system, which also subsumes sentence tableaux,

DEFINITION 121 (General free-variable tableau). *General free-variable tableaux* are defined as sentence tableaux are, but with the γ- and the δ-rule replaced by the following three rules. Let B be (the set of formulae on) the actual tableau branch and S the set of input sentences of the current tableau T.

($\gamma*$) $\quad \dfrac{\gamma}{\gamma(t)}$ $\quad$ where t is any term of the language $\mathcal{L}$ and $\{x/t\}$ is free for $\gamma(x)$

(δ^+) $\quad \dfrac{\delta}{\delta(f(x_1,\ldots,x_n))}$ $\quad$ where f is new to S and T, and $x_1,\ldots,x_n$ are the free variables in δ,

(S) $\quad$ Modify T to $T\sigma$ $\quad$ where σ is free for all formulae in T.

The δ^+-rule [Hähnle, 1994; Fitting, 1996] we use here is already an improvement of the original δ-rule used in [Fitting, 1996]. The difference between both rules will be discussed in Section 6. The additional *substitution rule* denoted with (S), which is now needed to achieve closure of certain branches, differs strongly from the tableau rules presented up to now. While all those rules were conservative in the sense that the initial tableau was not modified but just expanded, the substitution rule is *destructive*. This has severe consequences on free-variable tableaux, both proof-theoretically and concerning the functionality of the calculus, which will be discussed below.

But how do we know that the calculus of general free-variable tableau produces correct proofs? It is clear that the method of proving the correctness of sentence tableaux (using Proposition 76) will not work. In free-variable tableaux, branches cannot be treated separately, because they may share free variables. As an example, consider a tableau T with the two branches $P(x) \oplus \neg P(a)$ and $Q(x) \oplus \neg Q(b)$

which cannot be closed using the rules of general free-variable tableaux, although both branches are unsatisfiable. The notion of satisfiability is too coarse for free-variable tableaux. What will work here is the following finer notion which was developed in [Hähnle, 1994] and also used in [Fitting, 1996].

DEFINITION 122 ($\forall$-satisfiability). A collection $\mathcal{C}$ of sets of first-order formulae is called $\forall$-*satisfiable* if there is an interpretation $\mathcal{I}$ such that, for every variable assignment $\mathcal{A}$, $\mathcal{I}$ is an $\{\mathcal{A}\}$-model for some element of $\mathcal{C}$.

It is evident that, for closed first-order formulae, $\forall$-satisfiability of a collection coincides with ordinary satisfiability of some element of the collection. In order to illustrate the difference of this concept for formulae with free variables, consider the tableau T mentionend above. The collection consisting of the two sets of formulae $\{P(x), \neg P(a)\}$ and $\{Q(x), \neg Q(b)\}$ is $\forall$-satisfiable (set, e.g. $\mathcal{U} = \{0, 1\}$, $\mathcal{I}(P) = \{0\}, \mathcal{I}(Q) = \{1\}, \mathcal{I}(a) = 1$, and $\mathcal{I}(b) = 0$).

We now give a generalized version of Proposition 76 which can be used to prove correctness both of sentence tableaux and of general free-variable tableaux.

PROPOSITION 123. *Let $C' = C \cup \{S\}$ be a $\forall$-satisfiable collection of sets of first-order formulae.*

1. *If $\alpha \in S$, then $C \cup \{S \cup \{\alpha_i\}\}$ is $\forall$-satisfiable, for every α-subformula α_i of α.*

2. *If $\beta \in S$, then $C \cup \{S \cup \{\beta_1\}, \ldots, S \cup \{\beta_n\}\}$ is $\forall$-satisfiable where $\beta_1, \ldots, \beta_n$ is the β-subformula sequence of β.*

3. *If $\forall x F = \gamma \in S$, then $C \cup \{S \cup \{\gamma(t)\}\}$ is $\forall$-satisfiable, for any term t of the language $\mathcal{L}$ provided $\{x/t\}$ is free for F.*

4. *If $\delta \in S$, then $C \cup \{S \cup \{\delta(t)\}\}$ is $\forall$-satisfiable for any Skolemization $\delta(t)$ of δ w.r.t. $\bigcup C'$.*

5. *$C'\sigma$ is $\forall$-satisfiable, for any substitution σ which is free for all formulae in $\bigcup C'$.*

Proof. By the definition of $\forall$-satisfiability, there is an interpretation $\mathcal{I}$ such that, for every variable assignment $\mathcal{A}$, $\mathcal{I}$ is an $\{\mathcal{A}\}$-model for some member of C'. The non-trivial case for proving items (1) to (4) is the one in which S is $\{\mathcal{A}\}$-satisfied by $\mathcal{I}$ and no element of C is. Let A be the collection of all variable assignments, for which this holds. Items (1) and (2) are immediate from the definition of formula assignment. For (3), let $\mathcal{A}$ be an arbitrary element from A. Then, be item (8) of formula assignments, $\mathcal{I}^{\mathcal{A}'}(F) = \top$, for all x-variants of $\mathcal{A}$. Since $\mathcal{A}\sigma$ is an x-variant of $\mathcal{A}$, $\mathcal{I}^{\mathcal{A}\sigma}(F) = \top$. Now σ is free for F, therefore, Lemma 45 can be applied yielding that $\mathcal{I}^{\mathcal{A}}(F\sigma) = \mathcal{I}^{\mathcal{A}\sigma}(F)$. Item (4): since $\delta(t)$ is a Skolemization of δ w.r.t. $\bigcup C'$, by Proposition 55, there exists an A-model $\mathcal{I}'$ for $S \cup \{\delta(t)\}$ which is identical to $\mathcal{I}$ except for the interpretation of the new function symbol f in $\delta(t)$.

Since f does not occur in C, for all variable assignment $\mathcal{A} \notin A$, some element of C is $\{\mathcal{A}\}$-satisfied by $\mathcal{I}'$. Consequently, $\mathcal{I}'$ is a $\forall$-model for $C \cup \{S \cup \{\delta(t)\}\}$. For (5), let $\mathcal{A}$ be any variable assignment. Consider its modification $\mathcal{A}\sigma$. Since C' is assumed as $\forall$-satisfiable, $\mathcal{I}^{\mathcal{A}\sigma}(S') = \top$, for some $S' \in C'$. Now σ is free for F, therefore, again by Lemma 45, $\mathcal{I}^{\mathcal{A}}(S\sigma) = \mathcal{I}^{\mathcal{A}\sigma}(S)$. ■

PROPOSITION 124 (Soundness of general free-variable tableaux). *If a set of formulae S is satisfiable, then every general free-variable tableau for S has an open branch.*

Proof. First, note that the satisfiability of a set of formulae S entails the $\forall$-satisfiability of the collection $\{S\}$. Then the proof is by induction on the number of inference steps, using Proposition 123 on the collection of the sets of formulae on the branches of a general free-variable tableau. ■

The completeness of general free-variable tableaux is trivial, because the calculus is obviously a generalization of the calculus of sentence tableaux, So general free-variable tableaux are only relevant as a common framework but not as a calculus supporting the *finding* of proofs. What we are interested in is to apply a substitution only if this immediately leads to the closure of a branch, and we will even restrict this to an atomic closure.

DEFINITION 125 (Free-variable tableau). *Free-variable tableaux with atomic closure*, or just *free-variable tableaux*, are defined as general free-variable tableaux, but with the following two modifications. The γ^*-rule is replaced with the weaker γ'-rule and the substitution rule (S) is replaced with the weaker *closure rule* (C)

(γ')	$\dfrac{\gamma}{\gamma(x)}$	where x is a variable new to S and T,
(C)	Modify T to $T\sigma$	if two literals K and L are on a branch such that σ is a minimal unifier for $\{K, \sim L\}$.

Note that the applied substitution will be automatically free for the formulae in the tableau. This is because the γ'-rule guarantees that no variable occurs bound and free in formulae of the tableau and, since K and L are quantifier-free, the minimal unifier σ has only free variables in the terms of its range.

Let us now consider an example. It is apparent that the destructive modifications render it more difficult to represent a free-variable tableau deduction. We solve this problem by not applying the substitutions $\sigma_1, \ldots, \sigma_n$ explicitly to the tableau T, but by annotating them below the nodes at the respective leaves. The represented tableau then is $T\sigma_1 \cdots \sigma_n$. In Figure 6, a free-variable tableau for the same first-order sentence is displayed for which in Figure 3 a sentence tableau is displayed. Comparing both tableaux, we can observe that it is much easier to find the closed free-variable tableau than the closed sentence tableau. The substitutions that close the branches need not be blindly guessed, they can be automatically computed

$$(1)\ \neg\exists x(\forall y \forall z P(y, f(x,y,z)) \rightarrow (\forall y P(y, f(x,y,x)) \wedge \forall y \exists z P(g(y), z)))$$

$$\gamma'(1)$$

$$(2)\ \neg(\forall y \forall z P(y, f(x_1,y,z)) \rightarrow (\forall y P(y, f(x_1,y,x_1)) \wedge \forall y \exists z P(g(y), z)))$$

$$\alpha(2)$$

$$(3)\ \forall y \forall z P(y, f(x_1,y,z))$$

$$\alpha(2)$$

$$(4)\ \neg(\forall y P(y, f(x_1,y,x_1)) \wedge \forall y \exists z P(g(y), z))$$

$$\beta(4)$$

Left branch:

$$(5)\ \neg\forall y P(y, f(x_1,y,x_1))$$

$$\delta'(5)$$

$$(7)\ \neg P(h(x_1), f(x_1,h(x_1),x_1))$$

$$\gamma'(3)$$

$$(8)\ \forall z P(y_1, f(x_1,y_1,z))$$

$$\gamma'(8)$$

$$(9)\ P(y_1, f(x_1,y_1,z_1))$$

$$\sigma_1 = \{y_1/h(x_1), z_1/x_1\}$$

Right branch:

$$(6)\ \neg\forall y \exists z P(g(y), z)$$

$$\delta'(6)$$

$$(10)\ \neg\exists z P(g(b), z)$$

$$\gamma'(10)$$

$$(11)\ \neg P(g(b), z_2)$$

$$\gamma'(3)$$

$$(12)\ \forall z P(y_2, f(x_1,y_2,z))$$

$$\gamma'(12)$$

$$(13)\ P(y_2, f(x_1,y_2,z_3))$$

$$\sigma_2 = \{y_2/g(b), z_2/f(x_1,g(b),z_3)\}$$

Figure 6. Closed free-variable tableau

from the respective pairs of literals to be unified, viz. (7) and (9) on the left and (11) and (13) on the right branch.

4.3 Completeness of Free-variable Tableaux

The completeness of free-variable tableaux is not difficult to prove. For formulae in Skolem or weak Skolem form,[13] the construction of any atomically closed sentence tableau can be simulated step by step by the calculus of free-variable tableaux. This is evident, because only the γ-rule has a different effect for this class of formulae. The simulation then proceeds by simply delaying the instantiations of γ-formulae and performing the substitutions later by using the closure rule. Unfortunately, for general formulae, no identical simulation of sentence tableaux is possible, as becomes clear when comparing Figure 6 with Figure 3. The problem is that more complex Skolem functions may be necessary in free-variable tableaux.

[13] I.e. in which no tableau subformula of type γ has a tableau subformula of type δ.

But modulo such a modification, a so-called *Skolem variant*, a tree-isomorphic simulation exists.

DEFINITION 126 (Skolem variant of a sentence tableau). (inductive)

1. Any sentence tableau T is a *Skolem variant* of itself.
2. If c is a constant introduced by a δ-rule application on a branch B of a Skolem variant T' of a sentence tableau T and t is any ground term whose dominating function symbol is new to B, then the formula tree obtained from replacing any occurrence of c in T' by t is a Skolem variant of T.

It is clear that Skolem variants preserve the closedness of a formula tree.

LEMMA 127. *Any Skolem variant of a(n atomically) closed sentence tableau is (atomically) closed.*

Another problem is that the order in which a free-variable tableau is constructed can influence the arity of the Skolem functions in δ^+-rules. Consider, for example, a tableau consisting of a left branch $P(x) \oplus \neg P(a)$ and a right branch $\exists y(Q(x,y) \wedge \neg Q(a,y))$. If we decide to close the left branch first using the unifier $\{x/a\}$, then the performance of the δ^+-rule on the instantiated right branch will produce a Skolem constant. If the right branch is selected first, then we have to introduce a complex Skolem term $f(x)$, since x is still free. So, in the presence of δ-formulae, different orders of constructing a free-variable tableau can make a difference in the final tableau, as opposed to sentence tableaux which are completely independent of the construction order. As a matter of fact, we want completeness of free-variable tableaux independent of the construction order. The order of construction is formalized with the notion of a *branch selection function*.

DEFINITION 128 (Branch selection function). A *(branch) selection function* ϕ is a mapping assigning an open branch to every tableau T which is not atomically closed. Let ϕ be a branch selection function and let $T_1, \ldots, T_n$ be a sequence of *successive* tableaux, i.e., each T_{i+1}, can be obtained from T_i by a tableau inference step. The tableau T_n is said to be *(constructed) according to* ϕ if each T_{i+1} can be obtained from T_i by performing an inference on the branch $\phi(T_i)$.

LEMMA 129. *Let T' be any atomically closed sentence tableau. Then, for any branch selection function ϕ, there exists an atomically closed free-variable tableau T for S constructed according to ϕ such that T is more general than a Skolem variant of T'; and if every formula $F \in S$ is in weak Skolem form, then T is even more general than T'.*

Proof. Let T' be any atomically closed sentence tableau and ϕ any branch selection function. We define sequences $T_1, \ldots, T_m$ of free-variable tableaux which correspond to initial segments of T' as follows. T_1 is the one-node initial tableau of T'. Let T_i be the i-th element of such a sequence $T_1, \ldots, T_m$, $1 \leq i < m$, and B the inital segment of the branch in T' which corresponds to the selected branch $\phi(T_i)$ in T_i, i.e., B and $\phi(T_i)$ are paths from the root to the same tree position.

1. If B is atomically open, then some expansion step has been performed in the construction of T' to expand B. T_{i+1} is the result of performing a corresponding free-variable tableau expansion step on $\phi(T_i)$.

2. If B is atomically closed (and $\phi(T_i)$ is atomically open), then two complementary literals must be on B. Let K and L be the corresponding literals on $\phi(T_i)$. T_{i+1} is the result of applying a minimal unifier of $\{K, \sim L\}$ to T_i.

We show by induction on the sequence length that any of the tableaux in such a sequence is more general than an initial segment of some Skolem variant of T'. The induction base is evident. For the induction step, let T_i be more general than an initial segment T_i^{Sk} of a Skolem variant of T'. For case (1), we consider first the subcase in which B is not expanded by a δ-step. Then an expansion of $\phi(T_i)$ corresponding to the sentence tableau expansion of B is possible and produces a tableau T_{i+1} that is more general than the respective expansion of T_i^{Sk}, which is an initial segment of some Skolem variant of T'. The subcase of δ-expansion is the problematic one, since one (possibly) has to move to another Skolem variant of T'. Let $\delta(f(x_1, \ldots, x_n))$, $n \geq 0$, be the formula by which T_i was expanded. Each variable x_j, $1 \leq j \leq n$, has been introduced in T_i by a γ'-step using a node N_j. If $t_1, \ldots, t_n$ are the respective ground terms at the same term positions in T_i^{Sk}, then let T_{i+1}^{Sk} be the formula tree obtained by expanding the branch corresponding to B with $\delta(f(t_1, \ldots, t_n))$. By construction, T_{i+1} is more general than T_{i+1}^{Sk}, which is an initial segment of a Skolem variant of T'. In case (2), $\phi(T_i)$ is atomically open, but B is atomically closed. By the induction assumption, T_i is more general than T_i^{Sk}, which has the branch atomically closed. Therefore, there exists a minimal unifier for the literals K and the complement of L on $\phi(T_i)$, and $T_{i+1} = T_i\sigma$ is more general than T_i^{Sk}. Now any such sequence $T_1, \ldots, T_m$ must be of finite length, since in each simulation step either a different node position of T is expanded or closed, i.e m is less or equal to the number of nodes of T'. Consequently, $T = T_m$ is an atomically closed free-variable tableau for S that is more general than a Skolem variant of T'. Finally, if S is in weak Skolem form, no free variable can occur in a δ-formula in a free-variable tableau for S. In this case, one can always use the same Skolem constants in the construction of T and T' and never has to move to a proper Skolem variant of T'. Then T is more general than T'. ■

From this lemma immediately follows the refutational completeness of free-variable tableaux.

THEOREM 130 (Free-variable tableau completeness). *If S is an unsatisfiable set of first-order sentences, then there exists a finite atomically closed free-variable tableau for S.*

4.4 *Proof Procedures for Free-variable Tableaux*

We have proven the completeness of free-variable tableaux via a simulation of sentence tableaux instead of providing an independent completeness proof. The advantage of this approach is that we are assured that, for any atomic sentence tableau proof, there is a free-variable tableau proof of the same tree size. The disadvantage of this completeness proof, however, is that it is proof-theoretically weaker than the one given for sentence tableaux, since we do not specify *how to systematically construct* a closed free-variable tableau, as it is done with the systematic sentence tableau procedure. The simple reason for this is the following. Since the calculus of free-variable tableaux is destructive, in general, the (tree) union of the tableaux in a successive tableau sequence cannot be performed. The fundamental proof-theoretic difference from sentence tableaux is that with the application of substitutions to tableaux the paradigm of saturating a branch (possibly up to a Hintikka set) is lost. A notion of *saturated systematic free-variable tableau* can only be defined for the fragment of the calculus without the closure rule. Completeness could then be shown in the standard way by using any one-to-one association between the set of variables and the set of all ground terms which is then applied to the saturated tableaux at the end (cf. p. 195 in [Fitting, 1996]). This is proof-theoretically possible, but useless for efficient proof search, because the employment of a fixed association between variables and ground terms degrades free-variable tableaux to sentence tableaux. The question is whether there exists a systematic procedure for free-variable tableaux of the same type and functionality as for sentence tableaux but with variable instantiations guided by unification? The problem can at best be recognized with an example.

$$(1)\ \neg P(a,b,c) \wedge \neg P(c,a,b) \wedge \forall x \forall y \forall v \forall w (P(x,y,v) \vee P(y,x,w))$$

$\alpha(1)$

$$(2)\ \neg P(a,b,c)$$

$\alpha(1)$

$$(3)\ \neg P(c,a,b)$$

$\alpha(1)$

$$(4)\ \forall x \forall y \forall v \forall w (P(x,y,v) \vee P(y,x,w))$$

$4{*}\gamma'$

$$(8)\ P(x_1,y_1,v_1) \vee P(y_1,x_1,w_1)$$

$\beta(8)$

$$(9)\ P(x_1,y_1,v_1) \qquad (10)\ P(y_1,x_1,w_1)$$

Figure 7. Free-variable tableau for a datalogic formula (see Definition 103)

Consider the formula on top of Figure 7. Since the formula is a satisfiable datalogic formula, any regular Herbrand tableau construction will terminate with an open branch which is a Hintikka set. Let us contrast this with the behaviour of free-variable tableaux. Referring to the figure, after eight steps we have produced the displayed two-branch tableau. What shall we do next? If we close the left branch by unifying $P(x_1, y_1, v_1)$ and the complement of $\neg P(a, b, c)$ or $\neg P(c, a, b)$, the applied unifier blocks the immediate closure of the right branch. We could proceed and try another four γ'-steps, producing a similar situation than before. Since always new free variables are imported by the γ'-rule, the procedure never terminates, even if we only permit regular tableaux. How do we know when to stop and how can we produce a model? In fact, no systematic procedure for free-variable tableaux has been devised up to now that both is guided by unification and has the same functionality as sentence tableaux. It was only very recently that such a procedure has been proposed for the restricted class of formulae in clausal form. This procedure, which is based on a nondestructive variant of free-variable tableaux, is described in [Billon, 1996].

But we will further pursue the destructive line and discuss a radically different paradigm of searching for tableau proofs. Instead of saturation of a single tableau, one considers *all* tableaux that can be constructed. If all existing tableaux are enumerated in a fair manner, for any unsatisfiable input sentence, one will eventually find a closed free-variable tableau. The fair enumeration is facilitated by the fact that the set of all existing tableaux can be organized in the form of a tree.

DEFINITION 131 (Search tree). Let S be a set of sentences, C a tableau calculus, and ϕ a branch selection function. The *search tree of C and ϕ for S* is a tree $\mathcal{T}$ with its non-root nodes labelled with tableaux defined as follows.

1. The root of $\mathcal{T}$ consists of a single unlabelled node.
2. The successors of the root are labelled with all single-node tableaux for S.
3. Every non-leaf node $\mathcal{N}$ in $\mathcal{T}$ labelled with a tableau T has as many successor nodes as there are successful applications of a single inference step in C applied to the branch in T selected by ϕ, and the successor nodes of $\mathcal{N}$ in $\mathcal{T}$ are labelled with the respective resulting tableaux.

If the input set is finite, the search tree branches finitely, and a fair enumeration can be achieved by simply inspecting the search tree levelwise from the top to the leaves. Any closed tableau will eventually be found after finitely many steps. In practice, this could be implemented as a procedure which explicitly constructs competitive tableaux and thus investigates the search tree in a *breadth-first* manner. The *explicit* enumeration of tableaux, however, suffers from two severe disadvantages. The first one is that, due to the branching rate of the search tree, an enormous amount of space is needed to store all tableaux. The second disadvantage is that the cost for adding new tableaux increases during the proof process as the sizes of the proof objects increase. These weaknesses give sufficient reason why in practice no-one has succeeded with an explicit tableau enumeration approach up to now.

The customary and successful paradigm therefore is to perform tableau enumeration in an implicit manner, using *iterative deepening search* procedures. With this approach, iteratively increasing finite initial segments of a search tree are explored. Although, according to this methodology, initial parts of the search tree are explored several times, no significant efficiency is lost if the initial segments increase exponentially [Korf, 1985]. Due to the construction process of tableaux from the root to the leaves, many tableaux have identical or structurally identical subparts. This motivates one to explore finite initial segments of the search tree in a *depth-first* manner by strongly exploiting *structure sharing* techniques and *backtracking*. Using this approach, at each time only one tableau is kept in memory, which is extended following the branches of the search tree, and truncated when a leaf node of the respective inital segment of the search tree is reached. The advantage is that, due to the application of Prolog techniques, very high inference rates can be achieved this way (see [Stickel, 1988; Letz *et al.*, 1992], or [Beckert and Posegga, 1994]). The respective initial segments are determined by so-called *completeness bounds*.

DEFINITION 132 (Completeness bound). A *size bound* is a total mapping assigning to any tableau T a nonnegative integer n, the *size* of T. A size bound is called a *completeness bound* for a tableau calculus C if, for any finite set S of formulae and any $n \geq 0$, there are only finitely many C-tableaux with size less or equal to n.

A common methodology of developing completeness bounds for the strict (or strictly regular) free-variable tableau calculus C is to limit the application of γ'-steps in certain ways (see also [Fitting, 1996]). We give three concrete examples. First, one may simply limit the application of γ'-steps permitted in the tableau (1) or on each branch of the tableau (2). Another variant (3) is the so-called *multiplicity* bound which has also been used in other frameworks [Prawitz, 1960] and [Bibel, 1987]. The natural definition of this bound is for finite sets S of formulae in Skolem form and for tableaux in which every $F \in S$ is used in the tableau only once at the beginning. Then, with multiplicity n, to each γ-node in the tableau, at most n γ'-steps are permitted.

It is obvious that all mentioned size bounds are completeness bounds for the tableau calculus C, that is, for any finite input set of formulae S: for every n, there are only finitely many C-tableaux of size n for S, and if S is unsatisfiable, then, for some n, there is a closed C-tableau with size n.

Interestingly, one can make complexity assessments about the problem of determining whether a tableau with a certain limit exists. For example, for the bounds (1) and (3), one can demonstrate [Letz, 1998; Voronkov, 1997] that, for some finite input set S, the recognition problem of the existence of a closed tableau for S with a certain limit is complete for the complexity class Σ_2^p in the *polynomial hierarchy* [Garey and Johnson, 1978].

A general disadvantage of completeness bounds of the γ-type is that they are too uniform to be useful in practice. Normally, the first initial segment of the search tree containing a closed tableau with size n may have an astronomic size, with the obvious consequence that a proof will not be found. In the next section, we shall mention completeness bounds in which normally the first proof is in a much smaller initial segment.

We conclude this section with mentioning an obvious method for reducing the effort for finding closed free-variable tableaux. In fact, it is not necessary to consider *all* free-variable tableaux in an initial segment of a search tree. Since only the closure rule is destructive, we can work with the following refined calculus which, at least concerning the tableau expansion rules, is deterministic, similar to the systematic tableau procedure.

DEFINITION 133 (Expansion-deterministic free-variable tableau). The calculus of *expansion-deterministic free-variable tableaux* is defined as systematic sentence tableaux are but with the respective free-variable rules, plus the closure rule.

So the only way indeterminism can occur in this calculus is by the application of closure steps. In order to minimize the search effort, one should even prefer the closure rule (if applicable) to all expansion rules. The completeness of this refinement of free-variable tableaux immediately follows from Lemma 129, since the calculus can simulate the construction of any systematic sentence tableau.

As a final remark of this section, it should be emphasized that, from a search-theoretic perspective, tableau enumeration procedures are not optimally suited for confluent calculi (like the ones mentioned so far). This is because, for confluent tableau calculi, on *any* branch of the search tree there must be a closed tableau if the input set is unsatisfiable. A tableau enumeration procedure, however, does not take advantage of this proof-theoretic virtue of the calculus.

5 CLAUSAL TABLEAUX

The efforts in automated deduction for classical logic have been mainly devoted to the development of proof procedures for formulae in clausal form. This has two reasons. First, as discussed in Section 2, in classical logic, any first-order formula can be transformed into clausal form without affecting the satisfiability status and with only a polynomial increase of the formula size. Since, for formulae in clausal form, the tableau rules can be reduced and presented in a more condensed form, simpler and more efficient proof procedures can be achieved this way. Second and even more important, due to the uniform structure of formulae in clausal form, it is much easier to detect additional refinements and redundancy elimination techniques than for the full first-order format. This section will provide plenty of evidence for this fact.

Since in clause logic negations are only in front of atomic formulae, only atomic branch closures can occur. Accordingly, when a closed free-variable tableau for a set of clauses S is to be constructed and a clause in S has been selected for branch expansion, one can deterministically decompose it to the literal level. Such a macro step consists of an formula rule application, a sequence of γ'-steps and possibly a β-step. It is convenient to ignore the intermediate formulae and reformulate such a sequence of inference steps as a single condensed tableau expansion rule.

DEFINITION 134 (Variable renaming). Let F be a formulae, S a set of formula, and $\sigma = \{x_1/y_1, \ldots, x_n/y_n\}$ a substitution such that all $y_1, \ldots, y_n$ are distinct variables new to S. $F\sigma$ is called a *renaming of $x_1, \ldots, x_n$ in F w.r.t. S*.

Clausal tableaux are trees labelled with literals (and other control information) inductively defined as follows.

DEFINITION 135 (Clausal tableau). (inductive) Let S be any set of clauses. A tree consisting of just one unlabelled node is a *clausal tableau for S*. If B is a branch of a clausal tableau T for S, then the formula trees obtained from the following two inference rules are *clausal tableaux for S*:

(E) $B \oplus L_1 \mid \cdots \mid L_n$ where c is the matrix[14] of a clause in S and $L_1 \vee \cdots \vee L_n$ is a renaming of the free variables in c w.r.t. the formulae in T; the rule (E) is called *clausal expansion rule* or just *expansion rule*,

(C) the closure rule of free-variable tableaux, also called *reduction rule*.

DEFINITION 136 (Tableau clause). For any non-leaf node N in a clausal tableau, the set of nodes $N_1, \ldots, N_m$ immediately below N is called the node *family below* N; if the nodes $N_1, \ldots, N_m$ are labelled with the literals $L_1, \ldots, L_m$, respectively, then the clause $L_1 \vee \cdots \vee L_m$ is named the *tableau clause* below N; The tableau clause below the root node is called the *start* or *top clause* of T.

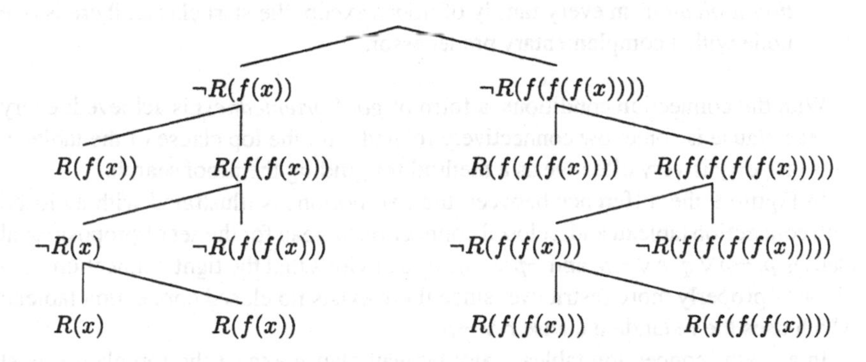

Figure 8. Closed clausal tableau

[14]Recall that the *matrix* of a prenex formula is the formula with the quantifier prefix removed.

In Figure 8, a closed clausal tableau is displayed, here the unifiers of closure steps were already applied. The input is the set of clauses $\forall x(R(x) \vee R(f(x)))$ and $\forall x(\neg R(x) \vee \neg R(f(f(x))))$ corresponding to the negation of the formula F presented in Example 12. So we have demonstrated that F is valid. Clausal tableau provide a relatively concise representation of proofs containing all relevant information. Note that a full free-variable tableau proof including all intermediate inference steps would have more than twice the size.

The completeness and confluence of clausal tableaux for clause formulae follow immediately from the fact that clausal tableaux can simulate free-variable tableaux, by simply omitting formula steps, β-steps, and γ'-steps and performing clausal expansion steps in place.

Clausal tableaux provide a large potential for refinements, i.e., for imposing additional restrictions on the tableau construction. For instance, one can integrate *ordering restrictions* (see [Klingenbeck and Hähnle, 1994]) as they are successfully used in resolution-based systems. The most influential structural refinement paradigm of clausal tableau, however, is the following.

5.1 Connection Tableaux

A closer look at the clausal tableau displayed in Figure 8 reveals an interesting structural property, which we call connectedness.

DEFINITION 137 (Path connectedness, connectedness).

1. A clausal tableau is said to be *path connected* or called a *path connection tableau* if, in every family of nodes except the start clause, there is one node with a complementary ancestor.

2. A clausal tableau is said to be *(tightly) connected* or called a *(tight) connection tableau* if, in every family of nodes except the start clause, there is one node with a complementary predecessor.

With the connection conditions, a form of *goal-orientedness* is achieved: every tableau clause is somehow connectively related with the top clause of the tableau. This property is very effective as a method for guiding the proof search.

In Figure 9 the difference between the two notions is illustrated with a closed path connection tableau and a closed connection tableau for the set of propositional clauses p, $\neg p \vee q$, $r \vee \neg p$, and $\neg p \vee \neg q$. It is obvious that the tight connection condition is properly more restrictive, since there exists no closed connection tableau which uses the redundant clause $r \vee \neg p$.

In a (path) connection tableau, any tableau clause except the top clause must have a literal complementary to some path literal. Therefore, we can require that immediately after attaching a new clause, one of the new branches must be closed using its leaf literal. This motivates to glue together two steps (an expansion and a closure step) into a single inference step.

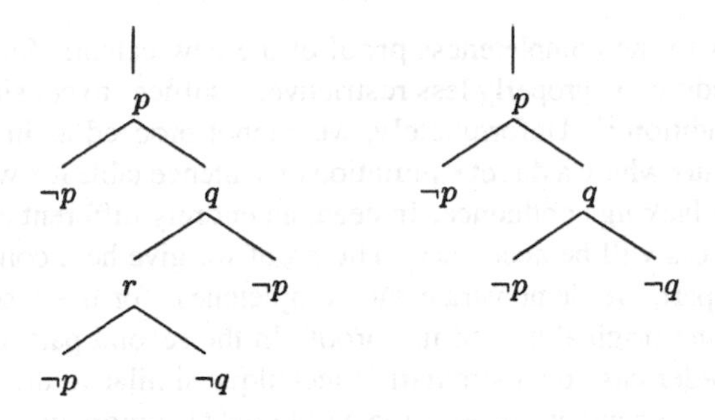

Figure 9. A path connection and a connection tableau

DEFINITION 138 ((Path) connection extension rule). The *path extension rule* (E_P) and the *(tight) extension rule* (E_C) are defined as follows: perform a clausal expansion step, immediately followed by a reduction step unifying one of the newly attached literals with the complement of the literal (E_P) at one of its ancestor nodes or (E_C) at its predecessor node, respectively.

Accordingly, the clausal inference rules can be reformulated in the following more procedurally oriented fashion.

DEFINITION 139 (Connection tableau calculi). The *path connection tableau calculus* and the *connection tableau calculus* consist of three inference rules:

(E_I) the *start rule*, which is simply a clausal expansion step performed only once at the root node,

(C) the closure or reduction rule, and

either (E_P) the path extension rule, or

(E_C) the extension rule, respectively.

Proof-theoretically, (path) connection tableaux strongly differ from all the tableau calculi presented so far.

PROPOSITION 140. *(Path) connection tableaux are not confluent.*

Proof. Consider the unsatisfiable set of clauses $S = \{p, q, \neg q\}$ and the (path) connection tableau T for S obtained by starting with the clause p. T is open and no (path) connection tableau rule can be applied to T, since no clause contains a literal that is complementary to p. ■

With (path) connection tableaux, we have encountered the first tableau calculi that lack confluence. The important consequence to be drawn from this property is that, for those tableau calculi, systematic branch saturation procedures *cannot exist*. Consequently, in order to find proofs, we must use tableau enumeration procedures.

Let us turn now to the completeness proof of the new calculi. Since the path connectedness condition is properly less restrictive, it suffices to consider the tight connectedness condition.[15] Unfortunately, we cannot proceed as in the case of free-variable tableaux where a direct simulation of sentence tableaux was possible, just because of the lacking confluence. Instead, an entirely different approach for proving completeness will be necessary. The proof we give here consists of two parts. In the first part, we demonstrate the completeness for the case of ground formulae—this is the original part of the proof. In the second part, this result is lifted to the first-order case by a simulation technique similar to the one used in the last section. Beforehand, we need some additional terminology.

DEFINITION 141 (Essentiality, relevance, minimal unsatisfiability). A formula F in a set of formulae S is called *essential in* S if S is unsatisfiable and $S \setminus \{F\}$ is satisfiable. A formula F in a set of formulae S is named *relevant in* S if there exists an unsatisfiable subset $S' \subseteq S$ such that F is essential in S'. An unsatisfiable set of formulae S is said to be *minimally unsatisfiable* if each formula in S is essential in S.

DEFINITION 142 (Strengthening). The *strengthening* of a set of clauses S *by* a set of literals $P = \{L_1, \ldots, L_n\}$, written $P \triangleright S$, is the set of clauses obtained by first removing all clauses from S containing literals from P and afterwards adding the n unit clauses $L_1, \ldots, L_n$.

Clearly, every strengthening of an unsatisfiable set of clauses is unsatisfiable. In the ground completeness proofs, we will make use of the following lemma.

LEMMA 143 (Mate Lemma). *Let S be an unsatisfiable set of ground clauses. For any literal L contained in any relevant clause c in S, there exists a clause c' in S such that*

1. *c' contains $\sim L$ and*
2. *c' is relevant in the strengthening $\{L\} \triangleright S$.*

Proof. From the relevance of c follows that S has a minimally unsatisfiable subset S_0 containing c. Every formula in S_0 is essential in S_0, hence there is an interpretation $\mathcal{I}$ for S_0 with $\mathcal{I}(S_0 \setminus \{c\}) = \top$ and $\mathcal{I}(c) = \bot$; Since $\mathcal{I}$ assigns $\bot$ to every literal contained in c, $\mathcal{I}(L) = \bot$. Consider the interpretation $\mathcal{I}'$ which is identical to $\mathcal{I}$ except that $\mathcal{I}'(L) = \top$. Clearly, $\mathcal{I}(c) = \top$. Now the unsatisfiability of S_0 guarantees the existence of another clause c' in S_0 with $\mathcal{I}'(c') = \bot$. We prove that c' meets the conditions (i) and (ii). On the one hand, the clause c' must contain the literal $\sim L$, otherwise $\mathcal{I}(c') = \bot$, which contradicts the selection of $\mathcal{I}$, hence (i). On the other hand, the essentiality of c' in S_0 entails that there exists an interpretation $\mathcal{I}''$ with $\mathcal{I}''(S_0 \setminus \{c'\}) = \top$ and $\mathcal{I}''(c') = \bot$. Since $\sim L$ is in c',

[15] The tight connectedness condition is also the favourable one, because it achieves a more effective pruning of the proof search.

$\mathcal{I}''(L) = \top$. Therefore, c' is essential in $S_0 \cup \{L\}$ and also in its subset $\{L\} \triangleright S_0$. From this and the fact that $\{L\} \triangleright S_0$ is a subset of $\{L\} \triangleright S$ it follows that c' is relevant in $\{L\} \triangleright S$. ■

We even will show completeness for *regular* connection tableaux. Interestingly, this refinement will help in the proof, since it guarantees termination.

PROPOSITION 144 (Ground completeness of regular connection tableaux). *For any finite unsatisfiable set of ground clauses S and any clause c that is relevant in S, there exists a closed regular connection tableau for S with start clause c, for any branch selection function ϕ.*

Proof. A closed regular connection tableau T for S with start clause c can be constructed as follows. Expand the initial one-node tableau with the start clause c. Then iterate the following non-deterministic procedure as long as the intermediate tableau is not yet closed.

> Suppose N be the leaf node of the open branch selected by ϕ in the current tableau and L the literal at N. Let furthermore c be the tableau clause of N and $P = \{L_1, \ldots, L_m, L\}, m \geq 0$, the set of literals on the branch from the root up to the node N. Select any clause c' which is relevant in $P \triangleright S$ and contains $\sim L$; perform an extension step (E_C) at the node N using c'.

First, evidently, the procedure admits solely the construction of regular connection tableaux, since in any expansion step the attached clause c' must contain the literal $\sim L$ (connectedness) and no literals from the path to its predecessor node (regularity). Because of the regularity restriction, there can be only branches of finite length. Consequently, the procedure must terminate either because every leaf is closed or because no clause c' exists for an extension step which meets the conditions stated in the procedure. We prove that the second alternative can never occur, since, for any open leaf node N with literal L, there exists such a clause c'. This will be shown by induction on the node depth. The induction base, $n = 1$, is evident from the Mate Lemma. For the induction step from depth n to $n+1, n \geq 1$, let N be an open leaf node of tableau depth $n+1$ with literal L, tableau clause c, and with the set P of path literals up to the predecessor of N. By the induction assumption, c is relevant in $P \triangleright S$. Let S_0 be any minimally unsatisfiable subset of $P \triangleright S$ containing c. By the Mate Lemma, S_0 contains a clause c' that contains $\sim L$. Since no literal in $P' = P \cup \{L\}$ is contained in a non-unit clause of $P' \triangleright S$ and because N was assumed to be open, no literal in P' is contained in c' (regularity). Finally, since S_0 is minimally unsatisfiable, c' is essential in S_0; therefore, c' is relevant in $P' \triangleright S$. ■

The second half of the completeness proof is a standard lifting argument.

DEFINITION 145 (Ground instance set). Let S be a set of clauses and S' a set of ground clauses. If, for any clause $c' \in S'$, there exists a clause $c \in S$ such that

c' is a substitution instance of the matrix of c, then S' is called a *ground instance set* of S.

PROPOSITION 146. *Every unsatisfiable set of clauses has a finite unsatisfiable ground instance set.*

Proof. Let S be an unsatisfiable set of clauses. Since the elements of S are in prenex form, there exists a finite closed quantifier preferring sentence tableau T for S. By Proposition 89, the initial ground set G of T is unsatisfiable. G must be finite and can consist only of clauses that are all instances of the matrices of clauses in S. Consequently, G is a ground instance set of S. ■

LEMMA 147. *Let T' be a closed (regular) connection tableau for a ground instance set S' of a set of clauses S. Then, for any branch selection function ϕ, there exists a closed (regular) connection tableau T for S constructed according to ϕ such that T is more general than T'.*

Proof. The proof is exactly as the proof of Lemma 129 except that here it is much simpler, since no δ-rule applications can occur. Whenever an expansion or extension step is performed in the construction of T' with a clause c', then a clause $c \in S$ is selected with c' being a ground instance of the matrix of c and a respective expansion or extension step with c is performed in the construction of T. Furthermore, as in the proof of Lemma 129, it may be necessary to perform additional closure steps, which obviously are not needed in the ground proof. ■

THEOREM 148 (Completeness of regular connection tableaux). *For any unsatisfiable set of clauses, any clause c that is relevant in S, and any branch selection function ϕ, there exists a closed regular connection tableau constructed according to ϕ and with a top clause that is a substitution instance of c.*

Proof. Immediate from Proposition 144 and Lemma 147. ■

5.2 Tableaux and Related Calculi

Due to the fact that tableau calculi work by building up tree structures whereas other calculi derive new formulae from old ones, the close relation of tableaux with other proof systems is not immediately evident. However, there are strong similarities between tableau proofs and deductions in other calculi. There is the well-known correspondence of tableau deductions with Gentzen's *sequent system* deductions, which was shown, e.g. in [Smullyan, 1968]. Here, we elaborate a correspondence that is motivated by achieving similarity to calculi from the area of automated deduction. In order to clarify the relation, it is helpful to reformulate the process of tableau construction in terms of formula generation procedures. There are two natural formula interpretations of tableaux which we shall mention and which have both its merits.

DEFINITION 149. The *branch formula* of a finite formula tree T is the disjunction of the conjunctions of the formulae on the branches of T.

Another finer view is preserving the underlying tree structure of the formula tree.

DEFINITION 150 (Formula of a formula tree). (inductive)

1. *The formula* of a one-node formula tree labelled with the formula F is simply F.

2. *The formula* of a complex formula tree with root N and label F, and immediate formula subtrees $T_1, \ldots, T_n$, in this order, is $F \wedge (F_1 \vee \cdots \vee F_n)$ where F_i is the formula of T_i, for every $1 \leq i \leq n$.

If the root is unlabelled like in clausal tableaux, the prefix '$F\wedge$' is omitted.

By the $\wedge$-distributivity (Proposition 34, item 5), the branch formula and the formula of a formula tree T are strongly equivalent. Furthermore, we have the following evident correspondence.

PROPOSITION 151. *The collection of the sets of formulae on the branches of a finite formula tree T is $\forall$-satisfiable if and only if the (branch) formula of T is satisfiable.*

For certain purposes, it is helpful to only consider the open branches of tableaux, which we call *goal trees.*

DEFINITION 152 (Goal tree). The *goal tree of* a tableau T is the formula tree obtained from T by deleting all closed nodes.

For proving the unsatisfiability of a set of formulae using the tableau framework, it is not necessary to *explicitly construct* a closed tableau, it is sufficient to *know* that the deduction *corresponds* to a closed tableau. The goal tree of a tableau contains only the open nodes of a tableau. For the continuation of the refutation process, all other parts of the tableau may be disregarded without any harm.

DEFINITION 153 (Goal formula).

1. The *goal formula* of any closed tableau is the logical falsum $\bot$, which is false under every interpretation.

2. The *goal formula* of any open tableau is the formula of the goal tree of T.

Using the goal formula interpretation, one can view the tableau construction as a linear deduction process in which always a new goal formula is deduced from the previous one until eventually the falsum is derived. In Example 154, we give a goal formula deduction that corresponds to a construction of the tableau in Figure 8, under a branch selection function ϕ that always selects the right-most branch.

EXAMPLE 154 (Goal formula deduction). The set $S = \{\forall x(R(x) \vee R(f(x))), \forall x(\neg R(x) \vee \neg R(f(f(x))))\}$ has the following goal formula refutation.

$$\neg R(x) \vee \neg R(f(f(x)))$$
$$\neg R(x) \vee (\neg R(f(f(x))) \wedge R(f(f(f(x)))))$$
$$\neg R(x) \vee (\neg R(f(f(x))) \wedge R(f(f(f(x)))) \wedge \neg R(f(x)))$$
$$\neg R(x) \vee (\neg R(f(f(x))) \wedge R(f(f(f(x)))) \wedge \neg R(f(x)) \wedge R(f(f(x))))$$
$$\neg R(x)$$
$$\neg R(f(x)) \wedge R(f(f(x)))$$
$$\neg R(f(x)) \wedge R(f(f(x))) \wedge \neg R(x)$$
$$\neg R(f(x)) \wedge R(f(f(x))) \wedge \neg R(x) \wedge R(f(x))$$
$$\bot$$

DEFINITION 155. The *open branch formula* of a formula tree T is either the falsum $\bot$ in case T is closed or otherwise the branch formula of the goal tree of T.

PROPOSITION 156. *The open branch formula and the goal formula of any formula tree T are strongly equivalent to the (branch) formula of T.*

Proof. Immediate from the fact that any formula of the form $\Phi \vee (F \wedge \cdots \wedge \neg F)$ is strongly equivalent to Φ. ■

Using the goal tree or goal formula notation, one can immediately identify a close similarity of clausal tableaux with two well-known calculi from automated deduction which were developed in a different conceptual framework. The first one is the connection calculus presented in [Bibel, 1987] Chapter III.6. It can be viewed as a version of path connection tableaux, restricted to depth-first branch selection functions, i.e., always an open branch with maximal length or *depth* has to be selected. See [Letz *et al.*, 1994] for a more detailed comparison.

Another framework is the *model elimination calculus* which was introduced in [Loveland, 1968] and improved in [Loveland, 1978]. Although model elimination became popular in the resolution-like style of Loveland [1978], the calculus is better viewed as a variant of the connection tableau calculus. The initial presentation in [Loveland, 1968] is indeed in a tableau-like fashion. This view has various advantages concerning generality, elegance, and the possibility of defining extensions and refinements.[16] Here, we treat a subsystem of the original model elimination calculus without factoring and lemmata, called *weak model elimination* in [Loveland, 1978], which is still refutation-complete. The fact that weak model elimination is indeed a specialized subsystem of the connection tableau calculus becomes apparent when considering the goal formula deductions of connection tableaux. The weak model elimination calculus consists of three inference rules which exactly correspond to the start, extension, and closure rule of connection tableaux. In model elimination, the selection of open branches is performed

[16]The soundness and completeness results for model elimination, for example, are immediate consequences of the soundness and completeness proofs of connection tableaux, which are short and simple if compared with the rather involved proofs in [Loveland, 1978]. Furthermore, only the tableau view can identify model elimination as a cut-free calculus.

in a *depth-first right-most* manner, i.e. always the right-most open branch has to be selected (there is also a variant with depth-first left-most selection). Due to the depth-first right-most (or left-most) restriction of the branch selection function, a one-dimensional 'chain' representation of goal formulae is possible (as used in Loveland, [1968; 1978]), in which no logical operators are used but only the two brackets '[' and ']'. The model elimination proof corresponding to the goal formula deduction given in Example 154 is depicted in Example 157.

EXAMPLE 157 (Model elimination deduction). $S = \{\forall x(R(x) \vee R(f(x))), \forall x(\neg R(x) \vee \neg R(f(f(x))))\}$ has the following model elimination refutation.

$$
\begin{array}{l}
\neg R(x)\ \neg R(f(f(x))) \\
\neg R(x)\ [\,\neg R(f(f(x)))\,]\ R(f(f(f(x)))) \\
\neg R(x)\ [\,\neg R(f(f(x)))\ R(f(f(f(x))))\,]\ \neg R(f(x))) \\
\neg R(x)\ [\,\neg R(f(f(x)))\ R(f(f(f(x))))\ \neg R(f(x)))\,]\ R(f(f(x))) \\
\neg R(x) \\
[\,\neg R(f(x))\,]\ R(f(f(x))) \\
[\,\neg R(f(x))\ R(f(f(x)))\,]\ \neg R(x) \\
[\,\neg R(f(x))\ R(f(f(x)))\ \neg R(x)\,]\ R(f(x)) \\
\bot
\end{array}
$$

The transformation from goal formulae with depth-first right-most selection function to model elimination chains works as follows. To any goal formula generated with a depth-first right-most branch selection function, apply the following operation: as long as logical operators are contained in the string, replace every conjunction $L \wedge F$ with $[L]F$ and every disjunction $L_1 \vee \cdots \vee L_n$ with $L_1 \cdots L_n$.

Accordingly, in a model elimination chain, any occurrence ${}^sL$ of an unbracketed literal corresponds to a leaf node N of the goal tree of the corresponding connection tableau T, and the occurrences of bracketed literals to the left of ${}^sL$ correspond to the ancestor nodes of N in T. From this it is evident that weak model elimination is just a variant of the connection tableau calculus.

5.3 *Proof Procedures for Connection Tableaux*

How can we find closed connection tableaux? The nonconfluence of the calculus entails that no systematic branch saturation procedures exist. Consequently, only tableau enumeration procedures can be used. Furthermore, in contrast to the case of free-variable tableaux, in which expansions steps can be performed in a deterministic manner and only the closure steps need to be backtracked, for connection tableaux, *any* inference step has to be backtracked. This is because certain expansion steps may move us into a satisfiable part of the formula, which we may not be able to leave because of the connectedness condition. The nonconfluence proof uses such an example.

On the other hand, connection tableau procedures have the advantage that, due to the connection condition, the detected proofs are often free of redundancies, in contrast to proofs generated with general tableau calculi, which are typically full of redundant parts. Since therefore connection tableau proofs are normally rather short, one can also employ finer and more restrictive completeness bounds than just limiting the γ'-rule. The most widely used completeness bounds for connection tableaux are the following two. On each iterative-deepening level n, consider only tableaux

1. that have a tree depth smaller than n,
2. that have less than n branches.

The first completeness bound, called *depth* bound is coarser than the second one. The second bound is also called *inference bound*, since in any connection tableau T, the number of closed tableau branches is about the number of inference steps to construct T. Concerning complexity, in the worst case, the number of connection tableaux with n branches/depth n is exponential/doubly exponential w.r.t. the number of literals in the input set. An experimental comparison of the two bounds is contained, for example, in [Letz *et al.*, 1992]. A combination of both bounds, the so-called *weighted-depth bound*, turned out to be particularly successful in practice [Ibens and Letz, 1997]. In this paper also advanced methods of branch selection are described.

Methods of Search Pruning

In general, there are two different methodologies for reducing the search effort of tableau enumeration procedures. On the one hand, one can attempt to *refine* the tableau calculus, that is, disallow certain inference steps if they produce tableaux of a certain *structure*—the regularity and the connectedness conditions are of this type. The effect on the tableau search tree is that the respective nodes together with the dominated subtrees vanish so that the branching rate of the tableau search tree decreases. These *tableau structural* methods of redundancy elimination are *local* pruning techniques in the sense that they can be performed by looking at single tableaux only.

The other approach is to improve the *proof search procedure* so that information coming from the proof search itself can be used to even eliminate proof attempts not excluded by the calculus. More specifically, these *global* methods compare *competitive* tableaux in the search tree, i.e., tableaux on different branches, and attempt to show that one tableau (together with its successors) is redundant in the presence of the other. A natural approach here is to exploit *subsumption* between tableaux, in a similar manner subsumption between clauses is used in formula saturation procedures like resolution [Robinson, 1965].

To this end, the notion of subsumption has to be generalized from clauses to formula trees. For a powerful definition of subsumption between formula trees, the following notion of formula tree *contractions* proves helpful.

DEFINITION 158 ((Formula) tree contraction). A (formula) tree T is called a *contraction* of a (formula) tree T' if T' can be obtained from T by attaching n (formula) trees to n non-leaf nodes of T, for some $n \geq 0$.

Figure 10. Illustration of the notion of tree contractions

In Figure 10, the tree on the left is a contraction of itself, of the second and the fourth tree but not a contraction of the third one. Furthermore, the third tree is a contraction of the fourth one. No-one of the other trees is a contraction of another one.

Now subsumption can be defined easily by building on the *'more general'*-relation between formula trees.

DEFINITION 159 (Formula tree subsumption). A formula tree T *subsumes* a formula tree T' if T is more general than a formula tree contraction of T'.

Since the exploitation of subsumption between entire tableaux has not enough reductive potential, we favour the following form of subsumption deletion.

PROCEDURE 160 (Subsumption deletion). For any pair of *competitive* nodes $\mathcal{N}$ and $\mathcal{N}'$ in a tableau search tree $\mathcal{T}$, if the goal tree of the tableau at $\mathcal{N}$ subsumes the goal tree of the tableau at $\mathcal{N}'$, then the whole subtree of the search tree with root $\mathcal{N}'$ is deleted from $\mathcal{T}$.

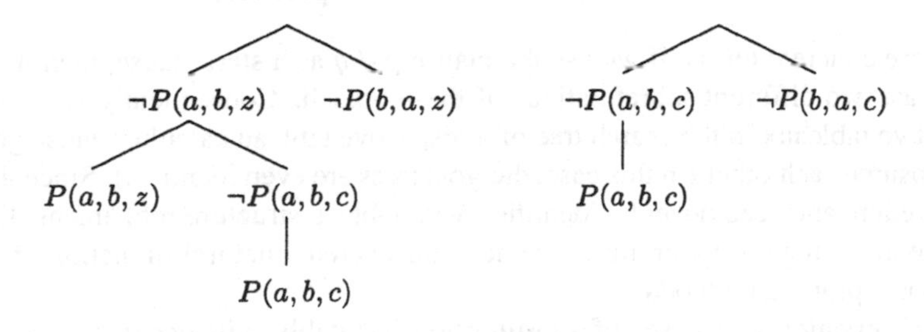

Figure 11. Subsumption of the goal formulae of two tableaux

This search pruning method preserves completeness of the proof search. Let us illustrate this with Figure 11 which displays two connection tableaux for the set of clauses S given in Example 161. Since z is a free variable, the goal tree of the left tableau ($\oplus\neg P(b,a,z)$) subsumes the goal tree of the right tableau ($\oplus\neg P(b,a,c)$). This entails that whenever the right tableau can be completed to a closed tableau, then the left one can also be closed. Consequently, the right tableau and all its possible expansions can be completely ignored.

EXAMPLE 161. The set of clauses $S = \{\forall x \forall y \forall z(\neg P(x,y,z) \vee \neg P(y,x,z)), \forall x \forall y \forall z \forall v(P(x,y,z) \vee P(x,y,v)), P(a,b,c)\}$.

With subsumption deletion, a form of *global* redundancy elimination is achieved which is complementary to the purely tableau *structural* pruning methods discussed so far. In order to illustrate that cases of goal tree subsumption inevitably must occur in the search for proofs—and not only for artificial examples—, we will now present a phenomenon of logic which sheds light on the following problematic property of proof search. Even for minimally unsatisfiable sets of input formulae (which are therefore free of redundancies), the strengthening process (which is implicitly performed in tableau calculi) may introduce redundancies. Let us formulate this more precisely, for the clausal case.

PROPOSITION 162. *If a set of clauses S is minimally unsatisfiable and L is a literal occurring in clauses of S, then the strengthening $\{L\} \triangleright S$ may contain more than one minimally unsatisfiable subset or, equivalently, not every relevant clause in $\{L\} \triangleright S$ may be essential.*

Proof. We use a set S consisting of the following propositional clauses

$$\begin{array}{llll} p \vee q, & \neg q \vee r, & \neg p \vee \neg q \vee \neg s, & \neg p \vee q \vee r, \\ p \vee \neg r \vee \neg s, & \neg q \vee s, & \neg p \vee \neg r \vee s, & \neg p \vee q \vee \neg r \vee \neg s. \end{array}$$

The set S is minimally unsatisfiable, but in the strengthening $\{p\} \triangleright S$ the clauses $\neg q \vee r$ and $\neg q \vee s$ both are relevant but no more essential, since the new unit clause p is also falsified by the two interpretations[17] $\{q, s\}$ and $\{q, r\}$ which have rendered the two mentioned clauses essential in S, respectively. ■

In more concrete terms, if we use the clause $p \vee q$ as a start clause, then there are at least two different subrefutations of the p-branch. Consequently, there are at least two tableaux in the search tree of a respective tableau calculus whose goal trees subsume each other (in this case, the goal trees are even identical). Since this type of redundancy can never be identified with tableau structure refinements like connectedness, regularity, or allies, we have uncovered a natural limitation of all purely local pruning methods.

The observation that cases of subsumption inevitably will occur in practice would motivate one to organize the enumeration of tableaux in such a way that cases of subsumption can really be detected. This could be achieved with a proof procedure which explicitly constructs competitive tableaux and thus investigates the search tree in a breadth-first manner. As already mentioned, the explicit enumeration of tableaux or goal trees, however, is practically impossible. The implicit

[17] We are using Notation 65 for representing interpretations.

enumeration of tableaux, which is performed with iterative-deepening search procedures, has the disadvantage that it is not easily possible to implement subsumption techniques in an adequate way, since at each time only one tableau is in memory. A restricted concept of subsumption deletion, however, can be achieved with the mechanism of memorization of so-called *failure substitutions*.[18] We present the method for depth-first branch selection functions only.

DEFINITION 163 (Solution substitution, failure substitution). Given an initial segment $\mathcal{T}$ of a tableau search tree for a tableau calculus and a depth-first branch selection function, let $\mathcal{N}$ be a node in $\mathcal{T}$ and T the tableau at $\mathcal{N}$. Let furthermore S denote the set of open branches of T and B the selected branch.

1. If $\mathcal{N}'$ with tableau $\mathcal{T}'$ is a node below $\mathcal{N}$ in the search tree $\mathcal{T}$ such that the set of open branches of $\mathcal{T}'$ is $(S \setminus \{B\})\sigma$, for some substitution σ (i.e., a subrefutation of the branch B has been achieved), then σ is called a *solution (substitution) of B at $\mathcal{N}$ via $\mathcal{N}'$*.

2. If the subtree of the search tree $\mathcal{T}$ with root $\mathcal{N}'$ does not contain a closed tableau, then σ is called a *failure substitution for $\mathcal{N}$*.

We describe how failure substitutions are applied in a search procedure which explores tableau search trees in a depth-first manner using backtracking.

PROCEDURE 164 (Local failure caching). Let $\mathcal{T}$ be an initial segment of a tableau search tree.

1. Whenever a branch B in a tableau at a node $\mathcal{N}$ in $\mathcal{T}$ has been closed *via* $\mathcal{N}'$, then the solution substitution σ is stored at the leaf node of B. If no closed tableau is found below $\mathcal{N}'$ and the proof procedure backtracks over $\mathcal{N}'$, then σ is turned into a failure substitution.

2. In any alternative solution attempt of the branch B below $\mathcal{N}$, if a substitution τ is computed such that one of the failure substitutions stored at the node N is more general than τ, then the proof procedure immediately backtracks.

3. When the search node $\mathcal{N}$ at which B was selected for solution is backtracked, then all failure substitutions at B are deleted.

The working of this method can be comprehended by considering the set of clauses S given in Example 161 and the corresponding Figure 11. Assume we are performing an iterative-deepening procedure and are exploring a certain finite initial segment $\mathcal{T}$ of the search tree. Suppose further that the tableau on the left is found first by the search procedure. Then, after the unsolvability of the branch $\oplus P(b, a, z)$ in the segment $\mathcal{T}$ has been detected (note that S is satisfiable), backtracking occurs. But the failure substitution $\sigma = \{x/a, y/b\}$ is annotated at the top

[18] In [Letz *et al.*, 1994] the term 'anti-lemmata' was used for this technique.

left node. In the alternative solution attempt of the branch displayed on the right, the substitution $\tau = \{x/a, y/b, z/c\}$ is computed which is subsumed by σ. This means that the right tableau is ignored, together with its possible expansions.

Whenever a failure substitution σ for a tableau branch B in a tableau T is more general than a substitution τ computed during an alternative solution attempt of B, then the goal tree of T subsumes the goal tree of the alternative tableau. This shows that local failure caching achieves a restricted form of subsumption deletion. The term 'local' has been used in order to distinguish this method from a *global* method of failure caching described in [Astrachan and Stickel, 1992]. In the global method, a permanent failure (and solution) cache is installed, which is used on every future branch B' in which the literals are instances of the literals on the branch B. This method is more powerful, but it is very expensive and needs an enormous amount of memory. Using the local method, failure substitutions do not survive the existence of a branch B. This significantly reduces the amount of memory.

The theorem prover SETHEO [Letz *et al.*, 1992] is an efficient implementation of the regular connection tableau calculus. In the version of the calculus described in [Letz *et al.*, 1994], also the local failure caching mechanism is integrated in an efficient way, by employing constraint technology. Experimental results confirm the theoretical conjecture that a significant search pruning effect can be achieved with the method. It should be noted that this mechanism is not limited to the clausal case, it works for any type of tableau enumeration procedure which is implemented using backtracking.

6 EXTENSIONS OF TABLEAU SYSTEMS

The analytic tableau approach has proven successful, both proof-theoretically and in the practice of automated deduction. It is well-known, however, since the work of Gentzen [1935] that the purely analytic paradigm suffers from a fundamental weakness, namely, the poor *deductive power*. That is, for very simple examples, the smallest tableau proof may be extremely large if compared with proofs in other calculi. In this section, we shall review methods which can remedy this weakness and lead to significantly shorter proofs.

The methods we mention are of three completely different sorts. First, we present mechanisms that amount to adding additional inference rules to tableau systems. The mechanisms are all centered around the (backward) cut rule, which, in its full form, may lead to nonelementarily smaller tableau proofs. Those mechanisms have the widest application, since they already improve the behaviour of tableaux for propositional logic. The propositional aspects of those extensions have already been elaborated in the previous chapter. Second, we consider so-called liberalizations of the δ-rule which may also lead to nonelementarily smaller tableau proofs. Their application, however, is restricted to formulae that are not in Skolem form. Since in automated deduction normally a transformation into

Skolem form is performed, the techniques seem mainly interesting as an improvement of this transformation. Finally, we consider in some more detail a line of improvement which is first-order by its very nature, since it can only be effective for free-variable tableaux. It is motivated by the fact that free variables in tableaux need not necessarily be treated as rigid by the closure rule. The generalization of the rule results in a calculus in which the complexity of proofs can be significantly smaller than the Herbrand complexity of the input formula, which normally is a lower bound to the length of any analytic tableau proof.

6.1 Controlled Integration of the Cut Rule

Gentzen's sequent calculus [Gentzen, 1935] contains the *cut rule* which in the tableau format can be formulated as follows.

DEFINITION 165 ((Tableau) cut rule). The *(tableau) cut rule* is the following tableau expansion rule

$$\text{(Cut)} \qquad \frac{}{F \mid \neg F} \qquad \text{where } F \text{ is any first-order formula.}$$

The formula F is called the *cut formula* of the cut step.

The cut rule is logically redundant, i.e., whenever there exists a closed tableau with cuts for an input set S, then there exists a closed cut-free tableau for S. Even the following stronger redundancy property holds. For this, note that the effect of the cut rule can be simulated by adding, for every applied cut with cut formula F, the special tautological formula $F \vee \neg F$ to the input set, since then the cuts can be performed by using the β-rule on those tautologies. So in a sense the power of the cut can already be contained in an input set if the right tautologies are contained. That tautologies need not be used as expansion formulae in a tableau is evident from the fact that, for every interpretation and variable assignment, one of the tableau subformulae of a tautology will become true.

Although tautologies and therefore the cut rule are redundant, they can lead to nonelementary reductions of the proof length [Orevkov, 1979; Statman, 1979]. While this qualifies the cut rule as one of the fundamental methods for representing proofs in a condensed format, obviously, the rule has the disadvantage that it violates the tableau subformula property. Consequently, from the perspective of proof *search*, an unrestricted use of the cut rule is highly detrimental, since it blows up the search space.

The problem therefore is to perform cuts in a controlled manner. A controlled application of the cut rule can be achieved, for instance, by performing a cut in combination with the β-rule only.

DEFINITION 166 (β-cut rule). Whenever a β-step is to be applied, first, perform a cut step with one of the formulae β_1 or β_2, afterwards perform the β-step on the new right branch. The entire operation is displayed in Figure 12.

Since one of the new branch is closed, only two open branches have been added,

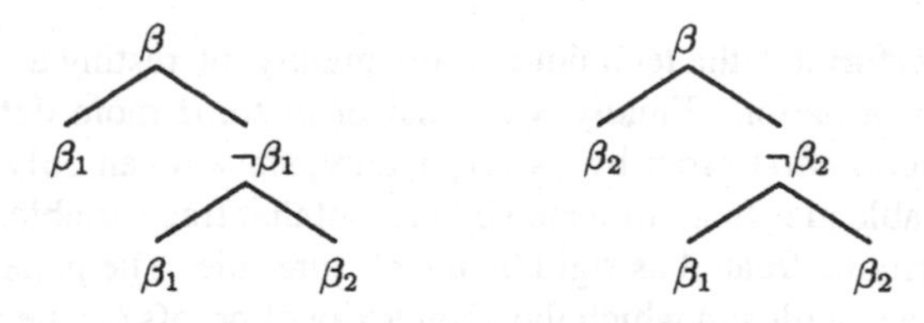

Figure 12. β-rule with cut

like in the standard β-rule, but one of the branches has one more formula on it which can additionally be used for closure. The β-cut rule fulfils the weaker tableau subformula property that any formula in a tableau is either a tableau subformula or the negation of a tableau subformula in the input set. This property suffices for guaranteeing that there exist no infinite decomposition sequences. In the previous chapter, the proof shortening effect of the β-cut rule is studied extensively, for the case of propositional logic (see also [d'Agostino and Mondadori, 1994] and [Letz *et al.*, 1994]). It is shown there that, for certain propositional formulae with only exponential cut-free tableau proofs, a shortening to a linear proof length can be obtained with the β-cut rule. In first-order logic, even a nonelementary proof length reduction can be achieved, as demonstrated in [Egly, 1997]. In [Letz *et al.*, 1994]) also the effect of the β-cut rule on connection tableaux is analyzed and some weaknesses of the rule for the free-variable case are identified. This leads to a generalization of the rule to the so-called *folding up* rule.

6.2 *Liberalizations of the δ-Rule*

Our completeness proof of free-variable tableaux has revealed that, for any atomic sentence tableau proof, there is a free-variable tableau proof of the same tree size. Interestingly, the converse, does not hold. The reason lies in the use of the δ^+-rule in free-variable tableaux taken from [Hähnle, 1994; Fitting, 1996], which can lead to significantly shorter tableau proofs [Baaz and Fermüller, 1995].

EXAMPLE 167. Any closed sentence tableau for the formula $\forall x(P(x) \wedge \exists y \neg P(y))$ requires two applications of the γ-rule whereas there is a closed free-variable tableau with only one application of the γ'-rule.[19]

Interestingly, in the first edition of Fitting's book [1996], a more restrictive variant of the δ^+-rule was given, in which the new Skolem term had to contain all variables on the branch and not only the ones contained in the respective δ-formula. Liberalization of the δ-rule means reducing the number of variables to be considered in the respective Skolem term. So, the δ^+-rule introduced in Sec-

[19] On should mention, that this weakness has already been recognized in [Smullyan, 1968], who identified a condition under which a Skolem term in a δ-rule application need not be new. With this liberalization, shorter sentence tableau proofs can be constructed.

tion 4 is already a liberalization of the original δ-rule in free-variable tableaux. In a sense, however, this older version of free-variable tableaux is conceptually cleaner with respect to the 'rigid' treatment of free-variables. The original idea of a rigid interpretation of the free variables in a tableau is that they may stand for arbitrary ground terms. Accordingly, the notion of ground satisfiability was introduced [Fitting, 1996].

DEFINITION 168 (Ground satisfiability). A collection $\mathcal{C}$ of sets of formulae is *ground satisfiable* if every ground instance of $\mathcal{C}$ has a satisfiable element.

Evidently, $\forall$-satisfiability of a collection $\mathcal{C}$ entails ground satisfiability, but the converse does not hold. As an example consider the ground satisfiable collection $\{\{P(x), \exists y \neg P(y)\}\}$ which is not $\forall$-satisfiable. The difference between the old and the new δ^+-rule is that the old one preserves ground satisfiability, but the new one does not. The closure rule (C), however, preserves ground satisfiability and hence subscribes to a rigid interpretation of the free variables. So, the system of free-variable tableaux (Definition 125) introduced in Section 4 is somewhat undecided in its treatment of free variables. One may argue, however, that with the new system, smaller tableau proofs can be formulated, and this is what counts. In the last part of this section, we will therefore draw the radical consequence and also liberalize the closure rule in such a manner that ground satisfiability is no more preserved. The gain is that with the new rule again a size-reduction of tableau proofs can be achieved.

The δ^+-rule is by far not the "best" Skolemization rule, in the sense that it includes a minimal number of variables in the Skolem term. Consider, for example, the δ-formula $\exists x(P(y) \wedge P(x))$. With the δ^+-rule a unary Skolem term $f(y)$ has to be used. But it is evident that y is irrelevant and a Skolem constant a could be used instead. There are a number of improvements of the δ^+-rule which shall briefly be mentioned here. In [Beckert *et al.*, 1993], it is shown that for each δ-formula stemming from the same occurrence in the input set and each number n of free variables in δ, always the same Skolem function symbol f_δ^n may be used without affecting the soundness of the rule. The thus improved δ-rule is named δ^{++}. In [Baaz and Fermüller, 1995], this method is further liberalized by identifying a *relevant* subset of the free variables in δ and excluding the other variables from the Skolem term. This method can identify the irrelevancy of the variable y in the aforementioned example. Furthermore, the notion of relevancy defined there can be decided in linear time. The corresponding rule is called δ^*. In the paper, a pairwise comparison between the four mentioned Skolemization methods (the one in [Fitting, 1996], δ^+, δ^{++}, and δ^*) is made w.r.t. the proof shortening effect that may be obtained. Interestingly, for each of the improvements, there are example formulae, for which a nonelementary proof length reduction can be achieved w.r.t. the previous rule in the sequence.

Note that, in general, there is no notion of relevant free variables in a δ-formula which is both *minimal* and can be computed efficiently. This can be illustrated with the following simple consideration. Consider a δ-formula of the form $\exists y(F \wedge G)$

containing a free variable x in G but not in F. Suppose further that $\exists y(F \wedge G)$ be strongly equivalent to $F \wedge \exists yG$. Then, obviously, the variable x is not relevant, but this might only be identifiable by proving the strong equivalence of two formulae, which is undecidable in general. So the constraint for any notion of relevant variables is that it be efficiently computable.

6.3 Liberalization of the Closure Rule

The final improvement of the tableau rules that we investigate is again of a quantificational nature. It deals with the problem that the rigid interpretation of free variables often leads to an unnecessary lengthening of tableau proofs.

DEFINITION 169 (Local variable). A variable x occurring free on an open tableau branch is called *local (to the branch)* if x does not occur free on other open branches of the tableau.

If a variable is local to a branch, then any formula containing x can be treated as universally quantified[20] in x. i.e., the universal closure of the formula w.r.t. the variable could be added to the branch. Let us formulate this as a tableau rule.

DEFINITION 170 (Generalization rule). The *generalization rule* is the following expansion rule which can be applied to any open branch of a tableau

(G) $\dfrac{F}{\forall x F}$ where x is a local variable.

PROPOSITION 171. *The generalization rule preserves ∀-satisfiability.*

Proof. Given a tableau T with a local variable x, assume F is any formula on an open branch B of T and T' is the tableau obtained by adding the formula $\forall x F$ to B. We work with the coincidence between ∀-satisfiability and the satisfiability of the open branch formula of a tableau. Let $\mathcal{B} = B_1 \vee \cdots \vee B \vee \cdots \vee B_n$ be the open branch formula of T with $B = F_1 \wedge \cdots \wedge F \wedge \cdots \wedge F_m$. Then $B_1 \vee \cdots \vee (B \wedge \forall x F) \vee \cdots \vee B_n$ is the open branch formula $\mathcal{B}'$ of T'. Now $\mathcal{B}$ is equivalent to $\forall x \mathcal{B}$. Since x does occur free in B only, $\forall x \mathcal{B}$ is strongly equivalent to $\mathcal{B}'$. Consequently, the satisfiability of $\mathcal{B}$ entails the satisfiability of $\mathcal{B}'$. ∎

However, it is apparent that the generalization rule does not preserve ground satisfiability. As a matter of fact, the generalization rule is just of a theoretical interest, since it violates the tableau subformula property. Since we are mainly interested in calculi performing atomic branch closure, it is clear that the new universal formula will be decomposed by the γ'-rule, thus producing a renaming of x in F. And, as instantiations are only performed in closure steps, we would perform the generalization implicitly, exactly at that moment. This naturally leads to a local version of the closure rule.

[20] In [Beckert and Hähnle, 1992], the term *universal* variable was used for a similar notion.

DEFINITION 172 (Local reduction rule). Let T be a tableau and S the set of formulae in T. Suppose K and L are two literals on a branch of T. Let $K\tau$ be a variable renaming of all local variables in K w.r.t. S, and $L\theta$ a variable renaming of all local variables in L w.r.t. $S \cup \{K\tau\}$. Then, the *local closure rule* is the following rule.

(C_L) Modify T to $T\sigma$ if σ is a minimal unifier for $\{K\tau, \sim L\theta\}$ and consider the branch as closed.

The soundness of the local closure rule follows from the fact that its effect can be simulated by a number of applications of the generalization rule, the γ'-rule, and the ordinary closure rule.

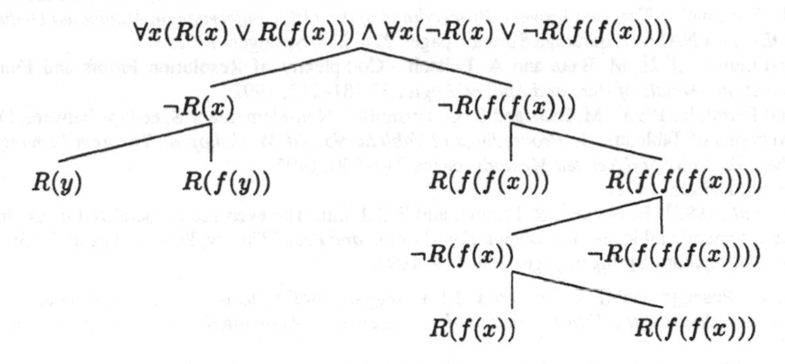

Figure 13. Closed clausal tableau with local reduction rule

Using the local closure rule instead of the standard closure rule, one can achieve a significant shortening of proofs, as illustrated with the following tableau which is smaller than the one given in Figure 8. Assume the tableau construction is performed using a right-most branch selection function. The crucial difference then occurs when the right part of the tableau is closed and a tableau clause of the form $R(y) \vee R(f(y))$ is attached on the left. Since the variable x is now local, it can be renamed and the two remanining branches can be closed using the local closure rule.

Although the displayed tableau has no unsatisfiable ground instance, the soundness of the local reduction rule assures that we have indeed refuted the input set. Note also, that tableau calculi containing the generalization rule or the local reduction rule are not independent of the branch selection function. As long as the right part of the tableau is not closed, the variable x is not local on the left branch and a renaming of x is not permitted. Consequently, the order in which branches are selected can strongly influence the size of the final tableau. The gain, however, is that the local reduction rule permits to build refutations that are smaller than the Herbrand complexity of the input.

ACKNOWLEDGEMENTS

I would like to thank Reiner Hähnle for his corrections and comments on a previous version of the text.

Technische Universität München, Germany.

REFERENCES

[Andrews, 1981] P. Andrews. Theorem Proving via General Matings. *Journal of the Association for Computing Machinery*, 28(2):193–214, 1981.

[Astrachan and Stickel, 1992] O. W. Astrachan and M. E. Stickel. Caching and Lemmaizing in Model Elimination Theorem Provers. *Proceedings of the 11th Conference on Automated Deduction (CADE-11)*, LNAI 607, Saratoga Springs, pages 224–238, Springer, 1992.

[Baaz and Leitsc, 1992] M. Baaz and A. Leitsch. Complexity of Resolution Proofs and Function Introduction. *Annals of Pure and Applied Logic*, 57:181–215, 1992.

[Baaz and Fermülle, 1995] M. Baaz and C. G. Fermüller. Non-elementary Speedups between Different Versions of Tableaux. In *Proceedings of Tableaux95, 4th Workshop on Theorem Proving with Analytic Tableaux and Related Methods*, pages 217–230, 1995.

[Beckert and Hähnle, 1992]

[Beckert *et al.*, 1993] B. Beckert, R. Hähnle, and P. Schmitt. The even more liberalized δ-rule in free variable semantic tableaux. In *Computational Logic and Proof Theory*, Proceedings of the 3rd Kurt Gödel Colloquium, Springer, pages 108–119, 1993.

[Beckert and Posegga, 1994] B. Beckert and J. Posegga. leanTAP: lean, tableau-based theorem proving. *Proceedings of the 12th International Conference on Automated Deduction*, pages 108–119, 1994.

[Bernays and Schönfinkel, 1928] P. Bernays and M. Schönfinkel. Zum Entscheidungsproblem der Mathematischen Logik. *Mathematische Annalen*, 99:342–372, 1928.

[Beth, 1955] E. W. Beth. Semantic Entailment and Formal Derivability. *Mededlingen der Koninklijke Nederlandse Akademie van Wetenschappen*, 18(13):309–342, 1955.

[Beth, 1959] E. W. Beth. *The Foundations of Mathematics*. North-Holland, Amsterdam, 1959.

[Bibel, 1981] W. Bibel. On Matrices with Connections. *Journal of the ACM*, 28:633–645, 1981.

[Bibel, 1987] W. Bibel. *Automated Theorem Proving*. Vieweg Verlag, Braunschweig. Second edition, 1987.

[Billon, 1996] J.-P. Billon. The Disconnection Method. In *Proceedings of Tableaux96, 5th Workshop on Theorem Proving with Analytic Tableaux and Related Methods*, 1996.

[Boy de la Tour, 1990] T. Boy de la Tour. Minimizing the Number of Clauses by Renaming. *Proceedings of the 10th International Conference on Automated Deduction*, pages 558–572, 1990.

[Church, 1936] A. Church. An Unsolvable Problem of Elementary Number Theory. *American Journal of Mathematics*, 58:345–363, 1936.

[Corbin and Bidoit, 1983] J. Corbin and M. Bidoit. A Rehabilitation of Robinson's Unification Algorithm. In *Information Processing*, pages 909–914. North-Holland, 1983.

[d'Agostino and Mondadori, 1994] M. d'Agostino and M. Mondadori. The taming of the cut. *Journal of Logic and Computation*, 4(3):285–319, 1994.

[Eder, 1985] E. Eder. An Implementation of a Theorem Prover based on the Connection Method. In W. Bibel and B. Petkoff, editors, *AIMSA: Artificial Intelligence Methodology Systems Applications*, pages 121–128. North-Holland, 1985.

[Egly, 1997] U. Egly. Non-elementary Speed-ups in Proof Length by Different Variants of Classical Analytic Calculi. In *Proceeding of TABLEAUX'97*, pages 158–172, Springer, 1997.

[Fitting, 1996] M. Fitting. *First-Order Logic and Automated Theorem Proving*, Springer, 1996. First edition, 1990.

[Fitting, 1996] M. Fitting. *First-Order Logic and Automated Theorem Proving*, Springer, 1996. Second revised edition of [Fitting, 1996].

[Gentzen, 1935] G. Gentzen. Untersuchungen über das logische Schließen. *Mathematische Zeitschrift*, 39:176–210 and 405–431, 1935. Engl. translation in [Szabo, 1969].

[Garey and Johnson, 1978] M.R. Garey und D.S. Johnson, *Computers and Intractability: A Guide to the Theory of NP-Completeness*, Freeman, New York, 1978.

[Gödel, 1930] K. Gödel. Die Vollständigkeit der Axiome des logischen Funktionenkalküls. *Monatshefte für Mathematik und Physik*, 37:349–360, 1930.

[Hähnle, 1994] R. Hähnle and P. Schmitt. The liberalized δ-rule in free variable semantic tableaux. *Journal of Automated Reasoning* 13(2):211–221, 1994.

[Herbrand, 1930] J. J. Herbrand. Recherches sur la théorie de la démonstration. *Travaux de la Société des Sciences et des Lettres de Varsovie, Cl. III, math.-phys.*, 33:33–160, 1930.

[Hilbert and Ackermann, 1928] D. Hilbert and W. Ackermann. *Grundzüge der theoretischen Logik.* Springer, 1928. Engl. translation: Mathematical Logic, Chelsea, 1950.

[Hilbert and Bernays, 1934] D. Hilbert and P. Bernays. *Grundlagen der Mathematik.* Vol. 1, Springer, 1934.

[Hintikka, 1955] K. J. J. Hintikka. Form and Content in Quantification Theory. *Acta Philosophica Fennica*, 8:7–55, 1955.

[Huet, 1976] G. Huet. *Resolution d'equations dans les languages d'ordre* $1, 2, \ldots, \omega$. PhD thesis, Université de Paris VII, 1976.

[Huet, 1980] G. Huet. Confluent Reductions: Abstract Properties and Applications to Term Rewriting Systems. *Journal of the Association for Computing Machinery*, 27(4):797–821, 1980.

[Ibens and Letz, 1997] O. Ibens and R. Letz. Subgoal Alternation in Model Elimination. In *Proceeding of TABLEAUX'97*, pages 201–215, Springer, 1997.

[Klingenbeck and Hähnle, 1994] S. Klingenbeck and R. Hähnle. Semantic tableaux with ordering restrictions. In *Proceedings of the 12th International Conference on Automated Deduction*, pages 708–722, 1994.

[Korf, 1985] R. E. Korf. Depth-First Iterative Deepening: an Optimal Admissible Tree Search. *Artificial Intelligence*, 27:97–109, 1985.

[Krivine, 1971] J.-L. Krivine. *Introduction to Axiomatic Set Theory*, Reidel, Dordrecht, 1971.

[Lassez *et al.*, 1988] J.-L. Lassez, M. J. Maher, and K. Marriott. Unification Revisited. *Foundations of Deductive Databases and Logic Programming* (ed. J. Minker), pages 587–625, Morgan Kaufmann Publishers, Los Altos, 1988.

[Letz *et al.*, 1992] R. Letz, J. Schumann, S. Bayerl, and W. Bibel. SETHEO: A High-Performance Theorem Prover. *Journal of Automated Reasoning*, 8(2):183–212, 1992.

[Letz, 1993] R. Letz. *First-Order Calculi and Proof Procedures for Automated Deduction.* PhD thesis, Technische Hochschule Darmstadt, 1993.

[Letz *et al.*, 1994] R. Letz, K. Mayr, and C. Goller. Controlled Integration of the Cut Rule into Connection Tableaux Calculi. *Journal of Automated Reasoning*, 13:297–337, 1994.

[Letz, 1998] R. Letz. *Structures and Complexities of First-order Calculi.* Forthcoming.

[Loveland, 1968] D. W. Loveland. Mechanical Theorem Proving by Model Elimination. *Journal of the Association for Computing Machinery*, 15(2):236–251, 1968.

[Loveland, 1978] D. W. Loveland. *Automated Theorem Proving: a Logical Basis.* North-Holland, 1978.

[Martelli and Montanari, 1976] A. Martelli and U. Montanari. Unification in Linear Time and Space: a Structured Presentation. Technical report. Internal Rep. No. B76-16, 1976.

[Martelli and Montanari, 1982] A. Martelli and U. Montanari. An Efficient Unification Algorithm. *ACM Transactions on Programming Languages and Systems*, Vol. 4, No. 2, pages 258–282, 1982.

[Ohlbach, 1991] H.-J. Ohlbach. Semantics Based Translation Methods for Modal Logics. *Journal of Logic and Computation*, 1(5):691–746, 1991.

[Orevkov, 1979] V.-P. Orevkov. Lower Bounds for Increasing Complexity of derivations after Cut Elimination. *Zapiski Nauchnykh Seminarov Leningradskogo Otdeleniya Matematicheskogo Instituta im V. A. Steklova AN USSR*, 88:137–161, 1979. English translation in *J. Soviet Mathematics*, 2337-2350, 1982.

[Paterson and Wegman, 1978] M. S. Paterson and M. N. Wegman. Linear Unification. *Journal of Computer and Systems Sciences*, 16:158–167, 1978.

[Prawitz, 1960] D. Prawitz. An Improved Proof Procedure. *Theoria*, 26:102–139, 1960.

[Prawitz, 1969] D. Prawitz. Advances and Problems in Mechanical Proof Procedures. In J. Siekmann and G. Wrightson (editors). *Automation of Reasoning. Classical Papers on Computational Logic*, Vol. 2, pages 285–297, Springer, 1983.

[Reckhow, 1976] R. A. Reckhow. *On the Lenghts of Proofs in the Propositional Calculus*. PhD thesis, University of Toronto, 1976.

[Reeves, 1987] S. Reeves. Semantic tableaux as a framework for automated theorem-proving. In C. S. Mellish and J. Hallam, editors, *Advances in Artificial Intelligence (Proceedings of AISB-87)*, pages 125–139, Wiley, 1987.

[Robinson, 1965] J. A. Robinson. A Machine-oriented Logic Based on the Resolution Principle. *Journal of the Association for Computing Machinery*, 12:23–41, 1965.

[Shostak, 1976] R. E. Shostak. Refutation Graphs. *Artificial Intelligence*, 7:51–64, 1976.

[Smullyan, 1968] R. M. Smullyan. *First Order Logic*. Springer, 1968.

[Statman, 1979] R. Statman. Lower Bounds on Herbrand's Theorem. In *Proceedings American Math. Soc.*, 75:104–107, 1979.

[Stickel, 1988] M. A. Stickel. A Prolog Technology Theorem Prover: Implementation by an Extended Prolog Compiler. *Journal of Automated Reasoning*, 4:353–380, 1988.

[Szabo, 1969] M. E. Szabo. *The Collected Papers of Gerhard Gentzen*. Studies in Logic and the Foundations of Mathematics. North-Holland, Amsterdam, 1969.

[Tarski, 1936] A. Tarski. Der Wahrheitsbegriff in den formalisierten Sprachen. *Studia Philosophica*, 1, 1936.

[Turing, 1936] A. M. Turing. On Computable Numbers, with an Application to the Entscheidungsproblem. *Proceedings of the London Mathematical Society*, 42:230–265, 1936.

[Voronkov, 1997] A. Voronkov. Personal communication, 1997.

[Wallen, 1989] L. Wallen. *Automated Deduction for Non-Classical Logic*. MIT Press, Cambridge, Mass., 1989.

BERNHARD BECKERT

EQUALITY AND OTHER THEORIES

1 INTRODUCTION

Theory reasoning is an important technique for increasing the efficiency of automated deduction systems. The knowledge from a given domain (or theory) is made use of by applying efficient methods for reasoning in that domain. The general purpose *foreground reasoner* calls a special purpose *background reasoner* to handle problems from a certain theory.

Theory reasoning is indispensable for automated deduction in real world domains. Efficient equality reasoning is essential, but most specifications of real world problems use other theories as well: algebraic theories in mathematical problems and specifications of abstract data types in software verification to name a few.

Following the pioneering work of M. Stickel, theory reasoning methods have been described for various calculi; e.g., resolution [Stickel, 1985; Policriti and Schwartz, 1995], path resolution [Murray and Rosenthal, 1987], the connection method [Petermann, 1992; Baumgartner and Petermann, 1998], model elimination [Baumgartner, 1992], connection tableaux [Baumgartner *et al.*, 1992; Furbach, 1994; Baumgartner, 1998], and the matrix method [Murray and Rosenthal, 1987a].

In this chapter, we describe how to combine background reasoners with the ground, the free variable, and the universal formula versions of semantic tableaux. All results and methods can be adapted to other tableau versions for first-order logic: calculi with signed formulae, with different δ-rules, with methods for restricting the search space such as connectedness or ordering restrictions, with lemma generation, etc. Difficulties can arise with adaptations to tableau calculi for other logics, in particular if the consequence relation is affected (e.g., non-monotonic logics and linear logic); and care has to be taken if theory links or theory connections have to be considered [Petermann, 1993; Baumgartner, 1998; Baumgartner and Petermann, 1998].

Background reasoners have been designed for various theories, in particular for equality reasoning; an overview can be found in [Baumgartner *et al.*, 1992; Furbach, 1994; Baumgartner, 1998], for set theory in [Cantone *et al.*, 1989]. Reasoning in single models, e.g. natural numbers, is discussed in [Bürckert, 1990].

One main focus of this chapter is efficient equality reasoning in semantic tableaux. Equality, however, is the only theory that is discussed in detail. There is no uniform way for handling theories, which is, after all, the reason for using a background reasoner but which makes it impossible to present good background

M. D'Agostino et al. (eds.), Handbook of Tableau Methods, 197–254.

reasoners for all possible theories. The second main focus of this chapter is therefore on the interaction between foreground and background reasoners, which plays a critical rôle for the efficiency of the combined system.

The chapter is organized as follows: in Section 2, the basic concepts of theory reasoning are introduced, and the main classifications of theory reasoning methods are discussed. The ground, the free variable, and the universal formula version of semantic tableaux, which are the versions that have to be distinguished for theory reasoning, are defined in Section 3, and methods are presented to add theory reasoning to these versions of tableaux. Soundness of these methods is proven in Section 4. In Section 5, completeness criteria for background reasoners are defined. Total and partial background reasoners for the equality theory are presented in Sections 6 and 7. Incremental theory reasoning, which is a method for improving the interaction between foreground and background reasoners, is introduced in Section 8. Finally, in Section 9, methods for handling equality are described that are based on modifying the input formulae.

2 THEORY REASONING

2.1 First-Order Logic: Syntax and Semantics

We use the logical connectives $\wedge$ (conjunction), $\vee$ (disjunction), $\supset$ (implication), $\leftrightarrow$ (equivalence), $\neg$ (negation), and the quantifier symbols $\forall$ and $\exists$.

NOTATION 1. A *first-order signature* $\Sigma = \langle P_\Sigma, F_\Sigma, \alpha_\Sigma \rangle$ consists of a set P_Σ of *predicate symbols*, a set F_Σ of *function symbols*, and a function α_Σ assigning an *arity* $n \geq 0$ to the predicate and function symbols; for each arity, there are infinitely many function and predicate symbols. Function symbols of arity 0 are called *constants*. In addition, there is an infinite set V of *object variables*.

$Term_\Sigma$ is the set of all terms and $Term^0_\Sigma \subset Term_\Sigma$ is the set of all ground terms built from Σ in the usual manner. $Form_\Sigma$ is the set of all first-order formulae over Σ; a formula $\phi \in Form_\Sigma$ must not contain a variable that is both bound and free in ϕ (see Section 1.1 in the chapter by Letz for formal definitions of $Term_\Sigma$ and $Form_\Sigma$). $Lit_\Sigma \subset Form_\Sigma$ is the set of all literals.

DEFINITION 2. A variable $x \in V$ is *free* in a first-order formula ϕ, if there is an occurrence of x in ϕ that is not inside the scope of a quantification $(\forall x)$ or $(\exists x)$; x is *bound* in ϕ if it occurs in ϕ inside the scope of a quantification $(\forall x)$ or $(\exists x)$.

A *sentence* is a formula $\phi \in Form_\Sigma$ not containing any free variables.

NOTATION 3. $Subst_\Sigma$ is the set of all substitutions, and $Subst^*_\Sigma \subset Subst_\Sigma$ is the set of all idempotent substitutions with finite domain.

A substitution $\sigma \in Subst_\Sigma$ with a finite domain $\{x_1, \ldots, x_n\}$ can be denoted by $\{x_1 \mapsto t_1, \ldots, x_n \mapsto t_n\}$, i.e. $\sigma(x_i) = t_i$ $(1 \leq i \leq n)$.

The restriction of σ to a set $W \subset V$ of variables is denoted by $\sigma_{|W}$.

A substitution σ may be applied to a quantified formula ϕ; however, to avoid undesired results, the bound variables in ϕ must neither occur in the domain nor the scope of σ. This is not a real restriction as the bound variables in ϕ can be renamed.

DEFINITION 4. A formula ϕ' is an *instance* of a formula ϕ if there is a substitution $\sigma = \{x_1 \mapsto t_1, \ldots, x_n \mapsto t_n\} \in Subst^*_\Sigma$ such that

1. $\phi' = \phi\sigma$,
2. none of the variables $x_1, \ldots, x_n$ is bound in ϕ, and none of the variables that are bound in ϕ occurs in the terms $t_1, \ldots, t_n$.

If an instance does not contain any variables, it is a *ground instance.*

DEFINITION 5. A formula $\phi \in Form_\Sigma$ is *universally quantified* if it is of the form $(\forall x_1) \cdots (\forall x_n)\psi$, $n \geq 0$, where ψ does not contain any quantifications.

In this case, if a formula ψ' is an instance of ψ (Definition 4), it is as well called an instance of ϕ.

DEFINITION 6. A *structure* $M = \langle D, I \rangle$ for a signature Σ consists of a non-empty domain D and an interpretation I which gives meaning to the function and predicate symbols of Σ.

A *variable assignment* is a mapping $\nu : V \to D$ from the set of variables to the domain D.

The combination of an interpretation I and an assignment ν associates (by structural recursion) with each term $t \in Term_\Sigma$ an element $t^{I,\nu}$ of D.

The *evaluation function* $val_{I,\nu}$ maps the formulae in $Form_\Sigma$ to the truth values *true* and *false* (in the usual way, see Section 1.2 in the chapter by Letz). If $val_{I,\nu}(\phi) =$ *true*, which is denoted by $(M, \nu) \models \phi$, holds for all assignments ν, then M *satisfies* the formula ϕ (is a *model* of ϕ); M satisfies a set Φ of formulae if it satisfies all elements of Φ.

A formula ϕ is a *tautology* if it is satisfied by all structures.

DEFINITION 7. A formula $\psi \in Form_\Sigma$ is a *(weak) consequence* of a set $\Phi \subset Form_\Sigma$ of formulae, denoted by $\Phi \models \psi$, if all structures that are models of Φ are models of ψ as well.

In addition to the normal (weak) consequence relation $\models$, we use the notion of strong consequence:

DEFINITION 8. A formula $\psi \in Form_\Sigma$ is a *strong consequence* of a set $\Phi \subset Form_\Sigma$ of formulae, denoted by $\Phi \models^\circ \psi$, if for all structures $M = \langle D, I \rangle$ and all variable assignments ν:

$$\text{If } (M, \nu) \models \phi \text{ for all } \phi \in \Phi, \text{ then } (M, \nu) \models \psi \ .$$

A difference between the strong consequence relation $\models^\circ$ and the weak consequence relation $\models$ is that the following holds for $\models^\circ$ (but not for $\models$):

LEMMA 9. *Given a set* $\Phi \subset Form_\Sigma$ *of formulae and a formula* $\psi \in Form_\Sigma$, *if* $\Phi \models^\circ \psi$, *then* $\Phi\sigma \models^\circ \psi\sigma$ *for all substitutions* $\sigma \in Subst^*_\Sigma$.

2.2 Theories

We define any satisfiable set of sentences to be a theory.

DEFINITION 10. A *theory* $\mathcal{T} \subset Form_\Sigma$ is a satisfiable set of sentences.

In the literature, often the additional condition (besides satisfiability) is imposed on theories that they are closed under the logical consequence relation. Without that restriction, we do not have to distinguish between a theory and its defining set of axioms.

EXAMPLE 11. The most important theory in practice is the equality theory $\mathcal{E}$.[1] It consists of the following axioms:

(1) $(\forall x)(x \approx x)$ (reflexivity),

(2) for all function symbols $f \in F_\Sigma$:

$$(\forall x_1)\cdots(\forall x_n)(\forall y_1)\cdots(\forall y_n)((x_1 \approx y_1 \wedge \ldots \wedge x_n \approx y_n) \supset f(x_1,\ldots,x_n) \approx f(y_1,\ldots,y_n))$$

where $n = \alpha_\Sigma(f)$ (monotonicity for function symbols),

(3) for all predicate symbols $p \in P_\Sigma$:

$$(\forall x_1)\cdots(\forall x_n)(\forall y_1)\cdots(\forall y_n)((x_1 \approx y_1 \wedge \ldots \wedge x_n \approx y_n) \supset (p(x_1,\ldots,x_n) \supset p(y_1,\ldots,y_n)))$$

where $n = \alpha_\Sigma(p)$ (monotonicity for predicate symbols),

Symmetry and transitivity of $\approx$ are implied by reflexivity (1) and monotonicity for predicate symbols (3) (observe that $\approx \in P_\Sigma$).

EXAMPLE 12. The theory $\mathcal{OP}$ of partial orderings consists of the axioms

(1) $(\forall x)\neg(x < x)$ (anti-reflexivity),

(2) $(\forall x)(\forall y)(\forall z)((x < y) \wedge (y < z) \supset (x < z))$ (transitivity).

$\mathcal{OP}$ is a finite theory; contrary to the equality theory, it does not contain monotonicity axioms.

An important class of theories, called *equational theories*, are extensions of the equality theory $\mathcal{E}$ by additional axioms that are universally quantified equalities.

[1] The equality predicate is denoted by $\approx \in P_\Sigma$ such that no confusion with the meta-level equality $=$ can arise.

An overview of important equational theories and their properties can be found in [Siekmann, 1989].

EXAMPLE 13. The AC-theory for the function symbol f contains (besides $\mathcal{E}$) the additional axioms

$$(\forall x)(\forall y)(\forall z)(f(f(x,y),z) \approx f(x,f(y,z)))$$

and

$$(\forall x)(\forall y)(f(x,y) \approx f(y,x)) \quad ,$$

which state associativity resp. commutativity of f; it is an equational theory.

Other typical examples for equational theories are specifications of algebraic structures:

EXAMPLE 14. Group theory can be defined using, in addition to $\mathcal{E}$, the equalities

$$\begin{array}{rcl} (\forall x)(\forall y)(\forall z)((x \circ y) \circ z & \approx & x \circ (y \circ z)) \\ (\forall x)(\quad x \circ e & \approx & x \qquad) \\ (\forall x)(\quad x \circ x^{-1} & \approx & e \qquad) \end{array}$$

The definitions of structure, satisfiability, tautology, and logical consequence are adapted to theory reasoning in a straightforward way:

DEFINITION 15. Let $\mathcal{T}$ be a theory. A $\mathcal{T}$*-structure* is a structure that satisfies all formulae in $\mathcal{T}$. A formula ϕ (a set Φ of formulae) is $\mathcal{T}$*-satisfiable* if there is a $\mathcal{T}$-structure satisfying ϕ (resp. Φ), else it is $\mathcal{T}$*-unsatisfiable*. A sentence ϕ is a $\mathcal{T}$*-tautology* if it is satisfied by all $\mathcal{T}$-structures.

A formula ϕ is a (weak) $\mathcal{T}$*-consequence* of a set Ψ of formulae, denoted by $\Psi \models_{\mathcal{T}} \phi$, if ϕ is satisfied by all $\mathcal{T}$-structures that satisfy Ψ. A formula ϕ is a *strong* $\mathcal{T}$*-consequence* of a set Ψ of formulae, denoted by $\Psi \models^{\circ}_{\mathcal{T}} \phi$, if for all $\mathcal{T}$-structures M and all variable assignments ν:

$$\text{If } (M,\nu) \models \psi \text{ for all } \psi \in \Psi, \text{ then } (M,\nu) \models \phi\,.$$

LEMMA 16. *Given a theory $\mathcal{T}$, a set Φ of sentences, and a sentence ψ, the following propositions are equivalent:*

1. $\Phi \models_{\mathcal{T}} \psi$.
2. $\Phi \cup \mathcal{T} \models \psi$.
3. $\Phi \cup \mathcal{T} \cup \{\neg\psi\}$ *is unsatisfiable.*
4. $\Phi \cup \{\neg\psi\}$ *is* $\mathcal{T}$*-unsatisfiable.*

2.3 Properties of Theories

The following definitions clarify which properties theories should have to be useful in practice:

DEFINITION 17. A theory $\mathcal{T}$ is *(finitely) axiomatizable* if there is a (finite) decidable set $\Psi \subset Form_\Sigma$ of sentences (the axioms) such that: $\phi \in Form_\Sigma$ is a $\mathcal{T}$-tautology if and only if $\Psi \models \phi$.

A theory $\mathcal{T}$ is *complete* if, for all sentences $\phi \in Form_\Sigma$, either ϕ or $\neg\phi$ is a $\mathcal{T}$-tautology.

All theories that we are concerned with, including equality, are axiomatizable. An example for a theory that is not axiomatizable is the set $\mathcal{T}$ of all satisfiable sentences.

If a theory $\mathcal{T}$ is axiomatizable, then the set of $\mathcal{T}$-tautologies is enumerable; it may, however, be undecidable (a simple example for this is the empty theory). If $\mathcal{T}$ is both axiomatizable and complete, then the set of $\mathcal{T}$-tautologies is decidable.

Another important method for characterizing a theory $\mathcal{T}$—besides axiomatization—is the model theoretic approach, where $\mathcal{T}$ is defined as the set of all formulae that are true in a given structure M. Theories defined this way are always complete, because $M \models \phi$ or $M \models \neg\phi$ for all sentences ϕ.

DEFINITION 18. A theory $\mathcal{T}$ is *universal* if it is axiomatizable using an axiom set consisting of universally quantified formulae (Definition 4).

THEOREM 19. *Let $\mathcal{T}$ be a universal theory. A set Φ of universally quantified formulae is $\mathcal{T}$-unsatisfiable if and only if there is a finite set of ground instances of formulae from Φ that is $\mathcal{T}$-unsatisfiable.*

In the literature on theory reasoning, all considerations are usually restricted to universal theories, because the Herbrand-type Theorem 19 holds exactly for universal theories [Petermann, 1992]. This theorem is essential for theory reasoning if the background reasoner can only provide formulae without variable quantifications (e.g., only literals or clauses); this is, for example, the case if theory reasoning is added to clausal tableaux or resolution.

EXAMPLE 20. The theory $\mathcal{T} = \{(\exists x)p(x)\}$ is not universal. Consequently, there are sets Φ of universally quantified formulae that are $\mathcal{T}$-unsatisfiable whereas all finite sets of ground instances of formulae from Φ are $\mathcal{T}$-satisfiable. An example is $\Phi = \{(\forall x)(\neg p(x))\}$; even the set $\{\neg p(t) \mid t \in Term^0_\Sigma\}$ of *all* ground instances of Φ is $\mathcal{T}$-satisfiable (using a $\mathcal{T}$-structure where not all elements of the domain are represented by ground terms).

The restriction to universal theories is not a problem in practice, because it is easy to get around using Skolemization.

EXAMPLE 21. An extension of $\mathcal{OP}$ that contains the density axiom

$$(\forall x)(\forall y)((x < y) \supset (\exists z)((x < z) \wedge (z < y)))$$

Key	$\mathcal{E}$-Refuter
$\{\neg(a \approx a)\}$	id
$\{\neg(x \approx a)\}$	$\{x \mapsto a\}$
$\{(\forall x)(\neg(x \approx a))\}$	id
$\{p(a), \neg p(b)\}$	$\langle id \quad , \{\neg(a \approx b)\}\rangle$
$\{p(f(a), f(b)), f(x) \approx x\}$	$\langle\{x \mapsto a\}, \{p(a, f(b))\}\rangle$
	$\langle\{x \mapsto b\}, \{p(f(a), b)\}\rangle$
$\{p(f(a), f(b)), (\forall x)(f(x) \approx x)\}$	$\langle id \quad , \{p(a, b)\}\rangle$

Table 1. Examples for $\mathcal{E}$-refuters

is not a universal theory. It can be made universal by replacing the above axiom with

$$(\forall x)(\forall y)((x < y) \supset ((x < between(x, y)) \wedge (between(x, y) < y))) .$$

2.4 *Basic Definitions for Theory Reasoning*

The following are the basic definitions for theory reasoning:

DEFINITION 22. Let $\Phi \subset Form_\Sigma$ be a finite set of formulae, called *key*. A finite set $R = \{\rho_1, \ldots, \rho_k\} \subset Form_\Sigma$ of formulae ($k \geq 0$) is a $\mathcal{T}$*-residue* of Φ if there is a substitution $\sigma \in Subst^*_\Sigma$ such that

1. $\Phi\sigma \models^\circ_\mathcal{T} \rho_1 \vee \ldots \vee \rho_k$ (in case R is empty: $\Phi\sigma \models^\circ_\mathcal{T}$ *false*);

2. $R = R\sigma$.

Then the pair $\langle\sigma, R\rangle$ is called a $\mathcal{T}$*-refuter* for Φ. If the residue R is empty, the substitution σ is called a $\mathcal{T}$-refuter for Φ (it is identified with $\langle\sigma, \emptyset\rangle$).

EXAMPLE 23. Table 1 shows some examples for $\mathcal{E}$-refuters.

DEFINITION 24. A set $\Phi \subset Form_\Sigma$ of formulae is $\mathcal{T}$*-complementary* if, for all $\mathcal{T}$-structures $\langle D, I\rangle$ and all variable assignments ν, $val_{I,\nu}(\Phi) = false$.

EXAMPLE 25. The set $\{\neg(x \approx y)\}$ is $\mathcal{E}$-unsatisfiable; it is, however, not $\mathcal{E}$-complementary because a variable assignment may assign different elements of the domain to x and y. The set $\{\neg(x \approx x)\}$ is both $\mathcal{E}$-unsatisfiable and $\mathcal{E}$-complementary.

In general, it is undecidable whether a formula set is $\mathcal{T}$-complementary; and, consequently, it is undecidable whether a pair $\langle\sigma, R\rangle$ is a refuter for a key Φ.

$\mathcal{T}$-complementarity generalizes the usual notion that formulae ϕ and $\neg\phi$ are complementary. The following lemmata are immediate consequences of the definitions:

LEMMA 26. *Given a theory $\mathcal{T}$, a substitution $\phi \in Subst^*_\Sigma$ is a $\mathcal{T}$-refuter for a set Φ of formulae if and only if the set $\Phi\sigma$ is $\mathcal{T}$-complementary.*

LEMMA 27. *Given a theory $\mathcal{T}$, a substitution σ and a set $R = \{\rho_1, \ldots, \rho_k\}$, $k \geq 0$, of formulae form a refuter $\langle\sigma, R\rangle$ for a set Φ of formulae if and only if*

1. *$\Phi\sigma \cup \{\neg\rho_1, \ldots, \neg\rho_k\}$ is $\mathcal{T}$-complementary;*
2. *$R = R\sigma$.*

There is an alternative characterization of $\mathcal{T}$-complementary sets that do not contain bound variables (e.g., sets of literals or clauses):

THEOREM 28. *Given a theory $\mathcal{T}$, a set Φ of formulae that does not contain any quantifiers is $\mathcal{T}$-complementary if and only if the existential closure $\exists\Phi$ of Φ is $\mathcal{T}$-unsatisfiable.*

Provided that the signature Σ contains enough function symbols not occurring in a universal theory $\mathcal{T}$, a quantifier-free formula set is $\mathcal{T}$-complementary if all its instances are $\mathcal{T}$-complementary:

THEOREM 29. *Given a universal theory $\mathcal{T}$ such that there are infinitely many function symbols of each arity $n \geq 0$ in F_Σ that do not occur in $\mathcal{T}$, then a set Φ of formulae that does not contain any bound variables is $\mathcal{T}$-complementary if and only if all ground instances of Φ are $\mathcal{T}$-unsatisfiable.*

EXAMPLE 30. Let $\mathcal{T}$ be the theory $\{p(t) \mid t \in Term^0_\Sigma\}$ that violates the pre-condition of Theorem 29, as all function symbols occur in $\mathcal{T}$. The formula $\neg p(x)$ is *not* $\mathcal{T}$-complementary because there may be elements in the domain of a $\mathcal{T}$-structure that are not represented by any ground term. Nevertheless, all instances of $\neg p(x)$ are $\mathcal{T}$-unsatisfiable, which shows that the pre-condition of Theorem 29 is indispensable.

By definition there is no restriction on what formulae may occur in keys or refuters. In practice, however, to restrict the search space, background reasoners do not compute refuters for all kinds of keys, and they do not compute all possible refuters (typically, keys are restricted to be sets of literals or universally quantified literals). To model this, we define background reasoners to be partial functions on the set of all possible keys:

DEFINITION 31. Let $\mathcal{T}$ be a theory; a *background reasoner* for $\mathcal{T}$ is a partial function

$$\mathcal{R} : 2^{Form_\Sigma} \longrightarrow Subst^*_\Sigma \times 2^{Form_\Sigma}$$

such that, for all keys $\Phi \subset Form_\Sigma$ for which $\mathcal{R}$ is defined, $\mathcal{R}(\Phi)$ is a set of $\mathcal{T}$-refuters for Φ.

A background reasoner $\mathcal{R}$ is *total* if, for all keys Φ for which $\mathcal{R}$ is defined, the residues of all refuters in $\mathcal{R}(\Phi)$ are empty, i.e. $\mathcal{R}(\Phi) \subset Subst^*_\Sigma$.

A background reasoner $\mathcal{R}$ is *monotonic* if, for all keys Φ and Ψ such that $\Phi \subset \Psi$: if $\mathcal{R}(\Phi)$ is defined, then $\mathcal{R}(\Psi)$ is defined and $\mathcal{R}(\Phi) \subset \mathcal{R}(\Psi)$.

EXAMPLE 32. A background reasoner for the theory $\mathcal{PO}$ of partial orderings can be defined as follows: For all keys Φ, let $\mathcal{R}(\Phi)$ be the smallest set such that:

1. for all terms $t, t', t'' \in Term_\Sigma$:
   if $t < t', t' < t'' \in \Phi$, then $\langle id, t < t'' \rangle \in \mathcal{R}(\Phi)$;
2. for all terms $t \in Term_\Sigma$: if $t < t \in \Phi$, then $id \in \mathcal{R}(\Phi)$;
3. for all literals $\phi \in Lit_\Sigma$: if $\phi, \neg\phi \in \Phi$, then $id \in \mathcal{R}(\Phi)$.

The combination of $\mathcal{R}$ and the ground version of tableaux leads to a complete calculus for $\mathcal{PO}$ (see Section 3.2).

A background reasoner has to compute refuters that are strong consequences of (an instance of) the key. In contrary to that, for tableau rules it is sufficient to preserve satisfiability. A tableau rule may deduce $p(c)$ from $(\exists x)p(x)$ where c is new, but $\langle id, \{p(c)\}\rangle$ is not a refuter for the key $\{(\exists x)p(x)\}$. A background reasoner may, however, do the opposite: $\langle id, \{(\exists x)p(x)\}\rangle$ is a refuter for the key $\{p(c)\}$ (this deduction usually does not help in finding a proof; see, however, Example 20).

2.5 *Total and Partial Theory Reasoning*

The central idea behind theory reasoning is the same for all calculi based in some way on Herbrand's theorem (tableau-like calculi, resolution, etc.): A key $\Phi \subset \Psi$ is chosen from the set Ψ of formulae already derived by the foreground reasoner and is passed to the background reasoner, which computes refuters $\langle \sigma, R\rangle$ for Φ.

There are two main approaches: if the background reasoner is total, i.e. only computes refuters with an empty residue R, we speak of *total* theory reasoning else of *partial* theory reasoning.

In the case of partial reasoning, where the residue $R = \{\rho_1, \ldots, \rho_k\}$ is not empty ($k \geq 1$), the formula $\rho_1 \vee \ldots \vee \rho_k$ is added to the set Ψ of derived formulae and the substitution σ is applied. If the foreground reasoner is then able to show that for some substitution τ the set $(\Psi\sigma \cup \{\rho_1 \vee \ldots \vee \rho_k\})\tau$ is $\mathcal{T}$-unsatisfiable, this proves that $\Psi\sigma\tau$ is $\mathcal{T}$-unsatisfiable.

Although total theory reasoning can be seen as a special case of partial theory reasoning, the way the foreground reasoner makes use of the refuter is quite different: no further derivations have to be made by the foreground reasoner; $\Phi\sigma$ and thus $\Psi\sigma$ have been proven to be $\mathcal{T}$-complementary by the background reasoner. In the tableau framework, where (usually) the key Φ is taken from a tableau branch B, this means that B is closed if the substitution σ is applied.

On the one hand, for total theory reasoning, more complex methods have to be employed to find refuters; the background reasoner has to make more complex deductions that, using partial reasoning, could be divided into several expansion steps followed by a simple closure step. On the other hand, the restriction to total theory reasoning leads to a much smaller search space for the foreground reasoner, because there are less refuters for each key and the search is more goal-directed.

2.6 Other Classifications of Theory Reasoning

Besides total and partial theory reasoning, there are several other ways to distinguish different types of background reasoners.

One possibility is to classify according to the information given to the background reasoner: (complex) formulae, literals, or terms [Baumgartner *et al.*, 1992]. Stickel distinguishes *narrow* theory reasoning, where all keys consist of literals, and *wide* theory reasoning, where keys consist of clauses [Stickel, 1985]. This type of classification is not used here, since all these types are subsumed by formula level theory reasoning. We will, however, restrict (nearly) all considerations to keys consisting of literals.

Another possibility is to classify background reasoners according to the type of calculus they use for deductions, the main divisions being *bottom up* and *top down* reasoning.

Local and *non-local* theory reasoning can be distinguished according to the effect that calling the background reasoner has on the tableau [Degtyarev and Voronkov, 1996]. In particular, the effect of calling the background reasoner is local if only local variables are instantiated by applying the theory expansion or closure rule to a tableau branch B, i.e. no variables occurring on other branches than B are instantiated.

3 THEORY REASONING FOR SEMANTIC TABLEAUX

3.1 Unifying Notation

Following Smullyan [Smullyan, 1995], the set of formulae that are not literals is divided into four classes: α for formulae of conjunctive type, β for formulae of disjunctive type, γ for quantified formulae of universal type, and δ for quantified formulae of existential type (unifying notation). This classification is motivated by the *tableau expansion rules* which are associated with each (non-literal) formula.

DEFINITION 33. The non-literal formulae in $Form_\Sigma$ are assigned a *type* according to Table 2. A formula of type $\xi \in \{\alpha, \beta, \gamma, \delta\}$ is called a *ξ-formula.*

NOTATION 34. The letters α, β, γ, and δ are used to denote formulae of (and only of) the appropriate type. In the case of γ- and δ-formulae the variable x bound by the (top-most) quantifier is made explicit by writing $\gamma(x)$ and $\gamma_1(x)$ (resp. $\delta(x)$ and $\delta_1(x)$); accordingly $\gamma_1(t)$ denotes the result of replacing all occurrences of x in γ_1 by t.

3.2 The Ground Version of Semantic Tableaux

We first present the classical *ground* version of tableaux for first-order logic. This version of tableaux is called 'ground', because universally quantified variables are replaced by *ground* terms when the γ-rule is applied.

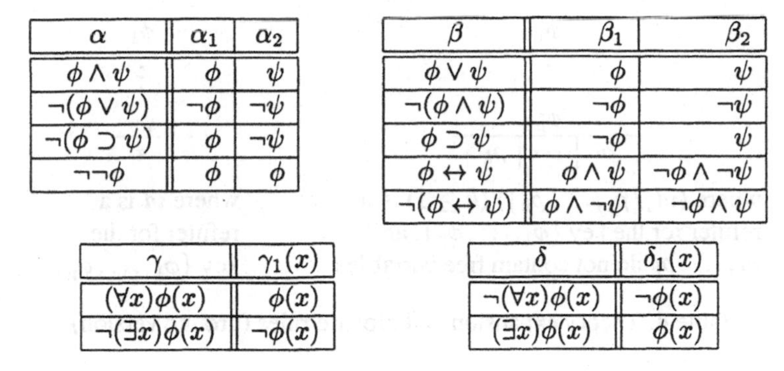

α	α_1	α_2
$\phi \wedge \psi$	ϕ	ψ
$\neg(\phi \vee \psi)$	$\neg\phi$	$\neg\psi$
$\neg(\phi \supset \psi)$	ϕ	$\neg\psi$
$\neg\neg\phi$	ϕ	ϕ

β	β_1	β_2
$\phi \vee \psi$	ϕ	ψ
$\neg(\phi \wedge \psi)$	$\neg\phi$	$\neg\psi$
$\phi \supset \psi$	$\neg\phi$	ψ
$\phi \leftrightarrow \psi$	$\phi \wedge \psi$	$\neg\phi \wedge \neg\psi$
$\neg(\phi \leftrightarrow \psi)$	$\phi \wedge \neg\psi$	$\neg\phi \wedge \psi$

γ	$\gamma_1(x)$
$(\forall x)\phi(x)$	$\phi(x)$
$\neg(\exists x)\phi(x)$	$\neg\phi(x)$

δ	$\delta_1(x)$
$\neg(\forall x)\phi(x)$	$\neg\phi(x)$
$(\exists x)\phi(x)$	$\phi(x)$

Table 2. Correspondence between formulae and rule types

$$\frac{\alpha}{\begin{matrix}\alpha_1\\ \alpha_2\end{matrix}} \qquad \frac{\beta}{\beta_1 \mid \beta_2} \qquad \frac{\gamma}{\gamma_1(t)} \qquad \frac{\delta}{\delta_1(t)}$$

where t is any ground term. (for γ)
where t is a ground term new to the tableau. (for δ)

Table 3. Rule schemata for the ground version of tableaux

The calculus is defined using a slightly non-standard representation of tableaux: a tableau is a multi-set of branches, which are multi-sets of first-order formulae; as usual, the branches of a tableau are implicitly disjunctively connected and the formulae on a branch are implicitly conjunctively connected. In graphical representations, tableaux are shown in their classical tree form.

DEFINITION 35. A *tableau* is a (finite) multi-set of tableau branches, where a tableau *branch* is a (finite) multi-set of first-order formulae.

In Table 3 the ground expansion rule schemata for the various formula types are given schematically. Premisses and conclusions are separated by a horizontal bar, while vertical bars in the conclusion denote different *extensions*. The formulae in an extension are implicitly conjunctively connected, and different extensions are implicitly disjunctively connected.

To prove a sentence ϕ to be a tautology, we apply the expansion rules starting from the initial tableau $\{\{\neg\phi\}\}$. A tableau T is expanded by choosing a branch B of T and a formula $\phi \in B$ and replacing B by as many updated branches as the rule corresponding to ϕ has extensions. Closed branches are removed from the tableau instead of just marking them as being closed; thus, a proof is found when the empty tableau has been derived.

There is a theory expansion and a theory closure rule. For both rules, a key $\Phi \subset B$ is chosen from a branch B, and a refuter $\langle\sigma, R\rangle$ for Φ is computed. Since formulae in ground tableaux do not contain free variables, the formulae in the residue, too, have to be sentences. The application of substitutions to formulae

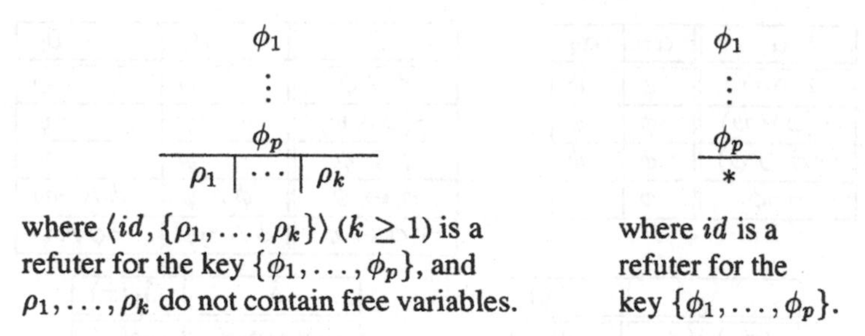

where $\langle id, \{\rho_1, \ldots, \rho_k\}\rangle$ $(k \geq 1)$ is a refuter for the key $\{\phi_1, \ldots, \phi_p\}$, and $\rho_1, \ldots, \rho_k$ do not contain free variables.

where id is a refuter for the key $\{\phi_1, \ldots, \phi_p\}$.

Table 4. Theory expansion and closure rules (ground version)

without free variables does not have any effect. Thus, if $\langle\sigma, R\rangle$ is a refuter for a key Φ taken from a ground tableau, then $\langle id, R\rangle$ is a refuter for Φ as well; thus, it is possible to use only refuters of this form for ground tableaux.

DEFINITION 36. A background reasoner $\mathcal{R}$ for a theory $\mathcal{T}$ is a *ground background reasoner* for $\mathcal{T}$ if, for all keys $\Phi \subset Form_\Sigma$ for which $\mathcal{R}$ is defined, all formulae in $\mathcal{R}(\Phi)$ are sentences, i.e. do not contain free variables.

Whether an expansion or a closure rule is to be applied depends on whether the residue $R = \{\rho_1, \ldots, \rho_k\}$ is empty or not. If $k \geq 1$, then the tableau is expanded. The old branch is replaced by k new branches, one for each ρ_i (since the ρ_i are implicitly disjunctively connected). The closure rule is applied if the residue is empty ($k = 0$); it can be seen as a special case of the expansion rule: the old branch is replaced by 0 new branches, i.e. it is removed. The rule schemata are shown in Table 4; in this and all following schematic representations of rules, the symbol $*$ is used to denote that a branch is closed if the rule is applied.

If the residue R is empty, the key is $\mathcal{T}$-complementary and the branch it has been taken from is $\mathcal{T}$-closed. This is a straightforward extension of the closure rule for tableaux without theory reasoning, where a branch is closed if it contains complementary formulae ϕ and $\neg\phi$, i.e. the $\emptyset$-complementary key $\{\phi, \neg\phi\}$ (therefore, the rule that a branch containing complementary formulae ϕ and $\neg\phi$ is closed does not have to be considered separately).

DEFINITION 37 (Ground tableau proof.). Given a theory $\mathcal{T}$ and a ground background reasoner $\mathcal{R}$ for $\mathcal{T}$ (Definition 36), a *ground tableau proof* for a first-order sentence $\phi \in Form_\Sigma$ consists of a sequence

$$\{\{\neg\phi\}\} = T_0, T_1, \ldots, T_{n-1}, T_n = \emptyset \qquad (n \geq 0)$$

of tableaux such that, for $1 \leq i \leq n$, the tableau T_i is constructed from T_{i-1}

1. by applying one of the expansion rules for ground tableaux from Table 3, i.e. there is a branch $B \in T_{i-1}$ and a formula $\phi \in B$ (that is not a literal)

$$\frac{\begin{array}{c} t < t' \\ t' < t'' \end{array}}{t < t''} \qquad \frac{t < t}{*} \qquad \frac{\begin{array}{c} \phi \\ \neg\phi \end{array}}{*}$$

Table 5. Expansion and closure rules for the theory $\mathcal{PO}$ of partial orderings

such that

$$T_i = (T_{i-1} \setminus \{B\}) \cup \begin{cases} \{(B \setminus \{\alpha\}) \cup \{\alpha_1, \alpha_2\}\} & \text{if } \phi = \alpha \\ \{(B \setminus \{\beta\}) \cup \{\beta_1\}, (B \setminus \{\beta\}) \cup \{\beta_2\}\} & \text{if } \phi = \beta \\ \{B \cup \{\gamma_1(s)\}\} & \text{if } \phi = \gamma(x) \\ \{(B \setminus \{\delta(x)\}) \cup \{\delta_1(t)\}\} & \text{if } \phi = \delta(x) \end{cases}$$

where $s \in Term^0_\Sigma$ is any ground term, and $t \in Term^0_\Sigma$ is a ground term not occurring in T_{i-1} nor in $\mathcal{T}$;

2. by applying the ground theory expansion rule, i.e. there is a branch $B \in T_{i-1}$ and, for a key $\Phi \subset B$, there is a $\mathcal{T}$-refuter $\langle id, \{\rho_1, \ldots, \rho_k\}\rangle$ $(k \geq 1)$ in $\mathcal{R}(\Phi)$ such that

$$T_i = (T_{i-1} \setminus \{B\}) \cup \{B \cup \{\rho_1\}, \ldots, B \cup \{\rho_k\}\} ;$$

3. or by closing a branch $B \in T_{i-1}$, i.e. $T_i = T_{i-1} \setminus \{B\}$ where B is $\mathcal{T}$-closed (i.e. $id \in \mathcal{R}(\Phi)$ for a key $\Phi \subset B$).

It is possible to describe a background reasoner using tableau rule schemata. The reasoner from Example 32 then takes the form that is shown in Table 5.

Even without theory reasoning, the construction of a closed tableau is a highly non-deterministic process, because at each step one is free to choose a branch B of the tableau and a formula $\phi \in B$ for expansion. If ϕ is a γ-formula, in addition, a term has to be chosen that is substituted for the bound variable.

There are two ways for resolving the non-determinism (actual implementations usually employ a combination of both): (1) fair strategies can be used such that, for example, each formula will finally be used to expand each branch on which it occurs. (2) Backtracking can be used; if a branch cannot be closed (observing a limit on its length), other possibilities are tried; for example, other terms are used in γ-rule applications. If no proof is found, the limit has to be increased (iterative deepening).

The theory expansion rule makes things even worse, because it is highly non-deterministic. In which way it has to be applied to be of any use, in particular when and how often the rule is applied, and which types of keys and refuters are used depends on the particular theory and is part of the domain knowledge (see Section 5.3).

$\dfrac{\alpha}{\alpha_1 \atop \alpha_2}$	$\dfrac{\beta}{\beta_1 \mid \beta_2}$	$\dfrac{\gamma}{\gamma_1(y)}$	$\dfrac{\delta}{\delta_1(f(x_1, \ldots, x_n))}$
		where y is a free variable.	where f is a new Skolem function symbol, and $x_1, \ldots, x_n$ are the free variables in δ.

Table 6. Tableau expansion rule schemata for free variable tableau

where $\langle\sigma, \{\rho_1, \ldots, \rho_k\}\rangle$ ($k \geq 1$) is a refuter for the key $\{\phi_1, \ldots, \phi_p\}$, and σ is applied to the whole tableau.

where σ is a refuter for the key $\{\phi_1, \ldots, \phi_p\}$, and σ is applied to the whole tableau.

Table 7. Theory expansion and closure rules (free variable version)

3.3 Free Variable Semantic Tableaux

Using free variable quantifier rules is crucial for efficient implementation—even more so if a theory has to be handled. They reduce the number of possibilities to proceed at each step in the construction of a tableau proof and thus the size of the search space. When γ-rules are applied, a new free variable is substituted for the quantified variable instead of replacing it by a ground term, which has to be 'guessed'. Free variables can later be instantiated 'on demand' when a tableau branch is closed or the theory expansion rule is applied to expand a branch.

To preserve correctness, the schema for δ-rules has to be changed as well: the Skolem terms introduced now contain the free variables occurring in the δ-formula (the free variable rule schemata are shown in Table 6).

Again, there is both a theory expansion and a theory closure rule. The difference to the ground version is that now there are free variables both in the tableau and in the refuter (the formulae that are added). When theory reasoning is used for expansion or for closing, the substitution σ of a refuter $\langle\sigma, R\rangle$ has to be applied to the whole tableau; the theory rule schemata are shown in Table 7.

In case there is a refuter σ with an empty residue for a key taken from a branch, that branch is $\mathcal{T}$-closed under the substitution σ, i.e. it is closed when σ is applied to the whole tableau.

It is often difficult to find a substitution σ that instantiates the variables in a tableau T such that *all* branches of T are $\mathcal{T}$-closed. The problem is simplified (as is usually done in practice) by closing the branches of T one after the other: if a substitution is found that closes a single branch B, it is applied (to the whole tableau) to close B before other branches are handled. This is not a restriction because, if a substitution is known to $\mathcal{T}$-close several branches, it can be applied to close one of them; after that the other branches are closed under the empty substitution.

DEFINITION 38 (Free variable tableau proof.). Let $\mathcal{T}$ be a theory and let $\mathcal{R}$ be a background reasoner for $\mathcal{T}$, a *free variable tableau proof* for a first-order sentence ϕ consists of a sequence

$$\{\{\neg\phi\}\} = T_0, T_1, \ldots, T_{n-1}, T_n = \emptyset \qquad (n \geq 0)$$

of tableaux such that, for $1 \leq i \leq n$, the tableau T_i is constructed from T_{i-1}

1. by applying one of the free variable expansion rules from Table 6, that is, there is a branch $B \in T_{i-1}$ and a formula $\phi \in B$ (that is not a literal) such that

$$\begin{array}{rcl} T_i & = & (T_{i-1} \setminus \{B\}) \\ & & \cup \begin{cases} \{(B \setminus \{\alpha\}) \cup \{\alpha_1, \alpha_2\}\} & \text{if } \phi = \alpha \\ \{(B \setminus \{\beta\}) \cup \{\beta_1\}, (B \setminus \{\beta\}) \cup \{\beta_2\}\} & \text{if } \phi = \beta \\ \{B \cup \{\gamma_1(y)\}\} & \text{if } \phi = \gamma(x) \\ \{(B \setminus \{\delta(x)\}) \cup \{\delta_1(f(x_1, \ldots, x_n))\}\} & \text{if } \phi = \delta(x) \end{cases} \end{array}$$

   where $y \in V$ is a new variable not occurring in T_{i-1} nor in $\mathcal{T}$, $f \in F_\Sigma$ is a Skolem function symbol not occurring in T_{i-1}, and $x_1, \ldots, x_n$ are the free variables in ϕ;

2. by applying the free variable theory expansion rule, that is, there is a branch $B \in T_{i-1}$, a key $\Phi \subset B$, and a $\mathcal{T}$-refuter $\langle\sigma, \{\rho_1, \ldots, \rho_k\}\rangle$ $(k \geq 1)$ in $\mathcal{R}(\Phi)$ such that

$$T_i = (T_{i-1} \setminus \{B\})\sigma \cup \{B\sigma \cup \{\rho_1\}, \ldots, B\sigma \cup \{\rho_k\}\} ;$$

3. or by closing a branch $B \in T_{i-1}$, that is, $T_i = (T_{i-1} \setminus \{B\})\sigma$ where the branch B is $\mathcal{T}$-closed under σ, i.e. $\sigma \in \mathcal{R}(\Phi)$ for a key $\Phi \subset B$.

3.4 Semantic Tableaux with Universal Formulae

Free variable semantic tableaux can be further improved by using the concept of universal formulae [Beckert and Hähnle, 1992]: γ-formulae (in particular axioms that extend the theory) have often to be used multiply in a tableau proof, with different instantiations for the free variables they contain. An example is the axiom $(\forall x)(\forall y)((x < y) \supset (p(x) \supset p(y)))$, that extends the theory of partial orderings by defining the predicate symbol p to be monotonous. The associativity axiom $(\forall x)(\forall y)(\forall z)((x \cdot y) \cdot z \approx x \cdot (y \cdot z))$ is another typical example. Usually, it has

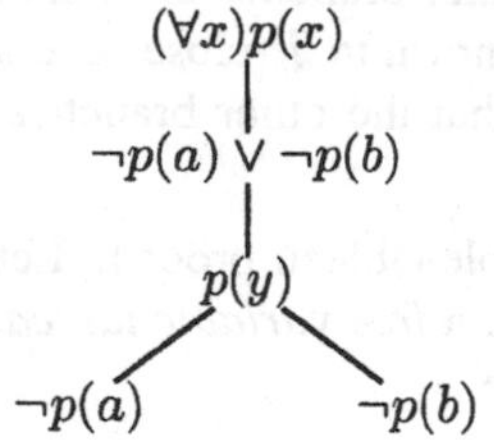

$(\forall x)p(x)$

$\neg p(a) \vee \neg p(b)$

$p(y)$

$\neg p(a)$ $\neg p(b)$

Figure 1. The advantage of using universal formulae

to be applied several times with different substitutions for x, y and z to prove even very simple theorems from, for example, group theory. In semantic tableaux, the γ-rule has to be applied repeatedly to generate several instances of the axiom each with different free variables substituted for x, y and z. Free variables in tableaux are *not* implicitly universally quantified (as it is, for instance, the case with variables in clauses when using a resolution calculus) but are *rigid*, which is the reason why a substitution must be applied to all occurrences of a free variable in the whole tableau.

Supposed a tableau branch B contains a formula $p(x)$, and the expansion of the tableau proceeds with creating new branches. Some of these branches contain occurrences of x; for closing the generated branches, the same substitution for x has to be used on all of them. Figure 1 gives an example for the situation: this tableau cannot be closed immediately as no single substitution closes both branches. To find a proof, the γ-rule has to be applied again to create another instance of $(\forall x)p(x)$.

In particular situations, a logical consequence of the formulae already on the tableau (in a sense made precise in Definition 39) may be that $(\forall y)p(y)$ can be added to B. This is trivially true in Figure 1. In such cases, different substitutions for y can be used without destroying soundness of the calculus. The tableau in Figure 1 then closes immediately. Recognizing such situations and exploiting them allows to use more general closing substitutions, yields shorter tableau proofs, and in most cases reduces the search space.

DEFINITION 39. Let $\mathcal{T}$ be a theory, and let ϕ be a formula on a branch B of a tableau T. Let T' be the tableau that results from adding $(\forall x)\phi$ to B for some $x \in V$. The formula ϕ is $\mathcal{T}$*-universal* on B with respect to x if $T \models_{\mathcal{T}} T'$, where T and T' are identified with the formulae that are the disjunctions of their branches, respectively, and a branch is the conjunction of the formulae it contains. Let $UVar(\phi) \subset V$ denote the universal variables of ϕ.[2]

[2] When the context is clear, a formula ϕ which is universal on a branch B w.r.t. a variable x is just referred to by 'the universal formula ϕ', and the variable x by 'the universal variable x'.

The above definition is an adaptation of the definition given in [Beckert and Hähnle, 1998] to theory reasoning.

The problem of recognizing universal formulae is of course undecidable in general. However, a wide and important class can be recognized quite easily (using this class can already shorten tableau proofs exponentially): If tableaux are seen as trees, a formula ϕ on a branch B of a tableau T is ($\mathcal{T}$-)universal w.r.t. x if all branches B' of T are closed which contain an occurrence of x that is not on B as well; this holds in particular if the branch B contains all occurrences of x in T.

Assume there is a sequence of tableau rule applications that introduces a variable x by a γ-rule application and does not contain a rule application distributing x over different subbranches; then the above criterion is obviously satisfied and all formulae that are generated by this sequence are universal w.r.t. x.

THEOREM 40. *Given a theory $\mathcal{T}$, a formula ϕ on a branch B of a tableau T is $\mathcal{T}$-universal w.r.t. x on B if in the construction of T the formula ϕ was added to B by either*

1. *applying a γ-rule and x is the free variable that was introduced by that application;*
2. *applying an α-, δ- or γ-rule to a formula ψ that is $\mathcal{T}$-universal w.r.t. x on B;*
3. *applying a β-rule to a formula ψ that is $\mathcal{T}$-universal w.r.t. x on B, and x does not occur in any formula except ϕ that has been added to the tableau by that β-rule application;*
4. *applying the theory expansion rule to a refuter $\{\rho_1, \ldots, \rho_k\}$ for a key $\Phi \subset B$ (i.e. $\phi = \rho_i$ for some $i \in \{1, \ldots, k\}$), and all formulae in Φ are $\mathcal{T}$-universal w.r.t. x on B, and x does not occur in any of the ρ_j for $j \neq i$.*

The knowledge that formulae are $\mathcal{T}$-universal w.r.t. variables they contain can be taken advantage of by universally quantifying the formulae in a key w.r.t. (some of) their universal variables. Similar to the ground and free variable cases, the closure rule can be seen as a special case of the expansion rule. The new theory rule schemata are shown in Table 8.

DEFINITION 41 (Universal formula version.). Let $\mathcal{T}$ be a theory, and let $\mathcal{R}$ be a background reasoner for $\mathcal{T}$. A *universal formula tableau proof* for a first-order sentence ϕ consists of a sequence

$$\{\{\neg\phi\}\} = T_0, T_1, \ldots, T_{n-1}, T_n = \emptyset \qquad (n \geq 0)$$

of tableaux such that, for $1 \leq i \leq n$, the tableau T_i is constructed from T_{i-1}

1. by applying one of the free variable expansion rules from Table 6 (see Definition 38 for a formal definition);

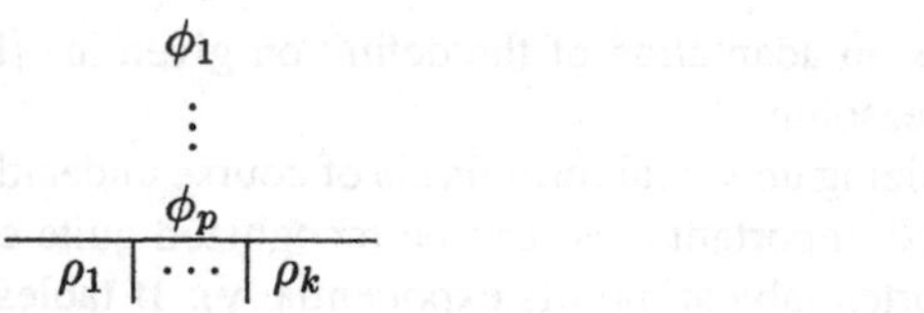

where $\langle\sigma, \{\rho_1, \ldots, \rho_k\}\rangle$ $(k \geq 1)$ is a refuter for the key $\{(\forall x_1^i) \cdots (\forall x_{m_i}^i)\phi_i \mid 1 \leq i \leq p\}$, the formula ϕ_i is $\mathcal{T}$-universal w.r.t. the variables $x_1^i, \ldots, x_{m_i}^i$; and σ is applied to the whole tableau.	where σ is a refuter for the key $\{(\forall x_1^i) \cdots (\forall x_{m_i}^i)\phi_i \mid 1 \leq i \leq p\}$, the formula ϕ_i is $\mathcal{T}$-universal w.r.t. the variables $x_1^i, \ldots, x_{m_i}^i$; and σ is applied to the whole tableau.

Table 8. Theory expansion and closure rules (universal formula version)

2. by applying the universal formula theory expansion rule to a branch $B \in T_{i-1}$, that is,

$$T_i = (T_{i-1} \setminus \{B\})\sigma \cup \{B\sigma \cup \{\rho_1\}, \ldots, B\sigma \cup \{\rho_k\}\}$$

where $\langle\sigma, \{\rho_1, \ldots, \rho_k\}\rangle$ is a $\mathcal{T}$-refuter in $\mathcal{R}(\Phi)$ for a key

$$\Phi = \{(\forall x_1^i) \cdots (\forall x_{m_i}^i)\phi_i \mid 1 \leq i \leq p\}$$

such that

(a) $\phi_1, \ldots, \phi_p \in B$,

(b) $\{x_1^i, \ldots, x_{m_i}^i\} \subset UVar(\phi_i)$ for $1 \leq i \leq p$,

3. or by closing a branch $B \in T_{i-1}$, that is,

$$T_i = (T_{i-1} \setminus \{B\})\sigma$$

where the branch B is $\mathcal{T}$-closed under σ, i.e. $\sigma \in \mathcal{R}(\Phi)$ for a key

$$\Phi = \{(\forall x_1^i) \cdots (\forall x_{m_i}^i)\phi_i \mid 1 \leq i \leq p\}$$

such that

(a) $\phi_1, \ldots, \phi_p \in B$,

(b) $\{x_1^i, \ldots, x_{m_i}^i\} \subset UVar(\phi_i)$ for $1 \leq i \leq p$.

Although it is easier to add theory reasoning to the ground version of tableau, to prove even simple theorems, free variable tableaux have to be used. These are sufficient for simple theories. If, however, finding a refuter requires complex deductions where the formulae both in the key and in the theory have to be used multiply with different instantiations, then universal formula tableaux have to be used (free variable tableaux are a special case of universal formula tableaux),

The following example illustrates the advantage of using the universal formula expansion rule as compared to the free variable rule:

EXAMPLE 42. Consider again the tableau T shown in Figure 1. The substitution $\sigma = \{y \mapsto a\}$ is a refute for the key $\{p(y), \neg p(a)\}$, which is taken from the left branch of T. If σ is used to close the left branch, then the variable y is instantiated with a in the whole tableau T. However, if the formula $p(y)$ is recognized to be universal w.r.t. y, then the key $\{(\forall y)p(y), \neg p(a)\}$ can be used instead, for which the empty substitution id is a refuter; thus using the universal formula closure rule, the left branch can be closed without instantiating y. Then the right branch can be closed, too, without generating a second free variable instance $p(y')$ of $(\forall x)p(x)$.

Using the universal formula technique is even more important in situations like the following:

EXAMPLE 43. Supposed the equality $f(x) \approx x$ and the literals $p(f(a), f(b))$ and $\neg p(a, b)$ are $\mathcal{E}$-universal w.r.t. x on a tableau branch. In that case, the key $\{(\forall x)(f(x) \approx x),\ p(f(a), f(b)),\ p(a, b)\}$ can be used, for which id is a refuter.

In free variable tableaux, the key $\{(f(x) \approx x),\ p(f(a), f(b)),\ p(a, b)\}$ has to be used, which allows to derive the refuters $\langle\{x \mapsto a\}, \{f(b) \approx b\}\rangle$ and $\langle\{x \mapsto b\}, \{f(a) \approx a\}\rangle$ only; a refuter with an empty residue, which closes the branch immediately, cannot be deduced anymore.

4 SOUNDNESS

In this section, soundness of semantic tableaux with theory reasoning is proven for the universal formula version. Soundness of the free variable version follows as a corollary because free variable tableaux are a special case of universal formula tableaux. For the ground version, soundness can be proven completely analogously.

First, satisfiability of tableaux is defined (Definition 44), then it is proven that satisfiability is preserved in a sequence of tableaux forming a tableau proof.

DEFINITION 44. Given a theory $\mathcal{T}$, a tableau T is *$\mathcal{T}$-satisfiable* if there is a $\mathcal{T}$-structure M such that, for every variable assignment ν, there is a branch $B \in T$ with

$$(M, \nu) \models B .$$

LEMMA 45. *If a tableau T is $\mathcal{T}$-satisfiable, then $T\sigma$ is $\mathcal{T}$-satisfiable for all substitutions $\sigma \in Subst^*_\Sigma$.*

Proof. By hypothesis there is a $\mathcal{T}$-structure $M = \langle D, I\rangle$ such that, for all variable assignments ν, there is a branch $B_\nu \in T$ with $(M, \nu) \models B_\nu$. We claim that, for the same structure M, we have also for all variable assignments ξ that there is a branch B with $(M, \xi) \models B\sigma$.

To prove the above claim, we consider a given variable assignment ξ. Let the variable assignment ν be defined by

$$\nu(x) = (x\sigma)^{I,\xi} \text{ for all } x \in V.$$

That implies for all terms $t \in Term_\Sigma$, and in particular for all terms t occurring in B_ν,

$$(t\sigma)^{I,\xi} = t^{I,\nu}$$

and therefore

$$val_{I,\xi}(B_\nu\sigma) = val_{I,\nu}(B_\nu) = \mathit{true}\ .$$

■

LEMMA 46. *Given a universal formula tableau proof $(T_j)_{0 \leq j \leq n}$, if the tableau T_i ($0 \leq i < n$) is $\mathcal{T}$-satisfiable, then the tableau T_{i+1} is $\mathcal{T}$-satisfiable as well.*

Proof. We use the notation from Definition 41. Let B be the branch in T_i to which one of the classical expansion rules or the theory expansion rule has been applied or that has been removed by applying the theory closure rule to derive the tableau T_{i+1}. Let $M = \langle D, I\rangle$ be a $\mathcal{T}$-structure satisfying T_i.

β-rule: Let ν be an arbitrary variable assignment. There has to be a branch B' in T_i such that $(M, \nu) \models B'$. If B' is different from B then $B' \in T_{i+1}$ and we are through.

If, on the other hand, $B' = B$, then $(M, \nu) \models B$. Let β be the formula in B to which the β-rule has been applied. By the property of β-formulae, $(M, \nu) \models \beta$ entails $(M, \nu) \models \beta_1$ or $(M, \nu) \models \beta_2$; and, therefore, $(M, \nu) \models (B \setminus \{\beta\}) \cup \{\beta_1\}$ or $(M, \nu) \models (B \setminus \{\beta\}) \cup \{\beta_2\}$. This concludes the proof for the case of a β-rule application, because $(B \setminus \{\beta\}) \cup \{\beta_1\}$ and $(B \setminus \{\beta\}) \cup \{\beta_2\}$ are branches in T_{i+1}.

α- and γ-rule: Similar to the β-rule.

δ-rule: Let δ be the δ-formula to which the δ-rule is applied to derive T_{i+1} from T_i; $\delta_1(f(x_1, \dots, x_m))$ is the formula added to the branch (f is a new Skolem function symbol and $x_1, \dots, x_m$ are the free variables in δ). We define a structure $M' = \langle D, I'\rangle$ that is identical to M, except that the new function symbol f is interpreted by I' in the following way: For every set $d_1, \dots, d_m$ of elements from the domain D, if there is an element d such that $(M, \xi) \models \delta_1(x)$ where $\xi(x_j) = d_j$ ($1 \leq j \leq m$) and $\xi(x) = d$, then $f^{I'}(d_1, \dots, d_m) = d$. If there are several such elements d, one of them may be chosen; and if there is no such element, an arbitrary element from the domain is chosen. It follows from this construction that for all variable assignments ν: if $(M, \nu) \models \delta$, then $(M', \nu) \models \delta_1(f(x_1, \dots, x_m))$. Since f does not occur in $\mathcal{T}$, M' is a $\mathcal{T}$-structure.

We proceed to show that M' satisfies T_{i+1}. Let ν be an arbitrary variable assignment. There has to be a branch B' in T_i with $(M, \nu) \models B'$. If B' is different from B, then we are done because $(M',\nu) \models B'$ (as f does not occur in B') and $B' \in T_{i+1}$.

In the interesting case where $\delta \in B' = B$, we have $(M,\nu) \models \delta$ which entails $(M',\nu) \models \delta_1(f(x_1,\ldots,x_m))$. Thus, $(B \setminus \{\delta\}) \cup \{\delta_1(f(x_1,\ldots,x_m))\}$, which is a branch in T_{i+1}, is satisfied by M'.

Theory Expansion Rule: Let $\langle\sigma, \{\rho_1,\ldots,\rho_k\}\rangle$ be the refuter used to expand the tableau. Since T_i is $\mathcal{T}$-satisfiable, the tableau $T_i\sigma$ is $\mathcal{T}$-satisfiable as well (Lemma 45). Let M be a $\mathcal{T}$-structure satisfying $T_i\sigma$, and let ν be an arbitrary variable assignment. There has to be a branch $B' \in T_i\sigma$ with $(M,\nu) \models B'$. Again, the only interesting case is where $B' = B\sigma$, and $B\sigma$ is the only branch in $T_i\sigma$ satisfied by (M,ν).

By definition of universal formulae, that implies $B \models^{\circ}_{\mathcal{T}} (\forall x_1^j)\cdots(\forall x_{m_j}^j)\,\phi_j$ $(1 \leq j \leq p)$ and, thus, $B\sigma \models^{\circ}_{\mathcal{T}} (\forall x_1^j)\cdots(\forall x_{m_j}^j)\,\phi_j\sigma$ (Lemma 9), which implies $(M,\nu) \models \Phi\sigma$ where $\Phi = \{(\forall x_1^j)\cdots(\forall x_{m_j}^j)\,\phi_j \mid 1 \leq j \leq p\}$, as $(M,\nu) \models B\sigma$. Because $\langle\sigma, \{\rho_1,\ldots,\rho_k\}\rangle$ is a refuter for Φ and, thus, $\Phi\sigma \models^{\circ}_{\mathcal{T}} \rho_1 \vee \ldots \vee \rho_k$, we have $(M,\nu) \models \rho_j$ for some $j \in \{1,\ldots,k\}$. This, finally, implies that M satisfies the branch $B\sigma \cup \{\rho_j\}$ in T_{i+1}.

Theory Closure Rule: In the same way as in the case of the theory expansion rule, we conclude that $(M,\nu) \models \Phi\sigma$. But now this leads to a contradiction: because the residue is empty, $val_{I,\nu}(\Phi\sigma) = \mathit{false}$ by definition. Thus the assumption that $B' = B\sigma$ has to be wrong, and the branch $B\sigma$ can be removed from the tableau. ■

Based on this lemma, soundness of semantic tableaux with theory reasoning can easily be proven:

THEOREM 47. *If there is a universal formula tableau proof*

$$\{\{\neg\phi\}\} = T_0, T_1, \ldots, T_{n-1}, T_n = \emptyset$$

for a sentence $\phi \in Form_\Sigma$ (Definition 41), then ϕ is a $\mathcal{T}$-tautology.

Proof. None of the tableaux in the sequence which the tableau proof consists of can be $\mathcal{T}$-satisfiable, otherwise the empty tableau $T_n = \emptyset$ had to be $\mathcal{T}$-satisfiable as well (according to Lemma 46); this, however, is impossible because the empty tableau has no branches.

Thus, the first tableau $\{\{\neg\phi\}\}$ in the sequence is $\mathcal{T}$-unsatisfiable, i.e. $\neg\phi$ is $\mathcal{T}$-unsatisfiable, which is equivalent to ϕ being a $\mathcal{T}$-tautology. ■

5 COMPLETENESS

5.1 Complete Background Reasoners

The most important feature of a background reasoner is completeness—besides soundness which is part of the definition of background reasoners. We define a background reasoner to be complete if its combination with the foreground reasoner leads to a complete calculus.

DEFINITION 48. A (ground) background reasoner for a theory $\mathcal{T}$ is

- a *complete ground* background reasoner for $\mathcal{T}$ if, for every $\mathcal{T}$-tautology ϕ, a ground tableau proof (Definition 37) can be built using $\mathcal{R}$.
- a *complete free variable* background reasoner for $\mathcal{T}$ if, for every $\mathcal{T}$-tautology ϕ, a free variable tableau proof (Definition 38) can be built using $\mathcal{R}$.
- a *complete universal formula* background reasoner for $\mathcal{T}$ if, for every $\mathcal{T}$-tautology ϕ, a universal formula tableau proof (Definition 41) can be built using $\mathcal{R}$.

Because free variable tableaux are a special case of universal formula tableaux, a complete universal variable background reasoner has to be a complete free variable background reasoner as well.

The existence of complete background reasoners is trivial, because an oracle-like background reasoner that detects all kinds of inconsistencies (and thus does all the work) is complete for all versions of tableau:

THEOREM 49. *Let $\mathcal{T}$ be a theory. If a background reasoner $\mathcal{R}$ satisfies the condition:*

$$id \in \mathcal{R}(\{\phi\})$$

for all $\mathcal{T}$-unsatisfiable sentences $\phi \in Form_\Sigma$, then $\mathcal{R}$ is a complete ground, free variable, and universal variable background reasoner.

Proof. If ϕ is a $\mathcal{T}$-tautology, then its negation $\neg\phi$ is $\mathcal{T}$-unsatisfiable and thus $id \in \mathcal{R}(\{\neg\phi\})$. By applying the theory closure rule using this refuter, the empty tableau $T_1 = \emptyset$ can be derived from the initial tableau $T_0 = \{\{\neg\phi\}\}$. ■

This completeness result is only of theoretical value. In practice, theory reasoners are needed that, on the one hand, lead to short tableau proofs and, on the other hand, can be computed easily (i.e. fast and at low cost). Of course, there is a trade off between these two goals.

There is a complete background reasoner for a theory $\mathcal{T}$ such that $\mathcal{R}(\Phi)$ is enumerable for all keys Φ if and only if the theory $\mathcal{T}$ is axioamatizable. Thus, it is not possible to implement a complete background reasoner for a non-axiomatizable theory.

5.2 *Completeness Criteria*

General completeness criteria that work for all theories such as 'a background reasoner that computes all existing refuters is complete' are not useful in practice. For many theories, and in particular for equality, highly specialized background reasoners have to be used to build an efficient prover. These exploit domain knowledge to restrict the number of refuters (and thus the search space); domain knowledge has to be used to prove such background reasoners to be complete.

Therefore, the completeness criteria presented in the following refer to the semantics of the particular theory, and there is no uniform way to prove that a background reasoner satisfies such a criterion. Nevertheless, the criteria give some insight in what has to be proven to show completeness of a background reasoner.

Fairness of the Foreground Reasoner

First, a characterization of fair tableau construction rules (i.e. fairness of the foreground reasoner) is given for the ground version. This notion is used in the proof that a background reasoner that satisfies a completeness criterion can be combined with a fair foreground reasoner to form a complete calculus.

For the multi-set representation of tableaux, the notion of fair tableau construction is somewhat more difficult to formalize than for the tree representation, but this has no effect on which construction rules are fair.

The condition for α-, β-, and δ-formulae is that the respective tableau rule is applied sooner or later. The γ-rule has to be applied to each γ-formula infinitely often, and—this is specific for the ground version—each ground term has to be used for one of these applications. The background reasoner has to be called with all keys for which it is defined and all refuters have to be used sooner or later.

DEFINITION 50. Given a ground background reasoner $\mathcal{R}$, a *ground tableau construction rule* for $\mathcal{R}$ is a rule that, when supplied with a formula ϕ, deterministically specifies in which way a sequence $(T_i)_{i\geq 0}$ of ground tableaux starting from $T_0 = \{\{\neg\phi\}\}$ is to be constructed. The rule is *fair* if for all Φ:

1. If there is a branch B that occurs in all tableaux T_i, $i > n$ for some $n \geq 0$, then B is exhausted, i.e. no expansion rule or closure rule can be applied to B.

2. For all infinite sequences $(B_i)_{i\geq 0}$ of branches such that $B_i \in T_i$ and either $B_{i+1} = B_i$ or the tableau T_{i+1} has been constructed from T_i by applying an expansion rule to B_i and B_{i+1} is one of the resulting new branches in T_{i+1} ($i \geq 0$):

    (a) for all α-, β-, and δ-formulae $\phi \in B_i$ ($i \geq 0$), there is a $j \geq i$ such that the tableau T_{j+1} has been constructed from T_j by applying the α-, β-, or δ-rule to $\phi \in B_j$.

    (b) for all γ-formulae $\phi \in B_i$ ($i \geq 0$) and all terms $t \in Term^0_\Sigma$, there is a $j \geq 0$ such that the tableau T_{j+1} has been constructed from T_j by applying the γ-rule to $\phi \in B_j$ and t is the ground term that has been substituted for the universally quantified variable in ϕ.

(c) for all keys $\Phi \subset B_i$ ($i \geq 0$) such that $\mathcal{R}(\Phi)$ is defined and all refuters $\langle id, R\rangle \in \mathcal{R}(\Phi)$, there is a $j \geq 0$ such that the tableau T_{j+1} has been constructed from T_j by applying the theory expansion or the theory closure rule to B_j using the key $\Phi \subset B_j$ and the refuter $\langle id, R\rangle$ (if the background reasoner is monotonic, a key $\Phi' \supset \Phi$ may be used as well).

A Completeness Criterion for Ground Background Reasoners

The criterion we are going to prove is that a background reasoner is complete if, for all $\mathcal{T}$-unsatisfiable downward saturated sets Ξ, it can either derive a residue consisting of new formulae that are not yet in Ξ, or it can detect the $\mathcal{T}$-unsatisfiability of Ξ.

DEFINITION 51. A set $\Phi \subset Form_\Sigma$ is downward saturated if the following conditions hold for all formulae $\phi \in \Phi$ that are not a literal:

1. if $\phi = \alpha$, then $\alpha_1, \alpha_2 \in \Phi$;
2. if $\phi = \beta$, then $\beta_1 \in \Phi$ or $\beta_2 \in \Phi$;
3. if $\phi = \delta(x)$, then $\delta_1(t) \in \Phi$ for some term $t \in Term^0_\Sigma$;
4. if $\phi = \gamma(x)$, then $\gamma_1(t) \in \Phi$ for all terms $t \in Term^0_\Sigma$.

THEOREM 52. *A ground background reasoner $\mathcal{R}$ for a theory $\mathcal{T}$ is complete if, for all (finite or infinite) $\mathcal{T}$-unsatisfiable downward saturated sets $\Xi \subset Form_\Sigma$ that do not contain free variables:*

1. there is a key $\Phi \subset \Xi$ such that $id \in \mathcal{R}(\Phi)$; or

2. there is a key $\Phi \subset \Xi$ such that there is a refuter $\langle id, \{\rho_1, \ldots, \rho_k\}\rangle$ in $\mathcal{R}(\Phi)$ with $\{\rho_1, \ldots, \rho_k\} \cap \Xi = \emptyset$ ($k \geq 1$).

Proof. Let $\mathcal{R}$ be a background reasoner satisfying the criterion of the theorem, and let ϕ be a $\mathcal{T}$-tautology. We prove that, using $\mathcal{R}$ and an arbitrary fair ground tableau construction rule (Definition 50), a tableau proof for ϕ is constructed. Let $(T_i)_{i\geq 0}$ be the sequence of ground tableaux that is constructed according to the fair rule starting from $T_0 = \{\{\neg\phi\}\}$.

Supposed $(T_i)_{i\geq 0}$ is not a tableau proof. If the sequence is finite, then there has to be at least one finite exhausted branch B^* in the final tableau T_n. If the sequence is infinite, then there is a sequence $(B_i)_{i\geq 0}$, $B_i \in T_i$, of branches as described in Condition 2 in the definition of fairness (Definition 50). In that case, $B^* = \bigcup_{i\geq 0} B_i$ is the union of these branches. We proceed to prove that the set B^* is $\mathcal{T}$-satisfiable. Because of the fairness conditions, B^* is downward saturated. If it were $\mathcal{T}$-unsatisfiable, then there had to be a key $\Phi \subset B^*$ such that

$id \in \mathcal{R}(\Phi)$ or such that there is a refuter $\langle id, \{\rho_1, \ldots, \rho_k\}\rangle \in \mathcal{R}(\Phi)$ $(k \geq 1)$ where $\{\rho_1, \ldots, \rho_k\} \cap B^* = \emptyset$. Because keys are finite, there then had to be an $i \geq 0$ such that $B_i \supset \Phi$; thus, for some $j \geq 0$, the tableau T_{j+1} had to be constructed from T_j applying the theory closure rule to B_j—which is according to the construction of the sequence $(B_i)_{i\geq 0}$ not the case—, or T_{j+1} had to be constructed from T_j applying the theory rule to B_j such that $\rho_m \in B_{j+1}$ for some $m \in \{1, \ldots, k\}$—which is impossible because $\{\rho_1, \ldots, \rho_k\} \cap B^* = \emptyset$.

We have shown B^* to be $\mathcal{T}$-satisfiable. Because $\neg\phi \in B^*$, $\neg\phi$ is $\mathcal{T}$-satisfiable as well. This, however, is a contradiction to ϕ being a $\mathcal{T}$-tautology, and the assumption that $(T_i)_{i\geq 0}$ is not a tableau proof has to be wrong. ∎

The criterion can be simplified if the residues a background reasoner computes for keys consisting of literals consist of literals as well. This is a reasonable assumption, which is usually satisfied in practice (and is true for all total background reasoners).

COROLLARY 53. *A ground background reasoner $\mathcal{R}$ for a theory $\mathcal{T}$ is complete if, for all keys $\Phi \subset Lit_\Sigma$ for which $\mathcal{R}$ is defined, $R \subset Lit_\Sigma$ for all $\langle \sigma, R\rangle \in \mathcal{R}(\Phi)$, and for all (finite or infinite) $\mathcal{T}$-unsatisfiable sets $\Xi \subset Form_\Sigma$ of* literals *that do not contain free variables:*

1. *there is a key $\Phi \subset \Xi$ such that $id \in \mathcal{R}(\Phi)$; or*

2. *there is a key $\Phi \subset \Xi$ such that there is a refuter $\langle id, \{\rho, \ldots, \rho_k\}\rangle$ in $\mathcal{R}(\Phi)$ with $\{\rho_1, \ldots, \rho_k\} \cap \Xi = \emptyset$ $(k \geq 1)$.*

Proof. Let Ξ be a downward saturated $\mathcal{T}$-unsatisfiable set of formulae and define $\Xi' = \Xi \cap Lit_\Sigma$ to be the set of literals in Ξ. Then Ξ' is $\mathcal{T}$-unsatisfiable, because a $\mathcal{T}$-structure satisfying Ξ' would satisfy Ξ as well.

Thus, there is a key $\Phi \subset \Xi' \subset \Xi$ such that $id \subset \mathcal{R}(\Phi)$—in which case we are done—, or there is a key $\Phi \subset \Xi'$ and a refuter $\langle id, \{\rho_1, \ldots, \rho_k\}\rangle \in \mathcal{R}(\Phi)$ with $\{\rho_1, \ldots, \rho_k\} \cap \Xi' = \emptyset$. In the latter case, the refuter satisfies the condition $\{\rho_1, \ldots, \rho_k\} \cap \Xi = \emptyset$, because by assumption $\{\rho_1, \ldots, \rho_k\} \subset Lit_\Sigma$. ∎

There is a strong relation between the criterion from Theorem 52 and the definition of Hintikka sets for theory reasoning: similar to classical first-order Hintikka sets (see Section 2.3 in the chapter by Fitting), any set is satisfiable that does not contain 'obvious' inconsistencies (inconsistencies that can be detected by the background reasoner) and is downward saturated (the background reasoner cannot add new formulae).

EXAMPLE 54. The complete background reasoner for the theory $\mathcal{OP}$ of partial orderings from Example 32 can be turned into a definition of Hintikka sets for $\mathcal{OP}$: A set Ξ that is downward saturated and, in addition, satisfies the following conditions is $\mathcal{OP}$-satisfiable:

1. For all terms $t, t', t'' \in Term_\Sigma$:
   if $(t < t'), (t' < t'') \in \Xi$, then $(t < t'') \in \Xi$.

2. There is no literal of the form $(t < t)$ in Ξ.

3. There are no literals $\phi, \neg\phi$ in Ξ.

A Completeness Criterion for Free Variable Background Reasoners

The criterion for free variable background reasoners is based on lifting completeness of a ground background reasoner. If the ground background reasoner computes a refuter $\langle id, R\rangle$ for a ground instance $\Phi\tau$ of a key Φ, then, to be complete, the free variable background reasoner has to compute a refuter for Φ that is more general than $\langle \tau, R\rangle$.

THEOREM 55. *Let $\mathcal{R}$ be a free variable background reasoner for a theory $\mathcal{T}$; $\mathcal{R}$ is complete if there is a complete ground background reasoner $\mathcal{R}^g$ for $\mathcal{T}$ such that, for all keys $\Phi \subset Form_\Sigma$, all ground substitutions τ, and all refuters $\langle id, R^g\rangle$ in $\mathcal{R}^g(\Phi\tau)$, there is a refuter $\langle \sigma, R\rangle \in \mathcal{R}(\Phi)$ and a substitution τ' with*

1. $\tau = \tau' \circ \sigma$,

2. $R\tau' = R^g$.

Proof. Let the sentence ϕ be a $\mathcal{T}$-tautology. Since $\mathcal{R}^g$ is a complete ground background reasoner, using $\mathcal{R}^g$ a ground tableau proof

$$\{\{\neg\phi\}\} = T_0^g, \ldots, T_n^g = \emptyset$$

can be constructed, where the new terms introduced by δ-rule applications have been chosen in an arbitrary way (see below).

By induction we prove that there is a free variable tableau proof

$$\{\{\neg\phi\}\} = T_0, \ldots, T_n = \emptyset$$

such that $T_i\tau_i = T_i^g$ for substitutions $\tau_i \in Subst^*_\Sigma$ $(0 \le i \le n)$.

$i = 0$: Since ϕ is a sentence, $T_0\tau_0 = T_0^g$ for $\tau_0 = id$.

$i \to i+1$: Depending on how T_{i+1}^g has been derived from T_i^g, there are the following subcases:

If T_{i+1}^g has been derived from T_i^g by applying an expansion rule to a formula ϕ^g on a branch $B_i^g \in T_i^g$, then there has to be a formula ϕ_i on a branch $B_i \in T_i$ such that $\phi_i\tau_i = \phi_i^g$ and $B_i\tau_i = B_i^g$. If an α- or β-rule has been applied, then apply the same rule to ϕ_i to derive T_{i+1} from T_i and set $\tau_{i+1} = \tau_i$. If a γ-rule has been applied and the term t has been substituted for the quantified variable in the ground tableau, then derive T_{i+1} from T_i by applying a γ-rule to ϕ_i and substituting a new free variable x for the quantified variable; set $\tau_{i+1} = \tau_i \cup \{x \mapsto t\}$. If a δ-rule has been applied, then apply the free variable δ-rule to ϕ_i to derive T_{i+1} from T_i and set $\tau_{i+1} = \tau_i$. In addition, the new ground term introduced in the ground tableau—that we are free to choose as long as it does not occur in T_i—shall be $f(x_1, \ldots, x_n)\tau_i$ where $f(x_1, \ldots, x_n)$ is the Skolem term that has been substituted for the existentially quantified variable in the free variable tableau.

If T^g_{i+1} has been derived from T^g_i by applying the theory closure or the theory expansion rule using a key Φ^g taken from a branch $B^g_i \in T^g_i$ and a refuter id or $\langle id, R^g\rangle$ in $\mathcal{R}^g(\Phi^g)$, then there has to be a key Φ_i on a branch $B_i \in T_i$ such that $\Phi_i\tau_i = \Phi^g_i$ and $B_i\tau_i = B^g_i$. Thus. there is a refuter $\langle\sigma, R\rangle \in \mathcal{R}(\Phi)$ and a substitution τ' such that $\tau_i = \tau' \circ \sigma$ and $R\tau' = R^g$. In that case, derive T_{i+1} from T_i by applying the theory expansion or closure rule using the key Φ_i and the refuter $\langle\sigma, R\rangle$, and set $\tau_{i+1} = \tau'$. ■

EXAMPLE 56. The criterion from Theorem 55 can be used to prove completeness of the free variable background reasoner $\mathcal{R}$ for the theory $\mathcal{OP}$ of partial orderings that satisfies the following conditions. The proof is based on the completeness of the ground background reasoner for $\mathcal{OP}$ from Example 32. The conditions for $\mathcal{R}$ are:

1. For all terms $s_1, t, t', s_2 \in Term_\Sigma$ where t and t' are unifiable:
   if $(s_1 < t), (t' < s_2) \in \Phi$, then $\langle\mu, \{(s_1 < s_2)\mu\}\rangle \in \mathcal{R}(\Phi)$ where μ is a most general unifier (MGU) of t and t'.

2. For all terms $t, t' \in Term_\Sigma$ that are unifiable:
   if $(t < t') \in \Phi$, then $\mu \in \mathcal{R}(\Phi)$ where μ is an MGU of t and t'.

3. For all atoms $\phi, \phi' \in Form_\Sigma$ that are unifiable:
   If $\phi, \neg\phi' \in \Phi$, then $\mu \in \mathcal{R}(\Phi)$ where μ is an MGU of ϕ and ϕ'.

In the ground case, a tableau proof can be constructed deterministically using a fair tableau construction rule. In the free variable case, however, there are additional choice points because there may be refuters with incompatible substitutions. Thus lifting the notion of fairness to the free variable case such that no backtracking at all is needed to construct a tableau proof is very difficult (though not impossible).

A Completeness Criterion for Universal Formula Background Reasoners

A criterion for the completeness of universal formula background reasoners can be defined based on completeness of free variable background reasoners (the proof of the theorem is similar to that of Thereom 55).

DEFINITION 57. Let $\Phi \subset Form_\Sigma$ be a key, and let $(\forall x_1)\cdots(\forall x_k)\phi$ be a universally quantified literal in Φ. If ϕ' is constructed from ϕ by replacing the variables $x_1, \ldots, x_k$ by free variables $y_1, \ldots, y_k$ that do not occur in Φ, then ϕ' is a *free variable instance* of $(\forall x_1)\cdots(\forall x_k)\phi$ (w.r.t. Φ).

THEOREM 58. *Let $\mathcal{R}$ be a universal formula background reasoner for a theory $\mathcal{T}$; $\mathcal{R}$ is complete if there is a complete free variable background reasoner $\mathcal{R}^{fv}$ for $\mathcal{T}$ such that, for all keys $\Phi \subset Form_\Sigma$, the following holds:*

Let the key Φ^{fv} be constructed from Φ by replacing all formulae $\phi \in \Phi$ of the form $(\forall x_1) \cdots (\forall x_n)\psi$ by a free variable instance of ϕ, and let F be the set of all the free variables occurring in Φ^{fv} but not in Φ. Then, for all refuters $\langle \sigma^{fv}, R^{fv} \rangle \in \mathcal{R}^{fv}(\Phi^{fv})$, there is a refuter $\langle \sigma, R \rangle \in \mathcal{R}(\Phi)$ where

1. $\sigma^{fv}{}_{|(V \setminus F)} = \sigma$,
2. $R^{fv} = R(\sigma^{fv}{}_{|F})$.

5.3 Completeness Preserving Refinements

Restrictions on Keys

In this section, additional refinements are discussed that are indispensable for an efficient implementation of theory reasoning.

An important simplification usually used in implementations is to impose the restriction on keys that they must consist of literals (universally quantified literals in the case of universal formula tableau). The proof for Theorem 52 shows that completeness is preserved if this restriction is combined with any complete background reasoner.

COROLLARY 59. *Let $\mathcal{R}$ be a complete ground, free variable, or universal formula background reasoner. Then the restriction $\mathcal{R}'$ of $\mathcal{R}$ to keys Φ that consist of (universally quantified) literals is complete ($\mathcal{R}'$ is undefined for other keys).*

The set of keys that have to be considered can be further restricted. The background reasoner has only to be defined for keys that contain a pair of complementary literals or at least one formula in which a symbol occurs that is defined by the theory:

DEFINITION 60. A set of function or predicate symbols is *defined* by a theory $\mathcal{T}$ if for all sets Φ of formulae that do not contain these symbols: Φ is satisfiable if and only if Φ is $\mathcal{T}$-satisfiable.

For example, the equality theory $\mathcal{E}$ defines the equality predicate $\approx$; the theory of partial orderings defines the predicate symbol $<$.

Similarly, only keys have to be considered that contain a pair of complementary literals or consist of formulae that *all* have a predicate symbol in common with the theory (which may or may not be defined by the theory). Thus, for the theory $\mathcal{OP}$, all formulae in keys have to contain the predicate symbol $<$ as it is the only predicate symbol in $\mathcal{OP}$. For the equality theory $\mathcal{E}$, however, this restriction is useless because $\mathcal{E}$ contains all predicate symbols.

COROLLARY 61. *Let $\mathcal{R}$ be a complete ground, free variable, or universal formula background reasoner for a theory $\mathcal{T}$. Then the restriction $\mathcal{R}'$ of $\mathcal{R}$ to keys Φ that*

1. *(a) contain at least one occurrence of a function or predicate symbol defined by $\mathcal{T}$, and*

   (b) consist of formulae that all have at least one predicate symbol in common with $\mathcal{T}$,

2. *or contain a pair ϕ and $\neg\psi$ (resp. $(\forall\overline{x})\phi$) and $(\forall\overline{y})\neg\psi$), where ϕ and ψ are unifiable,*

is complete ($\mathcal{R}'$ is undefined for other keys).

Most General Refuters

There is another important refinement that can be combined with all complete background reasoners: completeness is preserved if only most general refuters are computed (this is a corollary to Theorem 55). The subsumption relation on refuters may or may not take the theory $\mathcal{T}$ into account:

DEFINITION 62. Let $\mathcal{T}$ be a theory; and let $W \subset V$ be a set of variables. The subsumption relations $\leq^W$ and $\leq^W_{\mathcal{T}}$ on refuters are defined by:

- $\langle\sigma, R\rangle \leq^W \langle\sigma', R'\rangle$ if there is a substitution $\tau \in Subst^*_\Sigma$ such that
  1. $\sigma'(x) = \sigma(x)\tau$ for all $x \in W$, and
  2. $R'\tau \supset R$.
- $\langle\sigma, R\rangle \leq^W_{\mathcal{T}} \langle\sigma', R'\rangle$ if there is a refuter $\langle\sigma'', R''\rangle$ such that
  1. $\langle\sigma, R\rangle \leq^W \langle\sigma'', R''\rangle$, and
  2. $\Phi\sigma'' \cup \{\bigvee R''\} \models^\circ_{\mathcal{T}} \Phi\sigma' \cup \{\bigvee R'\}$ for all formula sets $\Phi \subset Form_\Sigma$ (including, in particular, the empty set).

In addition, we use the abbreviations $\leq = \leq^V$ and $\leq_{\mathcal{T}} = \leq^V_{\mathcal{T}}$ where V is the set of all variables.

The set W contains the 'relevant' variables, including *at least* those occurring in the two refuters that are compared. If, for example, the theory expansion rule is used to extend a tableau branch, then W contains all free variables occurring in the tableau. It is of advantage to keep the set W as small as possible; but, if the context is not known, the set $W = V$ of all variables has to be used.

The intuitive meaning of $\langle\sigma, R\rangle \leq^W_{\mathcal{T}} \langle\sigma', R'\rangle$ is that the effects of using the refuter $\langle\sigma', R'\rangle$ can be simulated by first applying a substitution τ and then using the resulting refuter $\langle\sigma'', R''\rangle$ of which the refuter $\langle\sigma', R'\rangle$ is a logical consequence.

EXAMPLE 63. The refuter $\langle id, \{p(x)\}\rangle$ subsumes $\langle\{x \mapsto a\}, \{p(a)\}\rangle$ w.r.t. the subsumption relation $\leq^W$ (and thus w.r.t. $\leq^W_\mathcal{T}$) for all variable sets W; however, it subsumes the refuter $\langle id, \{p(a)\}\rangle$ only if $x \notin W$.

The refuter $\langle\sigma, \{\phi\}\rangle$ is more general than $\langle\sigma, \{\phi, \psi\}\rangle$ w.r.t. all subsumption relations, i.e. only refuters with a minimal residue are most general.

Let $\mathcal{T}$ be the equational theory $\mathcal{E} \cup \{a \approx b\}$. Then $\langle id, p(a)\rangle$ and $\langle id, p(b)\rangle$ resp. $\{x \mapsto a\}$ and $\{x \mapsto b\}$ subsume each other w.r.t. $\leq_\mathcal{T}$.

COROLLARY 64. *Let $\mathcal{R}$ be a complete free variable or universal formula background reasoner for a theory $\mathcal{T}$. Then a background reasoner $\mathcal{R}'$ is complete as well if, for all keys Φ and refuters $\langle\sigma, R\rangle \in \mathcal{R}(\Phi)$, there is a refuter $\langle\sigma', R'\rangle$ in $\mathcal{R}'(\Phi)$ that subsumes $\langle\sigma, R\rangle$ w.r.t. $\leq$ or $\leq_\mathcal{T}$ (Definition 62).*

If the subsumption relations $\leq^W$ and $\leq^W_\mathcal{T}$ are used, the context in which a background reasoner is used has to be taken into consideration:

THEOREM 65. *Let $\mathcal{R}$ be a complete free variable or universal formula background reasoner for a theory $\mathcal{T}$. Then, for every $\mathcal{T}$-tautology ϕ, a free variable tableau proof resp. a universal formula tableau proof can be built using $\mathcal{R}$ observing the restriction that each $\mathcal{T}$-refuter that is used in a theory expansion or closure rule application is minimal in $\mathcal{R}(\Phi)$ w.r.t. $\leq^W$ or $\leq^W_\mathcal{T}$, where Φ is the key that has been chosen for that rule application and W is the set of free variables in the tableau to which the rule is applied.*

The number of refuters that have to be considered is closely related to the number of choice points when the theory expansion or closure rule is applied to a tableau. Therefore, it is desirable to compute a *minimal* set of refuters. Nevertheless, it is often not useful to ensure minimality since there is a trade-off between the gain of computing a minimal set and the extra cost for checking minimality and removing subsumed refuters. While it is relatively easy to decide whether $\langle\sigma, R\rangle \leq^W \langle\sigma', R'\rangle$, it can (depending on the theory $\mathcal{T}$) be difficult to decide and is in general undecidable whether $\langle\sigma, R\rangle \leq^W_\mathcal{T} \langle\sigma', R'\rangle$.

Other Search Space Restrictions

There are other useful restrictions that, however, cannot be imposed on an arbitrary background reasoner without destroying completeness. Nevertheless, for every theory, there are background reasoners that have at least some of the following features:

- To avoid branching when the theory expansion rule is applied, only refuters $\langle\sigma, R\rangle$ are computed where the residue is either empty or a singleton.
- Only total refuters are computed, i.e. the residues are empty.
- The sets of refuters computed for a key are restricted to be
  - finite (in which case their computation terminates);

$$\frac{t \approx s \quad \phi[t]}{\phi[s]} \qquad \frac{s \approx t \quad \phi[t]}{\phi[s]} \qquad \frac{\neg(t \approx t)}{*} \qquad \frac{\phi \quad \neg\phi}{*}$$

Table 9. Jeffrey's equality theory expansion and closure rules

- empty or a singleton (then theory expansion or closure rules are—at least for a single key—deterministic);

There is, of course, a trade-off between these desirable features, in particular between total and partial theory reasoning (see Section 2.5).

6 PARTIAL EQUALITY REASONING

6.1 *Partial Equality Reasoning for Ground Tableaux*

Virtually all approaches to handling equality can be regarded as a special case of the general methods for theory reasoning in semantic tableaux. Exception are, for example, the method of *equality elimination* [Degtyarev and Voronkov, 1996] and applying transformations from first-order logic with equality into first-order logic without equality to the input formulae [Brand, 1975; Bachmair *et al.*, 1997] (see Section 9).

The first methods for adding equality to the ground version of semantic tableaux have been developed in the 1960s [Jeffrey, 1967; Popplestone, 1967], following work by S. Kanger on how to add equality to sequent calculi [Kanger, 1963]. R. Jeffrey introduced the additional tableau expansion and closure rules shown in Table 9 (i.e. a partial reasoning method); a similar set of rules has been described by Z. Lis in [1960]. If a branch B contains a formula $\phi[t]$ and an equality $t \approx s$ or $s \approx t$ that can be 'applied' to $\phi[t]$ to derive a formula $\phi[s]$ (which is constructed by substituting one occurrence of t in $\phi[t]$ by s), then $\phi[s]$ may be added to B.

There are two closure rules. The first one is the usual closure rule for ground tableaux with and without theory reasoning: a branch B is closed if there are formulae ϕ and $\neg\phi$ in B. The second one is an additional equality theory closure rule: a branch is closed if it contains a formula of the form $\neg(t \approx t)$.

THEOREM 66 (Jeffrey, 1967). *A ground background reasoner $\mathcal{R}$ for the theory $\mathcal{E}$ of equality is complete if it satisfies the following conditions:*

1. *For all terms $t, s \in Term^0_\Sigma$ and sentences $\phi \in Form_\Sigma$: if $\phi[t], (t \approx s) \in \Phi$ or $\phi[t], (s \approx t) \in \Phi$, then $\langle id, \{\phi[s]\}\rangle \in \mathcal{R}(\Phi)$.*
2. *For all terms $t \in Term^0_\Sigma$: if $\neg(t \approx t) \in \Phi$, then $id \in \mathcal{R}(\Phi)$.*
3. *For all sentences $\phi \in Form_\Sigma$: if $\phi, \neg\phi \in \Phi$, then $id \in \mathcal{R}(\Phi)$.*

(1) $a \approx b$		(1) $a \approx b$		(1) $a \approx b$
(2) $p(a, a)$	$\rightsquigarrow$	(2) $p(a, a)$		(2) $p(a, a)$
(3) $\neg p(b, b)$		(3) $\neg p(b, b)$	$\rightsquigarrow$	(3) $\neg p(b, b)$
		(4) $p(a, b)$		(4) $p(a, b)$
				(5) $p(b, b)$
				$*$

Figure 2. The application of Jeffrey's additional rules to expand and close a tableau branch (Example 67)

EXAMPLE 67. Figure 2 shows an example for the application of Jeffrey's equality expansion and closure rules: The equality (1) is applied to the formula (2) to derive formula (4) and to (4) to derive (5). The branch is closed by the complementary formulae (3) and (5). Note that it is not possible to derive $p(b, b)$ in a single step.

The background reasoner is still complete if the formula ϕ to which an equality is applied is restricted to be (a) an inequality $\neg(s \approx t)$, or (b) a literal $p(t_1, \ldots, t_n)$ or $\neg p(t_1, \ldots, t_n)$ where $p \not\equiv \approx$; i.e. equalities do not have to be applied to complex formulae or to equalities.

Jeffrey's rules resemble paramodulation [Robinson and Wos, 1969] (see [Snyder, 1991] for an overview on various techniques for improving paramodulation).

Besides being based on the ground version of tableaux, the new expansion rules have a major disadvantage: they are symmetrical and their application is completely unrestricted. This leads to much non-determinism and a huge search space; an enormous number of irrelevant formulae (residues) can be derived. If, for example, a branch B contains the formulae $f(a) \approx a$ and $p(a)$, then all the formulae $p(f(a)), p(f(f(a))), \ldots$ can be added to B.

The rules presented by S. Reeves [1987] (see Table 10) generate a smaller search space. They are the tableau counterpart of RUE-resolution [Digricoli and Harrison, 1986] and are more goal-directed than Jeffrey's expansion rules: only literals that are potentially complementary are used for expansion. Like RUE-resolution, the rules are based upon the following fact: If an $\mathcal{E}$-structure M satisfies the inequality $\neg(f(a_1, \ldots, a_k) \approx f(b_1, \ldots, b_k))$ or it satisfies the formulae $p(a_1, \ldots, a_k)$ and $\neg p(b_1, \ldots, b_k)$, then at least one of the inequalities

$$\neg(a_1 \approx b_1), \ldots, \neg(a_k \approx b_k)$$

is satisfied by M. In addition, a rule is needed that implements the symmetry of equality, i.e. that allows to deduce $s \approx t$ from $t \approx s$. With these equality theory expansion rules, it is sufficient to use the same closure rules as in Theorem 66:

THEOREM 68 (Reeves, 1987). *If a ground background reasoner $\mathcal{R}$ for the the-*

$$\frac{\begin{array}{c} p(t_1,\ldots,t_k) \\ \neg p(s_1,\ldots,s_k) \end{array}}{\neg(t_1 \approx s_1) \mid \cdots \mid \neg(t_k \approx s_k)} \qquad \frac{\neg(f(t_1,\ldots,t_k) \approx f(s_1,\ldots,s_k))}{\neg(t_1 \approx s_1) \mid \cdots \mid \neg(t_k \approx s_k)}$$

$$\frac{t \approx s}{s \approx t} \qquad \frac{\neg(t \approx t)}{*} \qquad \frac{\begin{array}{c} \phi \\ \neg\phi \end{array}}{*}$$

Table 10. Reeves' equality expansion and closure rules

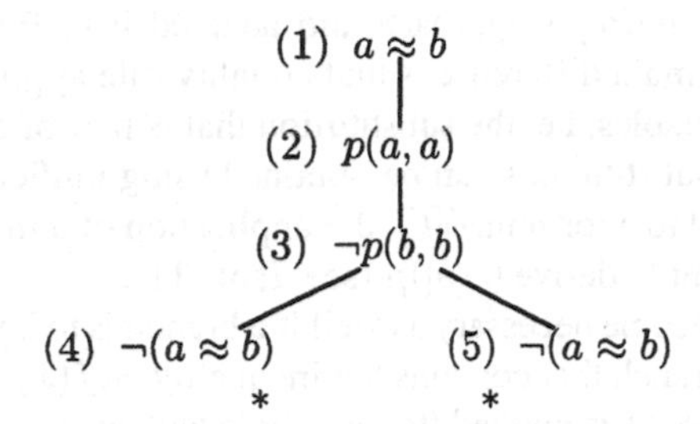

Figure 3. Applying Reeves's equality expansion rule (Example 69)

ory $\mathcal{E}$ of equality satisfies the following conditions, it is complete:

1. *For all terms $t = f(t_1, \ldots, t_k)$ and $s = f(s_1, \ldots, s_k)$ $(k \geq 1)$: if $\neg(t \approx s) \in \Phi$, then $\langle id, \{\neg(s_1 \approx t_1), \ldots, \neg(s_k \approx t_k)\}\rangle \in \mathcal{R}(\Phi)$.*
2. *For all literals $\psi = p(t_1, \ldots, t_k)$ and $\psi' = \neg p(s_1, \ldots, s_k)$ $(k \geq 1)$: if $\psi, \psi' \in \Phi$, then $\langle id, \{\neg(s_1 \approx t_1), \ldots, \neg(s_k \approx t_k)\}\rangle \in \mathcal{R}(\Phi)$.*
3. *For all terms $s, t \in Term^0_\Sigma$: if $(s \approx t) \in \Phi$, then $\langle id, \{t \approx s\}\rangle \in \mathcal{R}(\Phi)$.*
4. *For all terms $t \in Term^0_\Sigma$: if $\neg(t \approx t) \in \Phi$, then $id \in \mathcal{R}(\Phi)$.*
5. *For all literals $\phi \in Lit_\Sigma$: if $\phi, \neg\phi \in \Phi$, then $id \in \mathcal{R}(\Phi)$.*

EXAMPLE 69. Figure 3 shows the application of Reeves' rule to expand and close the same tableau branch as in Figure 2: It is applied to the atomic formulae (2) and (3) to generate the inequalities (4) and (5). The branches are closed by the formulae (1) and (4) and (1) and (5), respectively.

Reeves' approach, however, can lead to heavy branching, because the new expansion rules can as well be applied to pairs of equalities and inequalities. In the worst case, the number of branches generated is exponential in the number of equalities on the branch.

$$\frac{\begin{array}{c} t \approx s \\ \phi[t'] \end{array}}{(\phi[s])\mu} \qquad \frac{\begin{array}{c} s \approx t \\ \phi[t'] \end{array}}{(\phi[s])\mu} \qquad \frac{\neg(t \approx t')}{*} \qquad \frac{\begin{array}{c} \phi \\ \phi' \end{array}}{*}$$

where μ is an MGU of t and t' resp. ϕ and ϕ'
and μ is applied to the whole tableau.

Table 11. Fitting's equality reasoning rules for free variable tableaux

6.2 *Partial Equality Reasoning for Free Variable Tableaux*

M. Fitting extended Jeffrey's approach and adapted it to free variable tableaux [Fitting, 1996]. The main difference is that equality rule applications may require instantiating free variables, i.e. the substitution that is part of a refuter may not be the identity. These substitutions can be obtained using unification: If an equality $t \approx s$ is to be applied to a formula $\phi[t']$, the application of a most general unifier μ of t and t' is sufficient to derive $(\phi[s])\mu$ (see Table 11).

Unification can become necessary as well if a branch is to be closed using equality; for example, a branch that contains the inequality $\neg(f(x) \approx f(a))$ is closed if the substitution $\{x \mapsto a\}$ is applied (to the whole tableau):

THEOREM 70 (Fitting, 1990). *$\mathcal{R}$ is a complete free variable background reasoner for the equality theory $\mathcal{E}$ if it satisfies the following conditions:*

1. *For all terms $t, t' \in Term_\Sigma$ that are unifiable and all $\phi \in Form_\Sigma$:*
   if $(t \approx s), \phi[t'] \in \Phi$, then $\langle \mu, \{(\phi[s])\mu\} \rangle \in \mathcal{R}(\Phi)$ where μ is an MGU of t and t'.

2. *For all terms $t, t' \in Term_\Sigma$ that are unifiable:*
   if $\neg(t \approx t') \in \Phi$, then $\mu \in \mathcal{R}(\Phi)$ where μ is an MGU of t and t'.

3. *For all literals $\phi, \phi' \in Lit_\Sigma$ that are unifiable:*
   If $\phi, \neg\phi' \in \Phi$, then $\mu \in \mathcal{R}(\Phi)$ where μ is an MGU of ϕ and ϕ'.

EXAMPLE 71. Figure 4 shows a free variable tableau that proves the following set of formulae to be inconsistent:

$$\begin{array}{ll} (1) & (\forall x)(g(x) \approx f(x) \vee \neg(x \approx a)) \\ (2) & (\forall x)(g(f(x)) \approx x) \\ (3) & b \approx c \\ (4) & p(g(g(a)), b) \\ (5) & \neg p(a, c) \end{array}$$

By applying the standard free variable tableau rules, formula (6) is derived from formula (2), (7) from (1), and (8) and (9) from (7). The framed formulae are added to the left branch by applying Fitting's equality expansion rules. Formula (10) is derived by applying equality (8) to (4) (the substitution $\{x_2 \mapsto a\}$ has to be applied), formula (11) is derived by applying (6) to (10) (the substitution $\{x_1 \mapsto a\}$ has to be applied), and formula (12) is derived by applying (3) to (11). Formulae (12) and (5) close the left branch. The right branch is closed by the inequality (9) (the substitution $\{x_2 \mapsto a\}$ has already been applied).

The example demonstrates a difficulty involved in using free variable equality expansion rules: If equality (8) is applied to (4) in the wrong way, i.e. if the formula (10′) $p(f(g(a)), b)$ is derived instead of (10) $p(g(f(a)), b)$, then the term $g(a)$ is substituted for x_2 and the tableau cannot be closed. Either a new instance of (7), (8) and (9) has to be generated by applying the γ-rule to (1), or backtracking has to be initiated.

Completeness is preserved if the restriction is made that the formulae ϕ und ϕ' in Theorem 70 which the equality expansion rule is applied to have to be literals (similar to the ground case). However, the restriction that equalities must not be applied to equalities (that can be employed in the ground case) would destroy completeness, as the following example demonstrates.

EXAMPLE 72. Let the tableau branch B contain the formulae

$$a \approx b,\ f(h(a), h(b)) \approx g(h(a), h(b)),\ \neg(f(x, x) \approx g(x, x)) .$$

A refuter with the residue $\{f(h(a), h(a)) \approx g(h(a), h(a))\}$ can be derived, provided it is allowed to apply equalities to equalities. After this formula has been added to the branch, the closing refuter $\{x \mapsto h(a)\}$ can be found.

If the application of equalities to equalities is prohibited, completeness is lost: then the only possibility is to apply $a \approx b$ to the inequality in B. All refuters that can be derived that way instantiate the variable x either with a or with b, which in the sequel makes it impossible to close the branch. Note that the criterion from Theorem 55, which would guarantee completeness, is not satisfied.

6.3 *Partial Equality Reasoning for Tableaux with Universal Formulae*

Fitting's method can easily be extended to free variable tableaux *with universal formulae* [Beckert, 1997]. When equalities are used to derive new formulae, universality of both the equality $t \approx s$ (resp. $s \approx t$) and the formula $\phi[t']$ it is applied to has to be taken into consideration. The difference to the equality expansion rules from Section 6.2 is that, instead of the MGU μ of t and t', only its restriction μ' to variables is applied w.r.t. which *not* all formulae in the precondition of the rule are $\mathcal{E}$-universal (apart from that, the rule schemata are the same as the free variable schemata in Table 11). If an equality is universal with respect to a variable x, the variable x does not have to be instantiated to apply the equality. When branches are closed, the universality of formulae has to be taken into consideration as well.

$$(1)\ (\forall x)(g(x) \approx f(x) \vee \neg(x \approx a))$$
$$(2)\ (\forall x)(g(f(x)) \approx x)$$
$$(3)\ b \approx c$$
$$(4)\ p(g(g(a)), b)$$
$$(5)\ \neg p(a, c)$$
$$(6)\ g(f(x_1)) \approx x_1$$
$$(7)\ g(x_2) \approx f(x_2) \vee \neg(x_2 \approx a)$$
$$(8)\ g(x_2) \approx f(x_2) \qquad (9)\ \neg(x_2 \approx a)$$
$$(10)\ \boxed{p(g(f(a)), b)}$$
$$(11)\ \boxed{p(a, b)}$$
$$(12)\ \boxed{p(a, c)}$$

Figure 4. Using Fitting's expansion rules (Example 71)

EXAMPLE 73. If the method from Theorem 40 for recognizing universal formulae is used, the tableau in Figure 4 (without the framed formulae) can be closed using the substitution $\{x_2 \mapsto a\}$. The variable x_1 does not have to be instantiated, because equality (6) is recognized to be universal w.r.t. to x_1.

The background reasoner does not have to cope with the problem of recognizing universal formulae, because in keys the universal formulae are explicitly universally quantified (Definition 41).

THEOREM 74. *A background reasoner $\mathcal{R}$ that satisfies the following conditions is a complete universal formula background reasoner for the equality theory $\mathcal{E}$:*

1. *For all $s, t, t' \in Term_\Sigma$ such that t, t' are unifiable, and all $\phi \in Form_\Sigma$: if $(\forall \overline{x})(t \approx s), (\forall \overline{y})\phi[t'] \in \Phi$, then $\langle \mu_{|F}, \{(\phi[s])\mu\}\rangle \in \mathcal{R}(\Phi)$ where μ is an MGU of t and t' and F is the set of variables that are free in $(\forall \overline{x})(t \approx s)$ or $(\forall \overline{y})\phi[t']$.*[3]

2. *For all $t, t' \in Term_\Sigma$ that are unifiable: if $(\forall \overline{x})\neg(t \approx t') \in \Phi$, then $\mu_{|F} \in \mathcal{R}(\Phi)$ where μ is an MGU of t and t' and F is the set of variables that are free in $(\forall \overline{x})\neg(t \approx t')$.*

3. *For all atoms $\phi, \phi' \in Form_\Sigma$ that are unifiable: if $(\forall \overline{x})\phi, (\forall \overline{y})\neg\phi' \in \Phi$, then $\mu_{|F} \in \mathcal{R}(\Phi)$ where μ is an MGU of ϕ and ϕ' and F is the set of variables that are free in $(\forall \overline{x})\phi$ or $(\forall \overline{y})\phi'$.*

---

[3] $(\forall \overline{x})$ is an abbreviation for $(\forall x_1) \cdots (\forall x_m)$ $(m \geq 0)$. Without making a real restriction, we assume the sets of free and bound variables occurring in Φ to be disjoint.

7 TOTAL EQUALITY REASONING

7.1 *Total Equality Reasoning and E-unification*

The common problem of all the partial reasoning methods described in Section 6.1, which are based on additional tableau expansion rules, is that there are virtually no restrictions on the application of equalities. Because of their symmetry, this leads to a very large search space; even very simple problems cannot be solved in reasonable time.

It is difficult to transform more elaborate and efficient methods for handling equality, such as completion-based approaches, into (sufficiently) simple tableau expansion rules (i.e. partial background reasoners). A set of rules that implement a completion procedure for the ground version of tableaux has been described in [Browne, 1988]; however, these equality expansion rules are quite complicated, and the method cannot be extended to free variable tableaux.

If total equality reasoning is used, i.e. if no equality expansion rules are added, then the problem of finding refuters that close a tableau branch is equivalent to solving E-unification problems.

Depending on the version of semantic tableaux to which equality handling is added, different types of E-unification problems have to be solved. These are introduced in the following section.

7.2 *Universal, Rigid and Mixed E-unification*

The different versions of E-unification that are important for handling equality in semantic tableaux are: the classical 'universal' E-unification, 'rigid' E-unification, and 'mixed' E-unification, which is a combination of both. The different versions allow equalities to be used differently in an equational deduction: in the universal case, the equalities can be applied several times with different instantiations for the variables they contain; in the rigid case, they can be applied more than once but with only one instantiation for each variable; in the mixed case, there are both types of variables.

Which type of E-unification problems has to be solved to compute refuters, depends on the version of semantic tableaux that equality reasoning is to be added to. Universal E-unification can only be used in the ground case. For handling equality in free variable tableaux, rigid E-unification problems have to be solved. For tableaux with universal formulae, both versions have to be combined [Beckert, 1994]; then equalities contain two types of variables, namely universal (bound) and rigid (free) ones.

DEFINITION 75. A *mixed E-unification problem* $\langle E, s, t\rangle$ consists of a finite set E of universally quantified equalities $(\forall x_1)\cdots(\forall x_m)(l \approx r)$ and terms s and t.

E	s	t	MGUs	Type
$\{f(x) \approx x\}$	$f(x)$	a	$\{x \mapsto a\}$	rigid
$\{f(a) \approx a\}$	$f(a)$	a	id	ground
$\{(\forall x)(f(x) \approx x)\}$	$g(f(a), f(b))$	$g(a, b)$	id	universal
$\{f(x) \approx x\}$	$g(f(a), f(b))$	$g(a, b)$	—	rigid
$\{(\forall x)(f(x, y) \approx f(y, x))\}$	$f(a, b)$	$f(b, a)$	$\{y \mapsto b\}$	mixed

Table 12. Examples for the different versions of E-unification

A substitution $\sigma \in \mathit{Subst}^*_\Sigma$ is a *solution* to the problem $\langle E, s, t\rangle$ if

$$E\sigma \models^\circ_\mathcal{E} (s\sigma \approx t\sigma)\ .^4$$

The major differences between this definition and that generally given in the literature on (universal) E-unification are:

- The equalities in E are *explicitly* quantified (instead of considering all the variables in E to be *implicitly* universally quantified).
- The strong consequence relation $\models^\circ_\mathcal{E}$ is used instead of $\models_\mathcal{E}$.
- The substitution σ is applied not only to the terms s und t but also to the set E.

A mixed E-unification problem $\langle E, s, t\rangle$ is *universal* if there are no free variables in E, and it is *rigid* if there are no bound variables in E (if E is ground, the problem is both rigid and universal).

EXAMPLE 76. Table 12 shows some simple examples for the different versions of E-unification. The fourth problem has no solution, since the free variable x would have to be instantiated with both a and b. Contrary to that, the empty substitution id is a solution to the third problem where the variable x is universally quantified.

Syntactical unification is a special case of E-unification, namely the case where the set E of equalities is empty.

For handling equality in free variable tableaux, the problem of finding a simultaneous solution to several mixed E-unification problems plays an important rôle, as it corresponds to the problem of finding a substitution that allows to simultaneously close several tableau branches.

DEFINITION 77. A finite set $\{\langle E_1, s_1, t_1\rangle, \ldots, \langle E_n, s_n, t_n\rangle\}$ $(n \geq 1)$ of E-unification problems is called *simultaneous* E-unification problem. A substitution σ

[4] This is equivalent to $E\sigma \models_\mathcal{E} (s\sigma \approx t\sigma)$ where the free variables in $E\sigma$ are 'held rigid', i.e. treated as constants.

is a solution to the simultaneous problem if it is a solution to every component $\langle E_k, s_k, t_k\rangle$ $(1 \le k \le n)$.

7.3 *Extracting E-unification Problems from Keys*

The important formulae in a key from which E-unification problems are extracted are: equalities, inequalities, and pairs of potentially complementary literals:

DEFINITION 78. Literals

$$(\forall x_1)\cdots(\forall x_k)p(s_1,\ldots,s_n) \text{ and } (\forall y_1)\cdots(\forall y_l)\neg p(t_1,\ldots,t_n)\ ,$$

where $p \neq \approx$, are called a pair of *potentially complementary literals* ($n \ge 0$ and $k, l \ge 0$, i.e. the literals may or may not be universally quantified).

We now proceed to define the set of equalities and the set of E-unification problems of a key. All considerations here are restricted to keys consisting of universally quantified literals.

DEFINITION 79. Let $\Phi \subset Form_\Sigma$ be a key. The *set* $E(\Phi)$ *of equalities* consists of the universally quantified equalities in Φ, i.e. all formulae in Φ of the form $(\forall x_1)\cdots(\forall x_k)(s \approx t)$ $(k \ge 0)$.

EXAMPLE 80. As an example, we use the tableau from Figure 4. Its left branch is denoted by B_1 and its right branch by B_2. If the method for recognizing universal formulae from Theorem 40 is used and keys Φ_1 and Φ_2 are built from the literals on the branches B_1 and B_2, respectively (according to Theorem 41), then both $E(\Phi_1)$ and $E(\Phi_2)$ contain the equalities $b \approx c$ and $(\forall x)(g(f(x)) \approx x)$; $E(\Phi_1)$ contains, in addition, the equality $g(x_2) \approx f(x_2)$.

DEFINITION 81. Let $\Phi \subset Form_\Sigma$ be a key. The *set* $P(\Phi)$ *of E-unification problems* consists exactly of:

1. for each pair $\phi, \psi \in \Phi$ of potentially complementary literals, the problem

   $$\langle E(\Phi), \langle t_1,\ldots,t_n\rangle, \langle s_1,\ldots,s_n\rangle\rangle$$

   where $p(t_1,\ldots,t_n)$ and $\neg p(s_1,\ldots,s_n)$ are free variable instances of ϕ resp. ψ (Definition 57);

2. for each inequality $\phi = (\forall x_1)\cdots(\forall x_k)(\neg(t' \approx s'))$ in Φ $(k \ge 0)$, the problem

   $$\langle E(\Phi), t, s\rangle$$

   where $\neg(t \approx s)$ is a free variable instance of ϕ (Definition 57).

The problems in $P(\Phi)$ of the form $\langle E(\Phi), \langle s_1,\ldots,s_k\rangle, \langle t_1,\ldots,t_k\rangle\rangle$ are actually simultaneous E-unification problems (sharing the same set of equalities), since the non-simultaneous problems $\langle E(\Phi), s_i, t_i\rangle$ $(1 \le i \le k)$ have to be solved simultaneously.

LEMMA 82. *A substitution is a solution to a simultaneous mixed E-unification problem of the form* $\{\langle E, s_1, t_1\rangle, \ldots, \langle E, s_n, t_n\rangle\}$ $(n \ge 1)$ *iff*

- *it is a solution to the non-simultaneous mixed E-unification problem* $\langle E, f(s_1, \ldots, s_n), f(t_1, \ldots, t_n)\rangle$ *(the function symbol* f *must not occur in the original problem), and*
- *it does not instantiate variables with terms containing* f.

All substitutions that are $\mathcal{E}$-refuters, i.e. that close a tableau branch B, can be computed by extracting the set $P(\Phi)$ of mixed E-unification problems from a key $\Phi \subset B$ according to the above definition and solving the problems in $P(\Phi)$. If one of the problems in $P(\Phi)$ has a solution σ, all instances of $\Phi\sigma$ are $\mathcal{E}$-unsatisfiable; therefore, σ is a refuter for Φ. The pair of potentially complementary literals corresponding to the solved unification problem has been proven to actually be $\mathcal{E}$-complementary; or the corresponding inequality has been proven to be $\mathcal{E}$-complementary (provided the refuter is applied).

EXAMPLE 83. We continue Example 80. Again, B_1 denotes the left and B_2 the right branch of the tableau in Figure 4 (without the framed formulae), and Φ_1 and Φ_2 are keys extracted from these branches. Then both $P(\Phi_1)$ and $P(\Phi_2)$ contain the problem $\langle E(\Phi_i), \langle g(g(a)), b\rangle, \langle a, c\rangle\rangle$. $P(\Phi_2)$ contains, in addition, the problem $\langle E(\Phi_2), x_2, a\rangle$.

Apart from the version of E-unification problems that have to be solved, the way equality is handled is nearly the same for the different versions of semantic tableaux. Therefore, it is sufficient to only formulate one general soundness and completeness theorem:

THEOREM 84. *A total (universal formula, free variable, or ground) background reasoner* $\mathcal{R}$ *is complete for the equality theory* $\mathcal{E}$ *if it satisfies the following condition for all keys* $\Phi \subset Lit_\Sigma$*: If a substitution* σ *is a most general solution w.r.t. the subsumption relation* $\le^W$ *(or* $\le^W_\mathcal{E}$*) of one of the problems in* $P(\Phi)$*, then* $\mathcal{R}(\Phi)$ *contains the restriction of* σ *to the variables occurring in* Φ.

EXAMPLE 85. We continue from Examples 80 and 83: $\sigma = \{x_2 \mapsto a\}$ is a solution to the two mixed E-unification problems

$$\begin{array}{rcl} \langle E(\Phi_1), \langle g(g(a)), b\rangle, \langle a, c\rangle\rangle & \in & P(\Phi_1)\ , \\ \langle E(\Phi_2), x_2, a\rangle & \in & P(\Phi_2)\ . \end{array}$$

When the theory closure rule is used to close one of the branches (and thus σ is applied to the tableau), the other branch can then be closed using the empty substitution.

7.4 Solving Ground E-unification Problems

In [Shostak, 1978], it is proven that *ground* E-unification is decidable; consequently, by considering all variables to be constants, it is decidable whether the empty substitution *id* is a solution to a given *rigid* E-unification problem $\langle E, s, t\rangle$, i.e. whether $E \models^{\circ}_{\mathcal{E}} s \approx t$. This can be decided by computing a congruence closure, namely the equivalence classes of the terms (and subterms) occurring in $\langle E, s, t\rangle$ w.r.t. the equalities in E.

DEFINITION 86. Let $\langle E, s, t\rangle$ be a ground (or rigid) E-unification problem; and let $T_{\langle E,s,t\rangle} \subset \mathit{Term}_\Sigma$ be the set of all (sub-)terms occurring in $\langle E, s, t\rangle$. The equivalence class $[t]_{\langle E,s,t\rangle}$ of a term $t \in T_{\langle E,s,t\rangle}$ is defined by:

$$[t]_{\langle E,s,t\rangle} = \{s \in T_{\langle E,s,t\rangle} \mid E \models^{\circ}_{\mathcal{E}} s \approx t\} \ .$$

Since a ground E-unification problem $\langle E, s, t\rangle$ is solvable (and *id* a solution) if and only if $[s]_{\langle E,s,t\rangle} = [t]_{\langle E,s,t\rangle}$, one can decide whether $\langle E, s, t\rangle$ is solvable by computing these equivalence classes. Shostak proved that for computing the equivalence classes of all terms in $T_{\langle E,s,t\rangle}$, no terms that are not in $T_{\langle E,s,t\rangle}$ have to be considered: If s can be derived from t using the equalities in E, then this can be done without using an intermediate term that does not occur in the original problem, i.e. there is a sequence of terms $s = r_0, r_1, \ldots, r_k = t$, $k \geq 0$, all occurring in $\langle E, s, t\rangle$ such that r_i is derivable in one step from r_{i-1} using the equalities in E.

Since the number of subterms in a given problem is polynomial in its size, and the congruence closure can be computed in time polynomial in the number of subterms and the number of equalities, the solvability of a ground E-unification problem can be decided in polynomial time.

There are very efficient and sophisticated methods for computing the congruence closure, for example the algorithm described in [Nelson and Oppen, 1980], which is based on techniques from graph theory.

7.5 Solving Universal E-unification Problems

To solve a universal E-unification problem, one has to decide whether the equality of two given terms (or of instances of these terms) follows from E or, equivalently, whether the terms are equal in the free algebra of E. Overviews of methods for universal E-unification can be found in [Siekmann, 1989; Gallier and Snyder, 1990; Jouannaud and Kirchner, 1991; Snyder, 1991].

7.6 Solving Rigid E-unification Problems

Rigid E-unification and its significance for automated theorem proving was first described in [Gallier *et al.*, 1987]. It can be used for equality handling in semantic tableaux and other *rigid variable* calculi for first-order logic, including the mating method [Andrews, 1981], the connection method [Bibel, 1987], and model elimination [Loveland, 1969]; an overview of rigid E-unification can be found in [Beckert, 1998].

The solution to a rigid E-unification problem $\langle E, s, t\rangle$ is a substitution representing the instantiations of free variables that have been necessary to show that the two given terms are equal; it is an $\mathcal{E}$-refuter for the key $E \cup \{\neg(s \approx t)\}$. A single variable can only be instantiated once by a substitution and, accordingly, to solve a rigid E-unification problem, the equalities of the problem can only be used with (at most) one instantiation for each variable they contain; a variable is either instantiated or not, that is, uninstantiated variables have to be treated as constants.

Rigid E-unification does not provide an answer to the question of how many different instantiations of an equality are needed to solve a problem. If a single instance is not sufficient, then the answer is 'not unifiable'. If several different instances of an equality are needed, a sufficient number of copies of that equality (with different rigid variables) has to be provided for the rigid E-unification problem to be solvable.

The following theorem clarifies the basic properties of rigid E-unification by listing different characterizations of the set of solutions of a given problem:

THEOREM 87. *Given a rigid E-unification problem $\langle E, s, t\rangle$ and a substitution $\sigma = \{x_1 \mapsto t_1, \ldots, x_n \mapsto t_n\} \in Subst^*_\Sigma$, the following are equivalent conditions for σ being a solution to $\langle E, s, t\rangle$:*

1. *$E\sigma \models^\circ_\mathcal{E} s\sigma \approx t\sigma$, i.e. σ is by definition a solution to $\langle E, s, t\rangle$;*
2. *$E\sigma \models_\mathcal{E} s\sigma \approx t\sigma$ over a set V^0 of variables and a signature Σ^0 such that the variables occurring in $\langle E, s, t\rangle$ are constants, i.e. $V^0 = V \setminus W$ and $\Sigma^0 = \langle P_\Sigma, F_\Sigma \cup W, \alpha_\Sigma \cup \{x \mapsto 0 \mid x \in W\}\rangle$ where W is the set of variables occurring in $\langle E, s, t\rangle$.*
3. *$(E\sigma)\tau \models_\mathcal{E} (s\sigma)\tau \approx (t\sigma)\tau$ for all substitutions $\tau \in Subst^*_\Sigma$;*
4. *$E \cup \{x_1 \approx t_1, \ldots, x_n \approx t_n\} \models^\circ_\mathcal{E} s \approx t$; provided that none of the variables x_i occurs in any of the terms t_j $(1 \le i, j \le n)$;*
5. *σ is the restriction to the variables occurring in $\langle E, s, t\rangle$ of a substitution which is a solution to the rigid E-unification problem $\langle E', yes, no\rangle$ where $E' = E \cup \{eq(x, x) \approx yes,\ eq(s, t) \approx no\}$, and (a) the constants yes, no, the predicate eq, and the variable x do not occur in $\langle E, s, t\rangle$, and (b) the constants yes, no do not occur in the terms $t_1, \ldots, t_n$.*

The last characterization of solutions in the above theorem shows that it is always possible to solve a rigid E-unification problem by transforming it into a problem in which the terms to be unified are constants.

If a rigid E-unification problem is solvable, then it has infinitely many solutions. But there are, for each problem, *finite* sets of solutions w.r.t. the subsumption relation $\leq^W_{\mathcal{E},E}$ that is defined as follows:

DEFINITION 88. Let $E \subset Form_\Sigma$ be a set of rigid (i.e. quantifier-free) equalities; and let $W \subset V$ be a set of variables. Then the subsumption relations $\sqsubseteq^W_{\mathcal{E},E}$ and $\leq^W_{\mathcal{E},E}$ are on $Subst^*_\Sigma$ defined by:

- $\sigma \sqsubseteq^W_{\mathcal{E},E} \tau$ iff $E\sigma \models^\circ_\mathcal{E} \sigma(x) \approx \tau(x)$ for all $x \in W$;
- $\sigma \leq^W_{\mathcal{E},E} \tau$ iff there is a substitution $\sigma' \in Subst^*_\Sigma$ such that

$$\sigma \leq^W \sigma' \text{ and } \sigma' \sqsubseteq^W_{\mathcal{E},E} \tau .$$

The intuitive meaning of $\sigma \leq^W_{\mathcal{E},E} \tau$ is that the effects of applying τ to the set E of equalities can be simulated by first applying σ, then some other substitution ρ, and then equalities form $(E\sigma)\rho$.

LEMMA 89. *Let $E \subset Form_\Sigma$ be a set of rigid equalities, and let σ, τ be substitutions such that $\sigma \leq^W_{\mathcal{E},E} \tau$ where the set W contains all variables occurring in E. Then there is a substitution ρ such that $(E\sigma)\rho \models^\circ_\mathcal{E} E\tau$.*

It is possible to effectively compute a *finite* set $\mathcal{U}$ of solutions for a rigid E-unification problem $\langle E, s, t\rangle$ that is complete w.r.t. the subsumption relation $\leq^W_{\mathcal{E},E}$, i.e. for every solution σ of $\langle E, s, t\rangle$ there is a solution τ in $\mathcal{U}$ such that $\tau \leq^W_{\mathcal{E},E} \sigma$. This immediately implies the decidability of the question whether a given rigid E-unification problem $\langle E, s, t\rangle$ is solvable or not. On first sight this might be somewhat surprising since universal E-unification is undecidable; however, the additional restriction of rigid E-unification, that variables in E may only be instantiated once, is strong enough to turn an undecidable problem into a decidable one.

The problem of deciding whether a rigid E-unification problem has a solution is, in fact, NP-complete. This was first proven in [Gallier *et al.*, 1988] and then, more detailed, in [Gallier *et al.*, 1990; Gallier *et al.*, 1992]. The NP-hardness of the problem was already shown in [Kozen, 1981]. An alternative proof for the decidability of rigid E-unification was presented in [de Kogel, 1995], it is easy to understand but uses an inefficient decision procedure. More efficient methods using term rewriting techniques are described in [Gallier *et al.*, 1992; Becher and Petermann, 1994; Plaisted, 1995]. The procedure described in [Becher and Petermann, 1994] has been implemented and integrated into a prover for first-order logic with equality [Grieser, 1996].

7.7 Rigid Basic Superposition

In [Degtyarev and Voronkov, 1998], a method called *rigid basic superposition* has been presented for computing a *finite* (incomplete) set of solutions for rigid

E-unification problems that is 'sufficient' for handling equality in rigid variable calculi, i.e. can be used to build a complete free variable background reasoner for the equality theory $\mathcal{E}$. The procedure is an adaptation of basic superposition (in the formulation presented in [Nieuwenhuis and Rubio, 1995]) to rigid variables. It uses the concept of ordering constraints:

DEFINITION 90. An *(ordering) constraint* is a (finite) set of expressions of the form $s \simeq t$ or $s \succ t$ where s and t are terms. A substitution σ is a solution to a constraint C iff (a) $s\sigma = t\sigma$ for all $s \simeq t \in C$, i.e. σ is a unifier of s and t, (b) $s\sigma > t\sigma$ for all $s \succ t \in C$, where $>$ is an arbitrary but fixed term reduction ordering, and (c) σ instantiates all variables occurring in C with ground terms.

There are efficient methods for deciding the satisfiability of an ordering constraint C and for computing most general substitutions satisfying C in case the reduction ordering $>$ is a lexicographic path ordering (LPO) [Nieuwenhuis and Rubio, 1995].

The rigid basic superposition calculus consists of the two transformation rules shown below. They are applied to a rigid E-unification problem $\langle E, s, t\rangle \cdot C$ that has an ordering constraint C attached to it. The computation starts initially with the unification problem that is to be solved and the empty constraint. A transformation rule may be applied to $\langle E, s, t\rangle \cdot C$ only if the constraint is satisfiable before and after the application.

Left rigid basic superposition. If there are an equality $l \approx r$ or $r \approx l$ and an equality $u \approx v$ or $v \approx u$ in E and l' is a subterm of u, then replace the latter equality by $u[r] \approx v$ (where $u[r]$ is the result of replacing one occurrence of l' in u by r) and add $l \succ r$, $u \succ v$, and $l \simeq l'$ to C.

Right rigid basic superposition. If there is an equality $l \approx r$ or $r \approx l$ in E and l' is a subterm of s or of t, then replace s (resp. t) with $s[r]$ (resp. $t[r]$) and add $l \succ r$, $s \succ t$ (resp. $t \succ s$) and $l \simeq l'$ to C.

As the constraint expressions that are added by a rule application have to be satisfiable, they can be seen as a pre-condition for that application; for example, since $l \simeq l'$ is added to C, the terms l and l' have to be unifiable.

The two transformation rules are repeatedly applied, forming a non-deterministic procedure for transforming rigid E-unification problems. The process terminates when (a) the terms s and t become identical or (b) no further rule application is possible without making C inconsistent. Provided that no transformation is allowed that merely replaces an equality by itself, all transformation sequences are finite.

It is possible to only allow transformations where the term l' is *not* a variable, thus improving the efficiency of the procedure and reducing the number of solutions that are computed.

Let $\langle E, s, t\rangle \cdot C$ be any of the unification problems that are reachable by applying rigid basic superposition transformations to the original problem. Then, any solution to $C \cup \{s \simeq t\}$ is a solution to the original problem. Let $\mathcal{U}$ be the set of all such solutions that are most general w.r.t. $\leq^W$. The set $\mathcal{U}$ is finite because the application of rigid basic superposition rules always terminates.

EXAMPLE 91. Consider the rigid E-unification problem[5]

$$\langle E, s, t\rangle = \langle\{fa \approx a,\ g^2x \approx fa\},\ g^3x,\ x\rangle\ ,$$

and let $>$ be the LPO induced by the ordering $g > f > a$ on the function symbols.

The computation starts with

$$\langle E, s, t\rangle \cdot C = \langle\{fa \approx a,\ g^2x \approx fa\},\ g^3x,\ x\rangle \cdot \emptyset\ .$$

The only possible transformation is to use the right rigid basic superposition rule, applying the equality $(l \approx r) = (g^2x \approx fa)$ to reduce the term g^3x (all other transformations would lead to an inconsistent constraint). The result is the unification problem $\langle E, gfa, x\rangle \cdot \{g^2x \succ fa,\ g^3x \succ x,\ g^2x \simeq g^2x\}$; its constraint can be reduced to $C_1 = \{g^2x \succ fa\}$. A most general substitution satisfying $C_1 \cup \{gfa \simeq x\}$ is $\sigma_1 = \{x \mapsto gfa\}$.

A second application of the right rigid basic superposition rule leads to the unification problem $\langle E, ga, x\rangle \cdot \{g^2x \succ fa,\ fa \succ a,\ gfa \succ x,\ fa \simeq fa\}$; its constraint can be reduced to $C_2 = \{g^2x \succ fa,\ gfa \succ x\}$. A most general substitution satisfying $C_2 \cup \{ga \simeq x\}$ is $\sigma_2 = \{x \mapsto ga\}$.

At that point the process terminates because no further rule application is possible. Thus, σ_1 and σ_2 are the only solutions that are computed by rigid basic superposition for this example.

7.8 *Solving Mixed E-unification Problems*

Since universal E-unification is already undecidable, *mixed* E-unification is—in general—undecidable as well. It is, however, possible to enumerate a complete set of MGUs.

EXAMPLE 92. The following example requires only very little non-equality reasoning. A powerful equality handling technique is needed to find a closed tableau, and the universal formula version of tableaux has to be used to restrict the search space: If Γ consists of the axioms[6]

$$(\forall x)(i(tr, x) \approx x)$$
$$(\forall x)(\forall y)(\forall z)(i(i(x,y), i(i(y,z), i(x,z))) \approx tr)$$
$$(\forall x)(\forall y)(i(i(x,y),y) \approx i(i(y,x),x))$$

[5] In this example, we use g^2x as an abbreviation for $g(g(x))$, etc.

[6] This is an axiomatization of propositional logic, $i(x,y)$ stands for 'x implies y' and tr for 'true'.

$$\Gamma$$

$$(1)\ \neg(\forall x)(\forall y)(\forall z)(\exists w)(i(x,w) \approx tr \wedge w \approx i(y,i(z,y)))$$

$$(2)\ \neg(i(c_1,w_1) \approx tr \wedge w_1 \approx i(c_2,i(c_3,c_2)))$$

$$(3)\ \neg(i(c_1,w_1) \approx tr) \qquad (4)\ \neg(w_1 \approx i(c_2,i(c_3,c_2)))$$

Figure 5. The tableau that has to be closed to prove the theorem from Example 92

then

$$\Gamma \models_{\mathcal{E}} (\forall x)(\forall y)(\forall z)(\exists w)(i(x,w) \approx tr \wedge w \approx i(y,i(z,y))) .$$

To prove this, the tableau shown in Figure 5 has to be closed. Formula (2) is derived from the negated theorem (1) by three δ- and one γ-rule application; (3) and (4) are derived from (2).

To close the left branch, the E-unification problem

$$P_l = \langle \Gamma,\ i(c_1,w_1),\ tr \rangle$$

has to be solved, and the problem

$$P_r = \langle \Gamma,\ w_1,\ i(c_2,i(c_3,c_2)) \rangle$$

has to be solved to close the right branch.

The search for solutions performed by the tableau-based theorem prover $_3T\!AP$ [Beckert *et al.*, 1996], that uses a completion-based method for finding solutions of mixed E-unification problems, proceeds as follows. One of the first rules that are deduced from Γ is the reduction rule $(\forall x)(i(x,x) \rightarrow tr)$. Using this rule, the solution $\sigma = \{w_1 \mapsto c_1\}$ to the problem P_l is found and applied to the tableau. Then the Problem $P_r\sigma$ has to be solved to close the right branch; unfortunately, no solution exists. Thus, after a futile try to close the right branch, backtracking is initiated. More reduction rules are computed until finally the rule $(\forall x)(i(x,tr) \rightarrow tr)$ is applied to the problem P_l and the solution $\sigma' = \{w_1 \mapsto tr\}$ is found. Now the problem $P_r\sigma'$ has to be solved to close the right branch. It takes the computation of 48 critical pairs to deduce the rule $(\forall x)(\forall y)(i(y,i(x,y)) \rightarrow tr)$ which can be applied to show that the empty substitution is a solution to $P_r\sigma'$ and that therefore the right branch is closed.

7.9 *Simultaneous E-unification*

Instead of closing one branch after the other, one can search for a simultaneous refuter for all branches of a tableau. However, this is much more difficult than closing a single branch. Although (non-simultaneous) *rigid* E-unification is decidable, it is undecidable whether a simultaneous solution to several E-unification problems exists [Degtyarev and Voronkov, 1996a]. It is as well undecidable whether there is a substitution closing all branches of a given free variable tableau simultaneously after it has been expanded by a *fixed* number of copies of the universally quantified formulae it contains [Voda and Komara, 1995; Gurevich and Veanes, 1997].

In the same way as it may be surprising on first sight that simple rigid E-unification is decidable, it may be surprising that moving from simple to simultaneous problems destroys decidability—even more so considering that the simultaneous versions of other decidable types of unification (including syntactical unification and ground E-unification) are decidable. However, simultaneous rigid E-unification turns out to have a much higher expressiveness than simple rigid E-unification; it is even possible to encode Turing Machines into simultaneous rigid E-unification problems [Veanes, 1997]. For an overview of simultaneous rigid E-unification see [Degtyarev and Voronkov, 1998; Beckert, 1998].

Since simultaneous rigid E-unification is undecidable, sets of unifiers can only be enumerated; in general they are not finite. Solutions to a simultaneous problem can be computed combining solutions to its constituents $\langle E_i, s_i, t_i \rangle$; however, it is not possible to compute a finite complete set of unifiers of the simultaneous problem by combining solutions from finite sets of unifiers of the constituents that are complete w.r.t. the subsumption relation $\leq^W_{\mathcal{E},E}$, because they are complete w.r.t. different relations $\leq^W_{\mathcal{E},E_i}$. Thus, the subsumption relation $\leq^W_{\mathcal{E}}$ has to be used, which is the same for all i (but does not allow to construct *finite* complete sets of unifiers).

The undecidability of simultaneous rigid E-unification implies that, if a background reasoner produces only a *finite* number of solutions to any (non-simultaneous) rigid E-unification problem, then closing a tableau T may require to extend T by additional instances of equalities and terms even if there is a substitution that closes all branches of T simultaneously and there is, thus, a solution to a simultaneous rigid E-unification problem extracted from T. That notwithstanding, the background reasoner *may* be complete; and in that case the advantages of finite sets of solutions prevail. A complete background reasoner of this type can be built using rigid basic superposition (Section 7.7). It is not known whether the same can be achieved using (finite) sets of unifiers that are complete w.r.t. the subsumption relation $\leq^W_{\mathcal{E},E}$.

8 INCREMENTAL THEORY REASONING

Besides the efficiency of the foreground and the background reasoner, the interaction between them plays a critical rôle for the efficiency of the combined system:

It is a difficult problem to decide whether it is useful to call the background reasoner at a certain point or not, and how much time and other resources to spend for its computations. In general, giving a perfect answer to these questions is as difficult as the theory reasoning problem itself. Even with good heuristics at hand, one cannot avoid calling the background reasoner at the wrong point: either too early or too late.

This problem can (at least partially) be avoided by using incremental methods for background reasoning [Beckert and Pape, 1996], i.e. algorithms that—after a futile try to solve a theory reasoning problem—allow to save the results of the background reasoner's computations and to reuse this data for a later call.[7] Then, in case of doubt, the background reasoner can be called early without running the risk of doing useless computations. In addition, an incremental background reasoner can reuse data multiply if different extensions of a problem have to be handled. An important example are completion-based methods for equality reasoning, which are inherently incremental.

As already mentioned in Section 2.5, one of the main problems in using theorem reasoning techniques in practice is the efficient combination of foreground and background reasoner and their interaction—in particular if (a) the computation steps of the background reasoner are comparatively complex, and (b) in case calling the background reasoner may be useless because no refuter exists or can be found.

On the one hand, a late call to the background reasoner can lead to bigger tableaux and redundancy. Although several branches may share the same subbranch and thus contain the same key for which a refuter exists, the background reasoner is called separately for these branches and the refuter has to be computed repeatedly. On the other hand, an early call to the background reasoner may not be successful and time consuming; this is of particular disadvantage if the existence of a refuter is undecidable and, as a result, the background reasoner does not terminate although no refuter exists.

Both these phenomena may considerably decrease the performance of a prover, and it is very difficult to decide (resp. to develop good heuristics which decide)

1. when to call the background reasoner;

2. when to stop the background reasoner if it does not find a refuter.

EXAMPLE 93. The following example shows that earlier calls to the background reasoner can reduce the size of a tableau proof exponentially. Let $\Gamma \subset Form_\Sigma$ be a set of formulae and let $\phi_n \in Form_\Sigma$, $n \geq 0$, be formulae such that, for some theory $\mathcal{T}$, $\Gamma \models_\mathcal{T} \neg\phi_n$ $(n \geq 0)$. Figure 6 shows a proof for

$$\Gamma \models_\mathcal{T} \phi_0 \leftrightarrow \phi_1 \leftrightarrow \cdots \leftrightarrow \phi_n ,$$

[7] This should not be confused with deriving a refuter and handing it back to the foreground reasoner. The information derived by an incremental background reasoner cannot be used by the foreground reasoner, but only by the background reasoner during later calls.

$$\Gamma$$
$$\neg(\phi_0 \leftrightarrow \phi_1 \leftrightarrow \cdots \leftrightarrow \phi_n)$$
$$\phi_0 \wedge \neg(\phi_1 \leftrightarrow \cdots \leftrightarrow \phi_n) \qquad \neg\phi_0 \wedge (\phi_1 \leftrightarrow \cdots \leftrightarrow \phi_n)$$
$$\phi_0 \qquad \neg\phi_0$$
$$\neg(\phi_1 \leftrightarrow \cdots \leftrightarrow \phi_n) \qquad \phi_1 \leftrightarrow \cdots \leftrightarrow \phi_n$$
$$\neg\phi_{n-1}$$
$$\phi_{n-1} \leftrightarrow \phi_n$$
$$\phi_{n-1} \wedge \neg\phi_n \qquad \neg\phi_{n-1} \wedge \phi_n$$
$$\phi_{n-1} \qquad \neg\phi_{n-1}$$
$$\neg\phi_n \qquad \phi_n$$

Figure 6. Short tableau proof for $\Gamma \models_{\mathcal{T}} \phi_0 \leftrightarrow \cdots \leftrightarrow \phi_n$ (Example 93)

where the background reasoner is called when a literal of the form ϕ_n appears on a branch (with the key $\Phi = \Gamma \cup \{\phi_n\}$). As a result, all the left-hand branches are closed immediately and the tableau is of linear size in n.

If the background reasoner were only called when a branch is exhausted, i.e. when no further expansion is possible, then the tableau would have 2^n branches and the background reasoner would have to be called 2^n times (instead of n times).

An *incremental* background reasoner can be of additional advantage if the computations that are necessary to show that $\Gamma \models_{\mathcal{T}} \neg\phi_n$ are similar for all n. In that case, a single call to the background reasoner in the beginning may provide information that later can be reused to close all the branches with less effort.

Even the best heuristics cannot avoid calls to the background reasoner at the wrong time. However, under certain conditions, it is possible to avoid the adverse consequences of early calls: If the algorithm that the background reasoner uses is *incremental*, i.e. if the data produced by the background reasoner during a futile try to compute refuters can be reused for a later call.

If early calls have no negative effects, the disadvantages of late calls can easily be avoided by using heuristics that, in case of doubt, call the background reasoner at an early time. The problem of not knowing when to stop the background reasoner is solved by calling it more often with less resources (time, etc.) for each call.

An additional advantage of using incremental background reasoners in the tableau framework is that computations can be reused repeatedly for different extensions of a branch—even if the computation of refuters proceeds differently for these extensions.

8.1 Incremental Keys and Algorithms

Obviously, there has to be some strong relationship between the keys transferred to the background reasoner, to make it possible to reuse the information computed. Since, between calls to the background reasoner, (1) the tableau may be extended by new formulae and (2) substitutions (refuters) may be applied (to the tableau), these are the two operations we allow for changing the key:

DEFINITION 94. A sequence $(\Phi_i)_{i\geq 0}$ of keys is *incremental* if, for $i \geq 0$, there is a set $\Psi_i \subset Form_\Sigma$ of formulae and a substitution σ_i such that $\Phi_{i+1} = \Phi_i\sigma_i \cup \Psi_i$, where $\Psi_i = \Psi_i\sigma_i$.

In general, not all refuters of Φ_i are refuters of Φ_{i+1} (because a substitution is applied); nor are all refuters of Φ_{i+1} refuters of Φ_i (because new formulae are added).

To be able to formally denote the state the computation of a background reasoner has reached and the data generated, we use the following notion of incremental background reasoner:

DEFINITION 95. An *incremental background reasoner* $\mathcal{R}_{\mathcal{A},\mathcal{I},\mathcal{S}}$ is a background reasoner (Definition 31) that can be described using

1. an *algorithm* (a function) $\mathcal{A} : \mathcal{D} \to \mathcal{D}$ operating on a data structure $\mathcal{D}$,
2. an initialization function $\mathcal{I} : 2^{Form_\Sigma} \to \mathcal{D}$ that transforms a given key into the data structure format, and
3. an output function $\mathcal{S}$ that extracts computed refuters from the data structure,

such that for every key $\Phi \subset Form_\Sigma$ for which $\mathcal{R}_{\mathcal{A},\mathcal{I},\mathcal{S}}$ is defined

$$\mathcal{R}_{\mathcal{A},\mathcal{I},\mathcal{S}}(\Phi) = \bigcup_{i\geq 0} \mathcal{S}(\mathcal{A}^i(\mathcal{I}(\Phi))) \ .$$

The above definition does not restrict the type of algorithms that may be used; every background reasoner whose computations proceed in steps can be described this way. If a background reasoner applies different transformations to the data at each step of its computation, this can be modeled by adding the state of the reasoner to the data structure such that the right operation or sub-algorithm can be applied each time the background reasoner is invoked.

Of course, the input and output functions have to be reasonably easy to compute; in particular, the cost of their computation has to be much smaller than that of applying the algorithm $\mathcal{A}$, which is supposed to do the actual work.

The goal is to be able to stop the background reasoner when it has reached a certain state in its computations for a key Φ, and to proceed from that state with a new key $\Phi' = \Phi\sigma \cup \Psi$. For that purpose, an update function is needed that adapts the data structure representing the state of the computation to the new formulae Ψ and the substitution σ.

DEFINITION 96. Let $\mathcal{T}$ be a theory and $\mathcal{R}_{\mathcal{A},\mathcal{I},\mathcal{S}}$ a complete incremental background reasoner for $\mathcal{T}$. An update function

$$\mathcal{U} : \mathcal{D} \times 2^{Form_\Sigma} \times Subst^*_\Sigma \longrightarrow \mathcal{D}$$

is *correct* (for $\mathcal{R}_{\mathcal{A},\mathcal{I},\mathcal{S}}$) if a complete background reasoner $\mathcal{R}'_{\mathcal{A},\mathcal{I},\mathcal{S}}$ is defined by: for every key Φ

1. choose $\Phi' \subset Form_\Sigma$ and $\sigma \in Subst^*_\Sigma$ such that $\Phi = \Phi'\sigma \cup \Psi$ arbitrarily;
2. compute $D_n = \mathcal{U}(\mathcal{A}^n(\mathcal{I}(\Phi')),\ \Psi,\ \sigma)$ for an arbitrary $n \geq 0$;
3. set $\mathcal{R}'_{\mathcal{A},\mathcal{I},\mathcal{S}}(\Phi) = \bigcup_{i \geq 0} \mathcal{S}(\mathcal{A}^i(D_n))$.

According to the above definition, a correct update function behaves as expected when used for a single incremental step. Theorem 97 shows that this behavior extends to sequences of incremental steps. In addition, the algorithm can be applied arbitrarily often between incremental steps:

THEOREM 97. *Let $\mathcal{T}$ be a theory, $\mathcal{R}_{\mathcal{A},\mathcal{I},\mathcal{S}}$ a complete incremental background reasoner for $\mathcal{T}$, and $\mathcal{U}$ a correct update function for $\mathcal{R}_{\mathcal{A},\mathcal{I},\mathcal{S}}$.*

Then $\mathcal{R}^*_{\mathcal{A},\mathcal{I},\mathcal{S}}$ is a complete background reasoner for $\mathcal{T}$ that is defined by: for every key Φ

1. choose an arbitrary incremental sequence $(\Phi_i)_{i \geq 0}$ of keys where

$$\Phi_{i+1} = \Phi_i\sigma_i \cup \Psi_i \qquad (i \geq 0) ,$$

and $\Phi = \Phi_k$ for some $k \geq 0$;

2. let $(D_i)_{i \geq 0} \subset \mathcal{D}$ be defined by

(a) $D_0 = \mathcal{I}(\Phi_0)$,

(b) $D_{i+1} = \mathcal{U}(\mathcal{A}^{n_i}(D_i),\ \sigma_{i+1},\ \Psi_{i+1})$ for some $n_i \geq 0$;

3. set $\mathcal{R}^*_{\mathcal{A},\mathcal{I},\mathcal{S}}(\Phi) = \bigcup_{j \geq 0} \mathcal{S}(\mathcal{A}^j(D_k))$.

EXAMPLE 98. Let $(\Phi_i)_{i \geq 0}$ be an incremental sequence of keys such that $\Phi_{i+1} = \Phi_i\sigma_i \cup \Psi_i$ $(i \geq 0)$. Then, for every sound and complete incremental background reasoner $\mathcal{R}_{\mathcal{A},\mathcal{I},\mathcal{S}}$, the trivial update function defined by

$$\mathcal{U}(D, \Psi_i, \sigma_i) = \mathcal{I}(\Phi_i\sigma_i \cup \Psi_i)$$

is correct.

The above example shows that it is not sufficient to use any correct update function to achieve a better performance of the calculus, because using the trivial update function means that no information is reused. A useful update function has to preserve the information contained in the computed data.

Whether there actually is a useful and reasonably easy to compute update function depends on the theory $\mathcal{T}$, the background reasoner, and its data structure.

Such a useful update function exists for a background reasoner for completion-based equality handling [Beckert and Pape, 1996]. Another important example are background reasoners based on resolution: if a resolvent can be derived from a key Φ, then it is valid for all extensions $\Phi \cup \Psi$ of Φ; resolvents may be invalid for an instance $\Phi\sigma$ of the key, but to check this is much easier than to re-compute all resolvents. In [Baumgartner, 1996], a uniform translation from Horn theories to partial background reasoners based on unit-resulting positive hyper-resolution with input restriction is described. This procedure can be used to generate incremental background reasoners for a large class of theories.

8.2 Semantic Tableaux and Incremental Theory Reasoning

The incremental theory reasoning method presented in the previous section is easy to use for tableau-like calculi, because the definition of incremental sequences of keys matches the construction of tableau branches. The keys of a sequence are taken from an expanding branch, and the substitutions are those applied to the whole tableau.

The keys used in calls to the background reasoner as well as the information computed so far by the background reasoner have to be attached to the tableau branches:

DEFINITION 99. A *tableau for incremental theory reasoning* is a (finite) multi-set of tableau branches where a tableau *branch* is a triple $\langle \Theta, D, \Phi \rangle$; Θ is a (finite) multi-set of first-order formulae, $D \in \mathcal{D}$ (where $\mathcal{D}$ is the data structure used by the background reasoner), and $\Phi \subset Form_\Sigma$ is a key.

Now, the tableau calculus with theory reasoning introduced in Section 3.4 can be adapted to *incremental* theory reasoning: calling the background reasoner is added as a further possibility of changing the tableau (besides expanding and closing branches).

DEFINITION 100 (Incremental reasoning version). Given a theory $\mathcal{T}$, an incremental background reasoner $\mathcal{R}_{\mathcal{A},\mathcal{I},\mathcal{S}}$ for $\mathcal{T}$ (Definition 95), and a correct update function $\mathcal{U}$ for $\mathcal{R}_{\mathcal{A},\mathcal{I},\mathcal{S}}$ (Definition 96), an *incremental theory reasoning tableau proof* for a first-order sentence ϕ consists of a sequence

$$\{\{\neg\phi\}\} = T_0, T_1, \ldots, T_{n-1}, T_n = \emptyset \qquad (n \geq 0)$$

of tableaux such that, for $1 \leq i \leq n$, the tableau T_i is constructed from T_{i-1}

1. by applying one of the expansion rules from Table 6, i.e. there is a branch $B = \langle \Theta, D, \Phi \rangle \in T_{i-1}$ and a formula $\phi \in \Theta$ (that is not a literal) such that

$$T_i \;=\; (T_{i-1} \setminus \{B\})$$

$$\cup \begin{cases} \{\langle(\Theta \setminus \{\alpha\}) \cup \{\alpha_1, \alpha_2\}, D, \Phi\rangle\} & \text{if } \phi = \alpha \\ \{\langle(\Theta \setminus \{\phi\}) \cup \{\beta_1\}, D, \Phi\rangle, & \\ \quad \langle(\Theta \setminus \{\beta\}) \cup \{\beta_2\}, D, \Phi\rangle\} & \text{if } \phi = \beta \\ \{\langle\Theta \cup \{\gamma_1(y)\}, D, \Phi\rangle\} & \text{if } \phi = \gamma(x) \\ \{\langle(\Theta \setminus \{\phi\}) \cup \{\delta_1(f(x_1, \ldots, x_n))\}, D, \Phi\rangle\} & \text{if } \phi = \delta(x) \end{cases}$$

where $y \in V$ is a new variable not occurring in T_{i-1}, $f \in F_\Sigma$ is a Skolem function symbol not occurring in T_{i-1} nor in $\mathcal{T}$, and $x_1, \ldots, x_n$ are the free variables in ϕ;

2. by applying the incremental theory expansion rule, i.e. there is a branch $B = \langle\Theta, D, \Phi\rangle$ in T_{i-1} and a $\mathcal{T}$-refuter $\langle\sigma, \{\rho_1, \ldots, \rho_k\}\rangle$ ($k \geq 1$) in the set $\mathcal{S}(D)$, and

$$\begin{aligned} T_i \quad = \quad & \{\langle\Theta'\sigma, D', \Phi'\rangle \mid \langle\Theta', D', \Phi'\rangle \in (T_{i-1} \setminus \{B\})\} \cup \\ & \{\langle\Theta \cup \{\rho_j\}, D, \Phi\rangle \mid 1 \leq j \leq k\} \end{aligned}$$

3. by applying the incremental theory closure rule, i.e. there is a branch $B = \langle\Theta, D, \Phi\rangle$ in T_{i-1} that is $\mathcal{T}$-closed under σ, i.e. $\sigma \in \mathcal{S}(D)$, and

$$T_i = \{\langle\Theta'\sigma, D', \Phi'\rangle \mid \langle\Theta', D', \Phi'\rangle \in (T_{i-1} \setminus \{B\})\}$$

4. or by calling the background reasoner, i.e. there is a branch $B = \langle\Theta, D, \Phi\rangle$ in T_{i-1}, a number $c > 0$ of applications, and a key

$$\Phi' = \Phi\sigma \cup \Psi \subset \Theta = \{(\forall x_1^i) \cdots (\forall x_{m_i}^i)\phi_i \mid 1 \leq i \leq p\}$$

where

(a) $\phi_1, \ldots, \phi_p \in \Theta$,

(b) $\{x_1^i, \ldots, x_{m_i}^i\} \subset UVar(\phi_i)$ for $1 \leq i \leq p$.

and

$$T_i = (T_{i-1} \setminus \{B\}) \cup \{\langle\Theta, \mathcal{A}^c(\mathcal{U}(D, \Psi, \sigma)), \Phi'\rangle\} .$$

Soundness and completeness of the resulting calculus are a corollary of Theorems 47, 58, and 97:

THEOREM 101. *Let $\phi \in Form_\Sigma$ be a sentence. If there is an incremental tableau proof for ϕ (Definition 100), then ϕ is a $\mathcal{T}$-tautology.*

If $\mathcal{R}_{\mathcal{A},\mathcal{I},\mathcal{S}}$ is a complete incremental background reasoner for a theory $\mathcal{T}$ and the formula ϕ is a $\mathcal{T}$-tautology, then an incremental tableau proof for ϕ can be constructed using $\mathcal{R}_{\mathcal{A},\mathcal{I},\mathcal{S}}$.

The maximal cost reduction that can be achieved by using an incremental reasoner is reached if the costs are those of the non-incremental background reasoner called neither too early nor too late, i.e. if always the right key in the incremental sequence is chosen and the background reasoner is only called for that key (which is not possible in practice).

In practice, the costs of an incremental method are between the ideal value and the costs of calling a non-incremental reasoner for each of the keys in an incremental sequence (without reusing).

But even if the costs for one sequence, i.e. for closing one tableau branch, are higher than those of using a non-incremental method, the overall costs for closing the whole tableau can be small, because information is reused for more than one branch.

9 EQUALITY REASONING BY TRANSFORMING THE INPUT

Methods based on transforming the input are inherently not specific for semantic tableaux (although they might be more suitable for tableaux than for other calculi). They do not require the tableau calculus to be adopted to theory reasoning.

The simplest—however useless—method is to just add the theory axioms to the input formulae. A better way to 'incorporate' the equality axioms into the formulae to be proven is D. Brand's STE-modification [Brand, 1975], which is described below. An improved transformation using term orderings has been presented in [Bachmair *et al.*, 1997].

Usually, STE-modification is only defined for formulae in clausal form; but since a transformation to clausal form may be of disadvantage for non-normal form calculi like semantic tableaux, we present an adaptation of Brand's method for formulae in Skolemized negation normal form.

DEFINITION 102. Let ϕ be a formula in Skolemized negation normal form. The *E-modification* of ϕ is the result of applying the following transformations iteratively as often as possible:

1. If a literal of the form $p(\dots, s, \dots)$ or $\neg p(\dots, s, \dots)$ occurs in the formula where $s \notin V$, then replace it by

$$(\forall x)(\neg(s \approx x) \vee p(\dots, x, \dots)) \text{ resp.}$$
$$(\forall x)(\neg(s \approx x) \vee \neg p(\dots, x, \dots))$$

where x is a new variable.

2. If an equality of the form $f(\dots, s, \dots) \approx t$ or $t \approx f(\dots, s, \dots)$ occurs in the formula where $s \notin V$, then replace it by

$$(\forall x)(\neg(s \approx x) \vee f(\dots, x, \dots) \approx t)$$

where x is a new variable.

The *STE-modification* of ϕ is the result of (non-iteratively) replacing in the E-modification ϕ' of ϕ all equalities $s \approx t$ by

$$(\forall x)(\neg(t \approx x) \vee s \approx x) \wedge (\forall x)(\neg(s \approx x) \vee t \approx x)$$

where x is a new variable.

EXAMPLE 103. The STE-modification of $(\forall x)(p(f(a, g(x))))$ is

$$(\forall x)(\forall u)(\forall v)(\forall w)(\neg(a \approx u) \vee \neg(g(x) \approx v) \vee \neg(f(u,v) \approx w) \vee p(w)) .$$

The STE-modification of $(\forall x)(\forall y)(f(x,y) \approx g(a))$ is

$$\begin{aligned}(\forall x)(\forall y)(\forall z)(\neg(a \approx z) \vee ((\forall u)(\neg(f(x,y) \approx u) \vee g(z) \approx u) \wedge \\ (\forall u)(\neg(g(z) \approx u) \vee f(x,y) \approx u))) .\end{aligned}$$

To prove the $\mathcal{E}$-unsatisfiability of the STE-modification of a formula, it is sufficient to use the reflexivity axiom; symmetry, transitivity and monotonicity axioms are not needed any more.

THEOREM 104 (Brand, 1975). *Let ϕ be a sentence in Skolemized negation normal form, and let ϕ' be the STE-modification of ϕ. Then ϕ is $\mathcal{E}$-unsatisfiable if and only if*

$$\phi' \wedge (\forall x)(x \approx x)$$

is unsatisfiable.

10 CONCLUSION

We have given an overview of how to design the interface between semantic tableaux (the foreground reasoner) and a theory background reasoner. The problem of handling a certain theory has been reduced to finding an efficient background reasoner for that theory. The search for efficient methods has not come to an end, however, because there is no universal recipe for designing background reasoners. Nevertheless, some criteria have been presented that a background reasoner should satisfy and useful features it should have.

Specialized methods have been presented for handling equality; the most efficient of these are based on E-unification techniques. Similar to the design of background reasoners in general, the problem of developing E-unification procedures is difficult to solve in a uniform way. The research in the field of designing such procedures for certain equality theories has produced a huge amount of results, that is still rapidly growing, in particular for rigid and mixed E-unification.

ACKNOWLEDGEMENTS

I would like to thank Peter Baumgartner, Marcello D'Agostino, Paliath Narendran, and Christian Pape for fruitful comments on earlier versions of this chapter.

Universität Karlsruhe, Germany.

REFERENCES

[Andrews, 1981] P. B. Andrews. Theorem proving through general matings. *Journal of the ACM*, 28:193–214, 1981.

[Bachmair *et al.*, 1997] L. Bachmair, H. Ganzinger and A. Voronkov. Elimination of equality via transformation with ordering constraints. Technical Report MPI-I-97-2-012, MPI für Informatik, Saarbrücken, 1997.

[Baumgartner, 1992] P. Baumgartner. A model elimination calculus with built-in theories. In H.-J. Ohlbach, editor, *Proceedings, German Workshop on Artificial Intelligence (GWAI)*, LNCS 671, pages 30–42. Springer, 1992.

[Baumgartner, 1996] P. Baumgartner. Linear and unit-resulting refutations for Horn theories. *Journal of Automated Reasoning*, 16(3):241–319, 1996.

[Baumgartner, 1998] P. Baumgartner. *Theory Reasoning in Connection Calculi*. LNCS. Springer, 1998. To appear.

[Baumgartner *et al.*, 1992] P. Baumgartner, U. Furbach, and U. Petermann. A unified approach to theory reasoning. Forschungsbericht 15/92, University of Koblenz, 1992.

[Baumgartner and Petermann, 1998] P. Baumgartner and U. Petermann. Theory reasoning. In W. Bibel and P. H. Schmitt, editors, *Automated Deduction – A Basis for Applications*, volume I. Kluwer, 1998.

[Becher and Petermann, 1994] G. Becher and U. Petermann. Rigid unification by completion and rigid paramodulation. In B. Nebel and L. Dreschler-Fischer, editors, *Proceedings, 18th German Annual Conference on Artificial Intelligence (KI-94), Saarbrücken, Germany*, LNCS 861, pages 319–330. Springer, 1994.

[Beckert, 1994] B. Beckert. A completion-based method for mixed universal and rigid E-unification. In A. Bundy, editor, *Proceedings, 12th International Conference on Automated Deduction (CADE), Nancy, France*, LNCS 814, pages 678–692. Springer, 1994.

[Beckert, 1997] B. Beckert. Semantic tableaux with equality. *Journal of Logic and Computation*, 7(1):39–58, 1997.

[Beckert, 1998] B Beckert. Rigid E-unification. In W. Bibel and P. H. Schmitt, editors, *Automated Deduction – A Basis for Applications*, volume I. Kluwer, 1998.

[Beckert and Hähnle, 1992] B. Beckert and R. Hähnle. An improved method for adding equality to free variable semantic tableaux. In D. Kapur, editor, *Proceedings, 11th International Conference on Automated Deduction (CADE), Saratoga Springs, NY, USA*, LNCS 607, pages 507–521. Springer, 1992.

[Beckert and Hähnle, 1998] B. Beckert and R. Hähnle. Analytic tableaux. In W. Bibel and P. H. Schmitt, editors, *Automated Deduction – A Basis for Applications*, volume I. Kluwer, 1998.

[Beckert *et al.*, 1996] B. Beckert, R. Hähnle, P. Oel and M. Sulzmann. The tableau-based theorem prover $_3T^AP$, version 4.0. In *Proceedings, 13th International Conference on Automated Deduction (CADE), New Brunswick, NJ, USA*, LNCS 1104, pages 303–307. Springer, 1996.

[Beckert and Pape, 1996] B. Beckert and C. Pape. Incremental theory reasoning methods for semantic tableaux. In P. Miglioli, U. Moscato, D. Mundici, and M. Ornaghi, editors, *Proceedings, 5th Workshop on Theorem Proving with Analytic Tableaux and Related Methods, Palermo, Italy*, LNCS 1071, pages 93–109. Springer, 1996.

[Bibel, 1987] W. Bibel. *Automated Theorem Proving*. Vieweg, Braunschweig, second edition, 1987. First edition published in 1982.

[Brand, 1975] D. Brand. Proving theorems with the modification method. *SIAM Journal on Computing*, 4(4):412–430, 1975.

[Browne, 1988] R. J. Browne. Ground term rewriting in semantic tableaux systems for first-order logic with equality. Technical Report UMIACS-TR-88-44, College Park, MD, 1988.

[Bürckert, 1990] H. Bürckert. A resolution principle for clauses with constraints. In *Proceedings, 10th International Conference on Automated Deduction (CADE)*, LNCS 449, pages 178–192. Springer, 1990.

[Cantone *et al.*, 1989] D. Cantone, A. Ferro and E. Omodeo. *Computable Set Theory*, volume 6 of *International Series of Monographs on Computer Science*. Oxford University Press, 1989.

[de Kogel, 1995] E. de Kogel. Rigid E-unification simplified. In *Proceedings, 4th Workshop on Theorem Proving with Analytic Tableaux and Related Methods, St. Goar*, LNCS 918, pages 17–30. Springer, 1995.

[Degtyarev and Voronkov, 1996] A. Degtyarev and A. Voronkov. Equality elimination for the tableau method. In J. Calmet and C. Limongelli, editors, *Proceedings, International Symposium on Design and Implementation of Symbolic Computation Systems (DISCO), Karlsruhe, Germany*, LNCS 1128, pages 46–60, 1996.

[Degtyarev and Voronkov, 1996a] A. Degtyarev and A. Voronkov. Simultaneous rigid E-unification is undecidable. In H. Kleine Büning, editor, *Proceedings, Annual Conference of the European Association for Computer Science Logic (CSL'95)*, LNCS 1092, pages 178–190. Springer, 1996.

[Degtyarev and Voronkov, 1998] A. Degtyarev and A. Voronkov. What you always wanted to know about rigid E-unification. *Journal of Automated Reasoning*, 20(1):47–80, 1998.

[Digricoli and Harrison, 1986] V. J. Digricoli and M. C. Harrison. Equality-based binary resolution. *Journal of the ACM*, 33(2):253–289, April 1986.

[Fitting, 1996] M. C. Fitting. *First-Order Logic and Automated Theorem Proving*. Springer, 1996. First edition, 1990.

[Furbach, 1994] U. Furbach. Theory reasoning in first order calculi. In K. v. Luck and H. Marburger, editors, *Proceedings, Third Workshop on Information Systems and Artificial Intelligence, Hamburg, Germany*, LNCS 777, pages 139–156. Springer, 1994.

[Gallier *et al.*, 1988] J. H. Gallier, P. Narendran, D. Plaisted and W Snyder. Rigid E-unification is NP-complete. In *Proceedings, Symposium on Logic in Computer Science (LICS)*. IEEE Press, 1988.

[Gallier *et al.*, 1990] J. H. Gallier, P. Narendran, D. Plaisted, and W. Snyder. Rigid E-unification: NP-completeness and application to equational matings. *Information and Computation*, pages 129–195, 1990.

[Gallier *et al.*, 1992] J. H. Gallier, P. Narendran, S. Raatz, and W. Snyder. Theorem proving using equational matings and rigid E-unification. *Journal of the ACM*, 39(2):377–429, April 1992.

[Gallier *et al.*, 1987] J. H. Gallier, S. Raatz, and W. Snyder. Theorem proving using rigid E-unification, equational matings. In *Proceedings, Symposium on Logic in Computer Science (LICS), Ithaka, NY, USA*. IEEE Press, 1987.

[Gallier and Snyder, 1990] J. H. Gallier and W. Snyder. Designing unification procedures using transformations: A survey. *Bulletin of the EATCS*, 40:273–326, 1990.

[Grieser, 1996] G. Grieser. An implementation of rigid E-unification using completion and rigid paramodulation. Forschungsbericht FITL-96-4, FIT Leipzig e.V., 1996.

[Gurevich and Veanes, 1997] Y. Gurevich and M. Veanes. Some undecidable problems related to the Herbrand theorem. UPMAIL Technical Report 138, Uppsala University, 1997.

[Jeffrey, 1967] R. C. Jeffrey. *Formal Logic. Its Scope and Limits*. McGraw-Hill, New York, 1967.

[Jouannaud and Kirchner, 1991] J.-P. Jouannaud and C. Kirchner. Solving equations in abstract algebras: A rule-based survey of unification. In J. Lassez and G. Plotkin, editors, *Computational Logic – Essays in Honor of Alan Robinson*, pages 257–321. MIT Press, 1991.

[Kanger, 1963] S. Kanger. A simplified proof method for elementary logic. In P. Braffort and D. Hirschberg, editors, *Computer Programming and Formal Systems*, pages 87–94. North Holland, 1963. *Reprint as pages 364–371 of:* Siekmann, J., and Wrightson, G. (eds.), *Automation of Reasoning. Classical Papers on Computational Logic*, vol. 1. Springer, 1983.

[Kozen, 1981] D. Kozen. Positive first-order logic is NP-complete. *IBM Journal of Research and Development*, 25(4):327–332, 1981.

[Lis, 1960] Z. Lis. Wynikanie semantyczne a wynikanie formalne. *Studia Logica*, 10:39–60, 1960. In Polish with English summary.

[Loveland, 1969] D. W. Loveland. A simplified format for the model elimination procedure. *Journal of the ACM*, 16(3):233–248, 1969.

[Murray and Rosenthal, 1987] N. V. Murray and E. Rosenthal. Inference with path resolution and semantic graphs. *Journal of the ACM*, 34(2):225–254, April 1987.

[Murray and Rosenthal, 1987a] N. V. Murray and E. Rosenthal. Theory links: Applications to automated theorem proving. *Journal of Symbolic Computation*, 4:173–190, 1987.

[Nelson and Oppen, 1980] G. Nelson and D. C. Oppen. Fast decision procedures based on congruence closure. *Journal of the ACM*, 27(2):356–364, April 1980.

[Nieuwenhuis and Rubio, 1995] R. Nieuwenhuis and A. Rubio. Theorem proving with ordering and equality constrained clauses. *Journal of Symbolic Computation*, 19:321–351, 1995.

[Petermann, 1992] U. Petermann. How to build-in an open theory into connection calculi. *Journal on Computer and Artificial Intelligence*, 11(2):105–142, 1992.

[Petermann, 1993] U. Petermann. Completeness of the pool calculus with an open built-in theory. In G. Gottlob, A. Leitsch, and D. Mundici, editors, *Proceedings, 3rd Kurt Gödel Colloquium (KGC), Brno, Czech Republic*, LNCS 713, pages 264–277. Springer, 1993.

[Plaisted, 1995] D. A. Plaisted. Special cases and substitutes for rigid E-unification. Technical Report MPI-I-95-2-010, Max-Planck-Institut für Informatik, Saarbrücken, November 1995.

[Policriti and Schwartz, 1995] A. Policriti and J. T. Schwartz. T-theorem proving I. *Journal of Symbolic Computation*, 20:315–342, 1995.

[Popplestone, 1967] R. J. Popplestone. Beth-tree methods in automatic theorem proving. In N. Collins and D. Michie, editors, *Machine Intelligence*, volume 1, pages 31–46. Oliver and Boyd, 1967.

[Reeves, 1987] S. V. Reeves. Adding equality to semantic tableau. *Journal of Automated Reasoning*, 3:225–246, 1987.

[Robinson and Wos, 1969] J. A. Robinson and L. Wos. Paramodulation and theorem proving in first order theories with equality. In B. Meltzer and D. Michie, editors, *Machine Intelligence*. Edinburgh University Press, 1969.

[Shostak, 1978] R. E. Shostak. An algorithm for reasoning about equality. *Communications of the ACM*, 21(7):583–585, 1978.

[Siekmann, 1989] J. H. Siekmann. Universal unification. *Journal of Symbolic Computation*, 7(3/4):207–274, 1989. Earlier version in *Proceedings, 7th International Conference on Automated Deduction (CADE), Napa, FL. USA*, LNCS 170, Springer, 1984.

[Smullyan, 1995] R. M. Smullyan. *First-Order Logic*. Dover Publications, New York, second corrected edition, 1995. First published in 1968 by Springer.

[Snyder, 1991] W. Snyder. *A Proof Theory for General Unification*. Birkhäuser, Boston, 1991.

[Stickel, 1985] M. E. Stickel. Automated deduction by theory resolution. *Journal of Automated Reasoning*, 1:333–355, 1985.

[Veanes, 1997] M. Veanes. *On Simultaneous Rigid E-Unification*. PhD Thesis, Uppsala University, Sweden, 1997.

[Voda and Komara, 1995] P. Voda and J. Komara. On Herbrand skeletons. Technical Report mff-ii-02-1995, Institute of Informatics, Comenius University, Bratislava, Slovakia, 1995.

A. WAALER AND L. WALLEN

TABLEAUX FOR INTUITIONISTIC LOGICS

1 INTRODUCTION

Despite the fact that for many years intuitionistic logic has served its function primarily in relation to foundational questions in mathematics, there has been a significant revival of interest over the last couple of decades stimulated by the application of intuitionistic formalisms in computer science [1982]. It is beyond the scope of this chapter to comment on these applications in detail which, broadly speaking, either exploit formalisations of the intuitionistic meaning of general mathematical abstractions as programming logics [Martin-Löf, 1984; Martin-Löf, 1996; Constable *et al.*, 1986], or exploit the similarity of systems of formal intuitionistic proofs under cut-elimination to systems of typed lambda terms under various forms of reduction (e.g. [Howard, 1980; Girard, 1989; Coquand, 1990]).[1] Both types of application rely on the rich proof theory possessed by intuitionistic formalisms in comparison with their classical counterparts.[2]

The basic proof theory of intuitionistic logic was explicitly formulated in the early part of the thirties. The first complete formalisation of intuitionistic predicate logic was due to Heyting [1930]. Four years after Heyting's publication Gentzen [1935] published the system of natural deduction (NJ) and the calculus of sequents (LJ) thereby providing elegant formulations of the logic. At the same time, relying on ideas of Brouwer and Kolmogorov, Heyting presented the constructive semantics of intuitionistic logic, published again in his seminal book on intuitionism [1956]. We will refer to Heyting's rendition of intuitionistic semantics as the 'proof interpretation'.[3]

The purpose of this chapter is to view intuitionistic logic from the perspective of systems of tableaux. The rules of a tableau system are usually viewed as steps towards the construction of a countermodel. Indeed, the main distinction between tableau systems and other types of calculi might be said to be the former's *analytic* emphasis on systematic completeness arguments, where partial derivations play a role as finitary evidence of non-provability, as opposed to the latter's *synthetic* emphasis on sound inference and complete proofs. The emphasis on analysis and

[1] It is noteworthy that both types of application focus on higher-order rather than first-order formalisms. An interesting compendium of papers illustrating many of these ideas in detail is Odifreddi's book on Logic and Computer Science [1990]. Intuitionistic type theories also play a role as the internal languages of universes of sets (see for example [1990]).

[2] There are, however, ways of giving classical logic a more 'intuitionistic' or 'constructive' proof theory; see e.g. [1989; 1992; 1990; 1991]).

[3] For extensive information about the history of the subject the reader should consult the article by van Dalen [1986] and the volumes by Troelstra and van Dalen [1988]; these together contain a comprehensive list of references.

M. D'Agostino et al. (eds.), Handbook of Tableau Methods, 255–296.

refutation makes the tableau framework particularly suitable for the analysis of theorem-proving algorithms, an observation which gives this volume its special character.

Apart from being of current interest in computer science, intuitionistic logic is worthy of attention within computational logic itself precisely because of its rich proof theory and non-classical structure. Analysis of techniques for proof-search in intuitionistic logic sheds light on the basic techniques of the field. It is our aim to bring a number of these techniques into focus by giving a treatment of them in intuitionistic logic from the analysis/refutational perspective of tableaux.

In the remainder of this introduction we give Heyting's definition of the meaning of the intuitionistic connectives via his proof interpretation (Section 1.1) and present the proof system LJ as a formalisation of this interpretation (Section 1.2) in a form easily related to tableaux. The , a corollary to Gentzen's theorem (Section 1.3), restricts the formulae that need to be considered in a LJ-proof and supports a view of cut-free LJ as a (refutational) tableau system. In preparation for the development of this view we end the introduction by defining Kripke's alternative semantic scheme for intuitionistic logic (Section 1.4).

The theorems of intuitionistic predicate logic form a proper subset of the theorems of classical predicate logic, and LJ can be obtained from LK, the for classical logic, by a restriction of the latter to sequents whose succedents contain at most one formula [Gentzen, 1935]. This calculus bears a satisfactory relationship with Heyting's proof interpretation. From the refutational perspective of proof-search the specialisation of LK to LJ is far from unique, and intuitionistic systems can be obtained in a number of interesting ways, each inducing a different search space. The key concern is the extent to which the resultant inference rules inter-permute. Roughly speaking: the more the permutation properties, the smoother the search, and the more compact/uniform the search space.

The rules of LJ can indeed be inverted to form a standard [Smullyan, 1968], but unlike the classical sequent calculus whose inversion reflects the semantics of classical logic more or less directly, recasting the proof interpretation in a refutational framework is problematic. We can, however, maintain the idea of proof-search as a systematic attempt to construct evidence for non-provability by working purely formally with the notion of a *refutation tree* (Section 2.2). Refutation trees give rise most naturally to an alternative, more classical, semantics for the logic. There are in fact two kinds of semantics in this vein for intuitionistic logic, one due to Beth [1959], and one due to Kripke [1963]. A Beth countermodel to a classically valid sequent is infinite, a property not always shared by Kripke models. For this reason we shall work with the latter.

The above describes the technical heart of this Chapter (Section 2). We formulate a modified version of LJ called LB (after Beth), which is a notational variant of Fitting's tableau system for intuitionistic logic [1969]. The utility of LB for proof-search stems from the fact that it has fewer restrictions on rule permutation compared to LJ. A Schütte-type is proved by means of a refutation tree construction which allows discussion of search strategies.

Syntactically, refutation trees are a description of the search space induced by a particular calculus, in this case LB. Refutation trees have the property that, by definition, a 'successful' construction codes a proof in that calculus, and an 'unsuccessful' construction yields (in the limit) evidence for a counter-proof, or refutation. The observation that a Kripke model can be obtained *directly* from a refutation tree in the latter case establishes the completeness of Kripke models w.r.t. LB, and simultaneously justifies their form.

With the construction of Section 2 completed we turn our attention to various types of optimisation which have been suggested in the literature with the aim of obtaining less redundant, or more uniform, search spaces. Given the above, we should expect to see optimisations arise from restrictions on the refutation tree construction which take advantage of the structure of the logic. This is how such optimisations are presented.

We focus on two issues in particular: (i) restrictions on in propositional and predicate logic (Sections 3.2 and 3.3 respectively); and (ii) the treatment of first-order quantifiers using ideas from Herbrand, Skolem and Robinson. We present simple expositions of each of these techniques in turn, relating them to the construction in Section 2, and to classical logic.

A final methodological comment is in order before we begin the exposition. It is commonplace to express an optimisation in proof-search by modifying the rules of a calculus. As a specification for a revised *implementation* such an exposition is to be highly recommended. Our purpose here is different. We seek to explain optimisations in terms of their effect on the search space. This is related directly to the character of a calculus as a *refutation* system: i.e. how search using the calculus approximates evidence for non-provability. The close correspondence between these two notions in the case of classical logic has been a major obstacle to the development of computationally improved methods of proof-search for non-classical logics.

1.1 The proof interpretation

We use a formal language with the following logical symbols: $\bot, \wedge, \vee, \supset, \exists, \forall$. Formulae are defined in the usual way from a set of predicate symbols, a set of variables, and a set of terms called parameters; the sets are assumed to be non-empty and disjoint. Observe that there are no function symbols in the language.[4] An *atomic* formula is a formula without any occurrences of logical symbols and no free variables. Predicate symbols with arity 0 (i.e. propositional letters) are denoted p, q, r. P, Q denote predicate symbols with arity one; A, B, C denote arbitrary closed formulae. $A[t/x]$ denotes the result of replacing every occurrence of the variable x in A with the term t.

[4]Extension of many of the ideas that follow to a language with function symbols and constants is non-trivial and seems to require more global techniques equivalent to those involved in, say, [Wallen, 1990], or [Ohlbach, 1990].

We now turn to an informal presentation of the of intuitionistic logic. This proceeds in two stages; we present the (canonical) proofs of formulae first, then extend this notion to hypothetical proofs of formulae. The presentation is inspired by Martin-Löf's exposition of the intuitionistic system of natural deduction [1996].

First, there cannot be a clause for atomic formulae, since a (canonical) proof of a proposition represented by an atomic formula must be given relative to a particular theory. The clauses for the logical constants are as follows.

> There are no (canonical) proofs of $\perp$.
>
> A (canonical) proof of $A \wedge B$ is a pair (π_1, π_2), where π_1 is a proof of A and π_2 is a proof of B.
>
> A (canonical) proof of $A \vee B$ is a pair (n, π), where $n = 0$ and π is a proof of A or $n = 1$ and π is a proof of B.
>
> A (canonical) proof of $A \supset B$ is a construction f which maps any proof π of A into a proof $f(\pi)$ of B.
>
> A (canonical) proof of $\forall z A$ is a construction f which maps any term t into a proof $f(t)$ of $A[t/z]$.
>
> A (canonical) proof of $\exists z A$ is a pair (t, π), where t is a term, and π is a proof of $A[t/z]$.

A *hypothetical proof* of C from a collection of formulae Γ is a construction which, if provided with proofs of the formulae in Γ will become a proof of C. Here we take 'provided' and 'will become' as primitive, unexplained notions (though in a formal framework one might think in terms of textual juxtaposition or substitution).

1.2 Gentzen's system LJ

The standard formal presentation of the proof interpretation of intuitionistic logic is the system of natural deduction (or the typed λ-calculus). We will now introduce Gentzen's sequent calculus LJ and explain how it may be seen as a formalisation of the proof interpretation. Conceptually the main difference between Gentzen's and his system of natural deduction is that while the rules of the latter define relations between formulae (assertions), the rules of the former operate on objects that can naturally be taken to represent judgements of logical consequence.

The basic syntactic objects in the calculus are *sequents*: expressions of the form $\Gamma \rightarrow \Delta$, where Γ, the *antecedent*, and Δ, the *succedent*, are (possibly empty) finite multisets of closed formulae. We write A for the singleton multiset $\{A\}$, and let ',' denote multiset union.

In a formulation of intuitionistic logic, the succedent is restricted to contain at most one formula. Such sequents are usually referred to as *single-conclusioned*.

The intended interpretation of $\Gamma \to C$ is the judgement that there is a hypothetical proof of the succedent thesis C from the antecedent hypotheses Γ. That there is a hypothetical proof of C from Γ means that there is a construction which, given proofs of the hypotheses will produce a proof of C, as discussed above. In particular, a sequent may be used to represent judgements about atomic formulae, like the propositional sequent $p, q \to r$. If such sequents when added to the calculus are called *non-logical axioms*.

The rules of LJ come in three groups. The *identity* rules reflect basic properties of (hypothetical) proofs.

$$\frac{}{A \to A}\ \text{ID} \qquad\qquad \frac{\Gamma \to A \quad \Gamma, A \to C}{\Gamma \to C}\ \text{CUT}$$

The identity axiom ID states that there is a hypothetical proof of A from A. Such an 'identity' construction clearly exists since if we provide a proof of A we need do nothing to it to obtain a proof of A. An axiom is a zero-premiss rule. We will occasionally define axioms by introducing the sequent in the conclusion. In ID both occurrences of A are called *principal* formulae.

The rule CUT captures our ability to compose hypothetical proofs to form new ones. Given a hypothetical proof of C from Γ, A, say π, and a hypothetical proof of A from Γ, say π', the two hypothetical proofs can be combined into a hypothetical proof of C from Γ. The construction is easy to describe: if given proofs of the formulae in Γ we would provide these to π' and get a proof of A; we would then take this proof together with proofs of the formulae of Γ and provide them to the construction π yielding a proof of C. CUT therefore formalises composition of the constructions π and π' that justify the judgements represented by the premisses of the rule.

It will occasionally be convenient to use the CUT rule in the form

$$\frac{\Gamma_1 \to A \quad \Gamma_2, A \to C}{\Gamma_1, \Gamma_2 \to C}$$

which can easily be derived from CUT by . The *structural* rules are and on the left.

$$\frac{\Gamma \to \Delta}{\Gamma, A \to \Delta}\ \text{LT} \qquad\qquad \frac{\Gamma, A, A \to \Delta}{\Gamma, A \to \Delta}\ \text{LC}$$

The formula A in the conclusion of the thinning rule is said to be *introduced by thinning*. Thinning reflects the fact that a hypothetical proof may ignore proofs of a hypothesis in the construction of a proof of its conclusion; contraction reflects the fact that a hypothetical proof may make multiple use of the proof of one of its hypotheses in constructing the proof of its conclusion.

Except for the falsity rule the *logical* rules come in pairs. The rule for falsity states that there is a construction which can provide a proof of any formula if provided with a proof of $\bot$.

$$\frac{}{\bot \to C}\ \text{L}\bot$$

$$\frac{\Gamma, A, B \rightarrow C}{\Gamma, A \wedge B \rightarrow C} \text{L}\wedge \qquad \frac{\Gamma \rightarrow A \quad \Gamma \rightarrow B}{\Gamma \rightarrow A \wedge B} \text{R}\wedge$$

$$\frac{\Gamma, A \rightarrow C \quad \Gamma, B \rightarrow C}{\Gamma, A \vee B \rightarrow C} \text{L}\vee \qquad \frac{\Gamma \rightarrow A}{\Gamma \rightarrow A \vee B} \text{R}\vee_1 \qquad \frac{\Gamma \rightarrow B}{\Gamma \rightarrow A \vee B} \text{R}\vee_2$$

$$\frac{\Gamma \rightarrow A \quad \Gamma, B \rightarrow C}{\Gamma, A \supset B \rightarrow C} \text{L}\supset \qquad \frac{\Gamma, A \rightarrow B}{\Gamma \rightarrow A \supset B} \text{R}\supset$$

$$\frac{\Gamma, A[t/x] \rightarrow C}{\Gamma, \forall x A \rightarrow C} \text{L}\forall \qquad \frac{\Gamma \rightarrow A[a/x]}{\Gamma \rightarrow \forall x A} \text{R}\forall$$

$$\frac{\Gamma, A[a/x] \rightarrow C}{\Gamma, \exists x A \rightarrow C} \text{L}\exists \qquad \frac{\Gamma \rightarrow A[t/x]}{\Gamma \rightarrow \exists x A} \text{R}\exists$$

Restrictions: In R$\forall$ and L$\exists$ the parameter a cannot appear in the conclusion. C may be absent.

Figure 1. The logical rules of LJ

C may be absent in L$\perp$. This axiom is in accordance with the proof interpretation: as no proof can be given of $\perp$, every such proof can be mapped into a proof of any formula C. The logical rules of LJ are given in Figure 1. A *logical rule* is a specification of a relation between at most two sequents, called *premisses*, and one sequent, called the *conclusion*. If there are two premisses in a rule they are called the *left* and the *right* premiss respectively. The *principal* formula of a logical rule is the formula displayed in the conclusion with the logical symbol of the rule as the outermost one. The *side* formulae of a logical rule are the immediate subformulae of the principal formula, which occur in the premisses of an inference. The parameter a occurring in the side formula of L$\exists$ and R$\forall$ in Figure 1 is called the *eigenparameter* of the inferences.

Trees regulated by the rules are called *derivations*. An *LJ-proof* is a derivation whose leaves are all axioms. The *endsequent* of a proof π is the sequent at its root. π is said to prove its endsequent. When referring to a proof we will say that the premiss of an inference occurs *above* its conclusion, and that one inference occurs above another if the conclusion of the former either is a premiss of the latter, or occurs above one of the latter's premisses.

As usual in proof systems for intuitionistic logic we take negation as a defined symbol: $\neg A$ is shorthand for $A \supset \perp$. This gives rise to the following two derived rules:

$$\frac{\Gamma \rightarrow A}{\Gamma, \neg A \rightarrow C} \text{L}\neg \qquad \frac{\Gamma, A \rightarrow}{\Gamma \rightarrow \neg A} \text{R}\neg$$

The derivation of the left negation rule is trivial. To justify the right negation rule we first apply R$\supset$

$$\frac{\Gamma, A \to \bot}{\Gamma \to A \supset \bot}$$

Observe now that any proof of a sequent $\Gamma \to \bot$ can be transformed into a proof of $\Gamma \to$ simply by omitting the succedent occurrences of $\bot$ throughout the proof.

REMARK 1. We can obtain classical logic by adding a rule for *reductio ad absurdum*:

$$\frac{\Gamma, \neg A \to}{\Gamma \to A}\text{RAA}$$

This is not the standard sequent calculus formulation; the standard sequent calculus for classical logic, LK, is given in Figure 5 in Section 2.1.

The right rules can be seen as reflecting the meaning of the logical symbols. A left rule captures the function of the given logical symbol when it occurs as the outermost symbol in a hypothesis; i.e. the left rules formalise how the logical constants are used. We illustrate the relationship between the logical rules and the proof interpretation by considering the implication and universal quantifier; the reader is invited to check the other cases. We have to show that the rules in LJ can be naturally interpreted as properties of the constructions and proofs referred to in the proof interpretation.

Right implication rule. Assume that $\Gamma, A \to B$ holds; i.e. there is a hypothetical proof of B from Γ and A. We must argue that there is a hypothetical proof of $A \supset B$ from Γ. For this to be the case there must be a construction which, when provided with proofs of the formulae in Γ becomes a proof of $A \supset B$. A proof of $A \supset B$ is itself a construction which maps proofs of A into proofs of B. Suppose we provide proofs of the formulae in Γ to the hypothetical proof of our assumption. Then we are left with a hypothetical proof of B dependent only on A. That is to say, we have a construction which will yield a proof of B if provided with a proof of A. This is sufficient evidence to conclude that the construction given in our assumption can yield a construction which is a proof of $A \supset B$, when given proofs of Γ.[5]

Left implication rule. Assume that $\Gamma \to A$ and $\Gamma, B \to C$ both hold. Now assume we have a proof of $A \supset B$. This assumption is formally represented by the judgement $\to A \supset B$. By definition this proof is a construction which maps a proof of A into a proof of B; we are therefore justified in asserting the sequent $A \to B$. Two applications of cut give:

$$\frac{\dfrac{\Gamma \to A \quad A \to B}{\Gamma \to B} \qquad \Gamma, B \to C}{\Gamma \to C}$$

[5]Notice that in this argument we are in effect using an informal version of the S_{mn} theorem [Kleene, 1952] for constructions.

which shows that given a hypothetical proof of B from A, the sequent $\Gamma \rightarrow C$ holds. This is sufficient to conclude that there is a hypothetical proof of C from Γ and $A \supset B$, i.e. that $\Gamma, A \supset B \rightarrow C$ holds.

Right rule for the universal quantifier. Assume that there is a hypothetical proof π of $A[a/x]$ from Γ. Due to the eigenparameter condition we know that no premiss in Γ is dependent on a, which means that a can be taken to denote any object. We can therefore obtain a hypothetical proof of $A[t/x]$ from Γ, for any term t, simply by substituting t for a in π. We are therefore justified in concluding that there must be a hypothetical proof of $\forall x A$ from Γ.

Left rule for the universal quantifier. Assume that $\Gamma, A[t/x] \rightarrow C$ holds. Assume further that we have a proof of $\forall x A$. By definition of the proof interpretation, $\rightarrow A[t/x]$ must hold, and we can assert $\Gamma \rightarrow C$:

$$\frac{\rightarrow A[t/x] \qquad \Gamma, A[t/x] \rightarrow C}{\Gamma \rightarrow C}$$

Thus, if we have a proof of $\forall x A$, then $\Gamma \rightarrow C$ holds. This establishes the grounds on which we may assert the conclusion of the rule: $\Gamma, \forall x A \rightarrow C$.

1.3 Cut-elimination and the subformula property

The most important property of the sequent calculus is the eliminability of the CUT rule, Gentzen's so-called 'Hauptsatz' [1935]: any LJ-provable sequent has an LJ-proof without CUT. An immediate consequence of this result for LJ is that an LJ-provable sequent has an LJ-proof comprised only of subformulae of the sequent. This is an immediate consequence of the form of the rules. This result is central to tableau methods as it limits the formulae that need to be considered in the search.

We will now illustrate the two sources of complexity in proof-search for intuitionistic logic that go beyond that encountered in proof-search for classical logic: the non-eliminability of *contraction* on left implications, and the non-invertibility of certain inference rules. Both sources are present within intuitionistic propositional logic and ultimately explain the fact that the decision problem for the propositional fragment is PSPACE complete [Statman, 1979], while that for classical logic is NP-complete.

In the search for an LJ-proof we assume that derivations grow upwards. Whenever we apply a rule with principal formula C we say that we *expand* C to generate a new sequent, an instance of the premiss of the rule. If the rule has more than one premiss this expansion will generate two branches instead of one. When inference r_1 occurs above r_2 in a derivation we say that r_2 has been applied *before* r_1. We shall use this terminology throughout the chapter.

EXAMPLE 2. One of the first results in intuitionistic logic, due to Kolmogorov, was that the double negation of any theorem in classical propositional logic is intuitionistically valid. The result cannot be lifted to first-order logic. We prove $\neg\neg(p \vee \neg p)$. For the first step we are left no choices:

$$\frac{\neg(p \vee \neg p) \rightarrow}{\rightarrow \neg\neg(p \vee \neg p)}$$

If we now apply L$\neg$, we will generate the sequent $\rightarrow p \vee \neg p$, which is not intuitionistically valid. However, if we first apply contraction we get

$$\cfrac{\cfrac{\neg(p \vee \neg p) \rightarrow p \vee \neg p}{\neg(p \vee \neg p), \neg(p \vee \neg p) \rightarrow}}{\neg(p \vee \neg p) \rightarrow}$$

which the reader can easily expand into a proof (apply R$\vee_2$, R$\neg$, L$\neg$, and R$\vee_1$). If C is a theorem of classical propositional logic, but not intuitionistically valid, any proof in LJ of $\rightarrow \neg\neg C$ must use contraction on $\neg C$.

EXAMPLE 3. Backtracking will occur in LJ whenever a right disjunction is expanded, in that we must choose one right rule over another. If one choice does not lead to a proof, one must check whether this is the case for the other option as well. Since this particular instance of does not occur in the tableau proof system of Section 2.1, we consider the slightly more general case of choosing between L$\supset$ and R$\vee$.

Consider the sequent $\Gamma, p \supset (q \vee r) \rightarrow (p \supset q) \vee r$. This sequent is classically provable for any Γ. In general it is not intuitionistically provable. The reader is invited to check that if $\Gamma, p, r \rightarrow q$ is provable in LJ, and $\Gamma \rightarrow p$ is not, then we must expand the right disjunction first (followed by an instance of R$\supset$), while if $\Gamma \rightarrow p$ is provable and $\Gamma, p, r \rightarrow q$ is not, we must first expand the left implication.

A *permutation* is obtained from a derivation by interchanging two inferences. The significance of for proof-search is as follows. Assume that in the search for a proof of a sequent rules r_1 and r_2 both apply, and assume that r_1 does not permute over r_2. Then, other things being equal, we should apply r_2 before r_1. Kleene [1952b] enumerates the seven cases where permutation fails in general in LJ. The cases are:

$$\frac{\text{L}\forall}{\text{R}\forall} \qquad \frac{\text{L}\forall \text{ or R}\exists}{\text{L}\exists} \qquad \frac{\text{L}\supset}{\text{R}\supset} \qquad \frac{\text{R}\vee_1 \text{ or R}\vee_2 \text{ or R}\exists}{\text{L}\vee}$$

Permutation will also fail in LJ in the following cases:

$$\frac{\text{L}\supset \text{ or R}\forall \text{ or R}\supset \text{ or R}\wedge}{\text{L}\vee}$$

unless the two inferences immediately above the L$\vee$-inference have the same principal formula. In general this will not be the case.

EXAMPLE 4. Consider the task of showing that the set of identity axioms in LJ can be restricted to *atomic sequents*: sequents whose antecedents and succedents contain atomic formulae only. This is established by a simple inductive argument. The individual cases provide good illustrations of the permutability properties.

Conjunction. The task is to prove $A \wedge B \rightarrow A \wedge B$ using simpler axioms. We have a choice whether to start with a left or a right rule. Expanding the antecedent first gives the proof:

$$\dfrac{\dfrac{\dfrac{A \rightarrow A}{A, B \rightarrow A} \qquad \dfrac{B \rightarrow B}{A, B \rightarrow B}}{A, B \rightarrow A \wedge B}}{A \wedge B \rightarrow A \wedge B}$$

The reader can easily check that the inferences interchange; i.e. we may just as well expand the succedent first.

Disjunction. In this case we are left with no choices since we cannot commit ourselves to one disjunct before we apply the left disjunction rule. We therefore get the proof:

$$\dfrac{\dfrac{A \rightarrow A}{A \rightarrow A \vee B} \qquad \dfrac{B \rightarrow B}{B \rightarrow A \vee B}}{A \vee B \rightarrow A \vee B}$$

Implication. This case also shows a dependency among the rules: since the succedent side formula in the left premiss of the left implication inference must be matched by the antecedent side formula of the right implication inference, the right implication rule must be applied before the left one.

$$\dfrac{\dfrac{A \rightarrow A \qquad \dfrac{B \rightarrow B}{A, B \rightarrow B}}{A \supset B, A \rightarrow B}}{A \supset B \rightarrow A \supset B}$$

Quantifiers. These cases are also subject to dependencies, but this time due to the eigenparameter conditions (as in classical logic).

1.4 Kripke semantics

We end this introduction by introducing and proving the soundness of LJ with respect to this semantics. Since proofs and models are dual, the relationship between the proof systems and the Kripke semantics is most clearly seen in the construction of countermodels to non-provable sequents. We shall show in Section 2 how Kripke models arise naturally from searches by establishing a model existence theorem based on a systematic search procedure.

A *Kripke model* is a quadruple $(U, D, \leq, \Vdash')$; the *universe* U is non-empty set partially ordered by $\leq$; the *domain function* D maps each point in U to a non-empty set of terms in the language such that $D(x) \subseteq D(y)$ whenever $x \leq y$;[6] $\Vdash'$ is a binary relation between U and the set of atomic formulae such that if $x \Vdash' p$ and $x \leq y$, then $y \Vdash' p$, and $\forall x \in U, x \nVdash \bot$. The *forcing relation* $\Vdash$ is the weakest

[6]Note that we have opted for an interpretation of the quantifiers by substitution.

relation that contains $\Vdash'$ and is closed under the following rules.

$$
\begin{array}{lll}
x \Vdash A \wedge B & \text{if} & x \Vdash A \text{ and } x \Vdash B, \\
x \Vdash A \vee B & \text{if} & x \Vdash A, \\
z \Vdash A \vee B & \text{if} & x \Vdash B, \\
x \Vdash A \supset B & \text{if} & y \Vdash B \text{ whenever } y \nVdash A \text{ and } y \geq x, \\
x \Vdash \forall z A & \text{if} & y \Vdash A[t/z] \text{ for each } y \geq x \text{ and each } t \in D(y), \\
x \Vdash \exists z A & \text{if} & x \Vdash A[t/z] \text{ for some } t \in D(x).
\end{array}
$$

In the sequel, we do shall not indicate in the notation which model is being used as this should always be clear from the context. A sequent $\Gamma \rightarrow C$ is *valid* in a model if, for each point x which forces every formula in Γ, x forces C. A model is a *countermodel* to $\Gamma \rightarrow C$ if it contains a point at which every formula in Γ is forced, but C is not forced.

LEMMA 5. *If $x \Vdash A$ and $x \leq y$, then $y \Vdash A$.*

Proof. Induction on A. For atomic formulae the statement is immediate. We prove the statement for the case that A is $B \supset C$. Assume $x \Vdash B \supset C$ and $x \leq y$. By transitivity of $\leq$, for each z such that $y \leq z$, either $z \nVdash B$ or $z \Vdash C$. By definition, $y \Vdash B \supset C$. ■

We now establish the soundness of LJ with respect to Kripke semantics. Since LJ is a formalisation of the , the soundness result can be viewed as a formal variant of the correspondence between LJ and the proof interpretation sketched above.

PROPOSITION 6. *Let $\Gamma \rightarrow C$ be provable in LJ. Then $\Gamma \rightarrow C$ is valid in all Kripke models.*

Proof. Let π prove $\Gamma \rightarrow C$. By induction on the length of π we prove that for each point x in every model, if x forces every formula in Γ, x forces C. If the last rule of π is an axiom, this holds trivially. We consider only three cases as the other cases are similar.

Case 1. The last rule of π is a right implication:

$$
\frac{\begin{array}{c}\pi_1\vdots \\ \Gamma, A \rightarrow B\end{array}}{\Gamma \rightarrow A \supset B}
$$

Let $x \Vdash \Gamma$ and let y be any point such that $x \leq y$. Then $y \Vdash \Gamma$ by Lemma 5. If $y \Vdash A$, then $y \Vdash B$ by the induction hypothesis; otherwise $y \nVdash A$. In any case y meets the forcing condition for $A \supset B$.

Case 2. The last rule of π is a left implication:

$$
\frac{\begin{array}{c}\pi_1\vdots \\ \Gamma \rightarrow A\end{array} \qquad \begin{array}{c}\pi_2\vdots \\ \Gamma, B \rightarrow C\end{array}}{\Gamma, A \supset B \rightarrow C}
$$

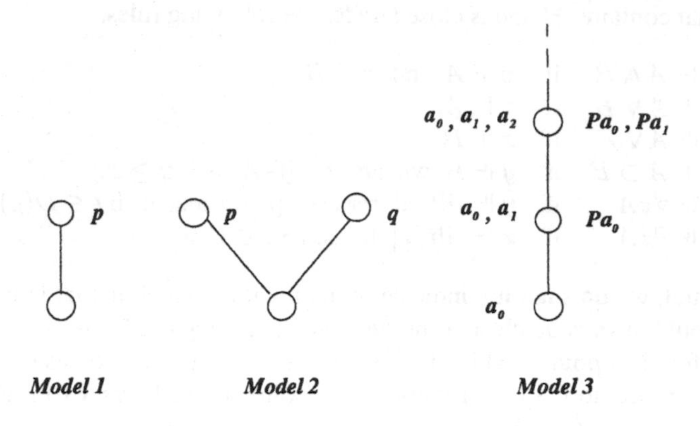

Figure 2. Example Kripke models

Let x be any point in a model that forces $\Gamma, A \supset B$. By the induction hypothesis applied to π_1, $x \Vdash A$. By the forcing condition for $A \supset B$, $x \Vdash B$. By the induction hypothesis applied to π_2, $x \Vdash C$.

Case 3. The last rule of π is R$\forall$:

$$\frac{\begin{array}{c}\vdots \\ \pi: \\ \Gamma \to A[a/x]\end{array}}{\Gamma \to \forall x A}$$

Let $x \Vdash \Gamma$ and $x \leq y$. The eigenparameter condition ensures that we can substitute t for a in π for any term $t \in D(y)$ and get a proof of $\Gamma \to A[t/x]$. By induction hypothesis and Lemma 5, $y \Vdash A[t/x]$. ■

REMARK 7. A simple corollary to Proposition 6 is obtained via the contrapositive proposition: the existence of a countermodel to a sequent implies that the sequent is not intuitionistically provable.

EXAMPLE 8. Models in propositional intuitionistic logic are restrictions of models for the first-order language in that the domain function D is redundant. A countermodel to $p \vee \neg p$ is shown in Figure 2 as model 1. The figure depicts a model in which $U = \{x, y\}$, $\Vdash'$ is $\{\langle y, p\rangle\}$, and $\leq$ is the reflexive, transitive closure of $\{\langle x, y\rangle\}$. Model 2 of the same figure is a countermodel to $(p \supset q) \vee (q \supset p)$, another formula which is classically valid.

EXAMPLE 9. Predicate symbols are restricted to unary predicates and prop ositional letters. Classical monadic predicate logic is decidable, but its intuitionistic version is undecidable. A countermodel to the classically valid formula $\forall x \neg\neg Px \supset \neg\neg\forall x Px$ is shown as model 3 in Figure 2 (domains are listed beside the circles marking the points). There exists no finite countermodel to this formula.

2 A SYSTEMATIC SEARCH PROCEDURE

The search space of a cut-free sequent calculus is in part determined by the permutation properties of the calculus. The existence of Kripke models for intuitionistic logic is also closely linked to the properties of permutation. This is as one should expect, given that a common way of viewing proof search is as a systematic attempt to construct a countermodel. One might say that the property of intuitionistic logic which gives rise to Kripke models is reflected syntactically in the properties of permutation.

As shown in Section 1.3 Gentzen's system LJ is quite asymmetrical w.r.t. permutation. To facilitate the formulation of search procedures for intuitionistic logic we will study a slightly modified formulation of LJ. The system appears under many labels in the literature. It seems to have been formulated independently by several authors; Takeuti [1975, p. 52] credits the invention of the system to Maehara [1954]. Since the system is a sequential formulation of the branch expansion rules of a system of Beth tableaux, we will refer to the system as LB.

2.1 LB: a multi-succedent variant of LJ

The syntax of LB differs from LJ syntax in that the LB system operates on sequents with (possibly) multiple succedent formulae; the latter are interpreted disjunctively. In semantic terms, $\Gamma \to \Delta$ is forced at x iff every point $y \geq x$ which forces every formula in Γ, forces at least one formula in Δ. Since there can be more than one formulae in the succedent the system contains thinning and contraction on both the left and the right.

The rules and axioms of LB are given in Figure 3. Observe that R$\supset$, R$\forall$ and the axiom involving $\perp$ are defined on sequents with at most one formula in the succedent; the other rules define relations between sequents with (possibly) multiple succedent formulae. The system admits negation rules as derived rules:

$$\frac{\Gamma \to A}{\Gamma, \neg A \to \Delta}\ \mathrm{L}\neg \qquad\qquad \frac{\Gamma, A \to}{\Gamma \to \neg A}\ \mathrm{R}\neg$$

CUT is admissible in LB.

EXAMPLE 10. Consider once again the task of showing that the set of identity axioms can be restricted to atomic sequents, cf. Example 4. We repeat two cases for LB.

Disjunction. In LB we can expand the succedent disjunction without committing ourselves to one disjunct, and hence expand the succedent before we apply the left disjunction rule.

$$\cfrac{\cfrac{\cfrac{A \to A}{A \to A, B} \qquad \cfrac{B \to B}{B \to A, B}}{A \vee B \to A, B}}{A \vee B \to A \vee B}$$

STRUCTURAL RULES

$$\frac{\Gamma \to \Delta}{\Gamma, A \to \Delta}\ \mathrm{LT} \quad \frac{\Gamma \to \Delta}{\Gamma \to A, \Delta}\ \mathrm{RT} \qquad \frac{\Gamma, A, A \to \Delta}{\Gamma, A \to \Delta}\ \mathrm{LC} \quad \frac{\Gamma \to A, A, \Delta}{\Gamma \to A, \Delta}\ \mathrm{RC}$$

AXIOMS

$$\bot \to C \qquad \Gamma, A \to A, \Delta$$

LOGICAL RULES

$$\frac{\Gamma, A, B \to \Delta}{\Gamma, A \wedge B \to \Delta}\ \mathrm{L}\wedge \qquad \frac{\Gamma \to A, \Delta \quad \Gamma \to B, \Delta}{\Gamma \to A \wedge B, \Delta}\ \mathrm{R}\wedge$$

$$\frac{\Gamma, A \to \Delta \quad \Gamma, B \to \Delta}{\Gamma, A \vee B \to \Delta}\ \mathrm{L}\vee \qquad \frac{\Gamma \to A, B, \Delta}{\Gamma \to A \vee B, \Delta}\ \mathrm{R}\vee$$

$$\frac{\Gamma \to A, \Delta \quad \Gamma, B \to \Delta}{\Gamma, A \supset B \to \Delta}\ \mathrm{L}\supset \qquad \frac{\Gamma, A \to B}{\Gamma \to A \supset B}\ \mathrm{R}\supset$$

$$\frac{\Gamma, A[t/x] \to \Delta}{\Gamma, \forall x A \to \Delta}\ \mathrm{L}\forall \qquad \frac{\Gamma \to A[a/x]}{\Gamma \to \forall x A}\ \mathrm{R}\forall$$

$$\frac{\Gamma, A[a/x] \to \Delta}{\Gamma, \exists x A \to \Delta}\ \mathrm{L}\exists \qquad \frac{\Gamma \to A[t/x], \Delta}{\Gamma \to \exists x A, \Delta}\ \mathrm{R}\exists$$

Restrictions: In R$\forall$ and L$\exists$ the parameter a cannot appear in the conclusion.

Figure 3. The logical rules of LB

Implication. A superficial look at the syntax gives the impression that since the L$\supset$ -rule sanctions classical sequents, L$\supset$ may permute with R$\supset$. Let us try this for the implication case.

$$\dfrac{\to A, A \supset B \qquad \dfrac{\dfrac{B \to B}{B, A \to B}}{B \to A \supset B}}{A \supset B \to A \supset B}$$

Observe that the succedent occurrence of A in the left leaf must be removed by thinning before the implication can be expanded. The resulting sequent is not, in general, provable. Therefore, in LB, as in LJ, we must expand the right implication below the left one.

LB has more symmetrical permutation properties than LJ. As well as the usual restrictions on permutation caused by the eigenparameter conditions, permutation will fail in the case of an inference with a succedent side formula occurring over a R$\supset$ or a R$\forall$. Since the premiss of the latter inferences must be intuitionistic, this situation can arise only when the inference just above is a L$\supset$. Therefore, in general, permutation will fail in the following five cases:

$$\frac{\text{L}\forall}{\text{R}\forall} \qquad \frac{\text{L}\forall \text{ or R}\exists}{\text{L}\exists} \qquad \frac{\text{L}\supset}{\text{R}\supset \text{ or R}\forall}$$

Consider now a sequent at a point in the search for an LB-proof. The task of a search procedure is to generate new sequents from a given one that better approximate axioms, so the problem is which formula to choose for expansion. The permutation properties impose certain constraints in that applications of L$\exists$, R$\forall$, and R$\supset$ should have priority. More precisely, eigenparameters should be introduced as early as possible in the search, and R$\supset$ should be applied to strengthen the antecedent prior to an application of L$\supset$. This latter constraint arises from the fact that a proof of the left premiss of the L$\supset$ inference may require a stronger antecedent, as is illustrated in Example 10. There is, however, a consideration which runs contrary to this. Observe that prior to applications of R$\forall$ and R$\supset$ we must in general apply thinning to the succedent. This suggests that we should *postpone* application of these two rules as long as possible.

The two conflicting considerations about R$\supset$ make the expansion of left implications particularly tricky. The non-permutabilities suggest that R$\supset$ should have priority over L$\supset$. However we will then run the danger of deleting succedent information by thinnings that is needed at a later stage in the search. Consider the following figure:

$$\frac{\Gamma \to A, D \supset B, C \qquad \Gamma, B \vee C \to D \supset B, C}{\Gamma, A \supset (B \vee C) \to D \supset B, C}$$

Here we have applied L$\supset$ and delayed the R$\supset$; the right premiss is provable. However, in the expansion of the right premiss the choice is between L$\vee$ and R$\supset$, which means we must opt for the L$\vee$. If $\Gamma \to A$ is provable, the figure above will lead to a proof. However, if we apply R$\supset$ before L$\supset$ we will not, in general, obtain a proof.

The discussion indicates that in a sequent $\Gamma, A \supset B \to \Delta$ one should expand the left implication (after first duplicating it by a contraction) if $\Gamma, A \supset B \to A$ is provable, otherwise not. This indeterminacy forces branching in the search space and this, together with the need for on certain subformulae of the endsequent, underlies the PSPACE-hardness of the propositional decision problem [Statman, 1979].[7] The strategy we adopt in this section is to apply L$\supset$ to every antecedent implication and test the left premiss for provability prior to expansion of a R$\supset$. Kripke models arise as natural objects from the syntax in a particularly direct way using this strategy, as we will see below.

The search will proceed in stages. At each stage we try to expand with all rules

[7] In formulations of intuitionistic logic without explicit contraction, contraction is eliminated in favour of additional branching in the search space [1992; 1993].

$$\frac{\Gamma, A \supset B \rightarrow A \qquad \Gamma, B \rightarrow \Delta}{\Gamma, A \supset B \rightarrow \Delta} \text{ LC}\supset$$

$$\frac{\Gamma, \forall x A, A[t/x] \rightarrow \Delta}{\Gamma, \forall x A \rightarrow \Delta} \text{ LC}\forall \qquad \frac{\Gamma \rightarrow A[t/x], \exists x A, \Delta}{\Gamma \rightarrow \exists x A, \Delta} \text{ RC}\exists$$

Figure 4. Derived rules in LB

except R⊃ and R∀. An application of R⊃ or R∀ indicates that new antecedent information is available in the resulting sequent, either directly by the left side formula of a R⊃, or indirectly in that R∀ introduces a new parameter with which we can instantiate a left universal quantifier. However, sometimes the left side formula of a R⊃ can be removed immediately by thinning, and even if a new term is available, the antecedent is not strengthened before a left universal quantifier has actually been instantiated with this term. What we want to say is that a new stage in the search is initiated when the antecedent has been strengthened.

We introduce the binary relation $\prec$ to make this idea precise. Let the sequent S be above the sequent T in a derivation. Then $T \prec S$ iff there is an intervening instance of R⊃ whose left side formula is not introduced by thinning, or an intervening instance of R∀ whose eigenparameter has been used as parameter in an intervening L∀ -inference.

We now introduce the search strategy in detail. The axioms are generalised to include sequents of the form: $\Gamma, A \rightarrow A, \Delta$ and $\Gamma, \bot \rightarrow A, \Delta$. Contraction is needed for left implications and universal quantifiers. For simplicity we use the derived rules of Figure 4. The 'C' in the names of the rules stands for 'contraction'. Observe that the implication need only be copied into the left branch of the left implication rule.

We order the logical rules into three distinct groups:

GROUP 1: R∧, L∧, R∨, L∨, LC⊃, L∃.

GROUP 2: LC∀, RC∃.

GROUP 3: R∀, R⊃.

The search proceeds by always opting for a rule in the group with the lowest number, should there be one that is applicable. To avoid vicious circles in the search we introduce some additional constraints. Let us say that a term is *available* at a sequent if there is an occurrence of this term below the sequent in the derivation. There are five constraints on application of the rules.

(S1) the left side formula of a R⊃ is immediately removed by thinning if there is an antecedent occurrence of the same formula in a sequent below in the derivation,

(S2) if LC$\supset$ has been used to expand a formula A in sequent T and A also occurs in the sequent S to be expanded, then A can be expanded only if $T \prec S$,

(S3) an antecedent universal quantifier $\forall x A$ can be instantiated with a term t if t is available and there is no antecedent occurrence of $A[t/x]$ in a sequent lower down in the derivation,

(S4) a succedent existential quantifier $\exists x A$ can be instantiated with a term t if t is available and $A[t/x]$ does not already occur in the succedent,

(S5) right thinning is only applied prior to R$\supset$ and R$\forall$.

The last condition defines the points of . If the removal of a succedent occurrence of A does not lead to a proof, and A is an implication or universal quantifier, we must return to this point in the search and expand A instead. We must then use thinnings on the other formulae in the succedent.

EXAMPLE 11. The identity axioms are restricted to atomic formulae for this example. We illustrate the procedure on the following identity judgement:

$$(A \supset B) \supset C \rightarrow (A \supset B) \supset C\,.$$

Assume that the letters denote distinct atomic formulae. According to Example 10 we should expand the succedent first in order to find a proof. The search procedure will, however, opt for the antecedent implication. There is one applicable rule in GROUP 1: LC$\supset$. This yields

$$\frac{(A \supset B) \supset C \rightarrow A \supset B, (A \supset B) \supset C \qquad C \rightarrow (A \supset B) \supset C}{(A \supset B) \supset C \rightarrow (A \supset B) \supset C}$$

The right premiss is obviously provable. Consider the left premiss. Condition (S2) blocks a new application of LC$\supset$ since the antecedent is the same as that in the sequent we started with. So we must expand the succedent. Let us first expand the leftmost succedent implication, $A \supset B$. This requires the elimination of the other succedent formula by thinning (*).

$$\cfrac{\cfrac{\cfrac{(A \supset B) \supset C, A \rightarrow A \supset B, B \qquad C, A \rightarrow B}{(A \supset B) \supset C, A \rightarrow B}}{(A \supset B) \supset C \rightarrow A \supset B}}{(A \supset B) \supset C \rightarrow A \supset B, (A \supset B) \supset C}\ (*)$$

Observe that the rightmost leaf ($C, A \rightarrow B$) is not provable and no rule applies, thus terminating the search in this branch. Observe also that if we nevertheless try to expand the leftmost branch, condition (S2) blocks application of LC$\supset$. Hence we must apply right thinning to B and R$\supset$ to $A \supset B$. But then condition (S1) applies to the resulting sequent, and condition (S2) once again blocks further expansion of the tree:

$$\cfrac{\cfrac{\cfrac{(A \supset B) \supset C, A \rightarrow B}{(A \supset B) \supset C, A, A \rightarrow B}}{(A \supset B) \supset C, A \rightarrow A \supset B}}{(A \supset B) \supset C, A \rightarrow A \supset B, B}$$

This illustrates how termination is secured for the propositional fragment. The next step in the search is to backtrack to the point $(*)$ and expand the other succedent implication. This yields:

$$\frac{(A \supset B) \supset C \rightarrow (A \supset B) \supset C}{(A \supset B) \supset C \rightarrow A \supset B, (A \supset B) \supset C} \ (**)$$

Note that we have generated the endsequent once again. However, contrary to what was the case when the search started, condition (S2) applies and blocks yet another instance of LC$\supset$. Hence we must apply R$\supset$. The reader is encouraged to continue the example until a proof is constructed.

Some optimisations of this strategy may be easily developed by taking only local information into consideration. For example, we can always postpone the left implication in the sequent $\Gamma, A \supset B \rightarrow C \supset D, \Delta$ whenever either Γ or Δ is empty. More substantial optimisations are possible once we take global considerations into account, i.e. information obtained from the entire proof search state (partial derivation). The definition of such strategies is quite complex and is beyond the scope of this chapter. Proof search on the basis of such global considerations is the fundamental idea behind connection driven proof methods for intuitionistic logic and certain based search strategies such as the intuitionistic resolution calculus of Mints [1990; 1993].

The search procedure is defined for a language without function symbols. The inclusion of function symbols complicates the definition of the search. For example, if a formula Pfa occurs in a sequent, where f is a unary function symbol, this formula must be given the same treatment as the formula $\exists x(Px \wedge Fxa)$ in a language without function symbols. For this reason we omit function symbols from the language, leaving the extension of the search procedure to function symbols as an exercise for the reader.

The right contraction rule, implicit in the rule RC$\exists$, is in fact redundant. Note that a succedent occurrence of an existentially quantified formula is always removed by thinning prior to applications of R$\supset$ and R$\forall$. The formulation in terms of right contraction has the advantage that backtracking in the search can be isolated to a single point: prior to instances of right thinning.

A search procedure for LJ can be designed along the same lines as the procedure for LB [Waaler, 1997a]. However, due to the non-permutabilities involving disjunction, a search procedure based on the LJ rules has to be more complex in its definition. From a search perspective the main difference between LB and LJ is that in LB succedent information is preserved longer in the development of a proof. However, commitments must be made before right implications or universal quantifications are expanded, and, as expected, there is thus no difference between LJ and LB as far as the complexity of the search space is concerned.

By shifting from LJ to LB we do, however, lose the tight connection between the formal system and the constructive semantics of intuitionistic logic. LB does enjoy , but the argument that LJ is a formalisation of the does not carry over to LB in a straightforward way. The connection between LB and the proof interpretation

$$\frac{\Gamma, A, B \rightarrow \Delta}{\Gamma, A \wedge B \rightarrow \Delta}\ \mathrm{L}\wedge \qquad \frac{\Gamma \rightarrow A, \Delta \quad \Gamma \rightarrow B, \Delta}{\Gamma \rightarrow A \wedge B, \Delta}\ \mathrm{R}\wedge$$

$$\frac{\Gamma, A \rightarrow \Delta \quad \Gamma, B \rightarrow \Delta}{\Gamma, A \vee B \rightarrow \Delta}\ \mathrm{L}\vee \qquad \frac{\Gamma \rightarrow A, B, \Delta}{\Gamma \rightarrow A \vee B, \Delta}\ \mathrm{R}\vee$$

$$\frac{\Gamma \rightarrow A, \Delta \quad \Gamma, B \rightarrow \Delta}{\Gamma, A \supset B \rightarrow \Delta}\ \mathrm{L}\supset \qquad \frac{\Gamma, A \rightarrow B, \Delta}{\Gamma \rightarrow A \supset B, \Delta}\ \mathrm{R}\supset$$

$$\frac{\Gamma, A[t/x] \rightarrow \Delta}{\Gamma, \forall x A \rightarrow \Delta}\ \mathrm{L}\forall \qquad \frac{\Gamma \rightarrow A[a/x], \Delta}{\Gamma \rightarrow \forall x A, \Delta}\ \mathrm{R}\forall$$

$$\frac{\Gamma, A[a/x] \rightarrow \Delta}{\Gamma, \exists x A \rightarrow \Delta}\ \mathrm{L}\exists \qquad \frac{\Gamma \rightarrow A[t/x], \Delta}{\Gamma \rightarrow \exists x A, \Delta}\ \mathrm{R}\exists$$

Restrictions: In R$\forall$ and L$\exists$ the parameter a cannot appear in the conclusion.

Figure 5. The logical rules of LK

can be made more explicit by a translation of LB-proofs into proofs in LJ. Such translations are defined by Fitting [1969] and Schmitt and Kreitz [1996] by means of the introduction of cuts, and by Waaler at the level of cut-free proofs [Waaler, 1997b].

REMARK 12. We close this section by a remark about classical logic. If we replace the R$\supset$ and R$\forall$ rules of LB with the rules

$$\frac{\Gamma, A \rightarrow B, \Delta}{\Gamma \rightarrow A \supset B, \Delta} \qquad \frac{\Gamma \rightarrow A[a/x], \Delta}{\Gamma \rightarrow \forall x A, \Delta}$$

the resulting system is Gentzen's system LK for classical logic. The logical rules of LK are shown in figure 5 for reference.

The search for LB-proofs mimics the search in classical logic to a certain extent, in that the search is performed classically for all rules except the two rules for implication and the right rule for universal quantifier. This is reflected in the truth conditions for Kripke models, which are the same as for classical logic except for the clauses for implication and the universal quantifier.

2.2 *Refutation trees*

A Schütte-type proof of the for *classical logic* [1960] proceeds by induction on an open branch in a complete tree. The construction for intuitionistic logic introduced in Section 2.3 generalises this procedure. In this section we give a formal presentation of the intuitionistic counterpart to an 'open branch'. Due to the definition will have to be more complex than that for classical logic. The method introduced

$$\frac{\Gamma, A, B \rightarrow \Delta}{\Gamma, A \wedge B \rightarrow \Delta}\ \mathrm{L}\wedge \qquad \frac{\Gamma \rightarrow A, \Delta}{\Gamma \rightarrow A \wedge B, \Delta}\ \mathrm{R}\wedge_1 \qquad \frac{\Gamma \rightarrow B, \Delta}{\Gamma \rightarrow A \wedge B, \Delta}\ \mathrm{R}\wedge_2$$

$$\frac{\Gamma, A \rightarrow \Delta}{\Gamma, A \vee B \rightarrow \Delta}\ \mathrm{L}\vee_1 \qquad \frac{\Gamma, B \rightarrow \Delta}{\Gamma, A \vee B \rightarrow \Delta}\ \mathrm{L}\vee_2 \qquad \frac{\Gamma \rightarrow A, B, \Delta}{\Gamma \rightarrow A \vee B, \Delta}\ \mathrm{R}\vee$$

$$\frac{\Gamma, A \supset B \rightarrow A, \Delta}{\Gamma, A \supset B \rightarrow \Delta}\ \mathrm{LC}\supset_1 \qquad \frac{\Gamma, B \rightarrow \Delta}{\Gamma, A \supset B \rightarrow \Delta}\ \mathrm{L}\supset_2 \qquad \frac{\Gamma, A \rightarrow B}{\Gamma \rightarrow A \supset B}\ \mathrm{R}\supset$$

$$\frac{\Gamma, \forall x A, A[t/x] \rightarrow \Delta}{\Gamma, \forall x A \rightarrow \Delta}\ \mathrm{LC}\forall \qquad \frac{\Gamma \rightarrow A[a/x]}{\Gamma \rightarrow \forall x A}\ \mathrm{R}\forall$$

$$\frac{\Gamma, A[a/x] \rightarrow \Delta}{\Gamma, \exists x A \rightarrow \Delta}\ \mathrm{L}\exists \qquad \frac{\Gamma \rightarrow A[t/x], \exists x A}{\Gamma \rightarrow \exists x A}\ \mathrm{RC}\exists$$

$$\frac{\Gamma, A \rightarrow \Delta}{\Gamma \rightarrow \Delta}\ \mathrm{LT} \qquad \frac{\Gamma \rightarrow C_1 \quad \cdots \quad \Gamma \rightarrow C_n}{\Gamma \rightarrow C_1, \ldots, C_n, \Delta}\ \text{Split}$$

L∃ and R∀ are subjected to the usual conditions on eigenparameters. In Split, $C_1, \ldots, C_n$ are all the implications and universally quantified formulae in the succedent of the conclusion, and $n > 0$.

Figure 6. Refutation rules for LB

below is an adaptation of one developed for LJ in [Waaler, 1997a].

The rules of this formal system are given in Figure 6. Observe that only Split, a structural rule which replaces right thinning, is branching: the function of this inference is to record the backtracking in a search. Once again we divide the logical rules into three distinct groups:

GROUP 1: R$\wedge_1$, R$\wedge_2$, L$\wedge$, R$\vee$, L$\vee_1$, L$\vee_2$, LC$\supset_1$, L$\supset_2$, L$\exists$.

GROUP 2: LC$\forall$, RC$\exists$.

GROUP 3: R$\forall$, R$\supset$.

The relation $\prec$ is defined on in an analogous manner to the definition on derivations trees.

We will be interested in trees generated from the rules in Figure 6 that satisfy the following conditions (compare these with conditions (S1)–(S5) in the previous section):

(R1) thinning only occurs on the left side formula of a R$\supset$, and it occurs whenever there is an antecedent occurrence of the same formula below this inference in the tree;

(R2) if there is an instance of $LC\supset_1$ with conclusion T, and there is an instance of $LC\supset_1$ above with identical principal formula and with conclusion S, then $T \prec S$;

(R3) the side formula of a $LC\forall$ does not have an antecedent occurrence below this inference in the tree, and the term used to instantiate the quantifier is available at the conclusion;

(R4) the term used to instantiate the quantifier in an instance of $RC\exists$ with premiss S is available at the conclusion of the inference, and if another instance of the side formula occurs in the succedent of a sequent T below S, then $T \prec S$;

(R5) every instance of Split has a $R\supset$ or $R\forall$ immediately above it in every branch.

A *refutation tree for a sequent* S is a tree with S as root, regulated by the rules (R1-R5), and satisfying the condition that for each rule occurence in the tree no rule from a lower group is applicable to its conclusion. A refutation tree may be thought of as being constructed by successive applications of the rules from conclusion to premisses, making sure to use any applicable rule from a lower group before an applicable rule from a higher group.

R is said to be a *complete refutation tree for a sequent* S if it is a refutation tree for S, R is not a proper subtree of any larger refutation tree for S, and, for every sequent $\Gamma \to \Delta$ in R, $\bot \notin \Gamma$ and $\Gamma \cap \Delta = \emptyset$.

EXAMPLE 13. The following is a refutation tree for the sequent $\to (p \supset q) \vee (q \supset p)$.

$$\dfrac{\dfrac{\dfrac{p \to q}{\to p \supset q} \qquad \dfrac{q \to p}{\to q \supset p}}{\to p \supset q, q \supset p}}{\to (p \supset q) \vee (q \supset p)}$$

The reader should check that the relevant conditions are met. The sequent is classically valid but intuitionistically unprovable.

EXAMPLE 14. This example illustrates termination. Consider the following tree.

$$\dfrac{\dfrac{\dfrac{\dfrac{\dfrac{\dfrac{(p \supset q) \supset r, p \to q}{(p \supset q) \supset r, p, p \to q}}{(p \supset q) \supset r, p \to p \supset q}}{(p \supset q) \supset r, p \to p \supset q, q}}{(p \supset q) \supset r, p \to q}}{(p \supset q) \supset r \to p \supset q}}{(p \supset q) \supset r \to}$$

The first step is to apply $LC\supset_1$. (R2) now blocks a repeated application of $LC\supset_1$ and we must hence apply $R\supset$. $LC\supset_1$ and $R\supset$ are then re-applied; observe that condition (R2) is met for the $LC\supset_1$ and that we must use Split before this application of $R\supset$. Finally, according to condition (R1), we must apply left thinning on p. Condition (R2) now blocks further expansion of the tree.

EXAMPLE 15. We construct a refutation tree for $\neg\forall x(Px \vee \neg Px) \rightarrow$. The first step in the construction is to apply $LC{\supset}_1$ and then expand the succedent, which yields the following block:

$$\cfrac{\cfrac{\cfrac{\cfrac{\cfrac{Pa_0, \neg\forall x(Px \vee \neg Px) \rightarrow}{\neg\forall x(Px \vee \neg Px) \rightarrow \neg Pa_0}}{\neg\forall x(Px \vee \neg Px) \rightarrow Pa_0, \neg Pa_0}}{\neg\forall x(Px \vee \neg Px) \rightarrow Pa_0 \vee \neg Pa_0}}{\neg\forall x(Px \vee \neg Px) \rightarrow \forall x(Px \vee \neg Px)}}{\neg\forall x(Px \vee \neg Px) \rightarrow}$$

Note that the leaf is similar to the endsequent except for the addition of Pa_0 in the antecedent, where a_0 is the eigenparameter of the R$\forall$-inference. This means that $LC{\supset}_1$ applies again, yielding a similar block above the leaf, only differing in the choice of a new eigenparameter a_1 and the addition of Pa_0 in the antecedent of its (i.e. the blocks) endsequent. The complete refutation tree hence consists of a countably infinite sequence of such blocks.

LEMMA 16. *If a sequent S is not provable in LB, then there exists a complete refutation tree with S as endsequent. Moreover, if S is restricted to propositional intuitionistic logic, then the complete refutation tree is finite.*

Proof. The existence of a complete refutation tree follows in the usual way via non-constructive reasoning. In the propositional case, the only rule that has the potential to give rise to an infinite refutation tree is $LC{\supset}_1$, owing to the contraction on the principal formula that is built into the rule. However, condition (R2) constrains the application of this rule as follows: in between every two instances of $LC{\supset}_1$ with identical principal formulae, must be a R$\supset$ whose left side formula is not removed by thinning. This obtains only when the left side formula strengthens the antecedent of the premiss compared to the antecedent of the conclusion. A sequent can only be strengthened by formulae that occur as subformulae of the endsequent S; and there are only a finite number of those. ■

Note that contraction has been built into one rule for left implication and the left rule for universal quantification, which in combination with condition (R1) restricts the copying process to the branches where a new copy of a formula can potentially contribute to the search. A search procedure for LJ based on the same idea is described and implemented by Sahlin, Franzén and Haridi [1992]. This procedure also contains various optimisations. For further results about in LB we refer to Sections 3.2 and 3.3, and (for LJ) to the study by Waaler [1997a].

2.3 The model existence theorem

In this section we show how a complete refutation tree with endsequent S defines a countermodel to S in a natural way.

We first define the basic construction which maps a complete refutation tree u into a Kripke model. The universe U of the Kripke model contains u and every subtree x of u whose endsequent is either

(a) a premiss of a R$\supset$ -inference whose left side formula is not introduced by thinning, or

(b) a premiss of a R$\forall$ -inference whose eigenparameter occurs in a side formula of a LC$\forall$ -inference in x.

When the left side formula of a R$\supset$ -inference r is introduced by thinning, we will say that the *premiss* of r is the sequent immediately above the thinning (so that the premiss and conclusion of r have equal antecedents). A sequent occurring in a tree $x \in U$ is called an *x-sequent* if it does not occur in any $y \in U$ such that y is a proper subtree of x. r is an *x-inference* if the conclusion of r is an x-sequent. Let x^- and x^+ be sets of formulae defined as follows.

$A \in x^-$ iff A occurs as an antecedent formula of an x-sequent;

$A \in x^+$ iff A occurs as a succedent formula of an x-sequent.

Let $x \leq y$ iff y is a subtree of x, $t \in D(x)$ iff there is an x-sequent at which t is available, with the proviso that (the dummy term) $0 \in D(x)$ only if $D(x)$ is otherwise empty. For any closed atomic formula p, $x \Vdash' p$ iff $p \in x^-$. It is easy to check that $(U, D, \leq, \Vdash')$ is a Kripke model.

EXAMPLE 17. Let us apply the above construction to the complete refutation tree in Example 13. The model will consist of 3 points: let x denote the refutation tree given in Example 13, y denote its subtree $p \to q$ and z denote the subtree $q \to p$ of x. Then $x \leq y$, $x \leq z$, and $\Vdash' = \{\langle y, p\rangle, \langle z, q\rangle\}$. The model is shown as model 2 in Figure 2. It is immediate that this is a countermodel to $\to (p \supset q) \vee (q \supset p)$.

EXAMPLE 18. The complete refutation tree in Example 15 gives rise to an infinite model depicted as model 3 in Figure 2. This is a countermodel to $\neg\forall x(Px \vee \neg Px) \to$.

In Section 1.4 we motivated Kripke models; to shed some light on the construction of the countermodel and provide a link to the syntax of refutation trees, observe first that $x \leq y$ if either $x = y$ or there is an x-sequent T and a y-sequent S such that $T \prec S$. $T \prec S$ implies that the antecedent of S is stronger than the antecedent of T. In terms of the this can be taken to mean that we are in possession of proofs of more formulae at the point y than at the point x.

THEOREM 19 (Model Existence). *Let S be an intuitionistic sequent not provable in LB. Then S has a counter-model.*

Proof. It is immediate from Lemma 16 that a complete refutation tree for S exists. Let the corresponding Kripke model $(U, D, \leq, \Vdash')$ be as defined above. We prove by induction on A that for each point x in the model, if $A \in x^-$, then $x \Vdash A$, and if $A \in x^+$, then $x \nVdash A$. The model existence theorem follows from this. There are six cases to consider.

Case 1: A is atomic or $\bot$. If $A \in x^-$, A must be distinct from $\bot$. By definition of $\Vdash'$, $x \Vdash A$. Let $A \in x^+$. Observe that the x-sequents form a block of non-branching inferences possibly with Split and right inferences immediately above. Let S be the uppermost x-sequent that is the premiss of a left inference; if there are no left x-inferences, take S to be the endsequent of x. Since $A \in x^+$, A must occur in the succedent of an x-sequent T such that the antecedent of S is equal to the antecedent of T. Since an atomic antecedent occurrence in a sequent will occur in every other sequent above, A will occur in the antecedent of T, i.e. if $A \in x^-$, A will occur in both the antecedent and the succedent of T. Hence $A \notin x^-$, and $x \nVdash A$.

Case 2: A is $B \wedge C$. Assume $A \in x^-$. Then A must be the principal formula of a L$\wedge$ -inference whose premiss is an x-sequent, giving $B \in x^-$ and $C \in x^-$. The induction hypothesis and the forcing condition yield $x \Vdash A$. Assume $A \in x^+$. By construction, A must be the principal formula of an inference of type R$\wedge_1$ or R$\wedge_2$ whose premiss is an x-sequent; thus either $B \in x^+$ or $C \in x^+$. Apply the induction hypothesis and the forcing condition for conjunction.

Case 3: A is $B \vee C$. This case is the dual of case 2.

Case 4: A is $B \supset C$. Let $A \in x^-$ and $y \geq x$. By construction A must be the principal formula of either a y-inference of type LC$\supset_1$, in which case $B \in y^+$, or a z-inference L$\supset_2$, $x \leq z \leq y$, in which case $C \in z^-$. By induction hypothesis and Lemma 5, either $y \nVdash B$ or $y \Vdash C$. Since this holds for every $y \geq x$, $x \Vdash B \supset C$. Let $A \in x^+$. Then A must be the principal formula of a R$\supset$ -inference r. Assume first that the left premiss of r is introduced by thinning, i.e. that r is an x-inference. Then $C \in x^+$ and there must be a point $u \leq x$ such that $B \in u^-$. By induction hypothesis and Lemma 5, $x \Vdash B$ and $x \nVdash C$. Assume that the left premiss of r is not introduced by thinning. Then the subtree of which the premiss of r is the root gives rise to a point y such that $x \leq y$. By induction hypothesis $y \Vdash B$ and $y \nVdash C$. In either case $x \nVdash B \supset C$.

Case 5: A is $\forall z B$. Let $A \in x^-$, $x \leq y$, and $t \in D(y)$. There must then be a point $v \leq y$ such that $t \in D(v)$ and for every point $w \leq v$, $t \in D(w)$ iff $v = w$. By construction of the refutation tree there must be a v-inference of type LC$\forall$ with side formula $B[t/z]$, i.e. $B[t/z] \in v^-$. By induction hypothesis, $v \Vdash B[t/z]$, hence, by Lemma 5, $y \Vdash A[t/z]$. Since this holds for any y and t, $x \Vdash \forall z B$. Let $A \in x^+$. A must be the principal formula of an inference r with eigenparameter a. If r is an x-inference, i.e. if a is not used in the instantiation of a LC$\forall$-inference above, then $a \in D(x)$ and $B[a/x] \in x^+$, and $x \nVdash B[a/x]$ by the induction hypothesis. If r is not an x-inference, there is a y-sequent, $y \geq x$, such that $B[a/x] \in y^+$. By induction hypothesis $y \nVdash B[a/x]$. In either case $x \nVdash \forall z B$.

Case 6: A is $\exists z B$. Let $A \in x^-$ and t be the eigenparameter of the inference in x introducing $\exists z B$. Then $t \in D(x)$ and $B[t/z] \in x^-$. Let $A \in x^+$. A must be introduced by an inference of type RC$\exists$; let the conclusion of this inference be S. By construction, every term $t \in D(x)$ must be available at S. Hence $B[t/z] \in x^-$ for each $t \in D(x)$. In both cases we conclude by induction hypothesis and the forcing condition for A. ∎

Observe that the proof of the existence of a countermodel is carried out constructively from the refutation tree, but the theorem itself is non-constructive. If we want to conclude that LB is complete with respect to Kripke models, i.e. that every sequent valid in all Kripke models is provable in LB, we must argue contrapositively (and non-constructively).

THEOREM 20 (Completeness). *LB is complete.*

Proof. If a sequent is not provable in LB, the construction of the Model Existence theorem gives us a countermodel. Hence the sequent is not valid. ■

REMARK 21. Refutation trees for propositional endsequents are finite (Lemma 16). Hence the search procedure will always terminate for propositional sequents. By completeness we can conclude that the search strategy defines a decision procedure for propositional intuitionistic logic.

3 REFINEMENTS

3.1 *Tableaux*

The Beth tableau system for intuitionistic logic is basically a notational variant of the system LB given above. Fitting presents an elegant formulation using ideas from both Beth and Hintikka in [Fitting, 1969]. We follow his presentation below.

We work with two signs T and F which can be informally understood as denoting 'proven' and 'not proven' respectively. A *signed* formula is a pair consisting of a sign and a formula, written TA or FA. Writing S for a multiset of signed formulae, the *reduction rules* of the tableau system can be written as in Figure 7. S, H is shorthand for $S \cup \{H\}$ (multiset union), S_T means $\{TA \mid TA \in S\}$ (as a multiset), and $S, H \mid S, H'$ means that the reduction produces two branches with these endpoints.

The presentation differs from Fitting's in two ways. First, since the rules are defined on multisets, the duplication of formulae is made explicit in the rules $T\supset$, $T\forall$ and $T\exists$. Second, we still treat negation as a defined symbol. A multiset is said to be *closed* if it contains occurrences of both TA and FA, or if it contains $T\bot$. A branch is closed if it contains a closed multiset.

Rather than repeating the technical definition of tableaux based on these multisets of signed formulae let us simply observe the following. Consider a mapping I from multisets of signed formulae to sequents defined by:

$$I(S) \quad = \quad \{A \mid TA \in S\} \rightarrow \{A \mid FA \in S\}.$$

It is easy to see that I maps the reduction rules of Figure 7 into slightly modified and inverted versions of the logical rules of LB (see Figure 3); $T\supset$, $T\forall$, and $F\exists$ map into inverted versions of the derived rules in figure 4; $F\supset$ and $F\forall$ when inverted map into the rules:

$$\frac{S, T(A \wedge B)}{S, TA, TB}\,T\wedge \qquad \frac{S, F(A \wedge B)}{S, FA \mid S, FB}\,F\wedge$$

$$\frac{S, T(A \vee B)}{S, TA \mid S, TB}\,T\vee \qquad \frac{S, F(A \vee B)}{S, FA, FB}\,F\vee$$

$$\frac{S, T(A \supset B)}{S, T(A \supset B), FA \mid S, TB}\,T\supset \qquad \frac{S, F(A \supset B)}{S_T, TA, FB}\,F\supset$$

$$\frac{S, T\forall x A}{S, T\forall x A, TA[t/x]}\,T\forall \qquad \frac{S, F\forall x A}{S_T, FA[a/x]}\,F\forall$$

$$\frac{S, T\exists x A}{S, TA[a/x]}\,T\exists \qquad \frac{S, F\exists x A}{S, F\exists x A, FA[t/x]}\,F\exists$$

Restrictions: The parameter a must be new.

Figure 7. Fitting/Beth Tableau rules [Fitting, 1969]

$$\frac{\Gamma, A \rightarrow B}{\Gamma \rightarrow A \supset B, \Delta} \qquad \frac{\Gamma \rightarrow A[a/x]}{\Gamma \rightarrow \forall x A, \Delta}$$

These two rules are admissible in LB. A multiset which closes a branch in a tableau corresponds to an axiom $\Gamma, A \rightarrow A, \Delta$ or $\Gamma, \bot \rightarrow \Delta$. This makes left and right thinning redundant. Observe, however, that left thinning plays an important role in the systematic search procedure for LB in deleting a side formula of a R$\supset$ whenever there is an occurrence of this formula below in the LB-derivation. To compensate for this we may introduce the tableau rule

$$\frac{S, F(A \supset B)}{S_T, FB}\,F\supset'$$

and use this rule in the search when there is already an occurrence of TA in the branch. The formulation of the systematic search procedure within the context of tableau rules is now straightforward.

Simple though this relationship is, the reader should not forget that the perspective of the tableau system is that of reduction and refutation, not (directly) proof. What is significant is that, whereas in (propositional) classical logic this perspective gives rise to a set of reduction rules which can be seen as working locally on single formulae, for intuitionistic logic the operation *must* be applied to a sequent. Hence the nodes of the reduction trees consist of multisets of formulae rather than single formulae. Our decision to base the earlier sections on sequent calculi is therefore justified.

3.2 Contraction-free systems of propositional logic

The idea behind contraction-free systems for the propositional fragment of intuitionisti logic can be traced back to a paper by Vorob'ev from 1952 [1952]. The idea has since been independently rediscovered by several logicians and computer scientists. The exposition below is inspired by Dyckhoff's treatment in [1992] and, for the semantical completeness proof, by the paper of Pinto and Dyckhoff [1995]. Variations of the same system are introduced by Hudelmaier [1992; 1993] and Lincoln *et al.* [1991]. We will present the idea in relation to LB, and leave the formulation of the associated tableau system to the reader.

As shown in Section 2 the only rule that needs contraction in the propositional fragment of LB is the left implication rule, a fact captured by the derived rule LC⊃ of Figure 4. The basic idea behind limiting contraction is to design rules that are sensitive to the structure of A in an antecedent occurrence of $A \supset B$. The rules are as follows.

$$\frac{\Gamma, p, B \to \Delta}{\Gamma, p, p \supset B \to \Delta}\ \mathrm{L}{\supset}_1^* \qquad \frac{\Gamma, C \supset (D \supset B) \to \Delta}{\Gamma, (C \wedge D) \supset B \to \Delta}\ \mathrm{L}{\supset}{\wedge}$$

$$\frac{\Gamma, C \supset B, D \supset B \to \Delta}{\Gamma, (C \vee D) \supset B \to \Delta}\ \mathrm{L}{\supset}{\vee} \qquad \frac{\Gamma, D \supset B \to C \supset D \quad \Gamma, B \to \Delta}{\Gamma, (C \supset D) \supset B \to \Delta}\ \mathrm{L}{\supset}{\supset}$$

The p in $\mathrm{L}{\supset}_1^*$ can be restricted to being atomic.

PROPOSITION 22. *The four rules above,* CUT *and contraction are all derived rules in LB.*

Proof. The derivation of $\mathrm{L}{\supset}_1^*$ is obtained by a simple application of L⊃. The three other cases are given by the following derivations.

L⊃∨:

$$\frac{\begin{matrix}\vdots\\ (C \vee D) \supset B \to D \supset B\end{matrix} \qquad \dfrac{\begin{matrix}\vdots\\ (C \vee D) \supset B \to C \supset B\end{matrix} \quad \Gamma, C \supset B, D \supset B \to \Delta}{\Gamma, (C \vee D) \supset B, D \supset B \to \Delta}}{\Gamma, (C \vee D) \supset B \to \Delta}$$

L⊃∧:

$$\frac{\begin{matrix}\vdots\\ (C \wedge D) \supset B \to C \supset (D \supset B)\end{matrix} \qquad \Gamma, C \supset (D \supset B) \to \Delta}{\Gamma, (C \wedge D) \supset B \to \Delta}$$

L⊃⊃:

$$\frac{\dfrac{\begin{matrix}\vdots\\ (C \supset D) \supset B \to D \supset B\end{matrix} \qquad \Gamma, D \supset B \to C \supset D}{\Gamma, (C \supset D) \supset B \to C \supset D, \Delta} \qquad \Gamma, B \to \Delta}{\Gamma, (C \supset D) \supset B \to \Delta}\ \mathrm{LC}{\supset}$$

■

EXAMPLE 23. By instantiating B by $\bot$ in L⊃∧, L⊃∨, and L⊃⊃ we can obtain the following instances of the rules above:

$$\frac{\Gamma, C \supset \neg D \rightarrow \Delta}{\Gamma, \neg(C \wedge D) \rightarrow \Delta} \qquad \frac{\Gamma, \neg C, \neg D \rightarrow \Delta}{\Gamma, \neg(C \vee D) \rightarrow \Delta} \qquad \frac{\Gamma, \neg D \rightarrow C \supset D}{\Gamma, \neg(C \supset D) \rightarrow \Delta}$$

In the same way we can obtain a double negation rule from L⊃⊃:

$$\frac{\Gamma \rightarrow \neg A}{\Gamma, \neg\neg A \rightarrow \Delta}$$

Let us now prove the sequent $\neg(p \vee \neg p) \rightarrow$. In Example 2 we showed that a proof of this sequent in LB requires contraction. The new rules do, however, permit a much simpler proof:

$$\cfrac{\cfrac{\neg p \rightarrow \neg p}{\neg p, \neg\neg p \rightarrow}}{\neg(p \vee \neg p) \rightarrow}$$

The reader is encouraged to redo Example 11 to see how the systematic search procedure will work using the new rules in place of L⊃.

Proposition 22 shows that every sequent derivable by the rules of LB and any of the rules $L\supset_1^*$, L⊃∨, L⊃∧, L⊃⊃ is derivable in LB. Since LB also satisfies a cut-elimination theorem, we can conclude that the sequent is provable in LB without CUT. The interest in these rules lies not in their soundness, but in the fact that together the four rules can replace L⊃, thus making explicit contraction redundant.

THEOREM 24. *The system obtained from propositional LB by deleting* LC *and* L⊃, *and adding the rules* $L\supset_1^*$, L⊃∨, L⊃∧, *and* L⊃⊃ *is complete.*

Proof. We prove a model existence theorem on the basis of the new set of rules. The proof follows the same line of argument as the proof of the model existence theorem for LB (theorem 19), restricted to a propositional language. Given that we replace L⊃ we must also modify the refutation rules of LB accordingly (figure 6). First, since L⊃⊃ is the only left implication rule with a right premiss, $L\supset_2$ must be modified to account for this, giving $L\supset_2^*$:

$$\frac{\Gamma, B \rightarrow \Delta}{\Gamma, (C \supset D) \supset B \rightarrow \Delta}\ L\supset_2^*$$

Second, the refutation rule $LC\supset_1$ is replaced by the four rules $L\supset_1^*$, L⊃∨, L⊃∧, and Split*. Split* is the more complex rule:

$$\frac{\Gamma_1, D_1 \supset B_1 \rightarrow C_1 \supset D_1 \quad \cdots \quad \Gamma_n, D_n \supset B_n \rightarrow C_n \supset D_n \qquad \Gamma \rightarrow \Delta}{\Gamma' \rightarrow \Delta}\ \text{Split}^*$$

in which the sets Γ, Γ_i and Γ' are given by:

Γ is $\Gamma', (C_1 \supset D_1) \supset B_1, \ldots, (C_n \supset D_n) \supset B_n$,

Γ_i is $\Gamma \setminus \{(C_i \supset D_i) \supset B_i\}$,

Γ' contains no formulae of the form $(C \supset D) \supset B$.

Once again the set of logical refutation rules is ordered into distinct groups:

GROUP 1: $R\wedge_1$, $R\wedge_2$, $L\wedge$, $R\vee$, $L\vee_1$, $L\vee_2$, $L\supset_1^*$, $L\supset\vee$, $L\supset\wedge$, $L\supset_2^*$.

GROUP 2: Split*.

GROUP 3: $R\supset$.

Since we work in the propositional fragment, the definition of the constraints is somewhat simpler than those required for LB in Section 2.2. First, $T \prec S$ iff there is an instance of $R\supset$ in between T and S whose left side formula is not introduced by thinning. Second, the refutation trees for the system under consideration must meet three conditions:

(RC1) thinning only occurs on the left side formula of a $R\supset$, and it occurs whenever there is an antecedent occurrence of the same formula below this inference in the tree;

(RC2) if there is an instance of Split* with conclusion T, and there is an instance of Split* above with conclusion S, then $T \prec S$;

(RC3) every instance of Split has a $R\supset$ immediately above it in every branch.

The definition of the model $(U, \leq, \Vdash')$ is exactly as in Section 2.3 (neglecting the domain function). Observe that $(U, \leq)$ is a finite tree. Let us say that the *degree* of a point x in U is the number of distinct points y in U such that $y \geq x$. y is *immediately above* x if $x \leq y$ and for every z such that $x \leq z \leq y$, either $z = x$ or $z = y$. Note that the ''immediately above' relation is reflexive.

The proof of the model existence theorem for LB proceeded by induction on the structure of a formula A. This was made possible by the fact that any side formula of a refutation rule of LB is simpler than the principal formula of the rule (and is in fact a subformula of it). Observe that this is not the case for $L\supset\wedge$, for example. Following Dyckhoff [1992] we therefore introduce the *weight* of a formula, 'wt' as follows:

$$\text{wt}(p)=\text{wt}(\bot)=1,$$

$$\text{wt}(A \vee B)=\text{wt}(A \supset B)=\text{wt}(A)+\text{wt}(B)+1,$$

$$\text{wt}(A \wedge B)=\text{wt}(A)+\text{wt}(B)+2.$$

The function 'wt' induces a well-founded relation over formulae. Observe that $\text{wt}((C \wedge D) \supset B) > \text{wt}(C \supset (D \supset B))$, which is what we need for the induction to go through. As in Theorem 19 we prove that for all $x \in U$ and every formula A, if $A \in x^-$, then $x \Vdash A$, and if $A \in x^+$, then $x \not\Vdash A$. The proof is by induction on the lexicographical ordering of all pairs (m, n), where m is wt(A) and n is the degree of x.

Cases 1–3 in the induction are identical to the proof of Theorem 19.

Case 4 needs modification. Let A be an implication. The case that $A \in x^+$ is exactly as in theorem 19. However, if $A \in x^-$, there are four subcases to consider.

Sub-case 1: A is $p \supset B$. Consider a point $y \geq x$. If $p \in y^-$, there must be a point z such that $x \leq z \leq y$, $p \in z^-$, and such that for every $u \leq z$, if $p \in u^-$, then $u = z$. There must then be a z-inference of type $\mathrm{L}\supset_1^*$ which applies, giving $B \in z^-$. By induction hypothesis, $z \Vdash B$, and by Lemma 5, $y \Vdash B$. Since this holds for every $y \geq x$, $x \Vdash p \supset B$.

Sub-case 2: A is $(C \wedge D) \supset B$. By construction there has to be an x-inference of type $\mathrm{L}\supset\wedge$ whose principal formula is A. Hence $C \supset (D \supset B) \in x^-$. Since the weight of this formula is less than the weight of A, the induction hypothesis applies and gives $x \Vdash C \supset (D \supset B)$. By the forcing condition, $x \Vdash (C \wedge D) \supset B$.

Sub-case 3: A is $(C \vee D) \supset B$. Same as previous case, referring to $\mathrm{L}\supset\vee$ instead of $\mathrm{L}\supset\wedge$.

Sub-case 4: A is $(C \supset D) \supset B$. In this case either Split* applies or there is an x-inference of type $\mathrm{L}\supset_2^*$ with principal formula A. In the latter case $B \in x^-$, and we can conclude by the induction hypothesis. Assume that Split* applies. Consider first the premiss $\Gamma, D \supset B \rightarrow C \supset D$. By induction hypothesis, x forces $D \supset B$. Furthermore, by condition (2) this premiss must be the conclusion of a $\mathrm{R}\supset$ with principal formula $C \supset D$. Hence there must be a point y immediately above x such that both $C \in y^-$ and $D \supset B \in y^-$. Consider now any point $z \geq y$. By induction hypothesis and Lemma 5, $z \Vdash C$ and $z \Vdash D \supset B$, which entails that if $z \Vdash C \supset D$, then $z \Vdash B$. Consider now any other point v immediately above x. The endsequent of v must be the premiss of a $\mathrm{R}\supset$ immediately above the Split* under consideration, possibly with a Split in between. By inspection of the rules we see that $A \in v^-$. Since the degree of v is less than the degree of x, the induction hypothesis can be used to conclude $v \Vdash A$. Hence, for any $z \geq x$, if $z \Vdash C \supset D$, then $z \Vdash B$, and we are done. ∎

EXAMPLE 25. To illustrate the construction in the proof given above we repeat Example 14 for the new set of rules. Consider the following tree.

$$\dfrac{\dfrac{p, q \supset r \rightarrow q}{q \supset r \rightarrow p \supset q} \qquad (p \supset q) \supset r \rightarrow}{(p \supset q) \supset r \rightarrow}$$

Observe that the conditions pertaining to a complete refutation tree are met. The countermodel defined by the tree is given as model 1 in Figure 2.

REMARK 26. The proof of Theorem 24 cannot be lifted to first-order intuitionistic logic. The reason is that in the first-order case the models defined by refutation trees may contain infinite branches, a situation which makes the degree of the points in the branch undefined. This situation is shown in Example 18.

$$\frac{}{\Gamma, A, \neg A \to \Delta}\ \text{ID}^*$$

$$\frac{\Gamma, \neg A \to \qquad \Gamma, \neg B \to}{\Gamma, \neg(A \wedge B) \to}\ \text{L}\neg\wedge \qquad \frac{\Gamma, \neg A, \neg B \to \Delta}{\Gamma, \neg(A \vee B) \to \Delta}\ \text{L}\neg\vee$$

$$\frac{\Gamma, A, \neg B \to}{\Gamma, \neg(A \supset B) \to}\ \text{L}\neg\supset \qquad \frac{\Gamma, A \to}{\Gamma, \neg\neg A \to}\ \text{L}\neg\neg$$

$$\frac{\Gamma, \neg\forall x A \to A[a/x]}{\Gamma, \neg\forall x A \to}\ \text{L}\neg\forall \qquad \frac{\Gamma, \neg\exists x A, \neg A[t/x] \to \Delta}{\Gamma, \neg\exists x A \to \Delta}\ \text{L}\neg\exists$$

Restrictions: In L$\neg\forall$ the parameter a cannot appear in the conclusion.

Figure 8. Sequential formulation of rules in Miglioli *et al.* [Miglioli *et al.*, 1994]

3.3 Limiting contraction in intuitionistic predicate logic

Owing to the interaction between right universal and left implication, i.e. the non-permutability of L$\supset$ over R$\forall$, contraction on left implications cannot be removed in the first-order case. In Example 18 we presented a countermodel to $\neg\forall x(Px \vee \neg Px) \to \bot$. It is easy to see that this formula has no finite countermodel, and the refutation tree given in Example 15 provides insight into why no finite countermodel can exist. This is sufficient to conclude that the method of the previous section cannot be lifted to the first-order case.

However, if we restrict attention to left negations, the method does transfer to first-order intuitionistic logic. This observation is due to Miglioli *et al.* [1994], who presented the idea within the framework of a tableau system. The idea can be formulated in terms of the sequent rules in Figure 8.

PROPOSITION 27. *The rules in Figure 8 are all derived rules in LB.*

Proof. L$\neg\wedge$ is derived by two instances of CUT:

$$\frac{\dfrac{\dfrac{\begin{matrix}\vdots\\ \neg(A \wedge B), B \to \neg A\end{matrix} \qquad \Gamma, \neg A \to}{\Gamma, \neg(A \wedge B), B \to}}{\Gamma, \neg(A \wedge B) \to \neg B} \qquad \Gamma, \neg B \to}{\Gamma, \neg(A \wedge B) \to}$$

The derivations of the other rules are left for the reader. Observe that either R$\neg$ (i.e. R$\supset$) or R$\forall$ is used in the arguments for the rules that operate on intuitionistic (single-conclusioned) sequents, and that they are not used in the arguments for the other rules. ∎

THEOREM 28. *The system obtained from LB by adding the rules in Figure 8, and*

$$\frac{\Gamma, \neg A \rightarrow \quad \Gamma, \neg(A \wedge B) \rightarrow \Delta}{\Gamma, \neg(A \wedge B) \rightarrow \Delta}\ \mathrm{L}\neg\wedge_1^* \qquad \frac{\Gamma, \neg B \rightarrow \quad \Gamma, \neg(A \wedge B) \rightarrow \Delta}{\Gamma, \neg(A \wedge B) \rightarrow \Delta}\ \mathrm{L}\neg\wedge_2^*$$

$$\frac{\Gamma, \neg A, \neg B \rightarrow \Delta}{\Gamma, \neg(A \vee B) \rightarrow \Delta}\ \mathrm{L}\neg\vee$$

$$\frac{\Gamma, A, \neg B \rightarrow \quad \Gamma, \neg(A \supset B) \rightarrow \Delta}{\Gamma, \neg(A \supset B) \rightarrow \Delta}\ \mathrm{L}\neg\supset^* \qquad \frac{\Gamma, A \rightarrow \quad \Gamma, \neg\neg A \rightarrow \Delta}{\Gamma, \neg\neg A \rightarrow \Delta}\ \mathrm{L}\neg\neg^*$$

$$\frac{\Gamma, \neg\forall x A \rightarrow A[a/x] \quad \Gamma, \neg\forall x A \rightarrow \Delta}{\Gamma, \neg\forall x A \rightarrow \Delta}\ \mathrm{L}\neg\forall^* \qquad \frac{\Gamma, \neg\exists x A, \neg A[t/x] \rightarrow \Delta}{\Gamma, \neg\exists x A \rightarrow \Delta}\ \mathrm{L}\neg\exists$$

$\mathrm{L}\neg\forall^*$ is subject to eigenparameter condition on the parameter a.

Figure 9. Refutation rules for the system in Miglioli *et al.* [Miglioli *et al.*, 1994]

forbidding L$\supset$ *on formulae of the form* $\neg A$*, is complete.*

Proof. As in the proof of Theorem 24, we prove a model existence theorem on the basis of the new set of rules by referring to the corresponding proof for LB, emphasising only the modifications of that construction.

Since we have added new inference rules, the set of refutation rules must also be modified. We add to the set of rules in figure 6 the rules in Figure 9; observe that in the latter figure, the rules that are branching are precisely those that define points of backtracking in the search. The points of backtracking can no longer be isolated to applications of Split, in contrast to the situation in LB.

The rules are ordered into four groups:

GROUP 1: $\mathrm{R}\wedge_1, \mathrm{R}\wedge_2, \mathrm{L}\wedge, \mathrm{R}\vee, \mathrm{L}\vee_1, \mathrm{L}\vee_2, \mathrm{LC}\supset_1, \mathrm{L}\supset_2, \mathrm{L}\exists, \mathrm{L}\neg\vee$.

GROUP 2: $\mathrm{LC}\forall, \mathrm{RC}\exists, \mathrm{L}\neg\exists$.

GROUP 3: $\mathrm{L}\neg\wedge_1^*, \mathrm{L}\neg\wedge_2^*, \mathrm{L}\neg\supset^*, \mathrm{L}\neg\neg^*, \mathrm{L}\neg\forall^*$.

GROUP 4: $\mathrm{R}\forall, \mathrm{R}\supset$.

The $\prec$ relation needs to be modified slightly. Let the sequent S be above the sequent T in a derivation. Then $T \prec S$ iff there is an intervening instance of R$\supset$ whose left side formula is not introduced by thinning, an intervening instance of one of the four propositional rules in GROUP 3 such that S occurs over the left premiss of the rule, or an intervening instance of R$\forall$ or L$\neg\forall^*$ whose eigenparameter has been used as parameter in an intervening inference of type L$\forall$. Using this relation we can formulate two conditions that a refutation tree must satisfy in addition to the five conditions introduced for LB in Section 2.2. First condition (1') replaces condition (1):

(R1') thinning only occurs on the left side formula of a R$\supset$ or the side formula of a propositional inference in GROUP 3, and it occurs whenever there is an antecedent occurrence of the same formula below this inference in the tree,

The new clauses are:

(R6) if there is an instance of a rule in GROUP 3 with conclusion T and there is an instance of the same rule above with identical principal formula, the conclusion of which is S, then $T \prec S$,

(R7) the side formula of a L$\neg\exists$ does not have an antecedent occurrence below this inference in the tree, and the term used to instantiate the quantifier is available at the conclusion.

Finally, we must modify slightly the definition of the Kripke model $(U, D, \leq, \Vdash')$ associated with a refutation tree u to account for the fact that, in addition to the rules in Group 4, the rules in GROUP 3 will give rise to new points in the model. We leave the details to the reader.

The new construction allows us to prove smoothly that for all $x \in U$ and every formula A, if $A \in x^-$, then $x \Vdash A$, and if $A \in x^+$, then $x \nVdash A$. The proof, which is by induction on A, adds new sub-cases to the proof of Theorem 19 for cases where the rules in Figure 9 apply. We consider one sub-case to Case 4 in the proof of theorem 19: A is $\neg(B \wedge C)$. Let $A \in x^-$ and $y \geq x$. By construction there must be z-inference r of type L$\neg\wedge_1^*$ or L$\neg\wedge_2^*$ whose principal formula is A, and which satisfies either of the two conditions:

(i) $y \leq z$, or

(ii) $x \leq z \leq y$ and every y-sequent occurs above the left premiss of r.

Assume that r is L$\neg\wedge_1^*$; the case that r is L$\neg\wedge_2^*$ is symmetrical. By inspecting this rule we see that the left premiss of r gives rise to a point u immediately above z such that $\neg B \in u^-$. By induction hypothesis we conclude that $u \Vdash \neg B$. In case (i) $y \leq u$; hence, by Lemma 5, $y \nVdash B$. In case (ii) induction hypothesis and Lemma 5 allow us to conclude that $y \Vdash \neg B$. Since one of these two cases must hold for every $y \geq x$ we conclude that $x \Vdash \neg(B \wedge C)$.

The other cases are similar (or simpler) and are left for the reader. ∎

In their tableau presentation Miglioli *et al.* [1994] introduce the sign F_c ('certain falsehood') in addition to the two signs used by Fitting. The symbol can informally be taken to stand for 'negation proven'. In order to present the set of tableau rules we will no longer treat negation as a defined symbol, i.e. we include explicit negation rules. S_T now means $\{TA \mid TA \in S\} \cup \{F_cA \mid F_cA \in S\}$. Keeping this in mind the tableau system consists of the rules in Figure 7 together with the rules in Figure 10.

$$\frac{S, T\neg A}{S, F_c A}\,T\neg \qquad \frac{S, F\neg A}{S_T, TA}\,F\neg \qquad \frac{S, F_c\neg A}{S_T, TA}\,F_c\neg$$

$$\frac{S, F_c A \wedge B}{S_T, F_c A \mid S_T, F_c B}\,F_c\wedge \qquad \frac{S, F_c A \vee B}{S, F_c A, F_c B}\,F_c\vee \qquad \frac{S, F_c A \supset B}{S_T, TA, FB}\,F_c\supset$$

$$\frac{S, F_c\forall x A}{S_T, FA[a/x], F_c\forall x A}\,F_c\forall \qquad \frac{S, F_c\exists x A}{S, FA[t/x], F_c\exists x A}\,F_c\exists$$

Restrictions: The parameter a in $F_c\forall$ must be new.

Figure 10. Tableau rules of Miglioli *et al.* [Miglioli *et al.*, 1994]

3.4 *The treatment of quantifiers*

One of the most important problems encountered in the formulation of methods of systematic proof-search is how to constrain the non-determinism that arises when choosing terms for use with quantifier rules. This problem was solved for the classical predicate calculus by Robinson [1965], drawing indirectly on ideas from Herbrand [1967]. There are in fact two separate problems to be addressed, one *algebraic* and the other *logical*. The algebraic problem is how to limit the class of terms considered in the search; the subformula property (section 1.3) does not serve to constrain term structure to the same degree as it constrains propositional structure. The logical problem is how to ensure that solutions to the algebraic problem are logically adequate given the meaning attributed to the quantifiers by a logic.

The adaptation of Robinson's solution to classical tableaux is discussed in other chapters in this volume (see Chapter 3). Here we shall be content to provide a brief description of the main elements of that solution as it relates to intuitionistic logic.[8]

Let us first address the algebraic problem of limiting the class of terms considered in a search. Terms normally used in expansion with the rules L∀ and R∃ are instead replaced by *indeterminates*, sometimes called 'free variables' or 'Skolem variables' (cf. Chapter 3). The L∀ and L∃ rules in LB are modified accordingly:

$$\frac{\Gamma, C[\beta/x] \rightarrow \Delta}{\Gamma, \forall x C \rightarrow \Delta}\ \mathrm{L}\forall i \qquad \frac{\Gamma \rightarrow C[\beta/x], \Delta}{\Gamma \rightarrow \exists x C, \Delta}\ \mathrm{R}\exists i$$

Here β is a fresh symbol selected from an alphabet of indeterminates assumed to be disjoint from the terms of the language. Without loss of generality we can assume that the identity axioms are atomic; they are further modified to require

[8]A comprehensive solution for intuitionistic predicate calculus was presented in [Wallen, 1990]; a formulation of that solution would take us beyond the scope of a volume on tableau methods. A Herbrand-Robinson treatment of quantification in an intuitionistic type theory can be found in [Pym and Wallen, 1990].

$$\dfrac{\dfrac{\dfrac{\dfrac{P\beta \to Pa}{P\beta \to \forall x Px} \quad q, P\beta \to q}{\forall x Px \supset q, P\beta \to q}}{\forall x Px \supset q \to P\beta \supset q}}{\forall x Px \supset q \to \exists y(Py \supset q)} \qquad \dfrac{\dfrac{\dfrac{\dfrac{Pa \to Pa}{Pa \to \forall x Px} \quad q, Pa \to q}{\forall x Px \supset q, Pa \to q}}{\forall x Px \supset q \to Pa \supset q}}{\forall x Px \supset q \to \exists y(Py \supset q)}$$

Figure 11. An LB-skeleton of $\forall x Px \supset q \to \exists y(Py \supset q)$ (to the left) and an instantiation of the skeleton with $\beta = a$ (to the right)

only that the predicate symbols of the principal formulae are identical. All other logical and structural rules are as in LB (see Figure 3).

In what follows we shall call trees constructed with the modified axioms and the L$\forall i$ and R$\exists i$ rules *LB-skeletons.* This is to reflect the fact that they do not necessarily carry logical force. The axioms of an LB-skeleton induce a system of simultaneous equations as follows. If the skeleton contains no axioms the induced system of equations is empty; otherwise, let $(P_i s_{i_1} \dots s_{i_m}, P_i t_{i_1} \dots t_{i_m})$, for $i = 1, \dots, n$, be the pairs of principal formulae of the n axioms of the skeleton. The system of simultaneous equations induced is:

$$s_{i_j} = t_{i_j}\,, \qquad i = 1, \dots, n\,;\; j = 1, \dots, m\,.$$

The solubility of such systems over the universe of terms augmented with the indeterminates as additional generating elements can be determined by , a method which also suffices to calculate 'most general' solutions. An example will help to illustrate these ideas.

EXAMPLE 29. Consider the classically valid, but intuitionistically unprovable sequent: $\forall x Px \supset q \to \exists y(Py \supset q)$. An LB-skeleton of the sequent is given in figure 11. The system of equations induced by the skeleton is the singleton

$$\beta \;=\; a$$

which is in solved form already.

The equation-solving perspective, in which the structure of the axioms of a skeleton is used to decide which equations require solution, provides a completely adequate algebraic solution to the problem of constraining the terms considered in a search. Only those terms that materially effect the axioms are ever considered. The fact that most general solutions exist, and can be feasibly computed, makes this one of the central results of automated deduction.

Notice, however, that the eigenparameter condition on R$\forall$ is not satisfied in the tree to the right in figure 11. This brings us to the second problem: how to distinguish the logically adequate solutions from the larger class of algebraically adequate ones. We should expect a solution to this problem to reflect the logical properties of the quantifiers, and therefore to differ systematically from one logic to the next.

Let us first briefly address classical logic. In the classical predicate calculus the quantificational structure is 'separable' from the propositional structure of the logic. This is expressed, variously, by the prenex normal form theorem, the of Smullyan [1968], or the of Gentzen [1935]. Herbrand's theorem[9] [1967] also reflects this property by expressing first-order provability in terms of the provability of quantifier free forms. A formulation of Herbrand's theorem for non-prenex formulae is given in [Bibel, 1982].

Proof-theoretically, this separability can be seen via the permutation properties of the calculus LK (see Figure 5). Recall that the only impermutable rule pairs are the following:

$$\frac{\text{L}\forall \text{ or R}\exists}{\text{R}\forall \text{ or L}\exists}$$

which stem from the eigenparameter conditions.

Following Herbrand, the identification (in classical logic) of those algebraic solutions which nevertheless will fail to satisfy eigenparameter conditions can be achieved quite simply. The eigenparameters in L∃ and R∀ are associated with information which eliminates those solutions which would result in an eigenparameter appearing in the derivation *below* the L∃ or R∀ rule with which it is associated. How this is done in practice is somewhat immaterial. It has become customary to introduce so-called 'Skolem' terms in place of the eigenparameters, and to give as arguments to the terms exactly those indeterminates which appear in the conclusion of the rule (cf. Chapter 3).

We shall first consider this solution as it applies to LB and then show that the classically sound generalisation of this method known as [10] fails intuitionistically. The central modification is to the rules L∃ and R∀ as follows:

$$\frac{\Gamma, A[f\vec{\beta}/x] \to \Delta}{\Gamma, \exists x A \to \Delta}\ \text{L}\exists i \qquad \frac{\Gamma \to A[f\vec{\beta}/x]}{\Gamma \to \forall x A}\ \text{R}\forall i$$

where in each case $\vec{\beta}$ is the list of indeterminates in the conclusion, and f is a 'function' symbol uniquely associated with the inference that introduces it. We shall call the LB system with modified quantifier rules and axioms as described above LB*i*. The 'logical' rules of LB*i* are summarised in Figure 12. The modified LB-skeleton of Figure 11 is shown in Figure 13. The equation induced by this LB*i*-skeleton is

$$\beta = f\beta$$

[9] Theorem 5, page 554 in van Heijenoort's collection [1967], though modern versions are much clearer; see for example Theorem 2.6.4 page 120 of [Girard, 1987].

[10] Skolemisation is discussed extensively in Chapter 3.

$$\frac{\Gamma, A, B \to \Delta}{\Gamma, A \wedge B \to \Delta}\ \mathrm{L}\wedge \qquad \frac{\Gamma \to A, \Delta \quad \Gamma \to B, \Delta}{\Gamma \to A \wedge B, \Delta}\ \mathrm{R}\wedge$$

$$\frac{\Gamma, A \to \Delta \quad \Gamma, B \to \Delta}{\Gamma, A \vee B \to \Delta}\ \mathrm{L}\vee \qquad \frac{\Gamma \to A, B, \Delta}{\Gamma \to A \vee B, \Delta}\ \mathrm{R}\vee$$

$$\frac{\Gamma \to A, \Delta \quad \Gamma, B \to \Delta}{\Gamma, A \supset B \to \Delta}\ \mathrm{L}\supset \qquad \frac{\Gamma, A \to B}{\Gamma \to A \supset B}\ \mathrm{R}\supset$$

$$\frac{\Gamma, A[\beta/x] \to \Delta}{\Gamma, \forall x A \to \Delta}\ \mathrm{L}\forall i \qquad \frac{\Gamma \to A[f\vec{\beta}/x]}{\Gamma \to \forall x A}\ \mathrm{R}\forall i$$

$$\frac{\Gamma, A[f\vec{\beta}/x] \to \Delta}{\Gamma, \exists x A \to \Delta}\ \mathrm{L}\exists i \qquad \frac{\Gamma \to A[\beta/x], \Delta}{\Gamma \to \exists x A, \Delta}\ \mathrm{R}\exists i$$

Restrictions: In L$\forall i$ and R$\exists i$, the indeterminate β must not have been introduced by any other rule in the skeleton. In L$\exists i$ and R$\forall i$, the Herbrand function f must not have been introduced by any other rule in the skeleton; $\vec{\beta}$ is the list of indeterminates appearing in the conclusion of the rule.

Figure 12. The 'logical' rules of LBi

which has no finite solution. This adequately reflects the fact that any solution for β which contains the parameter associated with the R$\forall$ rule will fail to produce a sound LB-derivation from the LBi-skeleton; the parameter will occur in the conclusion of the R$\forall$ rule and the skeleton will thereby fail to satisfy the eigenparameter condition.

The search procedure of Section 2.1 can be easily modified for LBi. The quantifier rules of LBi replace those of LB in GROUPS 1,2 and 3, and an LBi-skeleton is produced following the constraints of the procedure. The ordering on expansions imposed by the procedure ensures that if the system of equations induced by the LBi-skeleton is solvable, the resulting instantiation will produce an LB-proof.

Here is an example of an intuitionistically valid sequent proved by means of this search procedure.

EXAMPLE 30. The sequent we shall consider is $\exists y(Py \supset q) \to \forall x Px \supset q$. An LB$i$-skeleton for this sequent is shown in Figure 13. The induced equation, $\beta = a$, is already in solved form; the reader can easily check that the instantiated skeleton is an LB-proof.

REMARK 31. The method of refutation trees can be extended to LBi. This is left as an exercise for the reader.

The similarity of this essentially proof-theoretical technique to certain (seman-

$$\dfrac{\dfrac{\dfrac{\dfrac{P\beta \rightarrow Pf\beta}{P\beta \rightarrow \forall xPx} \qquad q, P\beta \rightarrow q}{\forall xPx \supset q, P\beta \rightarrow q}}{\forall xPx \supset q \rightarrow P\beta \supset q}}{\forall xPx \supset q \rightarrow \exists y(Py \supset q)} \qquad\qquad \dfrac{\dfrac{\dfrac{\dfrac{P\beta \rightarrow Pa}{\forall xPx \rightarrow Pa} \qquad \forall xPx, q \rightarrow q}{Pa \supset q, \forall xPx \rightarrow q}}{\exists y(Py \supset q), \forall xPx \rightarrow q}}{\exists y(Py \supset q) \rightarrow \forall xPx \supset q}$$

Figure 13. To the left an LB*i*-skeleton for $\forall xPx \supset q \rightarrow \exists y(Py \supset q)$. To the right an LB*i*-skeleton for $\exists y(Py \supset q) \rightarrow \forall xPx \supset q$

tic) methods of quantifier elimination in classical logic due to Skolem[11] has had both a positive and a negative influence on the adaptation of this technique to non-classical logics. On the positive side, skolemisation points to a generalisation of the above technique in the context of classical logic. The of our example endsequent is the sequent $Pa \supset q \rightarrow P\beta \supset q$. This formula is *classically* unsatisfiable if and only if the original formula is. The important point to note is that the parameter a does not depend in any way on the indeterminate. Semantically, this is justified by the fact that the $\forall$ is not in the scope of the $\exists$. Thus we can conclude that, for classical logic at least, the interaction we discovered when seeking to instantiate the skeleton of Figure 11 was a property of of that particular LB-skeleton and not an intrinsic dependency within the endsequent itself.

On the negative side, the generality of the technique as a way to express non-permutabilities in a sequent calculus or tableau system has not been widely appreciated, as it is assumed that the technique relies on properties of classical logic. (See [1990] for an alternative view.)

Proof-theoretically, again at least in classical logic, the independence of the parameter a from the indeterminate β can be justified by the fact that there is a skeleton in which the order of the quantifier inferences is reversed from that in the skeleton of Figure 11. The end-piece of such a derivation is shown below:

$$\dfrac{\forall xPx \supset q \rightarrow \forall xPx, \exists y(Py \supset q) \qquad \dfrac{\dfrac{q, Pa \rightarrow q}{q \rightarrow Pa \supset q}}{q \rightarrow \exists y(Py \supset q)}}{\forall xPx \supset q \rightarrow \exists y(Py \supset q)}$$

Within LB we must thin away the formula $\exists y(Py \supset q)$ in the right branch of this figure before we can apply R$\forall$. It is not hard to see that the resulting sequent $\forall xPx \supset q \rightarrow \forall xPx$ is unprovable in LB. The classical R$\forall$ rule of LK does, however, permit auxiliary formulae on the right. Recall that it has the form

$$\dfrac{\Gamma \rightarrow C[a/x], \Delta}{\Gamma \rightarrow \forall xC, \Delta}\ \text{R}\forall$$

[11] A comprehensive exposition of the relevant techniques arising from Skolem's work can be found in [Shoenfield, 1967].

$$\cfrac{\cfrac{\cfrac{q, Pa \to q}{q \to Pa \supset q}}{q \to \exists y(Py \supset q)} \quad \cfrac{\cfrac{\cfrac{\forall x Px \supset q, Pa \to Pa, q}{\forall x Px \supset q \to Pa, Pa \supset q}}{\forall x Px \supset q \to Pa, \exists y(Py \supset q)}}{\forall x Px \supset q \to \forall x Px, \exists y(Py \supset q)}}{\forall x Px \supset q \to \exists y(Py \supset q)}$$

Figure 14. An LK derivation of $\forall x Px \supset q \to \exists y(Py \supset q)$

which allows successful completion of the end-piece in LK. The completion of this derivation in LK is given in Figure 14.

The soundness of the skolemisation technique within classical logic can be seen to rest on the fact that the permutation constraints on the quantifier rules arise *only* from their axioms and mutual relationships in the endsequent.[12] Our example does not show directly that a similar generalisation is unavailable for intuitionistic logic, since neither the endsequent nor its Skolem normal form is intuitionistically provable. However, the example below shows that skolemisation is unsound intuitionistically.

EXAMPLE 32. The skolemised form of the intuitionistically unprovable sequent $\forall x(Px \vee q) \to \forall x Px \vee q$ is the intuitionistically provable sequent $P\beta \vee q \to Pa \vee q$.

This suggests that it might be difficult to extend skolemisation technique to intuitionistic logic. This difficulty is, however, simply a reflection of the (negative) influence of the semantic view of skolemisation. According to this view, a Herbrand function should be considered an extension to the first-order language.

The reader should immediately see that the problem lies not in the skolemisation technique itself, but in the assumption that the technique should be independent of the logical properties of the quantifiers. Once again we look to the permutation properties for our solution. Recall that for LB we have the following impermutabilities

$$\frac{\text{L}\forall}{\text{R}\forall} \qquad \frac{\text{L}\forall \text{ or R}\exists}{\text{L}\exists} \qquad \frac{\text{L}\supset}{\text{R}\supset \text{ or R}\forall}$$

We have simply to encode these impermutabilities in the Herbrand terms to extend skolemisation to intuitionistic logic. The first three cases are incorporated into the Herbrand function as currently constituted. The last two involving L$\supset$ are not. Note that the essential permutation required to reverse the order of the R$\forall$ and L$\exists$ in leftmost skeleton of Figure 13 is L$\supset$ over R$\supset$ and R$\forall$; these are the two

[12]Note that there are no quantifiers in a Skolem normal form, a reflection of the existence of a prenex normal form.

cases above which the classical formulation of the skolemisation technique fails to account for.

In fact, to extend the technique to intuitionistic logic requires a simple modification of the notion of 'Skolem normal form' to reflect the (permutation) properties of the intuitionistic connectives. Of course we should not necessarily expect to be able to internalise these functions in any way, though in fact they gain a suitable interpretation using . The technical details of such an approach are described in various formats in [Wallen, 1990; Pym and Wallen, 1990]. Shankar [1992] reformulated the technique for use with LJ. The examples above are inspired by his paper. Similar techniques have been introduced for by Fitting [1988] and, in the context of resolution, by Ohlbach [1990].

Arild Waaler

Department of Informatics, University of Oslo and Department of IT and Mathematics, Finnmark College, Norway.

Lincoln Wallen

Oxford University Computing Laboratory, Oxford.

REFERENCES

[Bell, 1990] J. L. Bell. *Toposes and Local Set Theories: an introduction*, volume 14 of *Oxford Logic Guides*. OUP, Oxford, 1990.

[Beth, 1959] E. W. Beth. *The Foundations of Mathematics*. North-Holland, Amsterdam, 1959.

[Bibel, 1982] W. Bibel. Computationally improved versions of Herbrand's Theorem. In J. Stern, editor, *Proceedings of the Herbrand Symposium, Logic Colloquium '81*, pages 11–28, Amsterdam, 1982. North-Holland Publishing Co.

[Constable *et al.*, 1986] R. L. Constable, S. F. Allen, H. M. Bromley, W. R. Cleaveland, F. F. Cremer, R. W. Harper, D. J. Howe, T. B. Knoblock, N. P. Mendler, P. Panandagen, J. T. Sasaki, and S. F. Smith. *Implementing Mathematics with the Nuprl Proof Development System*. Prentice-Hall, Englewood Cliffs, New Jersey, 1986.

[Coquand, 1990] T. Coquand. Metamathematical investigations of a calculus of constructions. In P. Odifreddi, editor, *Logic and Computer Science*, volume 31 of *APIC*, pages 91–122. Academic Press, London, 1990.

[Dyckhoff, 1992] R. Dyckhoff. Contraction-free sequent calculi for intuitionistic logic. *Journal of Symbolic Logic*, 57(3):795–807, 1992.

[Fitting, 1969] M. Fitting. *Intuitionistic Logic, Model Theory, and Forcing*. North-Holland, 1969.

[Fitting, 1988] M. Fitting. First-order modal tableaux. *Journal of Automated Reasoning*, 4:191–213, 1988.

[Gentzen, 1935] G. Gentzen. Untersuchungen über das logische Schließen. *Mathematische Zeitschrift*, 39:176–210, 405–431, 1934–1935, 1934–35. English translation in M.E. Szabo *The Collected Papers of Gerhard Gentzen*, North-Holland, Amsterdam, 1969.

[Girard, 1987] J.-Y. Girard. *Proof Theory and Logical Complexity*, volume 1 of *Studies in Proof Theory*. Bibliopolis, 1987.

[Girard, 1989] J.-Y. Girard. *Proofs and Types*. Cambridge University Press, Cambridge, 1989.

[Griffin, 1990] T. Griffin. A formulas-as-types notion of control. In *Proc. of the Seventeenth Annual Symp. on Principles of Programming Languages*, pages 47–58, 1990.

[Herbrand, 1967] J. Herbrand. Investigations in proof theory. In J. van Heijenoort, editor, *From Frege to Gödel: A Source Book of Mathematical Logic*, pages 525–581. Harvard University Press, Cambridge, MA, 1967.

[Heyting, 1930] A. Heyting. Die formalen Regeln der intuitionistischen Logik. Sitzungsberichte der Preussischen Akademie von Wissenschaften, 1930.

[Heyting, 1956] A. Heyting. *Intuitionism, An Introduction.* North-Holland, Amsterdam, rev. 2nd edition, 1956.

[Howard, 1980] W. A. Howard. The formulae-as-types notion of construction. In J. P. Hindley and J. R. Seldin, editors, *To H. B. Curry: Essays on Combinatory Logic, Lambda-calculus and Formalism*, pages 479–490. Academic Press, London, 1980.

[Hudelmaier, 1992] J. Hudelmaier. Bounds for cut-elimination in intuitionistic propositional logic. *Archive for Mathematical Logic*, 31:331–354, 1992.

[Hudelmaier, 1993] J. Hudelmaier. An $O(n\log n)$-space decision procedure for intuitionistic propositional logic. *Journal of Logic and Computation*, 3(1):63–76, 1993.

[Kleene, 1952] S. C. Kleene. *Introduction to Metamathematics.* North-Holland, Amsterdam, 1952.

[Kleene, 1952b] S. C. Kleene. Permutability of inferences in Gentzen's calculi LK and LJ. *Memoirs of the American Mathematical Society*, 10:1–26, 1952.

[Kripke, 1963] S. Kripke. Semantical analysis of intuitionistic logic I. In J. Crossley and M. Dummett, editors, *Formal Systems and Recursive Functions*, pages 92–129. North-Holland, Amsterdam, 1963.

[Lincoln, 1991] P. Lincoln, A. Scedrov, and N. Shankar. Linearizing intuitionistic implication. In *Sixth annual IEEE symposium on Logic in Computer Science: Proceedings, Amsterdam 1991*, pages 51–62. IEEE Computer Society Press, Los Alamitos, California, 1991.

[Maehara, 1954] S. Maehara. Eine Darstellung der intuitionistische Logik in der klassischen. *Nagoya Mathematical Journal*, 7:45–64, 1954.

[Mints, 1990] G. Mints. Gentzen-type systems and resolution rule, part i: Propositional logic. In *Proceedings International Conference on Computer Logic, COLOG'88, Tallinn, USSR*, volume 417 of *Lecture Notes in Computer Science*, pages 198–231, Berlin, 1990. Springer-Verlag.

[Mints, 1993] G. Mints. Gentzen-type systems and resolution rule, part ii: Predicate logic. In *Proceedings ASL Summer Meeting, Logic Colloquium '90 Helsinki, Finland*, volume 2 of *Lecture Notes in Logic*, pages 163–190, berlin, 1993. Springer-Verlag.

[Martin-Löf, 1982] P. Martin-Löf. Constructive mathematics and computer programming. In L.J. Cohen, editor, *Proceedings of the 6th International Congress for Logic, Methodology and the Philosophy of Science, 1979*, pages 153–175, Amsterdam, 1982. North-Holland.

[Martin-Löf, 1984] P. Martin-Löf. *Intuitionistic Type Theory.* Bibliopolis, Napoli, 1984. Notes by Giovanni Sambin of a Series of Lectures given in Padua, June 1980.

[Martin-Löf, 1996] P. Martin-Löf. On the meaning of the logical constants and the justification of the logical laws. *Nordic Journal of Philosophical Logic*, 1(1):11–60, 1996.

[Miglioli *et al.*, 1994] P. Miglioli, U. Moscato, and M. Ornaghi. An Improved Refutation System for Intuitionistic Predicate Logic. *Journal Automated Reasoning*, 13:361–373, 1994.

[Murthy, 1991] C. Murthy. An evaluation semantics for classical proofs. In *Proc. of Sixth Symp. on Logic in Comp. Sci.*, pages 96–109. IEEE, Amsterdam, The Netherlands, 1991.

[Odifreddi, 1990] P. Odifreddi, editor. *Logic and Computer Science*, volume 31 of *The Apic Studies in Data Processing*. Academic Press, London, 1990.

[Ohlbach, 1990] H.-J. Ohlbach. Semantics-based translation methods for modal logics. *Journal of Logic and Computation*, 1(5):691–746, 1990.

[Parigot, 1992] M. Parigot. $\lambda\mu$-calculus: an algorithmic interpretation of classical natural deduction. In *Proc. LPAR '92*, volume 624 of *LNCS*, pages 190–201, Berlin, 1992. Springer Verlag.

[Pinto and Dyckhoff, 1995] L. Pinto and R. Dyckhoff. Loop-free construction of counter-models for intuitionistic propositional logic. In Behara, Fritsch, and Lintz, editors, *Symposia Gaussiana, Conf. A*, pages 225–232. Walter de Gruyter & Co, 1995.

[Pym and Wallen, 1990] D. J. Pym and L. A. Wallen. Investigations into proof-search in a system of first-order dependent function types. In M.E. Stickel, editor, *Tenth Conference on Automated Deduction, Lecture Notes in Computer Science 449*, pages 236–250. Springer-Verlag, 1990.

[Robinson, 1965] J. A. Robinson. A machine oriented logic based on the resolution principle. *J. Assoc. Comput. Mach.*, 12:23–41, 1965.

[Schütte, 1960] K. Schütte. *Beweistheorie.* Springer Verlag, Berlin, 1960.

[Sahlin *et al.*, 1992] D. Sahlin, T. Franzén, and S. Haridi. An Intuitionistic Predicate Logic Theorem Prover. *Journal of Logic and Computation*, 2(5):619–656, 1992.

[Shankar, 1992] N. Shankar. Proof search in the intuitionistic sequent calculus. In D. Kapur, editor, *Proceedings of 11th International Conference on Automated Deduction*, volume 607 of *Lecture Notes in AI*, pages 522–536, Berlin, 1992. Springer Verlag.

[Shoenfield, 1967] J.R. Shoenfield. *Mathematical Logic*. Addison-Wesley, Reading, MA, 1967.

[Schmiktt and Kreitz, 1996] S. Schmitt and C. Kreitz. Converting non-classical matrix proofs into sequent-style systems. In M.A. McRobbie and J.K. Slaney, editors, *Proceedings of 13th International Conference on Automated Deduction*, volume 1104 of *Lecture Notes in AI*, pages 418–432, Berlin, 1996. Springer Verlag.

[Smullyan, 1968] R. M. Smullyan. *First-Order Logic*. Springer-Verlag, New York, 1968.

[Statman, 1979] R. Statman. Intuitionistic logic is polynomial-space complete. *Theor. Comp. Science*, 9:73–81, 1979.

[Takeuti, 1975] G. Takeuti. *Proof Theory*. North-Holland, Amsterdam, 1975.

[Troelstra and van Dalen, 1988] A. S. Troelstra and D. van Dalen. *Constructivism in Mathematics*, volume 121 of *Studies in Logic and the Foundations of Mathematics*. North-Holland, Amsterdam, 1988. Volumes I and II.

[van Dalen, 1986] D. van Dalen. *Intuitionistic Logic*, volume Handbook of Philosophical Logic III, chapter 4, pages 225–339. Reidel, Dordrecht, 1986.

[van Heijenoort, 1967] J. van Heijenoort. *From Frege to Gödel: A Source Book of Mathematical Logic*. Harvard University Press, Cambridge, MA, 1967.

[Vorob'ev, 1952] N. N. Vorob'ev. *The derivability problem in the constructive propositional calculus with strong negation*. Doklady Akademii Nauk SSSR, 85:689–692, 1952. (In Russian).

[Waaler, 1997a] A. Waaler. Essential contractions in intuitionistic logic. Technical Report PRG-TR-37-97, Oxford University Computing Laboratory, UK, 1997. Submitted for publication.

[Waaler, 1997b] A. Waaler. A sequent calculus for intuitionistic logic with classical permutability properties. Technical Report PRG-TR-38-97, Oxford University Computing Laboratory, UK, 1997. Submitted for publication.

[Wallen, 1990] L. A. Wallen. *Automated deduction in nonclassical logics*. MIT Press, 1990.

RAJEEV GORÉ

TABLEAU METHODS FOR MODAL AND TEMPORAL LOGICS

1 INTRODUCTION

Modal and temporal logics are finding new and varied applications in Computer Science in fields as diverse as Artificial Intelligence [Marek *et al.*, 1991], Models for Concurrency [Stirling, 1992] and Hardware Verification [Nakamura *et al.*, 1987]. Often the eventual use of these logics boils down to the task of deducing whether a certain formula of a logic is a logical consequence of a set of other formula of the same logic. The method of semantic tableaux is now well established in the field of Automated Deduction [Oppacher and Suen, 1988; Baumgartner *et al.*, 1995; Beckert and Possega, 1995] as a viable alternative to the more traditional methods based on resolution [Chang and Lee, 1973]. In this chapter we give a systematic and unified introduction to tableau methods for automating deduction in modal and temporal logics. We concentrate on the propositional fragments restricted to a two-valued (classical) basis and assume some prior knowledge of modal and temporal logic, but give a brief overview of the associated Kripke semantics to keep the chapter self-contained.

One of the best accounts of proof methods for modal logics is the book by Melvin Fitting [1983]. To obtain generality, Fitting uses Smullyan's idea of abstract consistency properties and the associated maximal consistent set approach for proving completeness. As Fitting notes, maximal consistent sets can also be used to determine decidability, but in general, they do not give information about the efficacy of the associated tableau method. Effectiveness however is of primary importance for automated deduction, and a more constructive approach using finite sets, due to Hintikka, is more appropriate. We therefore base our work on a method due to Hintikka [1955] and Rautenberg [1983].

In Section 2 we give the syntax and (Kripke) semantics for propositional modal logics, the traditional axiomatic methods for defining modal logics and the correspondences between axioms and certain conditions on frames.

In Section 3 we give a brief overview of the history of modal tableau systems.

Section 4 is the main part of the chapter and it can be split into two parts.

In Section 4.1 we motivate our study of modal tableau systems. In Section 4.2 we cover the syntax of modal tableau systems, explain tableau constructions and tableau closure. Section 4.3 covers the (Kripke) semantics of modal tableau systems and the notions of soundness and completeness with respect to these semantics. Sections 4.4–4.6 relate our tableau systems to the well-known systems of Fitting and Smullyan, and then cover proof theoretic issues like structural rules, admissible rules and derivable rules. Section 4.8 covers decidability issues like the

M. D'Agostino et al. (eds.), Handbook of Tableau Methods, 297–396.

subformula property, the analytic superformula property, and finiteness of proof search. Sections 4.9–4.12 explain the technical machinery we need to prove the soundness and completeness results, and their connections with decidability. The first half of Section 4 concludes with a summary of the techniques covered so far and sets up the specific examples of tableau systems covered in the second half.

The second half of Section 4 covers tableau systems for: the basic systems; modal logics with epistemic interpretations; modal logics with 'provability' interpretations and mono-modal logics with temporal interpretations. Sections 4.18–4.19 cover proof-theoretic issues again by highlighting some deficiencies of the tableau methods of Section 4. Section 4.20 closes the loop on the Kripke semantics by highlighting the finer characterisation results that are immediate from our constructive proofs of tableau completeness. Finally, Section 4.21 covers the connection between modal tableau systems and modal sequent systems, and the admissibility of the cut rule.

Section 5 is a very brief guide to tableau methods for multimodal logics, particularly linear and branching time logics over discrete frames with operators like 'next', 'until' and 'since'.

Section 6 gives a brief overview of labelled modal tableau systems where labels attached to formulae are used to explicitly keep track of the possible worlds in the tableau constructions.

2 PRELIMINARIES

2.1 Syntax and Notational Conventions

The sentences of modal logics are built from a denumerable non-empty set of primitive propositions $\mathcal{P} = \{p_1, p_2, \cdots\}$, the parentheses) and (, together with the classical connectives $\wedge$ ('and'), $\vee$ ('inclusive or'), $\neg$ ('not'), $\rightarrow$ ('implies'), and the non-classical unary modal connectives $\Box$ ('box') and $\Diamond$ ('diamond').

A well-formed formula, hereafter simply called a **formula**, is any sequence of these symbols obtained from the following rules: any $p_i \in \mathcal{P}$ is a formula and is usually called an **atomic formula**; and if A and B are formulae then so are $(\neg A)$, $(A \wedge B)$, $(A \vee B)$, $(A \rightarrow B)$, $(\Box A)$ and $(\Diamond A)$. For convenience we use $\bot$ to denote a constant false formula $(p_1 \wedge \neg p_1)$ (say) and then use $\top = (\neg\bot)$ to define a constant true formula.

Lower case letters like p and q denote members of $\mathcal{P}$. Upper case letters from the beginning of the alphabet like A and B together with P and Q (all possibly annotated) denote formulae. Upper case letters from the end of the alphabet like X, Y, Z (possibly annotated) denote *finite* (possibly empty) sets of formulae.

The symbols $\neg$, $\wedge$, $\vee$ and $\rightarrow$ respectively stand for logical negation, logical conjunction, logical disjunction and logical (material) implication. To enable us to omit parentheses, we adopt the convention that the connectives $\neg, \Box, \Diamond$ are of equal binding strength but bind tighter than $\wedge$ which binds tighter than $\vee$ which

Axiom	Defining Formula
K	$\Box(A \to B) \to (\Box A \to \Box B)$
T	$\Box A \to A$
D	$\Box A \to \Diamond A$
4	$\Box A \to \Box\Box A$
5	$\Diamond A \to \Box\Diamond A$
B	$A \to \Box\Diamond A$
2	$\Diamond\Box A \to \Box\Diamond A$
M	$\Box\Diamond A \to \Diamond\Box A$
L	$\Box((A \land \Box A) \to B) \lor \Box((B \land \Box B) \to A)$
3	$\Box(\Box A \to B) \lor \Box(\Box B \to A)$
X	$\Box\Box A \to \Box A$
F	$\Box(\Box A \to B) \lor (\Diamond\Box B \to A)$
R	$\Diamond\Box A \to (A \to \Box A)$
G	$\Box(\Box A \to A) \to \Box A$
Grz	$\Box(\Box(A \to \Box A) \to A) \to A$
Go	$\Box(\Box(A \to \Box A) \to A) \to \Box A$
Z	$\Box(\Box A \to A) \to (\Diamond\Box A \to \Box A)$
Zbr	$\Box(\Box A \to A) \to (\Box\Diamond\Box A \to \Box A)$
Zem	$\Box\Diamond\Box A \to (A \to \Box A)$
Dum	$\Box(\Box(A \to \Box A) \to A) \to (\Diamond\Box A \to \Box A)$
Dbr	$\Box(\Box(A \to \Box A) \to A) \to (\Box\Diamond\Box A \to \Box A)$

Figure 1. Axiom names and defining formulae

binds tighter than $\to$. So $\neg A \lor B \land C \to D$ should be read as $(((\neg A) \lor (B \land C)) \to D)$. The symbols $\Box$ and $\Diamond$ can take various meanings but traditionally stand for 'necessity' and 'possibility'. In the context of temporal logic, they stand for 'always' and 'eventually' so that $\Box A$ is read as 'A is always true' and $\Diamond A$ is read as 'A is eventually true'.

2.2 *Axiomatics of Modal Logics*

The logics we shall study are all normal extensions of the basic modal logic **K** and are traditionally axiomatised by taking the rule of necessitation **RN** (if A is a theorem then so is $\Box A$) and modus ponens **MP** (if A and $A \to B$ are theorems then so is B) as inference rules, and by taking the appropriate formulae from Figure 1 as axiom *schemas*. Thus the rule of uniform substitution **US** is built in so that any substitutional instance of an axiom schema or theorem, is also a theorem.

If a logic is axiomatised by adding axioms $A_1, A_2, \cdots, A_n$ to **K** then its name is written as $\mathbf{KA_1A_2 \cdots A_n}$. Sometimes however, the logic is so well known in the literature by another name that we revert to the traditional name. The logic

KT4, for example, is usually known as **S4**.

For an introduction to these notions see the introductory texts by Hughes and Cresswell [1968; 1984] or Chellas [1980], or the article by Fitting [1993].

We write $\vdash_{\mathbf{L}} A$ to denote that A is a theorem of an axiomatically formulated logic **L**. As with classical logic, the notion of theoremhood can be extended to the notion of 'deducibility' where we write $X \vdash_{\mathbf{L}} A$ to mean 'there is a deduction of A from the set of formulae X'. However, some care is needed when extending this notion to modal logics if we want to preserve the 'deduction theorem': $X \vdash_{\mathbf{L}} (A \to B)$ iff $X \cup \{A\} \vdash_{\mathbf{L}} B$, since it is well known that the deduction theorem fails if we use the notion of deducibility from classical Hilbert system formulations (due to the rule of necessitation). Fitting [1993] shows how to set up the notions of 'deducibility' so that the deduction theorem holds, but since axiomatics are of a secondary nature here, we omit details. The important point is that the notion of theoremhood $\vdash_{\mathbf{L}} A$ remains the same since it corresponds to 'deducibility' of A from the empty set viz: $\{\} \vdash_{\mathbf{L}} A$. We return to this point in Section 4.3.

2.3 Kripke Semantics for Modal Logics

A Kripke **frame** is a pair $\langle W, R\rangle$ where W is a non-empty set (of possible worlds) and R is a binary relation on W. We write wRw' iff $(w, w') \in R$ and we say that world w' is **accessible from** world w, or that w' is **reachable from** w, or w' is a **successor** of w, or even that w **sees** w'. We also write $w\not{R}w'$ to mean $(w, w') \notin R$.

A Kripke **model** is a triple $\langle W, R, V\rangle$ where V is a mapping from primitive propositions to sets of worlds; that is, $V : \mathcal{P} \mapsto 2^W$. Thus $V(p)$ is the set of worlds at which p is 'true' under the valuation V.

Given some model $\langle W, R, V\rangle$, and some $w \in W$, we write $w \models p$ iff $w \in V(p)$, and say that w satisfies p or p is true at w. We also write $w \not\models p$ to mean $w \notin V(p)$. This satisfaction relation $\models$ is then extended to more complex formulae according to the primary connective as below:

$w \models p$	iff	$w \in V(p)$;
$w \models \neg A$	iff	$w \not\models A$;
$w \models A \wedge B$	iff	$w \models A$ and $w \models B$;
$w \models A \vee B$	iff	$w \models A$ or $w \models B$;
$w \models A \to B$	iff	$w \not\models A$ or $w \models B$;
$w \models \Box A$	iff	for all $v \in W$, $w\not{R}v$ or $v \models A$;
$w \models \Diamond A$	iff	there exists some $v \in W$, with wRv and $v \models A$.

We say that w **satisfies** A iff $w \models A$ where the valuation is left as understood. If $w \models A$ we sometimes also say that A **is true at** w, or that w **makes** A **true**.

A formula A is **satisfiable in a model** $\langle W, R, V\rangle$ iff there exists some $w \in W$ such that $w \models A$. A formula A is **satisfiable on a frame** $\langle W, R\rangle$, iff there exists some valuation V and some world $w \in W$ such that $w \models A$. A formula A is **valid in a model** $\langle W, R, V\rangle$, written as $\langle W, R, V\rangle \models A$, iff it is true at every world in

Axiom	Condition	First-Order Formula
T	Reflexive	$\forall w : R(w, w)$
D	Serial	$\forall w \, \exists w' : R(w, w')$
4	Transitive	$\forall s, t, u : (R(s, t) \wedge R(t, u)) \rightarrow R(s, u)$
5	Euclidean	$\forall s, t, u : (R(s, t) \wedge R(s, u)) \rightarrow R(t, u)$
B	Symmetric	$\forall w, w' : R(w, w') \rightarrow R(w', w)$
2	Weakly-directed	$\forall s, t, u \, \exists v :$ $(R(s, t) \wedge R(s, u)) \rightarrow (R(t, v) \wedge R(u, v))$
L	Weakly-connected	$\forall s, t, u :$ $(R(s, t) \wedge R(s, u)) \rightarrow (R(t, u) \vee t = u \vee R(u, t))$
X	Dense	$\forall u, v \, \exists w : R(u, v) \rightarrow (R(u, w) \wedge R(w, v))$

Figure 2. Axioms and corresponding first-order conditions on R

W. A formula A is **valid in a frame** $\langle W, R \rangle$, written as $\langle W, R \rangle \models A$, iff it is valid in all models $\langle W, R, V \rangle$ (based on that frame). An axiom (schema) is valid in a frame iff all instances of that axiom (schema) are valid in all models based on that frame.

Given a class of frames $\mathcal{C}$, an axiomatically formulated logic $\mathbf{L}$ is **sound** with respect to $\mathcal{C}$ if for all formulae A:

$$\text{if } \vdash_{\mathbf{L}} A \text{ then, } \mathcal{F} \models A \text{ for all frames } \mathcal{F} \in \mathcal{C}.$$

Logic $\mathbf{L}$ is **complete** with respect to $\mathcal{C}$ if for all formulae A:

$$\text{if } \mathcal{F} \models A \text{ for all frames } \mathcal{F} \in \mathcal{C}, \text{ then } \vdash_{\mathbf{L}} A.$$

A logic $\mathbf{L}$ is **characterised** by a class of frames $\mathcal{C}$ iff $\mathbf{L}$ is sound and complete with respect to $\mathcal{C}$.

2.4 Known Correspondence and Completeness Results

The logics we study are known to be characterised by certain classes of frames because it is known that particular axioms correspond to particular restrictions on the reachability relation R. That is, suppose $\langle W, R \rangle$ is a frame, then a certain axiom A_1 will be valid on $\langle W, R \rangle$ *if and only if* the reachability relation R meets a certain condition. Many of the restrictions are definable as formulae of first-order logic where the binary predicate $R(x, y)$ represents the reachability relation, as shown in Figure 2, where the correspondences between certain axioms and certain conditions are also summarised. Some interesting properties of frames which cannot be captured by any one axiom are given in Figure 3; see [Goldblatt, 1987]. But some quite bizarre axioms, whose corresponding conditions cannot be expressed in first-order logic [van Benthem, 1984; van Benthem, 1983] are of particular interest precisely because of this 'higher order' nature. Some of these 'higher order' conditions are explained next.

Given a frame $\langle W, R\rangle$, an **R-chain** is a sequence of (not necessarily distinct) points from W with $w_1Rw_2Rw_3R\cdots Rw_n$. An **∞-R-chain** is an R-chain where n can be chosen arbitrarily large. A **proper R-chain** is an R-chain where the points are distinct. For example, a single reflexive point gives an (improper) ∞-R-chain: $wRwRwRw\cdots$.

Transitive frames are of particular interest when R is viewed as a flow of time. Informally, if $\langle W, R\rangle$ is a frame where R is transitive, then a **cluster** C is a maximal subset of W such that for all *distinct* worlds w and w' in C we have wRw' and $w'Rw$. A cluster is **degenerate** if it is a single irreflexive world, otherwise it is **nondegenerate**. A nondegenerate cluster is **proper** if it consists of two or more worlds. A nondegenerate cluster is **simple** if it consists of a single reflexive world. Note that in a nondegenerate cluster, R is reflexive, transitive *and symmetric*.

Because clusters are maximal we can order them with respect to R and speak of a cluster preceding another one. Similarly, a cluster C is **final** if no other cluster succeeds it and a cluster is **last** if *every* other cluster precedes it. For an introduction to Kripke frames, Kripke models and the notion of clusters see Goldblatt [1987] or Hughes and Cresswell [1984].

Figure 4 encapsulates the known characterisation results for each of our logics by listing the conditions on some class of frames that characterises each logic. The breaks in Figure 4 correspond to the grouping of the tableau systems for these logics under Sections 4.14–4.17. Thus we define a frame to be an **L-frame** iff it meets the restrictions of Figure 4. Then, a model $\langle W, R, V\rangle$ is an **L-model** iff $\langle W, R\rangle$ is an **L**-frame. A formula A is **L-valid** iff it is true in every world of every **L**-model. An **L**-model $\langle W, R, V\rangle$ is **an L-model for a finite set** X of formulae iff there exists some $w_0 \in W$ such that for all $A \in X$, $w_0 \models A$. A set X is **L-satisfiable** iff there is an **L**-model for X.

An axiomatically formulated logic **L** has the **finite model property** if every nontheorem A of **L** can be falsified at some world in some *finite* **L**-model. That is, if $\not\vdash_{\mathbf{L}} A$ implies that $\{\neg A\}$ has a *finite* **L**-model.

Property Name	Property of R
Irreflexive	$\forall w : \neg R(w, w)$
Intransitive	$\forall s, r, t : (R(s,t) \wedge R(t,r)) \rightarrow \neg R(s,r)$
Antisymmetric	$\forall s, t : (R(s,t) \wedge R(t,s)) \rightarrow (s = t)$
Asymmetric	$\forall w_1, w_2 : R(w_1, w_2) \rightarrow \neg R(w_2, w_1)$
Strict-order	$\forall w_1, w_2 : (w_1 \neq w_2) \rightarrow (R(w_1, w_2) \text{ exor } R(w_2, w_1))$

Figure 3. Names of some non-axiomatisable conditions on R

L	Axiomatic Basis	L-frame restriction
K	K	no restriction
KT	KT	reflexive
KD	KD	serial
K4	$K4$	transitive
K5	$K5$	Euclidean
KB	KB	symmetric
KDB	KDB	symmetric and serial
B	KTB	reflexive and symmetric
KD4	$KD4$	serial and transitive
K45	$K45$	transitive and Euclidean
KD5	$KD5$	serial and Euclidean
KD45	$KD45$	serial, transitive and Euclidean
S4	$KT4$	reflexive and transitive
KB4	$KB4$	symmetric and transitive
S5	$KT5$	reflexive, transitive and symmetric
S4R	$KT4R$	reflexive, transitive and $\forall x, y, z$: $(x \neq z \wedge R(x,z)) \rightarrow (R(x,y) \rightarrow R(y,z))$
S4F	$KT4F$	reflexive, transitive and $\forall x, y, z$: $(R(x,z) \wedge \neg R(z,x)) \rightarrow (R(x,y) \rightarrow R(y,z))$
S4.2	$KT4.2$	reflexive, transitive and weakly-directed
S4.3	$KT4.3$	reflexive, transitive and weakly-connected
S4.3.1	$KT4.3Dum$	reflexive, transitive, weakly-connected and no nonfinal proper clusters
S4Dbr	$KT4Dbr$	reflexive, transitive and no nonfinal proper clusters
K4DL	$KD4L$	serial, transitive and weakly-connected
K4DLX	$KD4LX$	serial, transitive, weakly-connected and dense
K4DLZ	$KD4LZ$	serial, transitive, weakly-connected and no nonfinal non-degenerate clusters
K4DZbr	$K4DZbr$	serial, transitive and no nonfinal nondegenerate clusters
G	KG	transitive and no ∞-R-chains (irreflexive)
Grz	$KGrz$	reflexive, transitive, no proper clusters and no proper ∞-R-chains
K4Go	$K4Go$	transitive, no proper clusters and no proper ∞-R-chains
GL	KGL	transitive, weakly-connected, no proper clusters and no ∞-R-chains (irreflexive)

Figure 4. Axiomatic Bases and L-frames

2.5 Logical Consequence

Suppose we are given some finite set of formulae Y, some formula A, and assume that the logic of interest is $\mathbf{L}$. We say that the formula A is a **local logical consequence** of the set Y iff: for every $\mathbf{L}$-model $\langle W, R, V \rangle$ and for every $w \in W$, if $w \models Y$ then $w \models A$. We write $Y \models_{\mathbf{L}} A$ whenever A is a local logical consequence of Y in logic $\mathbf{L}$; thus the subscript is for the logic, not for the word 'local'.

Since both Y and A are evaluated at the same world w in this definition, it is straightforward to show that $Y \models_{\mathbf{L}} A$ iff $\{\} \models_{\mathbf{L}} \widehat{Y} \rightarrow A$ where $\{\}$ is the empty set, and $\widehat{Y}$ is just the conjunction of the members of Y. Furthermore, a semantic version of the usual deduction theorem holds for local logical consequence viz: $Y, A \models_{\mathbf{L}} B$ iff $Y \models_{\mathbf{L}} A \rightarrow B$ where we write Y, A to mean $Y \cup \{A\}$.

As we saw in Section 2.2, the traditional axiomatically formulated logics obey the deduction theorem only if deducibility is defined in a special way. Fitting [1983] shows that a stronger version of logical consequence called global logical consequence corresponds to this notion of deducibility. Fitting also gives tableau systems that cater to both notions of logical consequence. We concentrate only on the local notion since Fitting's techniques can be used to extend our systems to cater for the global notion.

2.6 Summary

The semantic notion of validity $\models A$ and the axiomatic notion of theoremhood $\vdash A$ are tied to each other via the notions of soundness and completeness of the axiomatic deducibility relation $\vdash$ with respect to some class of Kripke frames. These notions can be generalised respectively to logical consequence $Y \models A$ and $Y \vdash A$. By careful definition we can maintain the soundness and completeness results intact for these generalisations. Unfortunately, axiomatic systems are notoriously bad for proof *search* because they give no guidance on how to look for a proof. Tableau systems also give rise to a syntactic notion of theoremhood but have the added benefit that they facilitate proof search in a straightforward way. Such systems are the subject of the rest of this chapter.

3 HISTORY OF MODAL TABLEAU SYSTEMS

The history of modal tableau systems can be traced back through two routes, one semantic and one syntactic.

The syntactic route began with the work of Gerhard Gentzen [1935] and the numerous attempts to extend Gentzen's results to modal logics. Curry [1952] appears to be the first to seek Gentzen systems for modal logics, soon followed by Ohnishi and Matsumoto [1957a; 1959; 1957]. Kanger [1957] is the first to use extra-logical devices to obtain Gentzen systems and is the precursor of what are now known as prefixed or labelled tableau systems. Once the basic method was worked out other authors tried to find similar systems for other logics, turning modal Gentzen systems into an industry for almost twenty years.

Not surprisingly, modal Gentzen systems involve a cut-elimination theorem. In many respects this early work on modal Gentzen systems was very difficult because these authors had no semantic intuitions to guide them and had to work quite hard to obtain a syntactic cut-elimination theorem. As we shall see, the task is much easier when we use the associated Kripke semantics.

The semantic route began with the work of Beth for classical propositional logic [Beth, 1955; Beth, 1953] but lay dormant for modal logics for almost twenty years until the advent of Kripke semantics [Kripke, 1959]. From then on, modal tableau systems, and in general modal logic, witnessed a resurgence.

The two routes began to meet in the late sixties when it was realised that classical semantic tableau systems and classical Gentzen systems were essentially the same thing. Zeman [1973] appears to be the first to give an account of both traditions simultaneously, although he is sometimes unable to relate his tableau systems to his Gentzen systems (c.f. his tableau system for **S4.3** is cut-free, yet his sequent system for **S4.3** is not). Rautenberg [1979] gives a rigorous account and covers many logics but has not received much attention as his book is written in German. Fitting's book [1983] is the most widely known and covers most of the basic logics.

During the eighties the two traditions were seen as two sides of the same coin, but more recently, the semantic tradition has assumed prominence in the field of automated deduction, while the syntactic tradition has gained prominence in the field of type theory [Masini, 1993; Borghuis, 1993]. In automated deduction, the primary emphasis is on finding a proof, whereas in type theory, the primary emphasis is on the ability to distinguish different proofs so as to put a computational interpretation on proofs.

Regardless of this historical basis, there are essentially two types of tableau systems which we shall call **explicit systems** and **implicit systems.** Recall that tableau systems are essentially semantic in nature, hence the reachability relation R plays a crucial part. In explicit systems, the reachability relation is represented *explicitly* by some device, and we are allowed to reason directly about the known properties of R, such as transitivity or reflexivity. In implicit systems, there is no explicit representation of the reachability relation, and these properties must be *built into the rules* in some way since we are not allowed to reason explicitly about R. We shall see that in some sense the two types of systems are dual in nature since implicit systems can be turned into explicit systems by giving a systematic method or strategy for the application of the implicit tableau rules.

Here is an outline of what follows. In the first few sections we introduce the syntax of implicit modal tableau systems by defining the form of the rules and tableau systems. These are all purely syntactic aspects of modal tableau systems allowing us to associate a syntactic deducibility relation with modal tableau systems. In the second part we introduce the semantics of modal tableau rules, and systems, and define the notions of soundness and completeness of modal tableau systems with respect to these semantics. In the last part we introduce the mathematical structures that we shall need to prove the soundness and completeness of the given tableau systems.

We then give tableau systems in decreasing detail for: the basic modal logics; the monotonic modal logics used to define nonmonotonic modal logics of knowledge and belief; modal logics with 'provability interpretations'; monomodal logics of linear and branching time; and multimodal logics of linear and branching time.

In the later sections of this chapter we introduce explicit tableau systems since they are an extension of implicit tableau systems. The extra power of explicit tableau systems comes from the labels which carry very specific semantic information about the (counter-)model under construction. Consequently we see that explicit tableau systems are better for the symmetric logics.

For the sake of brevity we do not consider quantified modal logics, but see Fitting [1983] for a treatment of quantified modal tableau systems.

4 MODAL TABLEAU SYSTEMS WITH IMPLICIT ACCESSIBILITY

4.1 Purpose of Modal Tableau Systems

As stated in the introduction, we concentrate on the use of modal tableau systems for performing deduction. In this context, modal tableau systems can be seen as refutation procedures that decompose a given set of formulae into a network of sets with each set representing a possible world in the associated Kripke model. Thus, our modal tableau systems are anchored to the semantics of the modal logic although they can be used in sequent form to obtain metamathetical results like interpolation theorems as well; see [Fitting, 1983] and [Rautenberg, 1983; Rautenberg, 1985].

The main features of semantic tableau systems carry over from classical propositional logic in that a set of formulae X is deemed consistent if and only if no tableau for X closes. Furthermore, from these open tableaux, we can construct a model to demonstrate that X is indeed satisfiable, thus tying the syntactic notion of consistency to the semantic notion of satisfiability.

Now, assume we are given some finite set of formulae $Y = \{A_1, \cdots, A_k\}$, and some formula A. Let $\widehat{Y} = (A_1 \land A_2 \land \cdots \land A_k)$ with $\widehat{Y} = \bot$ when $k = 0$. By definition, if the set $Y \cup \{\neg A\}$ is *not* **L**-satisfiable, then, in every **L**-model, each world that makes each member of Y true, must also make A true. That is, if the set $Y \cup \{\neg A\}$ is *not* **L**-satisfiable, then, the formula $\widehat{Y} \to A$ must be **L**-valid. Modal tableau systems give us a purely syntactic method of determining whether or not some given formula is **L**-valid. Thus, they give us a method of determining whether A is a *local logical consequence* of a set of formulae Y.

4.2 *Syntax of Modal Tableau Systems*

The most popular tableau formulation is due to Smullyan as expounded by Fitting [1983]. Following Hintikka [1955] and Rautenberg [1983; 1985], we use a slightly different formulation where formulae are carried from one tableau node to its child because the direct correspondence between sequent systems and tableau systems is easier to see using this formulation. To minimise the number of rules, we work with primitive notation, taking $\Box, \neg$ and $\wedge$ as primitives and defining all other connectives from these. Thus, for example, there are no explicit rules for $\vee$ and $\rightarrow$ but these can be obtained by rewriting $A \vee B$ as $\neg(\neg A \wedge \neg B)$ and $A \rightarrow B$ as $\neg(A \wedge \neg B)$. All our tableau systems work with *finite* sets of formulae.

We use the following notational conventions:

- $\bot$ denotes a constant false proposition and $\emptyset$ denotes the empty set;
- p, q denote primitive (atomic) propositions from $\mathcal{P}$;
- P, Q, Q_i and P_i denote (well formed) formulae;
- X, Y, Z denote *finite* (possibly empty) sets of (well formed) formulae;
- $(X;Y)$ stands for $X \cup Y$ and $(X;P)$ stands for $X \cup \{P\}$;
- $\Box X$ stands for $\{\Box P \mid P \in X\}$;
- $\neg\Box X$ stands for $\{\neg\Box P \mid P \in X\}$.

We use P and Q as formulae in the tableau rules and use A and B in the axioms to try to separate the two notions. Note that $(X;P;P) = (X;P)$ and also that $(X;P;Q) = (X;Q;P)$ so that the number of copies of the formulae and their order are immaterial as far as the notation is concerned.

A **tableau rule** ρ consists of a **numerator** $\mathcal{N}$ above the line and a (finite) list of **denominators** $\mathcal{D}_1, \mathcal{D}_2, \ldots, \mathcal{D}_k$ (below the line) separated by vertical bars:

$$(\rho) \quad \frac{\mathcal{N}}{\mathcal{D}_1 \mid \mathcal{D}_2 \mid \cdots \mid \mathcal{D}_k}$$

The numerator is a finite set of formulae and so is each denominator. We use the terms numerator and denominator rather than premiss and conclusion to avoid confusion with the sequent terminology. As we shall see later, each tableau rule is read downwards as 'if the numerator is **L**-satisfiable, then so is one of the denominators'.

The numerator of each tableau rule contains one or more distinguished formulae called the **principal formulae**. Each denominator usually contains one or more distinguished formulae called the **side formulae**. Each tableau rule is labelled with a name which usually consists of the main connective of the principal formula, in parentheses, but may consist of a more complex name. The rule name appears

$$(\wedge)\ \frac{X; P \wedge Q}{X; P; Q} \qquad (\bot)\ \frac{X; P; \neg P}{\bot} \qquad (\vee)\ \frac{X; \neg(P \wedge Q)}{X; \neg P \mid X; \neg Q}$$

$$(\neg)\ \frac{X; \neg\neg P}{X; P} \qquad (\theta)\ \frac{X; Y}{X} \qquad (K)\ \frac{\Box X; \neg\Box P}{X; \neg P}$$

Figure 5. Tableau rules for $\mathcal{C}\mathbf{K}$ where X, Y are sets and P, Q are formulae

at the left when the rule is being defined, and appears at the right when we use a particular instance of the rule.

For example, below at right is a tableau rule with:

1. a rule name $(\vee)$;
2. a numerator $X; \neg(P \wedge Q)$ with a principal formula $\neg(P \wedge Q)$; and
3. two denominators $X; \neg P$ and $X; \neg Q$ with respective side formulae $\neg P$ and $\neg Q$.

$$(\vee)\ \frac{X; \neg(P \wedge Q)}{X; \neg P \mid X; \neg Q}$$

A **tableau system** (or calculus) $\mathcal{C}\mathbf{L}$ is a finite collection of tableau rules $\rho_1, \rho_2, \cdots, \rho_m$ identified with the set of its rule names; thus $\mathcal{C}\mathbf{L} = \{\rho_1, \rho_2, \cdots, \rho_m\}$. Figure 5 contains some tableau rules which we shall later prove are those that capture the basic normal modal logic $\mathbf{K}$; thus $\mathcal{C}\mathbf{K} = \{(\bot), (\wedge), (\vee), (\neg), (K), (\theta)\}$.

A **$\mathcal{C}$L-tableau for** X is a finite tree with root X whose nodes carry *finite formula sets*. A tableau rule with numerator $\mathcal{N}$ is applicable to a node carrying set Y if Y is an instance of $\mathcal{N}$. The steps for extending the tableau are:

- choose a leaf node n carrying Y where n is not an end node, and choose a rule ρ which is applicable to n;
- if ρ has k denominators then create k successor nodes for n, with successor i carrying an appropriate instantiation of denominator D_i;
- all with the proviso that if a successor s carries a set Z and Z has already appeared on the branch from the root to s then s is an end node.

A branch in a tableau is **closed** if its end node is $\{\bot\}$; otherwise it is **open**. A tableau is **closed** if all its branches are closed; otherwise it is **open**.

The rule $(\bot)$ is really a check for inconsistency, therefore, we say that a set X is **$\mathcal{C}$L-consistent** if no $\mathcal{C}$L-tableau for X is closed. Conversely we say that a formula A is a **theorem** of $\mathcal{C}\mathbf{L}$ iff there is a closed tableau for the set $\{\neg A\}$. We write $\vdash_{\mathcal{C}\mathbf{L}} A$ if A is a theorem of $\mathcal{C}\mathbf{L}$ and write $Y \vdash_{\mathcal{C}\mathbf{L}} A$ if $Y \cup \{\neg A\}$ is $\mathcal{C}\mathbf{L}$-inconsistent.

EXAMPLE 1. The formula $\Box(p \to q) \to (\Box p \to \Box q)$ is an instance of the axiom **K**. Its negation can be written in primitive notation and simplified to $\Box(\neg(p \wedge \neg q)) \wedge \Box p \wedge \neg\Box q$. Below at left is a closed $\mathcal{C}\mathbf{K}$-tableau for the (singleton) set $X = \{\Box(\neg(p \wedge \neg q)) \wedge \Box p \wedge \neg\Box q\}$ where each node is labelled at the right by the rule that produces its successor(s). Below at right is a more succinct version of the same $\mathcal{C}\mathbf{K}$-tableau. Hence $\Box(p \to q) \to (\Box p \to \Box q)$ is a theorem of $\mathcal{C}\mathbf{K}$.

$$\{\Box\neg(p \wedge \neg q) \wedge \Box p \wedge \neg\Box q\}\ (\wedge)$$
$$|$$
$$\{\Box\neg(p \wedge \neg q) \wedge \Box p, \neg\Box q\}\ (\wedge)$$
$$|$$
$$\{\Box\neg(p \wedge \neg q), \Box p, \neg\Box q\}\ (K)$$
$$|$$
$$\{\neg(p \wedge \neg q), p, \neg q\}\ (\vee)$$
$$\swarrow \qquad \searrow$$
$$\{\neg p, p, \neg q\}\ (\bot) \qquad \{\neg\neg q, p, \neg q\}\ (\bot)$$
$$| \qquad\qquad\qquad |$$
$$\bot \qquad\qquad\qquad \bot$$

$$\cfrac{\cfrac{\cfrac{\cfrac{\Box\neg(p \wedge \neg q) \wedge \Box p \wedge \neg\Box q}{\Box\neg(p \wedge \neg q) \wedge \Box p; \neg\Box q}\ (\wedge)}{\Box\neg(p \wedge \neg q); \Box p; \neg\Box q}\ (\wedge)}{\neg(p \wedge \neg q); p; \neg q}\ (K)}{\neg p; p; \neg q\ (\bot) \quad | \quad \neg\neg q; p; \neg q\ (\bot)}\ (\vee)$$

4.3 *Soundness and Completeness*

Tableau systems give us a syntactic way to define consistency, and hence theoremhood. As with the axiomatic versions of these notions, the notions of soundness and completeness relate these syntactic notions to the semantic notions of satisfiability and validity as follows.

Soundness: We say that $\mathcal{C}\mathbf{L}$ is **sound with respect to L-frames** (the Kripke semantics of **L**) if: $Y \vdash_{\mathcal{C}\mathbf{L}} A$ implies $Y \models_{\mathbf{L}} A$. In words, if there is a closed $\mathcal{C}\mathbf{L}$-tableau for $Y \cup \{\neg A\}$ then any **L**-model that makes Y true at world w must make A true at world w.

Completeness: We say that $\mathcal{C}\mathbf{L}$ is **complete with respect to L-frames** (the Kripke semantics of **L**) if: $Y \models_{\mathbf{L}} A$ implies $Y \vdash_{\mathcal{C}\mathbf{L}} A$. In words, if every **L**-model that makes Y true at world w also makes A true at world w, then some $\mathcal{C}\mathbf{L}$-tableau for $Y \cup \{\neg A\}$ must close.

We already know that axiomatically formulated **L** is also sound and complete with respect to **L**-frames. If we can show that $\mathcal{C}\mathbf{L}$ is also sound and complete with respect to **L**-frames then we can complete the link between $\mathcal{C}\mathbf{L}$ and **L** via: $Y \vdash_{\mathcal{C}\mathbf{L}} A$ iff $Y \models_{\mathbf{L}} A$ iff $\models_{\mathbf{L}} \widehat{Y} \to A$ iff $\vdash_{\mathbf{L}} \widehat{Y} \to A$. Thus our tableau systems, as given, capture axiomatically formulated theoremhood only. As stated previously, they can be easily extended to handle the stronger notion of 'deducibility' using techniques for handling global logical consequence from Fitting [1983].

4.4 *Relationship to Smullyan Tableau Systems*

Tableau systems are often presented using trees where each node is labelled by a single (possibly signed) formula [Fitting, 1983]. The associated tableau rules then allow us to choose some formula on the current branch as the principal formula of the rule, and then to extend all branches below this formula by adding other formulae onto the end of these branches. For modal logics, some of the tableau rules demand the deletion of formulae from the current branch, as well as the addition of new formulae. In fact, the tableau rules are often summarised using set notation by collecting into a numerator all the formulae on the branch prior to a tableau rule application, and collecting into one or more denominators all the formulae that remain after the tableau rule application. Such summarised rules correspond exactly to the tableau rules we use. In particular, the thinning rule (θ) allows us to capture the desired deletion rules.

4.5 *Structural Rules*

Tableau systems are closely related to Gentzen systems and both often contain three rules known as structural rules; so called because they do not affect a particular formula in the numerator but the whole of the numerator itself.

Exchange

Since we use sets of formulae, the order of the formulae in the set is immaterial. Thus a commonly used rule called the 'exchange' rule that simply swaps the order of formulae is implicit in our formulation.

Contraction

The $(\wedge)$ rule is shown below left. Consider the two applications of the $(\wedge)$ rule shown at right:

$$(\wedge)\ \frac{X; P \wedge Q}{X; P; Q} \qquad\qquad \frac{p \wedge q}{p\,;\,q}\ (\wedge) \qquad\qquad \frac{p \wedge q}{p \wedge q\,;\,p\,;\,q}\ (\wedge)$$

The left hand application is intuitive, corresponding to putting $X = \emptyset$, $P = p$, and $Q = q$ giving a numerator

$$\mathcal{N} = (X; P \wedge Q) = (\emptyset; p \wedge q) = \{p \wedge q\}$$

and hence obtaining the denominator

$$\mathcal{D} = (X; P; Q) = (\emptyset; p; q) = \{p, q\}.$$

However, the right-hand derivation is also legal since we can put $X = \{p \wedge q\}$, $P = p$, and $Q = q$ to give the numerator

$$\mathcal{N} = (X; P \wedge Q) = (p \wedge q; p \wedge q) = \{p \wedge q\}$$

and hence obtain the denominator

$$\mathcal{D} = (X; P; Q) = (p \wedge q; p; q) = \{p \wedge q, p, q\}.$$

Thus, although our tableau rules seem to delete the principal formulae in a rule application, they also allow us to carry that formula into the denominator if we so choose.

Now, in classical propositional logic, it can be shown that the deletion of the principal formula does no harm. However, in certain modal logics, the deletion of the principal formula leads to incompleteness. That is, a tableau for X may not close if we always delete the principal formula, and yet, a similar tableau for X may close if we carry a copy of the principal formula into the denominator. For an example, see Example 8 on page 324.

Completeness is essential if our tableau systems are to be used as decision procedures, thus we need a way to duplicate formulae. It is tempting to add a rule called the contraction rule (ctn) as shown below left. And below at right is an application of it where we duplicate the formula $\Box p$ in $\mathcal{N} = \{p \wedge q, \Box p\}$:

$$(\text{ctn}) \ \frac{X; P}{X; P; P} \qquad\qquad \frac{p \wedge q; \Box p}{p \wedge q; \Box p; \Box p} \ (\text{ctn})$$

But now we have a problem, for the definition of a tableau is in terms of nodes carrying *sets* and the two nodes of the right-hand tableau carry identical sets since $(p \wedge q; \Box p) = (p \wedge q; \Box p; \Box p) = \{p \wedge q, \Box p\}$. Thus, any explicit application of the contraction rule immediately gives a cycle and stops the tableau construction. An explicit contraction rule is not what we want.

In order to avoid these complexities we shall omit an explicit contraction rule from our tableau systems and make no assumptions about the deletion or copying of formulae when moving from the numerator to the denominator. However, when we wish to copy the principal formula into the denominator we shall explicitly show it in the denominator. So for example, the rule below at left explicitly stipulates that a copy of the principal formula $P \wedge Q$ must be carried into the denominator, whereas the rule below at right allows us to choose for ourselves:

$$(\wedge') \ \frac{X; P \wedge Q}{X; P \wedge Q; P; Q} \qquad\qquad (\wedge) \ \frac{X; P \wedge Q}{X; P; Q}$$

Thinning

The thinning rule (θ) allows us to convert any tableau for a given set Y into a tableau for a bigger set $(X; Y)$ simply by adding $(X; Y)$ as a new root node. It encodes the monotonicity of a logic since it encodes the principle that if $Y \vdash_{\mathcal{C}\mathbf{L}} A$ then $X \cup Y \vdash_{\mathcal{C}\mathbf{L}} A$. In tableau systems for classical logic it can be built into the basic consistency check by using a base rule like our $(\bot)$ (shown below right) since all formulae that are not necessary to obtain closure can be stashed in the set X. Alternatively it becomes necessary if we use a base rule like the one shown at below left:

$$\frac{P; \neg P}{\bot} \qquad\qquad \frac{X; P; \neg P}{\bot}\ (\bot)$$

Consequently, our tableau system $\mathcal{C}\mathbf{K}$ is complete for classical propositional logic without (θ) and the thinning rule is required only for the modal aspects. The thinning rule can also be built into the modal rules as we shall show, but we choose to make it explicit because it helps to keep the modal rules simpler.

Cut

The cut rule shown below encodes the law of the excluded middle but suffers the disadvantage that the new formulae P and $\neg P$ are totally arbitrary, bearing no relationship to the numerator X. To use the (cut) rule we have to guess the correct P (although note that modal tableau systems based on Mondadori's system **KE** [D'Agostino and Mondadori, 1994] can use cut sensibly):

$$(cut)\ \frac{X}{X; P \mid X; \neg P}$$

The redundancy of the cut rule is therefore very desirable and can be proved in two ways. The first is to allow the cut rule and show syntactically that whenever there is a closed $\mathcal{C}\mathbf{L}$-tableau for X containing uses of the cut rule, there is another closed $\mathcal{C}\mathbf{L}$-tableau for X containing no uses of the cut rule. This is the cut-elimination theorem of Gentzen. The alternative is to omit the cut rule from the beginning and show that the cut-free tableau system $\mathcal{C}\mathbf{L}$ is nevertheless sound and complete with respect to the semantics of the logic **L**. For most of our systems, we follow this latter route.

A more practical version of the cut rule, known as **analytical cut**, is one where P is a subformula of some formula in X. Thus the formulae that appear in the denominator are not totally arbitrary. Some of our systems require such an analytic cut rule for completeness. The use of analytic cut is not as bad as it may seem since it can lead to exponentially shorter proofs.

4.6 Derived Rules and Admissible Rules

Our rules are couched in terms of (set) variables like X, which denote sets of formulae, and formulae variables like $\neg\Box P$ which denote formulae with a particular structure. Thus our rules are really rule schemata which we instantiate by instantiating X to a set of formulae, and instantiating $\neg\Box P$ to a particular formula like $\neg\Box q$ say. And up till now, we have always applied the rules to sets of formulae. But if a sequence of rule applications is used often then it is worth defining a new rule as a macro or derived rule. And in defining a macro, we apply rules to set variables and to formula variables, not to actual sets of formulae.

More formally, a rule (ρ) with numerator $\mathcal{N}$ and denominators $\mathcal{D}_1, \mathcal{D}_2, \cdots, \mathcal{D}_k$ is **derivable in $\mathcal{C}\mathbf{L}$** iff there is a finite $\mathcal{C}\mathbf{L}$-tableau that begins with the schema $\mathcal{N}$ itself, and has leaves labelled with the schemata $\mathcal{D}_1, \mathcal{D}_2, \cdots, \mathcal{D}_k$, but where the rules are applied to schema rather than to actual sets of formulae. The addition of derived rules does not affect soundness and completeness of $\mathcal{C}\mathbf{L}$ since their applications can be replaced by the macro-expansion.

For example, in order to apply the (K) rule, the numerator (schema) $\Box X; \neg\Box P$ is not allowed to contain nonmodal formulae like $p \wedge q$. Before applying the (K) rule, these undesirable elements have to be 'thinned out' via the set Y as shown below left. But notice that here we have applied the (θ) rule, not to a set of formulae, but to a schema which represents a set of formulae. And similarly, the subsequent application of the (K) rule is also applied to a schema rather than an actual set of formulae. Since such an application of (θ) may be necessary before every application of (K) it may be worth defining a 'derived rule' $(K\theta)$ which builds in this thinning step as shown below right. In fact, if we replace (K) by $(K\theta)$ in $\mathcal{C}\mathbf{K}$ then (θ) becomes superfluous since these are the only necessary applications of (θ).

$$\cfrac{\cfrac{Y; \Box X; \neg\Box P}{\Box X; \neg\Box P}\ (\theta)}{X; \neg P}\ (K) \qquad\qquad (K\theta)\ \frac{Y; \Box X; \neg\Box P}{X; \neg P}$$

On the other hand, it is often possible (and useful) to add extra rules even though these rules are not derivable. For example, the cut rule is not derivable in $\mathcal{C}\mathbf{K}$ since the denominators of each rule of $\mathcal{C}\mathbf{K}$ are always related to the numerator of that rule, whereas (cut) breaks this property since the P in the denominator is arbitrary.

We can ensure that the new rules do not add to the deductive power of the system as follows. Let (ρ) be an arbitrary tableau rule with a numerator $\mathcal{N}$ and n denominators $\mathcal{D}_1, \mathcal{D}_2, \cdots, \mathcal{D}_n$ and let $\mathcal{C}\mathbf{L}\rho$ be the tableau system $\mathcal{C}\mathbf{L} \cup \{(\rho)\}$. Then the rule (ρ) is said to be **admissible** in $\mathcal{C}\mathbf{L}$ if: X is $\mathcal{C}\mathbf{L}$-consistent iff X is $\mathcal{C}\mathbf{L}\rho$-consistent. That is, if: a $\mathcal{C}\mathbf{L}$-tableau for X is closed iff a $\mathcal{C}\mathbf{L}\rho$-tableau for X is closed.

LEMMA 2. *If $\mathcal{C}\mathbf{L}$ is sound and complete with respect to $\mathbf{L}$-frames and (ρ) is sound with respect to $\mathbf{L}$-frames then (ρ) is admissible in $\mathcal{C}\mathbf{L}$.*

Proof. Since $\mathcal{C}\mathbf{L} \subseteq \mathcal{C}\mathbf{L}\rho$ we know that if X is $\mathcal{C}\mathbf{L}\rho$-consistent then X is $\mathcal{C}\mathbf{L}$-consistent. To prove the other direction suppose that $\mathcal{C}\mathbf{L}$ is sound and complete with respect to **L**-frames, that (ρ) is sound with respect to **L**-frames, and that X is $\mathcal{C}\mathbf{L}$-consistent. By the completeness of $\mathcal{C}\mathbf{L}$, the set X must be **L**-satisfiable. Since (ρ) is sound with respect to **L**-frames, so is $\mathcal{C}\mathbf{L}\rho$. Suppose X is not $\mathcal{C}\mathbf{L}\rho$-consistent. Then there is a closed $\mathcal{C}\mathbf{L}\rho$-tableau for X which must utilise the rule (ρ) since this is the only difference between $\mathcal{C}\mathbf{L}$ and $\mathcal{C}\mathbf{L}\rho$. But, by the soundness of $\mathcal{C}\mathbf{L}\rho$ this implies that X must be **L**-unsatisfiable; contradiction. Hence X must be $\mathcal{C}\mathbf{L}\rho$-consistent. ■

For example, there is no rule for $A \to B$ in our tableau system since we always use primitive notation and rewrite $A \to B$ as $\neg(A \land \neg B)$. But the following rules are clearly sound with respect to the semantics of classical logic, and hence are admissible for $\mathcal{C}\mathbf{PC}$ (the calculus $\mathcal{C}\mathbf{K}$ minus the (K) rule) since $\mathcal{C}\mathbf{PC}$ is sound and complete with respect to the same semantics:

$$(\to)\ \frac{X; P \to Q}{X; \neg P \mid X; Q} \qquad\qquad (\neg\to)\ \frac{X; \neg(P \to Q)}{X; P; \neg Q}$$

4.7 Invertible Rules

A tableau rule (ρ) is **invertible** in $\mathcal{C}\mathbf{L}$ iff: if there is a closed $\mathcal{C}\mathbf{L}$-tableau for (an instance of) the numerator $\mathcal{N}$ then there are closed $\mathcal{C}\mathbf{L}$-tableaux for (appropriate instances of) the denominators $\mathcal{D}_i$.

LEMMA 3. *The rules* $(\land)$, $(\lor)$ *and* $(\neg)$ *are invertible in* $\mathcal{C}\mathbf{PC}$.

Proof. The assumption is that we are given a closed $\mathcal{C}\mathbf{PC}$-tableau for some set X that matches the numerator $\mathcal{N}$ of rule (ρ), where (ρ) is one of $(\land)$, $(\lor)$ and $(\neg)$. We have to prove that there is a closed $\mathcal{C}\mathbf{PC}$-tableau for the corresponding instantiations of the denominators of (ρ).

We prove this simultaneously for all three rules by induction on the length of the given closed $\mathcal{C}\mathbf{PC}$-tableau for X. The induction argument requires slight modifications to our $\mathcal{C}\mathbf{PC}$-tableaux: we assume that all applications of the rule $(\bot)$ are restricted to atomic formulae since every closed $\mathcal{C}\mathbf{PC}$-tableau can be extended to meet this condition, and we also ignore the rule (θ) since any closed $\mathcal{C}\mathbf{PC}$-tableau that uses (θ) can be converted into one that does not use (θ).

The base case for the induction proof is when the length of the given closed $\mathcal{C}\mathbf{PC}$-tableau for X is 1; that is, there is some *atomic formula* p such that $\{p, \neg p\} \subseteq X$. The corresponding denominators of (ρ) must also contain $\{p, \neg p\}$ since neither p nor $\neg p$ can be the principal formulae of (ρ). So these denominator instances are also closed.

The induction hypothesis is that the lemma holds for all closed $\mathcal{C}\mathbf{PC}$-tableaux of lengths less than n. Suppose now that the given closed $\mathcal{C}\mathbf{PC}$-tableau for X is of length n. We argue by cases, but only give the case for the $(\land)$ rule in detail.

($\wedge$) The numerator is of the form $\mathcal{N} = (Z; P \wedge Q)$ and we have to provide a closed $\mathcal{C}\mathbf{PC}$-tableau for the corresponding denominator $(Z; P; Q)$. Consider the actual first rule application (τ) in the given closed $\mathcal{C}\mathbf{PC}$-tableau for $(Z; P \wedge Q)$.

If $P \wedge Q$ is not the principal formula A of (τ) then the denominators of (τ) are of the form $(Z_i'; P \wedge Q)$, $1 \leq i \leq 2$, since A must be some formula from Z. The given $\mathcal{C}\mathbf{PC}$-tableau for $(Z; P \wedge Q)$ is closed, so each $(Z_i'; P \wedge Q)$, $1 \leq i \leq 2$, must have a closed $\mathcal{C}\mathbf{PC}$-tableau of length less than n. Then, by the induction hypothesis, there are closed $\mathcal{C}\mathbf{PC}$-tableaux of length less than n for each $(Z_i'; P; Q)$, $1 \leq i \leq 2$.

If we now start a separate $\mathcal{C}\mathbf{PC}$-tableau for $(Z; P; Q)$ and use (τ) with the same $A \in Z$ as the principal formula, we obtain the sets $(Z_i'; P; Q)$ as the denominators of (τ). Since we already have closed $\mathcal{C}\mathbf{PC}$-tableaux for these sets, we have a closed $\mathcal{C}\mathbf{PC}$-tableau for $(Z; P; Q)$, as desired. It is crucial that the length of the new $\mathcal{C}\mathbf{PC}$-tableau is also n.

If $P \wedge Q$ is the principal formula A of (τ) then $(\tau) = (\wedge)$ has only one denominator $(Z; P; Q)$, and the $\mathcal{C}\mathbf{PC}$-tableau for it closes. But this is the closed $\mathcal{C}\mathbf{PC}$-tableau we had to provide. In this case, the length of the 'new' $\mathcal{C}\mathbf{PC}$-tableau is actually $n - 1$.

($\vee$) Similar to above.

($\neg$) Similar to above. ■

4.8 *Subformula Property and Analytic Superformula Property*

For a formula A, the **degree** $deg(A)$ counts the maximum depth of nesting while the **modal degree** $mdeg(A)$ counts the maximum depth of modal nesting. Their definitions are:

$deg(A) = 0$ when A is atomic
$deg(\neg A) = 1 + deg(A)$
$deg(A \wedge B) = 1 + max(deg(A), deg(B))$
$deg(\Box A) = 1 + deg(A)$

$mdeg(A) = 0$ when A is atomic
$mdeg(\neg A) = mdeg(A)$
$mdeg(A \wedge B) = max(mdeg(A), mdeg(B))$
$mdeg(\Box A) = 1 + mdeg(A)$

For a finite set X:

$deg(X) = \max\{deg(A) \mid A \in X\}$
$mdeg(X) = \max\{mdeg(A) \mid A \in X\}$

The set of all **subformulae** of a formula, or of a set of formulae, is used extensively. For a formula A, the *finite* set of all subformulae $Sf(A)$ is defined inductively as [Goldblatt, 1987]:

$$Sf(p) = \{p\} \text{ where } p \text{ is an atomic formula} \quad Sf(\neg A) = \{\neg A\} \cup Sf(A)$$
$$Sf(A \wedge B) = \{A \wedge B\} \cup Sf(A) \cup Sf(B) \quad Sf(\Box A) = \{\Box A\} \cup Sf(A)$$

Note that under this definition, a formula must be in primitive notation to obtain its subformulae; for example:

$$Sf(p \vee q) = Sf(\neg(\neg p \wedge \neg q)) = \{\neg(\neg p \wedge \neg q), \neg p \wedge \neg q, \neg p, \neg q, p, q\}$$
$$Sf(\Diamond p) = Sf(\neg\Box\neg p) = \{\neg\Box\neg p, \Box\neg p, \neg p, p\}$$

For a finite set of formulae X, the set of all subformulae Sf(X) consists of all subformulae of all members of X; that is, $Sf(X) = \bigcup_{A \in X} Sf(A)$. The set of **strict subformulae** of A is $Sf(A) \setminus \{A\}$.

A tableau rule has the **subformula property** iff every formula in the denominators is a subformula of some formula in the numerator. A tableau system $\mathcal{C}\mathbf{L}$ has the subformula property iff every rule in $\mathcal{C}\mathbf{L}$ has it.

If $\mathcal{C}\mathbf{L}$ has the subformula property then each rule can be seen to 'break down' its principal formula(e) into its subformulae. Furthermore, if the principal formula is not copied into the numerator, then termination is guaranteed without cycles since every rule application is guaranteed to give a denominator of lower degree, eventually leading to a node with degree zero.

Notice that the rules of $\mathcal{C}\mathbf{K}$ do not have the subformula property, for both the $(\vee)$ and (K) rule denominators contain formulae which are *negations* of a subformula of the principal formula. But clearly this is not a disaster since the degree is not actually *increased*, but may remain the same.

The modal tableau rules for more complex logics, however, introduce quite complex 'superformulae' into their denominators, thereby *increasing* both the degree and the modal degree. Nevertheless, all is not lost, for every tableau will be guaranteed to terminate, possibly with a cycle.

In order to prove this claim we need to introduce the idea of an **analytic superformula**. The intuition is simple: rules will be allowed to 'build up' formulae so long as the rules cannot conspire to give an infinite chain of 'building up' operations.

A tableau system $\mathcal{C}\mathbf{L}$ has the **analytical superformula** property iff to every finite set X we can assign, *a priori*, a *finite* set $X^*_{\mathcal{C}\mathbf{L}}$ such that $X^*_{\mathcal{C}\mathbf{L}}$ contains all formulae that may appear in *any* tableau for X.

LEMMA 4. *If* $\mathcal{C}\mathbf{L}$ *has the analytic superformula property then there are (only) a finite number of* $\mathcal{C}\mathbf{L}$*-tableaux for the finite set* X.

Proof. Since $\mathcal{C}\mathbf{L}$ has the analytical superformula property the only $\mathcal{C}\mathbf{L}$-tableaux we need consider are those whose nodes carry subsets of the set $X^*_{\mathcal{C}\mathbf{L}}$. Since $X^*_{\mathcal{C}\mathbf{L}}$

is finite, the number of subsets of $X^*_{\mathcal{C}\mathbf{L}}$ is also finite. ■

For example, the calculus $\mathcal{C}\mathbf{K}$ has the analytic superformula property because for any given finite X we can put $X^*_{\mathcal{C}\mathbf{K}} = Sf(X) \cup \neg Sf(X) \cup \{\bot\}$.

4.9 *Proving Soundness*

By definition, a tableau system $\mathcal{C}\mathbf{L}$ is sound with respect to $\mathbf{L}$-frames if $Y \vdash_{\mathcal{C}\mathbf{L}} A$ implies $Y \models_{\mathbf{L}} A$.

Proof Outline. To prove this claim we assume that $Y \vdash_{\mathcal{C}\mathbf{L}} A$; that is, that we have a closed $\mathcal{C}\mathbf{L}$-tableau for $X = (Y; \neg A)$. Then we use induction on the structure of this tableau to show that X is $\mathbf{L}$-unsatisfiable; that is, that $Y \models_{\mathbf{L}} A$.

The base case is when the tableau consists of just one application of the $(\bot)$ rule. In this case, the set X must contain some P and also $\neg P$ and is clearly $\mathbf{L}$-unsatisfiable (since our valuations are always classical two-valued ones).

Now suppose that the (closed) $\mathcal{C}\mathbf{L}$-tableau is some finite but arbitrary tree. We know that all leaves of this (closed) tableau end in $\{\bot\}$. So all we have to show is that for each $\mathcal{C}\mathbf{L}$-tableau rule: if all the denominators are $\mathbf{L}$-unsatisfiable then the numerator is $\mathbf{L}$-unsatisfiable. This would allow us to lift the $\mathbf{L}$-unsatisfiability of the leaves up the tree to conclude that the root X is $\mathbf{L}$-unsatisfiable. Instead, we show the contrapositive; that is, for each $\mathcal{C}\mathbf{L}$-tableau rule we show that if the numerator is $\mathbf{L}$-satisfiable then at least one of the denominators is $\mathbf{L}$-satisfiable. ■

Thus proving the soundness of a tableau system is possible on a rule by rule basis. For example, the $(\wedge)$ rule is sound with respect to $\mathbf{K}$-models because if we are given some $\mathbf{K}$-model $\langle W, R, V\rangle$ with some $w \in W$ such that $w \models X; p \wedge q$, then we can always find a $\mathbf{K}$-model $\langle W', R', V'\rangle$ with some $w' \in W'$ such that $w' \models X; p; q$ by simply putting $\langle W, R, V\rangle = \langle W', R', V'\rangle$ and putting $w = w'$.

As another example the (K) rule is sound with respect to $\mathbf{K}$-models because if we are given some $\mathbf{K}$-model $\langle W, R, V\rangle$ with some $w \in W$ such that $w \models \Box X; \neg\Box P$ then we know that w has some successor $w' \in W$ such that wRw' and $w' \models \neg P$ (by the definition of $w \models \neg\Box P$). Furthermore, since $w \models \Box X$ and wRw' we know that $w' \models X$ (by the definition of $w \models \Box P$). Thus we can find a $w' \in W$ such that $w' \models X; \neg P$. In this case, although the underlying model has remained the same, the world w' may be different from w.

4.10 *Static Rules, Dynamic Rules and Invertibility*

The previous two examples show that, in general, the numerator and denominators of a tableau rule either represent the same world in the same model as in the $(\wedge)$ example, or they represent different worlds in the same model as in the (K) example. We therefore categorise each rule as either a **static rule** or as a **transitional rule**.

The intuition behind this sorting is that in the static rules, the numerator and denominator represent the same world (in the same model), whereas in the transitional rules, the numerator and denominator represent different worlds (in the same model).

For example, the tableau rules for $\mathcal{C}\mathbf{K}$ are categorised as follows:

$\mathcal{C}\mathbf{L}$	Static Rules	Transitional Rules
$\mathcal{C}\mathbf{K}$	$(\theta), (\bot), (\neg), (\wedge), (\vee)$	(K)

The division of rules into static or transitional ones is based purely on the semantic arguments outlined above. But there is a proof-theoretic reason behind this sorting as captured by the following lemma.

LEMMA 5. *The static rules of $\mathcal{C}\mathbf{L}$, except (θ), are precisely the rules that are invertible in $\mathcal{C}\mathbf{L}$.*

Proof. We shall have to prove this lemma for each $\mathcal{C}\mathbf{L}$ by extending Lemma 3. And it is precisely the requirement of invertibility that sometimes requires us to copy the principal formula into the numerator; see Section 4.14. ■

4.11 Proving Completeness via Model-Graphs

By definition, $\mathcal{C}\mathbf{L}$ is complete with respect to $\mathbf{L}$-frames iff: $Y \models_{\mathbf{L}} A$ implies $Y \vdash_{\mathcal{C}\mathbf{L}} A$.

Proof Outline. We prove the contrapositive. That is, we assume $Y \nvdash_{\mathcal{C}\mathbf{L}} A$ by assuming that *no* $\mathcal{C}\mathbf{L}$-tableau for $X = (Y; \neg A)$ is closed. Then we pick and choose sets with certain special properties from *possibly different* open tableaux for X, and use them as possible worlds to construct an $\mathbf{L}$-model $\mathcal{M}$ for X, safe in the knowledge that each of these sets is $\mathcal{C}\mathbf{L}$-consistent. The model $\mathcal{M}$ is deliberately constructed so as to contain a world w_0 such that $w_0 \models Y$ and $w_0 \models \neg A$. Hence we demonstrate by construction that $Y \not\models A$. The basic idea is due to Hintikka [Hintikka, 1955]. ■

In order to do so we first need some technical machinery.

Downward Saturated Sets

A set X is **closed with respect to a tableau rule** if, whenever (an instantiation of) the numerator of the rule is in X, so is (a corresponding instantiation of) at least one of the denominators of the rule. A set X is $\mathcal{C}\mathbf{L}$**-saturated** if it is $\mathcal{C}\mathbf{L}$-consistent and closed with respect to the static rules of $\mathcal{C}\mathbf{L}$ excluding (θ).

LEMMA 6. *For each $\mathcal{C}\mathbf{L}$ with the analytic superformula property and each finite $\mathcal{C}\mathbf{L}$-consistent X there is an effective procedure to construct some finite $\mathcal{C}\mathbf{L}$-saturated (and $\mathcal{C}\mathbf{L}$-consistent) X^s with $X \subseteq X^s \subseteq X^*_{\mathcal{C}\mathbf{L}}$.*

Proof. Since X is $\mathcal{C}\mathbf{L}$-consistent, we know that no $\mathcal{C}\mathbf{L}$-tableau for X closes and hence that the $(\perp)$ rule is not applicable.

Let $X_0 = X$, let $i = 0$ and let $(\rho) \neq (\theta)$ be a static rule of $\mathcal{C}\mathbf{L}$ with respect to which X_0 is not closed. If there are none, then we are done.

Given a $\mathcal{C}\mathbf{L}$-consistent set X_i which is not closed with respect to the static rule $(\rho) \neq (\theta)$, apply (ρ) to (the numerator) X_i to obtain the corresponding denominators. At least one of these denominators must have only open $\mathcal{C}\mathbf{L}$-tableaux. So choose a denominator for which no $\mathcal{C}\mathbf{L}$-tableau closes and let Y_i be the $\mathcal{C}\mathbf{L}$-consistent set carried by it.

Suppose that this application of (ρ) has a principal formula $A \in X_i$ and side formulae $\{B_1, \cdots, B_k\} \subseteq Y_i$. Put $X_{i+1} = (Y_i; A)$ by adding A to Y_i, thereby making X_{i+1} closed with respect to this particular application of (ρ).

For a contradiction, assume that X_{i+1} is $\mathcal{C}\mathbf{L}$-inconsistent; that is, assume that there is a closed $\mathcal{C}\mathbf{L}$-tableau for $(Y_i; A)$. Since (ρ) was applicable to A, putting $\mathcal{N} = (Y_i; A)$ and $\mathcal{D} = (Y_i; B_1; \cdots; B_k)$ gives a part of an instance of (ρ); 'part of' because (ρ) may have more than one denominator and $\mathcal{D}$ is an instance of only one of them. But (ρ) is invertible in $\mathcal{C}\mathbf{L}$, so if there is a closed $\mathcal{C}\mathbf{L}$-tableau for $(Y_i; A)$, then there is a closed $\mathcal{C}\mathbf{L}$-tableau for $(Y_i; B_1; \cdots; B_k)$. Since $\{B_1, \cdots, B_k\} \subseteq Y_i$, this means that there is a closed $\mathcal{C}\mathbf{L}$-tableau for Y_i. Contradiction, hence X_{i+1} is $\mathcal{C}\mathbf{L}$-consistent; that is, no $\mathcal{C}\mathbf{L}$-tableau for X_{i+1} closes.

Now repeat this procedure on X_{i+1}. Since X_{i+1} is closed with respect to at least one more rule application, the number of choices for (ρ) is one less. Furthermore, the resulting set X_{i+2} is guaranteed to be $\mathcal{C}\mathbf{L}$-consistent.

By always iterating on the new set we obtain a sequence of finite $\mathcal{C}\mathbf{L}$-consistent sets $X_0 \subseteq X_{i+1} \subseteq \cdots$, terminating with some final X_n because X_n is closed with respect to every static rule of $\mathcal{C}\mathbf{L}$, except (θ), and is $\mathcal{C}\mathbf{L}$-consistent, as desired. Let $X^s = X_n$.

Since each rule carries subsets of $X^*_{\mathcal{C}\mathbf{L}}$ to subsets of $X^*_{\mathcal{C}\mathbf{L}}$ and we start with $X \subseteq X^*_{\mathcal{C}\mathbf{L}}$, we have $X \subseteq X^s \subseteq X^*_{\mathcal{C}\mathbf{L}}$. ■

In classic logic, such sets are called downward saturated sets and form the basis of Hintikka's [1955] method for proving completeness of classical tableau systems. In the next section we introduce the technical machinery necessary to extend this method to modal logics.

Model Graphs and Satisfiability Lemma

The following definition from Rautenberg [1983] is central for the model constructions. A **model graph** for some finite fixed set of formulae X is a finite $\mathbf{L}$-frame $\langle W, R \rangle$ such that all $w \in W$ are $\mathcal{C}\mathbf{L}$-saturated sets with $w \subseteq X^*_{\mathcal{C}\mathbf{L}}$ and

(i) $X \subseteq w_0$ for some $w_0 \in W$;

(ii) if $\neg\Box P \in w$ then there exists some $w' \in W$ with wRw' and $\neg P \in w'$;

(iii) if wRw' and $\Box P \in w$ then $P \in w'$.

LEMMA 7. *If $\langle W, R \rangle$ is a model graph for X then there exists an* **L***-model for X [Rautenberg, 1983].*

Proof. For every $p \in \mathcal{P}$, let $\vartheta(p) = \{w \in W : p \in w\}$. Using simultaneous induction on the degree of an arbitrary formula $A \in w$, it is easy to show that (a) $A \in w$ implies $w \models A$; and (b) $\neg A \in w$ implies $w \not\models A$. By (a), $w_0 \models X$ hence the model $\mathcal{M} = \langle W, R, \vartheta \rangle$ is an **L**-model for X [Rautenberg, 1983]. ■

This model graph construction is similar to the subordinate frames construction of Hughes and Cresswell [1984] except that Hughes and Cresswell use maximal consistent sets and do not consider cycles, giving infinite models rather than finite models.

4.12 Finite Model Property and Decidability

In the above procedure, if $\mathcal{M}$ can be chosen finite (for finite X) then the logic **L** has the finite model property (fmp). It is known that a finitely axiomatisable normal modal logic with the finite model property must be decidable; see Hughes and Cresswell [1984, page 154]. Hence $\mathcal{C}\mathbf{L}$ provides a decision procedure for determining whether $Y \models_{\mathbf{L}} A$.

4.13 Summary

In the rest of this section we present tableau systems for many propositional normal modal logics based on the work of Rautenberg [1983], Fitting [1983], Shvarts [1989], Hanson [1966], Goré [1992; 1991; 1994] and Amerbauer [1993]. Most of the systems are cut-free but even those that are not use only an analytical cut rule. Each tableau system immediately gives an analogous (cut-free) sequent system. The presentation is based on the basis laid down in the previous subsections and is therefore rather repetitive. The procedure for each tableau system $\mathcal{C}\mathbf{L}$ is:

1. define the tableau rules for $\mathcal{C}\mathbf{L}$;

2. define $X^*_{\mathcal{C}\mathbf{L}}$ for a given fixed X;

3. prove that the $\mathcal{C}\mathbf{L}$ rules are sound with respect to **L**-frames;

4. prove that each $\mathcal{C}\mathbf{L}$-consistent X can be extended (effectively) to a $\mathcal{C}\mathbf{L}$-saturated X^s with $X \subseteq X^s \subseteq X^*_{\mathcal{C}\mathbf{L}}$;

5. prove that the $\mathcal{C}\mathbf{L}$ rules are complete with respect to **L**-frames by giving a procedure to construct a *finite* **L**-model for any finite $\mathcal{C}\mathbf{L}$-consistent X and hence prove that **L** has the finite model property, that **L** is decidable and that $\mathcal{C}\mathbf{L}$ is a decision procedure for deciding local logical consequence ($Y \models_{\mathbf{L}} A$) in **L**.

4.14 *The Basic Normal Systems*

In this section we study the tableau systems which capture the basic normal modal logics obtained from various combinations of the five basic axioms of reflexivity, transitivity, seriality, Euclideaness, and symmetry. We shall see that implicit tableau systems can handle certain combinations of the first four properties with ease, but require an analytic cut rule to handle symmetry. In each case, we give the tableau calculi and prove them sound and complete with respect to the appropriate semantics. We shall also see that some of the basic logics have no known implicit tableau systems, leaving an avenue for further work.

The following notational conventions are useful for defining $X^*_{\mathcal{C}\mathbf{L}}$ for each X and each $\mathcal{C}\mathbf{L}$. For any finite set X :

- let $Sf(X)$ denote the set of all subformulae of all formulae in X ;
- let $\neg Sf(X)$ denote $\{\neg P \mid P \in Sf(X)\}$;
- let $\widetilde{X}$ denote the set $Sf(X) \cup \neg Sf(X) \cup \{\bot\}$;
- let $\Box(\widetilde{X} \to \Box\widetilde{X})$ denote the set $\{\Box(P \to \Box P) \mid P \in \widetilde{X}\}$.

We sometimes write SfX instead of $Sf(X)$ whence $\widetilde{X} = (Sf\neg SfX) \cup \{\bot\}$.

Tableau Calculi

All the tableau calculi contain the rules of $\mathcal{C}\mathbf{PC}$ and one or more logical rules from Figure 6 on page 322. The tableau systems are shown in Figure 7 on page 323 and the only structural rule is (θ). The calculi marked with a superscript † require analytic cut whilst the others are cut-free. The entries marked by question-marks are open questions.

The rules are categorised into two sorts, static rules and transitional rules as explained on page 318. This sorting should become even clearer once we prove soundness.

The semantic and sometimes axiomatic intuitions behind these rules are as follows.

Intuitions for (K) **:** if the numerator represents a world w where $\Box X$ and $\neg\Box P$ are true, then since $\neg\Box P = \Diamond\neg P$, there must be a world w' reachable from w such that w' makes P false and makes all the formulae in X true. The denominator of the (K) rule represents w'.

Intuitions for (T) **:** if the numerator represents a world w where X and $\Box P$ are true, then every successor of w must make P true. By reflexivity of R the world w itself must be one of these successors.

Intuitions for (D) **:** if the numerator represents a world w where X and $\Box P$ are true, then by seriality of R there must exist some w' such that wRw'. Then the definition of $\Box P$ forces P to be true at w'. Hence $\neg\Box\neg P$, that is $\Diamond P$, must be

$$(K)\ \frac{\Box X; \neg\Box P}{X; \neg P} \qquad (T)\ \frac{X; \Box P}{X; \Box P; P} \qquad (D)\ \frac{X; \Box P}{X; \Box P; \neg\Box\neg P}$$

$$(KD)\ \frac{\Box X; \neg\Box P}{X; \neg P} \text{ where } \{\neg\Box P, \neg P\} \text{ may be empty}$$

$$(K4)\ \frac{\Box X; \neg\Box P}{X; \Box X; \neg P} \qquad (S4)\ \frac{\Box X; \neg\Box P}{\Box X; \neg P}$$

$$(45)\ \frac{\Box X; \neg\Box Y; \neg\Box P}{X; \Box X; \neg\Box Y; \neg\Box P; \neg P}$$

$$(45D)\ \frac{\Box X; \neg\Box Y; \neg\Box P}{X; \Box X; \neg\Box Y; \neg\Box P; \neg P} \text{ where } Y \cup \{P\} \cup \{\neg P\} \text{ may be empty}$$

$$(B)\ \frac{X; \neg\Box P}{X; \neg\Box P; P \mid X; \neg\Box P; \neg P; \Box\neg\Box P} \qquad (T_\Diamond)\ \frac{X; \Box P; \neg\Box Q}{X; \Box P; P; \neg\Box Q}$$

$$(5)\ \frac{X; \neg\Box P}{X; \neg\Box P; \Box\neg\Box P} \qquad (S5)\ \frac{\Box X; \neg\Box Y; \neg\Box P}{\Box X; \neg\Box Y; \neg\Box P; \neg P}$$

$$(sfc\Box)\frac{X; \Box P}{X; \Box P; P \mid X; \Box P; \neg P}$$

$$(sfc\Diamond)\frac{X; \neg\Box P}{X; \neg\Box P; P \mid X; \neg\Box P; \neg P}$$

$$(sfc\vee)\frac{X; \neg(P \wedge Q)}{X; \neg P; \neg Q \mid X; \neg P; Q \mid X; P; \neg Q}$$

$$(sfc) = \{(sfc\Box), (sfc\Diamond), (sfc\vee)\} \qquad (sfcT) = \{(sfc\vee), (sfc\Diamond)\}$$

Figure 6. Tableau Rules for Basic Systems

$\mathcal{C}$L	Static Rules	Transitional Rules	$X^*_{\mathcal{C}\mathbf{L}}$
$\mathcal{C}$PC	$(\theta), (\bot), (\neg), (\wedge), (\vee)$	—	$\widetilde{X}$
$\mathcal{C}$K	$\mathcal{C}$PC	(K)	$\widetilde{X}$
$\mathcal{C}$T	$\mathcal{C}$PC, (T)	(K)	$\widetilde{X}$
$\mathcal{C}$D	$\mathcal{C}$PC, (D)	(K)	$Sf\neg Sf\Box\widetilde{X}$
$\mathcal{C}$D′	$\mathcal{C}$PC	(KD)	$\widetilde{X}$
$\mathcal{C}$KB	?	?	?
$\mathcal{C}$K4	$\mathcal{C}$PC	$(K4)$	$\widetilde{X}$
$\mathcal{C}$K5	?	?	?
$\mathcal{C}$KDB	?	?	?
$\mathcal{C}$KD5	?	?	?
$\mathcal{C}$K4D	$\mathcal{C}$PC, (D)	$(K4)$	$Sf\neg Sf\Box\widetilde{X}$
$\mathcal{C}$K45	$\mathcal{C}$PC	(45)	$\widetilde{X}$
$\mathcal{C}$K45D	$\mathcal{C}$PC	$(45D)$	$\widetilde{X}$
$\mathcal{C}$S4	$\mathcal{C}$PC, (T)	$(S4)$	$\widetilde{X}$
$\mathcal{C}$S5π^-	$\mathcal{C}$PC, (T)	$(S5)$	$\widetilde{X}$
$\mathcal{C}^\dagger$K45	$\mathcal{C}$PC, (sfc)	(45)	$\widetilde{X}$
$\mathcal{C}^\dagger$K45D	$\mathcal{C}$PC, (sfc)	$(45D)$	$\widetilde{X}$
$\mathcal{C}^\dagger$K4B	$\mathcal{C}$PC, $(sfc), (T_\Diamond), (5)$	$(K4)$	$Sf\neg Sf\Box\widetilde{X}$
$\mathcal{C}^\dagger$S4	$\mathcal{C}$PC, $(T), (sfcT)$	$(S4)$	$\widetilde{X}$
$\mathcal{C}^\dagger$B	$\mathcal{C}$PC, $(T), (B), (sfcT)$	(K)	$Sf\neg Sf\Box\widetilde{X}$
$\mathcal{C}^\dagger$S5	$\mathcal{C}$PC, $(T), (5), (sfcT)$	$(S4)$	$Sf\neg Sf\Box\widetilde{X}$
$\mathcal{C}^\dagger$S5′	$\mathcal{C}$PC, $(T), (sfcT)$	$(S5)$	$\widetilde{X}$

Figure 7. Tableau Calculi for Basic Systems

true at w *itself.* Note that (D) is a static rule since its numerator and denominator represent the same world, and also that (D) creates a superformula $\neg\Box\neg P$.

Intuitions for (KD) **:** if the numerator represents a world w where $\Box X$ is true, then the seriality of R guarantees a successor w' for this world, and the definition of $\Box X$ forces X to be true at w'. So we can apply the (KD) rule even when the numerator contains no formulae of the form $\neg\Box P$. Of course, if such a formula is present then the intuitions for the (K) rule suffice. Note that (KD) is a transitional rule since the numerator and denominator represent different worlds, and also that it has the subformula property.

Intuitions for $(K4)$ **:** if the numerator represents a world w where $\Box X$ and $\neg\Box P$ are true, there must be a world w' representing the denominator, with wRw', such that w' makes X true and makes P false. Then by transitivity of R, any and all successors of w' must also make X true, hence w' makes $\Box X$ true. If w' does not have successors then it makes $\Box X$ true vacuously.

Intuitions for $(S4)$ **:** if the numerator represents a world w where $\Box X$ and $\neg\Box P$ are true, then by transitivity of R there must be a world w' representing the denominator, with wRw', such that w' makes $\Box X$ true and makes P false.

Intuitions for (B) **:** if R is symmetric and reflexive and the numerator represents a world w where X and $\neg\Box P$ are true, we know that this world either makes P true or makes P false. If w makes P true then we have the left denominator. If w makes P false, then we have the right denominator which also contains $\Box\neg\Box P$ since $A \rightarrow \Box\Diamond A$ is a theorem of **B**.

Intuitions for (5) **:** Suppose R is Euclidean and the numerator represents a world w where X and $\neg\Box P$ are true. Then we immediately have that w also makes $\Box\neg\Box P$ true since $\neg\Box A \rightarrow \Box\neg\Box A$ is just another way of writing the axiom 5 which we know must be valid in all Euclidean Kripke frames.

Intuitions for (sfc) **:** if the numerator represents a world w where $\neg(P \wedge Q)$ is true, then we know that w either makes both P and Q false; or makes P false and Q true; or makes P true and Q false. The other cases use similar intuitions.

Intuitions for $(sfcT)$ **:** as for the (sfc) rule except that by reflexivity we cannot have both $\Box P$ and $\neg P$ true at w so one of the cases cannot occur.

EXAMPLE 8. The following example shows that copying the principal formula into the denominator is crucial since the left $\mathcal{C}$**KT**-tableau, using a non-copying application of a rule (T'), does not close but the right one, using (T), does close.

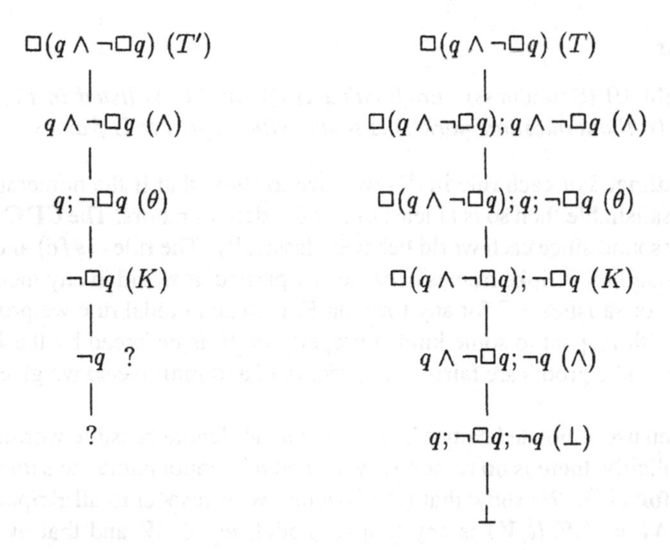

EXAMPLE 9. The following example shows that the order of the modal rule applications is important, since the $\mathcal{C}\mathbf{KT}$-tableau below does not close precisely because (K) (and hence (θ)) is applied at the start. If we apply the $(\vee)$ rule first then the tableau can be closed:

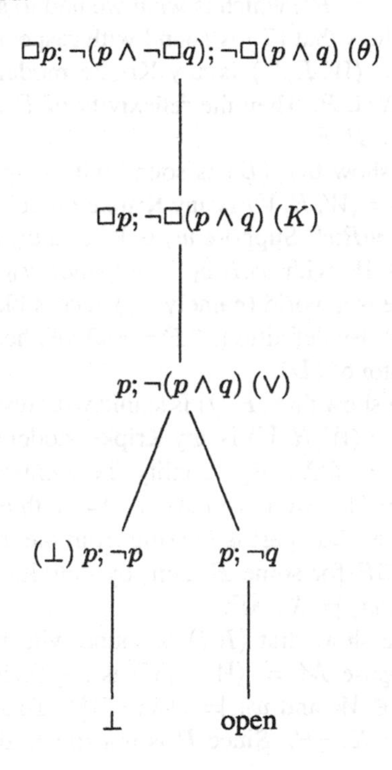

Soundness

THEOREM 10 (Soundness). *Each calculus $\mathcal{C}\mathbf{L}$ and $\mathcal{C}^{\dagger}\mathbf{L}$ listed in Figure 7 on page 323 (without question marks!) is sound with respect to* $\mathbf{L}$*-frames.*

Proof Outline. For each rule in $\mathcal{C}\mathbf{L}$ we have to show that if the numerator of the rule is $\mathbf{L}$-satisfiable then so is at least one of the denominators. The $\mathcal{C}\mathbf{PC}$ rules are obviously sound since each world behaves classically. The rules (sfc) and $(sfcT)$ are also sound for Kripke frames because any particular world in any model either satisfies P or satisfies $\neg P$ for any formula P. For each modal rule we prove that it is sound with respect to some known property of R as enforced by the $\mathbf{L}$-frames restrictions. The proofs are fairly straightforward and intuitive so we give a sketch only.

We often use annotated names like w_1 and w' to denote possible worlds. Unless stated explicitly, there is no reason why w_1 and w' cannot name the same world.

Proof for (K)**:** We show that (K) is sound with respect to all Kripke frames. Suppose $\mathcal{M} = \langle W, R, V\rangle$ is any Kripke model, $w_0 \in W$ and that w_0 satisfies the numerator of (K). That is, suppose $w_0 \models \Box X; \neg\Box P$. We have to show that there exists some world that satisfies the denominator of (K). By definition of the satisfaction relation, $w_0 \models \neg\Box P$ implies that there exists a $w_1 \in W$ with $w_0 R w_1$ and $w_1 \models \neg P$. Since $w_0 \models \Box X$ and $w_0 R w_1$, the definition of $\models$ implies that $w_1 \models X$, hence $w_1 \models (X; \neg P)$, which is what we had to show.

Proof for (T)**:** We show that (T) is sound with respect to all reflexive Kripke frames. Suppose $\mathcal{M} = \langle W, R, V\rangle$ is any Kripke model where R is reflexive, $w_0 \in W$ and $w_0 \models \Box X; \Box P$. Then the reflexivity of R and the definition of $\models$ implies that $w_0 \models \Box X; \Box P; P$.

Proof for (D)**:** We show that (D) is sound with respect to all serial Kripke frames. So suppose $\mathcal{M} = \langle W, R, V\rangle$ is any Kripke model where R is serial. That is, $\forall w \in W, \exists w' \in W : wRw'$. Suppose $w_0 \in W$ and $w_0 \models X; \Box P$. By seriality there exists some $w_1 \in W$ with $w_0 R w_1$. And since $w_0 \models \Box P$ we must have $w_1 \models P$. But then there is a world (namely w_1) accessible from w_0 that satisfies P, and hence $w_0 \models \Diamond P$. By definition, $\Diamond P = \neg\Box\neg P$, hence $w_0 \models \neg\Box\neg P$, thus satisfying the denominator of (D).

Proof for (KD)**:** We show that (KD) is sound with respect to all serial Kripke frames. So suppose $\mathcal{M} = \langle W, R, V\rangle$ is any Kripke model where R is serial. Suppose $w_0 \in W$ and $w_0 \models \Box X$. By seriality there exists some $w_1 \in W$ with $w_0 R w_1$, and since $w_0 \models \Box X$ we must have $w_1 \models X$ thus satisfying the denominator of (KD) when the $\neg\Box P$ part is missing from the numerator. On the other hand, if $w_0 \models \Box X; \neg\Box P$ for some P then, by definition, there is a world w_2 accessible from w_0 with $w_2 \models X; \neg P$.

Proof for $(K4)$**:** We show that $(K4)$ is sound with respect to all transitive Kripke frames. So suppose $\mathcal{M} = \langle W, R, V\rangle$ is any Kripke model where R is transitive. Suppose $w_0 \in W$ and $w_0 \models \Box X; \neg\Box P$. Thus there exists $w_1 \in W$ with $w_0 R w_1$ and $w_1 \models X; \neg P$. Since R is transitive, all successors of w_1 are reachable from w_0, hence $w_0 \models \Box X$ implies that every successor of w_1, if there are any, must also satisfy X. By the definition of $\models$ this gives $w_1 \models X; \Box X; \neg P$. If w_1 has no successors then it vacuously satisfies $\Box A$ for any formula A, hence it vacuously satisfies $\Box X$, and we are done.

Proof for $(S4)$**:** The proof for $(K4)$ also shows that $(S4)$ is sound with respect to all transitive Kripke models.

Proof for (45)**:** Let $\mathcal{M} = \langle W, R, V\rangle$ be any Kripke model where R is transitive and Euclidean. Suppose that $w_0 \in W$ and $w_0 \models \Box X; \neg\Box Y; \neg\Box P$. We have to show that there exists a $w' \in W$ such that $w' \models X; \Box X; \neg\Box Y; \neg\Box P; \neg P$.

We need only prove that there exists a $w' \in W$ such that $w' \models \neg\Box Y; \neg\Box P; \neg P$ since the $X; \Box X$ part will follow from the transitivity of R. Since $w_0 \models \neg\Box P$ we know that there exists some w' with $w_0 R w'$ and $w' \models \neg P$. By the definition of Euclideaness $w_0 R w'$ and $w_0 R w'$ (sic) implies $w' R w'$. Hence w' is reflexive and we have $w' \models \neg\Box P$. Now, if Y is empty then we are done; otherwise if $Y = \{Q_1, Q_2, \cdots, Q_n\}$, $n \geq 1$, there will be worlds $w_1, w_2, \ldots, w_n$ (not necessarily distinct) where $w_0 R w_i$ for each $1 \leq i \leq n$ and such that $w_i \models \neg Q_i$. Since R is Euclidean, $w_0 R w'$ and $w_0 R w_i$ implies that $w' R w_i$ for each $1 \leq i \leq n$. But then $w' \models \neg\Box Y$ and we are done.

Proof for $(45D)$**:** Let $\mathcal{M} = \langle W, R, V\rangle$ be any Kripke model where R is serial, transitive and Euclidean, and suppose that $w_0 \in W$ and $w_0 \models \Box X; \neg\Box Y; \neg\Box P$. We have to show that there exists a $w' \in W$ such that $w' \models X; \Box X; \neg\Box Y; \neg\Box P; \neg P$ allowing for the case where the $\neg\Box Y; \neg\Box P$ part is missing. Since R is transitive and Euclidean the proof for $\mathcal{C}$**K45** applies when the $\neg\Box Y; \neg\Box P$ part is present. If there are no formulae of the form $\neg\Box P$ in w_0 then seriality guarantees that there is some world w' with wRw', and then transitivity of R ensures that $w' \models X; \Box X$.

Proof for (B)**:** We show that (B) is sound with respect to all symmetric Kripke frames. Suppose $\mathcal{M} = \langle W, R, V\rangle$ is any Kripke model where R is symmetric, $w_0 \in W$ and $w_0 \models X; \neg\Box P$. We show that $w_0 \models P$ or $w_0 \models \neg P; \Box\neg\Box P$. If $w_0 \models P$ then $w_0 \models X; \neg\Box P; P$ and we are done. Otherwise $w_0 \models \neg P$. In this latter case, suppose $w_0 \not\models \Box\neg\Box P$. Then $w_0 \models \neg\Box\neg\Box P$, that is $w_0 \models \Diamond\Box P$, so there exists some $w_1 \in W$ with $w_0 R w_1$ and $w_1 \models \Box P$. Since R is symmetric, $w_0 R w_1$ implies $w_1 R w_0$ which together with $w_1 \models \Box P$ gives $w_0 \models P$. But this contradicts the supposition that $w_0 \models \neg P$. Hence $w_0 \models X; \neg\Box P; P$ or $w_0 \models X; \neg\Box P; \neg P; \Box\neg\Box P$ and we are done.

Proof for $(T_\Diamond)$**:** We show that $(T_\Diamond)$ is sound with respect to all Kripke frames that are symmetric and transitive. Suppose $\mathcal{M} = \langle W, R, V\rangle$ is any Kripke model where R is symmetric and transitive, $w_0 \in W$ and $w_0 \models X; \Box P; \neg\Box Q$. Then there exists some $w_1 \in W$ with $w_0 R w_1$ and $w_1 \models \neg Q$. By symmetry, $w_0 R w_1$ implies $w_1 R w_0$. By transitivity, $w_0 R w_1$ and $w_1 R w_0$ implies $w_0 R w_0$. Therefore $w_0 \models P$ and we are done.

Proof for (5)**:** We show that (5) is sound with respect to all Euclidean Kripke frames. Suppose $\mathcal{M} = \langle W, R, V\rangle$ is any Kripke model where R is Euclidean, and suppose $w_0 \in W$ with $w_0 \models X; \neg\Box P$. We have to show that $w_0 \models \Box\neg\Box P$.

Assume for a contradiction that $w_0 \not\models \Box\neg\Box P$; that is, $w_0 \models \neg\Box\neg\Box P$, which is the same as $w_0 \models \Diamond\Box P$. Thus there exists some $w_1 \in W$ with $w_0 R w_1$ and $w_1 \models \Box P$. Since $w_0 \models \neg\Box P$ there is also some w_2 with $w_0 R w_2$ and $w_2 \models \neg P$. Since R is Euclidean, $w_0 R w_1$ and $w_0 R w_2$ implies $w_1 R w_2$. And since $w_1 \models \Box P$ we must have $w_2 \models P$. Contradiction; hence $w_0 \models \Box\neg\Box P$ as desired.

Proof for $(S5)$**:** We show that $(S5)$ is sound with respect to all Kripke frames that are transitive and Euclidean. Suppose $\mathcal{M} = \langle W, R, V \rangle$ is any Kripke model where R is transitive and Euclidean. Suppose $w_0 \in W$ and $w_0 \models \Box X; \neg\Box Y; \neg\Box P$. Thus there exists some world $w' \in W$ with $w_0 R w'$ and $w' \models \neg P$. Suppose $Y = \{Q_1, Q_2, \cdots, Q_n\}, n \geq 1$. Thus there exist (not necessarily distinct) worlds $w_1, w_2, \cdots, w_n$ such that $w_0 R w_i$ and $w_i \models \neg Q_i$, for $1 \leq i \leq n$. Since R is Euclidean, $w' R w'$ and $w' R w_i$ for each i. The first gives $w' \models \neg\Box P$, and the second gives $w' \models \neg\Box Q_i$, $1 \leq i \leq n$. Hence $w' \models \neg P; \neg\Box P; \neg\Box Y$. If Y is empty then we just get $w' \models \neg P; \neg\Box P$. Now choose any arbitrary world w' such that $w' R w'$ (there is at least one since w' is a reflexive world). By transitivity of R, $w_0 R w'$, hence $w' \models X$. Since w' was an arbitrary successor for w' this holds for all successors of w'. Hence $w' \models \Box X$ as well giving $w' \models \Box X; \neg P; \neg\Box P; \neg\Box Y$. ■

Invertibility Again

Before moving on to completeness, we return to the relationship between static rules and invertible rules.

LEMMA 11. *For every* $\mathcal{C}\mathbf{L}$*, the static rules of* $\mathcal{C}\mathbf{L}$*, except* (θ)*, are invertible in* $\mathcal{C}\mathbf{L}$*.*

Proof. We have to extend the proof of Lemma 3 to each $\mathcal{C}\mathbf{L}$. We consider only the case of $\mathcal{C}\mathbf{KT}$ since the proofs for other calculi are similar. The main point is to highlight the need for copying the principal formula $\Box P$ of the (T) rule into the denominator.

Proof for $\mathcal{C}\mathbf{KT}$**:** As stated already, the induction argument requires slight modifications to our $\mathcal{C}\mathbf{L}$-tableaux: we assume that all applications of the rule $(\bot)$ are restricted to atomic formulae since every closed $\mathcal{C}\mathbf{L}$-tableau can be extended to meet this condition. The rule (θ) interferes with the induction argument so we proceed in two steps. We prove the lemma for the calculus $\mathcal{C}\mathbf{K}\theta\mathbf{T}$ in which the (K) and (θ) rules are replaced by the rule $(K\theta)$. We then leave it to the reader to prove that a finite set X has a closed $\mathcal{C}\mathbf{KT}$-tableau iff it has a closed $\mathcal{C}\mathbf{K}\theta\mathbf{T}$-tableau but give some hints at the end of the proof.

The assumption is that we are given a closed $\mathcal{C}\mathbf{K}\theta\mathbf{T}$-tableau for some set X that matches the numerator $\mathcal{N}$ of a static rule (ρ) of $\mathcal{C}\mathbf{K}\theta\mathbf{T}$; that is, (ρ) is one of $(\wedge)$, $(\vee)$, $(\neg)$ and (T). Our task is to provide a closed $\mathcal{C}\mathbf{K}\theta\mathbf{T}$-tableau for the appropriate instance of the denominators of (ρ).

We again proceed by induction on the length of the given closed $\mathcal{C}\mathbf{K}\theta\mathbf{T}$-tableau for X. The base case for the induction proof is when the length of the given closed

$\mathcal{C}\mathbf{K}\theta\mathbf{T}$-tableau for X is 1; and the argument of Lemma 3 suffices. The induction hypothesis is that the lemma holds for all closed $\mathcal{C}\mathbf{K}\theta\mathbf{T}$-tableaux of lengths less than n. Suppose now that the given closed $\mathcal{C}\mathbf{K}\theta\mathbf{T}$-tableau for X is of length n. We argue by cases, but only give the case $(\rho) = (T)$ in detail since the cases for the static $\mathcal{C}\mathbf{PC}$ rules are similar.

$(\rho) = (T)$ The set X of the given closed $\mathcal{C}\mathbf{K}\theta\mathbf{T}$-tableau of length n is of the form $\mathcal{N} = (Z; \Box P)$ and we have to provide a closed $\mathcal{C}\mathbf{K}\theta\mathbf{T}$-tableau for $(Z; \Box P; P)$, the denominator corresponding to $(\rho) = (T)$.

Consider the first rule application (τ) in the given closed $\mathcal{C}\mathbf{K}\theta\mathbf{T}$-tableau for $(Z; \Box P)$. If $\Box P$ is not the principal formula A of (τ) then there are two subcases:

(i) If (τ) is a static (logical) rule of $\mathcal{C}\mathbf{K}\theta\mathbf{T}$ then the denominators of (τ) are of the form $(Z'_i; \Box P)$, $1 \leq i \leq 2$, since A must be some formula from Z. The given $\mathcal{C}\mathbf{K}\theta\mathbf{T}$-tableau for $(Z; \Box P)$ is closed, so each $(Z'_i; \Box P)$, $1 \leq i \leq 2$, must have a closed $\mathcal{C}\mathbf{K}\theta\mathbf{T}$-tableau of length less than n. Then, by the induction hypothesis, there are closed $\mathcal{C}\mathbf{K}\theta\mathbf{T}$-tableaux of length less than n for each $(Z'_i; \Box P; P)$, $1 \leq i \leq 2$.

If we now start a separate $\mathcal{C}\mathbf{K}\theta\mathbf{T}$-tableau for $(Z; \Box P; P)$ and use (τ) with the same $A \in Z$ as the principal formula, we obtain the set $(Z'_i; \Box P; P)$. Since we already have closed $\mathcal{C}\mathbf{K}\theta\mathbf{T}$-tableaux for these sets, we have a closed $\mathcal{C}\mathbf{K}\theta\mathbf{T}$-tableau for $(Z; \Box P; P)$, as desired. It is crucial that the length of the new $\mathcal{C}\mathbf{K}\theta\mathbf{T}$-tableau is also n.

(ii) If (τ) is $(K\theta)$ then $(Z; \Box P)$ is of the form $(Y; \Box W; \neg\Box Q; \Box P)$ and the denominator of (τ) is $(W; \neg Q; P)$. Furthermore, the $\mathcal{C}\mathbf{K}\theta\mathbf{T}$-tableau for $(W; \neg Q; P)$ is closed.

In this subcase, $(Z; \Box P; P)$ is of the form $(Y; \Box W; \neg\Box Q; \Box P; P)$. If we start a new $\mathcal{C}\mathbf{K}\theta\mathbf{T}$-tableau for the set $(Y; \Box W; \neg\Box Q; \Box P; P)$, then we can obtain the same set $(W; \neg Q; P)$ using $(K\theta)$. Since we already have a closed $\mathcal{C}\mathbf{K}\theta\mathbf{T}$-tableau for $(W; \neg Q; P)$ this is a closed $\mathcal{C}\mathbf{K}\theta\mathbf{T}$-tableau for $(Z; \Box P; P)$, also of length n. This is the closed $\mathcal{C}\mathbf{K}\theta\mathbf{T}$-tableau (of length n) we had to provide.

If $\Box P$ is the principal formula of (τ) then $(\tau) = (T)$ and (τ) has a denominator $(Z; \Box P; P)$. Furthermore, the $\mathcal{C}\mathbf{K}\theta\mathbf{T}$-tableau for $(Z; \Box P; P)$ closes. But this is exactly the closed $\mathcal{C}\mathbf{K}\theta\mathbf{T}$-tableau we had to provide.

$(\rho) = (\wedge)$, $(\rho) = (\vee)$, $(\rho) = (\neg)$: Similar to above.

In order to lift this proof to $\mathcal{C}\mathbf{KT}$ we have to show that X has a closed $\mathcal{C}\mathbf{KT}$-tableau iff it has a closed $\mathcal{C}\mathbf{K}\theta\mathbf{T}$-tableau. A closed $\mathcal{C}\mathbf{K}\theta\mathbf{T}$-tableau can be converted to a closed $\mathcal{C}\mathbf{KT}$-tableau simply by replacing the rule $(K\theta)$ with the appropriate application of (θ) immediately followed by an application of (K), see Section 4.6. Conversely, a closed $\mathcal{C}\mathbf{KT}$-tableau can be converted to a closed $\mathcal{C}\mathbf{K}\theta\mathbf{T}$-tableau by first moving every application of (θ) so that it immediately precedes an application of (K), and then replacing these pairs by an application of $(K\theta)$. ■

In Example 8 we saw the importance of copying the principal formula of the (T) rule into its denominator. We can now explain this in more proof-theoretic terms: the rule (T) is invertible in $\mathcal{C}\mathbf{KT}$, but the rule (T') is not invertible in $\mathcal{C}\mathbf{KT}'$. To see that (T') is not invertible in $\mathcal{C}\mathbf{KT}'$ consider the set $(\neg\Box p; \Box p)$:

- this set as the numerator of (T') has a corresponding denominator $(\neg\Box p; p)$,
- $(\neg\Box p; \Box p)$ has a closed $\mathcal{C}\mathbf{KT}'$-tableau, just apply the (K) rule once,
- but $(\neg\Box p; p)$ has no closed $\mathcal{C}\mathbf{KT}'$-tableau (try it).

The curious reader may be wondering why the proof of Lemma 11 fails for $\mathcal{C}\mathbf{KT}'$. In the above example, $\mathcal{N} = (\neg\Box p; \Box p)$ and (τ) is the transitional rule (K). If we had used $\mathcal{C}\mathbf{K}\theta\mathbf{T}'$ it would be $(K\theta)$, so we enter case (ii) of the proof with a known closed $\mathcal{C}\mathbf{K}\theta\mathbf{T}'$-tableau for $(\neg p; p)$. Our task is to provide a closed $\mathcal{C}\mathbf{K}\theta\mathbf{T}'$-tableau for $\mathcal{D} = (\neg\Box p; p)$, the denominator of the (T') rule corresponding to $\mathcal{N}$. But if we start a new $\mathcal{C}\mathbf{K}\theta\mathbf{T}'$-tableau for $(\neg\Box p; p)$, we cannot use the $(K\theta)$ rule to obtain the set $(\neg p; p)$. In fact, there is no rule which allows us to do this in $\mathcal{C}\mathbf{K}\theta\mathbf{T}'$.

Completeness

As we saw in Subsection 4.11 (page 318) , proving completeness boils down to proving the following: if X is a finite set of formulae and no $\mathcal{C}\mathbf{L}$-tableau for X is closed then there is an $\mathbf{L}$-model for X on an $\mathbf{L}$-frame $\langle W, R\rangle$.

We call a formula $\neg\Box P$ an **eventuality** since it entails that eventually $\neg P$ must hold. A world w is said to **fulfill** an eventuality $\neg\Box P$ when $w \models \neg P$. A sequence of worlds $w_1 R w_2 R \cdots R w_m$ is said to fulfill an eventuality $\neg\Box P$ when $w_i \models \neg P$ for some w_i in the sequence.

As expected we shall associate sets of formulae with possible worlds and use an explicit immediate successor relation $\prec$ from which we will obtain R. We abuse notation slightly by using w, w' and w_1 to sometimes denote worlds in a model, and sometimes to denote sets of formulae (in a model under construction). Thus, a *set* w is said to fulfill an eventuality $\neg\Box P$ when $\neg P \in w$. A sequence $w_1 \prec w_2 \prec \cdots \prec w_m$ of sets is said to fulfill an eventuality $\neg\Box P$ when $\neg P \in w_i$ for some w_i in the sequence.

Recall that a set X is **$\mathcal{C}\mathbf{L}$-saturated** iff it is $\mathcal{C}\mathbf{L}$-consistent and closed with respect to the static rules of $\mathcal{C}\mathbf{L}$ (excluding (θ)). We now have to check that the $X^*_{\mathcal{C}\mathbf{L}}$ defined in Figure 7 on page 323 allow (the Saturation) Lemma 6 (page 318) to go through.

LEMMA 12. *If there is a closed $\mathcal{C}\mathbf{L}$-tableau for X then there is a closed $\mathcal{C}\mathbf{L}$-tableau for X with all nodes in the finite set $X^*_{\mathcal{C}\mathbf{L}}$.*

Proof. Obvious from the fact that all rules for $C\mathbf{L}$ operate with subsets of $X^*_{C\mathbf{L}}$ only. ■

LEMMA 13. *For each $C\mathbf{L}$-consistent X there is an effective procedure to construct some finite $C\mathbf{L}$-saturated X^s with $X \subseteq X^s \subseteq X^*_{C\mathbf{L}}$.*

Proof. Same as on page 318. ■

A set X is **subformula-complete** if $P \in Sf(X)$ implies either $P \in X$ or $\neg P \in X$. Some of the completeness proofs make extensive use of the following lemma.

LEMMA 14 (subformula-complete). *If X is closed with respect to (the static rules) $\{ (\bot), (\neg), (\wedge), (\vee), (sfc) \}$, or $\{ (\bot), (\neg), (\wedge), (\vee), (sfc), (T_\Diamond) \}$ or $\{ (\bot), (\neg), (\wedge), (\vee), (T), (sfcT) \}$ then X is subformula-complete.*

Proof. The first case is obvious. The $(sfcT)$ rule is just a special case of (sfc) and always appears with (T). Thus, the lemma also holds if we have both $(sfcT)$ and (T) instead of (sfc). ■

THEOREM 15 (Completeness). *If X is a finite set of formulae and X is $C\mathbf{L}$-consistent then there is an $\mathbf{L}$-model for X on a finite $\mathbf{L}$-frame.*

Proof Outline: For each $C\mathbf{L}$ we give a way to construct a finite model graph $\langle W_0, R\rangle$ for X. Recall that a model graph for some finite fixed set of formulae X is a finite $\mathbf{L}$-frame $\langle W_0, R\rangle$ such that all $w \in W_0$ are $C\mathbf{L}$-saturated sets with $w \subseteq X^*_{C\mathbf{L}}$ and

(i) $X \subseteq w_0$ for some $w_0 \in W_0$;

(ii) if $\neg\Box P \in w$ then there exists some $w' \in W_0$ with wRw' and $\neg P \in w'$;

(iii) if wRw' and $\Box P \in w$ then $P \in w'$.

The first step is to create a $C\mathbf{L}$-saturated w_0 from X with $X \subseteq w_0 \subseteq X^*_{C\mathbf{L}}$. This is possible via Lemma 6 (page 318). So w_0, and in general w (possibly annotated) stands for a finite $C\mathbf{L}$-saturated set of formulae (that corresponds to a world of W_0). Since w_0 is $C\mathbf{L}$-consistent, we know that *no* $C\mathbf{L}$-tableau for w_0 closes. We use this fact to construct a graph of $C\mathbf{L}$-saturated worlds, always bearing in mind that the resulting model graph must be based on an $\mathbf{L}$-frame. The construction is a meta-level one since we are free to inspect all $C\mathbf{L}$-tableaux for w_0, choosing nodes at will, since all such $C\mathbf{L}$-tableaux are open. We use a successor relation $\prec$ while building this graph and then form R from $\prec$. Also, if w is a set of formulae in this construction then $w^\Box = \{P : \Box P \in w\}$.

By Lemma 7 (page 320), $w_0 \models X$ under the truth valuation $\vartheta : p \mapsto \{w \in W_0 : p \in w\}$, giving an $\mathbf{L}$-model for X at w_0 as desired.

Proof for $C\mathbf{K}$: If no $\neg\Box P$ occurs in w_0 then $\langle W, R\rangle = \langle \{w_0\}, \emptyset\rangle$ is the desired model graph since this is a $\mathbf{K}$-frame and (i)-(iii) are satisfied.

Otherwise, let $Q_1, Q_2, \cdots, Q_m$ be all the formulae such that $\neg\Box Q_i \in w_0$. Since w_0 is $\mathcal{C}\mathbf{K}$-consistent, no application of (θ) can lead to a closed $\mathcal{C}\mathbf{K}$-tableau; in particular, the set $\{\Box A : \Box A \in w_0\} \cup \{\neg\Box Q_i\}$ must be $\mathcal{C}\mathbf{K}$-consistent for each $1 \leq i \leq m$. Each of these sets matches the numerator of (K) so (K) is applicable to each of them. But we know that an application of (K) to any of these sets could not have led to a closed $\mathcal{C}\mathbf{K}$-tableau either, so each of their respective denominators $(w_0^\Box; \neg Q_i)$ for $i = 1, \cdots, m$ must be $\mathcal{C}\mathbf{K}$-consistent (by (θ) and (K)). Note that these nodes come from different $\mathcal{C}\mathbf{K}$-tableaux.

Create a $\mathcal{C}\mathbf{K}$-saturated $v_i \subseteq X^*_{\mathcal{C}\mathbf{K}}$ from each $(w_0^\Box; \neg Q_i)$ for $i = 1, \cdots, m$, by using the static rules, and (the Saturation) Lemma 6. Put $w_0 \prec v_i$ for $i = 1, \cdots, m$, giving the nodes of level 1. Continue to create the nodes of further levels using (θ) and (K) as above.

Note that the denominator of the (K) rule has a maximum modal degree which is strictly less than that of its numerator, and that the $\mathcal{C}\mathbf{K}$-saturation process does not increase the maximum modal degree. Hence a path $w_0 \prec w_1 \prec w_2 \cdots$ must terminate (without cycles) because each successor created by (K) has a maximum modal degree strictly lower than that of the parent node.

Let R be $\prec$ and let W_0 consist of all the nodes created in this process, then $\langle W_0, R\rangle$ is a finite, irreflexive and intransitive tree and a model graph for X. Hence by Lemma 7, there is a $\mathbf{K}$-model for X with root w_0.

Proof for $\mathcal{C}\mathbf{T}$: If no $\neg\Box P$ occurs in w_0 then $\langle W, R\rangle = \langle\{w_0\}, \{(w_0, w_0)\}\rangle$ is the desired model graph since (i)-(iii) are satisfied. Otherwise, let $Q_1, Q_2, \cdots, Q_m$ be all the formulae such that $\neg\Box Q_i \in w_0$ and $\neg Q_i \notin w_0$. Proceed as for $\mathcal{C}\mathbf{K}$, noting that $\mathcal{C}\mathbf{T}$-saturation now involves (T) as well, but ignoring the successor for $\neg\Box Q \in w$ if $\neg Q \in w$. Let R be the reflexive closure of $\prec$; that is, put wRw for all worlds in the tree and also put wRw' if $w \prec w'$. Termination is as for $\mathcal{C}\mathbf{K}$.

Proof for $\mathcal{C}\mathbf{D}$: If no $\neg\Box P$ occurs in w_0 then $\langle\{w_0\}, \{(w_0, w_0)\}\rangle$ is the desired model graph since (i)-(iii) are satisfied. Otherwise, proceed as for $\mathcal{C}\mathbf{K}$, except that $\mathcal{C}\mathbf{D}$-saturation now involves (D) as well, and let W_{end} be the nodes of (the resulting tree) W_0 that have no successors. For each $w, w' \in W_0$, put wRw' if $w \prec w'$ and put wRw if $w \in W_{end}$. We have to show that (i)-(iii) are satisfied by this R. The only interesting case is to show that $\Box P \in w$ implies $P \in w$ for $w \in W_{end}$. This is true since $w \in W_{end}$ implies that w contains no $\Box P$, as otherwise, w would contain $\neg\Box\neg P$ by (D) and hence would have a successor node by (K), contradicting that $w \in W_{end}$. Termination is as for $\mathcal{C}\mathbf{K}$.

Proof for $\mathcal{C}\mathbf{D}'$: If no $\neg\Box Q$ occurs in w_0 and no $\Box P$ occurs in w_0 then $\langle\{w_0\}, \{(w_0, w_0)\}\rangle$ is the desired model graph. Otherwise, let $Z = \{Q_1, \cdots, Q_m\}$ be all the formulae such that $\neg\Box Q_i \in w_0, 1 \leq i \leq m$, and let $Y = \{P_1, P_2, \cdots, P_n\}$ be all the formulae such that $\Box P_j \in w_0$, $1 \leq j \leq n$. We know $m + n \geq 1$. Since w_0 is $\mathcal{C}\mathbf{D}'$-consistent each $\neg Q_i; Y$ is $\mathcal{C}\mathbf{D}'$-consistent, for $i = 1, 2, \cdots, m$ by (θ) and (KD). Also, Y itself is $\mathcal{C}\mathbf{D}'$-consistent by (θ) and (KD). If Z is non-empty then create a Q_i-successor v_i using (KD) containing $(\neg Q_i; Y)$ for each Q_i. But if Z is empty then create a single P-successor y using (KD) containing Y. Put $w_0 \prec v_i$ for each $i = 1 \cdots m$, or $w_0 \prec y$, as the case may be, giving the node(s) of level one. Continuing in this way obtain the node(s) of level two etc. Again, a sequence $w_0 \prec w_1 \prec w_2 \cdots$ must terminate since (KD) reduces the maximum modal degree and $\mathcal{C}\mathbf{D}'$-saturation does not increase it. As in the first proof for $\mathcal{C}\mathbf{D}$ put wRw if $w \in W_{end}$ and put wRw' if $w \prec w'$. Property (iii) holds for $w \in W_{end}$ as end nodes do not contain any $\Box P$, as otherwise, w would have a successor by (KD), contradicting that $w \in W_{end}$.

Proof for $\mathcal{C}\mathbf{K4}$: If no $\neg\Box P$ occurs in w_0 then $\langle\{w_0\}, \emptyset\rangle$ is the desired model graph since it is an $\mathbf{K4}$-frame and (i)-(iii) are satisfied. Otherwise, let $Q_1, Q_2, \cdots$, Q_m be all the formulae such that $\neg\Box Q_i \in w_0$.

We can form the sets $\{\Box A : \Box A \in w_0\} \cup \neg\Box Q_i$ for $1 \leq i \leq m$, by (θ), each of which is a numerator for $(K4)$. Hence by $(K4)$ each denominator $X_i = \{A : \Box A \in w_0\} \cup \{\Box A : \Box A \in w_0\} \cup \neg Q_i$ for $1 \leq i \leq m$, is also $\mathcal{C}\mathbf{K4}$-consistent.

Clearly for each X_i we can find some $\mathcal{C}\mathbf{K4}$-saturated $v_i \supseteq X_i$, with $v_i \subseteq X^*_{\mathcal{C}\mathbf{K4}}$. Put $w_0 \prec v_i$, $i = 1, \cdots, m$ and call v_i the Q_i-successor of w_0. These are the immediate successors of w_0. Now repeat the construction with each v_i thus obtaining the nodes of level 2 and so on.

In general, the above construction of $\langle W_0, \prec\rangle$ runs ad infinitum. However, since $w \in W_0$ implies $w \subseteq X^*_{\mathcal{C}\mathbf{K4}}$, (a finite set), a sequence $w_0 \prec w_1 \prec \cdots$ in $\langle W_0, \prec\rangle$ either terminates, or a node repeats. If in the latter case $n > m$ are minimal with $w_n = w_m$ we stop the construction and identify w_n and w_m in $\langle W_0, \prec\rangle$ thus obtaining a circle instead of an infinite path. One readily confirms that $\langle W_0, R\rangle$ is a model graph for X where R is the transitive closure of $\prec$. It is obvious that clusters in $\langle W_0, R\rangle$ form a tree.

Proof for $\mathcal{C}\mathbf{K4D}$: If no $\neg\Box P$ occurs in w_0 then $\langle\{w_0\}, \{(w_0, w_0)\}\rangle$ is the desired model graph. Otherwise, proceed as for $\mathcal{C}\mathbf{K4}$, except that $\mathcal{C}\mathbf{K4D}$-saturation also involves (D). A sequence either terminates or cycles since $X^*_{\mathcal{C}\mathbf{K4D}}$ is finite. Put $w \prec w$ for all $w \in W_{end}$ and let R be the transitive closure of $\prec$. Property (iii) is satisfied by $w \in W_{end}$ just as in the proof for $\mathcal{C}\mathbf{D}$.

Proof for $\mathcal{C}\mathbf{K45}$: Suppose X is $\mathcal{C}\mathbf{K45}$-consistent and create a $\mathcal{C}\mathbf{K45}$-saturated superset $w_0 \subseteq X^*_{\mathcal{C}\mathbf{K45}}$ of X as usual. If no $\neg\Box P$ occurs in w_0 then $\langle\{w_0\}, \emptyset\rangle$ is the desired model graph since (i)-(iii) are satisfied.

Otherwise let $Q_i, Q_2, \cdots, Q_k$ be all the formulae such that $\neg\Box Q_i \in w_0$ and create a Q_i-successor for each Q_i using (θ) and the (45) rule. Continue construction of one such sequence $S = w_0 \prec w_1 \prec \cdots$ always choosing a successor that is new to the current sequence. Note that a successor may be new either because it fulfills an eventuality that is not fulfilled by the current sequence, or because it contains formulae that do not appear in previous nodes that fulfill the same eventuality. Since $X^*_{\mathcal{C}\mathbf{K45}}$ is finite, we must sooner or later come to a node w_m such that the sequence $S = w_0 \prec w_1 \prec \cdots \prec w_m$ already contains *all* the successors of w_m. That is, it is not possible to choose a new successor.

Now, the $(K45)$ rule guarantees that if $\neg\Box P \in w_0$ then $\neg\Box P \in w_i$, $i > 0$, so one of the successors of w_m must fulfill $\neg\Box P$, and furthermore, this successor must already appear in the sequence. However, there is no guarantee that this successor is w_1. So, choose the successor w_x of w_m that fulfills some eventuality in w_m, but that appears earliest in S and put $w_m \prec w_x$ giving $S = w_0 \prec w_1 \prec \cdots \prec w_x \prec \cdots \prec w_m \prec w_x$. There are two cases to consider depending on whether $x = 0$ or $x \neq 0$.

Case 1: If $x = 0$, put R as the reflexive, transitive and symmetric closure of $\prec$ over $W_0 = \{w_0, w_1, \cdots, w_m\}$. This gives a frame $\langle W_0, R\rangle$ which is a nondegenerate cluster.

Case 2: If $x \neq 0$, put $W_0 = \{w_0, w_x, w_{x+1}, \cdots, w_m\}$, discarding w_1, w_2, $\cdots$, w_{x-1}, and let R' be the reflexive, transitive and *symmetric* closure of $\prec$ over $W_0 \setminus \{w_0\}$. That is, $R' = \{(w_i, w_j) | w_i \in W_0, w_j \in W_0, i \geq x, j \geq x\}$. Now put $R' = R' \cup \{(w_0, w_x)\}$ and let R be the transitive closure of R'. The frame $\langle W_0, R\rangle$ now consists of a degenerate cluster w_0 followed by a nondegenerate cluster $w_x R w_{x+1} R \cdots R w_m R w_x$ where R is transitive and Euclidean.

Property (i) is satisfied by $\langle W_0, R\rangle$ by construction. We show that (ii) and (iii) are satisfied as follows.

Proof of (ii): The (45) rule also carries *all* eventualities from the numerator to the denominator, including the one it fulfills. Therefore, for all $w_i \in W_0$ we have: $\neg\Box P \in w_i$ implies $\neg\Box P \in w_m$. But we stopped the construction at w_m because no new Q_i-successors for w_m could be found. Hence there is a Q_i-successor for each eventuality of w_m. Since we have a cycle, and eventualities cannot disappear, these are all the eventualities that appear in the cycle. Furthermore, we chose w_x to be the successor of w_m that was earliest in the sequence S. Hence all of the eventualities of w_m are fulfilled by the sequence $w_x R \cdots R w_m$. All the eventualities of w_0 are also in w_m, hence (ii) holds.

Proof of (iii): The (45) rule carries all formulae of the form $\Box P$ from its numerator to its denominator. Hence $\Box P \in w$ and $w \prec v$ implies that $P \in v$ and $\Box P \in v$. But we know that $w_x \prec \cdots \prec w_m \prec w_x$ forms a cycle, hence (iii) holds as well.

Proof for $\mathcal{C}$K45D: Based on the previous proof. If the $(45D)$ rule is ever used with no eventualities present then this can only happen when w_0 contains no eventualities. For if w_0 contained an eventuality then so would all successors.

So if w_0 contains no eventualities and no formulae of the form $\Box P$ then $\langle \{w_0\}, \{(w_0, w_0)\}\rangle$ is the desired model graph. This gives a frame which is a simple (nondegenerate) cluster.

Otherwise, let $Q_1, \cdots, Q_k$ be all the formulae such that $\neg\Box Q_i \in w_0$ and let $P_1, \cdots, P_m$ be all the formulae such that $\Box P_j \in w_0$. Create a successor w_1 for w_0 using $(45D)$ for some Q_i or P_j and continue creating successors using $(45D)$, always choosing a successor new to the sequence until no new successors are possible. Choose w_x as the successor nearest to w_0 giving a cycle $w_0 \prec \cdots \prec w_x \prec \cdots \prec w_m \prec w_x$ and discard $w_1, w_2, \cdots w_{x-1}$ as in the previous proof.

Form R as in the proof for $\mathcal{C}$K45 where $x = 0$ gives a frame which is a simple cluster and $x \neq 0$ gives a frame which is a degenerate cluster followed by a nondegenerate cluster.

Properties (i)–(iii) can be proved in a similar manner.

Note that the requirement to continually choose a new successor is tantamount to following an infinite path in Shvarts' formulation [Shvarts, 1989]. That is, the inevitable cycle that we encounter constitutes an infinite branch if it is unfolded.

Proof for $\mathcal{C}$S4: If no $\neg\Box P$ occurs in w_0 then $\langle\{w_0\}, \{(w_0, w_0)\}\rangle$ is the desired model graph. Otherwise proceed as for $\mathcal{C}$**K4** except create a successor for eventuality $\neg\Box P \in w$ only if $\neg P \notin w$, and use $(S4)$ to create successors instead of $(K4)$. Then, a successor for w will be based on $\{\Box A : \Box A \in w\} \cup \neg P$. Let R be the reflexive and transitive closure of $\prec$ (instead of the transitive closure of $\prec$). We can add reflexivity because of closure with respect to (T).

Proof for $\mathcal{C}$S5π^-: see page 341.

Proof for $\mathcal{C}^\dagger$K45: Suppose X is $\mathcal{C}^\dagger$**K45**-consistent and create a $\mathcal{C}^\dagger$**K45**-saturated superset w_0 with $X \subseteq w_0 \subseteq X^*_{\mathcal{C}^\dagger\mathbf{K45}}$ as usual. If no $\neg\Box P$ occurs in w_0 then $\langle\{w_0\}, \emptyset\rangle$ is the desired model graph since (i)–(iii) are satisfied.

Otherwise, let $Q_i, Q_2, \cdots, Q_m$ be all the formulae such that $\neg\Box Q_i \in w_0$ and create a Q_i-successor v_i for each Q_i using the (45) rule. This gives all the nodes of level 1, so put $w_0 \prec v_i$, for each $i = 1 \cdots m$, and stop!

Consider any two nodes v_i and v_j with $i \neq j$. Using the facts that each node is subformula-complete and there are no building up rules, we show that

(a) $\Box P \in v_i$ implies $\Box P \in w_0$ implies $P \in v_j$, $P \in v_i$ and $\Box P \in v_j$;

(b) $\neg\Box P \in v_i$ implies $\neg\Box P \in w_0$ implies there exists a v_k such that $\neg P \in v_k$.

Proof of (a): Suppose $\Box P \in v_i$. Then $\Box P \in Sf(w_0)$ since there are no building up rules, and so $\Box P \in w_0$ or $\neg\Box P \in w_0$ since w_0 is subformula-complete. If $\neg\Box P \in w_0$ then $\neg\Box P \in v_i$ by (45), contradicting the $\mathcal{C}^\dagger$**K45**-consistency of v_i. Hence $\Box P \in w_0$. Note that this holds only because the (45) rule carries $\neg\Box P$ into its denominator along with $\neg\Box Y$.

Proof of (b): As for (a) except uniformly replace $\neg\Box P$ by $\Box P$ and vice-versa. The crux of the proof is that (45) preserves all formulae of the form $\Box P$ and $\neg\Box P$.

Hence we can put $v_i R v_j R v_i$ for all v_i and v_j giving a reflexive, transitive and symmetric nondegenerate cluster. If we also put $w_0 R v_i$ for all $i = 1 \cdots m$, and take the transitive closure, then we obtain a degenerate cluster followed by a nondegenerate cluster. If some $v_k = w_0$ then we obtain a lone nondegenerate cluster. In each case the frame is a **K45**-frame.

In either case, (i)–(iii) are satisfied giving a model graph and hence a **K45**-model for X.

Proof for $\mathcal{C}^\dagger$K45D: Similar to the proofs for $\mathcal{C}^\dagger$**K45** and $\mathcal{C}$**KD**.

Proof for $\mathcal{C}^\dagger$KB4: Suppose no $\mathcal{C}^\dagger$**KB4**-tableau for X is closed. Construct a $\mathcal{C}^\dagger$**KB4**-saturated w_0 from X as usual. If no $\neg\Box P$ occurs in w_0 then $\langle\{w_0\}, \emptyset\rangle$ is the desired model graph as (i)–(iii) are satisfied. Otherwise, create a successor v_i for each eventuality in w_0 using (θ) and $(K4)$ giving the nodes of level one, put $w_0 \prec v_i$ and stop. Since w_0 contains at least one eventuality, w_0 must be closed with respect to $(T_\Diamond)$, hence $\Box Q \in w_0$ implies $Q \in w_0$. We show that

(a) $\neg\Box P \in v_i$ implies $\neg\Box P \in w_0$; and

(b) $\Box P \in v_i$ implies $\Box P \in w_0$

from which properties (i)–(iii) follow.

(a) Suppose $\neg\Box P \in v_i$ and $\neg\Box P \notin w_0$. The only super-formulae are of the form $\Box A$ hence $\neg\Box P \in Sf(w_0)$ or $\neg\Box P \in \neg Sf(w_0)$ whence $\Box P \in Sf(w_0)$. Since w_0 is subformula-complete we must have $\Box P \in w_0$ and hence $\Box P \in v_i$ by $(K4)$; contradiction.

(b) Suppose $\Box P \in v_i$.

(i) If $\Box P \in Sf(w_0)$ then $\Box P \in w_0$ or $\neg\Box P \in w_0$. The latter implies $\Box\neg\Box P \in w_0$ by (5) which implies $\neg\Box P \in v_i$; contradiction. Hence if $\Box P \in v_i$ and $\Box P \in Sf(w_0)$ then $\Box P \in w_0$ whence $P \in v_i$ by $(K4)$ and $P \in w_0$ by $(T_\Diamond)$.

(ii) If $\Box P \notin Sf(w_0)$ then $\Box P = \Box\neg\Box Q$ for some eventuality $\neg\Box Q$ of v_i. Hence $\neg\Box Q \in v_i$. By (a) we then have $\neg\Box Q \in w_0$, which by (5) gives $\Box\neg\Box Q \in w_0$. But $\Box\neg\Box Q$ is $\Box P$, hence $\Box P \in Sf(w_0)$; contradiction. Hence case $\Box P \notin Sf(w_0)$ is impossible.

Now let R be the reflexive, transitive and symmetric closure of $\prec$. Note that reflexivity for w_0 comes from saturation with respect to $(T_\Diamond)$ and reflexivity for v_i comes from property (b) via $(K4)$. Thus when w_0 contains at least one eventuality, we get an **S5**-frame (showing that **K4B** is 'almost' **S5**).

Proof for $\mathcal{C}^\dagger$S4: If no $\neg\Box P$ occurs in w_0 then $\langle\{w_0\}, \{(w_0, w_0)\}\rangle$ is the desired model graph. Otherwise, let $Q_1, Q_2, \cdots, Q_k$ be all the formulae such that $\neg\Box Q_i \in w_0$ and $\neg Q_i \notin w_0$. Create a Q_i-successor v_i of level 1 for each Q_i using the (θ) and $(S4)$ rules, and continue in this way to obtain the nodes of level 2 and so on with the following termination condition:

(*) if $w_0 \prec w_1 \prec \cdots \prec w_{i-1} \prec w_i$ is a path in this construction and $i \geq 1$ is the least index such that $\Box A \in w_i$ implies $\Box A \in w_{i-1}$, then put $w_i \prec w_{i-1}$ giving a cycle on this path and stop!

First of all, this termination condition is satisfactory since $(S4)$ ensures that $\Box A \in w_j$ implies $\Box A \in w_{j+1}$ so that $\Box$-formulae accumulate and we eventually run out of new $\Box$-formulae since $X^*_{\mathcal{C}^\dagger\mathbf{S4}}$ is finite.

Second, note that $\mathcal{C}^\dagger$**S4** contains $(sfcT)$ and hence each w_i is subformula-complete. Since there are no building up rules, the only new formulae that may appear by saturating with the $(sfcT)$ rules are the negations of subformulae from the predecessor. Therefore, each $w_{n+1} \subseteq Sf(\widetilde{w_n})$ where $\widetilde{w} = Sf(w) \cup \neg Sf(w)$.

Let R be the reflexive and transitive closure of $\prec$. It is obvious that clusters of R form a tree. To prove that $\langle W_0, R\rangle$ is a model graph for X we have to prove (i)–(iii).

(i) Clearly (i) holds by construction;

(ii) Suppose $\neg\Box P \in w_j$ where w_j is some arbitrary world of some arbitrary path of our construction. If the termination condition was not applied to w_j, then either $\neg P \in w_j$ or w_j has a P-successor fulfilling $\neg\Box P$ by $(S4)$ and so (ii) is satisfied. That is (ii) holds for any world to which the termination condition was not applied.

If the termination condition was applied to w_j, then it could not have been applied to w_{j-1}. Hence (ii) holds for w_{j-1}. So all we have to show is that $\neg\Box P \in w_{j-1}$ because, in this case, (ii) would then hold for w_j from the fact that $w_j R w_{j-1}$ and the transitivity of R.

Suppose to the contrary that $\neg\Box P \notin w_{j-1}$. Since $\neg\Box P \in w_j$ by supposition, we must have $\Box P \in Sf(w_{j-1})$ by the second point we noted above. Then $\Box P \in w_{j-1}$ by (the subformula-completeness) Lemma 14, and $\Box P \in w_j$ by $(S4)$ contradicting the $\mathcal{C}^\dagger\mathbf{S4}$-consistency of w_j since $\neg\Box P \in w_j$. Hence (ii) also holds.

(iii) Suppose $\Box P \in w_j$. If (*) was not applied to w_j then (iii) holds as for $\mathcal{C}\mathbf{S4}$ by (T) since $(S4)$ preserves $\Box$-formulae. If (*) was applied to w_j then (iii) would follow from $\Box P \in w_{j-1}$ by $(S4)$ and (T). But this is exactly what (*) guarantees. Hence (iii) holds as well.

Proof for $\mathcal{C}^\dagger\mathbf{B}$: If no $\neg\Box P$ occurs in w_0 then $\langle\{w_0\}, \{(w_0, w_0)\}\rangle$ is the desired model graph as (i)–(iii) are satisfied. Otherwise, let $Q_1, Q_2, \cdots, Q_m$ be all the formulae such that $\neg\Box Q_i \in w_0$ and $\neg Q_i \notin w_0$. Since w_0 is $\mathcal{C}^\dagger\mathbf{B}$-saturated, w_0 is subformula-complete, hence $Q_i \in w_0$ for each Q_i. Create a Q_i-successor for each Q_i using (θ) and (K) giving the nodes of level one. Repeat this procedure to give the nodes of level two and so on. For any node w in this construction let $s(w)$ be the number of formulae with $P \in w$ and $\neg\Box P \in w$. Let $t(w) = s(w) + mdeg(w)$. To quote Rautenberg *'It is easily seen that $w \prec v \to t(v) < t(w)$, so that W_0 is finite.'*, but as shown in [Goré, 1992] Rautenberg's definition of *mdeg* is not sufficient. We accept Rautenberg's claim for the moment and return to this issue after completing the model construction.

Let R be the reflexive and symmetric closure of $\prec$ so that $\langle W_0, R\rangle$ is a $\mathbf{B}$-frame. We have to show that (i)–(iii) hold. The only difficulty is to show symmetry: that is, $\Box P \in w_{i+1}$ and $w_i \prec w_{i+1}$ implies $P \in w_i$. So suppose that $w_i \prec w_{i+1}$ and $\Box P \in w_{i+1}$. We have to show that $P \in w_i$. There are two cases: $\Box P \in Sf(w_i)$ or $\Box P \notin Sf(w_i)$.

Case 1: If $\Box P \in Sf(w_i)$, then $\Box P \in w_i$ or $\neg\Box P \in w_i$ since w_i is subformula-complete. If $\Box P \in w_i$ then $P \in w_i$ by (T) and we are done. Otherwise, if $\neg\Box P \in w_i$ and $P \notin w_i$ then $\neg P \in w_i$ and $\Box\neg\Box P \in w_i$ by (B) and so $\neg\Box P \in w_{i+1}$ contradicting the consistency of w_{i+1} since $\Box P \in w_{i+1}$ by supposition. Hence $\neg\Box P \in w_i$ also implies that $P \in w_i$.

Case 2: If $\Box P \notin Sf(w_i)$ then $\Box P = \Box\neg\Box Q$ for some $\neg\Box Q \in w_{i+1}$ and $\neg Q \in w_{i+1}$. Hence $\neg\Box Q \in Sf(w_i)$ or $\neg\Box Q \in \neg Sf(w_i)$ whence $\Box Q \in Sf(w_i)$. By subformula-completeness we then have $\Box Q \in w_i$ or $\neg\Box Q \in w_i$. If $\Box Q \in w_i$, then $Q \in w_{i+1}$ contradicting the $\mathcal{C}^{\dagger}\mathbf{B}$-consistency of w_{i+1} since $\neg Q \in w_{i+1}$. Hence $\neg\Box Q \in w_i$. But then $P \in w_i$ since P is $\neg\Box Q$ and we are done.

Now, we still have to show that this construction terminates. The crux of the matter is to use a definition of a metric mdg say, which is like our $mdeg$ but where $mdg(A \wedge B) = mdg(A) + mdg(B)$ rather than $max\{\,mdg(A), mdg(B)\}$ [Massacci, 1995]. Similarly, for a set X, we use $mdg(X) = \Sigma_{A \in X} mdg(A)$ rather than $max\{mdg(A) \mid A \in X\}$. Then, a rather tedious counting exercise, which we omit for brevity, suffices to show that if $w \prec v$ then $t(v) < t(w)$, which is enough to show termination. We have retained our version of $mdeg$ because it is useful for other purposes.

Proof for $\mathcal{C}^{\dagger}\mathbf{S5}$: If no $\neg\Box P$ occurs in w_0 then $\langle\{w_0\}, \{(w_0, w_0)\}\rangle$ is the desired model graph as (i)–(iii) are satisfied. Otherwise, let $Q_1, Q_2, \cdots, Q_m$ be all the formulae such that $\neg\Box Q_i \in w_0$. Since w_0 is $\mathcal{C}^{\dagger}\mathbf{S5}$-saturated, $\Box\neg\Box Q_i \in w_0$ for each Q_i by (5). Create a Q_i-successor for each Q_i using (θ) and $(S4)$ giving the nodes v_i of level one, put $w_0 \prec v_i$, for each $i = 1, 2, \cdots, m$ and stop! Let R be the reflexive, transitive and symmetric closure of $\prec$. By construction, $\langle W_0, R\rangle$ is an $\mathbf{S5}$-frame. We have to show that (i)–(iii) hold.

For any k, with $1 \leq k \leq m$, and $w_0 \prec v_k$, we show that:

(a) $\neg\Box P \in v_k$ implies $\neg\Box P \in w_0$; and

(b) $\Box P \in v_k$ implies $\Box P \in w_0$

from which (i)–(iii) follow.

(a) Suppose $w_0 \prec v_k$, $\neg\Box P \in v_k$ and $\neg\Box P \notin w_0$. Since $\neg\Box P \in Sf(w_0)$, and w_0 is subformula-complete, we have $\Box P \in w_0$. But then, by $(S4)$, $\Box P \in v_k$, contradicting the $\mathcal{C}^{\dagger}\mathbf{S5}$-consistency of v_k. Hence $\neg\Box P \in w_0$.

(b) Suppose $w_0 \prec v_k$ and $\Box P \in v_k$, then $\Box P \in Sf(w_0)$ or $\Box P \notin Sf(w_0)$.

(b1) If $\Box P \in Sf(w_0)$ and $\Box P \notin w_0$, then $\neg\Box P \in w_0$ since w_0 is subformula-complete. Then $\Box\neg\Box P \in w_0$ by (5) and $\neg\Box P \in v_k$ by $(S4)$, contradicting the $\mathcal{C}^{\dagger}\mathbf{S5}$-consistency of v_k. Hence, if $\Box P \in v_k$ and $\Box P \in Sf(w_0)$ then $\Box P \in w_0$.

(b2) If $\Box P \notin Sf(w_0)$ then $\Box P = \Box\neg\Box Q$ for some $\neg\Box Q \in v_k$ since this is the only way that formulae from outside $Sf(w_0)$ can appear in v_k. By (a), $\neg\Box Q \in v_k$ implies $\neg\Box Q \in w_0$ which by (5) implies $\Box\neg\Box Q \in w_0$. Since $\Box\neg\Box Q$ *is* $\Box P$, we have $\Box P \in w_0$. But this is absurd since it implies that $\Box P \in Sf(w_0)$ and our supposition was that $\Box P \notin Sf(w_0)$. Hence the subcase (b2) cannot occur.

Proof for $\mathcal{C}^{\dagger}\mathbf{S5}'$: For completeness suppose X is $\mathcal{C}\mathbf{S5}'$-consistent and create a $\mathcal{C}^{\dagger}\mathbf{S5}'$-saturated superset w_0 with $X \subseteq w_0 \subseteq X^*_{\mathcal{C}\mathbf{S5}'}$ as usual.

If no $\neg\Box P$ occurs in w_0 then $\langle\{w_0\}, \{(w_0, w_0)\}\rangle$ is the desired model graph. Otherwise, let $Q_1, Q_2, \cdots, Q_m$ be all the formulae such that $\neg\Box Q_i \in w_0$ and $\neg Q_i \notin w_0$. Create a Q_i-successor v_i of level 1 for each Q_i using the $(S5)$ rule and stop!

Let $W_0 = \{w_0, v_1, v_2, \cdots, v_m\}$. Consider any two nodes v_i and v_j of level 1 so that $w_0 \prec v_i$ and $w_0 \prec v_j$ with $i \neq j$. We claim that:

(a) $\Box P \in v_i$ implies $\Box P \in w_0$ implies $\Box P \in v_j$; and

(b) $\neg\Box P \in v_i$ implies $\neg\Box P \in w_0$ implies there exists a $w \in W_0$ with $\neg P \in w$.

Proof of (a): Suppose $\Box P \in v_i$, then $P \in v_i$ by (T). Also, $\Box P \in Sf(w_0)$ as there are no building up rules, hence $\Box P \in w_0$ or $\neg\Box P \in w_0$ by $(sfcT)$. If $\neg\Box P \in w_0$ then either $\neg P \in v_i$ or $\neg\Box P \in v_i$ by $(S5)$. The first contradicts the $\mathcal{C}^\dagger\mathbf{S5}'$-consistency of v_i since $P \in v_i$ and so does the second since $\Box P \in v_i$. Hence $\Box P \in w_0$. And then $\Box P \in v_j$ by $(S5)$ and $P \in v_j$ by (T).

Proof of (b): Suppose $\neg\Box P \in v_i$. Then as there are no building up rules, $\neg\Box P \in Sf(w_0)$. Hence $\Box P \in w_0$ or $\neg\Box P \in w_0$ since w_0 is subformula-complete. If $\Box P \in w_0$ then $\Box P \in v_i$ by $(S5)$, contradicting the $\mathcal{C}^\dagger\mathbf{S5}'$-consistency of v_i since $\neg\Box P \in v_i$ by supposition. Hence $\neg\Box P \in w_0$. And then either $\neg P \in w_0$, or there is some v_k such that $\neg P \in v_k$ by $(S5)$. That is, the w we seek is either w_0 itself, or one of the nodes of level 1.

Putting R equal to the reflexive, symmetric and transitive closure of $\prec$ gives an **S5**-model graph since (i)–(iii) follow from (a) and (b). ∎

Bibliographic Remarks and Discussion

The cut-free calculi $\mathcal{C}\mathbf{K}$, $\mathcal{C}\mathbf{T}$, $\mathcal{C}\mathbf{D}$, $\mathcal{C}\mathbf{D}'$, $\mathcal{C}\mathbf{K4}$, $\mathcal{C}\mathbf{K4D}$ and $\mathcal{C}\mathbf{S4}$ can all be traced back to Fitting [1973] via Fitting [1983] although our presentation is based on the work of Hintikka [1955] and Rautenberg [1983]. The system $\mathcal{C}\mathbf{K4D}$ is an obvious extension of Rautenberg's system $\mathcal{C}\mathbf{D}$, and $\mathcal{C}\mathbf{D}'$ is lifted straight from Fitting [1983]. The advantage of $\mathcal{C}\mathbf{D}'$ is that it has the subformula property whereas $\mathcal{C}\mathbf{D}$ does not. Clearly, the $(K4)$ rule can be extended to handle seriality as done in the (KD) rule to give a $(K4D)$ rule, but we omit details. The tableau systems $\mathcal{C}\mathbf{K45}$ and $\mathcal{C}\mathbf{K45D}$ are based on the work of Shvarts [1989] (also known as Schwarz), while $\mathcal{C}\mathbf{K4B}$ and the $(T_\Diamond)$ rule come from the work of Amerbauer [1993].

Some of the desired properties of R can be obtained in two different ways. For example, Rautenberg encodes the seriality of **D**-frames by the *static* (D) rule which adds an eventuality $\Diamond P$ for every formula of the form $\Box P$. The transitional (K) rule then fulfills that eventuality. On the other hand Shvarts [1989] and Fitting [1983] use the *transitional* rule (KD). Similarly, the $(S5)$ *transitional* rule due to Fitting builds in the effect of Rautenberg's *static* rule (5) by carrying $\neg\Box P$ and $\neg\Box Y$ from the numerator into the denominator.

Rautenberg [1983] does not explicitly distinguish transitional and static modal rules. Hence his rules for (T), (D), (B), (sfc) and $(sfcT)$ do not carry all the numerator formulae into their denominators. For example, Rautenberg's (T) rule is shown below left whereas ours is shown below right:

$$\frac{X; \Box P}{X; P} \qquad (T)\ \frac{X; \Box P}{X; \Box P; P}$$

Thus contraction is implicit in his systems and as we saw in Example 8 (page 324), contraction is necessary for some modal systems.

The $\mathcal{C}^{\dagger}\mathbf{S4}$ system is based on ideas of Hanson [1966] where he gives Kripke-like tableau systems for **S4** and **S5** using a form of $(sfcT)$ as early as 1966. The tableau system $\mathcal{C}^{\dagger}\mathbf{S4}$ is not exactly Hanson's system but the ideas are his. The advantage of adding $(sfcT)$ is that the termination condition in the completeness proof is much easier to check than the one for $\mathcal{C}\mathbf{S4}$ where we have to look at all predecessors in order to detect a cycle. However, the overheads associated with any sort of cut rule are significant, and a more detailed analysis shows that $\mathcal{C}^{\dagger}\mathbf{S4}$ performs much useless work. Hanson also suggests a tableau system for **S5** along these lines, but in it he uses a rule which explicitly adds a formula to the parent node to obtain symmetry. This is forbidden for our tableau systems since we cannot return to previous nodes.

The tableau systems of Heuerding *et al.* [1996] are further refinements of our tableau systems which allow for a more efficient check for cyclic branches. However, they are nonstandard in that the denominators and numerators carry extra sets to store the necessary information.

Notice that the effects of $(sfcT)$ on w_0 when R is to be transitive and there are no building up rules like (5) is to flush out all the eventualities that could possibly appear in any successor. That is, if $\neg\Box P$ is going to appear in a successor, it must be in $Sf(w_0)$. But then it must be in w_0 since otherwise by $(sfcT)$, we would have $\Box P \in w_0$ contradicting the appearance of $\neg\Box P$ in any consistent successor. Hence the number of eventualities never increases as all the eventualities that will ever appear are already in w_0. Indeed this fact may actually make things worse since we will have to fulfill $\neg\Box P$ at the first level of the model construction as well as at deeper levels where $\neg\Box P$ reappears. The refinements of Heuerding *et al.* [1996] may be useful in such cases since one of their ideas addresses exactly this point.

The idea behind (sfc) and $(sfcT)$ is to put extra information into a node before leaving it for good. That is, once we leave a node in our tableau procedure, we can never return to it. Also, the transitional rules usually lose information in the transition from the numerator to the denominator. The (sfc) and $(sfcT)$ rules are used to make up for this 'destructive' aspect of our transitional rules.

The completeness proofs in this section go through unchanged [Massacci, 1995] if we replace the (sfc) and $(sfcT)$ rules by the 'modal cut' rule (mc) shown below:

$$(mc)\ \frac{X}{\Box P; X \mid \neg\Box P; X} \text{ where } \Box P \in Sf(X)$$

Also, many of the rule combinations can be further refined. For example, the (B) rule subsumes the modal aspects of the $(sfcT)$ rule so that only the non-modal part is necessary in $\mathcal{C}^{\dagger}\mathbf{B}$; see also [Amerbauer, 1993] for further refinements.

The tableau systems $\mathcal{C}^{\dagger}\mathbf{B}$ and $\mathcal{C}^{\dagger}\mathbf{S5}$ are due to Rautenberg while $\mathcal{C}^{\dagger}\mathbf{S5}'$, $\mathcal{C}^{\dagger}\mathbf{K45}$ and $\mathcal{C}^{\dagger}\mathbf{K45D}$ are an amalgamation of ideas of Fitting, Hanson and Rautenberg. Note that in the latter, we add (sfc), not $(sfcT)$ since **K45**-frames and **K45**-frames are not reflexive. The advantage over the cut-free counterparts $\mathcal{C}\mathbf{K45}$ and

$\mathcal{C}$**K45D** is that the completeness proofs, and hence the satisfiability tests based upon them, are much simpler. Note that $\mathcal{C}^{\dagger}$**S5** does not have the subformula property, but $\mathcal{C}^{\dagger}$**S5′** does.

Fitting [1983, page 201] gives tableau calculi for the symmetric logics **KB**, **KDB**, **KTB**, and **S5** using a **semi-analytic cut** rule (sac), which he attributes to Osamu Sonabe. The (sac) rule is allowed to cut on subformulae of formulae that are in the numerator, and also on superformulae obtained by repeatedly prefixing modalities $\Box$, $\neg\Box$, $\Diamond$ and $\neg\Diamond$, to these subformulae. Since the superformulae are not bounded, as they are in Rautenberg's systems, the semi-analytic cut rule cannot give a decision procedure.

Fitting's semi-analytic system for **S5** is essentially $\mathcal{C}\mathbf{T} + (S5) + (sac)$. Fitting [1983, page 226] replaces the semi-analytic cut rule with a (static) building up rule of the form

$$(\pi)\ \frac{X; P}{X; \Diamond P; P}$$

and proves that his system $\mathcal{C}\mathbf{S5}\pi = \mathcal{C}\mathbf{T} + (S5) + (\pi)$ is sound and (weakly) complete with respect to **S5**-frames. But note that the (π) rule is not 'once off' since it can lead to an infinite chain $A \in w, \Diamond A \in w, \Diamond\Diamond A \in w, \cdots$ so this system cannot give a decision procedure for **S5** either. That is, we have merely traded one non-analytic rule for another.

Fitting then proves the curious fact that a single formula A is an **S5**-theorem if and only if a $\mathcal{C}\mathbf{S5}\pi$-tableau for $\{\neg A\}$ closes, and furthermore, that the (π) rule needs to be used only *once* at the beginning of the $\mathcal{C}\mathbf{S5}\pi$-tableau to lift $\neg A$ to $\neg\Box A$ [Fitting, 1983, page 229]. That is, the system $\mathcal{C}\mathbf{S5}\pi^{-}$ *without* the (π) rule is (weakly) complete for **S5** in the sense that A is an **S5**-theorem if and only if a $\mathcal{C}\mathbf{S5}\pi^{-}$-tableau for $\{\neg\Box A\}$ closes. Fitting gives a completeness proof in terms of maximal consistent sets, but a constructive completeness for this system is also easy as given below.

Completeness Proof for $\mathcal{C}\mathbf{S5}\pi^{-}$: Suppose no $\mathcal{C}\mathbf{S5}\pi^{-}$-tableau for the singleton set $\{\neg\Box A\}$ closes. Construct some $\mathcal{C}\mathbf{S5}\pi^{-}$-saturated set w_0 from $\neg\Box A$ by applying all the non-structural static rules; obtaining $w_0 = \{\neg\Box A\}$! Now construct a tree of $\prec$-successors as in the $\mathcal{C}\mathbf{S4}$ completeness proof except that we use the transitional rule $(S5)$ instead of $(S4)$ to create $\prec$-successors. Let R be the reflexive and transitive closure of $\prec$ to obtain a finite tree of finite clusters as in the $\mathcal{C}\mathbf{S4}$ case. Consider some final cluster C of this tree. Since C is final, any eventuality in any of its sets must be fulfilled by some set of C itself, as otherwise, C could not be final. But note that the $(S5)$ rule carries *all* its eventualities from its numerator into its denominator. Thus, in this case, $\neg\Box A$ is in every member of C, and hence some set $w_1 \in C$ has $\{\neg\Box A, \neg A\} \subseteq w_1$. But a *final* cluster is also symmetric, hence C is an **S5**-frame and hence an **S5**-model for $\{\neg A, \neg\Box A\}$ at w_1 under the usual valuation $\vartheta(p) = \{w : p \in w\}$. This completes the unusual proof for $\mathcal{C}\mathbf{S5}\pi^{-}$ that: if there is no closed $\mathcal{C}\mathbf{S5}\pi^{-}$-tableau for $\{\neg\Box A\}$ then $\neg A$ is **S5**-satisfiable. That is, if $\models_{\mathbf{S5}} A$ then $\vdash_{\mathcal{C}\mathbf{S5}\pi^{-}} \Box A$.

$$(R)\ \frac{X; \neg\Box P}{X; \neg\Box P; \neg P \mid X; \neg\Box P; \Box\neg\Box P; P}$$

$$(S4F)\ \frac{U; \Box X; \neg\Box P; \neg\Box Y}{U; \Box X; \neg\Box P; \neg\Box Y; \Box\neg\Box P \mid \Box X; \neg\Box P; \neg\Box Y; \neg P}$$

$$(S4.2)\ \frac{X; \neg\Box P}{X; \neg\Box P; \Box\neg\Box P \mid X; \neg\Box P; \Box(\neg\Box\neg\Box P)^*}\ \neg\Box P \text{ not starred}$$

Figure 8. Tableau rules for **S4R**, **S4F** and **S4.2**

For the logics with a symmetric R we seem to need analytic cut, either as (sfc) or as $(sfcT)$. The subformula property can be regained for some logics by changing the transitional rules to carry more information from the numerator to the denominator. But note that a building up rule seems essential for $\mathcal{C}\mathbf{B}$, so not all the systems are amenable to this trick.

4.15 Modal Logics of Knowledge and Belief

In this section we give a brief overview of tableau systems for the modal logics **S4R**, **S4F** and **S4.2**. These logics, together with the logics **K45** and **K45D**, have proved useful as nonmonotonic modal logics where the formula $\Box A$ is read as 'A is believed' or as 'A is known' [Moore, 1985; Schwarz, 1992; Fagin *et al.*, 1995; Schwarz and Truszczynski, 1992; Marek *et al.*, 1991]. In these logics, the reflexivity axiom, $\Box A \to A$, is deliberately omitted on the grounds that believing A should not imply that A is true. The logic **K45D** is another candidate for such logics of belief because its extra axiom, $\Box A \to \Diamond A$, which can be written as $\Box A \to \neg\Box\neg A$, encodes the intuition that 'if A is believed then $\neg A$ is not believed'.

Figure 8 shows the tableau rules we require. The tableau calculi we consider are shown below:

$\mathcal{C}$L	Static Rules	Transitional Rules	$X^*_{\mathcal{C}\mathrm{L}}$
$\mathcal{C}$**S4R**	$\mathcal{C}$**PC**, (T), (R)	$(S4)$	$Sf\neg Sf\Box\widetilde{X}$
$\mathcal{C}^\dagger$**S4.2**	$\mathcal{C}$**PC**, $(sfcT)$, (T), $(S4.2)$	$(S4)$	$Sf\neg Sf\Box X^*_{\mathcal{C}\mathbf{S4R}}$
$\mathcal{C}^\dagger$**S4F**	$\mathcal{C}$**PC**, $(sfcT)$, (T), $(S4.2)$	$(S4F)$, $(S4)$	$Sf\neg Sf\Box X^*_{\mathcal{C}\mathbf{S4R}}$

The $(S4F)$ rule is odd in that its left denominator is static whilst its right denominator is transitional. The $(S4.2)$ rule is the only potentially dangerous rule since its denominator contains a formula to which the rule can be applied in an endless fashion. To forbid this the new formula is marked with a star and the $(S4.2)$ rule is restricted to apply only to non-starred formulae. All other rules must treat starred formula as if they were non-starred.

The soundness and completeness of these calculi is proved in detail by Goré [Goré, 1991]. Goré actually proves soundness and completeness with respect to a class of finite frames, each of which is an **L**-frame as defined here. Consequently, these logics are also characterised by the classes of finite-**L**-frames shown in Figure 13. Note that the values of $X^*_{C\mathbf{L}}$ are different from those in [Goré, 1991] but it is easy to see that the new ones are the correct ones due to the effect of $(sfcT)$.

Tableau systems for the logics **K4.2** and **K4.2G** can be found in Amerbauer's dissertation [Amerbauer, 1993].

EXAMPLE 16. The formula $\Diamond\Box p \to \Box\Diamond p$ is an instance of the axiom 2, and hence is a theorem of **S4.2**. The following closed $C^\dagger\mathbf{S4.2}$-tableau for its negation $(\Diamond\Box p \wedge \neg\Box\Diamond p)$ which in primitive notation is $(\neg\Box\neg\Box p) \wedge (\neg\Box\neg\Box\neg p)$ illustrates the use of starred formulae.

$$(\neg\Box\neg\Box p) \wedge (\neg\Box\neg\Box\neg p)\ (\wedge)$$
$$\neg\Box\neg\Box p; \neg\Box\neg\Box\neg p\ (S4.2)$$

Left branch:

$$(S4)\ \neg\Box\neg\Box p; \neg\Box\neg\Box\neg p; \Box\neg\Box\neg\Box p$$
$$(\neg\neg)\ \neg\neg\Box\neg p; \Box\neg\Box\neg\Box p$$
$$(T)\ \Box\neg p; \Box\neg\Box\neg\Box p$$
$$(S4)\ \Box\neg p; \neg\Box\neg\Box p; \Box\neg\Box\neg\Box p$$
$$(\neg\neg)\ \Box\neg p; \neg\neg\Box p; \Box\neg\Box\neg\Box p$$
$$(T)\ \Box\neg p; \Box p; \Box\neg\Box\neg\Box p$$
$$(T)\ \neg p; \Box\neg p; \Box p; \Box\neg\Box\neg\Box p$$
$$(\bot)\ \neg p; \Box\neg p; p; \Box p; \Box\neg\Box\neg\Box p$$
$$\bot$$

Right branch:

$$\neg\Box\neg\Box p; \neg\Box\neg\Box\neg p; \Box(\neg\Box\neg\Box\neg\Box p)*\ (S4)$$
$$\neg\neg\Box p; \Box(\neg\Box\neg\Box\neg\Box p)*\ (\neg\neg)$$
$$\Box p; \Box(\neg\Box\neg\Box\neg\Box p)*\ (T)$$
$$\Box p; (\neg\Box\neg\Box\neg\Box p)*; \Box(\neg\Box\neg\Box\neg\Box p)*\ (S4)$$
$$\Box p; \neg\neg\Box\neg\Box p; \Box(\neg\Box\neg\Box\neg\Box p)*\ (\neg\neg)$$
$$\Box p; \Box\neg\Box p; \Box(\neg\Box\neg\Box\neg\Box p)*\ (T)$$
$$\Box p; \neg\Box p; \Box\neg\Box p; \Box(\neg\Box\neg\Box\neg\Box p)*\ (S4)$$
$$\Box p; \neg p; \Box\neg\Box p; \Box(\neg\Box\neg\Box\neg\Box p)*\ (T)$$
$$p; \Box p; \neg p; \Box\neg\Box p; \Box(\neg\Box\neg\Box\neg\Box p)*\ (\bot)$$
$$\bot$$

$$(G)\ \frac{\Box X; \neg\Box P}{X; \Box X; \neg P; \Box P} \qquad (Grz)\ \frac{\Box X; \neg\Box P}{X; \Box X; \neg P; \Box(P \to \Box P)}$$

Figure 9. Tableau Rules for logics of provability

4.16 Modal Logics with Provability Interpretations

In this section we give tableau calculi for the modal logics that have important readings as logics of 'provability' where $\Box A$ is read as 'it is provable in Peano Arithmetic that A holds'; see Fitting [1983, page 241] and Boolos [1979]. These systems are obtained either by adding the axiom G:$\Box(\Box A \to A) \to \Box A$, named after Gödel-Löb and sometimes called GL, or adding the axiom Grz:$\Box(\Box(A \to \Box A) \to A) \to A$, named after Grzegorczyk, or adding the axiom 4 and the axiom G_o:$\Box(\Box(A \to \Box A) \to A) \to \Box A$, to **K**.

It is known that both G and Grz imply the transitivity axiom 4 when they are respectively added to **K** [van Benthem and Blok, 1978]. But the logic $\mathbf{K4G_o}$ whose frames share some of the properties of **G**-frames and **Grz**-frames, explicitly contains 4 as an axiom. It is also known that Grz implies reflexivity.

Once again, all the tableau calculi contain the rules of $\mathcal{C}\mathbf{PC}$ and one or more logical rules from Figure 9 on page 344 as shown below:

$\mathcal{C}\mathbf{L}$	Static Rules	Transitional Rules	$X^*_{\mathcal{C}\mathbf{L}}$
$\mathcal{C}\mathbf{G}$	$\mathcal{C}\mathbf{PC}$	(G)	$\widetilde{X}$
$\mathcal{C}\mathbf{K4G_o}$	$\mathcal{C}\mathbf{PC}$	(Grz)	$Sf\Box(\widetilde{X} \to \Box\widetilde{X})$
$\mathcal{C}\mathbf{Grz}$	$\mathcal{C}\mathbf{PC}$, (T)	(Grz)	$Sf\Box(\widetilde{X} \to \Box\widetilde{X})$

The semantic and axiomatic intuitions behind these rules are more enlightening than any technical proof (of soundness) so we present these as well.

Intuitions for (G) **:** We know that axiomatically formulated logic **G** is characterised by **G**-frames. Therefore, axiom G must be valid on any **G**-frame; hence true in any world of any **G**-model. The axiom G is

$$\Box(\Box A \to A) \to \Box A.$$

Its contrapositive is

$$\neg\Box A \to \neg(\Box(\Box A \to A))$$

which is the same as

$$\neg\Box A \to \Diamond(\Box A \wedge \neg A).$$

Thus, if the numerator represents a world w where $\neg\Box P$ is true, then there exists another world w' where $\Box P$ is true and P is false, and w' is reachable from w.

The denominator represents this world.

Intuitions for (Grz) **:** The axiom Grz is

$$\Box(\Box(A \to \Box A) \to A) \to A.$$

It is known that 4 and T are theorems of **Grz** [Hughes and Cresswell, 1984, page 111], hence **S4** $\subseteq$ **Grz**. Segerberg [1971, page 107], and more recently Goré et al [1995], show that **Grz** = **S4Grz** = **S4G$_o$** where G_o is

$$\Box(\Box(A \to \Box A) \to A) \to \Box A$$

which gives the following (contraposed formulae) as theorems of **Grz**:

$$\begin{array}{lcl} \neg\Box A & \to & \neg\Box(\Box(A \to \Box A) \to A) \\ \neg\Box A & \to & \Diamond(\Box(A \to \Box A) \wedge \neg A). \end{array}$$

Thus, if $\neg\Box P$ is true at the numerator, then there exists some world where $\Box(P \to \Box P) \wedge \neg P$ eventually becomes true. The denominator of (Grz) represents this world.

THEOREM 17 (Soundness). *The calculi* $\mathcal{C}$**G**, $\mathcal{C}$**Grz** *and* $\mathcal{C}$**K4G$_o$** *are sound with respect to* **G**-*frames,* **Grz**-*frames and* **K4G$_o$**-*frames respectively.*

Proof Outline: For each rule in $\mathcal{C}$**L** we have to show that if the numerator of the rule is **L**-satisfiable then so is at least one of the denominators.

Proof of $\mathcal{C}$**G:** Suppose $\mathcal{M} = \langle W, R, V \rangle$ is a **G**-model, $w_0 \in W$ and $w_0 \models \Box X; \neg\Box P$. Thus there exists some $w_1 \in W$ with $w_0 R w_1$ and $w_1 \models X; \Box X; \neg P$ by the transitivity of R. Since R is irreflexive, $w_0 \neq w_1$. Suppose $w_1 \not\models \Box P$. Then $w_1 \models \neg\Box P$ and there exists some $w_2 \in W$ with $w_1 R w_2$ and $w_2 \models X; \Box X; \neg P$ by transitivity of R. Since R is irreflexive, $w_1 \neq w_2$. Since R is transitive, $w_2 = w_0$ would give $w_1 R w_0 R w_1$ implying $w_1 R w_1$ and contradicting the irreflexivity of R, hence $w_0 \neq w_2$. Suppose $w_2 \not\models \Box P$ then ... Continuing in this way, it is possible to obtain an infinite path of distinct worlds in $\mathcal{M}$ contradicting the **G**-frame condition on $\mathcal{M}$. Thus there must exist some $w_i \in W$ with $w_0 R w_i$ and $w_i \models X; \Box X; \neg P; \Box P$ and we are done.

Proof of (T) **for** $\mathcal{C}$**Grz:** The (T) rule is sound for **Grz**-frames since every **Grz**-frame is reflexive.

Proof of (Grz) **for** $\mathcal{C}$**K4G$_o$:** Suppose $\mathcal{M} = \langle W, R, V \rangle$ is a **K4G$_o$**-model, then R is transitive, there are no proper clusters, and there are no proper ∞-R-chains. Suppose $w_0 \in W$ is such that $w_0 \models \Box X; \neg\Box P$. We have to show that there exists some $w_n \in W$ with $w_0 R w_n$ and $w_n \models X; \Box X; \neg P; \Box(P \to \Box P)$. Since R is transitive, $w_0 \models \Box X$ means that $\forall w \in W, w_0 R w$ implies $w \models X; \Box X$. Thus our task is reduced to showing that there exists some $w_n \in W$ such that $w_0 R w_n$ and $w_n \models \neg P; \Box(P \to \Box P)$. Suppose for a contradiction that no such world exists in W. That is,

(a) $\forall w \in W, w_0 R w$ implies $w \not\models \neg P; \Box(P \to \Box P)$.

Since $w_0 \models \neg\Box P$, there exists some $w_1 \in W$ with $w_0 R w_1$ and $w_1 \models \neg P$. By (a), $w_1 \not\models \Box(P \to \Box P)$ and hence $w_1 \models \neg\Box(P \to \Box P)$. Thus there exists some $w_2 \in W$ with $w_1 R w_2$ and $w_2 \models \neg(P \to \Box P)$, that is, $w_2 \models P \wedge \neg\Box P$. Since $w_1 \models \neg P$, $w_1 \neq w_2$ and since **K4G$_o$**-models cannot contain proper clusters, $w_0 \neq w_2$. Since $w_2 \models \neg\Box P$ there exists some $w_3 \in W$ with $w_3 \models \neg P$. Since $w_2 \models P$, $w_3 \neq w_2$. And $w_3 \neq w_0$ and $w_3 \neq w_1$ as either would give a proper cluster. By (a), $w_3 \not\models \Box(P \to \Box P)$ and hence $w_3 \models \neg\Box(P \to \Box P)$. Continuing in this way, we either obtain an infinite path of distinct points, giving a proper ∞-R-chain, or we obtain a cycle, giving a proper cluster. Both are forbidden in **K4G$_o$**-frames. Hence (a) cannot hold and $\exists w \in W, w_0 R w$ and $w \models \neg P; \Box(P \to \Box P)$. That is, the desired w_n exists.

Proof of (Grz) **for** $\mathcal{C}$**Grz :** Every **Grz**-frame is a **K4G$_o$**-frame, hence the proof above suffices. ■

As we saw in Subsection 4.11, proving completeness boils down to proving the following: if X is a finite set of formulae and no $\mathcal{C}$**L**-tableau for X is closed then there is an **L**-model for X on an **L**-frame $\langle W, R\rangle$.

LEMMA 18. *If there is a closed* $\mathcal{C}$**L***-tableau for* X *then there is a closed* $\mathcal{C}$**L***-tableau for* X *with all nodes in the finite set* $X^*_{\mathcal{C}\mathbf{L}}$.

Proof. Obvious from the fact that all rules for $\mathcal{C}$**L** operate with subsets of $X^*_{\mathcal{C}\mathbf{L}}$ only. ■

LEMMA 19. *For each* $\mathcal{C}$**L***-consistent* X *there is an effective procedure to construct some finite* $\mathcal{C}$**L***-saturated* X^s *with* $X \subseteq X^s \subseteq X^*_{\mathcal{C}\mathbf{L}}$.

THEOREM 20 (Completeness). *If* X *is a finite set of formulae and* X *is* $\mathcal{C}$**L***-consistent then there is an* **L***-model for* X *on a finite* **L***-frame.*

As usual we construct some $\mathcal{C}$**L**-saturated w_0 from X with $X \subseteq w_0 \subseteq X^*_{\mathcal{C}\mathbf{L}}$.

Proof for $\mathcal{C}$**G:** If no $\neg\Box P$ occurs in w_0 then $\langle\{w_0\}, \emptyset\rangle$ is the desired model graph as (i)–(iii) are satisfied. Otherwise, let $Q_1, Q_2, \cdots, Q_m$ be all the formulae such that $\neg\Box Q_i \in w_0$. Create a $\mathcal{C}$**G**-saturated Q_i-successor for each Q_i using (θ) and (G) giving the nodes v_i of level one. Repeating this construction on the nodes of level one gives the nodes of level two, and so on for other levels. Consider any sequence $w_i \prec w_{i+1} \prec w_{i+2} \cdots$. Since w_i has a successor, there is some $\neg\Box Q \in w_i$ and $\Box Q \in w_{i+j}$ for all $j \geq 1$ by (G). Thus $w_i \neq w_{i+j}$ for any $j \geq 1$ and each such sequence must terminate since $X^*_{\mathcal{C}\mathbf{G}}$ is finite. Let R be the transitive closure of $\prec$; that is put wRw' if $w \prec w'$ and put wRv if $w \prec w' \prec v$. The resulting tree is a model graph $\langle W_0, R\rangle$ for X which is also a **G**-frame.

Proof for $\mathcal{C}$**Grz:** If no $\neg\Box P$ occurs in w_0 then $\langle\{w_0\}, \{(w_0, w_0)\}\rangle$ is the desired model graph as (i)–(iii) are satisfied. Otherwise, let $Q_1, Q_2, \cdots, Q_m$ be all the formulae such that $\neg\Box Q_i \in w_0$ and $\neg Q_i \notin w_0$. Create a $\mathcal{C}$**Grz**-saturated Q_i-successor for each Q_i using (θ) and (Grz) giving the nodes v_i of level one, and so on for other levels. Consider any sequence $w_i \prec w_{i+1} \prec w_{i+2} \cdots$. Since w_i has a successor, there is some Q such that $\neg\Box Q \in w_i$, $\neg Q \notin w_i$, and by (Grz), $\Box(Q \to \Box Q) \in w_{i+j}$ for all $j \geq 1$. Suppose $w_{i+j} = w_i$, then $\Box(Q \to \Box Q) \in w_i$ and hence $Q \to \Box Q \in w_i$ by (T). Since $Q \to \Box Q$ is just abbreviation for $\neg(Q \wedge \neg\Box Q)$, we know that $\neg Q \in w_i$ or $\neg\neg\Box Q \in w_i$. We created a successor w_{i+1} for w_i precisely because $\neg Q \notin w_i$ and so the first case is impossible. And if $\neg\neg\Box Q \in w_i$ then $\Box Q \in w_i$ by $(\neg)$, contradicting the **Grz**-consistency of w_i since $\neg\Box Q \in w_i$ by supposition. Thus each such sequence must terminate (without cycles). Let R be the reflexive and transitive closure of $\prec$ to obtain a model graph $\langle W_0, R\rangle$ for X which is also a **Grz**-frame.

Proof for $\mathcal{C}$K4G$_o$: If no $\neg\Box P$ occurs in w_0 then $\langle\{w_0\}, \{\emptyset\}\rangle$ is the desired model graph as (i)–(iii) are satisfied. Otherwise, let $Q_1, Q_2, \cdots, Q_m$ be all the formulae such that $\neg\Box Q_i \in w_0$. A $\mathcal{C}$**K4G$_o$**-saturated set v is reflexive iff $\Box A \in v$ implies $A \in v$. If v is non-reflexive then there exists some $\Box B \in v$ but $B \notin v$.

If w_0 is reflexive then create a $\mathcal{C}$**K4G$_o$**-saturated Q_i-successor for each $\neg\Box Q_i$ with $\neg Q_i \notin w_0$, otherwise if w_0 is non-reflexive then create a $\mathcal{C}$**K4G$_o$**-saturated Q_i-successor for each $\neg\Box Q_i$, $1 \leq i \leq m$. This gives the nodes of level one. Continue creating successors in this fashion for these nodes using (θ) and (Grz).

Consider any sequence $w_i \prec w_{i+1} \prec w_{i+2} \cdots$. Since w_i has a successor, there is some $\neg\Box Q \in w_i$ that gives rise to w_{i+1}. Also, $\Box(Q \to \Box Q) \in w_{i+j}$ for all $j \geq 1$.

If w_i is reflexive then $\neg Q \notin w_i$, and yet $\neg Q \in w_{i+1}$ by (Grz); hence $w_i \neq w_{i+1}$. Suppose $w_{i+j} = w_i$, $j \geq 2$. That $j \geq 2$ is crucial! Then $\Box(Q \to \Box Q) \in w_i$ and $Q \to \Box Q \in w_i$ by (Grz). Since $Q \to \Box Q$ is just abbreviation for $\neg(Q \wedge \neg\Box Q)$, we know that $\neg Q \in w_i$ or $\neg\neg\Box Q \in w_i$. Since w_i is reflexive, we created a successor w_{i+1} for w_i precisely because $\neg Q \notin w_i$ and so the first case is impossible. And if $\neg\neg\Box Q \in w_i$ then $\Box Q \in w_i$ by $(\neg)$, contradicting the **K4G$_o$**-consistency of w_i since $\neg\Box Q \in w_i$ by supposition.

If w_i is non-reflexive then there is some $\Box B \in w_i$, with $B \notin w_i$, and yet both $\Box B$ and B are in w_{i+j} by (Grz), for all $j \geq 1$; hence $w_i \neq w_{i+j}$, $j \geq 1$.

Thus each such sequence must terminate (without cycles). Let R be the transitive closure of $\prec$ and also put wRw if w is reflexive to obtain a model graph $\langle W_0, R\rangle$ for X which is also a **K4G$_o$**-frame.

As Amerbauer [1993] points out, this means that **K4G$_o$** is characterised by finite transitive trees of non-proper clusters refuting the conjecture of Goré [1992] that **K4G$_o$** is characterised by finite transitive trees of degenerate non-final clusters and simple final clusters.

Bibliographic Remarks and Related Systems

The tableau system $\mathcal{C}$**G** is from Fitting [1983] who attributes it to [Boolos, 1979], while $\mathcal{C}$**Grz** is from Rautenberg [1983]. Rautenberg gives a hint on how to extend these to handle $\mathcal{C}\mathbf{K4G_o}$ but Goré [1992] is unable to give an adequate system for $\mathcal{C}\mathbf{K4G_o}$, leaving it as further work. The given $\mathcal{C}\mathbf{K4G_o}$ is due to Martin Amerbauer [1993] who following suggestions of Rautenberg and Goré also gives systems for **KG.2** and **KGL** (which Amerbauer calls **K4.3G**).

Provability logics have also been studied using Gentzen systems, and appropriate cut-elimination proofs have been given by Avron [1984], Bellin [1985], Borga [1983], Borga and Gentilini [1986], Sambin and Valentini [1980; 1983; 1982], and Valentini [1983; 1986].

4.17 Monomodal Temporal Logics

In this section, which is based heavily on [Goré, 1994], we give tableau systems for normal modal logics with natural temporal interpretations where $\Box A$ is read as 'A is true always in the future' and $\Diamond A$ is read as 'A is true some time in the future'. All logics are 'monomodal' in that the reverse analogues of these operators, namely 'always in the past' and 'some time in the past', are not available. That is, the reachability relation R is taken to model the flow of time in a forward direction, and each possible world represents a point in this flow with some point deemed to be 'now'. We are allowed to look forwards but not backwards. In all cases time is taken to be transitive and the variations between the logics comes about depending on whether we view time as linear or branching; as dense or discrete; and as reflexive or non-reflexive (which is not the same as irreflexive). We explain these notions below.

Reflexive Monomodal Temporal Logics

The logics **S4.3**, **S4.3.1** and **S4Dbr** are all normal extensions of **S4** and are axiomatised by taking the appropriate formulae from Figure 1 as axiom schemas. Their respective axiomatisations are: **S4** is $KT4$; **S4.3** is $KT43$; **S4.3.1** is $KT43Dum$; and **S4Dbr** is $KT4Dbr$.

The Diodorean modal logics **S4.3** and **S4.3.1** have received much attention in the literature because of their interpretation as logics of dense and discrete *linear* time [Bull, 1965]. That is, it can be shown that $\langle \mathcal{I}, \leq \rangle \models A$ iff $\vdash_{\mathbf{S4.3}} A$ where $\mathcal{I}$ is either the set of real numbers or the set of rational numbers and $\leq$ is the usual (reflexive and transitive) ordering on numbers [Goldblatt, 1987, page57]. Consequently, between any two points there is always a third and **S4.3** is said to model **linear dense** time. It can be shown that $\langle \omega, \leq \rangle \models A$ iff $\vdash_{\mathbf{S4.3.1}} A$ where ω is the set of natural numbers [Goldblatt, 1987]. Hence, between any two points there is always a finite number (possibly none) of other points and **S4.3.1** is said to model **linear discrete** time. The formal correspondence between $\langle \mathcal{I}, \leq \rangle$ and **S4.3**-frames, and between $\langle \omega, \leq \rangle$ and **S4.3.1**-frames can be obtained by using a technique known as bulldozing and defining an appropriate mapping called a p-morphism [Goldblatt, 1987; Hughes and Cresswell, 1984].

The logics **S4** and **S4Dbr** can be given interpretations as logics of dense and discrete *branching* time. That is, it can be shown that **S4** is also characterised by the class of all reflexive transitive (and possibly infinite) trees [Hughes and Cresswell, 1984, page 120]. That is, by bulldozing each proper cluster of an **S4**-frame we can obtain an infinite dense sequence so that **S4** is the logic that models branching dense time. The axiomatic system **S4Dbr** is proposed by Zeman [1973, page 249] as the temporal logic for branching discrete time, but Zeman and Goré [1994] call this logic **S4.14**.

Therefore, the logics **S4**, **S4.3**, **S4.3.1** and **S4Dbr** cover the four possible combinations of discreteness and density paired with linearity and branching.

Figure 10 on page 350 shows the rules we need to add to $\mathcal{C}$**S4** in order to obtain tableau systems for **S4.3**, **S4.3.1** and **S4Dbr**. The tableau calculi $\mathcal{C}$**S4.3**, $\mathcal{C}$**S4.3.1** and $\mathcal{C}$**S4Dbr** are respectively the calculi for the logics **S4.3**, **S4.3.1** and **S4Dbr** as shown below:

$\mathcal{C}\mathbf{L}$	Static Rules	Transitional Rules	$X^*_{\mathcal{C}\mathbf{L}}$
$\mathcal{C}\mathbf{S4.3}$	$\mathcal{C}\mathbf{PC}, (T)$	$(S4.3)$	$\widetilde{X}$
$\mathcal{C}\mathbf{S4.3.1}$	$\mathcal{C}\mathbf{PC}, (T)$	$(S4), (S4.3.1)$	$Sf(\Box(\widetilde{X} \to \Box\widetilde{X}); \Box\widetilde{X})$
$\mathcal{C}\mathbf{S4Dbr}$	$\mathcal{C}\mathbf{PC}, (T)$	$(S4), (S4Dbr)$	$Sf(\Box(\widetilde{X} \to \Box\widetilde{X}); \Box\widetilde{X})$

Note that $\mathcal{C}$**S4.3** does not contain the rule $(S4)$ and that $\mathcal{C}$**S4.3.1** does not contain the rule $(S4.3)$ but does contain the rule $(S4)$. Also note that the $(S4.3.1)$ rule contains some static denominators and some transitional denominators.

LEMMA 21. *If there is a closed* $\mathcal{C}\mathbf{L}$ *tableau for the finite set* X *then there is a closed* $\mathcal{C}\mathbf{L}$ *tableau for* X *with all nodes in the finite set* $X^*_{\mathcal{C}\mathbf{L}}$.

Proof. Obvious from the fact that all rules for $\mathcal{C}\mathbf{L}$ operate with subsets of $X^*_{\mathcal{C}\mathbf{L}}$ only. ∎

LEMMA 22. *For each* $\mathcal{C}\mathbf{L}$*-consistent* X *there is an effective procedure to construct some finite* $\mathcal{C}\mathbf{L}$*-saturated* X^s *with* $X \subseteq X^s \subseteq X^*_{\mathcal{C}\mathbf{L}}$.

Proof. As on page 318. ∎

THEOREM 23. *The* $\mathcal{C}\mathbf{L}$ *rules are sound with respect to* $\mathbf{L}$*-frames.*

Proof. We omit details since the proofs can be found in [Goré, 1994], although note that there, the definition of **L**-frames is slightly different.

The intuition behind the $(S4.3)$ rule is based on a consequence of the characteristic **S4.3** axiom 3. Adding 3 to **S4** gives a weakly-connected R for **S4.3** so that eventualities can be weakly-ordered. If there are k eventualities, one of them must be fulfilled first. The $(S4.3)$ rule can be seen as a disjunctive choice between which one of the k eventualities is fulfilled first and an appropriate 'jump' to the corresponding world.

$$(S4Dbr)\quad \frac{\Box X; \neg\Box P}{\Box X; \Box\neg\Box P \mid \Box X; \neg P; \Box(P \to \Box P)}$$

$$(S4.3)\quad \frac{\Box X; \neg\Box\{P_1, \cdots, P_k\}}{\Box X; \neg\Box\overline{Y_1}; \neg P_1 \mid \cdots \mid \Box X; \neg\Box\overline{Y_k}; \neg P_k}$$

where $Y = \{P_1, \cdots, P_k\}$ and $\overline{Y_i} = Y \setminus \{P_i\}$

$$(S4.3.1)\quad \frac{U; \Box X; \neg\Box\{Q_1, \cdots, Q_k\}}{S_1 \mid S_2 \mid \cdots \mid S_k \mid S_{k+1} \mid S_{k+2} \mid \cdots \mid S_{2k}}$$

where

$Y = \{Q_1, \cdots, Q_k\}$;

$\overline{Y_j} = Y \setminus \{Q_j\}$;

$S_j = U; \Box X; \neg\Box\overline{Y_j}; \Box\neg\Box Q_j$

$S_{k+j} = \Box X; \neg Q_j; \Box(Q_j \to \Box Q_j); \neg\Box\overline{Y_j}$

for $1 \leq j \leq k$

Figure 10. Tableau rules ($S4Dbr$), ($S4.3$) and ($S4.3.1$)

The intuition behind the $(S4.3.1)$ rule is that each eventuality is either 'eternal', because it is fulfilled an infinite number of times in the sequence of worlds that constitute an **S4.3.1**-model, or 'non-eternal'. If the eventuality $\neg\Box P$ is 'eternal' then it can be stashed away (statically) as $\Box\neg\Box P$ and ignored until 'later'. Otherwise it must be dealt with immediately by fulfilling it via a transition. But there may be many such eventualities and since R is weakly-connected, they must be ordered.

THEOREM 24. *If X is a finite set of formulae and X is $\mathcal{C}$**L**-consistent then there is an **L**-model for X on a finite **L**-frame $\langle W, R\rangle$.*

Again we omit details since they can be found in [Goré, 1994] but note that there we used **S4.14** for **S4Dbr**. However, the proof for $\mathcal{C}$**S4.3** is reproduced below to give an idea of how to handle linearity.

Proof sketch for $\mathcal{C}$S4.3: The completeness proof of $\mathcal{C}$**S4.3** is similar to the completeness proof for $\mathcal{C}$**S4**. The differences are that only *one* sequence is constructed, and that in doing so, the $(S4.3)$ rule is used instead of the $(S4)$ rule. Note that the $(S4.3)$ rule guarantees only that *at least one* eventuality gives a $\mathcal{C}$**S4.3**-consistent successor whereas $(S4)$ guarantees that *every* eventuality gives a $\mathcal{C}$**S4**-consistent successor. And this crucial difference is why thinning seems essential. The basic idea is to follow one sequence, always attempting to choose a successor new to the sequence. Sooner or later, no such successor will be possible giving a sequence $S = w_0 \prec w_1 \prec w_2 \prec \cdots \prec w_m \prec w_{m+1} \prec \cdots \prec w_{n-1} \prec w_m$ containing a cycle $C = w_m \prec w_{m+1} \prec \cdots \prec w_{n-1} \prec w_m$ which we write pictorially as

$$S = w_0 \prec w_1 \prec w_2 \prec \cdots \prec \overline{w_m \prec w_{m+1} \cdots \prec w_{n-1}}.$$

The cycle C fulfills at least one of the eventualities in w_{n-1}, namely the $\neg\Box Q$ that gave the duplicated Q-successor w_m of w_{n-1}. But C may not fulfill *all* the eventualities in w_{n-1}.

Let $Y = \{P \mid \neg\Box P \in w_{n-1}$ and $\neg P \notin w_j,\ m \le j \le n-1\}$, so that $\neg\Box Y$ is the set of eventualities in w_{n-1} that remain unfulfilled by C. Let $w' = \{P \mid \Box P \in w_{n-1}\}$. Since $(\Box w'; \neg\Box Y) \subseteq w_{n-1}$ is $\mathcal{C}$**S4.3**-consistent by (θ), so is at least *one* of

$$X_j = \Box w' \cup \{\neg P_j\} \cup \neg\Box\overline{Y_j}, \text{ for } j = 1, \cdots, k$$

by $(S4.3)$. As before, choose the $\mathcal{C}$**S4.3**-consistent X_i that gives a **S4.3**-saturated P_i-successor for w_{n-1} which is new to S to sprout a continuation of the sequence, thus escaping out of the cycle. If no such new successor is possible then choose the successor $w_{m'}$ that appears earliest in S. This successor *must* precede w_m, as otherwise, C would already fulfill the eventuality that gives this successor. That is, we can extend C by putting $w_{n-1} \prec w_{m'}$. Recomputing Y using m' instead of m must decrease the size of Y since w_{n-1} has remained fixed. Repeating this procedure will eventually lead either to an empty Y or to a new successor. In the latter case we carry on the construction of S. In the former case we form a final cycle that fulfills all the eventualities of w_{n-1} and stop.

Sooner or later we must run out of new successors since $X^*_{C\mathbf{S4.3}}$ is finite and so only the former case is available to us. Let R be the reflexive and transitive closure of $\prec$ so that the overlapping clusters of $\prec$ become maximal disjoint clusters of R. It should be clear that $\langle W, R\rangle$ is a linear order of maximal, disjoint clusters that satisfies properties (i)–(iii), and hence that $\langle W, R\rangle$ is a model-graph for X.

Note that thinning seems essential. That is, in computing Y, we *have* to exclude the eventualities that are already fulfilled by the current cycle C in order to escape out of the cycle that they cause. We return to this point later. ■

Non-reflexive Monomodal Temporal Logics

The logics **S4.3** and **S4.3.1** respectively have counterparts called **K4DLX** and **K4DLZ** [Goldblatt, 1987] that omit reflexivity where the new axiom schemata are D, L, X, Z, and Zbr; see Figure 1 on page 299.

It is known that $\langle \mathcal{I}, <\rangle \models A$ iff $\vdash_{\mathbf{K4DLX}} A$ and $\langle \omega, <\rangle \models A$ iff $\vdash_{\mathbf{K4DLZ}} A$ where $\mathcal{I}$ is either the set of real numbers or the set of rational numbers and ω is the set of natural numbers [Goldblatt, 1987]. Hence these logics model transitive non-reflexive linear dense, and transitive non-reflexive linear discrete time respectively. I am not aware of a proof of completeness for the non-reflexive counterpart of **S4Dbr** but it seems reasonable to conjecture that **K4DZbr** is this counterpart.

The simplest way to handle the seriality axiom D is to use the static (D) rule of Rautenberg even though it breaks the subformula property. But (D) and $(K4Zbr)$ can conspire to give an infinite sequence of building up operations,[1] so we use the transitional $(KD4)$ and $(KD4L)$ rules instead; see Figure 11.

Another minor complication is the need for an explicit tableau rule to capture density (no consecutive degenerate clusters, see [Goldblatt, 1987]) for **K4DLX** but this is handled by the transitional rule $(K4DX)$, which is sound for **K4DLX**-frames.

The non-reflexive analogue of the $(S4.3)$ rule becomes very clumsy since it is based on the **K4LX**-theorem:

$$\Diamond P \wedge \Diamond Q \to \Diamond(P \wedge \Diamond Q) \vee \Diamond(Q \wedge \Diamond P) \vee \Diamond(P \vee Q)$$

and it is easier to use the rule $(K4L)$ which makes explicit use of subsets. The $(K4L)$ rule is similar to a rule given by Valentini [1986]. By using rules from Figure 11 it is possible to obtain cut-free tableau calculi possessing the analytic superformula property for these logics as:

[1] I missed this aspect in [Goré, 1994]

$$(K4D)\ \frac{\Box X; \neg\Box P}{X; \Box X; \neg P} \text{ where } \{\neg\Box P, \neg P\} \text{ may be empty}$$

$$(K4DX)\ \frac{\Box X; \neg\Box Y}{X; \Box X; \neg\Box Y} \text{ where } \neg\Box Y \text{ may be empty}$$

$$(K4Zbr)\ \frac{\Box X; \neg\Box P}{X; \Box X; \Box\neg\Box P \mid X; \Box X; \neg P; \Box P}$$

$$(K4L)\ \frac{\Box X; \neg\Box\{P_1, \cdots, P_k\}}{S_1 \mid S_2 \mid \cdots \mid S_m}$$

where $m = 2^k - 1, \quad 1 \leq i \leq m$;
$Y^1, \cdots, Y^m$ is an enumeration of the non-empty subsets of Y;
$\overline{Y^i} = Y \setminus Y^i$
$S_i = (X; \Box X; \neg\Box\overline{Y^i}; \neg Y^i)$

$$(K4LZ)\ \frac{U; \Box X; \neg\Box\{Q_1, \cdots, Q_k\}}{S_1 \mid S_2 \mid \cdots \mid S_k \mid S_{k+1} \mid S_{k+2} \mid \cdots \mid S_{k+m}}$$

where :
$Y = \{Q_1, \cdots, Q_k\}; m = 2^k - 1$;
$Y^1, \cdots, Y^m$ is an enumeration of the non-empty subsets of Y;
$\overline{Y_j} = Y \setminus \{Q_j\}$ for $1 \leq j \leq k$;
$\overline{Y^i} = Y \setminus Y^i$ for $1 \leq i \leq m$;
$S_j = U; \Box X; \neg\Box\overline{Y_j}; \Box\neg\Box Q_j$ for $1 \leq j \leq k$;
$S_{k+i} = X; \Box X; \neg Y^i; \Box Y^i; \neg\Box\overline{Y^i}$ for $1 \leq i \leq m$

Figure 11. Tableau rules for non-reflexive Diodorean logics

$\mathcal{C}\mathbf{L}$	Static Rules	Transitional Rules	$X^*_{\mathcal{C}\mathbf{L}}$
$\mathcal{C}\mathbf{K4DLX}$	$\mathcal{C}\mathbf{PC}$	$(K4DX), (K4L)$	$\widetilde{X}$
$\mathcal{C}\mathbf{K4DLZ}$	$\mathcal{C}\mathbf{PC}$	$(K4D), (K4LZ)$	$Sf\neg Sf\Box\widetilde{X}$
$\mathcal{C}\mathbf{K4DZbr}$	$\mathcal{C}\mathbf{PC}$	$(K4D), (K4Zbr)$	$Sf\neg Sf\Box\widetilde{X}$

First of all note that $(K4DX)$ is a transitional rule, not a static rule.

Now, it may appear as if the explicit subset notation would allow us to dispense with (θ) but this is not so. For (θ) allows us to *ignore* certain eventualities, whereas $(K4L)$ and $(K4LZ)$ only allow us to *delay* them. Thus using the reflexive analogues of these rules for **S4.3** and **S4.3.1** does not help to eliminate (θ).

The Saturation Lemma (Lemma 6 on page 318) will go through as for the other logics since the tableau systems have the analytic superformula property.

THEOREM 25. *The* $\mathcal{C}\mathbf{L}$ *rules are sound with respect to* $\mathbf{L}$*-frames.*

Proof. We omit details since the proofs are similar to the ones for the reflexive temporal logics and are not difficult.

THEOREM 26. *If X is a finite set of formulae and X is $\mathcal{C}\mathbf{L}$-consistent then there is an $\mathbf{L}$-model for X on a finite $\mathbf{L}$-frame $\langle W, R\rangle$.*

Again we omit details since they are similar to the proofs given in [Goré, 1994] but note that *there* we used the name Z_{14} for the axiom we here dub Zbr. However, the proof for $\mathcal{C}\mathbf{K4DLX}$ is reproduced below to give an idea of how to handle the density requirement.

Proof sketch for $\mathcal{C}\mathbf{K4DLX}$: The construction of the model graph is similar to the construction for $\mathcal{C}\mathbf{S4.3}$ except that we now know that every eventuality gives rise to two $\mathcal{C}\mathbf{K4DLX}$-consistent successors; one from $(K4DX)$ and at least one from $(K4L)$ (and (θ)). We again construct just one sequence but with the following twist.

A $\mathcal{C}\mathbf{K4DLX}$-saturated set v is reflexive iff $\Box A \in v$ implies $A \in v$. If v is non-reflexive then there exists some $\Box Q \in v$ but $Q \notin v$. If v is non-reflexive then create a successor v_1 for v using $(K4DX)$. If v_1 is non-reflexive then create a successor v_2 for v_1 using $(K4DX)$. Repeating this procedure must eventually give a $(K4DX)$-successor v_n that *is* reflexive. Note that $v \subseteq v_1 \subseteq v_2 \subseteq \cdots \subseteq v_n$ hence the sole purpose of v_n is to carry v *and* be reflexive; thus it need not fulfill any eventualities. Now discard $v_1, v_2, \cdots, v_{n-1}$ and put $v \prec v_n$.

So in the general $\mathcal{C}\mathbf{K4DLX}$ construction, if we are constructing a successor for w and w is reflexive then create a possibly non-reflexive $(K4L)$-successor, else create a reflexive $(K4DX)$-successor (like v_n) as shown above. In either case the sequence produced using $\prec$ satisfies the following criterion: there are no consecutive non-reflexive sets in the sequence.

Once again, this procedure may produce a cycle, and we may need thinning to escape from the cycle if it does not fulfill all its own eventualities as in the case for $\mathcal{C}\mathbf{S4.3}$. Nevertheless, eventually we will produce a sequence, possibly containing cycles, that fulfills all its eventualities, and furthermore that has no consecutive non-reflexive worlds in the sequence. Let R be the transitive closure of $\prec$ but also put wRw if w is reflexive. The resulting model is a finite reflexive and transitive linear sequence of R-clusters with no consecutive degenerate R-clusters. The density condition is met because if we have $w_1 R w_2$ then one of them must be reflexive, as otherwise they would form two consecutive degenerate R-clusters. Hence between any w_1 and w_2 we can always put a third world w which is a copy of the one that is reflexive.

The observation that we can detect reflexive worlds is due to Martin Amerbauer [1993].

A Note on **S4Dbr**

In a chapter on modal logic by Segerberg and Bull [1984, page 51], it is claimed that the logic **S4Dum** 'is characterised by the finite reflexive-and-transitive frames in which all but the final clusters are simple'. We show that this second claim is not correct by giving a finite reflexive-and-transitive model in which all but the final clusters are simple, but in which Dum is false. The model is pictured in Figure 12.

The explanation rests on the fact that $\Box(\Box(P \to \Box P) \to P)$ can be written as $\Box(\neg P \to \Diamond(P \land \Diamond\neg P))$. Thus Dum can be written as: $\Box(\neg P \to \Diamond(\neg P \land \Diamond P)) \land \Diamond\Box P \to P$.

This is just as well because we have just shown that **S4Dbr** characterises this class and Dum and *Dbr* are different. But note that the extra $\Box$ modality in *Dbr* is exactly what is needed since, in the counter-example of Figure 12, $w_0 \not\models \Box\Diamond\Box p$. That is, the counter-example does not falsify *Dbr* because the extra modality handles the branching inherent in **S4Dbr**-models which is absent in **S4.3.1**-models.

Related Work and Extensions

Zeman [1973] appears to have been the first to give a tableau system for **S4.3** but he is unable to extract the corresponding cut-free sequent system [Zeman, 1973, page 232]. Shimura [1991] has given a syntactic proof of cut-elimination for the corresponding sequent system for **S4.3**, whereas we give a semantic proof. Apparently, Serebriannikov has also obtained this system for **S4.3** but I have been unable to trace this paper. Rautenberg [1983] refers to 'a simple tableau' system for **S4.3** but does not give details since his main interest is in proving interpolation, and **S4.3** lacks interpolation. In subsequent personal communications I have been unable to ascertain the **S4.3** system to which Rautenberg refers [Rautenberg, 1990]. Bull [1985] states that *'Zeman's* **Modal Logic** *(XLII 581), gives tableau systems for* **S4.3** *and* **D** *in its Chapter 15, ...'*. The **D** mentioned by Bull is **S4.3.1** but Zeman [1973, page 245] merely shows that his tableau procedure for **S4.3** goes into unavoidable cycles when attempting to prove Dum. Zeman does not investigate remedies and consequently does *not* give a tableau system for **S4.3.1**. In fact, Bull [1965] mentions that Kripke used semantic tableau for **S4.3.1**, in 1963,

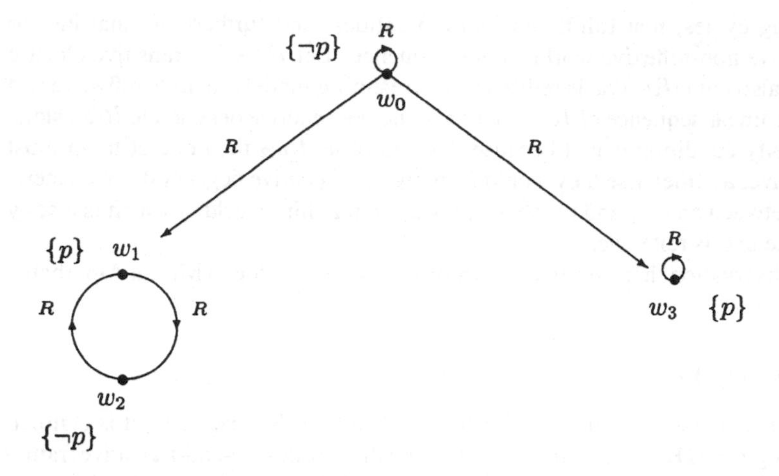

Dum can be written as: $\Box(\neg p \to \Diamond(p \wedge \Diamond\neg p)) \wedge \Diamond\Box p \to p$;

$w_0 \models \Diamond\Box p$ because $w_3 \models \Box p$;

$w_0 \models \neg p \to \Diamond(p \wedge \Diamond\neg p)$ because of w_1 and w_2;

$w_0 \models \Box(\neg p \to \Diamond(p \wedge \Diamond\neg p))$

but $w_0 \not\models p$.

Figure 12. A finite reflexive-and-transitive model in which all but the final clusters are simple in which Dum is false at w_0

but he gives no reference and subsequent texts that use semantic tableau do not mention this work [Zeman, 1973]. Presumably Kripke would have used tableaux where an explicit auxiliary relation is used to mimic the desired properties (like linearity) of R as is done in the semantic diagrams of Hughes and Cresswell [1968, page 290]. Note that no such explicit representation of R is required in our systems where the desired properties of R are obtained by appropriate tableau rules. Until very recently, I had thought that there were no other cut-free Gentzen systems for the logics S4.14 and S4.3.1 in the literature. Guram Bezhanishvili has recently informed me that Shimura [Shimura, 1992] has (independently) given almost identical systems for these logics. Shimura also gives interpolation theorems for these, and many other logics containing **S4**. At first sight, there seems to be some mismatch between the rules in [Shimura, 1992] and those given in [Goré, 1994], since the former are non-branching, while the latter are branching. However, the rules in [Shimura, 1992] contain an extra side-condition corresponding to an oracle, and also take a logic as a parameter. These side conditions correspond exactly to the missing second branches of the rules from [Goré, 1994] when we take into account that Shimura actually uses the logic **S5** as a parameter to his rules for **S4.14** and **S4.3.1**; see [Shimura, 1992, Lemma 2.6]. Thus, Shimura's oracles contain the nondeterminism inherent in these rules, and implicitly use an embedding of **S4** into **S5**.

Finally, these techniques extend easily to give a cut-free tableau system for **S4.3Grz** = **KGrz.3** [van Benthem and Blok, 1978] which is axiomatised as $KGrz.3$ where Grz is the Grzegorczyk axiom schema Grz:$\Box(\Box(A \to \Box A) \to A) \to A$. This logic is characterised by finite linear sequences of simple clusters but note that Shimura [1991] has already given a sequent system for this logic, and it is easy to turn his system into a tableau system.

The non-reflexive counterpart of **S4.3Grz** is **KLG** (sometimes called **G.3** or $\mathbf{GL_{lin}}$ or **K4.3W**) where L is as above and G is the Gödel-Löb axiom $\Box(\Box A \to A) \to \Box A$. Rautenberg [1983] shows that **KG** is characterised by the class of finite transitive trees of irreflexive worlds. Thus **KLG** is characterised by finite linear sequences of irreflexive worlds, but note that Valentini [1986] has already given a cut-free sequent system for this logic.

4.18 Eliminating Thinning

The structural rule (θ) corresponds to the sequent rule of weakening which explicitly enforces monotonicity; see page 312. From a theorem proving perspective, (θ) introduces a form of nondeterminism into each $C\mathbf{L}$ since we have to guess which formulae are really necessary for a proof. It is therefore desirable to eliminate (θ). There are two places where we resort to applications of (θ) in our completeness proofs. We consider each in turn.

The main applications of (θ) in our completeness proofs are the ones used to eliminate the formulae that do not match elements of the numerator, prior to an application of a transitional rule; see page 313. These applications of (θ) can be eliminated by building thinning in a deterministic way into the transitional rules. For example, we can change the $(S5)$ rule shown below left to the $(S5\theta)$ rule shown below right:

$$(S5)\ \frac{\Box X; \neg\Box Y; \neg\Box P}{\Box X; \neg\Box Y; \neg\Box P; \neg P} \qquad\qquad (S5\theta)\ \frac{X; \neg\Box P}{X'; \neg P}$$

where $X' = \{\Box A : \Box A \in X\} \cup \{\neg\Box B : \neg\Box B \in X\} \cup \{\neg\Box P\}$; see Fitting [Fitting, 1983]. The new transitional rule $(S5\theta)$ does the work of (θ) and $(S5)$. The crucial point is that we can specify X' exactly because we know exactly which formulae to throw away: namely, the ones that do not match the numerator of $(S5)$.

In some completeness proofs we also avoid creating a successor for $\neg\Box Q \in w$ if $\neg Q \in w$, thus pre-empting the reflexivity of R. This is *not* an application of (θ) when the transitional rule in question is non-branching like $(S4)$, because a consistent successor also exists for these eventualities, it is just that we are not interested in these successors.

However, (θ) appears essential for some of the branching transitional rules like $(S4.3)$, $(K4L)$ and $(S4.3.1)$ etc. even though we can also build thinning into these rules as well. For in the counter-model construction for $\mathcal{C}\mathbf{S4.3}$, we may reach a stage where all $\mathcal{C}\mathbf{S4.3}$-consistent successors already appear in S but no such cycle fulfills all the eventualities of the last node. At this stage it is essential to invoke applications of (θ) on subsets of the eventualities. That is, we must be able to *ignore* some of the eventualities in w_{n-1} using (θ) and this means that (θ) is now an essential rule of $\mathcal{C}\mathbf{S4.3}$.

The crucial difference between the branching transitional rules like $(S4.3)$ and the non-branching transitional rules like $(S4)$ is that the former guarantee only that *at least one* denominator is consistent, whereas the non-branching rules guarantee that every denominator is consistent (since they only have one denominator). But note that not all branching transitional rules are bad, for the $(S4Dbr)$ rule also branches, but the completeness proof (see [Goré, 1994]) goes through without recourse to (θ) because we can make a second pass of the initial model graph to obtain the desired frame.

It may be possible to eliminate thinning by using cleverer completeness proofs. For example, an alternate proof for $\mathcal{C}\mathbf{S4.3}$ may be possible by considering all $(S4.3)$-successors for every node, giving a tree of nondegenerate clusters, and then showing that any two worlds in this tree can be ordered as is done by Hughes and Cresswell [1984, page 30–31]. Note however that this seems to require a cut rule since Hughes and Cresswell use maximal consistent sets rather than saturated sets as we do.

Clearly the intuitions inherent in our semantic methods are no longer sufficient to prove that weakening is eliminable. We have obtained a syntactic proof of elimination of weakening in the sequent system containing the sequent analogues of the modified tableau rule $(S4.3\theta)$, but this is beyond the scope of this chapter.

4.19 Eliminating Contraction

As we have seen, contraction is built into our tableau rules by the ability to carry a copy of the principal formula into the denominator. But we believe it can be limited to the explicit contractions we have shown in our modal rules. Unfortunately, our set-based rules and completeness proofs are not sophisticated enough to *prove* this since (the saturation) Lemma 6 on page 318 requires that we copy the principal formula into the denominator. It is possible to rework all of our work using multisets instead of sets, but the proofs become very messy. For a more detailed study of contraction in modal tableau systems see the work of Hudelmaier [1994] and Miglioli *et al.* [1995].

4.20 Finite $\mathbf{L}$*-frames*

In all our completeness proofs we construct *finite* model graphs, hence our logics are also characterised by the *finite* frames shown in Figure 13. The frames in

Figure 13 are all based on trees of clusters or trees of worlds where we assume that clusters immediately imply transitivity. Consequently, each logic has the finite model property, and is decidable. These finer-grained results are not always obtainable when using other tableau methods.

4.21 Admissibility of Cut and Gentzen Systems

The cut rule is sound with respect to all our **L**-frames and each $\mathcal{C}\mathbf{L}$ is sound and complete with respect to the appropriate **L**-frames. Thus, putting (ρ) equal to (cut) in Lemma 2 (page 313) gives:

THEOREM 27. *The rule* (cut) *is admissible in each* $\mathcal{C}\mathbf{L}$.

Tableau systems are (upside down) cousins of proof systems called Gentzen systems or sequent systems; see Fitting [1983]. For example, the Gentzen system $\mathcal{G}K$ shown in Figure 14 is a proof system for modal logic **K**. That is, a formula A is valid in all **K**-frames (and hence a theorem of **K**) iff the sequent $\longrightarrow A$ is provable in $\mathcal{G}K$. Each of our tableau rules has a sequent analogue so it is possible to convert each tableau system $\mathcal{C}\mathbf{L}$ into a sequent system $\mathcal{G}L$. Then, $\mathcal{G}L$ is cut-free as long as $\mathcal{C}\mathbf{L}$ does not use (sfc) or $(sfcT)$. By induction it is straightforward to show that the sequent $X \longrightarrow Y$ is provable in $\mathcal{G}L$ iff there is a closed $\mathcal{C}\mathbf{L}$-tableau for $X; \neg Y$.

Our sequent systems do not possess all the elegant properties usually demanded of (Gentzen) sequent systems. For example, not only do some of our systems break the subformula property, but most do not possess separate rules for introducing modalities into the right and left sides of sequents.

Elegant modal sequent systems respecting these ideals of Gentzen have proved elusive although the very recent work of Avron [1994], Cerrato [1993], Masini [1992; 1991] and Wansing [1994] are attempts to redress this dearth. However, some of these methods have their own disadvantages. The systems of Cerrato enjoy the subformula property and separate introduction rules but do not enjoy cut-elimination in general (although the systems for **K** do so). The systems of Masini enjoy cut-elimination and give direct proofs of decidability but (currently) apply only to the logics **K** and **KD**. The systems of Wansing enjoy cut-elimination and clear introduction rules but do not immediately give decision procedures, and cannot handle logics like **S4.3.1** and **S4Dbr** [Kracht, 1996]. The hypersequents of Pottinger [1983] and Avron [1994] seem to retain most of the desired properties since they give cut-free systems with the subformula property for most of the basic modal logics including **S5**. It would be interesting to see if they can be extended to handle the Diodorean or provability logics.

5 TABLEAU SYSTEMS FOR MULTIMODAL TEMPORAL LOGICS

In this section we briefly survey tableau systems for multimodal temporal logics with future and past time connectives which have proved useful in Computer Sci-

L	finite-L-frames
K	finite intransitive tree of irreflexive worlds
T	finite intransitive tree of reflexive worlds
D	finite intransitive tree of worlds with reflexive final worlds
K4	finite tree of finite clusters
KDB	a single reflexive world; or a finite intransitive and symmetric tree of at least two worlds
K4D	finite tree of finite clusters with finite nondegenerate final clusters
K45	a single finite cluster; or a degenerate cluster followed by a finite nondegenerate cluster
K45D	a single finite nondegenerate cluster; or a degenerate cluster followed by a finite nondegenerate cluster
S4	finite tree of finite nondegenerate clusters
KB4	single finite cluster
S5	single finite nondegenerate cluster
B	finite symmetric tree of reflexive worlds
S4R S4.3Zem	a single finite nondegenerate cluster; or a simple cluster followed by a finite nondegenerate cluster
S4F	a sequence of at most two finite nondegenerate clusters
S4.2	a finite tree of finite nondegenerate clusters with one last cluster
S4.3	finite sequence of finite nondegenerate clusters
S4.3.1	finite sequence of finite nondegenerate clusters with no proper non-final clusters
S4Dbr	finite tree of finite nondegenerate clusters with no proper non-final clusters
K4L	finite sequence of finite clusters
K4DL	finite sequence of finite clusters with a nondegenerate final cluster
K4DLX	finite sequence of finite clusters with a nondegenerate final cluster, and no consecutive degenerate clusters
K4DLZ	finite sequence of degenerate clusters with a final simple cluster
K4DLZbr	finite tree of degenerate clusters with final simple clusters
G	finite transitive tree of irreflexive worlds
Grz S4Grz S4MDum	finite transitive tree of reflexive worlds
$K4G_o$	finite transitive tree of worlds
GL	finite transitive sequence of irreflexive worlds

Figure 13. Definition of finite-L-frames

$$X, P \longrightarrow P, Y \quad \text{(Ax)}$$

$$\frac{X, P, Q \longrightarrow Y}{X, P \wedge Q \longrightarrow Y} (\wedge \rightarrow) \qquad \frac{X \longrightarrow P, Y \quad X \longrightarrow Q, Y}{X \longrightarrow P \wedge Q, Y} (\rightarrow \wedge)$$

$$\frac{X \longrightarrow P, Y}{X, \neg P \longrightarrow Y} (\neg \rightarrow) \qquad \frac{X, P \longrightarrow Y}{X \longrightarrow \neg P, Y} (\rightarrow \neg)$$

$$\frac{X \longrightarrow Y}{X, U \longrightarrow V, Y} (\theta) \qquad \frac{X \longrightarrow P}{\Box X \longrightarrow \Box P} (\rightarrow \Box P : K)$$

Figure 14. Sequent rules for $\mathcal{G}K$

ence. The brevity is justified since the survey by Emerson [Emerson, 1990] covers tableau methods for these logics. Here we just try to show how these logics and their tableau methods relate to the methods we have seen so far.

In Computer Science the term 'temporal logic' is used to describe logics where the frames are discrete in the sense of **S4.3.1**-frames and **S4Dbr**-frames. The term 'linear temporal logic' is used when the frames are linear (discrete) sequences and the term 'branching temporal logic' is used when the frames are (discrete and) branching. If we wish to refer to the past then we can use a multimodal tense logic where $\blacksquare A$ is read as 'A is true at all points in the past' and $\blacklozenge A$ is read as 'A is true at some point in the past' [Burgess, 1984]. However, certain *binary* modal connectives have proved more useful.

The impetus for studying linear binary modal operators started with the seminal results of Kamp [Kamp, 1968]. Kamp showed that linear tense logic equipped with monomodal tense connectives like $\blacksquare$, $\blacklozenge$, $\Diamond$ and $\Box$ are 'expressively incomplete' because there are simple properties of linear orders that cannot be expressed using only these connectives together with the usual boolean connectives. One example is the property 'A is true now and remains true until B becomes true'. Kamp then showed that certain *binary* modal connectives are 'expressively complete' in that they capture *any* property expressible in the first-order theory of linear orders; that is, expressible using time point variables like t_1, t_2, the quantifiers $\forall$, $\exists$, the boolean connectives and the predicate $\leq$ familiar from number theory. Wolper then showed that even these connectives could not express all desirable properties of sequences [Wolper, 1983]; for example, properties that correspond to regular expressions from automata theory like 'A is true in every second state'. Wolper introduced extra connectives corresponding to regular expressions but these are beyond the scope of this article; see [Wolper, 1983].

5.1 Linear Temporal Logics

Syntax of Linear Temporal Logics

We add the unary modal connectives $\bullet$, $\circ$, $\blacklozenge$ and $\blacksquare$, and the binary modal connectives $\mathcal{U}$, $\mathcal{W}$, $\mathcal{S}$ and $\mathcal{Z}$. Any primitive proposition p is a formula, and if A and B are formulae, then so are: $(\neg A)$, $(A \wedge B)$, $(A \vee B)$, $(\Box A)$, $(\Diamond A)$, $(\blacksquare A)$, $(\blacklozenge A)$, $(\circ A)$, $(\bullet A)$, $(A\,\mathcal{U}B)$, $(A\,\mathcal{W}B)$, $(A\,\mathcal{S}B)$ and $(A\,\mathcal{Z}B)$.

Intuitively, $\circ A$ means 'A is true in the next state', $\bullet A$ means 'A is true in the previous state', $A\,\mathcal{U}B$ means 'A is true until B becomes true', and $A\,\mathcal{S}B$ means 'A has been true since B became true'. The others are explained shortly.

Semantics of Linear Temporal Logics

For brevity we concentrate on the linear temporal logic with future connectives only and dub it **PLTL** for propositional linear temporal logic, and follow Goldblatt [1987].

A **state sequence** is a pair $\langle S, \sigma\rangle$ where σ is a function from the natural numbers ω onto S enumerating the members of S as an infinite sequence $\sigma_0, \sigma_1, \cdots, \sigma_n \cdots$ (with repetitions when S is finite). A **model** $\mathcal{M} = \langle S, \sigma, V\rangle$ is a state sequence together with a valuation V that maps every primitive proposition onto a subset of S as usual. A model **satisfies** a formula at state σ_i according to:

$$(\mathcal{M}, \sigma_i) \models p \quad \text{iff} \quad \sigma_i \in V(p);$$

$$(\mathcal{M}, \sigma_i) \models \neg A \quad \text{iff} \quad (\mathcal{M}, \sigma_i) \not\models A;$$

$$(\mathcal{M}, \sigma_i) \models A \wedge B \quad \text{iff} \quad (\mathcal{M}, \sigma_i) \models A \text{ and } (\mathcal{M}, \sigma_i) \models B;$$

$$(\mathcal{M}, \sigma_i) \models A \vee B \quad \text{iff} \quad (\mathcal{M}, \sigma_i) \models A \text{ or } (\mathcal{M}, \sigma_i) \models B;$$

$$(\mathcal{M}, \sigma_i) \models \circ A \quad \text{iff} \quad (\mathcal{M}, \sigma_{i+1}) \models A;$$

$$(\mathcal{M}, \sigma_i) \models \Box A \quad \text{iff} \quad \forall j, j \geq i, (\mathcal{M}, \sigma_j) \models A;$$

$$(\mathcal{M}, \sigma_i) \models \Diamond A \quad \text{iff} \quad \exists j, j \geq i, (\mathcal{M}, \sigma_j) \models A;$$

$$(\mathcal{M}, \sigma_i) \models A\,\mathcal{U}B \quad \text{iff} \quad \exists k, k \geq i, (\mathcal{M}, \sigma_k) \models B \text{ and } \forall j, i \leq j < k, (\mathcal{M}, \sigma_j) \models A;$$

$$(\mathcal{M}, \sigma_i) \models A\,\mathcal{W}B \quad \text{iff} \quad (\mathcal{M}, \sigma_i) \models A\,\mathcal{U}B \text{ or } (\mathcal{M}, \sigma_i) \models \Box A.$$

Intuitively imagine the states $\sigma_0, \sigma_1, \cdots$ to form an infinite sequence where

$\sigma_i R \sigma_j$ iff $j = i + 1$ and R is functional. Now if we let $\leq$ be the reflexive and transitive closure of R, then $\Box$ is interpreted using $\leq$ while $\bigcirc$ is interpreted using R. For example, the formula $\bigcirc A$ is true at some state σ_i if A is true at *the* successor state σ_{i+1}. Note that the clause for $A \mathcal{U} B$ demands that there is some future state σ_k where B becomes true but does not specify a value for A at this state. A weaker version of $\mathcal{U}$ called $\mathcal{W}$ (for weak until) drops the first demand by allowing for the possibility that there is no future state where B is true as long as $\Box A$ is true at σ_i.

Note that we could also obtain $\Box$ and $\Diamond$ by defining $\Box A$ as $A \mathcal{W} \bot$ and $\Diamond A$ as $\top \mathcal{U} A$, and still maintain that $\Box A$ is $\neg\Diamond\neg A$.

If we wish to allow reasoning about the past we can also allow backward looking operators. The function σ must now map the set of integers onto S. Some care is needed to ensure the correct behaviour of the definitions below if time does not extend ad infinitum in the past [Fisher, 1991]:

$$(\mathcal{M}, \sigma_i) \models \bullet A \quad \text{iff} \quad (\mathcal{M}, \sigma_{i-1}) \models A;$$

$$(\mathcal{M}, \sigma_i) \models \blacksquare A \quad \text{iff} \quad \forall j, j \leq i, (\mathcal{M}, \sigma_j) \models A;$$

$$(\mathcal{M}, \sigma_i) \models \blacklozenge A \quad \text{iff} \quad \exists j, j \leq i, (\mathcal{M}, \sigma_j) \models A;$$

$$(\mathcal{M}, \sigma_i) \models A \mathcal{S} B \quad \text{iff} \quad \exists k, k \leq i, (\mathcal{M}, \sigma_k) \models B \text{ and } \forall j, k \leq j < i, (\mathcal{M}, \sigma_j) \models A;$$

$$(\mathcal{M}, \sigma_i) \models A \mathcal{Z} B \quad \text{iff} \quad (\mathcal{M}, \sigma_i) \models A \mathcal{S} B \text{ or } (\mathcal{M}, \sigma_i) \models \blacksquare A.$$

Axiomatisations

A Hilbert system for **PLTL** taken from Goldblatt [1987] is given below:

$$\begin{aligned}
&K : \Box(A \to B) \to (\Box A \to \Box B) \\
&K_o : \bigcirc(A \to B) \to (\bigcirc A \to \bigcirc B) \\
&Fun : \bigcirc\neg A \leftrightarrow \neg\bigcirc A \\
&Mix : \Box A \to (A \land \bigcirc\Box A) \\
&Ind : \Box(A \to \bigcirc A) \to (A \to \Box A) \\
&U1 : A \mathcal{U} B \to \Diamond B \\
&U2 : A \mathcal{U} B \leftrightarrow B \lor (A \land \bigcirc(A \mathcal{U} B))
\end{aligned}$$

We also need the inference rules of universal substitution **US**, modus ponens **MP** and an extended rule of necessitation **RN** viz: if $A \in \mathbf{L}$ then both $\Box A \in \mathbf{L}$ and $\bigcirc A \in \mathbf{L}$; see page 299.

The recursive nature of the Mix and $U2$ axioms gives rise to a fix-point characterisation of these operators which is the key to the tableau procedures for these logics; see [Wolper, 1983; Banieqbal and Barringer, 1987]. Notice also that the axiom Ind encodes an induction principle: if it is always the case that A being true now implies A is true in the next state, then A being true now implies A is true always in the future. It is this property that makes Gentzen systems for these logics difficult to obtain; see Section 5.1.

Finite Model Property, Decidability and Complexity

Wolper [1983] shows that although our models are infinite state sequences, linear temporal logic is also characterised by a class of finite frames. In fact, it is characterised by our finite-**S4.3.1**-frames; see [Goldblatt, 1987]. A tableau procedure is given by Wolper where he also shows that the problem of deciding satisfiability in **PLTL** is PSPACE-complete. Further complexity results for linear and branching time logics can be found in [Emerson and Sistla, 1984; Sistla and Clarke, 1985]. Decidability and incompleteness results for first-order linear temporal logics have been studied by Merz [1992].

Tableau Systems

Tableau systems for the fragment of linear temporal logic containing only future connectives have been studied by Wolper. He gives a tableau-based decision procedure for this logic, and extensions involving regular operators; see Wolper [1983; 1985].

The linear temporal logic including both future and past modalities has been extensively studied by Gough [1984]. Gough uses (the appropriately defined analogues of downward saturated) Hintikka sets to build a model graph for a given formula of this logic. A second phase then prunes nodes from this model graph to check that all eventualities can be fulfilled on a linear sequence. If this is not possible then the graph is pruned by removing the nodes that contain unfulfillable eventualities. If the initial node is removed by this pruning procedure then the initial formula is unsatisfiable on a linear model hence its negation is a theorem of this logic. The procedure has been automated and the resulting prover called dp is available by anonymous ftp from Graham Gough (gdg@cs.man.ac.uk) at the University of Manchester, England.

A system for temporal logic has also been implemented in the MGTP theorem prover by Koshimura and Hasegawa [1994].

The (informal) gist of any tableau procedure for linear temporal logics involving next-time modalities is to use the fix-point nature of the modalities to create a cyclic graph of (state) nodes. This graph is then pruned by deleting nodes that contain unfulfillable eventualities. For example, the following logical equivalences hold in linear temporal logic:

$$(A\,\mathcal{U}B) \equiv (B \vee \bigcirc(A\,\mathcal{U}B)) \qquad \Diamond B \equiv (\top\,\mathcal{U}B)$$
$$(A\,\mathcal{W}B) \equiv (A\,\mathcal{U}B) \vee \Box A \qquad \Box A \equiv (A\,\mathcal{W}\bot)$$

Suppose we are given an initial node n node containing a set of formulae X. For every formula in n that is an instance of the left hand of the above equivalences, we can add the appropriate instance of the right-hand side formula and mark the left hand instance as 'processed'. We can use the usual boolean rules for $\neg\neg$ and $\wedge$ to saturate this node by adding the appropriate subformulae to node n, again marking all parent formulae as 'processed'. For $\vee$ we put one disjunct in n and create a copy of the old n containing the other disjunct giving a branch in the tableau. Repeating this process on the new formulae means that n contains 'processed' formulae and unprocessed formulae. But all unprocessed formulae begin with $\bigcirc$ since these are the only formulae not touched by the above procedure. That is, all unprocessed formulae are in outermost-$\bigcirc$-form. For each node x we then create a successor node y and fill it with $\{A : \bigcirc A \in x\}$. Repeating this procedure on such successors produces a graph because the number of different formulae that can be generated from this process is finite, hence some nodes reappear. Note that we now allow arbitrary cycles whereas in the completeness proofs of Section 4 we confined cycles to nodes on the same branch. Some of these nodes contain eventualities like $\Diamond B$ or $A\,\mathcal{U}B$ since each of these demands the existence of some node that fulfills B. Now we make a second pass and delete nodes that contain both P and $\neg P$ for some formula; delete any node s whose eventualities cannot *all* be jointly fulfilled by some linear path through the graph beginning at s; and delete any nodes without successors. If the initial node ever gets deleted by this procedure then it can be shown that the initial set of formulae cannot be satisfied on a linear discrete model Wolper [1983; 1985]. Otherwise there will be a linear sequence of nodes that satisfies all the formulae in the initial node, thus demonstrating a linear discrete model for X.

Gentzen Systems

Gentzen systems for temporal logics have been given by various authors but almost all require either a cut rule or an infinitary rule for completeness Kawai [1987; 1988]. The exceptions appear to be the work of Gudzhinskas [1982] and Pliuskevicius [1991] but these articles are extremely difficult to read.

5.2 *Branching Temporal Logics*

Just as **S4Dbr** and **S4.3.1** are branching and linear respectively, there are branching analogues of the linear temporal logics we have seen using $\bigcirc$, $\mathcal{U}$ and even $\mathcal{S}$. We briefly cover the syntax and semantics of one of the most powerful of these branching time logics called **CTL***, and point to the abundant literature for tableau methods for these logics.

Syntax of Branching Temporal Logics

We again concentrate on the future fragment only and follow Emerson and Srinivasan [1988] using new modal connectives E and X in addition to $\mathcal{U}$. The syntax of branching time logics is given in terms of 'state' formulae and 'path' formulae where 'state' formulae are true or false at some state (world) and where 'path' formulae are true or false of (rather than on) a linear sequence of states (worlds). More formally:

any atomic formula p is a *state* formula;

if P and Q are *state* formulae then so are $P \wedge Q$ and $\neg P$;

if P is a *path* formula then EP is a *state* formula;

any *state* formula P is also a *path* formula ;

if P and Q are *path* formulae then so are $P \wedge Q$ and $\neg P$;

if P and Q are *path* formulae then so are XP and $(P \mathcal{U} Q)$;

The other boolean connectives are introduced in the usual way while AP abbreviates $\neg E \neg P$, and FP abbreviates $\top \mathcal{U} P$, and GP abbreviates $\neg F \neg P$. Note the absence of ○, ●, □, ◇, ■, and ◆.

Semantics of Branching Temporal Logics

The semantics of **CTL*** are again in terms of a Kripke structure $\mathcal{M} = \langle S, R, L \rangle$ where S is a non-empty set of states or worlds; R is a binary relation on S such that each state has at least one successor; and L is a function which assigns to each state a set of atomic propositions (those that are intended to be true at that state). Note that L is a slight variation on our usual V since the latter assigns atomic propositions to sets of worlds, but the two are equivalent in our classical two-valued setting.

A **fullpath** $x = s_0, s_1, s_2, \ldots$ in $\mathcal{M}$ is an infinite sequence of states such that $s_i R s_{i+1}$ for each i, $i \geq 0$. By $(\mathcal{M}, s) \models P$ and $(\mathcal{M}, x) \models P$ we mean that the state formula P is true at state s in model $\mathcal{M}$, and the path formula P is true of the path x in model $\mathcal{M}$, respectively. If $\mathcal{M}$ is understood then we just write $s \models P$ or $x \models P$. The formal definition of $\models$ is as below where s is an arbitrary state of some $\mathcal{M}$, where $x = s_0, s_1, s_2, \ldots$ is a fullpath in $\mathcal{M}$, and where x^i denotes the suffix fullpath $s_i, s_{i+1}, s_{i+2} \ldots$ of x:

$s \models p$ iff $p \in L(s)$;

$s \models P \wedge Q$ iff $s \models P$ and $s \models Q$;

$s \models \neg P$ iff $s \not\models P$;

$s \models EP$ iff for some fullpath y starting at s, $y \models P$;

$x \models P$ iff $s_0 \models P$ for any state formula P;

$x \models P \wedge Q$ iff $x \models P$ and $x \models Q$;

$x \models \neg P$ iff $x \not\models P$;

$x \models XP$ iff $x^1 \models P$;

$x \models P\,\mathcal{U}Q$ iff $\exists i \geq 0, x^i \models Q$ and $\forall j, 0 \leq j < i, x^i \models P$.

These definitions are enough to give a semantics for the modalities obtained via definitions: AP is true at state s if P is true of all paths beginning at s; FP is true of a path x if P is true of some suffix fullpath x^i $(i \geq 0)$ of x; and GP is true of a path x if P is true of all suffix fullpaths x^i, $(i \geq 0)$.

The notions of satisfiability and validity are the same as before for state formulae. A path formula P is **satisfiable** if there is some model $\mathcal{M}$ containing some path x such that $x \models P$, and is **valid** if for every model $\mathcal{M}$ and every fullpath path x in $\mathcal{M}$ we have $x \models P$.

As Emerson and Srinivasan note, a menagerie of branching time temporal logics can be obtained by restricting or extending these definitions [Emerson and Srinivasan, 1988].

Note that path formulae cannot be evaluated at states since there are no clauses in the definition of $\models$ for evaluating XP or $P\,\mathcal{U}Q$ at a state. But a state formula P can be evaluated on a fullpath simply by checking if the first state of the fullpath satisfies P. Hence, if P is a state formula, then formula XP cannot be evaluated at some state s, it must be evaluated with respect to a path x. Once a path is chosen however, it is just the same as the linear time formula $\bigcirc P$ since XP is true on path x if the second state of the path satisfies (state formula) P. Similarly, if P is a state formula, then FP and GP are just $\Diamond P$ and $\Box P$, except that they are evaluated over a *linear* sequence. But note that $\mathbf{CTL}^*$ is strictly more expressive than $\mathbf{PLTL}$.

Tableau Systems

The logic $\mathbf{CTL}^*$ is known to have the finite model property, in fact, it is characterised by finite-$\mathbf{S4Dbr}$-frames, but once again, note the presence of the extra modalities. Emerson and Srinivasan [1988] compare the expressiveness of various such branching time logics. Tableau methods for branching time logics can be found in Emerson [1985]. Once again, these tableau methods are based on the appropriate analogues of Hintikka-structures (see [Emerson and Halpern, 1985]) and use the following logical equivalences to expand formulae that match the left hand sides into an 'outermost-EX-normal' or 'outermost-AX-normal' form [Emerson and Srinivasan, 1988]:

$$E(P \vee Q) \equiv EP \vee EQ \qquad A(P \wedge Q) \equiv AP \wedge AQ$$
$$EGP \equiv P \wedge EXEGP \qquad AGP \equiv P \wedge AXAGP$$
$$EFP \equiv P \vee EXEFP \qquad AFP \equiv P \vee AXAFP$$
$$E(P\mathcal{U}Q) \equiv Q \vee (P \wedge EXE(P\mathcal{U}Q))$$
$$A(P\mathcal{U}Q) \equiv Q \vee (P \wedge AXA(P\mathcal{U}Q))$$

There are no clauses to 'expand' formulae beginning with EX or AX in some given set w. Each formula of the form EXP_i gives us reason to create a successor state w_i containing P_i just as $\neg\Box P \in w$ gave rise to a successor v containing $\neg P$ in the modal tableau completeness proofs of Section 4. Now any formula of the form $AXQ \in w$ allows us to put Q into each next state since XQ must be true of all paths that begin at w.

Again the procedure gives a cyclic graph since only a finite number of different sets can be built in this manner and some set reappears. Again, we form arbitrary graphs, not cyclic trees. And once again, a second phase prunes nodes that are inconsistent, that contain unfulfillable eventualities, or have no successor.

Note that all these tableau methods break the subformula property in a weak way since they introduce superformulae of the form $\bigcirc P$ or AXP or EXP where P is built from subformulae of the initial set. But we never apply an expansion rule to these superformulae, thus, EX and AXP act like 'wrappers' to keep this building up procedure in check, just as $\bigcirc$ acted as a wrapper for the **PLTL** procedure. These 'wrappers' are removed by creating successor state(s) and filling these with the 'unwrapped' (sub)formulae.

Gentzen Systems

Once we start to use graphs rather than trees, the connection with Gentzen systems becomes very tenuous. Gentzen systems for some branching time logics (without A and E) have been studied by Paech [1988]. Unfortunately, these systems require a (partly hidden) cut rule which means that they are not the proof-theoretic analogues of the tableau procedures mentioned above.

5.3 Bibliographic Remarks and Related Systems

The complexity of the decision problem for branching time logics has been studied by Emerson and Sistla [1983]. Temporal logics are known to be related to Büchi automata [Vardi and Wolper, 1986] and so their decision problems can be studied from an automata-theoretic perspective as well [Muller *et al.*, 1988].

All these branching time logics exclude past-time operators, but they can be added. The work of Gabbay [1987] and Gabbay *et al.* [1994] is particularly interesting because many temporal logics have the 'separation property': that is, any complicated formula A has a *logically equivalent* form A' where A' is a conjunction $B \wedge C \wedge D$ such that B involves only past-time modalities, C involves no modalities, and D involves only future-time modalities. Thus the decision problems for these logics can often be handled by separate routines for just past-time modalities, just future-time modalities and just pure propositional reasoning.

6 MODAL TABLEAU SYSTEMS WITH EXPLICIT ACCESSIBILITY

We now turn to tableau systems where the reachability relation R is represented explicitly. There are essentially two ways to represent R. One is to maintain a network of named nodes, where each node contains a set of formulae, and also maintain a separate relation $R(x, y)$ to represent that the node named y is reachable from the node named x. The names x and y are merely indices to allow cross-reference between these two 'data-structures'. The second is to incorporate *complex or structured* world names into the syntax, attaching the label l_1 to every formula that belongs to the world named l_1 and attaching l_2 to every formula that belong to the world named l_2. No separate reachability relation is kept since the reachability relation is built into the *structure* of the labels.

6.1 *History of Explicit Tableau Systems*

The most celebrated work is of course that of Kripke [1959] where possible worlds related by an accessibility relation are first proposed as a semantics for modal logics. Bull and Segerberg [1984] give an account of the genesis of the possible worlds approach and suggest that credit is also due to Hintikka and Kanger. Zeman [1973] even credits C. S. Pierce with the idea of 'a book of possible worlds' as far back as 1911!

Kripke follows Beth [1955] and divides each tableau into a left hand side and a right-hand side where the left side is for formulae that must be assigned 'true' and the right side is for formula that must be assigned 'false'; see [Fitting, 1993] for examples using this style of tableau. Thus it is clear that this is a refutation procedure and we are attempting to obtain a falsifying model of possible worlds for the given formula. To handle the added complexities of modal formulae like $\Box A$ and $\neg\Box A$, Kripke uses auxiliary tableaux, where a new tableau is used for each possible world and these auxiliary tableaux are interrelated by an auxiliary reachability relation R. Auxiliary tableaux may have tableaux auxiliary to them and so on, obtaining a complex web of tableaux.

Kripke uses two basic rules to handle modal formulae: one to handle $\Box A$ on the left of a tableau and one to handle $\Box A$ on the right of a tableau. They are,

Yl: If $\Box A$ appears on the left of a tableau t, then for every tableau t' such that tRt', put A on the left of t';

Yr: If $\Box A$ appears on the right of a tableau t, then start out a new tableau t', with A on the right, and such that tRt'.

Different constraints on this auxiliary relation give different tableau systems. That is, the definition of the auxiliary relation R changes with each logic, so that the auxiliary relation directly mimics the required accessibility relation. For example, the auxiliary relation R for **S4** is defined to be reflexive and transitive, so for any tableau t we have tRt *by definition*. These constraints form an extra theory about R that must be taken into account at each rule application.

Note also that the application of the Yl rule can have delayed consequences. For example, if a new auxiliary tableau t' is created and it happens to be auxiliary to the tableau t in which the Yl rule has already been applied, then we have to keep track of this previous application of Yl and add A to the left of t'. Thus, the meaning of 'every tableau t' such that tRt'' includes tableaux that may come into existence via the Yr rule at any later point of the construction. The rules are therefore like constraints that may be activated at a later time.

This is essentially a way to keep track of all worlds in the counter model being sought. When a new world comes into existence, it is immediately linked into this counter-model according to the constraints on R. That is, Kripke's method is a refutation procedure where extra modal information is kept in the auxiliary relation between tableaux. The construction is on a global level in that we can return to previous nodes of the tableau construction at will. In our tableau systems $\mathcal{C}\mathbf{L}$ we cannot return to nodes higher up in the tree.

The semantic diagrams of Hughes and Cresswell [1968] and the tableau systems of Zeman [1973] use essentially the same ideas except that Hughes and Cresswell use annotations of ones and zeros instead of using a left and right side. Slaght [1977] goes one step further than usual and adds rules for quantifiers and also incorporates a form of negated normal form by translating $\neg\Box P$ into $\Diamond\neg P$, $\neg\Diamond P$ into $\Box\neg P$, $\neg\exists x(\cdots)$ into $\forall x\neg(\cdots)$ and $\neg\Box x(\cdots)$ into $\exists x\neg(\cdots)$.

These ideas have been implemented by Catach in his TABLEAUX theorem prover [Catach, 1991; Catach, 1988]. Although labels are used in the TABLEAUX prover, they are used only as indices into an *explicit and separate* representation of the reachability relation. Indeed, Catach even laments the lack of modularity in this method [Catach, 1991, page 503].

Kanger's spotted formulae [Kanger, 1957], which precede Kripke's work, are the precursors of the second explicit approach which we call the labelled tableau method. In this method, each formula is prefixed with a label to retain its modal context and the reachability relation is encoded in the structure of the labels. Given two labels we can tell whether they are related by the reachability relation simply by inspecting their structure. Fitting's prefixed tableaux are direct applications of Kanger's idea to handle many different modal logics [Fitting, 1983, chapter 8]. And as we shall soon see, Massacci [1994; 1995a] has refined these ideas even further to give modular prefixed tableau systems for many modal logics. If we permit labels to contain variables then specialised 'string unification' methods can be used to detect closed tableau branches as is done by Wallen [1989], and Artosi and Governatori [1994]. The principle of using labels to 'bring some of the semantics into the syntax' is also the basis of Gabbay's Labelled Deductive

Systems [Gabbay, 1997].

Prefixes are one way to separate the modal component from the classical component. Another is to explicitly translate the modalities into a restricted subset of first-order logic. Specialised routines for first-order deduction, like resolution, can then be applied to this restricted subset. Such 'translation methods' have been investigated by Morgan [1976], Ohlbach [1990; 1993], Auffray and Enjalbert [1989], Frisch and Scherl [1991], and Gent [1991a; 1993; 1991].

In all these translational methods, the modal logics **K**, **T**, **K4**, **S4** and **S5** are easily handled and Gent has also obtained systems for **B** and **S4.3**. The most striking feature of Gent's work is that he is unable to give a system for **S4.3.1** and this is essentially due to the fact that the reachability relation R for **S4.3.1**-frames is not first-order definable. It is known that a formula of second order logic is required to express the reachability relation for **S4.3.1** [van Benthem, 1983]. This deficiency of translational methods is also mentioned by Auffray and Enjalbert [1989] while the method of Frisch and Scherl [1991] is limited to serial logics.

The biggest disadvantage of the translational methods is that first-order logic is known to be only semi-decidable, thus the translated system may not be decidable even though the original modal logic is decidable. Clearly it must be possible to identify decidable classes of first-order logic into which these translations will fall, but I am not aware of any such detailed investigations.

In all fairness, it must be mentioned that the translational methods seem to be much better for automated deduction in *first-order modal logics* where various domain restrictions can complicate matters for the first-order versions of our implicit tableau systems $\mathcal{C}$**L**; see [Ohlbach, 1990]. At the first-order level, all modal logics are only semi-decidable since they all include classical first-order logic. Then, decidability is no longer an important issue.

There is a subtle but deep significance to the use of labels which explains their increased power over implicit tableau methods. Our implicit tableaux were *local* in that, at all times, we worked with a set of formulae (denoting one particular world), with no explicit reference to the particular properties of the reachability relation since these properties were built into the rules. Labelled tableaux are *global* in that the labels allow us to 'see' the reachability relation and hence allows us to keep a picture of the whole model under construction.

6.2 Labelled Tableau Systems Without Unification

As stated previously the idea of labelled tableau systems goes back to at least Kripke and Kanger. The most attractive feature of labels for modal tableau systems is the ability to handle the symmetric logics like **S5** which require some form of analytic cut rule in the implicit systems we have studied so far, and also logics like **KB** for which I know no implicit tableau system formulation. We now review in some detail recent work of Massacci [1994] which gives simple labelled tableau systems for all the 15 distinct basic normal modal logics obtainable from **K** by the addition of any combination of the axioms T, D, 4, 5, and B in a *modular way*.

The prefixed tableaux of Fitting can be obtained as derived rules in this method. Hence our labelled tableau systems are a mixture of the methods of Fitting and Massacci.

The irony is that this method is essentially Kripke's 'reformulated method' based on his observation [1963, page 80] that:

> 'These considerations suggest that the rules, which we have stated in terms of R, could instead be stated in terms of the basic tree relation S defined in the preceding paragraph (letting R drop out of the picture altogether).'

Using trees it is possible to isolate the individual atomic aspects of reflexivity, transitivity, symmetry etc. To model combinations of these properties both Kripke and Fitting merge the respective atomic aspects into new rules. Fitting goes one step further by building in the *closure* of these properties as side-conditions, thereby requiring explicit reference to the underlying reachability relation. Massacci, on the other hand, merely adds the individual atomic rules as they are, and thereby obtains modularity. The closure is obtained by repeated applications of the atomic rules.

As an aside, note that [Massacci, 1994] contains some minor errors; for example, the system given there for **K5** is incomplete. Massacci has reworked, corrected, and extended his work into a journal version [1995a], but most of this section was written independently of [Massacci, 1995a]. Thus there is a lot of overlap between this section and [Massacci, 1995a], but there are also some subtle differences. In particular, we do not use an empty label at any stage, whereas Massacci sometimes uses an empty label to capture the **L**-accessibility conditions between labels.

We now switch to the tableau formulation of Fitting and Smullyan [1983] rather than sticking to the formulation of Rautenberg because the labels allow us to distinguish formulae that belong to one world (label) from those that belong to another (label), so there is no need to delete formulae when 'traversing' from one world to another. Consequently, we can work with a *single* set of *labelled* formulae.

A **label** is a nonempty sequence of positive integers separated by dots. We use lowercase Greek letters like σ, τ for labels and often omit the dots using σn instead of $\sigma.n$ if no confusion can arise. We use Γ to denote a set of labels. The **length** of a label σ is the number of integers it contains (or the number of dots plus one), and is denoted by $|\sigma|$. For example, 1, 1.21, and 1.2.1 are three labels respectively of lengths 1, 2 and 3. A label τ is a **simple extension** of a label σ if $\tau = \sigma.n$ for some $n \geq 1$. A label τ is an **extension** of a label σ if $\tau = \sigma.n_1.n_2.\cdots.n_k$ for some $k \geq 1$ with each $n_i \geq 1$.

A set of labels Γ is **strongly generated** (with root ρ) if:

1. there is some (root) label $\rho \in \Gamma$ such that every other label in Γ is an extension of ρ; and
2. $\sigma.n \in \Gamma$ implies $\sigma \in \Gamma$.

In what follows, we always assume that $\rho = 1$ as it simplifies some technical details.

As we shall soon see, the labels capture a basic reachability relation between the worlds they name where the world named by $\sigma.n$ is accessible from the world named by σ. A set of strongly generated labels can be viewed as a tree with root ρ where $\sigma.n$ is an immediate child of σ (whence the name 'strongly generated').

A **labelled formula** is a structure of the form $\sigma :: A$ where σ is a label and A is a formula. A **labelled tableau rule** has a numerator and one or more denominators as before, except that each numerator is comprised of a *single* labelled formula, and each denominator is comprised of at most two labelled formulae. There may be side conditions on the labels that appear in the rule. A **labelled tableau calculus** is simply a collection of labelled tableau rules.

A **labelled tableau** for a finite set of formulae $X = \{A_1, A_2, \cdots, A_n\}$ is a tree, where each node contains a single labelled formula, constructed by the systematic construction described in Figure 15. A **tableau branch** is any path from the root downwards in such a tree. A **branch is closed** if it contains some labelled formula $\sigma :: P$ and also contains $\sigma :: \neg P$. Otherwise it is **open**. A **tableau is closed** if every branch is closed, otherwise it is open.

A label σ is **used** on a branch if there is some labelled formula $\sigma :: P$ on that branch. A label σ is **new** to a branch if there is no labelled formula $\sigma :: P$ on that branch.

If $\mathcal{X}$ is a set of labelled formulae then we let $lab(\mathcal{X}) = \{\sigma | \sigma :: P \in \mathcal{X}\}$ be the set of all labels that appear in $\mathcal{X}$. Although a branch $\mathcal{B}$ of a tableau is defined as a set of nodes, each of which contains a formula, we often drop this pedantic distinction and use $\mathcal{B}$ to mean the set of labelled formulae on the branch. Then $lab(\mathcal{B})$ is just the set of labels that are used on branch $\mathcal{B}$.

In Figure 16 we list the rules we need, and in Figure 17 we show how they can be used to give labelled tableau systems for many basic modal logics including some symmetric logics that proved elusive using implicit tableau systems. All are based on those of Massacci [1994].

The rules are categorised into three types: the **PC**-rules are just the usual ones needed for classical propositional logic; the ν-rules are all the rules applicable to formulae of the form $\sigma :: \Box P$ (such formulae are called ν-formulae in many tableau formulations); and the single π-rule is the only rule applicable to formulae of the form $\sigma :: \neg\Box P$ (such formulae are called π-formulae in many tableau formulations).

As expected, there is no modal aspect to the **PC**-rules since the labels in the numerator and denominator(s) are identical. The π-rule is a 'successor creator' since it is the only rule allowed to create new labels. Each ν-rule is a licence to add the formula in the denominator to the already existing world named by the label of the denominator. It is the power to look *backwards* along the reachability relation (in rules like (lB) and $(l4^r)$ that allows us to handle the symmetric and Euclidean logics with such ease.

Notice that none of the rules explicitly mention the reachability relation between

Stage 1: Put the labelled formulae $1 :: A_i$, $1 \leq i \leq n$, in a vertical linear sequence of nodes, one beneath the other, in some order and mark them all as awake.

While the tableau is open and some formula is awake do:

Begin Stage n+1: Choose an awake labelled formula $\sigma :: A$ as close to the root as possible. If there are several awake formulae at the same level then choose the one on the leftmost branch. If $\sigma :: A$ is atomic then mark this formula as finished and stop stage $n+1$. Otherwise update the tableau as follows where 'updating a branch with a labelled formula' means adding the formula to the end of the branch and marking it as awake if it does not already appear on the branch (with any mark), but doing nothing if the formula already appears on the branch (with any mark). For every *open* branch $\mathcal{B}$ which passes through $\sigma :: A$, do:

($\wedge$) if $\sigma :: A$ is of the form $\sigma :: P \wedge Q$ then update $\mathcal{B}$ with $\sigma :: P$ and then update the new $\mathcal{B}$ with $\sigma :: Q$;

($\vee$) if $\sigma :: A$ is of the form $\sigma :: \neg(P \wedge Q)$ then split the end of $\mathcal{B}$ and update the left fork with $\sigma :: \neg P$ and update the right fork with $\sigma :: \neg Q$. If any of these updates fails to add the corresponding formula then delete that fork, possibly leaving $\mathcal{B}$ unaltered or with no fork;

($\neg$) if $\sigma :: A$ is of the form $\sigma :: \neg\neg P$ then update $\mathcal{B}$ with $\sigma :: P$;

(ν) if $\sigma :: A$ is of the form $\sigma :: \Box P$ then, for every ν-rule rule in the calculus which is applicable to $\sigma :: \Box P$, update $\mathcal{B}$ with the corresponding denominator;

(π) if $\sigma :: A$ is of the form $\sigma :: \neg\Box P$ then let k be the smallest integer such that the label σk is new on branch $\mathcal{B}$, update $\mathcal{B}$ with $\sigma k :: \neg P$, and mark all formula on $\mathcal{B}$ of the form $\sigma :: \Box Q$ as awake;

End Stage n+1: Once this has been done for every open branch that passes through $\sigma :: A$, if $\sigma :: A$ is of the form $\sigma :: \Box P$ then mark it as asleep, otherwise mark $\sigma :: A$ as finished, and terminate Stage n+1.

Figure 15. Systematic tableau construction for $X = \{A_1, A_2, \cdots, A_n\}$

$$(l\neg)\ \frac{\sigma :: \neg\neg P}{\sigma :: P} \qquad (l\wedge)\ \frac{\sigma :: P \wedge Q}{\begin{array}{c}\sigma :: P\\ \sigma :: Q\end{array}} \qquad (l\vee)\ \frac{\sigma :: \neg(P \wedge Q)}{\sigma :: \neg P \mid \sigma :: \neg Q}$$

$$(l\pi)\ \frac{\sigma :: \neg\Box P}{\sigma.n :: \neg P} \text{ where } \sigma.n \text{ is new to the current branch}$$

$$(lK)\ \frac{\sigma :: \Box P}{\sigma.n :: P} \qquad (lD)\ \frac{\sigma :: \Box P}{\sigma :: \neg\Box\neg P} \qquad (lT)\ \frac{\sigma :: \Box P}{\sigma :: P}$$

$$(lB)\ \frac{\sigma.n :: \Box P}{\sigma :: P} \qquad (l4)\ \frac{\sigma :: \Box P}{\sigma.n :: \Box P} \qquad (l5)\ \frac{1.n :: \Box P}{1 :: \Box\Box P}$$

$$(l4^r)\ \frac{\sigma.n :: \Box P}{\sigma :: \Box P} \qquad (l4^d)\ \frac{\sigma.n :: \Box P}{\sigma.n.m :: \Box P}$$

Note: except for σn in the rule $(l\pi)$, each label in the numerator and denominator must already exist on the branch.

Figure 16. Single Step Rules for the Basic Modal Logics

$\mathcal{LCL}$	PC-Rules	ν-Rules	π-Rule
$\mathcal{LCPC}$	$(l\neg), (l\wedge), (l\vee)$	—	—
$\mathcal{LCK}$	$\mathcal{LCPC}$	(lK)	$(l\pi)$
$\mathcal{LCT}$	$\mathcal{LCPC}$	$(lK), (lT)$	$(l\pi)$
$\mathcal{LCD}$	$\mathcal{LCPC}$	$(lK), (lD)$	$(l\pi)$
$\mathcal{LCKB}$	$\mathcal{LCPC}$	$(lK), (lB)$	$(l\pi)$
$\mathcal{LCK}4$	$\mathcal{LCPC}$	$(lK), (l4)$	$(l\pi)$
$\mathcal{LCK}5$	$\mathcal{LCPC}$	$(lK), (l4^d), (l4^r), (l5)$	$(l\pi)$
$\mathcal{LCKDB}$	$\mathcal{LCPC}$	$(lK), (lB), (lD)$	$(l\pi)$
$\mathcal{LCKD}5$	$\mathcal{LCPC}$	$(lK), (lD), (l4^d), (l4^r), (l5)$	$(l\pi)$
$\mathcal{LCK}4D$	$\mathcal{LCPC}$	$(lK), (lD), (l4)$	$(l\pi)$
$\mathcal{LCK}45$	$\mathcal{LCPC}$	$(lK), (l4), (l4^r), (l5)$	$(l\pi)$
$\mathcal{LCK}45D$	$\mathcal{LCPC}$	$(lK), (l4), (l4^r), (l5), (lD)$	$(l\pi)$
$\mathcal{LCK}4B$	$\mathcal{LCPC}$	$(lK), (lB), (l4), (l4^r)$	$(l\pi)$
$\mathcal{LCB}$	$\mathcal{LCPC}$	$(lK), (lT), (lB)$	$(l\pi)$
$\mathcal{LCS}4$	$\mathcal{LCPC}$	$(lK), (lT), (l4)$	$(l\pi)$
$\mathcal{LCS}5$	$\mathcal{LCPC}$	$(lK), (lT), (l4), (l4^r)$	$(l\pi)$

Figure 17. Labelled Tableau Systems for the Basic Logics

labels in their side-conditions. Furthermore, in all rules, the world named by the label in the denominator is at most *one step away* from the world named by the label in the numerator. For example, the (lT) rule adds the formula P to the same world, whereas the (lK) and (lB) rules add P to a successor and predecessor respectively.

At first sight, the 'single step' nature of the ν-rules seems a drawback since we know that a ν-formula can affect *all* successors, regardless of how many primitive steps it takes to reach them. One is immediately tempted to add side conditions that explicitly mention the reachability relation to capture this notion as is done by Fitting [1983]. But it is precisely this 'single step' nature that allows the rules to ignore the reachability relation and which gives us the modularity apparent in the calculi of Figure 17.

A particular rule may not capture a property of accessibility completely, but some combination of the rules will do so. For example, for transitivity we require $\sigma :: \Box P$ to be able to give $\sigma.\theta :: P$, for any $|\theta| \geq 1$, assuming that both these labels (worlds) σ and $\sigma.\theta$ exist. As Massacci [1994] points out, instead of building this transitive closure property into a side condition for $(l4)$, it is obtained by the combination of $(l4)$ and (lK), one step at a time, as shown below extreme left where we assume that $\theta = n.m$. That is, we cannot derive Fitting's actual rule for transitivity since that rule captures the *closure* of the transitivity property by referring to $\mathbf{L}$-accessibility in the side condition. But we can derive *every instance* of transitivity, thereby computing the closure by repeated applications of the single step rules. We can also derive other useful rules. For example, the rule of 'delayed reflexivity' (lT^d) below centre says something like 'all worlds $(\sigma.n)$ that have a predecessor (σ) are reflexive'. It can be derived in $\mathcal{LC}K5$ and $\mathcal{LC}K4B$ as shown below extreme right:

$$\dfrac{\dfrac{\sigma :: \Box P}{\sigma.n :: \Box P}\,(l4)}{\sigma.n.m :: P}\,(lK) \qquad (lT^d)\ \dfrac{\sigma.n :: \Box P}{\sigma.n :: P} \qquad \dfrac{\dfrac{\sigma.n :: \Box P}{\sigma :: \Box P}\,(l4^r)}{\sigma.n :: P}\,(lK)$$

derivation of transitivity $\qquad\qquad$ derivation of (lT^d)

As an aside, note that in a symmetric frame, like those for $\mathbf{K4B}$, any world that has a predecessor also has a successor, hence (lT^d) captures the essence of the $(T_\Diamond)$ rule of $\mathcal{C}^\dagger\mathbf{K4B}$ on page 322.

The systematic construction is based on the one given by Fitting [1983, page 402] for his prefixed tableau, and the one given by Massacci [1994], except that we have amalgamated two of Fitting's procedures in one here. Fitting first works with *occurrences* of labelled formulae in order to mark them as finished, adding fresh unfinished *occurrences* to handle necessary repetitions. Later he refines the procedure to stop explicit repetitions since this is just a form of contraction where such formulae may have to be used more than once for completeness.

We work with labelled formulae *per se*, avoiding repetitions right from the beginning, and mark most formulae as finished once we have dealt with them. But we do not mark ν-formulae as finished since they may need to be used again and again. Because we always start a stage at the highest awake formulae, these formulae get considered over and over again as desired.

Notice that the systematic procedure constructs only *one* tableau and that it traverses this tableau in a *breadth-first* manner (except that some formulae may change from asleep to awake and temporarily interrupt this traversal). Massacci [1994] gives an alternative systematic procedure where the formulae on a branch are processed using a different strategy; all formulae of the form $\sigma :: \neg\Box P$ on a branch are processed before all formulae of the form $\tau :: \Box Q$ for example. Space forbids us from comparing these strategies in more detail.

EXAMPLE 28. Below we show a closed systematic $\mathcal{LCK}$-tableau for $X = \{\Box(p \to q), \neg(\Box p \to \Box q)\}$. We assume that $A \to B$ is written as $\neg(A \wedge \neg B)$ and that $\neg(A \to B)$ is rewritten and simplified to $A \wedge \neg B$. We use 'a', 's' and 'f' for awake, asleep and finished respectively. The notation s/a indicates that the formula was asleep but was woken up during the stage.

Systematic $\mathcal{LCK}$-tableau for $X = \{\Box(p \to q), \neg(\Box p \to \Box q)\}$

Extant Tableau at Stage End	Marks at Stage End 1	2	3	4	5	6	7	8	9
$1 :: \Box(p \to q)$	a	s	s	s	s/a	s	s	s	s
$1 :: \neg(\Box p \to \Box q)$	a	a	f	f	f	f	f	f	f
$1 :: \Box p$			a	s	s/a	a	s	s	s
$1 :: \neg\Box q$			a	a	f	f	f	f	f
$1.1 :: \neg q$					a	a	a	f	f
$1.1 :: p \to q$						a	a	a	f
$1.1 :: p$							a	a	a
$1.1 :: \neg p$ (left branch)									a
$1.1 :: \neg\neg q$ (right branch)									a
(closed) (closed)									

EXAMPLE 29. The formula $(\Box\Diamond p) \wedge (\Diamond p)$ can be written in primitive notation as $(\Box\neg\Box\neg p) \wedge (\neg\Box\neg p)$. As the reader can verify, the systematic **S4**-tableau for $X = \{(\Box\neg\Box\neg p) \wedge (\neg\Box\neg p)\}$ neither terminates nor closes.

6.3 Soundness of Single Step Tableau Rules

The soundness of the tableau rules is proved using a method from Fitting [1983], but modified to cater for the strongly generated property. We first extend the primitive notion of reachability between labels σ and $\sigma.n$ into a general notion of **L**-accessibility *between labels* σ and τ, and show that it captures the conditions on **L**-frames.

Recall that label ρ is the root of a strongly generated set of labels if every other label in the set is an extension of ρ.

A set $\mathcal{X}$ of labelled formulae is **strongly generated** if $lab(\mathcal{X})$ is strongly generated. For any two labels σ and τ from some strongly generated set Γ of labels with root $\rho = 1$ we define an **L**-accessibility relation $\rhd$ according to Figure 18. These conditions are calculated by taking the appropriate closure of the underlying basic reachability relation between σ and $\sigma.n$. (Thanks to Nicolette Bonnette for many simplifications.) For example, the condition on **K45**-frames is calculated by computing the transitive and Euclidean closure of the basic reachability relation. It is here that our assumption that the root $\rho = 1$ simplifies the conditions for **L**-accessibility, but there is still a slight complication for the serial logics.

For any nonserial logic $\mathbf{L_1}$ we say that σ is an $\mathbf{L_1}$**-deadend** if there is no τ that is $\mathbf{L_1}$-accessible from σ. Now we can express the seriality condition for the serial counterpart $\mathbf{L} = \mathbf{L_1 D}$ by demanding that all $\mathbf{L_1}$-deadends be reflexive. In particular, we say that $\sigma \in \Gamma$ is a **K**-deadend if no label in Γ is a simple extension of σ. In Figure 18 we have computed the forms of the $\mathbf{L_1}$-deadends and added an extra condition to make them reflexive for each logic $\mathbf{L_1 D}$. The notation $|\Gamma|$ means the number of labels in Γ.

We leave it to the reader to generalise these conditions to account for the case where ρ is an arbitrary label. Note that the conditions on **L**-accessibility in Figure 18 and the conditions on accessibility in the finite-**L**-frames of Figure 13 on page 360 are closely related. We return to this point later.

But first we relate **L**-accessibility to the **L**-frames of Figure 4 on page 303.

THEOREM 30. *If* Γ *is a strongly generated set of labels with root* $\rho = 1$ *then* $\mathcal{F} = \langle \Gamma, \rhd \rangle$ *is an* **L***-frame.*

Proof. It is obvious that **KT**-accessibility, **K4**-accessibility and **KB**-accessibility forces $\mathcal{F}$ to be respectively reflexive, transitive and symmetric. We consider only the case for **K45** in detail.

We have to show that **K45**-accessibility forces $\mathcal{F}$ to be Euclidean and transitive. **K45**-accessibility $\rhd$ is Euclidean if $\sigma_0 \rhd \sigma_1$ and $\sigma_0 \rhd \sigma_2$ implies $\sigma_1 \rhd \sigma_2$, where **K45**-accessibility $\rhd$ is defined as:

$$\sigma \rhd \tau \text{ iff } (\tau = \sigma.\theta \text{ and } |\theta| \geq 1) \text{ or } (|\sigma| \geq 2 \text{ and } |\tau| \geq 2)$$

By substitution we get:

Definition of $\sigma \rhd \tau$ where σ and τ are nonempty and drawn from a strongly generated set of labels Γ with root $\rho = 1$	
Logics	for all $\tau, \sigma \in \Gamma$, τ is **L**-accessible from σ iff
K	$\tau = \sigma.n$ for some $n \geq 1$
KT	$\tau = \sigma.n$ or $\tau = \sigma$
KB	$\tau = \sigma.n$ or $\sigma = \tau.m$
K4	$\tau = \sigma.\theta$ and $\lvert\theta\rvert \geq 1$
K5	$\tau = \sigma.n$ or ($\lvert\sigma\rvert \geq 2$ and $\lvert\tau\rvert \geq 2$)
K45	($\tau = \sigma.\theta$ and $\lvert\theta\rvert \geq 1$) or ($\lvert\sigma\rvert \geq 2$ and $\lvert\tau\rvert \geq 2$)
KD	**K**-condition or (σ is a **K**-deadend and $\sigma = \tau$)
KDB	**KB**-condition or ($\lvert\Gamma\rvert = 1$ and $\sigma = \tau = 1$)
KD4	**K4**-condition or (σ is a **K**-deadend and $\sigma = \tau$)
KD5	**K5**-condition or ($\lvert\Gamma\rvert = 1$ and $\sigma = \tau = 1$)
KD45	**K45**-condition or ($\lvert\Gamma\rvert = 1$ and $\sigma = \tau = 1$)
KB4	$\lvert\Gamma\rvert \geq 2$
B	$\tau = \sigma$ or $\tau = \sigma.n$ or $\sigma = \tau.m$
S4	($\tau = \sigma.\theta$ and $\lvert\theta\rvert \geq 1$) or ($\tau = \sigma$)
S5	$\lvert\Gamma\rvert \geq 1$

Figure 18. Definition of **L**-accessibility $\rhd$

Hypotheses	Expanded Hypothesis
$\sigma_0 \rhd \sigma_1$ and $\sigma_0 \rhd \sigma_2$	$(\sigma_1 = \sigma_0.\theta_1$ and $\|\theta_1\| \geq 1)$ or $(\|\sigma_0\| \geq 2$ and $\|\sigma_1\| \geq 2)$ and $(\sigma_2 = \sigma_0.\theta_2$ and $\|\theta_2\| \geq 1)$ or $(\|\sigma_0\| \geq 2$ and $\|\sigma_2\| \geq 2)$
Goal	**Expanded Goal**
$\sigma_1 \rhd \sigma_2$	$(\sigma_2 = \sigma_1.\theta_3$ and $\|\theta_3\| \geq 1)$ or $(\|\sigma_1\| \geq 2$ and $\|\sigma_2\| \geq 2)$

Now, we know that σ_0 is nonempty, hence $|\sigma_0| \geq 1$. But this together with $(\sigma_1 = \sigma_0.\theta_1$ and $|\theta_1| \geq 1)$ in the left disjunct of the first hypothesis immediately gives $|\sigma_1| \geq 2$. Thus both disjuncts of the first line of the hypothesis imply $|\sigma_1| \geq 2$.

Similarly, $|\sigma_0| \geq 1$ together with $(\sigma_2 = \sigma_0.\theta_2$ and $|\theta_2| \geq 1)$ in the left disjunct of the second hypothesis gives $|\sigma_2| \geq 2$. Thus both disjuncts of the second hypothesis imply $|\sigma_2| \geq 2)$.

And the conjunction of these two gives the second disjunct of the goal showing that **K45**-accessibility relation $\rhd$ is indeed Euclidean.

To show that **K45**-accessibility is also transitive, we must show that $\sigma_0 \rhd \sigma_1$ and $\sigma_1 \rhd \sigma_2$ implies $\sigma_0 \rhd \sigma_2$. The same expansions can be used but the roles of hypotheses and goal are slightly altered. The argument is almost identical, except for one subcase which relies on the fact that $|\sigma_0| = 1$ implies $\sigma_0 = 1$. ■

Let $\mathcal{X}$ be a strongly generated set of labelled formulae, let $lab(\mathcal{X})$ be the set of labels that appear in $\mathcal{X}$ and let $\mathcal{M} = \langle W, R, V \rangle$ be some **L**-model where **L** is any one of the 15 distinct basic normal modal logics obtainable by adding any combination of the axioms T, D, B, 4 and 5 to logic **K**. Call a world in $\mathcal{M}$ **idealisable** iff it has an R-successor in $\mathcal{M}$.

An **L-interpretation of** (a strongly generated set of labelled formulae) $\mathcal{X}$ **in** $\mathcal{M}$ is a mapping $I : lab(\mathcal{X}) \mapsto W$ that satisfies: if $\sigma \rhd \tau$ and $I(\sigma)$ is idealisable then $I(\sigma)RI(\tau)$, where $\rhd$ is the appropriate **L**-accessibility relation from Figure 18 [Fitting, 1983].

A strongly generated set $\mathcal{X}$ of labelled formulae is **L-satisfiable under the L-interpretation** I if $I(\sigma) \models A$ for each $\sigma :: A$ in $\mathcal{X}$; and is **L-satisfiable** if it is **L**-satisfiable under some **L**-interpretation. A branch of a labelled tableau is **L**-satisfiable if the set of labelled formulae on it is **L**-satisfiable, and a tableau is **L**-satisfiable if some branch of the tableau is **L**-satisfiable.

PROPOSITION 31. *The set of labelled formulae $lab(\mathcal{B})$ from any branch $\mathcal{B}$ of a labelled tableau is a strongly generated set.*

Proof. By the fact that the initial label is always $\rho = 1$, and the fact that the only new labels that may be created are labels of the form $\sigma.n$, $n \geq 1$, which are all simple extensions of some $\sigma \in lab(\mathcal{B})$. ■

We now prove soundness of some of the rules leaving the others to the reader. Since the systematic procedure updates all branches that pass through the chosen formula, the soundness theorem states the following: if a tableau $\mathcal{T}$ is **L**-satisfiable and we apply rule $(l\rho)$ to get tableau $\mathcal{T}'$, then $\mathcal{T}'$ is also **L**-satisfiable. Since every rule has at most two denominators, a rule can cause a given branch to split into at most branches. Consequently we have to prove that if a branch $\mathcal{B}$ is **L**-satisfiable, and applying rule $(l\rho)$ causes it to be updated into branches $\mathcal{C}$ and $\mathcal{D}$, then at least one of the new branches is also **L**-satisfiable.

Soundness of $(l\pi)$ for L-frames: Suppose $\mathcal{B}$ is an **L**-satisfiable branch and that we apply the $(l\pi)$ rule to some awake $\sigma :: \neg\Box P$ on $\mathcal{B}$ to obtain branch $\mathcal{C}$ containing $\sigma n :: \neg P$ where σn is a simple extension of σ that is new to $\mathcal{B}$. We have to show that $\mathcal{C}$ is **L**-satisfiable.

Since $\mathcal{B}$ is **L**-satisfiable, there is some **L**-model $\mathcal{M} = \langle W, R, V\rangle$ and some **L**-interpretation I in $\mathcal{M}$ such that $I(\sigma) \in W$ and $I(\sigma) \models \neg\Box P$. Hence $I(\sigma)$ is idealisable as there is some $w \in W$ with $I(\sigma)Rw$ and $w \models \neg P$. Since σn is new, it does not appear in $\mathcal{B}$ and hence has no image under I. Extend I by putting $I(\sigma n) = w$. We then have $\sigma \rhd \sigma n$, $I(\sigma)RI(\sigma n)$, and $I(\sigma n) \models \neg P$ meaning that $\mathcal{C}$ is indeed **L**-satisfiable under the extended I in $\mathcal{M}$. ■

Soundness of $(l4^d)$ for K5-frames: Suppose $\mathcal{B}$ is a **K5**-satisfiable branch and that we apply the $(l4^d)$ rule to some $\sigma n :: \Box P$ to get a branch $\mathcal{C}$ containing $\sigma nm :: \Box P$. We have to show that $\mathcal{C}$ is also **K5**-satisfiable.

Since $\mathcal{B}$ is **K5**-satisfiable and the labels σn and σnm must already exist on $\mathcal{B}$, there is some **K5**-model $\mathcal{M} = \langle W, R, V\rangle$ and some **K5**-interpretation I in $\mathcal{M}$ such that $I(\sigma n) \in W$, $I(\sigma nm) \in W$ and $I(\sigma n) \models \Box P$. The label σn can exist on $\mathcal{B}$ only if σ also exists on $\mathcal{B}$ since $\mathcal{B}$ is strongly generated. Hence there is some $I(\sigma) \in W$. The configuration $\sigma \rhd \sigma n \rhd \sigma nm$ immediately implies $I(\sigma)RI(\sigma n)RI(\sigma nm)$ by the definition of I. Because R is Euclidean we know that $I(\sigma n)RI(\sigma n)$; that is $I(\sigma n)$ is reflexive. Then $I(\sigma n)RI(\sigma nm)$ and $I(\sigma n)RI(\sigma n)$ gives $I(\sigma nm)RI(\sigma n)$. Hence $I(\sigma nm) \models \Diamond\Box P$. Euclidean frames must validate axiom 5 ($\Diamond\Box A \to \Box A$) hence $I(\sigma nm) \models \Box P$. We have not altered I in any way, so by definition, $\mathcal{C}$ is **K5**-satisfiable under I in $\mathcal{M}$. ■

Soundness of $(l5)$ for K5-frames: Suppose $\mathcal{B}$ is a **K5**-satisfiable branch and that we apply the $(l5)$ rule to some $1.n :: \Box P$ to get a branch $\mathcal{C}$ containing $1 :: \Box\Box P$. We have to show that $\mathcal{C}$ is also **K5**-satisfiable.

As before there is some **K5**-model $\mathcal{M} = \langle W, R, V\rangle$ and some **K5**-interpretation I in $\mathcal{M}$ such that $I(1.n) \in W$ and $I(1.n) \models \Box P$. Since 1 is used on $\mathcal{B}$ and $1 \rhd 1.n$, there must be some $I(1) \in W$ with $I(1)RI(1.n)$.

Now suppose for a contradiction that $I(1) \models \neg\Box\Box P$; then there is some $w \in W$ such that $I(1)Rw$ and $w \models \neg\Box P$, which in turn implies that there is some $w' \in W$ such that wRw' and $w' \models \neg P$. Since R is Euclidean, $I(1)RI(1.n)$ and $I(1)Rw$ gives $wRI(1.n)$, and then wRw' gives $I(1.n)Rw'$. But then $I(1.n) \models$

$\Box P$ implies $w' \models P$; contradiction. Hence $I(1) \models \Box\Box P$ and $\mathcal{C}$ is **K5**-satisfiable under I in $\mathcal{M}$. ■

THEOREM 32. *If the systematic tableau for X closes then X is* **L**-*unsatisfiable.*

Proof. For a contradiction, suppose the tableau for X is closed and that X is **L**-satisfiable. The latter means that there is some **L**-model $\mathcal{M} = \langle W, R, V\rangle$ and some world $w \in W$ such that $w \models X$. Our tableau begins with nodes $1 :: A_i$, for each $A_i \in X$ so define an **L**-interpretation I in $\mathcal{M}$ such that $I(1) = w$. Then the initial tableau comprising the linear sequence of these nodes $1 :: A_i$ is **L**-satisfiable (under I in $\mathcal{M}$). Since each of our tableau rules is sound, any tableau obtained from this initial tableau by these rules is also **L**-satisfiable. Hence our tableau is **L**-satisfiable.

Suppose $\mathcal{B}$ is some branch of this closed tableau. Then $\mathcal{B}$ itself is closed and hence contains some labelled formula $\sigma :: P$ and also contains $\sigma :: \neg P$. Now any **L**-interpretation I' for $\mathcal{B}$ in any **L**-model $\mathcal{M}'$ would entail that $I'(\sigma) \models P$ and also that $I'(\sigma) \models \neg P$, which is clearly impossible. Hence $\mathcal{B}$ is not **L**-satisfiable. Since $\mathcal{B}$ was an arbitrary branch this must be true for all branches of this closed tableau. Then, by definition, our tableau is not **L**-satisfiable. Contradiction, hence if the tableau for X closes then X is **L**-unsatisfiable. ■

COROLLARY 33 (Soundness). *If the systematic tableau for $\{\neg A\}$ is closed then A is* **L**-*valid.*

6.4 Fairness, Infinite Tableaux, Chains and Periodicity

The systematic tableau construction may go on ad infinitum in some cases. We now prove some useful properties of our systematic labelled tableau procedure giving some insight into its behaviour.

We have already noted that the systematic procedure is essentially a breadth-first traversal of the tableau under construction except that certain formulae may awaken to interrupt this traversal. In what follows we refer to the *uninterrupted* sequence of node traversal as the **visit sequence**. That is, the visit sequence is the sequence in which the systematic procedure would visit the nodes if no ν-formula is reawakened. It has little to do with the sequence of nodes on a *particular* branch.

The systematic tableau is a finitely generated tree in that each node has at most two immediate children (since branches are caused only by the $(\vee)$ rule). By Königs lemma, an infinite but finitely generated tree must contain an infinite branch (see Fitting [1983, pages 404–407]). Hence there are four ways in which the systematic procedure can go on ad infinitum:

1. by constructing an *infinite branch* containing a sequence of distinct labelled formulae $\sigma :: P_1, \sigma :: P_2, \sigma :: P_3, \cdots, \sigma :: P_n, \cdots$ all with the same label σ;
2. by constructing an *infinite branch* containing a sequence of labelled formulae $\sigma.1 :: P_1, \sigma.2 :: P_2, \sigma.3 :: P_3, \cdots, \sigma.n :: P_n, \cdots$ all simple extensions of some common σ;

3. by constructing an *infinite branch* containing a sequence of labelled formulae $\sigma_1 :: P_1, \sigma_2 :: P_2, \sigma_3 :: P_3, \cdots, \sigma_n :: P_n, \cdots$ all with different labels ; and
4. by traversing a set of formulae that repeatedly switch from asleep to awake and vice-versa on the *visit sequence*.

We show that items (1), (2) and (4) cannot occur.

LEMMA 34. *In any branch of a systematic tableau for the finite set of formulae X, the maximum number of formulae with some given label σ is finite.*

Proof. By induction on the length of σ. If $|\sigma| = 1$ then $\sigma = 1$ and the only possible formulae with this label are either subformulae of X, negations of a subformula of X, or are obtained from some subformulae of X by the building up rules $(l5)$ and (lD). But no infinite sequence of building up rules is possible. If $|\sigma| \geq 1$ then σ must have been created by $(l\pi)$ which adds only the negation of a *subformula* of its numerator. For details see Fitting [1983, page 411]. ■

Item 1 above is then impossible since any branch has but a finite number of formulae with label σ and we do not permit the branch to contain repetitions. We leave it to the reader to compute actual bounds noting the presence of the 'building up rules' (lD) and $(l5)$; see Massacci [1994]

LEMMA 35. *In any branch of a systematic tableau for the finite set of formulae X, the number N_k of different labels of length k is finite.*

Proof. Proof by induction on k and the fact that the systematic tableau construction avoids repetitions. See Fitting [1983, pages 410–412] and Massacci [1994] for more exact bounds but once again beware that these need to be adjusted for the 'building up rules'. ■

Thus no branch can contain an infinite number of labels all of the same length k for any k, and item 2 above is also impossible.

We now turn to item 4 in some detail since these details cannot be found elsewhere. First note that although a branch does not contain repetitions, the visit sequence may do so.

LEMMA 36. *A particular labelled formula occurrence $\sigma :: \Box Q$ on the visit sequence can be awakened only a finite number of times.*

Proof. The only way to awaken a ν-formula occurrence $\sigma :: \Box Q$ is to visit some π-formula occurrence $\sigma :: \neg\Box P$ that appears on the *same branch* as $\sigma :: \Box Q$. Since the systematic tableau is finitely branching, the number of such branches is finite. A branch can contain $\sigma :: \neg\Box P$ at most once, hence the number of *occurrences* of $\sigma :: \neg\Box P$ on the visit sequence is (also) finite. Since σ must be of finite length, Lemma 34 guarantees that there are only a finite number of formulae with label

σ on any branch of the tableau. Hence there are a finite number of π-formulae occurrences that can awaken $\sigma :: \Box Q$.

If none of these π-formulae occurrences is visited then $\sigma :: \Box Q$ is never awakened. On the other hand, whenever one of these π-formulae occurrences is visited, it is marked as finished, and π-formulae are never reawakened, hence $\sigma :: \Box Q$ can be awakened only a finite number of times. Since this formula occurrence was an arbitrary ν-formula occurrence we know that *every* ν-formula occurrence can be awakened only a finite number of times. ∎

LEMMA 37 (Fairness). *If a labelled formula occurrence $\sigma :: A$ on the visit sequence is awake at the end of Stage n, the systematic procedure is guaranteed to visit it at some later stage.*

Proof: By induction on the number of π-formulae occurrences that precede $\sigma :: A$ in the visit sequence. Clearly, if $\sigma :: A$ is the root then it is immediately visited at Stage $n + 1$. Similarly, if there are no π-formulae occurrences between the root and $\sigma :: A$ on the visit sequence then every subsequent stage will visit the next intervening formulae occurrence in the visit sequence and mark it as asleep or finished. The absence of intervening π-formulae occurrences means that no formulae occurrences can awaken until after $\sigma :: A$ is visited. Hence there must come a stage that visits $\sigma :: A$.

Suppose the lemma holds for any labelled formula occurrence with j π-formulae occurrences preceding it in the visit sequence.

Consider some $\sigma :: A$ occurrence that is awake at the end of stage n but that has $j + 1$ π-formulae occurrences preceding it in the visit sequence. Let $\tau :: \neg\Box B$ be the last π-formula occurrence in the visit sequence that precedes $\sigma :: A$.

If $\tau :: \neg\Box B$ is not awake at the end of stage n then it must be finished, meaning that all π-formula occurrences preceding $\sigma :: A$ in the visit sequence must be finished. Each subsequent stage must visit one of the awake ν-formulae occurrences preceding $\sigma :: A$ and mark each one as asleep. No ν-formulae occurrences can awaken during this process since there are no awake π-formula occurrences preceding $\sigma :: A$. Hence there must come a stage that visits $\sigma :: A$.

If $\tau :: \neg\Box B$ is awake at the end of stage n then it satisfies the induction hypothesis, so it will eventually be visited at some later stage, and marked as finished, meaning that no π-formula occurrences preceding $\sigma :: A$ in the visit sequence are awake. Some ν-formulae occurrences preceding $\sigma :: A$ may be awakened by the visit to $\tau :: \neg\Box B$ but each of these will be visited in turn and put to sleep in the stages that follow. Again, no formulae occurrences will be awakened in this process. Hence there must come a stage when we visit the formula occurrence immediately after $\tau :: \neg\Box B$ in the visit sequence. If this is $\sigma :: A$ then we are done. Otherwise this stage and subsequent stages must bring us closer and closer to $\sigma :: A$ since none of these intervening formulae occurrences is a π-formula. ∎

LEMMA 38. *No labelled formula occurrence on the visit sequence can remain awake for ever.*

Proof. Suppose the occurrence $\sigma :: A$ is awake at stage n. Lemma 37 guarantees that $\sigma :: A$ will be visited at some later stage m with $m > n$. If $\sigma :: A$ is not a ν-formula then it will be marked as finished and will remain so hereafter. Else $\sigma :: A$ is a ν-formula and it will be marked as asleep at the end of stage m. If $\sigma :: A$ ever awakens at some later stage k then Lemma 37 again guarantees that it will be visited and put back to sleep. But this can happen only a finite number of times since Lemma 36 guarantees that $\sigma :: A$ can awaken only a finite number of times. Hence there must come a stage when $\sigma :: A$ is put to sleep, never to awaken again. ■

Thus the systematic procedure is 'fair' in that item 4 is also impossible. The only way the systematic procedure can go ad infinitum is for some branch to have at least one infinite sequence of longer and longer labels of the form σ, $\sigma.n_1$, $\sigma.n_1.n_2 \cdots$ where each label is a simple extension of its predecessor. In fact, since every label starts with a 1 we can be more precise as below (again following Fitting [Fitting, 1983]).

A **chain** is a sequence of labels $1, \sigma_1, \sigma_2 \cdots$ where each label in the sequence is a simple extension of its predecessor [Fitting, 1983]. A chain of labels $1, \sigma_1, \sigma_2 \cdots$ from branch $\mathcal{B}$ is **periodic** if there exist distinct labels σ_i and σ_j in the chain $(i < j)$ such that $\sigma_i :: A$ is on $\mathcal{B}$ iff $\sigma_j :: A$ is on $\mathcal{B}$; that is if $\{A | \sigma_i :: A \text{ on } \mathcal{B}\} = \{B | \sigma_j :: B \text{ on } \mathcal{B}\}$. A branch is **periodic** if every infinite chain (of labels) on $\mathcal{B}$ is periodic.

LEMMA 39. *If any branch of a systematic tableau for the finite set of formulae X is infinite, then it must be periodic [Fitting, 1983].*

Proof. Basically, given a finite X, there is a limit to the number of different (unlabelled) formulae we can play with, even with the building up rules. Thus any infinite chain of prefixed formulae from any one branch must repeat formulae at some stage. Since this is true for every chain on an infinite branch, the branch must become periodic. ■

We thus have a handle on the systematic construction since an infinite branch is not as bad as it first seemed. If we could keep track of cycles then we could obtain a decision procedure. We briefly return to this point later.

6.5 Completeness

Again we follow Fitting [1983, pages 408–410] but make adjustments for the strongly generated property. A strongly generated set $\mathcal{X}$ of labelled formulae is **L-downward-saturated** if it satisfies the following conditions, where $\triangleright$ is the appropriate **L**-accessibility relation between labels from Figure 18 (page 379):

0) there is no formula A such that both $\sigma :: A$ and $\sigma :: \neg A$ are in $\mathcal{X}$;

1) if $\sigma :: \neg\neg A \in \mathcal{X}$ then $\sigma :: A \in \mathcal{X}$

2) if $\sigma :: A \wedge B \in \mathcal{X}$ then $\sigma :: A \in \mathcal{X}$ and $\sigma :: B \in \mathcal{X}$;

3) if $\sigma :: \neg(A \wedge B) \in \mathcal{X}$ then $\sigma :: \neg A \in \mathcal{X}$ or $\sigma :: \neg B \in \mathcal{X}$;

4) if $\sigma :: \Box A \in \mathcal{X}$ then $\tau :: A \in \mathcal{X}$ for every $\tau \in lab(\mathcal{X})$ such that $\sigma \rhd \tau$;

5) if $\sigma :: \neg\Box A \in \mathcal{X}$ then $\tau :: \neg A \in \mathcal{X}$ for some $\tau \in lab(\mathcal{X})$ such that $\sigma \rhd \tau$.

LEMMA 40. *If $\mathcal{X}$ is a strongly generated set of labelled formulae that is* **L***-downward-saturated and $lab(\mathcal{X})$ has root $\rho = 1$, then $\mathcal{X}$ is* **L***-satisfiable in a model whose possible worlds are the labels that appear in $\mathcal{X}$.*

Proof. Suppose $\mathcal{X}$ is **L**-downward-saturated and let $lab(\mathcal{X})$ be the set of labels that appear in $\mathcal{X}$. Since $\mathcal{X}$ is strongly generated, so is $lab(\mathcal{X})$. Now define a model $\langle W, R, V\rangle$ as follows:

1. let $W = lab(\mathcal{X})$;
2. let $\sigma R \tau$ iff $\sigma \rhd \tau$ (that is, iff τ is **L**-accessible from σ);
3. for each primitive proposition p let $V(p) = \{\sigma | \sigma :: p \in \mathcal{X}\}$.

It is then easy to show by induction on the degree of a formula A and the **L**-downward-saturated property that: if $\sigma :: A \in \mathcal{X}$ then $\sigma \models A$ in the model $\langle W, R, V\rangle$. The identity mapping $I(\sigma) = \sigma$ is then an **L**-interpretation for $\mathcal{X}$ in the model $\langle W, R, V\rangle$.

Once again, the condition that $\rho = 1$ is forced upon us by our reliance on this condition in the definitions of **L**-accessibility. ■

We have already noted that the systematic procedure is essentially a breadth-first traversal of the tableau under construction. We have also identified the mode in which this procedure can go ad infinitum. Keeping these in mind, we say that an open tableau is **completed** if it is infinite or if no formulae in it is awake. But before we can prove the completeness theorem we need to show that our systematic procedure 'does everything that is necessary' in the following sense.

LEMMA 41. *If $\mathcal{B}$ is an open branch of a completed systematic tableau then $\mathcal{B}$ is closed with respect to every tableau rule in the calculus in that: every rule that could have been applied to a formula in $\mathcal{B}$ must have been applied at some stage.*

Proof. By fairness, every formula is visited at least once. Thus the **PC**-rules and the π-rules must have been applied whenever it was possible. For the ν-rules, suppose $\sigma :: \Box P$ is some ν-formula on $\mathcal{B}$ and suppose some instance of a ν-rule $(l\rho)$ is applicable to it because some label τ of the required form is used on $\mathcal{B}$.

Now, when $\sigma :: \Box P$ was first visited, if τ was used on the extant part of $\mathcal{B}$ then we are done for the given instance of rule $(l\rho)$ must have been applied then.

Else, τ must be $\sigma.n$ and must have been created at some later stage by some awake π-formula $\sigma :: \neg\Box Q$ on $\mathcal{B}$. The creation of $\sigma.n$ must have awakened σ ::

$\Box P$. Since $\mathcal{B}$ is completed, and our systematic procedure is fair, the procedure must have visited $\sigma :: \Box P$ at some later stage still. The given instance of rule $(l\rho)$ must have been applied at that later stage since τ was used on the extant part of $\mathcal{B}$. ∎

LEMMA 42. *If $\mathcal{B}$ is an open branch of a completed systematic tableau then $\mathcal{B}$ is* **L**-*downward-saturated.*

Proof. By Lemma 41, $\mathcal{B}$ is closed with respect to every rule of the calculus (in the appropriate sense). We now have to go through the necessary clauses (see page 385) to show that $\mathcal{B}$ is **L**-downward-saturated.

Clause **0)** is satisfied since $\mathcal{B}$ is open. Clauses **1)**, **2)** and **3)** are satisfied since $\mathcal{B}$ must be closed with respect to the classical propositional rules. Clause **5)** must be satisfied because $\mathcal{B}$ is closed with respect to $(l\pi)$. For clause **4)** assume that $\sigma :: \Box A \in \mathcal{B}$ and that $\sigma \rhd \tau$ for some τ in $lab(\mathcal{B})$. We have to show that $\tau :: A \in \mathcal{B}$ for each definition of **L**-accessibility $\rhd$ from Figure 18.

We give the proof for **K5** only. By the definition of **K5**-accessibility, $\sigma \rhd \tau$ means that

$$\tau = \sigma.n \text{ or } (|\sigma| \geq 2 \text{ and } |\tau| \geq 2)$$

Case 1: If $\tau = \sigma.n$ then $\sigma :: \Box A \in \mathcal{B}$ implies $\sigma n :: A \in \mathcal{B}$ by the fact that $\mathcal{B}$ is closed with respect to the rule (lK).
Case 2: Otherwise, if $(|\sigma| \geq 2$ and $|\tau| \geq 2)$ then $\sigma = 1.n_1.n_2 \cdots n_k$ for some $k \geq 1$ and $\tau = 1.m_1.m_2 \cdots m_l$ for some $l \geq 1$. Then starting from $(\sigma :: \Box A) = (1.n_1.n_2 \cdots n_k :: \Box A)$ we can obtain $1.n_1 :: \Box A \in \mathcal{B}$ and $1 :: \Box A \in \mathcal{B}$ by closure of $\mathcal{B}$ with respect to $(l4^r)$. From the first we can obtain $1 :: \Box\Box A \in \mathcal{B}$ by closure of $\mathcal{B}$ with respect to $(l5)$, and from this we obtain $1.m_1 :: \Box A \in \mathcal{B}$ by (lK). Now, if $l = 1$ then $\tau = 1.m_1$ and $1 :: \Box A \in \mathcal{B}$ immediately implies $\tau :: A \in \mathcal{B}$ by (lK). Otherwise, if $l \geq 2$ then $1.m_1 :: \Box A \in \mathcal{B}$ and closure with respect to $(l4^d)$ guarantee that $1.m_1.m_2 \cdots m_{l-1} :: \Box A \in \mathcal{B}$ from which we get $(1.m_1.m_2 \cdots m_l :: A) = (\tau :: A) \in \mathcal{B}$ by (lK) as desired. ∎

THEOREM 43. *If the systematic tableau for X does not close then X is* **L**-*satisfiable.*

Proof. Suppose the systematic tableau for X does not close. Then the tableau must be completed, and must contain some open branch $\mathcal{B}$ by definition. Lemma 42 guarantees that $\mathcal{B}$ is an **L**-downward-saturated set. Since $lab(\mathcal{B})$ must have root $\rho = 1$, Lemma 40 then guarantees that $\mathcal{B}$ is **L**-satisfiable (under the identity **L**-interpretation $I(\sigma) = \sigma$) in an **L**-model $\mathcal{M} = \langle lab(\mathcal{B}), \rhd, V\rangle$. Furthermore, if $\sigma :: A \in \mathcal{B}$ then $\sigma \models A$ in $\mathcal{M}$. The tableau started with a linear sequence of nodes $1 :: A_i$ for every $A_i \in X$, hence $1 :: A_i \in \mathcal{B}$ for every $A_i \in X$. But then $1 \models X$ in $\mathcal{M}$. ∎

COROLLARY 44 (Completeness). *If A is* **L**-*valid then the systematic tableau for $\{\neg A\}$ must close.*

These methods extend easily to cater for 'strong completeness' where we are allowed both 'global' and 'local' assumptions; see Fitting [1983] and Massacci [1994].

6.6 *Cycles, Termination and Decidability*

In the previous sections we have seen how an infinite tableau must give rise to a counter-model. But it is also possible to modify the systematic procedure to identify potential periodic chains and keep tabs on them during the systematic procedure. That is, once a chain of labels becomes periodic because σ_i and σ_j label identical sets of formulae, all formulae with the longer label are put to sleep. They are awakened only when periodicity for this chain is broken by the appearance of some new formula with a label σ_i or σ_j; see Massacci [1994]. Lemma 39 guarantees that every infinite branch will eventually become periodic, hence the modified systematic procedure will terminate for finite X. If the tableau has not closed then we are still guaranteed the same model as if we had allowed it to run ad infinitum. Thus these labelled tableaux can be used as decision procedures for the 15 basic logics. By keeping tabs on cycles we can also prove the finite model property for these logics since the resulting $\mathbf{L}$-frames are exactly the finite-$\mathbf{L}$-frames of Figure 13 (page 360).

The details are considerably more intricate than the preceding paragraph suggests since we have to preserve 'fairness' and completeness. But there simply is no space. Massacci [1994; 1995a] gives alternative proofs of decidability for his systematic procedure based on an interpretation of the tableau rules as term rewriting rules. But a check for periodicity cannot be avoided for the transitive logics.

6.7 *Extensions and Further Work*

The most obvious extensions of this approach are to multi-modal logics where different sorts of labels are used to model the different reachability relations.

An alternative extension is to change the π-rule, thereby obtaining systems for the provability logics, as shown in Figure 19. Note that first-order definability is not a hurdle for these labelled tableau systems since the class of $\mathbf{G}$-frames and $\mathbf{Grz}$-frames are *not* first-order definable. It may also be possible to extend these systems to handle some of the Diodorean modal logics.

We noted on page 378 that the $\mathbf{L}$-accessibility relation $\rhd$ and the finite-$\mathbf{L}$-frames of Figure 13 (page 360) are closely related. We also mentioned on page 305 that there is a duality between the explicit tableau methods and the implicit tableau methods. We now briefly explain these comments by way of an alternative labelled tableau system $\mathcal{LC}^*K45$ for logic $\mathbf{K45}$.

Consider the system $\mathcal{LC}^*K45 = \mathcal{LCPC} \cup \{ (l\pi_1) , (l4^r\pi) , (l4^r_1) , (lK) \}$ where the new rules are as given below:

$$(l\pi G) \quad \frac{\sigma :: \neg\Box P}{\sigma.n :: \neg P \quad \sigma.n :: \Box P} \text{ where } \sigma.n \text{ is new to the current branch}$$

$$(l\pi Grz) \quad \frac{\sigma :: \neg\Box P}{\sigma.n :: \neg P \quad \sigma.n :: \Box(P \to \Box P)} \text{ where } \sigma.n \text{ is new to the current branch}$$

$\mathcal{LCL}$	PC-Rules	ν-Rules	π-Rule	L-accessibility $\triangleright$
$\mathcal{LCG}$	$\mathcal{LCPC}$	$(lK), (l4)$	$(l\pi G)$	**K**
$\mathcal{LCK}4G_o$	$\mathcal{LCPC}$	$(lK), (l4)$	$(l\pi Grz)$	**K4**
$\mathcal{LCG}rz$	$\mathcal{LCPC}$	$(lK), (l4), (lT)$	$(l\pi Grz)$	**S4**

Figure 19. Labelled Tableau Systems for Provability Logics

$$(l4^r\pi) \quad \frac{1.n :: \neg\Box P}{1 :: \neg\Box P} \qquad (l4^r_1) \quad \frac{1.n :: \Box P}{1 :: \Box P} \qquad (lK) \quad \frac{\sigma :: \Box P}{\sigma.n :: P}$$

$$(l\pi_1) \quad \frac{1 :: \neg\Box P}{1.n :: \neg P} \text{ where } 1.n \text{ is new to the current branch}$$

Note: except for $1.n$ in the rule $(l\pi_1)$, each label in the numerator and denominator must already exist on the branch.

The system $\mathcal{LC}^*K45$ does not fit into the mould of our other labelled systems since: it has *two* π-rules, neither of which is the usual (π) rule; the $(l4^r\pi)$ rule does *not* create a successor but merely moves a π-formula from world $1.n$ to the root world 1; and the $(l\pi_1)$ rule is a special case of the usual (π) rule, and creates a successor for a π formula only if its label is the root label 1. We therefore need to modify the systematic procedure slightly so that one of the mutually exclusive rules $(l4^r\pi)$ or $(l\pi_1)$ is applied to the chosen (awake) π-formula as is appropriate. Then a π-formula with a label $\sigma \neq 1$ cannot cause the creation of a successor and a systematic $\mathcal{LC}^*K45$-tableau for a finite X will contain labels of length at most 2. Furthermore, even though the logic is transitive, we do not need any check for periodicity since every systematic tableau is guaranteed to terminate for finite X.

THEOREM 45. *The rules of $\mathcal{LC}^*K45$ are sound for* **K45***-frames.*

Proof. We have to show that if the numerator is **K45**-satisfiable then so is each denominator. So as in Section 6.3 (page 377), suppose there is some **K45**-model $\mathcal{M}$ and an **L**-interpretation I under which each numerator is **K45**-satisfiable in $\mathcal{M}$.

Proof for $(l4^r\pi)$ **:** If $I(1.n) \models \neg\Box P$ then $1 \triangleright 1.n$ gives $I(1)RI(1.n)$ which gives $I(1) \models \Diamond\neg\Box P$ which is $I(1) \models \Diamond\Diamond\neg P$. Then, by the variant $\Diamond\Diamond A \to \Diamond A$ of the transitivity axiom 4 we have $I(1) \models \Diamond\neg P$, that is, $I(1) \models \neg\Box P$ as required.

Proof for $(l4^r_1)$ **:** If $I(1.n) \models \Box P$ then $1 \rhd 1.n$ gives $I(1)RI(1.n)$, giving $I(1) \models \Diamond\Box P$, which by the Euclidean axiom $\Diamond\Box A \to \Box A$ gives $I(1) \models \Box P$, as required.

Proof for $(l\pi_1)$ **:** The rule $(l\pi_1)$ is just an instance of $(l\pi)$ and we know the latter is sound for all Kripke frames. ■

THEOREM 46. *The calculus $\mathcal{LC}^*K45$ is complete with respect to* **K45***-frames.*

Proof. We have to show that if the systematic tableau for X is open then some open branch $\mathcal{B}$ gives an **K45**-downward-saturated set of labelled formulae (see page 385).

Very well, suppose the systematic tableau for X is open. Choose an open branch $\mathcal{B}$. The branch must be closed with respect to all the rules of $\mathcal{LC}^*K45$ in the appropriate sense (page 386) since this is a consequence of the systematic procedure itself rather than the form of the rules. The clauses **0)** to **3)** of the definition of **K45**-downward-saturated go through as before. For clause **4)** note that $1 \rhd 1.n$ and $1.n \rhd 1.m$ for all n and m, where n and m are integers, captures **K45**-accessibility over $lab(\mathcal{B})$ completely since $\mathcal{B}$ contains labels of length at most 2. The derivation below left shows that clause **4)** must be satisfied while the derivation below right shows that clause **5)** must also be satisfied

$$\dfrac{\dfrac{1.n :: \Box P}{1 :: \Box P}\,(l4^r_1)}{1.m :: P}\,(lK) \qquad\qquad \dfrac{\dfrac{1.n :: \neg\Box P}{1 :: \neg\Box P}\,(l4^r\pi)}{1.m :: \neg P}\,(l\pi_1)$$

Thus X is **K45**-satisfiable under the identity **L**-interpretation $I(\sigma) = \sigma$ in the **K45**-model $\langle lab(\mathcal{B}), \rhd, V\rangle$ as defined in Lemma 40 on page 386. ■

The new rules of $\mathcal{LC}^*K45$ are essentially the operations that we required in the completeness proofs for $\mathcal{C}^\dagger$**K45** on page 335. Thus $\mathcal{LC}^*K45$ *implements* the completeness proof for $\mathcal{C}^\dagger$**K45**, but $\mathcal{LC}^*K45$ is cut-free! Furthermore, the **K45**-model created by the completeness proof for $\mathcal{LC}^*K45$ (above) is also a finite-**K45**-frame as defined on page 360. The extra power of rules that look backward against R, like $(l4^r_1)$ and $(l4^r\pi)$, have allowed us to eliminate even analytic cut.

For most cases, $\mathcal{LC}^*K45$ will be more efficient than $\mathcal{LC}K45$ due to the restriction that labels be at most length 2. Given a finite X, the number of prefixes of length 2 on any branch of a systematic tableau for X can be bounded by extending Lemma 35; see Massacci [1994] or Fitting [1983]. Hence, as pointed out to me by Massacci, we may even be able to determine the complexity of the decision and satisfiability problems for **K45** using this system, although such results are already known for most of the basic logics; see [Ladner, 1977; Halpern and Moses, 1985].

The system **KE** of Mondadori [D'Agostino and Mondadori, 1994] has already been described in another Chapter in this *Handbook*. Clearly, it should be possible to extend all our modal tableau systems by modifying our tableau rules to incorporate the rule (PB). The only work along these lines that I know of is the work of Artosi, Governatori and coworkers [Artosi and Governatori, 1994] who use both (PB) and labelled tableaux, but where the labels are allowed to contain *variables* as well as constants. A branch is now closed if it contains some $\sigma :: A$ and some $\tau :: \neg A$ as long as the labels σ and τ are unifiable as strings with different string unification algorithms for different modal logics. The rule (PB) is also driven by string unification of labels. That is, the restrictions on the reachability relation are not built into a notion like **L**-accessibility, but into the unification algorithms. The main advantage is that we can now 'detect' closure subject to a constraint that two given labels unify.

Ohlbach [1993] has also studied such systems but in a different guise, for Ohlbach literally translates modal logics into classical first-order logic.

Any method that uses labels is really translating the modal logic into classical first-order logic since all these methods use a label of 'universal force' for $\Box$-formulae and use a label of 'existential force' for $\Diamond$-formulae. The recent work of Russo [1995] makes these intuitions explicit.

ACKNOWLEDGEMENTS

I would like to thank Melvin Fitting, Jean Goubault, Alain Heuerding, Bob Meyer and Minh Ha Quang for their comments on earlier drafts. Particular thanks to Fabio Massacci for many useful comments and fruitful discussions, and Nicolette Bonnette for numerous corrections.

Australian National University, Canberra.

REFERENCES

[Auffray and Enjalbert, 1989] Y. Auffray and P. Enjalbert. Modal theorem proving: An equational viewpoint. In *11th International Joint Conference on Artificial Intelligence*, pages 441–445, 1989.

[Artosi and Governatori, 1994] A. Artosi and G. Governatori. Labelled model modal logic. In *Proceedings of the CADE-12 Workshop on Automated Model Building*, pages 11–17, 1994.

[Amerbauer, 1993] M. Amerbauer. *Schnittfreie Tableau- und Sequenzenkalküle für Normale Modale Aussagenlogiken*. PhD thesis, Naturwissenschaftliche Fakultät der Universität Salzburg, 1993.

[Avron, 1984] A. Avron. On modal systems having arithmetical interpretations. *Journal of Symbolic Logic*, 49:935–942, 1984.

[Avron, 1994] A. Avron. The method of hypersequents in proof theory of propositional non-classical logics. Technical Report 294-94, Institute of Computer Science, Tel Aviv University, Israel, 1994.

[Banieqbal and Barringer, 1987] B. Banieqbal and H. Barringer. Temporal logic with fixed points. In *Proc. Workshop on Temporal Logic in Specification, LNCS 398*, 1987.

[Bellin, 1985] G. Bellin. A system of natural deduction for GL. *Theoria*, 51:89–114, 1985.

[Beth, 1953] E. W. Beth. On Padoa's method in the theory of definition. *Indag. Math.*, 15:330–339, 1953.

[Beth, 1955] E. W. Beth. Semantic entailment and formal derivability. *Mededelingen der Koninklijke Nederlandse Akademie van Wetenschappen, Afd. Letterkunde*, 18:309–342, 1955.

[Borga and Gentilini, 1986] M. Borga and P. Gentilini. On the proof theory of the modal logic Grz. *ZML (now called Mathematical Logic Quarterly)*, 32:145–148, 1986.

[Baumgartner *et al.*, 1995] P. Baumgartner, R. Hähnle, and J. Posegga (Ed.). Proceedings of the fourth workshop on theorem proving with analytic tableaux and related methods. LNAI 918, 1995.

[Boolos, 1979] G. Boolos. *The Unprovability of Consistency*. Cambridge University Press, 1979.

[Borga, 1983] M. Borga. On some proof theoretical properties of the modal logic GL. *Studia Logica*, 42:453–459, 1983.

[Borghuis, 1993] T. Borghuis. Interpreting modal natural deduction in type theory. In Maarten de Rijke, editor, *Diamonds and Defaults*, pages 67–102. Kluwer Academic Publishers, 1993.

[Beckert and Possega, 1995] B. Beckert and J. Posegga. leanTAP: Lean tableau-based deduction. *Journal of Automated Reasoning*, 15(3):339–358, 1995.

[Bull and Segerberg, 1984] R. A. Bull and K. Segerberg. Basic modal logic. In D. Gabbay and F. Guenthner, editors, *Handbook of Philosophical Logic, Volume II: Extensions of Classical Logic*, pages 1–88. D. Reidel, 1984.

[Bull, 1965] R. A. Bull. An algebraic study of Diodorean modal systems. *Journal of Symbolic Logic*, 30(1):58–64, 1965.

[Bull, 1985] R. A. Bull. Review of 'Melvin Fitting, Proof Methods for Modal and Intuitionistic Logics, Synthese Library, Vol. 169, Reidel, 1983'. *Journal of Symbolic Logic*, 50:855–856, 1985.

[Burgess, 1984] J. Burgess. Basic tense logic. In D. Gabbay and F. Guenthner, editors, *Extensions of Classical Logic*, volume II of *Handbook of Philosophical Logic*, pages 1–88. Reidel, Dordrecht, 1984.

[Catach, 1988] L Catach. TABLEAUX: A general theorem prover for modal logics. In *Proc. International Computer Science Conference: Artificial Intelligence: Theory and Applications*, pages 249–256, 1988.

[Catach, 1991] L. Catach. TABLEAUX: A general theorem prover for modal logics. *Journal of Automated Reasoning*, 7:489–510, 1991.

[Cerrato, 1993] C. Cerrato. Modal sequents for normal modal logics. *Mathematical Logic Quarterly (previously ZML)*, 39:231–240, 1993.

[Chellas, 1980] B. F. Chellas. *Modal Logic: An Introduction.* Cambridge University Press, 1980.

[Chang and Lee, 1973] G. L. Chang and R. G. T. Lee. *Symbolic logic and mechanical theorem proving*. Academic Press, New York, 1973.

[Curry, 1952] H. B. Curry. The elimination theorem when modality is present. *Journal of Symbolic Logic*, 17:249–265, 1952.

[D'Agostino and Mondadori, 1994] M. D'Agostino and M. Mondadori. The taming of the cut: classical refutations with analytic cut. *Journal of Logic and Computation*, 4:285–319, 1994.

[Gudzhinskas, 1982] É. Gudzhinskas. Syntactical proof of the elimination theorem for von Wright's temporal logic. *Mat. Logika Primenen*, 2:113–130, 1982.

[Emerson and Halpern, 1985] E. A. Emerson and J. Y. Halpern. Decision procedures and expressiveness in the temporal logic of branching time. *Journal of Computer and System Sciences*, 30:1–24, 1985.

[Emerson, 1985] E. A. Emerson. Automata, tableaux, and temporal logics. In *Proc. Logics of Programs 1985, LNCS 193*, pages 79–87, 1985.

[Emerson, 1990] E. A. Emerson. Temporal and modal logic. In J. van Leeuwen, editor, *Handbook of Theoretical Computer Science*, volume Volume B: Formal Models and Semantics, chapter 16. MIT Press, 1990.

[Emerson and Sistla, 1983] E. A. Emerson and A. P. Sistla. Deciding branching time logic: A triple exponential decision procedure for CTL^*. *Proc. Logics of Programs, LNCS 164*, pages 176–192, 1983.

[Emerson and Sistla, 1984] E. A. Emerson and A. P. Sistla. Deciding branching time logic. In *Proc. 16th ACM Symposium Theory of Computing*, pages 14–24, 1984.

[Emerson and Srinivasan, 1988] E. A. Emerson and J. Srinivasan. Branching time temporal logic. In *Proc. School/Workshop on Linear Time, Branching Time and Partial Order In Logics and Models of Concurrency, The Netherlands, 1988, LNCS 354*, pages 123–172, 1988.

[Fagin *et al.*, 1995] R. Fagin, J. Halpern, Y. Moses and M. Vardi. *Reasoning about Knowledge*. The MIT Press, Cambridge, MA, 1995.

[Fisher, 1991] M. Fisher. A resolution method for temporal logic. In *Proc. 12th International Joint Conference on Artificial Intelligence 1991*, pages 99–104. Morgan-Kaufmann, 1991.

[Fitting, 1973] M. Fitting. Model existence theorems for modal and intuitionistic logics. *Journal of Symbolic Logic*, 38:613–627, 1973.

[Fitting, 1983] M. Fitting. *Proof Methods for Modal and Intuitionistic Logics*, volume 169 of *Synthese Library*. D. Reidel, Dordrecht, Holland, 1983.

[Fitting, 1993] M. Fitting. Basic modal logic. In D. Gabbay et al, editor, *Handbook of Logic in Artificial Intelligence and Logic Programming: Logical Foundations*, volume 1, pages 365–448. Oxford University Press, 1993.

[Frisch and Scherl, 1991] A. M. Frisch and R. B. Scherl. A general framework for modal deduction. In J. Allen, R. Fikes, and E. Sandewall, editors, *Proc. 2nd Conference on Principles of Knowledge Representation and Reasoning*. Morgan-Kaufmann, 1991.

[Gabbay, 1987] D. Gabbay. The declarative past and imperative future: Executable temporal logic for interactive systems. In *Proc. Workshop on Temporal Logic in Specification, LNCS 398*, pages 409–448. Springer-Verlag, 1987.

[Gabbay, 1997] D. Gabbay. *Labelled Deductive Systems*. Oxford University Press, 1997.

[Gentzen, 1935] G. Gentzen. Untersuchungen über das logische Schliessen. *Mathematische Zeitschrift*, 39:176–210 and 405–431, 1935. English translation: Investigations into logical deduction, in The Collected Papers of Gerhard Gentzen, edited by M. E. Szabo, pp 68-131, North-Holland, 1969.

[Gent, 1991] I. Gent. *Analytic Proof Systems for Classical and Modal Logics of Restricted Quantification*. PhD thesis, Dept. of Computer Science, University of Warwick, Coventry, England, 1991.

[Gent, 1991a] I. Gent. Theory tableaux. Technical Report 91-62, Mathematical Sciences Institute, Cornell University, 1991.

[Gent, 1993] I. Gent. Theory matrices (for modal logics) using alphabetical monotonicity. *Studia Logica*, 52(2):233–257, 1993.

[Goré *et al.*, 1995] R. Goré, W. Heinle, and A. Heuerding. Relations between propositional normal modal logics: an overview. Technical Report TR-16-95, Automated Reasoning Project, Australian National University, 1995. Available via `http://arp.anu.edu.au/` on WWW.

[Gabbay *et al.*, 1994] D M Gabbay, I. M. Hodkinson, and M A Reynolds. *Temporal logic - mathematical foundations and computational aspects, Volume 1*. Oxford University Press, 1994.

[Goldblatt, 1987] R. I. Goldblatt. *Logics of Time and Computation*. CSLI Lecture Notes Number 7, Center for the Study of Language and Information, Stanford, 1987.

[Goré, 1991] R. Goré. Semi-analytic tableaux for modal logics with applications to nonmonotonicity. *Logique et Analyse*, 133-134:73–104, 1991. (printed in 1994).

[Goré, 1992] R. Goré. Cut-free sequent and tableau systems for propositional normal modal logics. Technical Report 257, University of Cambridge, England, June, 1992.

[Goré, 1994] R. Goré. Cut-free sequent and tableau systems for propositional Diodorean modal logics. *Studia Logica*, 53:433–457, 1994.

[Gough, 1984] G. Gough. Decision procedures for temporal logics. Master's thesis, Dept. of Computer Science, University of Manchester, England, 1984.

[Hanson, 1966] W. Hanson. Termination conditions for modal decision procedures (abstract only). *Journal of Symbolic Logic*, 31:687–688, 1966.

[Hughes and Cresswell, 1968] G. E. Hughes and M. J. Cresswell. *Introduction to Modal Logic*. Methuen, London, 1968.

[Hughes and Cresswell, 1984] G. E. Hughes and M. J. Cresswell. *A Companion to Modal Logic*. Methuen, London, 1984.

[Hintikka, 1955] K. J. J. Hintikka. Form and content in quantification theory. *Acta Philosophica Fennica*, 8:3–55, 1955.

[Halpern and Moses, 1985] J. Y. Halpern and Y. Moses. A guide to the modal logics of knowledge and belief: Preliminary draft. In *Proc. International Joint Conference on Artificial Intelligence*, pages 480–490, 1985.

[Heuerding *et al.*, 1996] Alain Heuerding, Michael Seyfried, and Heinrich Zimmermann. Efficient loop-check for backward proof search in some non-classical logics. Submitted to Tableaux 96.

[Hudelmaier, 1994] J. Hudelmaier. On a contraction free sequent calculus for the modal logic S4. In *Proceedings of the 3rd Workshop on Theorem Proving with Analytic Tableaux and Related Methods*. Technical Report TR-94/5, Department of Computing, Imperial College, London, 1994.

[Kamp, 1968] J. A. W. Kamp. *Tense Logic and the Theory of Linear Order*. PhD thesis, Dept. of Philosophy, University of California, USA, 1968.

[Kanger, 1957] S. Kanger. *Provability in Logic*. Stockholm Studies in Philosophy, University of Stockholm, Almqvist and Wiksell, Sweden, 1957.

[Kawai, 1987] H. Kawai. Sequential calculus for a first-order infinitary temporal logic. *ZML (now Mathematical Logic Quarterly)*, 33:423–432, 1987.

[Kawai, 1988] H. Kawai. Completeness theorems for temporal logics T_{Ω} and $\Box T_{\Omega}$. *ZML (now Mathematical Logic Quarterly)*, 34:393–398, 1988.

[Koshimura and Hasegawa, 1994] M. Koshimura and R. Hasegawa. Modal propositional tableaux in a model generation theorem prover. In *Proceedings of the 3rd Workshop on Theorem Proving with Analytic Tableaux and Related Methods*. Technical Report TR-94/5, Department of Computing, Imperial College, London, 1994.

[Kracht, 1996] M. Kracht. Power and weakness of the modal display calculus. In H. Wansing, editor, *Proof Theory of Modal Logics*. Kluwer, 1996. to appear.

[Kripke, 1959] S. Kripke. A completeness theorem in modal logic. *Journal of Symbolic Logic*, 24(1):1–14, March 1959.

[Kripke, 1963] S. Kripke. Semantical analysis of modal logic I: Normal modal propositional calculi. *Zeitschrift für Mathematik Logik und Grundlagen der Mathematik*, 9:67–96, 1963.

[Ladner, 1977] R. Ladner. The computational complexity of provability in systems of modal propositional logic. *SIAM Journal of Computing*, 6(3):467–480, 1977.

[Masini, 1991] A. Masini. 2-sequent calculus: Classical modal logic. Technical Report TR-13/91, Universita Degli Studi Di Pisa, 1991.

[Masini, 1992] A. Masini. 2-sequent calculus: A proof theory of modalities. *Annals of Pure and Applied Logic*, 58:229–246, 1992.

[Masini, 1993] A. Masini. 2-sequent calculus: Intuitionism and natural deduction. *Journal of Logic and Computation*, 3(5):533–562, 1993.

[Massacci, 1994] F. Massacci. Strongly analytic tableaux for normal modal logics. In Alan Bundy, editor, *Proc. CADE-12, LNAI 814*, pages 723–737. Springer, 1994.

[Massacci, 1995] F. Massacci. Personal communication, December 1995.

[Massacci, 1995a] F. Massacci. Simple steps tableaux for modal logics. Submitted for publication, 1995.

[Merz, 1992] S. Merz. Decidability and incompleteness results for first-order temporal logics of linear time. *Journal of Applied Non-Classical Logic*, 2(2), 1992.

[Miglioli *et al.*, 1995] P. Miglioli, U. Moscato, and M. Ornaghi. Refutation systems for propositional modal logics. In P. Baumgartner, R. Hähnle, and J. Posegga, editors, *Proceedings of the 4th Workshop on Theorem Proving with Analytic Tableaux and Related Methods*, volume LNAI 918, pages 95–105. Springer-Verlag, 1995.

[Moore, 1985] R. C. Moore. Semantical considerations on nonmonotonic logic. *Artificial Intelligence*, 25:272–279, 1985.

[Morgan, 1976] C. G. Morgan. Methods for automated theorem proving in nonclassical logics. *IEEE Transactions on Computers*, C-25(8):852–862, 1976.

[Muller *et al.*, 1988] D. E Muller, A. Saoudi, and P. E Schupp. Weak alternating automata give a simple explanation of why most temporal and dynamic logics are decidable in exponential time. In *Proc. Logics in Computer Science*, pages 422–427, 1988.

[Marek *et al.*, 1991] W. Marek, G. Schwarz, and M. Truszczynski. Modal nonmonotonic logics: ranges, characterisation, computation. Technical Report 187-91, Dept. of Computer Science, University of Kentucky, USA, 1991.

[Nakamura *et al.*, 1987] H. Nakamura, M. Fujita, S. Kono, and H. Tanaka. Temporal logic based fast verification system using cover expressions. In C H Séquin, editor, *VLSI '87, Proceedings of the IFIP TC 10/WG 10.5 International Conference on VLSI, Vancouver*, pages 101–111, 1987.

[Ohlbach, 1990] H. J. Ohlbach. Semantics based translation methods for modal logics. Technical Report SEKI Report SR-90-11, Universität Kaiserslautern, Postfach, 3049, D-6750, Kaiserslautern, Germany, 1990.

[Ohlbach, 1993] H. J. Ohlbach. Translation methods for non-classical logics: An overview. *Bulletin of the Interest Group in Pure and Applied Logics*, 1(1):69–89, 1993.

[Ohnishi and Matsumoto, 1957] M. Ohnishi and K. Matsumoto. Corrections to our paper 'Gentzen method in modal calculi I'. *Osaka Mathematical Journal*, 10:147, 1957.

[Ohnishi and Matsumoto, 1957a] M. Ohnishi and K. Matsumoto. Gentzen method in modal calculi I. *Osaka Mathematical Journal*, 9:113–130, 1957.

[Ohnishi and Matsumoto, 1959] M. Ohnishi and K. Matsumoto. Gentzen method in modal calculi II. *Osaka Mathematical Journal*, 11:115–120, 1959.

[Oppacher and Suen, 1988] F. Oppacher and E. Suen. HARP: A tableau-based theorem prover. *Journal of Automated Reasoning*, 4:69–100, 1988.

[Paech, 1988] B. Paech. Gentzen-systems for propositional temporal logics. In *Proc. 2nd Workshop on Computer Science Logics, LNCS 385*, pages 240–253, 1988.

[Pliuskevicius, 1991] R. Pliuskevicius. Investigations of finitary calculus for a discrete linear time logic by means of finitary calculus. In *LNCS vol 502*, pages 504–528. Springer-Verlag, 1991.

[Pottinger, 1983] G. Pottinger. Uniform, cut-free formulations of T, S4 and S5. *Abstract in JSL*, 48:900–901, 1983.

[Rautenberg, 1979] W. Rautenberg. *Klassische und Nichtklassische Aussagenlogik*. Vieweg, Wiesbaden, 1979.

[Rautenberg, 1983] W. Rautenberg. Modal tableau calculi and interpolation. *Journal of Philosophical Logic*, 12:403–423, 1983.

[Rautenberg, 1985] W. Rautenberg. Corrections for modal tableau calculi and interpolation by W. Rautenberg, JPL 12 (1983). *Journal of Philosophical Logic*, 14:229, 1985.

[Rautenberg, 1990] W. Rautenberg. Personal communication, December 5th, 1990.

[Russo, 1995] A. Russo. Modal labelled deductive systems. Technical Report TR-95/7, Dept. of Computing, Imperial College, London, 1995.

[Sistla and Clarke, 1985] A. P. Sistla and E. M. Clarke. The complexity of propositional linear temporal logics. *Journal of the Association for Computing Machinery*, 32(3):733–749, 1985.

[Schwarz, 1992] G. Schwarz. Minimal model semantics for nonmonotonic modal logics. In *Proc. Logics in Computer Science*, 1992.

[Segerberg, 1971] K. Segerberg. An essay in classical modal logic (3 vols.). Technical Report Filosofiska Studier, nr 13, Uppsala Universitet, Uppsala, 1971.

[Shimura, 1991] T. Shimura. Cut-free systems for the modal logic S4.3 and S4.3GRZ. *Reports on Mathematical Logic*, 25:57–73, 1991.

[Shvarts, 1989] G. F. Shvarts. Gentzen style systems for K45 and K45D. In A. R. Meyer and M. A. Taitslin, editors, *Logic at Botik '89, Symposium on Logical Foundations of Computer Science, LNCS 363*, pages 245–256. Springer-Verlag, 1989.

[Slaght, 1977] R. L. Slaght. Modal tree constructions. *Notre Dame Journal of Formal Logic*, 18(4):517–526, 1977.

[Schwarz and Truszczynski, 1992] G. E. Schwarz and M. Truszczynski. Modal logic S4F and the minimal knowledge paradigm. In *Proc. of Theoretical Aspects of Reasoning About Knowledge*, 1992.

[Stirling, 1992] C. Stirling. Modal and temporal logics for processes. Technical report, Dept of Computer Science, Edinburgh University, 1992. ECS-LFCS-92-221.

[Sambin and Valentini, 1980] G. Sambin and S. Valentini. A modal sequent calculus for a fragment of arithmetic. *Studia Logica*, 34:245–256, 1980.

[Sambin and Valentini, 1982] G. Sambin and S. Valentini. The modal logic of provability: the sequential approach. *Journal of Philosophical Logic*, 11:311–342, 1982.

[Shimura, 1992] T. Shimura. Cut-free systems for some modal logics containing S4. *Reports on Mathematical Logic*, 26:39–65, 1992.

[Valentini, 1983] S. Valentini. The modal logic of provability: Cut-elimination. *Journal of Philosophical Logic*, 12:471–476, 1983.

[Valentini, 1986] S. Valentini. A syntactic proof of cut elimination for GL_{lin}. *Zeitschrift für Mathematische Logik und Grundlagen der Mathematik*, 32:137–144, 1986.

[van Benthem and Blok, 1978] J. F. A. K. van Benthem and W. Blok. Transitivity follows from Dummett's axiom. *Theoria*, 44:117–118, 1978.

[van Benthem, 1983] J. F. A. K. van Benthem. *The Logic of Time: a model-theoretic investigation into the varieties of temporal ontology and temporal discourse*. Synthese library; vol. 156, Dordrecht: Reidel, 1983.

[van Benthem, 1984] J. F. A. K. van Benthem. Correspondence theory. In D. Gabbay and F. Guenthner, editors, *Handbook of Philosophical Logic*, volume II, pages 167–247. D. Reidel, 1984.

[Valentini and Solitro, 1983] S. Valentini and U. Solitro. The modal logic of consistency assertions of Peano arithmetic. *ZML (now Mathematical Logic Quarterly)*, 29:25–32, 1983.

[Vardi and Wolper, 1986] M. Vardi and P. Wolper. Automata-theoretic techniques for modal logics of programs. *Journal of Computer and System Sciences*, 32, 1986.

[Wallen, 1989] L. A. Wallen. *Automated Deduction in Nonclassical Logics: Efficient Matrix Proof Methods for Modal and Intuitionistic Logics*. MIT Press, 1989.

[Wansing, 1994] Heinrich Wansing. Sequent calculi for normal modal propositional logics. *Journal of Logic and Computation*, 4, 1994.

[Wolper, 1983] P. Wolper. Temporal logic can be more expressive. *Information and Control*, 56:72–99, 1983.

[Wolper, 1985] P. Wolper. The tableau method for temporal logic: an overview. *Logique et Analyse*, 110-111:119–136, 1985.

[Zeman, 1973] J. J. Zeman. *Modal Logic: The Lewis-Modal Systems*. Oxford University Press, 1973.

MARCELLO D'AGOSTINO, DOV GABBAY, KRYSIA BRODA

TABLEAU METHODS FOR SUBSTRUCTURAL LOGICS

1 INTRODUCTION

Over the last few decades a good deal of research in logic has been prompted by the realization that logical systems can be successfully employed to formalize and solve a variety of computational problems. Traditionally, the theoretical framework for most applications was assumed to be *classical* logic. However, this assumption often turned out to clash with researchers' intuitions even in well-established areas of application. Let us consider, for example, what is probably the most representative of these application areas: logic programming. The idea that the execution of a Prolog program is to be understood as a derivation in classical logic has played a key role in the development of the area. This interpretation is the leitmotiv of Kowalski's well-known [1979], whose aim is described as an attempt 'to apply the traditional methods of [classical] logic to contemporary theories of problem solving and computer programming'. However, here are some quotations which are clearly in conflict with the received view (and with each other) as to the correct interpretation of logic programs:

> Relevance Logic not only shows promise as a standard for modular reasoning systems, but it has, in a sense, been already adopted by artificial intelligence researchers. The resolution method for Horn clauses *appears* to be based on classical logic, but *procedural derivation* (see [Kowalski, 1979]), the method actually used for logic programming, is not complete for classical logic, and is in fact equivalent to Relevance Logic.[...] the systems of modules which are actually used in computer science have the feature that relevance, not classical logic, provides a theory of their behaviour. [Garson, 1989, p.214]

> According to the standard view, a logic program is a definite set of Horn clauses. Thus logic programs are regarded as syntactically restricted first-order theories within the framework of classical logic. Correspondingly, the proof-theory of logic programs is considered as a specialized version of classical resolution, known as SLD-resolution. This view, however, neglects the fact that a program clause $a_0 \leftarrow a_1, a_2, \ldots, a_n$ is an expression of a fragment of positive logic [a subsystem of Intuitionistic Logic] rather than an implication formula of classical logic. The logical behaviour of such clauses is in no way related to any negation or complement operation. So (positive) logic programs are 'sub-classical'. The classical interpretation seems to be a semantical overkill' [Wagner, 1991, p.835].

This is just an example of a *general* phenomenon which arises in the application of logic: we start from a logic L (e.g. classical logic) and develop a deduction system

M. D'Agostino et al. (eds.), Handbook of Tableau Methods, 397–467.

for L. Then, in order to adapt this system to a given application problem, we often introduce *ad hoc procedural* 'perturbations' which restrict the original logic L and turn it into another logic L^*. Now we have two alternatives. We may regard L^* as the result of application-dependent constraints which belong to the 'pragmatics' of L. If we adopt this conservative option, the behaviour of L^* belongs to the domain of *applied logic*, the effects of the perturbations and the overall behaviour of the perturbed system can be regarded as typical engineering problems. Alternatively, instead of imposing some theoretical system borrowed from the literature on pure logic, we may decide to recognize L^* as a first-class citizen, i.e. a logical system in its own sake, which can become an independent object of investigation. In this way the behaviour of L^* becomes a new theoretical problem and belongs to the domain of *pure logic*. We can then try to provide a theoretical characterization of the new system by exploiting its analogy with more familiar ones (for instance, we can axiomatize L^*, produce Gentzen-style proof systems, develop algebraic and relational semantics for it).

In this chapter we shall focus on an important family of logical systems, which arise from 'perturbations' of classical and intuitionistic logic and are known as *substructural* or *resource* logics. Historically, the subject of substructural logics arises from the combination of four main components:

- the tradition of Relevance Logic (Anderson, Belnap, Meyer);
- the work on BCK Logic (Fitch, Nelson, Meredith, Prior);
- the work on Categorial Logic, (Curry, Howard, Lambek, Van Benthem);
- the recent developments in Linear Logic, (Girard, Lafont).

We shall not attempt here to provide an introduction to the subject and shall assume the reader to be familiar at least with the basic ideas involved. The reader is referred to [Dôsen, 1993] for the necessary historical and conceptual background. We shall, however, try and give a flavour of the subject by briefly discussing its impact on the traditional notion of logical deduction.

1.1 Background on substructural logics

Substructural logics ultimately stem from Gentzen's *Investigations into Logical Deduction* [Gentzen, 1935], and in particular from his characterization of deductive inference in terms of calculi of sequents. Such calculi can be seen as axiomatizations of the notion of logical consequence, expressed by the turnstile $\vdash$, which characterize classical and intuitionistic inferences. Gentzen's analysis brought to light that the properties of $\vdash$ fall in two categories:

- Properties characterizing the behaviour of the logical operators $\rightarrow, \wedge, \vee, \neg$, fixed by the *operational* rules.

- Properties pertaining to the interpretation of $\vdash$, fixed by the *structural* rules (including the cut rule).

Gentzen's approach already contained the possibility of *heuristic variations*. The operational rules of his sequent calculus can be seen as expressing the meaning of the logical operators, while the structural properties of the turnstile as expressing the properties of the underlying *information system*, *i.e.* the rules that govern the information-processing mechanism. From the viewpoint of the present chapter, this heuristic is central. Many perturbations of classical and intuitionistic logic, prompted by their interaction with computational, philosophical or linguistic problems, are captured by re-interpreting the relation $\vdash$, via variations of its structural properties.

Such 'perturbed' systems may have a decent proof-theory and even an intuitive semantics, but can we call them 'logics'? An answer to this question depends, of course, on our definition of 'logic'. Very seldom definitions are completely stable, and this is no exception. A 'logic' is usually identified with a *consequence relation*, i.e. a binary relation formalizing the intuitive notion of logical consequence. According to the traditional definition, which was first formulated explicitly by Tarski [1930a; 1930b], a consequence relation is a relation $\vdash$ between sets of formulae and formulae satisfying the following conditions:

(Identity) $$A \vdash A$$

(Monotonicity) $$\Gamma \vdash B \Longrightarrow \Gamma, A \vdash B$$

(Transitivity) $$\Gamma, A \vdash B \text{ and } \Delta \vdash A \Longrightarrow \Gamma, \Delta \vdash B$$

where Γ and Δ are *sets* of formulae (we write, as usual, 'A' instead of '$\{A\}$' and 'Γ, A' instead of $\Gamma \cup \{A\}$').

For instance, the system of intuitionistic implication can be shown to correspond to the smallest consequence relation closed under the following additional condition concerning the $\rightarrow$ operator:

(Cond) $$\Gamma, A \vdash B \Longleftrightarrow \Gamma \vdash A \rightarrow B$$

The closure conditions in the definition of consequence relation are all *structural* conditions, i.e. they do not involve any specific logical operator, but express basic properties of the notion of logical consequence[1]. The emergence of Relevance Logic (see the monumental [Anderson and Belnap, 1975] and [Anderson *et al.*, 1992]; for an overview see [Dunn, 1986]) showed the inadequacy of this traditional definition. If the system **R** proposed by Anderson and Belnap had to be called 'logic', the definition had to be amended. Let's see why.

---

[1]This terminology goes back to Gentzen [1935] and his distinction between structural and operational rules in the sequent calculi.

The whole idea of Relevance Logic is that in a 'proper' deduction all the premisses *have to be used at least once* to establish the conclusion, so as to stop the validity of the notorious 'fallacies of relevance' such as the so-called 'positive paradox' $A \to (B \to A)$. This criterion of use is ultimately sufficient to prevent the fallacies from being provable. For instance, in the typical natural deduction proof of the positive paradox the assumption B is discharged 'vacuously' by the application of the $\to$-introduction rule, i.e. it is not used in obtaining the conclusion of the subproof constituting the premiss of the rule application. If we restrict our notion of proof in such a way that all the assumptions have to be *used* in order to obtain the conclusion of the proof, such 'vacuous' applications of the $\to$-introduction rule are not allowed, since the subproof which occurs as premiss would not be a 'proper proof'. So, the standard proof is no longer an acceptable proof of the positive paradox, and it can be shown that no alternative proofs can be found.

The criterion of use is clearly of a 'metalevel' nature. It takes the form of a side-condition on the application of the natural deduction rules. Let us consider the restricted deducibility relation $\vdash'_{ND}$ which incorporates this side-condition. Is it a consequence relation in the traditional sense? The answer is obviously 'no', because it does not satisfy (Monotonicity), in that this condition would allow the addition of 'irrelevant' assumptions which are not used in deducing the conclusion. So, if we want to consider the system of relevant implication as a logic, we have to drop (Monotonicity) from our definition of a consequence relation.

But this is not the whole story. In the system **R**, the definition of a 'relevant' deduction requires that every single *occurrence* of an assumption is used to obtain the conclusion. Now, let us write $\Gamma \vdash_{\mathbf{R}} A$, where Γ is a finite sequence of formulae, to mean that there is an **R**-deduction of A using all the elements of Γ (which are *occurrences* of formulae). Consider the statement $A, A \vdash_{\mathbf{R}} A$. This is not provable because there is no way of using both occurrences of A in the antecedent in order to obtain the conclusion, i.e. one of these two occurrences is 'irrelevant'. Therefore, while $A \vdash_{\mathbf{R}} A$ is trivially provable, $A, A \vdash_{\mathbf{R}} A$ is not. It becomes provable if we 'dilute' the criterion of use to the effect that *at least one* occurrence of each assumption needs to be used, as in the 'mingle' system (see [Dunn, 1986]). The trouble is that the distinction between these two different approaches to the notion of relevance cannot be expressed by the traditional notion of consequence relation. Indeed, according to this notion, a consequence relation is taken to be a relation between *sets* of formulae and formulae. But there is no way to distinguish the set $\{A, A\}$ from the set $\{A\}$ and, therefore, between $A, A \vdash_R A$ and $A \vdash_R A$. To make this sort of distinction we need a finer-grained notion of consequence relation.

In fact, such a finer-grained notion was already contained in Gentzen's calculus of sequents [Gentzen, 1935]. In this approach, a (single-conclusion[2]) consequence

[2]Gentzen also considered *multiple-conclusion* consequence relations and showed that this more general notion was more convenient for the formalization of classical logic. See [Gentzen, 1935].

relation is axiomatized as a relation between a finite *sequence* of formulae and a formula. That such a relation holds between a sequence Γ and a formula A is expressed by the *sequent* $\Gamma \vdash A$ (Gentzen used the notation $\Gamma \Longrightarrow A$, but we prefer the 'turnstile' notation for reasons of uniformity). The axioms of Gentzen's system are given by all the sequents of the form $A \vdash A$. Gentzen specified also two sets of rules to derive new sequents from given ones, that he called *operational rules* and *structural rules*. While the first type of rules embodied, in his view, the meaning of the logical operators, the second type embodied the meaning of $\vdash$. These structural rules are the following:

Weakening
$$\frac{\Gamma, \Delta \vdash B}{\Gamma, A, \Delta \vdash B}$$

Exchange
$$\frac{\Gamma, A, B, \Delta \vdash C}{\Gamma, B, A, \Delta \vdash C}$$

Contraction
$$\frac{\Gamma, A, A, \Delta \vdash B}{\Gamma, A, \Delta \vdash B}$$

Cut
$$\frac{\Gamma \vdash A \qquad \Delta, A, \Lambda \vdash B}{\Delta, \Gamma, \Lambda \vdash B}$$

Here the Weakening rule corresponds to the Monotonicity condition of a consequence relation, and Cut corresponds to Transitivity, except that, in the sequent formulation, the structure on the left of $\vdash$ is a *sequence*, rather than a *set*, of formulae. Moreover, in Gentzen's system the role of the Identity condition is played by the assumption that every axiom $A \vdash A$ can be used at any step of a sequent proof. The remaining rules of Contraction and Exchange have the effect of making Gentzen's relations $\vdash$ deductively equivalent to the corresponding consequence relations with *sets* as first argument.

Gentzen's richer axiomatization provides the means of characterizing systems like **R** as 'substructural' consequence relations, i.e. logics for which the standard structural rules of Gentzen's axiomatization may not hold. In the case of **R**, the rule which is dropped is the Weakening rule, responsible for the arbitrary introduction of 'irrelevant' items in the antecedent of a sequent. After removing Weakening, one can consider a rule symmetric to the Contraction rule:

Expansion
$$\frac{\Gamma, A, \Delta \vdash B}{\Gamma, A, A, \Delta \vdash B}$$

The more 'liberal' mingle system is then distinguished from **R** by the fact that it allows for this weaker version of the Weakening rule.

The discussion of Relevance Logic clearly brings out the general idea that variations in the notion of logical consequence correspond to variations in the allowed *structural rules* of a suitable sequent-based system, leaving the basic *operational* rules unchanged. With Girard's Linear Logic [Girard, 1987; Avron, 1988a] the 'substructural movement' reached its climax. Linear Logic completely rejects the 'vagueness' of traditional proof-theory concerning the use and manipulation of assumptions in a deductive process. A 'proper' proof is one in which every assumption is used *exactly once*. If a particular assumption A can be used *ad libitum*, this has to be made *explicit* by prefixing it with the 'storage' operator !. This means that the Contraction rule is not sound in Linear Logic, since it says, informally, that a proof of B from two occurrences of A can be turned into a proof of B from one occurrence only of A. But this is impossible, unless A is one of those assumptions which can be used *ad libitum*, in which case we should prefix it with the storage operator. In the 'non-commutative' variant of Linear Logic — which was anticipated in 1958 by Lambek [1958] as a system intended for applications to mathematical linguistics (see Abrusci [1990; 1991] and [van Benthem, 1991] for further developments — also the *order* in which assumptions are used becomes crucial, and therefore the Exchange rule is also disallowed.

In this chapter we shall focus on tableau methods for substructural logics and shall discuss two main lines of research: the approach based on 'proof-theoretic' tableaux developed by McRobbie, Belnap and Meyer, which is motivated by the work done in the tradition of Relevance Logic; and the approach based on 'labelled tableaux', an outgrowth of Gabbay's research program on *Labelled Deductive Systems*.

1.2 Substructural Consequence Relations

In this chapter we take a *consequence relation* as a relation $\vdash$ between *sequences* of formulae and formulae, satisfying:

Identity $$A \vdash A$$

Surgical Cut $$\frac{\Gamma \vdash A \qquad \Delta, A, \Lambda \vdash B}{\Delta, \Gamma, \Lambda \vdash B}$$

Since the Γ, Δ and Λ range over sequences, an application of the cut rule replaces an *occurrence* of the formula A with the sequence Γ exactly in the position of A. This is why we call the cut 'surgical'.

Structural rules. Apart from the cut rule, which is part of their definition, consequence relations may or may not obey any of the following conditions describing their structural properties:

Exchange	$\dfrac{\Gamma, A, B, \Delta \vdash C}{\Gamma, B, A, \Delta \vdash C}$	Contraction	$\dfrac{\Gamma, A, A, \Delta \vdash C}{\Gamma, A, \Delta \vdash C}$
Expansion	$\dfrac{\Gamma, A, \Delta \vdash C}{\Gamma, A, A, \Delta \vdash C}$	Weakening	$\dfrac{\Gamma, \Delta \vdash C}{\Gamma, A, \Delta \vdash C}$

We can think of a sequent $\Gamma \vdash A$ as stating that the piece of information expressed by A is 'contained' in the data structure expressed by Γ. Then different combinations of the above structural rules can be seen as different ways of defining the properties of the 'information flow' expressed by the turnstile, depending on the structure of the data and on the allowed ways of manipulating it. For instance, if we disallow the Weakening rule, the consequence relation becomes sensitive to the *relevance* of the data to the conclusion: *all* the data has to be used in the derivation process. If also Expansion is disallowed, this notion of relevance extends to the single occurrences of the formulae in the data (as in Anderson and Belnap's system of Relevance Logic, each *occurrence* of a formula in the data has to be used). If Contraction is disallowed, each item of data can be used only *once* and has to be replicated if it is used more than once (as in Girard's Linear Logic and its satellites). In this way the process of deriving a formula becomes more similar to a physical process and the consequence relation becomes sensitive to the resources employed. Finally, if Exchange is disallowed, the 'chronology' of this process — i.e. the order in which formulae occur in the data — becomes significant (as in the Lambek calculus).

Notice that if Exchange is allowed, the antecedent of a sequent can be regarded as a *multiset*.[3] If Contraction and Expansion are also allowed (notice that the Weakening rule implies the Expansion rule) the antecedent of a sequent can be regarded as a *set* of formulae, as with the standard Gentzen systems, i.e. the number of occurrences of formulae does not count, nor does the order in which they occur.

The operator $\rightarrow$. The conditional operator, or 'implication' as is often improperly called, is usually characterized by the following equivalence:

$$C_{\rightarrow} \qquad \Gamma \vdash A \rightarrow B \Longleftrightarrow \Gamma, A \vdash B$$

which, under the assumption of Identity and Surgical cut is equivalent to the pair of sequent rules:

$$(1) \qquad \frac{\Gamma, A \vdash B}{\Gamma \vdash A \rightarrow B} \qquad \frac{\Gamma \vdash A \quad \Delta, B \vdash C}{\Delta, A \rightarrow B, \Gamma \vdash C}$$

[3]For multiset-based consequence relations, the reader is referred to [Avron, 1991].

For instance, we can derive the only-if part of $C_{\rightarrow}$ from the Gentzen rules (the if-part is a primitive rule) as follows:

$$\dfrac{\Gamma \vdash A \rightarrow B \qquad \dfrac{A \vdash A \quad B \vdash B}{A \rightarrow B, A \vdash B}}{\Gamma, A \vdash B}\text{[CUT]}$$

Notice that if *no* structural rule, except the cut rule, is to be used in the derivation, the shape of the latter depends crucially on the format of the operational rules and of the cut rule. Notice also that in systems which do not allow *Exchange* the conditional operator splits into two operators defined as follows:

$C_{\rightarrow_1}$ $\qquad \Gamma, A \vdash B \Longleftrightarrow \Gamma \vdash A \rightarrow_1 B$

$C_{\rightarrow_2}$ $\qquad A, \Gamma \vdash B \Longleftrightarrow \Gamma \vdash A \rightarrow_2 B.$

Of course, if *Exchange* is allowed $\rightarrow_1$ and $\rightarrow_2$ collapse into each other. In this context we shall use the symbol $\rightarrow$ without subscripts.

The operators $\otimes$ and $\wedge$. In the classical and intuitionistic sequent calculus the comma occurring in the left-hand side of a sequent is associated with classical conjunction. This operator represents a particular way of combining pieces of information. Its inferential role depends on two components: the operational rules defining its meaning, and the structural rules defining the meaning of the turnstile. In the new setting, in which the only conditions which are required to hold are Identity and Surgical cut, the comma is no longer equivalent to classical conjunction. However, the new consequence relations may still be closed under the standard condition defining classical conjunction. Such condition characterizes a new type of 'substructural' conjunction, denoted by $\otimes$ and sometimes called 'tensor' which is no longer classical (because of the failure of some or all of the structural rules), and yet is defined in the same way as its classical version. Therefore, in a sense, it has the same 'meaning', namely[4]:

$C_{\otimes}$ $\qquad \Gamma, A \otimes B, \Delta \vdash C \Longleftrightarrow \Gamma, A, B, \Delta \vdash C$

Clearly, a sequent $\Gamma \vdash A$ becomes equivalent to $\otimes\Gamma \vdash A$ where $\otimes\Gamma$ denotes the $\otimes$-concatenation of the formulae in Γ.

Given Identity and Surgical cut, $C_{\otimes}$ can be easily shown to be equivalent to the pair of sequent rules:

(2) $$\dfrac{\Gamma \vdash A \quad \Delta \vdash B}{\Gamma, \Delta \vdash A \otimes B} \qquad\qquad \dfrac{\Gamma, A, B, \Delta \vdash C}{\Gamma, A \otimes B, \Delta \vdash C}$$

[4]For the time being we are considering only single-conclusion sequents. Later on we shall consider also multi-conclusion sequents in the context of logical systems with classical (involutive) negation.

It is not difficult to see that $C_\otimes$ together with Identity and Surgical Cut, implies:

(3) $$\frac{A \vdash B \quad C \vdash D}{A \otimes C \vdash B \otimes D}$$

Moreover, let us consider the following restricted form of cut:

Transitivity $$\frac{A \vdash B \quad B \vdash C}{A \vdash C}$$

Then, it is easy to see that Transitivity, together with Identity, $C_\otimes$ and (3), imply Surgical Cut.

The rules for $\otimes$, as well as its definition $C_\otimes$, describe a type of conjunction which is sometimes called 'context-free' since its characterization does not impose any special condition on the 'context', i.e. on the structures of side formulae (the Γ and Δ). A 'context-dependent' type of conjunction is expressed by the following equivalence:

$C_\wedge$ $$\Gamma \vdash A \wedge B \Longleftrightarrow \Gamma \vdash A \text{ and } \Gamma \vdash B$$

which corresponds to the the following sequent rules:

(4) $$\frac{\Gamma \vdash A \quad \Gamma \vdash B}{\Gamma \vdash A \wedge B} \qquad \frac{\Gamma, A, \Delta \vdash C}{\Gamma, A \wedge B, \Delta \vdash C} \qquad \frac{\Gamma, B, \Delta \vdash C}{\Gamma, A \wedge B, \Delta \vdash C}$$

Here, in the two-premiss rule, the premisses must share the *same* structure of side formulae, that is the two proofs of A and B must depend exactly on the same structure of assumptions, namely Γ. The sequent $\Gamma \vdash A \wedge B$ expresses the fact that from this Γ we can derive *either* A *or* B at our choice. Of course, if we allowed Contraction, we could derive *both* A and B and then 'shrink' the two copies of Γ used in this derivation into one. If we allowed Weakening, on the other hand, we could derive A and B from different data-structures, say Γ and Δ and then expand both to Γ, Δ so as to satisfy the condition of the rule. Thus, under the assumption of Weakening and Contraction, $C_\wedge$ and the corresponding sequent rules would not define any new operator different from $\otimes$. If either of these rules is disallowed, they do define a new operator which is called, in Girard's terminology, *additive conjunction*.

The operator $\neg$. Negation can be defined *á la* Johansson, in terms of implication and the 'falsum' constant $\bot$. In this approach $\neg A$ is defined as $A \rightarrow \bot$. Then the condition $C_\rightarrow$ above immediately yields:

$C_\neg$ $$\Gamma, A \vdash \bot \Longleftrightarrow \Gamma \vdash \neg A.$$

Again, it is easy to see that, given Identity and Surgical cut, the condition $C_\neg$ is equivalent to the following pair of sequent rules:

(5) $\dfrac{\Gamma, A \vdash \bot}{\Gamma \vdash \neg A} \qquad \dfrac{\Gamma \vdash A}{\neg A, \Gamma \vdash \bot}$

The operator $\vee$ The disjunction operator can be defined by the following equivalence:

$\mathrm{C}_\vee \qquad \Gamma, A \vee B \vdash C \iff \Gamma, A \vdash C$ and $\Gamma, B \vdash C$

which corresponds to the the following sequent rules:

(6) $\dfrac{\Gamma, A \vdash C \quad \Gamma, B \vdash C}{\Gamma, A \vee B \vdash C} \qquad \dfrac{\Gamma \vdash A}{\Gamma \vdash A \vee B} \qquad \dfrac{\Gamma \vdash B}{\Gamma \vdash A \vee B}$

Notice that the disjunction operator defined by these rules is 'context-dependent' in a sense similar to the sense in which $\wedge$ is: a crucial condition in the two-premiss rule is that the *context* of the inference step, namely the structure of side formulae Γ, is the same in both premisses. This type of disjunction is called 'additive' to distinguish it from a context-free type of disjunction that arises when we consider logics with an involutive negation operator (see below).

Involutive negation and the operator $⅋$. Involutive negation is characterized by the additional condition expressed by the 'double negation law':

(7) $\neg\neg A \vdash A$

which, given $\mathrm{C}_\neg$ — or equivalently the sequent rules in (5) —, Identity and Surgical Cut, is equivalent to the rule:

(8) $\dfrac{\Gamma, \neg A \vdash \bot}{\Gamma \vdash A}$

Let us now define a binary operator $⅋$ as follows:

(9) $A ⅋ B =_{\mathrm{def}} \neg A \to B$

If (7) holds *and* Exchange is allowed, $⅋$ is commutative and associative. Moreover the following equivalences also hold:

(10) $\Gamma, A \vdash B \iff \Gamma \vdash \neg A ⅋ B \qquad \Gamma, \neg A \vdash B \iff \Gamma \vdash A ⅋ B$

Thus, in the logics satisfying the Exchange property and the double negation law (7) we can naturally introduce *multi-conclusion* sequents of the form

$$\Gamma \vdash \Delta$$

where Δ is (like Γ) a list of formulae, with the intended meaning $\otimes\Gamma \vdash ⅋\Delta$ (by $\otimes\Gamma$ and $⅋\Delta$ we denote, respectively, the $\otimes$-concatenation of the formulae in Γ and the $\parr$-concatenation of the formulae in Δ). The properties of the (classical) negation operator allow us to translate back and forth from the single-conclusion formulation to the multi-conclusion one. The operator $\parr$ corresponds to the comma in the right-hand side of a multi-conclusion sequent, just as the operator $\otimes$ corresponds to the comma in the left-hand side. Under these circumstances the following equivalences can also be easily shown:

(11) $A \otimes B \dashv\vdash \neg(A \rightarrow \neg B)$

(12) $A \otimes B \dashv\vdash \neg(\neg A \parr \neg B)$

When all the usual structural rules are allowed, $\parr$ is clearly equivalent to classical disjunction, just as $\otimes$ is equivalent to classical conjunction.

Notice that in the logics satisfying (7) and Exchange both $\otimes$ and $\parr$ are commutative, so the antecedent and the succedent of a sequent can be regarded as *multisets* rather than just sequences, and Surgical Cut can be replaced by the more familiar (though not necessarily more natural):

$$\frac{\Gamma \vdash A, \Delta \qquad \Lambda, A \vdash \Pi}{\Gamma, \Lambda \vdash \Delta, \Pi}$$

The reader can easily derive multi-conclusion versions of $C_\otimes$, $C_\rightarrow$ and $C_\neg$, and the corresponding sequent rules. Clearly $\parr$ can be defined directly by the following condition:

$C_\parr \qquad \Gamma \vdash \Delta, A, B, \Lambda \Longleftrightarrow \Gamma \vdash \Delta, A \parr B, \Lambda$

This condition is equivalent to the following pair of rules:

$$(13)\quad \frac{\Gamma, A \vdash \Delta \qquad B, \Lambda \vdash \Pi}{\Gamma, A \parr B, \Lambda \vdash \Delta, \Pi} \qquad \frac{\Gamma \vdash \Delta, A, B, \Lambda}{\Gamma \vdash \Delta, A \parr B, \Lambda}$$

which bring up the 'context-free' character of this type of disjunction as opposed to the the context-dependent character of $\vee$.

We conclude this section with a terminological remark. Following Girard [1987] the logical operators we have been considering so far can be classified into two categories, the *multiplicatives* and the *additives*. The multiplicative operators are the 'context-free' ones, i.e. those which can be defined via rules of inference with no conditions on the context of the inference step (see [Avron, 1988b] for this characterization), and include $\rightarrow, \neg, \otimes$ and $\parr$. The 'additive' operators, are the context-dependent conjunction and disjunction $\wedge$ and $\vee$.

2 PROOF-THEORETIC TABLEAUX

An early example of tableau methods for substructural logics is Dunn's method of 'coupled trees' for first degree entailment, described in Fiutting's introduction.

After that, the first work dealing *explicitly* with tableau methods for substructural logics was done by Michael McRobbie and Nuel Belnap Jr. [1979; 1984] and was concerned with systems belonging to the family of *Relevance Logic*. This approach has been recently extended by the same authors with the collaboration of Robert Meyer to the system of *Linear Logic* [1995]. As the authors remark, their method is 'purely proof-theoretic in character, as opposed to that of Smullyan which is *semantical*' [McRobbie and Belnap, 1979, p. 187], meaning by this that the justification of their approach relies entirely on the corresponding substructural sequent calculi. Indeed, their tableau rules can be seen as a rewriting of (suitable variants) of the sequent rules. As remarked in Chapter 2, this kind of 'proof-theoretic' interpretation of the tableau rules can be given also in the case of classical tableaux (and was in fact given by Smullyan himself). However, in this case a 'semantic' interpretation, regarding a closed tableau as a failed attempt to construct a countermodel, is not immediately available.

The link between the 'official' sequent-based formulation of the logics under consideration and the proof-theoretic tableau methods for them, is given by so-called *left-handed sequent system*. By exploiting the properties of classical negation, Gentzen's system **LK** can be reformulated in terms of sequents in which the succedent is always *empty*. This is an immediate consequence of the equivalence

$$\Gamma \vdash A, \Delta \Longleftrightarrow \Gamma, \neg A \vdash \Delta$$

which allows us to move any formula from the succedent to the antecedent. Each left-handed sequent means that the sequence of formulae on the left of the turnstile is *inconsistent*. So, to prove that $\Gamma \vdash A$ we prove, as in classical tableaux, that $\Gamma, \neg A \vdash$, i.e. that $\Gamma, \neg A$ is inconsistent. This implies that $\Gamma \vdash \neg\neg A$ and, given the double negation law, that $\Gamma \vdash A$. The rules of a left-handed sequent system for classical logic are given in Table 1.

In the case of classical logic, where all the three structural rules are allowed, we can replace the two-premiss rules (for conditional, disjunction and negated conjunction) with the following equivalent version where the sequences of 'side-formulae' Γ and Δ are identical in both premiss:

$$(14) \quad \frac{\Gamma, A \vdash \qquad \Gamma, B \vdash}{\Gamma, A \vee B \vdash} \quad \frac{\Gamma, \neg A \vdash \qquad \Gamma, \neg B \vdash}{\Gamma, \neg(A \wedge B) \vdash} \quad \frac{\Gamma, \neg A \vdash \qquad \Gamma, B \vdash}{\Gamma, A \rightarrow B \vdash}$$

It is not difficult to see that, given this 'context-dependent' version of the two-premiss rules, the rules of Smullyan's tableaux are obtained simply by reading the sequent rules upside-down and omitting the side-formulae. We have already observed that when some of the structural rules are disallowed, the context-dependent and the context-free version of the rules are by no means identical and, indeed, give rise to different logical operators, the so-called 'multiplicatives' being associated with the context-free rules and the 'additives' with the context-dependent versions.

AXIOM

$$A, \neg A \vdash$$

STRUCTURAL RULES

Exchange	*Contraction*	*Weakening*
$\dfrac{\Gamma, A, B, \Delta \vdash}{\Gamma, B, A, \Delta \vdash}$	$\dfrac{\Gamma, A, A \vdash}{\Gamma, A \vdash}$	$\dfrac{\Gamma \vdash}{\Gamma, A \vdash}$

OPERATIONAL RULES

Double negation	*Conditional*	*Negated conditional*
$\dfrac{\Gamma, A \vdash}{\Gamma, \neg\neg A \vdash}$	$\dfrac{\Gamma, \neg A \vdash \qquad \Delta, B \vdash}{\Gamma, \Delta, A \to B \vdash}$	$\dfrac{\Gamma, A, \neg B \vdash}{\Gamma, \neg(A \to B) \vdash}$
Conjunction1	*Conjunction2*	*Negated conjunction*
$\dfrac{\Gamma, A \vdash}{\Gamma, A \wedge B \vdash}$	$\dfrac{\Gamma, B \vdash}{\Gamma, A \wedge B \vdash}$	$\dfrac{\Gamma, \neg A \vdash \qquad \Delta, \neg B \vdash}{\Gamma, \Delta, \neg(A \wedge B) \vdash}$
Disjunction	*Negated disjunction1*	*Negated disjunction2*
$\dfrac{\Gamma, A \vdash \qquad \Delta, B \vdash}{\Gamma, \Delta, A \vee B \vdash}$	$\dfrac{\Gamma, \neg A \vdash}{\Gamma \neg(A \vee B) \vdash}$	$\dfrac{\Gamma, \neg B \vdash}{\Gamma \neg(A \vee B) \vdash}$

Table 1. A left-handed version of Gentzen's **LK**

2.1 *Proof-theoretic Tableaux for Relevance Logics*

The first paper on tableaux for Relevance Logic [McRobbie and Belnap, 1979] dealt with the multiplicative operators. As remarked above, given the involutive property of negation which holds in the most popular systems of relevance logics, these operators can all be defined in terms of $\to$ and $\neg$. A suitable left-handed Gentzen's system for the $\{\to, \neg\}$ fragment of Anderson and Belnap's main system of relevance logic, called **R**, is obtained by restricting the operational rules of Table 1 to the rules for $\to$ and $\neg$ and by dropping the Weakening rule, which is responsible for the introduction of irrelevancies in classical (and intuitionistic) reasoning, i.e. formulae which are not *used* in the derivation of the conclusion. It is therefore quite natural to think of relevant tableaux as special cases of classical tableaux with the global requirement that all formula-occurences are *used at least once* as premisses of some rule-application (including the closure rule). This basic observation leads straight to McRobbie and Belnap's 'relevant tableaux'. Consider, for instance, the classical tableau for the implicational formula corresponding to the Weakening rule (displayed in Figure 1). In the figure the $\surd$ sign indicates that the corresponding node has been *used*, namely a tableau rule has been applied to the formula-occurrence with which it is labelled. The tableau is

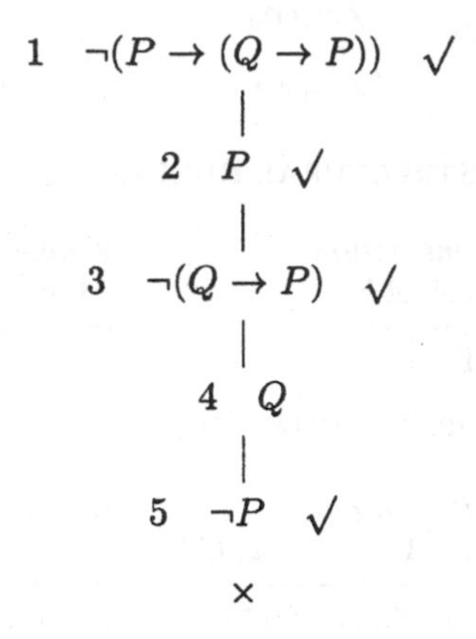

Figure 1. A tableau for the Weakening axiom. This tableau is closed classically but not relevantly

classically closed, but it is not closed relevantly, since it does not satisfy the global requirement of use (the formula-occurence in node 4 is not used).

To stick to pure implicational formulae, let us instead consider the formula corresponding to the structural rule of Contraction. In Figure 2 we show a tableau for this formula which is closed both classically and relevantly, since it satisfies the use requirement, i.e. each node is used at least once (notice that node 4 is used twice as premiss of the closure rule on different branches). In Figure 3 we show another example of a relevantly closed tableau involving the negation operator. So, a relevant tableau for the $\{\to, \neg\}$ fragment of **R** is nothing but a classical tableau satisfying the use requirement. The equivalence between such a tableau method and the left-handed Gentzen system for the same fragment is rather straightorward (a rigorous proof can be found in [McRobbie and Belnap, 1979]): the sequence of formulae on each branch correspond to the antecedent of a sequent and each tableau-expansion rule applied to a node m corresponds to an application of the corresponding sequent rule for the formula labelling m. Notice that in this tableau method the allowed structural rules of the sequent calculus, namely Exchange and Contraction, are 'built in'. The Exchange rule corresponds to the freedom of using nodes occurring anywhere in the branch, and the Contraction rule to the freedom of using a node more than once (as we do in Figure 2). It must be stressed that each node is required to be used at least once *in the whole tree*, and not on every branch. This corresponds to the context-free format of the conditional rule where the sequence of side formulae in the conclusion of the sequent rule is split into the two sequences (Γ and Δ) which occur in the premisses. Observe that in the relevant tableaux of McRobbie and Belnap, there is no guarantee that each tableau-construction is good, i.e. that every tableau for the initial sequence of formulae can

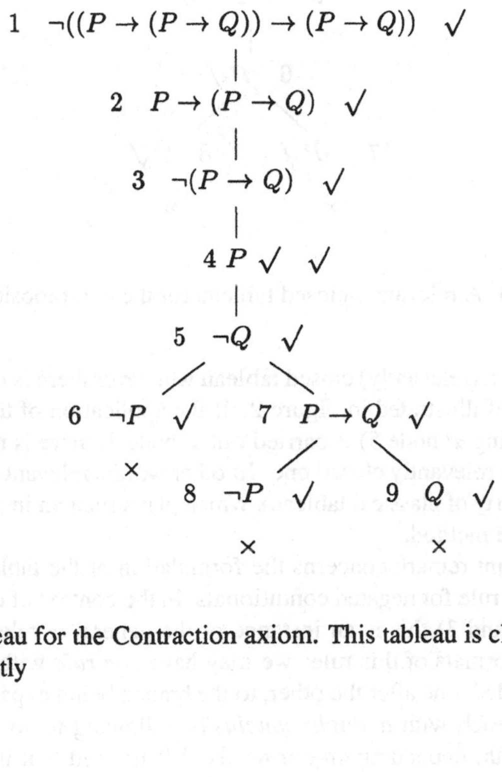

1 $\neg((P \to (P \to Q)) \to (P \to Q))$ √

2 $P \to (P \to Q)$ √

3 $\neg(P \to Q)$ √

4 P √ √

5 $\neg Q$ √

6 $\neg P$ √ ×

7 $P \to Q$ √

8 $\neg P$ √ ×

9 Q √ ×

Figure 2. A tableau for the Contraction axiom. This tableau is closed both classically and relevantly

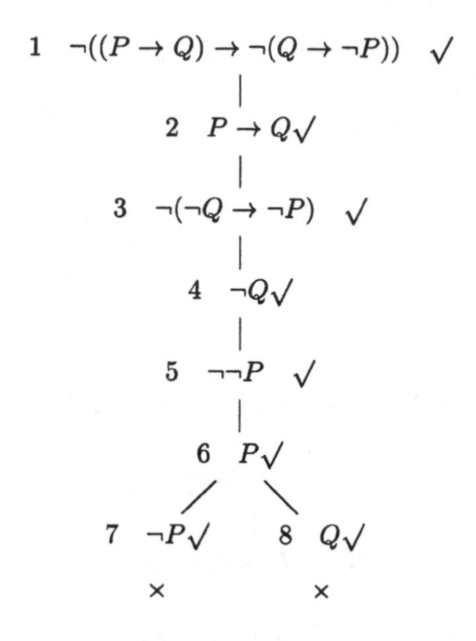

Figure 3. A relevantly closed tableau for the contraposition law

be expanded into a (relevantly) closed tableau whenever there is one. For instance, consider the proof illustrated in Figure 2. If the application of the branching rule to node 2 (occuring *at* node 5) is carried out at node 3, there is no way to expand the tableau into a relevantly closed one. In other words, relevant tableaux lack the *confluence* property of classical tableaux which plays such an important role in the automation of the method.

A last important remark concerns the formulation of the tableau rules. Let us first consider the rule for negated conditionals. In the context of classical tableaux (see Chapters 2 and 3) this is an instance of the general α rule. We considered two alternative formats of this rule: we may have *one rule* with *two conclusions* which are appended, one after the other, to the branch being expanded; or, we may have *two rules*, each with *a single conclusion*, allowing us to append only one conclusion or both, depending on our needs. We noticed that the second format may be preferable for certain refinements of tableaux (such as *regular tableaux*), in order to avoid repetition of identical formulae in the same branch. In the same connection we observed that both the α and the β rules may be interpreted as allowing the addition of certain formulae to a given branch, provided that they do not already occur in the branch itself. It is important to notice that in the context of relevant tableaux we have to stick to the *two-conclusion* version of the α-rule and that repetition of formulae *must not* be avoided. The reason is easy to see. For instance, if we applied the one-conclusion version of the α-rule to node 3 in the tableau contained in Figure 1, so as to obtain only $\neg P$ as conclusion and omitting the conclusion Q in node 4, we would be able to generate a 'relevantly' closed tableau for the positive paradox.

As mentioned in the previous section, the other 'multiplicative' operators, and so their rules, can be defined in terms of $\neg$ and $\rightarrow$. The 'fusion', of A and B, that the relevantists write as $A \circ B$, ($\circ$ is just the same as the operator $\otimes$ discussed in Section 1.2 where we followed the notation introduced by Girard) is defined as $\neg(A \rightarrow \neg B)$, while the 'fission' of A and B, written as $A + B$ (that in Section 1.2 we wrote as $A ⅋ B$ again following Girard's fancier notation) is defined as $\neg A \rightarrow B$. Thus, the rules for fusion and fission are the following:

Fusion

$$\frac{\Gamma, A, B \vdash}{\Gamma, A \circ B \vdash}$$

Negated fusion

$$\frac{\Gamma, \neg A \vdash \quad \Delta, \neg B \vdash}{\Gamma, \Delta, \neg(A \circ B) \vdash}$$

Fission

$$\frac{\Gamma, A \vdash \quad \Delta, B \vdash}{\Gamma, \Delta, A + B \vdash}$$

Negated fission

$$\frac{\Gamma, \neg A, \neg B \vdash}{\Gamma \neg(A + B) \vdash}$$

The reader should notice that the two-premiss rules have the same form as the context-free rules for disjunction and conjunction given in Table 1, while the one-premiss rules are different. For instance the rule for fusion requires *both A and B* to occur in the premiss. The justification for this choice emerges immediately if one translates $A \circ B$ into $\neg(A \rightarrow \neg B)$. Similarly, for the negated fission rule both $\neg A$ and $\neg B$ must occur in the premiss. Again, this difference would not be sufficient to distinguish fusion from conjunction and fission from disjunction if all the structural rules were allowed, but in the absence of Weakening, we are faced with two pairs of operators: on the one side we have the 'additive' disjunction and conjunction defined by the one-premiss rules in Table 1 and by the context-dependent two-premiss rules in (14); on the other, we have the 'multiplicative' versions of these operators, called 'fusion' and 'fission', which are defined by the rules listed above, where the two-premiss rules are the same as the ones in Table 1, once $\wedge$ is replaced with $\circ$ and $\vee$ with $+$. Tableau rules for fusion and fission are obtained immediately by reversing the sequent rules:

(15) $$\frac{A \circ B}{\begin{array}{c} A \\ B \end{array}} \qquad \frac{\neg(A \circ B)}{\neg A \mid \neg B} \qquad \frac{\neg(A + B)}{\begin{array}{c} \neg A \\ \neg B \end{array}} \qquad \frac{A + B}{A \mid B}$$

Let us now turn our attention to the rules for the 'additives', i.e. the operators defined by the context-dependent two-premiss rules for conjunction and disjunction given in (14) and by the one-premiss rules given in Table 1. Tableau methods for relevance logics including these operators were formulated for the first time in [McRobbie and Belnap, 1984]. The one-premiss sequent rules given above yield the following tableau rules:

(16) $\dfrac{A \wedge B}{A}$ $\dfrac{A \wedge B}{B}$ $\dfrac{\neg(A \vee B)}{\neg A}$ $\dfrac{\neg(A \vee B)}{\neg B}$

These rules correspond to the one-conclusion version of the α-rule and clearly bring out the difference between multiplicative and additive operators with respect to the use requirement. While using a formula such as $A \circ B$ generates two conclusions in one go which have *both* to be used to close the tableau, using a formula such as $A \wedge B$ generates one conclusion at a time, so that a formula of this form can be marked as used even if only one of the two conclusions has been generated (and used later on to close the tableau).

The tableau rules corresponding to the two-premiss sequent rules for the additives are more complicated. The sequence of side-formulae that appear in the conclusion must be the same as the one which appears in both premisses. So, when we try to prove a sequent, say, $\Gamma, A \vee B \vdash$ we must prove both sequents $\Gamma, A \vdash$ and $\Gamma, B \vdash$, and therefore *use* all the formula-occurrences in Γ on both branches. What does this mean for the tableau reading of the rules? We observed that when a branch is split by an application of one of the branching rules for the multiplicatives, each formula-occurrence which appears in the branch above this application must be used in *at least one* of the two branches generated by it. In the case of the additives, each formula-occurrence must be used in *both* the branches generated. So, while in the additive-free fragment a tableau is relevantly closed if every node is used at least once in the whole tree, in the multiplicative-free fragment it is necessary that each formula-occurrence is used at least once in each branch. A way of combining these two contrasting requirements for the mixed system containing both the multiplicatives and the additives is described in [McRobbie and Belnap, 1984] and consists in the following modifications of the classical rules:

(17)
$$\frac{A \vee B}{\begin{array}{c|c} A & B \\ C_1 & C_1 \\ \vdots & \vdots \\ C_m & C_m \end{array}} \qquad \frac{\neg(A \wedge B)}{\begin{array}{c|c} \neg A & \neg B \\ C_1 & C_1 \\ \vdots & \vdots \\ C_m & C_m \end{array}}$$

where $C_1, \ldots, C_m$ are all the *unused* formula occurrences appearing in the path up to the node *at which* the rule is applied (i.e. the last node in the branch before the rule application) as opposed to the node *to which* the rule is applied (i.e. the node labelled with the formula that instantiates the premiss of the rule). In fact, these rules are meant as $m + 1$-premiss rules with the 'old' occurrences of $C_1, \ldots, C_m$ as additional premisses which can therefore be marked as 'used' as well as the occurrence of the main premiss. In other words, applying either of these rules amounts to (i) applying the standard β-rule of Smullyan's tableaux to a given formula-occurrence, (ii) appending to both the new branches thereby generated fresh copies of all the unused formula-occurrences appearing in the branch before the splitting, and (iii) marking all the 'old' copies (as well as the formula-occurrence to which the rule is applied) as used. In Figure 4 we show a simple example of a closed tableau for a formula containing both additive and multiplicative operators (notice that the order of application of the rules is crucial), while Figure 5 shows an example of an open tableau for a formula which is a classical theorem but not a theorem of **R**. The reader will notice that the tableau is open as a result of the negated conjunction rule which requires to copy all the unused formula-occurrences in both branches generated by its application. While the

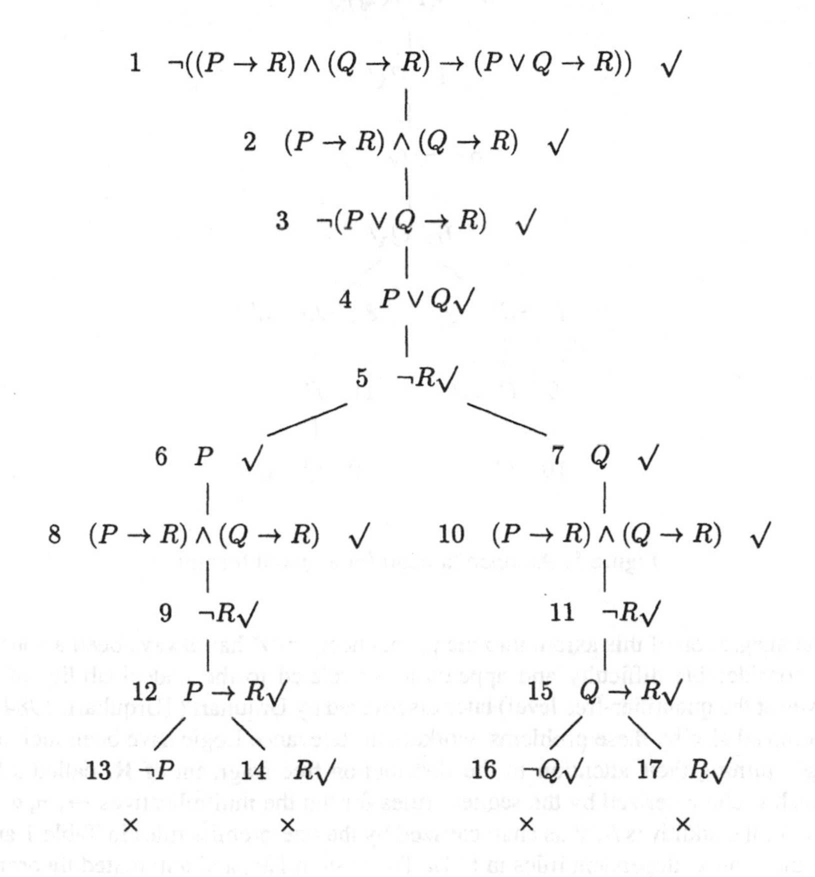

Figure 4. A closed tableau for a mixed formula

left-handed Getzen system for the multiplicative operators yield the multiplicative fragment of Anderson and Belnap's main system of relevance logic **R**, if we augment this system with the left-handed sequent rules for additive conjunction and disjunction we *do not* obtain a system equivalent to the full **R**. Indeed, the resulting system lacks a basic axiom of **R**, namely the following *distributivity principle*:

$$A \wedge (B \vee C) \rightarrow (A \wedge B) \vee (A \wedge C)$$

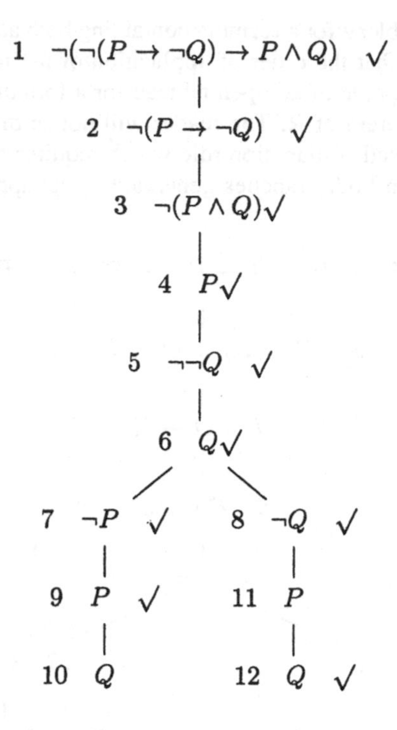

Figure 5. An open tableau for a mixed formula

The integration of this axiom into the proof-theory of **R** has always been a source of considerable difficulty and appears to be related to the undecidability of **R** (even at the quantifier-free level) later discovered by Urquhart ([Urquhart, 1984]). Prompted also by these problems, workers in Relevance Logic have been increasingly turning their attention to the distribution-free fragment of **R**, called **LR**, which is characterized by the sequent rules for the the multiplicatives $\rightarrow, \neg, \circ, +$ and for the additives $\wedge, \vee$ as characterized by the one-premiss rules in Table 1 and by the context-dependent rules in (14). The system **LR**, and automated theorem-proving techniques based on it, are studied extensively in [Thistlewaite *et al.*, 1988].

We conclude this section by mentioning McRobbie and Belnap's treatment of (the distribution-free fragment of) **RM**, i.e. the system of relevance logic which is obtained from any axiom system for **R** by adding the 'mingle' axiom $A \rightarrow (A \rightarrow A)$ or, equivalently, from the left-handed sequent system for **R**, by adding the following structural rule:

$$\frac{\Gamma, A \vdash}{\Gamma, A, A \vdash} \textit{Expansion}$$

This rule amounts to diluting the use requirement as follows: we do not insist that all formula-*occurrences* are used, but we are happy if at least one occurrence of each formula is used. The reader can easily check that this weaker criterion allows us to construct a 'mingle-closed' tableau for the Mingle axiom $A \to (A \to A)$.

The Expansion rule is subtly related to the following derived rule of mingle systems, that we express in the left-handed notation:

$$(18) \quad \frac{\Gamma \vdash \qquad \Delta \vdash}{\Gamma, \Delta \vdash}$$

The derivation of this rule is easier in a language with the $\bot$ operator. In such a language $\bot \vdash \bot$ is clearly an axiom, and we can write $\Gamma \vdash$ as $\Gamma \vdash \bot$. Then the rule is easily derived as follows:

$$\dfrac{\dfrac{\dfrac{\bot \vdash \bot}{\bot, \bot \vdash \bot}\text{ Expansion} \qquad \Gamma \vdash \bot}{\Gamma, \bot \vdash \bot}\text{ Cut} \qquad \Delta \vdash \bot}{\Gamma, \Delta \vdash \bot}\text{ Cut}$$

Notice that in this derivation we have made use of the left-handed version of the Cut rule, namely:

$$\frac{\Gamma, A \vdash \qquad \Delta, \neg A \vdash}{\Gamma, \Delta \vdash}$$

On the other hand, given this rule as primitive, Expansion can be derived as follows:

$$\dfrac{\dfrac{\dfrac{\Gamma, A \vdash \bot \qquad \Gamma, A \vdash \bot}{\Gamma, A, \Gamma, A \vdash \bot}}{\Gamma, \Gamma, A, A \vdash \bot}\text{ Exchange}}{\Gamma, A, A, \vdash \bot}\text{ Contraction}$$

Where the double lines mean that several steps of the indicated rule have been applied to reach the conclusion. (Notice that this derivation does not make any use of the cut rule.)

The above derivation suggests that if we incorporate the rule in question as primitive in our tableau method, we can simulate the role played by Expansion without any need to relax our use requirement, and so stick to the relevant policy that every *node* (formula-occurrence) has to be used. One way of incorporating this rule into the method consists in observing that its effect can be obtained by replacing the standard axioms of the form $A, \neg A \vdash$ with more general ones, of the form $\Gamma, \Gamma' \vdash$ where Γ and Γ' are complementary sequences of formulae, in the sense that their elements are pairwise complementary. To be more precise we can say that two sequences Γ and Γ' are *mingle-complementary* if they satisfy the condition that every element of Γ is complemented in Γ' and viceversa. So, the new more general axioms are all the sequents of the form $\Gamma, \Gamma' \vdash$ where Γ and Γ' are mingle-complementary sequences of formulae. This observation, yields the following more general closure rules:

> *Mingle closure*: a branch ϕ is mingle-closed if it contains two sequences of formulae Γ and Γ' which are mingle-complementary. The closure rule is applied *to all* the formula-occurrences in Γ and Γ' (which are therefore all marked as used).

As an exercise, the reader can try to use this more general rule to prove the formula $(P \to Q) \to ((Q \to R) \to (P \to (Q \to R)))$.

2.2 *Proof-theoretic Tableaux for Linear Logic*

A quite natural approach to understanding Linear Logic, although not the one preferred by the 'linear orthodoxy', consists in regarding it as a special case of Relevance Logic. This was the route taken in [Avron, 1988b], which showed how the funny connectives introduced in Girard's paper [1987] could be understood by analogy with well-known connectives which had been studied thouroughly by relevance logicians. Indeed, according to this interpretation, the difference between Relevance and Linear Logic does not lie in the definition of the logical operators but in their different approach to the manipulation of assumptions in a proof. Relevance logics, like **R**, are already much stricter than classical and intuitionistic logic, in that they do not allow for the introduction of unused assumptions in a proof, so banning irrelevance from our logical world. Linear Logic can be seen as combining this horror for irrelevance with a resource-concerned deductive policy. The Contraction rule embodies the traditional (classical, intuitionistic and relevantist) careless approach which does not distinguish between using a formula once and using it any number of times. By disallowing Contraction, Linear Logic eliminates this residual degree of vagueness from traditional proof-theory: if a formula is to be used n times it must be assumed n times. In terms of left-handed Gentzen systems, and according to this 'reductionist' point of view, the exponential-free fragment of Linear Logic can be seen as arising from the sequent system for **LR**, the distribution-free fragment of **R**, by removing Contraction, all the definitions of the logical operators — as well as the distinction between multiplicatives and additives — remaining the same.

It is straightforward to translate the Linear Logic deductive policy into a stricter *criterion of use*: a Linear proof is a Relevant proof in which each formula is used *exactly once*. This suggests that a tableau method for Linear Logic can be easily obtained from the relevant tableau method of the previous section by simply imposing such a stricter use requirement. This conjecture has been shown to be correct in [Meyer *et al.*, 1995]. The reader can check that the tableau for the Contraction axiom given in Figure 2, though being relevantly closed, is not closed according to this stricter use requirement, since the formula-occurrence at node 4 is used twice. On the other hand, the tableau in Figure 3 is closed linearly, as well as relevantly, since each formula-occurrence is used exactly once.

One of the main features of Linear Logic, which is not present in the relevance tradition, consists in bringing back the whole power of the lost structural rules by means of new *unary operators*, namely Girard's *exponentials*: the 'of course' operator, written as '!', and its dual 'why not', written as '?'. As mentioned in Section 1, $!A$ means that the formula A can be used as many times as needed (including no times at all). So the ! operator provides the means for simulating the effect of both the Weakening and the Contraction rule. In the cited [Meyer *et al.*, 1995] Meyer, McRobbie and Belnap present a treatment of Girard's exponentials in the context of their approach to Linear tableaux outlined above. The main difference between their exposition and ours is that they assume that all formulae are reduced to *De Morgan Normal Form* (DMNF), i.e. to equivalent ones which do not contain $\rightarrow$ and such that the negation operator is applied only to atomic formulae. Assuming such a reduction to DMNF, the left-handed sequent rules for the exponentials are the following ones:

$$(19)\quad \frac{\Gamma, !A, !A \vdash}{\Gamma, !A \vdash} \quad \frac{\Gamma \vdash}{\Gamma, !A \vdash} \quad \frac{\Gamma, A \vdash}{\Gamma, !A \vdash} \quad \frac{!\Gamma, A \vdash}{!\Gamma, ?A \vdash}$$

where $!\Gamma$ means that all the elements of Γ must be of the form $!A$. In the context of Linear tableaux, the first two rules are implemented by relaxing the use requirement so as to exclude the formulae of the form $!A$, i.e. each formula of this form can be used any number n, with $n \geq 0$, of times. The third rule can be directly translated into the following tableau rule:

$$\frac{!A}{\mathrm{A}}$$

As for the last rule in (19), Meyer, McRobbie and Belnap, drawing on its analogy with the **S4** rule for the necessitation operator which can be found in several Gentzen formulations of this logic and on their own the treatment of strict implication in [McRobbie and Belnap, 1979], offer the following translation into their tableau setting:

$$\frac{?A}{\overline{A}}$$

where the double line means that a *barrier* has been introduced between the node *at* which the rule is applied and the new node added by the rule. The meaning of such a barrier is that only formulae of the form $!A$ can cross it, i.e. if there is a barrier between two nodes m and n, where m precedes n, then the only rule that can be applied *to* m *at* n, is the 'necessitation' rule for !. The reader is referred to [Meyer *et al.*, 1995] for the details.

3 LABELLED TABLEAUX

In this section we shall describe a different approach, developed by the present authors, which may be regarded as being closer to the 'semantic' interpretation of classical tableaux. The following exposition is loosely based on [D'Agostino and Gabbay, 1994] to which we refer the reader for further details.

LDS

The methodology of *Labelled Deductive Systems*— or simply LDS — is a unifying framework for the study of logics and their interactions. It was proposed by Dov Gabbay[1996] in response to conceptual pressure arising from application areas, and has now become a research programme which aims to provide logicians, both pure and applied, with a common language and a common set of basic principles in which to express and to solve their problems.

For the theoretician LDS provides the foundations for a *taxonomy* of logics and brings out the common structure underlying different logical systems proposed for a variety of different purposes — whether philosophical, or mathematical, or practical. For the applied logician (e.g. the computer scientist or the engineer) it provides a powerful technique to develop logical systems tailored to the needs of a specific application (see [Gabbay, 1992]), so as to maximize the role of logic in applied research.

In the LDS approach the basic declarative units of a deductive process are not just formulae but *labelled formulae*, i.e. expressions of the form $A : x$ where A is a formula of a standard logical language and x is a term of a given labelling language. The labels refer to elements of a suitable algebraic structure that we call *the algebra of the labels* (or *labelling algebra*). In general, a labelled formula $A : x$ expresses a relation between a formula A and an element x of the labelling algebra. It is not necessary to commit to any particular assumption on the nature of this relation. On some occasions we can read $A : x$ 'semantically' as 'A holds *relative to* x', where x is an element of an appropriate space to which it makes sense to relativize the truth of the proposition A[5]. On others, the label x may identify a region of a *structured database* — whose structure is modelled by the labelling algebra —, and the labelled formula may record metalevel information about the dependence of A on x. According to some interpretations (originating in [William, 1980]) A may also represent a *type* and the label x a *term* of type A. What counts here is that in a labelled deductive system the deduction rules act on the labels as well as on the formulae, according to certain fixed rules of propagation which depend on the intepretation of the labelled formulae.

For example, given a language whose only connective is $\rightarrow$, the *modus ponens* (or $\rightarrow$-elimination) rule of natural deduction may be given a general formulation

[5]For instance a Kripke frame for modal logics; for an LDS approach to normal modal logics see [Russo, 1996].

as

$$\frac{\begin{array}{ll} A \to B : x \\ A & : y \end{array}}{B \quad : f(x, y)}$$

where x and y are terms of a given labelling algebra and f is a function (associated with modus ponens) giving the new label after the rule has been applied.

We can obtain an appropriate $\to$-introduction rule by inverting the elimination rule, as follows:

$$\frac{\begin{array}{ll} A & : t \\ B & : f(x, t) \end{array}}{A \to B : x} \text{ with } t \text{ atomic.}$$

To show $A \to B : x$ we assume $A : t$ (with a new atomic label t) and we must prove $B : z$, for a z such that $z = f(x, t)$.

Different f's or different labelling algebras yield different variants of modus ponens and, possibly, different logics. For example, if we take an arbitrary semigroup as the algebra of the labels, with $\circ$ as multiplication and $f(x, y) = x \circ y$, we have the rules:

$$\frac{\begin{array}{ll} A \to B : x \\ A & : y \end{array}}{B \quad : x \circ y} \qquad \frac{\begin{array}{ll} A & : t \\ B & : x \circ t \end{array}}{A \to B : x}$$

The concrete interpretation of these rules depends on the interpretation of the labelling algebra.

In the sequel we shall first describe a uniform Kripke-style semantic framework for the family of substructural logics under consideration and then we shall turn it into a *labelled refutation system* where the 'semantics' is brought into the 'syntax' via a suitable labelling discipline. The resulting method is a generalization of the classical tableau-like system **KE** described in Chapter 2 that covers the whole family of substructural logics in a unified framework where different logics are associated with different 'labelling algebras'. Variants of the Kripke-sytle semantics outlined in this section can be found in a large number of papers. We mention, in particular, [Dôsen, 1988; Dôsen, 1989; Ono and Komori, 1985; Ono, 1993] and [D'Agostino and Gabbay, 1994]. All these semantics ultimately stem from Urquhart's semantics of relevant implication described in [Urquhart, 1972]. For related semantical investigations into substructural logics see also [Avron, 1988a; Wansing, 1993; Sambin, 1993; Allwein and Dunn, 1993; Abrusci, 1991; MacCaull, 1996].

The LDS approach works both for the substructural logics arising from the sequent calculus **LJ**, and for those arising from **LK**. We shall call the former, 'intuitionistic substructural logics' and the latter, 'classical substructural logics'. Observe that, according to the exposition given above (see Section 1.2), each proper substructural logic of the classical type can be characterized as arising from an intuitionistic substructural logic simply by assuming the involution property of negation $\neg\neg A \rightarrow A$.

3.1 Substructural Implication Logics

In this section we deal with the pure implication fragments of substructural logics. Here the distinction between classical and intuitionistic systems is immaterial, in that all *proper* substructural systems of the classical type are also of the intuitionistic type, i.e. the strongest substructural implication logic of the classical type is intuitionistic implication. Since the labelled system described below is nothing but a labelled version of the classical tableau-like system **KE** (see Chapter 2), the logic of classical implication will be trivially obtained by ignoring the labels.

Uniform Semantics for Substructural Implication

As is well-known, intuitionistic implication can be characterized by means of Kripke models. The other substructural implication operators can also be given Kripke-style semantics of some sort. So, a crucial preliminary step towards a uniform treatment of all these implication systems consists in reducing their semantics to the same format. In the next definition we shall define a class of models which allows us to provide a uniform characterization of all the substructural implication logics.

DEFINITION 1. A *s.o.m. model* (s.o.m. stands for 'semilattice ordered monoid') is a 5-tuple $\mathbf{m} = \langle W, \circ, 1, \leq, V \rangle$, where

- $\langle W, \circ, 1 \rangle$ is a monoid with identity 1
- $\leq$ is a (lower) semilattice ordering on W, i.e. if a and b are in W, $a \sqcap b$ also belongs to W, where $a \sqcap b$ is the greatest lower bound of a and b.
- $\circ$ distributes over $\sqcap$, i.e.

$$(x \sqcap y) \circ z = (x \circ z) \sqcap (y \circ z)$$
$$z \circ (x \sqcap y) = (z \circ x) \sqcap (z \circ y)$$

  Notice that this property implies that $\circ$ is order-preserving on both arguments, i.e.

$$a_1 \leq b_1 \text{ and } a_2 \leq b_2 \Longrightarrow a_1 \circ a_2 \leq b_1 \circ b_2$$

- V is a monotonic valuation, i.e. a function of type $P \mapsto 2^W$, where P is the set of propositional letters, satisfying the following *persistence* condition (we write '$a \in \mathbf{m}$' to mean that a belongs to the *domain* of $\mathbf{m}$:

$$(\forall a, b \in \mathbf{m})(\forall p \in P)(a \leq b \text{ and } a \in V(p) \Longrightarrow b \in V(p))$$

and the following (downward) *continuity* condition :

$$(\forall a, b \in \mathbf{m})(\forall p \in P)(a \in V(p) \text{ and } b \in V(p) \Longrightarrow a \sqcap b \in V(p)).$$

The 'forcing' relation $a \Vdash_{\mathbf{m}} A$ (read 'a forces A in $\mathbf{m}$', 'A is true at a in $\mathbf{m}$' or '$\mathbf{m}$ verifies A at a'), where $\mathbf{m}$ is a *s.o.m.* model, is defined as follows:

1. $(\forall p \in P)(\forall a \in \mathbf{m})(a \Vdash_{\mathbf{m}} p \Longleftrightarrow a \in V(p))$
2. $(\forall a \in \mathbf{m})(a \Vdash_{\mathbf{m}} A \to B \Longleftrightarrow (\forall b \in \mathbf{m})(b \Vdash_{\mathbf{m}} A \Longrightarrow (a \circ b \Vdash_{\mathbf{m}} B)))$

We say that a formula A is *true* in a *s.o.m.* model $\mathbf{m}$, and write $\mathbf{m} \models A$, if and only if $1 \Vdash_{\mathbf{m}} A$.

Observe that the Kripke-style semantics described above is a generalization of Urquhart's semantics for relevant implication introduced in [Urquhart, 1972]. It is easy to show, from the above definition, that both the persistence and the continuity condition must hold also for arbitrary formulae, that is for all $a, b \in \mathbf{m}$ and all formulae A:

$$a \leq b \text{ and } a \Vdash_{\mathbf{m}} A \Longrightarrow b \Vdash_{\mathbf{m}} A$$

and

$$a \Vdash_{\mathbf{m}} A \text{ and } b \Vdash_{\mathbf{m}} A \Longrightarrow a \sqcap b \Vdash_{\mathbf{m}} A.$$

We can now construe all the substructural implication systems as characterized by similar structures, that is different classes of *s.o.m.* models, each class being identified by a different subset of the following additional constraints:

C1 $(\forall a, b \in \mathbf{m})(a \circ b \leq b \circ a)$

C2 $(\forall a, b \in \mathbf{m})(a \leq a \circ b)$

C3 $(\forall a \in \mathbf{m})(a \circ a \leq a)$

C4 $(\forall a \in \mathbf{m})(a \leq a \circ a)$.

For instance, Linear implication is characterized by the class of all *s.o.m.* models $\mathbf{m}$ such that $\leq$ and $\circ$ satisfy Condition C1 above, while for intuitionistic implication we take the class of all *s.o.m.* models satisfying C1, C2 and C3. Relevant implication corresponds to the constraints C1 and C3, mingle Implication to C1, C3 and C4, etc.

There is a close correspondence between Conditions C1–C4 above and the structural rules of p. 402. Condition C1 corresponds to the Exchange rule, Condition C2 to the Weakening rule, Condition C3 to the Contraction rule, and Condition C4 to the Expansion rule. To stress this correspondence we shall call the conditions C1–C4 *structural constraints*. Just as the sequent calculs for a substructural implication system is obtained from the sequent calculus for intuitionistic implication by disallowing some of the structural rules, the associated class of models is obtained from the class of all *s.o.m.* models by waiving the corresponding structural constraints.

Soundness and Completeness

Let us use the following abbreviations for the structural rules: E for the Exchange rule, C for Contraction, X for eXpansion and K for weaKening. Let us denote by E°, C°, X° and K° the corresponding structural constraints on *s.o.m.* models. By saying that a model is *of type* τ, where τ is a subset of $\{E^\circ, C^\circ, X^\circ, K^\circ\}$, we mean that it satisfies the constraints in τ.

Now, let $\mathbf{S}$ be an arbitrary subset of $\{E, C, X, K\}$. We say that an implication formula A is an $\mathbf{S}$-*theorem*, and write $\vdash_{\mathbf{S}} A$, if the sequent $\vdash A$ can be derived by using only the operational rules and the structural rules in $\mathbf{S}$. Let us denote by $\tau[\mathbf{S}]$ the set of structural constraints corresponding to the structural rules in $\mathbf{S}$ (the correspondence is summarized in Table 2).

$$x \circ y \leq y \circ x \qquad \frac{\Gamma, A, B, \Delta \vdash C}{\Gamma, B, A, \Delta \vdash C}$$

$$x \circ x \leq x \qquad \frac{\Gamma, A, A, \Delta \vdash C}{\Gamma, A, \Delta \vdash C}$$

$$x \leq x \circ y \qquad \frac{\Gamma, \Delta \vdash C}{\Gamma, A, \Delta \vdash C}$$

$$x \leq x \circ x \qquad \frac{\Gamma, A, \Delta \vdash C}{\Gamma, A, A, \Delta \vdash C}$$

Table 2. Correspondence between structural constraints and structural rules

Then, we can prove the following proposition:

PROPOSITION 2. *A formula A is an* $\mathbf{S}$*-theorem if and only if A is true in all models of type* $\tau[\mathbf{S}]$.

Proof. Let us say that a sequent $A_1, \ldots, A_n \vdash B$ is $\tau[\mathbf{S}]$-valid if the formula $A_1 \rightarrow (A_2 \rightarrow \cdots \rightarrow (A_n \rightarrow B) \cdots)$ is true in all models of type $\tau[\mathbf{S}]$. Then it can be easily checked that (i) all the sequents of the form $A \vdash A$ are $\tau[\mathbf{S}]$-valid, and (ii) the sequent rules for $\rightarrow$ and the structural rules in $\mathbf{S}$ preserve $\tau[\mathbf{S}]$-validity. It follows that if A is an $\mathbf{S}$-theorem, it must be $\tau[\mathbf{S}]$-valid. This proves the only-if direction.

For the if-direction, let us say that a set of formulae $\mathbf{U}$ is $\mathbf{S}$-decreasing if it satisfies the following condition: $A \in \mathbf{U}$ and $A \rightarrow B$ is an $\mathbf{S}$-theorem imply that $B \in \mathbf{U}$. Now, consider the 'canonical model' $\mathbf{m} = \langle W, \circ, 1, \leq, V \rangle$ where

- W is the set of all **S**-decreasing sets of formulae;
- $\circ$ is a binary operation defined as follows:

$$a \circ b =_{\text{def}} \{C \mid \exists A \in a, \exists B \in b, \vdash_{\mathbf{S}} A \to (B \to C)\}$$

- 1 is the set $\{A \mid \vdash_{\mathbf{S}} A\}$;
- the relation $\leq$ is set-inclusion;
- V is defined as follows:

$$a \in V(p) \Longleftrightarrow p \in a$$

  for all atomic formulae p and all a in W.

The reader can verify that the canonical model $\mathbf{m}$ is indeed a *s.o.m.* model of type $\tau[\mathbf{S}]$. For instance, the definition of $\circ$ immediately implies its associativity. To see that 1 is an identity element, reason as follows. First, $1 \circ a$ is equal, by definition, to $\{C \mid \exists A \in 1, \exists B \in a, \vdash_{\mathbf{S}} A \to (B \to C)\}$. Now, by the operational rules of the sequent calclus, $A \to (B \to C)$ is an **S**-theorem if and only if the sequent $A, B \vdash C$ is **S**-provable. So $1 \circ a$ is equal to the set of all C's such that $A, B \vdash C$ is **S**-provable, for some $A \in 1$ and some $B \in a$. Since 1 is the set of all **S**-theorems and the cut rule is admissible, $1 \circ a$ is equal to the set of all C's such that $B \vdash C$ is **S**-provable, i.e. $B \to C$ is an **S**-theorem, for some $B \in a$. But, since a is decreasing, such a set is equal to a itself. Hence, $1 \circ a = a$. A similar argument shows that $a \circ 1 = a$. All the other properties of models are easily seen to be satisfied by our canonical model. Moreover, it can be shown, by induction on the structure of A, that:

$$a \Vdash_{\mathbf{m}} A \Longleftrightarrow A \in a$$

holds true for every formula A. For the atomic case, this follows by the definition of V above and the standard definition of $\Vdash_{\mathbf{m}}$. Suppose that $A = B \to C$. Then, by definition of $\Vdash_{\mathbf{m}}$:

$$a \Vdash_{\mathbf{m}} B \to C \Longleftrightarrow (\forall b)(b \Vdash_{\mathbf{m}} B \Longrightarrow a \circ b \Vdash_{\mathbf{m}} C).$$

Let us denote by $[B]$ the set $\{D \mid \vdash_{\mathbf{S}} B \to D\}$. Clearly $[B] \in W$ and $B \in [B]$. So, by inductive hypothesis, $[B] \Vdash_{\mathbf{m}} B$ and, therefore, $a \circ [B] \Vdash_{\mathbf{m}} C$. Again by inductive hypothesis, $C \in a \circ [B]$ which implies, by definition of $\circ$, that $A \to (E \to C)$ is an **S**-theorem, i.e. the sequent $A, E \vdash C$ is **S**-provable, for some $A \in a$ and some $E \in [B]$. Since $B \vdash E$ is **S**-provable for all E in $[B]$, it follows by cut that $A, B \vdash C$ is **S**-provable, and therefore $A \to (B \to C)$ is an **S**-theorem, for some $A \in a$, which means that $B \to C \in a$. Thus we have shown that $a \Vdash_{\mathbf{m}} B \to C$ implies that $B \to C \in a$. The converse is easily shown by a similar argument.

Now, suppose that A is not an $\mathbf{S}$-theorem. Then $A \not\leq 1$ and therefore $1 \not\Vdash_{\mathbf{m}} A$ in the canonical model $\mathbf{m}$. So the canonical model is a countermodel to A of type $\tau[\mathbf{S}]$. This concludes our proof. ■

The above argument shows that every substructural sequent system for implication is sound and complete with respect to the corresponding class of models. So, our task of formulating a uniform semantics for substructural implication systems is completed.

All the possible types of models, which can be defined by imposing different combinations of our four structural constraints $E^\circ, C^\circ, X^\circ$, and K° — and correspond to all the possible substructural implication systems contained in Getnzen's intuitionistic sequent calculus **LJ** — are summarized in Table 3, where each row corresponds to a type of models, while the first column indicates the logic characterized by each type.

	$x \circ y \leq y \circ x$	$x \circ x \leq x$	$x \leq x \circ x$	$x \leq x \circ y$
Lambek's implications				
Linear implication	•			
Relevant implication	•	•		
Mingle implication	•	•	•	
Intuitionistic implication	•	•	•	•
BCK implication	•		•	•

Table 3. Correspondence between implication systems and sets of structural constraints

LKE *for Substructural Implication*

An inferential characterization of substructural implication logics can be obtained by turning the 'semantics' described in the previous section into the rules of a labelled deductive system (in the sense of [Gabbay, 1996])[6]. This takes the form of a labelled generalization of the classical system **KE** discussed in Chapter 2. We call this generalization **LKE**. The rules of **LKE** are tree-expansion rules which are immediately justified by — and are indeed equivalent to — our previous definitions (see [D'Agostino and Gabbay, 1994] for a systematic introduction to the **LKE** system; see also [Broda *et al.*, 1998] for a detailed discussion of the route which leads from sequent-based presentation of substructural consequence relations to **LKE** via algebraic and Kripke-style semantics).

In this section we introduce **KE**-type *elimination* rules for substructural implications. In these rules the declarative units are not just signed formulae as in the

[6]A similar approach is used by Fitting in his 'prefixed' tableaux for classical modal logics [Fitting, 1983]. For an extension to Intuitionistic Logic, see Lincoln Wallen's chapter in this *Handbook*.

classical **KE** system (or in the system of analytic tableaux) but *labelled signed formulae*, or *LS-formulae* for short. The points of our *s.o.m.* models are turned into 'labels', while signs play the usual role, so that $T\ A : x$ is interpreted as 'A is true at point x' and $F\ A : x$ is interpreted as 'A is *not true* at point x'. Different types of *s.o.m.* models — defined in terms of the structural constraints which are imposed — correspond to different *labelling algebras*, i.e. sets of rules that can be used in manipulating the labelling terms to verify whether or not the condition for the application of the closure rule is satisfied.

The rules for the implication fragment are listed in Table 4 and are discussed in more detail below. Notice that the use of *signed* formulae is crucial in this type of system.

$$\frac{\begin{array}{l} T\,A \to B : x \\ T\,A : y \end{array}}{T\,B : x \circ y} \qquad \frac{\begin{array}{l} T\,A \to B : x \\ F\,B : x \circ y \end{array}}{F\,A : y}$$

$$\frac{F\,A \to B : x}{\begin{array}{l} T\,A : a \\ F\,B : x \circ a \end{array}} \quad \text{where } a \text{ is a } \textit{new} \text{ atomic label}$$

$$\frac{}{T\,A : x \mid F\,A : x} \qquad \frac{\begin{array}{l} T\,A : x_1 \\ \vdots \\ T\,A : x_n \\ F\,A : y \end{array}}{\times} \text{ if } \sqcap(x_1, \ldots, x_n) \leq y$$

Table 4. The **LKE** rules for substructural implication

Generalized Bivalence and Non-contradiction The classical notions of truth and falsity are governed by two basic principles: the Principle of Bivalence (every proposition is either true or false) and the Principle of Non-Contradiction (no proposition is both true and false). In the **KE** system (see Chapter 2) these principles are expressed directly as *rules*. The Principle of Bivalence is expressed by the only branching rule of the system,[7] the rule PB, and the Principle of Non-Contradiction by the rule for closing a branch (as in the standard tableau method).

These principles still apply to our generalized framework except that they are expressed in terms of labelled propositions. Given an arbitrary proposition A and an arbitrary point x of a *s.o.m.* model, either it is true that x verifies A or it is false

[7]This fundamental principle of classical semantics is not directly expressed by the rules of the standard tableau method. On this point see Chapter 2.

that x verifies A (i.e. $x \Vdash A$ or $x \nVdash A$). So the classical rule of bivalence can be generalized as follows:

(20)
$$\frac{}{T\,A : x \mid F\,A : x}$$

for every label x.

Notice that the rule PB can be seen as a 'lemma introduction' rule: if A can be derived with label x (i.e. if the right-hand subtree is closed), this fact can be used as a 'lemma' (in the left-hand subtree).

A similar generalization applies to the Principle of Non-Contradiction: it is impossible that the same formula is verified and not verified by the same information token. Moreover, if a formula is verified by a token x it is also verified by every token y such that $x \leq y$. Hence, the following generalized closure rule is justified:

(21)
$$\frac{\begin{array}{c} T\,A : x \\ F\,A : y \end{array}}{\times}$$

provided that $x \leq y$.

In fact, to cover all the logics in the family we are investigating, we need an even more general closure rule. Let us consider models which are expansive but *not* monotonic. Suppose a formula A is verified by each of the points $x_1, \ldots, x_n$. By definition of valuation A is true at $a = x_1 \sqcap \cdots \sqcap x_n$. If the model is expansive we have

$$a \leq \overbrace{a \circ \cdots \circ a}^{n \text{ times}}.$$

Therefore, since $a \leq x_i$ for $i = 1, \ldots, n$, A is verified also by the token $x_1 \circ \cdots \circ x_n$. Hence, a branch containing all the $T\ A : x_i$ and $F\ A : x_1 \circ \cdots x_n$ should be considered closed. This agrees with the fact that in any logic which allows the Expansion rule, the following rule can be derived:

$$\frac{\Gamma \vdash A \qquad \Delta \vdash A}{\Gamma, \Delta \vdash A}.$$

This problem can be overcome by introducing a more general closure rule of which the previous one is a special case:

(22)
$$\frac{\begin{array}{c} T\,A : x_1 \\ \vdots \\ T\,A : x_n \\ F\,A : y \end{array}}{\times}$$

provided that $\bigsqcap\{x_1,\ldots,x_n\} \leq y$. So, in the previous example, a branch containing all $T\ A : x_i$ for $i = 1,\ldots,n$ and $F\ A : x_1 \circ \cdots x_n$, is closed by the above rule because, denoting with a the token $\bigsqcap\{x_1,\ldots,x_n\}$, we have that $a \leq x_i$ for $i = 1,\ldots,n$. Moreover, $a \leq a \circ \cdots \circ a$, and thus $a \leq x_1 \circ \cdots \circ x_n$. However this more general closure rule is needed only for the logics characterized by models which are expansive, but not monotonic. For all the other logics the simple special case (21) will suffice.

Elimination rules for $\rightarrow$. It is natural to assume that if a 'piece of information' x verifies a complex sentence of the form $A \rightarrow B$, and another 'piece of information' y verifies A, then the composition of the two will verify B, i.e. :

$$(23)\quad (\forall x \in \mathbf{m})(x \Vdash_{\mathbf{m}} A \rightarrow B \Longrightarrow (\forall y \in \mathbf{m})(y \Vdash_{\mathbf{m}} A \Longrightarrow (x \circ y \Vdash_{\mathbf{m}} B))).$$

This is half of the semantic clause given in Definition 1 and justifies immediately the following generalizations of the **KE** elimination rules for $T\ A \rightarrow B$:

$$(24)\quad \frac{\begin{array}{l} T\ A \rightarrow B : x \\ T\ A : y \end{array}}{T\ B : x \circ y} \qquad \frac{\begin{array}{l} T\ A \rightarrow B : x \\ F\ B : x \circ y \end{array}}{F\ A : y}\ .$$

If we assume that the converse of (23) also holds, as we do in Definition 1 we can justify also the following generalization of the **KE** elimination rule for $FA \rightarrow B$:

$$(25)\quad \frac{FA \rightarrow B : x}{\begin{array}{l} TA : a \\ FB : x \circ a \end{array}} \quad \boxed{\text{where } a \text{ is a } \textit{new atomic} \text{ label}}\ .$$

The new atomic label a stands for a 'piece of information' which instantiates the y in the existentially quantified expression which is obtained by contraposing the converse of (23).

The reader can check that the implication rules (together with our PB rule) are equivalent to the if-and-only-if valuation clause in Definition 1. This condition is formally identical to Urquhart's semantics of relevant implication [Urquhart, 1972]. However, here $\circ$ is not a semilattice join, but a monoid operation. Thus, the above condition can be used to characterize the whole family of substructural implication systems and not just the implication fragment of **R**.

The labelling algebra. For each implication system, the *labelling language* consists of a denumerable set of atomic labels denoted by 'a, b, c', etc. (possibly with subscripts), a distinguished atomic label '1', called *the unit label*, the identity symbol '=', and two binary operation symbols '$\circ$' and '$\sqcap$'. Complex labels are built up from the atomic ones by means of the binary operation symbols. The *basic labelling algebra*, which applies to all *s.o.m.* models, consists of the following ingredients:

1. the axioms expressing the fact that $\circ$ is a monoid operation with 1 as identity, i.e.

$$x \circ (y \circ z) = (x \circ y) \circ z \tag{26}$$

$$1 \circ x = x \circ 1 = x \tag{27}$$

2. The axioms expressing the fact that $\sqcap$ is a semilattice meet:

$$x \sqcap (y \sqcap z) = (x \sqcap y) \sqcap z \tag{28}$$

$$x \sqcap y = y \sqcap x \tag{29}$$

$$x \sqcap x = x \tag{30}$$

3. The axioms expressing the distributivity of $\sqcap$ over $\circ$:

$$(x \sqcap y) \circ z = (x \circ z) \sqcap (y \circ z) \tag{31}$$

$$z \circ (x \sqcap y) = (z \circ x) \sqcap (z \circ y) \tag{32}$$

Clearly all the variables x, y, z are intended as being universally quantified. The partial ordering $\leq$ can be defined, as usual, by putting $x \leq y$ if and only if $x \sqcap y = x$. Observe that the above axioms imply that $\circ$ is order-preserving, i.e.

(33) $x \leq y \Longrightarrow x \circ z \leq y \circ z \quad x \leq y \Longrightarrow z \circ x \leq z \circ y$

This basic labelling algebra augmented with a set τ of structural constraints, provides a specific labelling algebra sufficient to charaterize the notion of validity in the class $\mathcal{M}_{\langle\tau\rangle}$, i.e. the class of all models satisfying the constraints in τ. We shall denote by $\mathbf{LKE}\langle\tau\rangle$, the $\mathbf{LKE}$ system equipped with the labelling algebra obtained by adding the constraints in τ to the basic labelling algebra.

LKE-proofs. A proof of the validity of a formula A for the class $\mathcal{M}_{\langle\tau\rangle}$ consists in a refutation of the assumption that A is false at the identity element 1 of some model in $\mathcal{M}_{\langle\tau\rangle}$. Such a refutation is represented by a closed $\mathbf{LKE}\langle\tau\rangle$-tree starting with the LS-formula $F\ A : 1$, where the constraints in τ may be used together with the basic labelling algebra in order to close a branch. This means that the closure condition $\sqcap x_1, \ldots, x_n \leq y$ may be shown to hold on the basis of the labelling algebra. Whenever such a closed tree can be constructed, we say that A is an $\mathbf{LKE}\langle\tau\rangle$-theorem.

We shall also denote by $\vdash_{\mathbf{LKE}\langle\tau\rangle}$ the (finitary) deducibility relation of $\mathbf{LKE}\langle\tau\rangle$ defined as follows (where [] represents the empty list of formulae):

1. $[] \vdash_{\mathbf{LKE}\langle\tau\rangle} A$ iff A is an $\mathbf{LKE}\langle\tau\rangle$-theorem;

2. $A_1, \ldots, A_n \vdash_{\mathbf{LKE}\langle\tau\rangle} B$ iff $[] \vdash_{\mathbf{LKE}\langle\tau\rangle} A_1 \rightarrow (A_2 \rightarrow \cdots \rightarrow (A_n \rightarrow B) \cdots)$.

We shall also write simply $\vdash_{\mathbf{LKE}\langle\mathcal{T}\rangle}$ instead of $[] \vdash_{\mathbf{LKE}\langle\mathcal{T}\rangle} A$. This definition implies that $A_1, \ldots, A_n \vdash_{\mathbf{LKE}\langle\mathcal{T}\rangle} B$ if and only if there is a closed $\mathbf{LKE}\langle\mathcal{T}\rangle$-tree starting with the sequence of LS-formulae

$$T\,A_1 : a_1, \ldots, T\,A_n : a_n, F\,B : a_1 \circ \cdots \circ a_n$$

where $a_1, \ldots, a_n$ are all *distinct atomic* labels.

In this approach the whole family of substructural implication logics is, therefore, characterised by *the same* tree-expansion rules, and different members of the family are identified by the different labelling algebras that can be employed to check branch-closure. This is what we call the *separation-by-closure* property of the **LKE** system. Another property of **LKE** is the *atomic closure property*: if there is a closed tree starting from a given set of initial LS-formulae, then there is also one starting from the same set of LS-formulae such that the closure rule is applied only with *atomic* LS-formulae, i.e. LS-formulae of the form $\mathbf{s}p : x$ (where $\mathbf{s} = T$ or F) with p atomic.

Two examples of **LKE**-proofs are given in Figures 6 and 7. The first example shows a proof of the Contraction axiom, while the second shows a proof of the Weakening axiom:

$$\begin{array}{l}
F\,(P \to (P \to Q)) \to (P \to Q) : 1 \\
T\,P \to (P \to Q) : a \\
F\,P \to Q : 1 \circ a (= a) \\
T\,P : b \\
F\,Q : a \circ b \\
T\,P \to B : a \circ b \\
T\,Q : a \circ b \circ b
\end{array}$$

Figure 6. The Contraction axiom is valid for the class of contractive models (i.e. those satisfying the condition $x \circ x \leq x$, for all x). For, $b \circ b \leq b$ and, since $\circ$ is order-preserving, $a \circ b \circ b \leq a \circ b$. Therefore, this one-branch tree is closed. The axiom is not valid in Linear Logic which is characterised by a class of models which are not contractive

Liberalized rules. As we have seen, the F$\to$-rule introduces a *new* atomic label. In fact, we can formulate a 'liberalized' version of this rule to the effect that the atomic label introduced by it does not have to be *new*.

One of the crucial properties of our models ensures that whenever there are several 'pieces of information' which verify the same proposition A, their lattice meet still verifies A. Therefore, we can always assume, without loss of soundness, that the new label a introduced by the F$\to$-rule refers to the meet of all points b such that the labelled signed formula $TA : b$ occurs in the branch as a result of an application of the F$\to$ rule to conditionals sharing the same antecedent. Consider,

$$F\,(P \to (Q \to P) : 1$$
$$T\,P : a$$
$$F\,Q \to P : 1 \circ a (= a)$$
$$T\,Q : b$$
$$F\,P : a \circ b$$

Figure 7. The Weakening axiom is valid for the class of monotonic models (i.e. those satisfying the condition $x \leq x \circ y$, for all x and y). The axiom is not valid in Linear Logic and in Relevance Logic which are characterised by classes of models which are not monotonic

for example, the rule (25). The forcing clause for $\to$ (see Definition 1) implies that if x does not force $A \to B$, then there exists a y such that y forces A and $x \circ y$ does not force B. Let a be such a y. Now, suppose that after analysing $FA \to B : x$ we come across, in the same branch, a signed formula of the form $FA \to C : z$. If we apply the rule to $FA \to C : z$, we obtain $TA : b$ with a *new* atomic label b, and $FC : z \circ b$. Observe that both inferences remain sound, with respect to our semantics, if we replace a and b with their meet. For, $a \sqcap b \leq a$ and $a \sqcap b \leq b$, so if B is not forced by $x \circ a$, *a fortiori* it will not be forced by $a \circ (a \sqcap b)$. Similarly, if C is not forced by $z \circ b$, it will not be forced $z \circ (a \sqcap b)$. Therefore, we can assume, without loss of soundness, that the two new atomic labels introduced by the two applications of F$\to$ denote the same 'piece of information'. We can generalize this argument to any finite number of formulae of the form $TA : a$ which are introduced in a branch by the F$\to$ rule. Therefore:

FACT 3. If $T\ A : a_1, \ldots, T\ A : a_n$, with $a_1, \ldots, a_n$ atomic, occur in the same branch as a result of applications of the F$\to$ rule, then we can assume that $a_1 = \cdots = a_n$.

The above fact ensures that the domain of atomic labels introduced in a branch is not extended unnecessarily.

We can formulate the F$\to$ rule in a more convenient way so as to incorporate the content of Fact 3.

DEFINITION 4. We say that an *atomic label* a is *A-characteristic in a branch* ϕ if $TA : a$ occurs in ϕ as a result of an application of the F$\to$ rule.

Exploiting this terminology, we can formulate a 'liberalized' version of the rule as follows:

(34) $$\frac{FA \to B : x}{\begin{array}{l} TA : a \\ FB : x \circ a \end{array}}$$ for some *A-characteristic* atomic label a

In this formulation the atomic label does not have to be *new*. For instance, this version of the rule would allow for the following expansion step:

$$(35) \quad \frac{FA \rightarrow B : x}{FB : x \circ a}$$

whenever a is an atomic label such that $TA : a$ occurs above in the branch as a result of an application of the F→-rule (not of the PB rule). Only if no A-characteristic atomic label occurs in the branch does the rule force us to introduce a *new* atomic label a. The justification of this reformulation of the F→-rule is somewhat similar to the justification of the 'liberalized' δ rule in first-order analytic tableaux (on this point see [Beckert *et al.*, 1993] and Chapter 3 of this *Handbook*). In Figure 8 we show how the liberalized F→-rule simplifies the proof of the 'mingle' axiom $A \rightarrow (A \rightarrow A)$. The axiom is valid for the class of expansive models (i.e. those satisfying the condition $x \leq x \circ x$, for all x), so it is not valid in Linear Logic and in the relevance logic **R** which are characterised by classes of models which are not expansive. Contrast the proof on the left-hand side which makes use of the liberalized F→ rule with the proof on the right-hand side which does not and requires some further reasoning on the algebra of the labels While the first tree is immediately seen to be closed because of the two complementary formulae $TA : a$ and $FA : a \circ a$, for the second tree we have to apply the general closure rule to the formulae $TA : a, TA : b$ and $FA : a \circ b$. To see that this is a sound application of the closure rule, we need the following sequence of inequations which are justified by the algebra of the labels:

$$\begin{aligned} a \sqcap b &\leq (a \sqcap b) \circ (a \sqcap b) \\ a \sqcap b &\leq a \\ a \sqcap b &\leq b \\ a \sqcap b &\leq a \circ b. \end{aligned}$$

	$F\,(P \rightarrow (P \rightarrow P) : 1$
$F\,(P \rightarrow (P \rightarrow P) : 1$	$T\,P : a$
$T\,P : a$	$F\,P \rightarrow P : 1 \circ a (= a)$
$F\,P \rightarrow P : 1 \circ a (= a)$	$T\,P : b$
$F\,P : a \circ a$	$F\,P : a \circ b$

Figure 8. Two proofs of the mingle axiom

Free variables in the labels. So far, our examples have not made any use of the branching rule of 'generalized bivalence'. Indeed, this rule introduces a good deal of non-determinism into the system, in that it allows for the use of (a) arbitrary formulae and (b) arbitrary labels in each rule application. However, this non-determinism can be tamed to some extent. As for (a), it can be shown that the applications of PB can be restricted to *analytic* ones, i.e. involving only subformulae of formulae previously occurring in the branch. Let us say that an **LKE**-tree enjoys the *subformula property* if for every LS-formula $\mathsf{s}A : x$ occuring in it, where s is one of the two signs T or F, the formula A is a subformula of some formula B such that $\mathsf{s}B : y$, for some s and some y, occurs above in the same branch. Let us now say that two closed **LKE**-trees are *equivalent* if they are constructed starting from the same set of initial LS-formulae. Then it can be shown that:

PROPOSITION 5. *There is a procedure to turn every closed* **LKE***-tree into an equivalent one which enjoys the subformula property.*

The proof of this proposition (that we omit here) involves showing that all the applications of the labelled PB rule which violate the subformula property can be eliminated without loss. Indeed, it can also be shown that the applications of the PB rule can be restricted even further to *canonical* applications as in the canonical procedure for the classical **KE** system outlined in Chapter 2.

As for (b), it must be remarked that in order to obtain a fully mechanical refutation procedure we need also to 'tame' the non-determinism related to the use of labels in the generalized PB rule, i.e. we have to develop a procedure which either terminates with a closed tree or gives us enough information to construct a countermodel. To solve this problem, the best strategy is to apply the PB rule with a *variable* as label. This means that branches may be closed under appropriate substitutions of the variables with terms of the labelling algebra. An example of this use of variable labels is given in Figure 9. So, the closure of a branch ultimately depends on the solution of an inequation in the given algebra of the labels and the closure of the whole tree on the simultaneous solution of a system of inequations. This is a well defined algebraic problem which can be addressed via unification-like techniques. The solution to this problem crucially depends on what structural constraints are allowed in the algebra of the labels and on the development of suitable algorithms for solving the systems of inequations associated with **LKE**-trees with variables in the labels. It seems plausible that efficient decision algorithms based on our method will have to be logic-specific, exploiting the computational properties of each given labelling algebra (see [Broda *et al.*, 1997]).

3.2 Full Substructural Logics of the Intuitionistic Type

We shall now consider the other operators defined in Section 1.2 and show how they can be characterized by suitable elimination rules in our labelled **KE** system. For the time being, we shall restrict ourselves to intuitionistic substructural logics, namely those with a non-involutive negation. Again, these rules can be seen as importing the 'semantics' of the given operators into the syntax.

$$F\ ((P \to P) \to Q) \to ((Q \to R) \to R) : 1$$
$$T\ (P \to P) \to Q : a$$
$$F\ (Q \to R) \to R : a$$
$$T\ Q \to R : b$$
$$F R : a \circ b$$

$T\ P \to P : x$	$F\ P \to P : x$
$T\ Q : a \circ x$	$T\ P : c$
$T\ R : b \circ a \circ x$	$F\ P : x \circ c$

Figure 9. This tree is closed for all commutative models under the substitution $x = 1$

Semantics and **LKE**-*rules for* $\otimes$ *and* $\wedge$

The forcing relation over *s.o.m.* models defined in Section 3.1 can be extended to deal with the operator $\otimes$ as follows:

(36) $(\forall a \in \mathbf{m})(a \Vdash_{\mathbf{m}} A \otimes B \iff (\exists b \in \mathbf{m})(\exists c \in \mathbf{m})(b \circ c \leq a$ and $b \Vdash_{\mathbf{m}} A$ and $c \Vdash_{\mathbf{m}} B$.

Adding this clause to the definition of the forcing relation is sufficient to characterize the $\{\to, \otimes\}$ fragments of substructural logics. (It is not difficult to adapt the argument based on the canonical model given in the proof of Proposition 2 to show the equivalence between this extended semantics and the corresponding substructural sequent calculi.)

From (36) we can immediately derive the following **LKE**-rules:

(37) $\dfrac{T\ A \otimes B : x}{\begin{array}{l} T\ A : a \\ T\ B : b \end{array}}$ $(*)$. $\dfrac{\begin{array}{l} F\ A \otimes B : x \\ T\ A : y \end{array}}{F\ B : z}$ $(**)$ $\dfrac{\begin{array}{l} F\ A \otimes B : x \\ T\ B : z \end{array}}{F\ A : y}$ $(**)$

with the following side-conditions:

(*) a and b are *new atomic* labels satisfying the *local* constraint $a \circ b \leq x$;
(**) for all y and z such that $y \circ z \leq x$

We call the constraint in (*) *local* because it holds for the particular tentative countermodel under construction, as opposed to the *global* constraints, such as Monotonicity or Contraction, wich hold for *all* the models characterizing a given logic. Local constraints provide inequations that, together with the global constraints, may allow us to close a branch. The reader can check that our elimination rules (together with PB) are equivalent to the forcing clause given above.

We can formulate a liberalized version of the T$\otimes$ rule as we did for the F$\rightarrow$ rule as follows. First we extend the definition of A-characteristic label (see Definition 4):

DEFINITION 6. We say that an *atomic label* a is *A-characteristic in a branch* ϕ if $TA : a$ occurs in ϕ as a result of an application of the F$\rightarrow$ rule or of the T$\otimes$ rule.

Then, we can formulate our liberalized rule for $\otimes$:

(38) $$\frac{T\,A \otimes B : x}{\begin{array}{l} T\,A : a \\ T\,B : b \end{array}} \quad (***).$$

(***) where a and b are, respectively, A-characteristic and B-charac-. teristic atomic labels satisfying the *local* constraint $a \circ b \leq x$

Notice that if the local constraint $a \circ b \leq x$ is satisfied by two elements a and b, such that a forces A and b forces B, then it is also satisfied if a and b are, respectively, A-characteristic and B-characteristic. This means that, working with the liberalized T$\otimes$ rule we can always assume that the characteristic labels satisfy the local constraint (see [Broda *et al.*, 1997]).

A characterization of the additive conjunction $\wedge$ is obtained by adding a suitable semantic clause for $\wedge$:

(39) $(\forall a \in \mathbf{m})(a \Vdash_{\mathbf{m}} A \wedge B \Longleftrightarrow a \Vdash_{\mathbf{m}} A \text{ and } a \Vdash_{\mathbf{m}} B$

which can be immediately turned into the following $\mathbf{LKE}$-rules:

(40) $$\frac{T\,A \wedge B : x}{\begin{array}{l} T\,A : x \\ T\,B : x \end{array}} \qquad \frac{\begin{array}{l} F\,A \wedge B : x \\ T\,A : x \end{array}}{F\,B : x} \qquad \frac{\begin{array}{l} F\,A \wedge B : x \\ T\,B : x \end{array}}{F\,B : x}$$

Semantics and $\mathbf{LKE}$*-rules for* $\vee$

We want now to extend the forcing relation to formulae of the form $A \vee B$ where $\vee$ denotes the additive disjunction. This extension is not nearly as smooth as it was for the additive conjunction. Let us therefore attempt an intuitive analysis of the situation in terms of our intended model.

What does it mean that a piece of information x verifies a disjunction $A \vee B$? It does *not* necessarily mean that x verifies A or x verifies B, since the information carried by each of the propositions A and B is clearly greater than the information carried by their disjunction $A \vee B$. As we have seen above, in the sequent-based presentation of substructural logics additive disjunction is characterized by the following triple of rules:

$$\frac{\Gamma \vdash A}{\Gamma \vdash A \vee B} \qquad \frac{\Gamma \vdash B}{\Gamma \vdash A \vee B} \qquad \frac{\Gamma, A, \Delta \vdash C \qquad \Gamma, B, \Delta \vdash C}{\Gamma, A \vee B, \Delta \vdash C}$$

Let $[A] = \{B \mid A \vdash B\}$. A concise explanation of the meaning of $A \vee B$ can then be given as follows:

(41) $x \Vdash A \vee B \iff ((\forall C) C \in [A] \cap [B] \implies x \Vdash C)$

In other words, x verifies a disjunction $A \vee B$ if and only if x verifies all the propositions that can be inferred by both A and B (separately). Consider two pieces of information a and b such that a verifies A, b verifies B, and a and b are incomparable (neither $a \leq b$ nor $b \leq a$). Let us take their meet $a \sqcap b$. According to (41), $a \sqcap b$ verifies $A \vee B$, but it may well be that it does not verify either A or B.

It is not difficult to show that this explanation of the meaning of $\vee$ is equivalent to the combination of the sequent rules given above, as well as to the following semantic presentation (given in [Dôsen, 1989] and [Wansing, 1993]):

(42) $x \Vdash A \vee B \iff x \Vdash A$ or
$x \Vdash B$ or $(\exists z_1)(\exists z_2)(z_1 \sqcap z_2 \leq x$ and
$z_1 \Vdash A$ and $z_2 \Vdash B)$

Let us now see an alternative, but equivalent, definition of $x \Vdash A \vee B$ (see [Ono and Komori, 1985] and [Ono, 1993]).

(43) $x \Vdash A \vee B \iff (\exists z_1)(\exists z_2) z_1 \sqcap z_2 \leq x$ and
$(z_1 \Vdash A$ or $z_1 \Vdash B)$ and $(z_2 \Vdash A$ or $z_2 \Vdash B)$.

PROPOSITION 7. *The definitions in (42) and (43) are equivalent.*

Proof. We show that (42) implies (43) and leave it to the reader to show the converse. The if-direction is trivial. For the only-if direction, suppose $x \Vdash A \vee B$. Now, suppose that $x \Vdash A$ or $x \Vdash B$. Then, trivially, there are z_1 and z_2 (both equal to x), satisfying the required condition, since obviously $x \sqcap x \leq x$. If neither $x \Vdash A$, nor $x \Vdash B$, (42) implies that there are z_1 and z_2 such that $z_1 \sqcap z_2 \leq x$ and $z_1 \Vdash A$ and $z_2 \Vdash B$ and the required condition is again satisfied. ■

We have now to translate the semantics in (43) into suitable **LKE**-rules for disjunction. Unfortunately this extension turns out to be a source of inelegancies and complications. This is however no surprise, since similar difficulties are shared by other proof-theoretical approaches and seem inevitable when one tries to deal with the multiplicative and additive operators in the same proof system.

Let us associate with each branch ϕ of an **LKE**-tree and each LS-formula of the form $T\ A \vee B : x$ in ϕ a pair of new atomic labels c_1 and c_2 satisfying the condition in the right-hand side of (43). The equivalence in (43) justifies the following elimination rules:

$$\frac{\begin{array}{l}TA \vee B : x \quad (*)\\ FA : c_i\end{array}}{TB : c_i} \qquad \frac{\begin{array}{l}TA \vee B : x \quad (*)\\ FB : c_i\end{array}}{TA : c_i}$$

$$\frac{FA \vee B : x}{\begin{array}{l|l}FA : x_1 & FA : x_2\\ FB : x_1 & FB : x_2\end{array}} \quad (**)$$

(*) where (i) $i = 1, 2$ and (ii) c_1 and c_2 are the atomic labels associated with $T\ A \vee B : x$ in the branch, satisfying the constraint $c_1 \sqcap c_2 \leq x$.

(**) where x_1 and x_2 are *any* labels satisfying the constraint $x_1 \sqcap x_2 \leq x$. (Notice that in the application of this branching rule we can make use of variables as we did with the other branching rule PB.)

Observe that, for every x, $x \sqcap x \leq x$ and thus, the following is an instance of the rule scheme E$F\vee$:

$$\frac{FA \vee B : x}{\begin{array}{l}FA : x\\ FB : x\end{array}}$$

We now turn our attention to the distribitivity of $\wedge$ and $\vee$. Let us consider the following derived rules for conjunction:

$$(44) \quad \frac{\begin{array}{l}F\ A \wedge B : x\\ T\ A : y\end{array}}{F\ B : x} \qquad \frac{\begin{array}{l}F\ A \wedge B : x\\ T\ B : y\end{array}}{F\ A : x}$$

with the side-condition that $y \leq x$. These can be easily derived as follows:

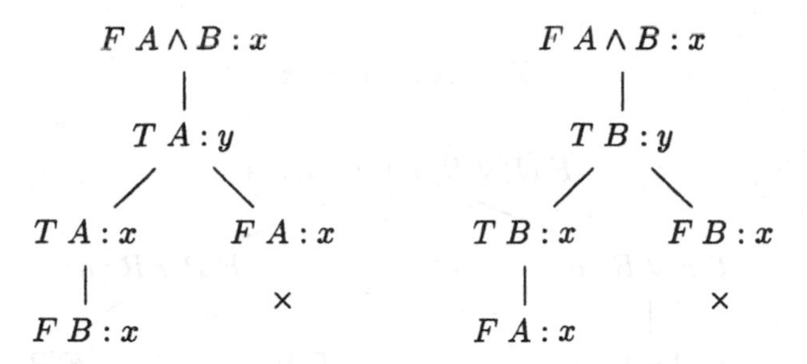

Consider also the following derived rules for disjunction:

(45) $\dfrac{T\ A \vee B : x \quad F\ A : y}{T\ B : c_i}$ $\qquad$ $\dfrac{T\ A \vee B : x \quad F\ B : y}{T\ A : c_i}$

where c_i is one of the two atomic labels associated with $T\ A \vee B : x$, and satisfies the side-condition that $c_i \leq y$. Again, it is easy to derive these rules by means of the official rules. These rules are 'logic-dependent' in that they depend on the specific algebra of the labels used to close one of the two branches.

So, we can replace each application of any of these derived rules with a suitable sequence of applications of the official rules. The tree in Figure 10 shows a proof of the non-critical distributivity law, namely the one which is sound in all substructural logics. The tree in Figure 11 shows a proof of the critical distributivity law which is sound for every system allowing for Monotonicity and Contraction. In the latter we make use of the logic-dependent derived rules described above. However, the application of these derived rules can be eliminated in favour of the official logic-independent rules.

LKE *rules for* $\neg$

Since we define $\neg A$ as $A \rightarrow \bot$, the elimination rules for $\neg$ can be immediately derived from the elimination rules for $\rightarrow$. So, the rule for eliminating $T\ \neg A : x$ will be:

(46) $\dfrac{T\ \neg A : x \quad T\ A : y}{T\ \bot : x \circ y}$

If we are dealing with the intuitionistic-like, non-involutive, negation, completeness is achieved by adding the following rule:

(47) $\dfrac{F\ \neg A : x}{T\ A : a \quad F\ \bot : x \circ a}$ $\quad$ where a is a *new atomic* label

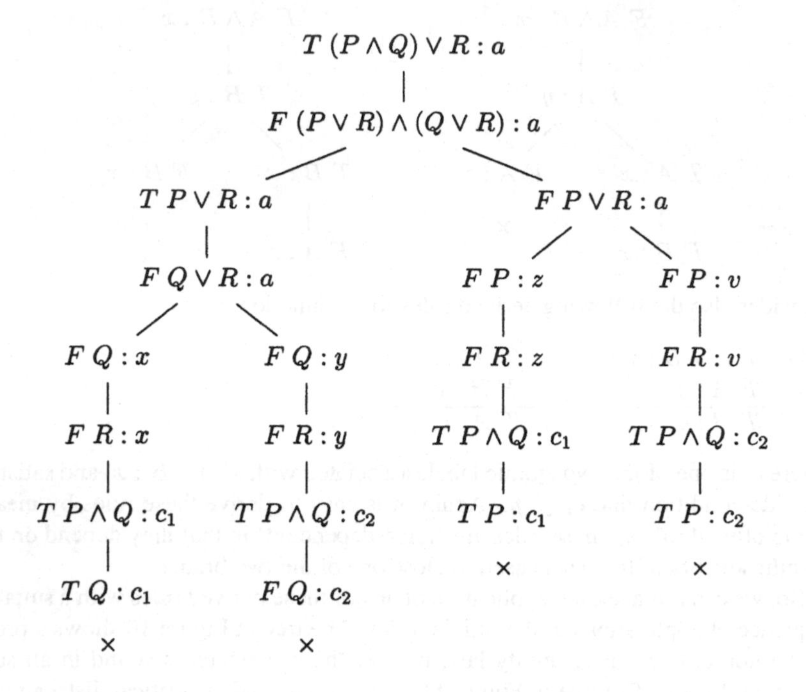

Figure 10. A proof of the non-critical distributivity law, making use of the local constraint $c_1 \sqcap c_2 \leq a$. The tree is sound and closed under the substitution $(x := c_1, y := c_2, z := c_1, v := c_2)$.

In order to characterize logics with the *ex-falso rule*

$$\frac{\Gamma \vdash \bot}{\Gamma \vdash A}$$

we need the following additional closure rule:

$$\text{(48)} \quad \frac{\begin{array}{l} T\,\bot : x \\ F\,A : x \end{array}}{\times}$$

In the next subsection, we shall show how to provide a more elegant and uniform treatment of the negation operator within the **LKE** system.

Constrained Variables

Some of our elimination rules are associated with *local constraints*, namely additional inequalities which are assumed to hold for the tentative countermodel under

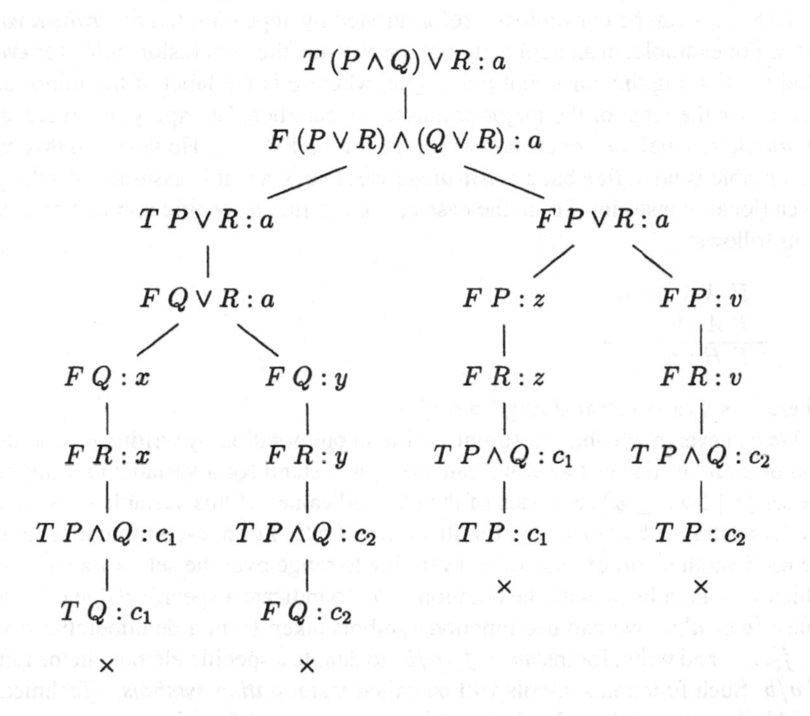

Figure 11. This is a closed tree under the substitution $(x := a \circ c_1, y := a \circ c_2)$ for all systems satisfying Monotonicity and Contraction. Monotonicity is needed to close the branches and for the correct application of the derived rules for conjunction and disjunction; Contraction is needed to show that the constraint $x \sqcap y \leq a$, required for a sound application of the $F\vee$ rule, is also satisfied. For, $(a \circ c_1) \sqcap (a \circ c_2) = a \circ (c_1 \sqcap c_2)$, by definition, and $c_1 \sqcap c_2 \leq a$ since c_1 and c_2 are the atomic labels associated with $T\,(P \wedge Q) \vee R : a$. Thus, $a \circ (c_1 \sqcap c_2) \leq a \circ a$ and, by Contraction, $a \circ (c_1 \sqcap c_2) \leq a$.

construction in a branch ϕ and may be used to close ϕ together with the global properties and constraints of the given algebra of the labels. Let us consider, for instance, the rules for $\otimes$. The T$\otimes$ rule, when applied to a formula $T\ A \otimes B : x$ introduces two atomic labels a and b which are assumed to satisfy the constraint $a \circ b \leq x$. The F$\otimes$ rule can be applied to a formula $F\ A \otimes B : x$ with the minor premiss $T\ A : y$, to yield $F\ A : z$ with the constraint that $y \circ z \leq x$.

Such rules can be conveniently reformulated by appealing to *constrained variables*. For example, in an application of the F$\otimes$ rule the conclusion holds for *every* label z satisfying the constraint $y \circ z \leq x$, where y is the label of the minor premiss and x the label of the major premiss. We can therefore apply this rule using a *variable* to label the conclusion as we do for the PB rule. However, in this case the variable is not a *free* but a *constrained* variable, since it is assumed to satisfy a given (local) constraint. So, in the case of the F$\otimes$ rule a specific application could be as follows:

$$\frac{\begin{array}{l} F\ A \otimes B : a \\ T\ A : b \end{array}}{F\ B : \alpha}$$

where α is a variable satisfying $b \circ \alpha \leq a$.

We can even make this constraint visible in our notation by writing α as a function of a and b. For instance, we can use a/b to stand for a variable ranging over the set $\{z \mid b \circ z \leq a\}$. So, each of the allowed values of this variable satisfies the given constraint: $b \circ (a/b) \leq a$ for all values of a/b. For non-commutative logics, we need another sort of constrained variable to range over the set $\{z \mid z \circ b \leq a\}$, which we can indicate with the notation $a\backslash b$. To indicate a specific value of a variable a/b or $a\backslash b$, we can use function symbols taken from a denumerable stock, $f_1, f_2, \ldots$, and write, for instance, $f_1(a/b)$ to denote a specific element in the range of a/b. Such function symbols will be called *instantiation symbols*. (Technically speaking we could describe them as denoting constant functions mapping every element of the range of the constrained variable to a fixed element in it.) Whether or not we want to consider constrained variables and instantiation symbols as part of our labelling language or as metalanguage abbreviations is, to a large extent, a matter of taste. If we allow them into the labelling language, then we must extend our notion of *atomic label* to include expressions of the form $f(x/y)$ where f is an instantiation operator.

Exploiting this notation the rules for $\otimes$ can be rewritten as follows:

$$(49) \quad \frac{T\ A \otimes B : x}{\begin{array}{l} T\ A : a \\ T\ B : f(x/a) \end{array}} \ (*). \qquad \frac{\begin{array}{l} F\ A \otimes B : x \\ T\ A : y \end{array}}{F\ B : x/y} \qquad \frac{\begin{array}{l} F\ A \otimes B : x \\ T\ B : y \end{array}}{F\ A : x\backslash y}$$

where x/y ranges over the set $\{z \mid y \circ z \leq x\}$. The side-condition (*) on the T$\otimes$ rule is the usual one, namely that a and $f(x/a)$ are, respectively, A-characteristic and B-characteristic atomic labels (in the new extended meaning of 'atomic').

This means that either (i) $f(x/a)$ is equal to a B-characteristic atomic label already occurring in the branch, or (ii) the instantiation symbol f is *new*.

Similar constrained variables can be helpful also to reformulate the negation and the disjunction rules. Let us start with negation. For every model $\mathfrak{m}$ and every element x of its domain, let us define:

$$Cons(x) = \{y \mid x \circ y \not\Vdash_{\mathfrak{m}} \bot\}.$$

In other words, $Cons(x)$ is the set of all pieces of information which are consistent with x. Observe that $Cons(1)$ is the set of all consistent points of the domain, since $1 \circ x = x$ for all x. We shall use the notation x^* for a constrained variable ranging over $Cons(x)$. Then, we can reformulate the F$\neg$ rule as follows:

(50) $$\frac{F\ \neg A : x}{T\ A : f(x^*)}$$

where $f(x^*)$ is an A-characteristic atomic label.

The rule (50) is equivalent to the original one provided that the following 0-premiss rule is also available:

(51) $$\frac{}{F\bot : x \circ f(x^*)}$$

for *every* function symbol f. This rule is clearly sound and it is easy to check that the pair of rules (50) and (51) are equivalent to the rule (47). Notice that the 0-premiss rule for $\bot$ given above is equivalent to the following additional closure rule:

(52) $$\frac{\begin{array}{c} T\ \bot : y_1 \\ \vdots \\ T\ \bot : y_n \end{array}}{\times}$$

provided that $\bigsqcap\{y_i\} \leq x \circ f(x^*)$ for some x and some f. The special case:

(53) $$\frac{T\ \bot : y}{\times}$$

provided that $y \leq x \circ f(x^*)$ for some x and some f, is sufficient for completeness except for the class of logics characterized by models which are expansive but non-monotonic.

When using constrained variables we must make sure that we do not 'create' elements whose existence is not guaranteed by the information available in the branch. For instance, suppose we are working with a monotonic labelling algebra, i.e. the constraint $x \leq x \circ y$ is satisfied for all x and y in the domain of the model. Suppose also that we have obtained a formula of the form $T\ \bot : x$. Now, we may be tempted to conclude that the branch is closed, because $x \leq x \circ f(x^*)$. Since

$x \circ f(x^*)$ is consistent by definition, so should be x. Thus monotonicity would be sufficient to guarantee that all the elements in the domain are consistent and, therefore, that the *ex-falso* rule is sound in every monotonic logic. But Minimal Logic, which can be characterized by removing the *ex-falso* rule from intuitionistic logic, is clearly a counterexample. However, the mistake is easily found: if x is an inconsistent element of the domain (it verifies $\bot$) and the logic is monotonic, the set $\{z \mid x \circ z \not\Vdash_{\mathbf{m}} \bot\}$ is *empty* and therefore $f(x^*)$ simply cannot exist. So, before introducing elements of the form $f(x^*)$, we must make sure that they exist in the model under consideration. Similar considerations hold for the labels of the form $f(x/y)$.

Luckily there are simple sufficient conditions to guarantee that the range of the constrained variables is not empty and, therefore, the expressions constructed by means of the instantiation symbols do have a denotation. First of all, the rules which introduce instantiation symbols take care of themselves, since the existence of a denotation is guaranteed by their premiss. Second, existence conditions depend on the logic under consideration. For instance, let us consider a logic with the *ex-falso* rule. In such a logic, an inconsistent point forces every formula. So, the occurrence of a formula of the form $F\ A : x$ in a branch is sufficient to ensure that $Cons(x)$ is non-empty, because x is certainly consistent (otherwise A would be true at x) and 1 belongs to $Cons(x)$ (since $x \circ 1 = x$). Hence, if $F\ A : x$ occurs in the branch under consideration, the range of x^* is non-empty and the expression $f(x^*)$ is denoting. This means that whenever we have two formulae of the form $T\ \bot : x$ and $F\ A : x$ in the same branch and $x \leq x \circ f(x^*)$, we can declare the branch closed. The logics with the *ex-falso* rule can then be characterized in terms of models satisfying the *global* constraint that $x \leq x \circ f(x^*)$, for all x, whenever $f(x^*)$ is defined, i.e. whenever $Cons(x)$ is non-empty (an LS-formula of the form $FA : x$ occurs in the branch. (This can be seen as a sort of restricted monotonicity which mirrors the fact that the logics in this class are chacterized by sequent systems with Weakening on the *right*.) Using this characterization the rule (48) can be easily derived. The advantage of this approach is that it allows us to cover logics with the *ex-falso* rule withouth any need for a special closure rule, but by imposing additional conditions on the labelling algebra.

Constrained variables can be used also to simplify the formulation of the disjunction rules. Here we have *two* constrained variables, that we denote by $\pi_1 x$ and $\pi_2 x$ respectively, satisfying the constraint:

$$\pi_1 x \sqcap \pi_2 x \leq x.$$

The values that these variables can take depend on each other. To express this dependency we can use the same instantiation symbol and stipulate that $f(\pi_1 x)$ depends on $f(\pi_2 x)$ as required by the constraint and, therefore,

$$f(\pi_1 x) \sqcap f(\pi_2 x) \leq x.$$

The disjunction rules can then be reformulated as follows:

$$\frac{\begin{array}{l}TA \vee B : x \quad (*)\\ FA : f(\pi_i x)\end{array}}{TB : f(\pi_i x)} \qquad \frac{\begin{array}{l}TA \vee B : x \quad (*)\\ FB : f(\pi_i x)\end{array}}{TA : f(\pi_i x)}$$

$$\frac{FA \vee B : x}{\begin{array}{l|l}FA : \pi_1 x & FA : \pi_2 x\\ FB : \pi_1 x & FB : \pi_2 x\end{array}}$$

(*) where (i) $i = 1, 2$ and (ii) $f(\pi_1 x)$ and $f(\pi_2 x)$ are the atomic labels associated with $T\ A \vee B : x$ in the branch.

Soundness and Completeness of **LKE**

Our claim is that the **LKE** systems which arise from augmenting the basic labelling algebra with sets of global constraints are equivalent to the corresponding subsystems of **LJ**, according to the correspondence between global constraints and structural rules.

Using the terminology of Section 3.1, let us say that A is a **S**-theorem, and write $\vdash_{\mathbf{S}} A$, if the sequent $\vdash A$ can be derived by using only the operational rules and the structural rules in **S**. Let us denote by $\tau[\mathbf{S}]$ the set of structural constraints corresponding to the structural rules in **S** (the correspondence is summarized in Table 2). Finally, we say that A is an $\mathbf{LKE}\langle\tau[\mathbf{S}]\rangle$-theorem if there is a closed **LKE**-tree starting with $FA : 1$, in which the the global constraints in $\tau[\mathbf{S}]$ may be used together with the basic labelling algebra in order to close a branch. Then we can prove the following result:

PROPOSITION 8. *A formula A is an* **S**-*theorem if and only if A is an* $\mathbf{LKE}\langle\tau[\mathbf{S}]\rangle$-*theorem.*

Proof. *(Outline)* To show the if-direction we can use the semantics as an intermediary. First extend Proposition 2 above to cover the new operators. The discussion of the **LKE**-rules and of the labelling algebra shows that $\mathbf{LKE}\langle\tau[\mathbf{S}]\rangle$ is sound with respect to the corresponding semantics (based on *s.o.m.* models of the $\tau[\mathbf{S}]$-type and on the forcing relations for all the operators) and therefore, also with respect to the sequent calculi of type **S**. The converse can be shown directly by simulating the sequent rules. Given a provable sequent $A_1, \ldots, A_n \vdash B$, we say that it is $\mathbf{LKE}\langle\tau[\mathbf{S}]\rangle$-provable if there is a closed $\mathbf{LKE}\langle\tau[\mathbf{S}]\rangle$-tree starting with the sequence $T\ A_1 : a_1, \ldots, T\ A_n : a_n \vdash F\ B : a_1 \circ \cdots \circ a_n$, where $a_1, \ldots, a_n$ are all atomic labels such that distinct labels are assigned to distinct formulae. Then the sequent rules can be simulated by showing that if their premisses are provable so are their conclusions. (The argument is not entirely trivial and involves renaming of the labels.) This implies the completeness of $\mathbf{LKE}\langle\tau[\mathbf{S}]\rangle$ with respect to the sequent calculi of type **S**. Since the sequent calculi are, in view of the extension of Proposition 2, complete with respect to the semantics, it follows that

LKE$\langle\tau[\mathbf{S}]\rangle$ is also complete with respect to the corresponding semantics based on *s.o.m.* models of the $\tau[\mathbf{S}]$-type. ■

3.3 Substructural Logics of the Classical Type

We now turn our attention to the substructural logics with involutive negation, that we have called 'classical substructural logics'. A trivial way of characterizing them within our framework would consist in adding a double negation rule of the form

$$\frac{T \neg\neg A : x}{T\, A : x}$$

or some other equivalent rule to our stock of **LKE**-rules. But this *ad hoc* solution is not particularly interesting, and we shall instead explore a different approach which aims to incorporate the involutive property of classical negation into the Kripke-style semantics of substructural logics that we have developed so far. This approach stems quite naturally from our discussion of constrained variables in Section 3.2.

Information Frames

In Section 3.1 we defined a *s.o.m.* model as a 5-tuple $\mathfrak{m} = \langle W, \circ, 1, \leq, V\rangle$. We can distinguish between two components of such a model: the underlying *frame*, namely the 4-tuple $\langle W, \circ, 1, \leq\rangle$ and the *valuation* V. In this way the same frame can support a variety of models depending on the valuation V. By a *s.o.m. frame* we shall therefore mean the underlying frame of a *s.o.m.* model.

The semantics based on *s.o.m.* models is adequate to deal with intuitionistic substructural logics. In order to add involutive negation and other nice properties to this semantics it is convenient to enrich it by moving from *s.o.m.* frames to what we shall call 'information frames' to stress the fact that they appear to have enough structure to model a large variety of information processes (including those involving modalities).

DEFINITION 9. An *information frame* is a structure $\mathcal{F} = (W, \circ, 1, \leq)$ such that:

1. W is a non-empty set of elements called *pieces of information* or *information tokens*;

2. $\leq$ is a partial ordering which makes W into a complete lattice;

3. $\circ$ is a binary associative operation on W which is

   (a) distributive over $\bigsqcup$: for every non-empty family $\{c_i\} \subseteq W$,

   $$\bigsqcup\{c_i \circ a\} = \bigsqcup\{c_i\} \circ a \text{ and } \bigsqcup\{a \circ c_i\} = a \circ \bigsqcup\{c_i\}$$

(b) distributive over $\bigsqcap$: for every non-empty family $\{c_i\} \subseteq W$,

$$\bigsqcap\{c_i \circ a\} = \bigsqcap\{c_i\} \circ a \text{ and } \bigsqcap\{a \circ c_i\} = a \circ \bigsqcap\{c_i\}.$$

4. $1 \in W$ and for every $a \in W$, $a \circ 1 = 1 \circ a = a$.

In other words, an information frame is a *s.o.m.* frame such that (i) the partial ordering $\leq$ is a complete lattice, and (ii) the monoid operation $\circ$ is fully distributive over $\bigsqcup$ and $\bigsqcap$. Recall that the properties of $\circ$ imply that this operation is order-preserving, i.e.

$$a_1 \leq a_2 \Longrightarrow a_1 \circ b \leq a_2 \circ b \text{ and } b \circ a_1 \leq b \circ a_2.$$

In analogy with what we did for *s.o.m.* models, we can define classes of information frames which satisfy additional structural constraints on the ordering $\leq$:

DEFINITION 10. We say that an information frame is:

commutative	if	$a \circ b \leq b \circ a$
contractive	if	$a \circ a \leq a$
expansive	if	$a \leq a \circ a$
monotonic	if	$a \leq a \circ b$

In this new context, a *model* will be defined as follows:

DEFINITION 11. A *model* $\mathbf{m}$ is a pair $\langle \mathcal{F}, V \rangle$ where $\mathcal{F}$ is an information frame and V is a valuation of atomic formulae satisfying the persistence condition of Definition 1 and the following continuity conditions:

1. *(downward continuity)* For every atomic formula p and every non-empty set S of information tokens:

$$(\forall a \in S) a \in V(p) \Longrightarrow \bigsqcap S \in V(p)$$

2. *(upward continuity)* For every atomic formula p and every non-empty set S of information tokens:

$$(\forall a \in S) a \notin V(p) \Longrightarrow \bigsqcup S \notin V(p).$$

The 'meaning' of the conditional operator (as expressed by the forcing relation) is the same as before. The reader can check that the continuity properties, as well as the persistence condition, hold also for arbitrary formulae, i.e.

(54) $a \leq b$ and $a \Vdash_{\mathbf{m}} A \Longrightarrow b \Vdash_{\mathbf{m}} A$

and, for every non-empty set S of information tokens:

(55) $(\forall a \in S) a \Vdash_{\mathbf{m}} A \Longrightarrow \bigsqcap S \Vdash_{\mathbf{m}} A$

(56) $(\forall a \in S) a \not\Vdash_{\mathbf{m}} A \Longrightarrow \bigsqcup S \not\Vdash_{\mathbf{m}} A.$

It can also be shown that the move from *s.o.m.* models to information frames does not increase the stock of valid formulae formulae and, therefore, allows us to preserve completeness without strengthening our labelling algebras. However, information frames will prove convenient for an elegant treatment of involutive negation.

Involutive Negation

The rules for negation given above in Section 3.2 are *intuitionistically* valid also in the new setting where models are based on information frames. The good news, is that now we can get rid of the cumbersome instantiation symbols, because we can take x^* to denote the join of $Cons(x)$. So we can treat x^* not as a constrained variable, but as a real point of the model. It is easy to check that the rules remain valid, under this interpretation, even if we remove the instantiation symbols to obtain the following simpler rules:

(57) $\dfrac{\begin{array}{c} T\,\neg A : x \\ T\,A : y \end{array}}{T\bot : x \circ y} \qquad \dfrac{F\,\neg A : x}{T\,A : x^*} \qquad \dfrac{T\,\bot : y}{\times}\ \boxed{y \leq x \circ x^*}$

In fact, if A is true at $f(x^*)$, i.e. at some point in $Cons(x)$, then it is true also at $\bigsqcup Cons(x)$. On the other hand, since $\bot$ is false at $x \circ y$ for every $y \in Cons(x)$, then, the distributivity of $\circ$ over $\bigsqcup$, together with the (upward) continuity property of valuations, imply that $\bot$ is false at $\bigsqcup Cons(x)$. So, if we treat * as a unary operation of the labelling algebra, the above rules are justified under the interpretation $x^* = \bigsqcup Cons(x)$, and still characterize the non-involutive negation of intuitionistic substructural logics.

Now, the involutive negation typical of classical substructural logics can be characterized simply by assuming that, in the labelling algebra, the operation * is an involution, i.e. $x^{**} = x$.

PROPOSITION 12. *For every frame $\mathcal{L}$ and every model $\mathbf{m}$ over $\mathcal{L}$, the following two statements are equivalent:*

1. *For all information tokens $x \in \mathbf{m}$ and all sentences A:*

$$x \Vdash_{\mathbf{m}} A \Longleftrightarrow x^{**} \Vdash_{\mathbf{m}} A;$$

2. *For all information tokens $x \in \mathbf{m}$ and all sentences A:*

$$x \Vdash_{\mathbf{m}} \neg\neg A \Longleftrightarrow x \Vdash_{\mathbf{m}} A.$$

Proof. The proposition follows immediately from observing that, according to our definitions:

$$x \Vdash_{\mathbf{m}} \neg A \iff x^* \not\Vdash_{\mathbf{m}} A$$

and therefore

$$x \Vdash_{\mathbf{m}} \neg\neg A \iff x^{**} \Vdash_{\mathbf{m}} A.$$

■

Let us consider two information tokens *equivalent* if they verify exactly the same propositions. Then, given a frame $\mathcal{L}$, we can restrict our attention to a subframe containing only a representative for each equivalence class of tokens. So we can assume without loss of generality that information frames do not contain distinct equivalent tokens that force exactly the same formulae. In this case, if any of the two properties in Proposition 12 holds, the operation * is an involution, i.e. $x^{**} = x$. We call a model *involutive* if the operation * is an involution. An example of a proof exploiting this property of the * operation is given in Figure 12, where we prove the formula $(\neg P \to \neg Q) \to (Q \to P)$. This is a theorem of Linear and Relevance Logic (as well as, of course, Classical Logic) but not a theorem of Intuitionistic Logic. Accordingly, the tree in the figure is closed under the assumption that the frame is commutative and involutive (therefore is not intuitionistically closed).

The reader familiar with the literature on Relevance Logic will be reminded of the $*$-operation used by Routley and Routley in their semantics of first-degree entailment [Routley and Routley, 1972] (see also [Routley and Meyer, 1973]). (Similar operations pervade a good deal of work on the semantics of negation in non-classical logics.) Here the $*$-operation is defined and explained in terms of our basic structures.

LKE *for Classical Substructural Logics*

In classical substructural logics, the multiplicative operators can all be defined in terms of $\to$ and $\neg$. It is therefore routine to work out suitable rules for them from the given rules for $\to$ and $\neg$. As for $\otimes$, we can exploit the following equivalence:

(58) $A \otimes B \dashv\vdash \neg(A \to \neg B)$

and turn the constrained variable x/y of Section 3.2 into a real point of the information frame, by taking / as denoting a new binary operator defined as follows:

(59) $x/y = (x^* \circ y)^*$

It can be easily shown that, given this definition, we obtain the following rules:

(60) $\dfrac{T\,A \otimes B : x}{\begin{array}{l} T\,A : a \\ T\,B : x/a \end{array}}$ $(*)$. $\qquad \dfrac{\begin{array}{l} F\,A \otimes B : x \\ T\,A : y \end{array}}{F\,B : x/y} \qquad \dfrac{\begin{array}{l} F\,A \otimes B : x \\ T\,B : y \end{array}}{F\,A : x/y}$

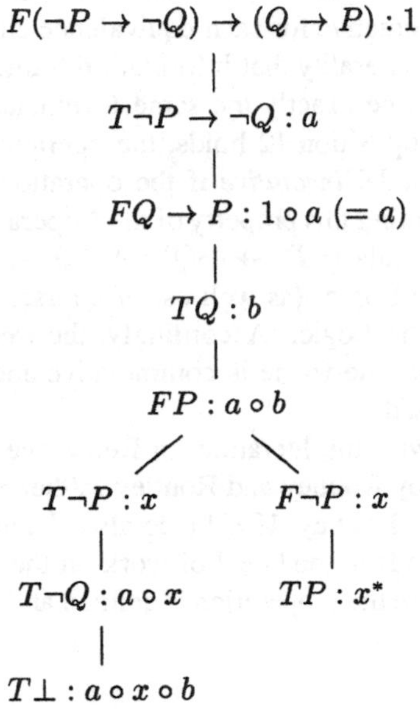

$$F(\neg P \to \neg Q) \to (Q \to P) : 1$$

$$T\neg P \to \neg Q : a$$

$$FQ \to P : 1 \circ a \; (= a)$$

$$TQ : b$$

$$FP : a \circ b$$

Left branch	Right branch
$T\neg P : x$	$F\neg P : x$
$T\neg Q : a \circ x$	$TP : x^{*}$
$T\bot : a \circ x \circ b$	

Figure 12. In this tree, the left-hand branch is closed for all commutative labelling algebras under the substitution $x = (a \circ b)^{*}$. Under this substitution the rigth-hand branch is closed only for *classical* models, since $(a \circ b)^{**} = a \circ b$.

(In the case of non-commutative logics we need, of course, to distinguish / from \. It is left to the reader to work out how.) As mentioned in Section 1.2, the operator ⅋ arises naturally in logical systems with an involutive negation and closed under Exchange. It can be defined in terms of $\neg$ and $\rightarrow$ as follows:

$$A⅋B =_{\text{def}} \neg A \rightarrow B.$$

So, its elimination rules can be derived from the rules for $\rightarrow$ and $\neg$ together with the assumption that models are restricted to the involutive ones (i.e. those in which the operation * is an involution):

(61) $$\frac{T\ A⅋B : x \quad F\ A : y^*}{T\ B : x \circ y} \qquad \frac{T\ A⅋B : x \quad F\ B : x \circ y}{T\ A : y^*}$$

and

(62) $$\frac{F\ A⅋B : x}{F\ A : a^* \quad F\ B : x \circ a}$$ where a is a *new atomic* label

The validity of these rules can be seen by deriving them from the rules for $\rightarrow$ and $\neg$. As for the additives, the good news is that we can get rid of the instantiation symbols in the $\vee$ rules, as we did for the other rules, by defining new unary operators π_1 and π_2 in a similar way. The bad news is that the simple rules for $\wedge$ become unsound under the new interpretation. This is due to the fact that now labels can stand for points which are *joins*. Now, if a forces A and b forces B, clearly $a \sqcup b$ forces *both* A and B, but does not necessarily force $A \wedge B$ (which is stronger than either A or B). This is easily seen if one thinks in terms of the canonical model, where $a \sqcup b$ is simply the set-theoretic union of a and b seens as decreasing sets of formulae. So the rule F$\wedge$ becomes unsound and new rules for conjunction have to be derived from the rules for $\vee$ and $\neg$, by exploiting the usual De Morgan laws (which hold in all classical substructural logics). Space prevents us from discussing the subtleties involved in the treatment of the classical additives for which we refer the reader to [D'Agostino and Gabbay, 1999].

3.4 Modal Substructural Logics

When logical systems are used to formalise some application area, they may require the addition of modality to the language for a variety of reasons: to cater for changes of the system in time, or perhaps for the dependency of the system on the context, or even to bring metalevel notions into the object level. Be the reasons and motivations what they may, we must develop the logical capability of incorporating modalities into these systems, so adding a new dimension which can explicitly and naturally account for 'accessibility' relations involved in the processes that are being modelled. This leads us to an important area of research: the hybrid systems which result from grafting the modal operators $\Box$ and $\Diamond$ onto the variety of substructural logics. The case-study of modal implication logics is discussed in [D'Agostino *et al.*, 1997] on which we shall base the following exposition. Previous work on the specific topic of this section has been concentrating on intuitionistic modal logics. The implication fragments of these logics belong to the wide family of substructural modal systems presented here (i.e. those satisfying *all* the structural rules of Table 2). We mention in particular the works of Božić and Došen [1984], and Plotkin and Stirling [1986] (see also [Sympson, 1993] for an overview).

Formally, we have a system with the conditional operator $\rightarrow$ and we want to add unary operators $\Box$ and $\Diamond$ which behave like modalities. As we have seen, substructural logics can be characterized by structures of 'objects' which can be understood as 'pieces of information' or, sometimes, as 'resources'. A natural way of defining modalities within such semantics consists therefore in adding an accessibility relation between pieces of information. This is the approach adopted in [D'Agostino *et al.*, 1997] that we shall briefly review in this section. By analogy with classical modal logics the intuitive idea is that the verification of a proposition of the form '$\Box A$' or '$\Diamond A$' by means of a given piece of information (or resource) a, depends on what is verified by other pieces of information (or resources) 'accessible' from a. Unlike classical modal logics, however, the verification of non-modal formulae, such as a conditional formula, by a given piece of information a also depends on what is or is not verified elsewhere in the structure (as with the intuitionistic conditional). We shall not pursue any particular intuitive interpretation, because it may change from one logic to the other, and shall leave it to the reader to envisage his or her favourite application contexts.

In the next section we investigate the semantics of the logical systems obtained by introducing the modalities $\Box$ and $\Diamond$ into the family of substructural implication logics. This can be seen as extending the semantic analysis of modal intuitionistic logics carried out in [Fischer-Servi, 1977; Božić and Došen, 1984; Plotkin and Stirling, 1986; Amati and Pirri, 1993; Sympson, 1993]. In Section 3.4, we show how to extend the **LKE** method presented in the previous sections to deal with the modal operators.

The Modal Operators

We start from the information frames defined in Section 3.3 and augment them with a binary relation R between points, called the *accessibility relation*. We assume that this relation is 'closed' under arbitrary $\bigsqcup$ and $\bigsqcap$, namely:

(63) $(\forall b \in S)aRb \Longrightarrow aR\bigsqcup S$ and $(\forall b \in S)aRb \Longrightarrow aR\bigsqcap S$.

We then introduce into our language the two unary operators $\Box$ and $\Diamond$ intended as the usual modalities. The forcing relation extends to these modal operators in the expected way. For $\Box$ we have the following clause:

(64) $a \Vdash_{\mathbf{m}} \Box A$ iff $b \Vdash_{\mathbf{m}} A$ for all b, such that aRb

and for $\Diamond$:

(65) $a \Vdash_{\mathbf{m}} \Diamond A$ iff $b \Vdash_{\mathbf{m}} A$ for some b, such that aRb

In order to preserve the persistence property expressed in (54) in the new setting, R must also satisfy the following conditions (see [Božić and Došen, 1984] and [Plotkin and Stirling, 1986]):

(66) $a \leq b$ and $aRc \Longrightarrow (\exists c')(bRc'$ and $c \leq c')$

and

(67) $a \leq b$ and $bRc \Longrightarrow (\exists c')(aRc'$ and $c' \leq c)$.

DEFINITION 13. A *modal information frame* is a pair $(\mathcal{F}, R)$ where $\mathcal{F}$ is an information frame and R is a binary accessibility relation defined on the domain of $\mathcal{F}$ and satisfying conditions (63), (66) and (67).

If R is defined as above, any atomic valuation V over an information frame can be extended to a forcing relation over a modal information frame by means of the forcing clauses for $\Box$ and $\Diamond$ preserving the persistence property.

DEFINITION 14. A *modal implication model* is a triple $(\mathcal{F}, R, V)$, where $(\mathcal{F}, R)$ is a modal information frame and V is a valuation. The *forcing* relation over a modal implication model $\mathbf{m}$ is a relation between elements of $\mathbf{m}$ and formulae, satisfying the following conditions:

1. $(\forall p \in P)(\forall a \in \mathbf{m})(a \Vdash_{\mathbf{m}} p \Longleftrightarrow a \in V(p))$
2. $(\forall a \in \mathbf{m})(a \Vdash_{\mathbf{m}} A \rightarrow B \Longleftrightarrow (\forall b \in \mathbf{m})(b \Vdash_{\mathbf{m}} A \Longrightarrow (a \circ b \Vdash_{\mathbf{m}} B)))$
3. $a \Vdash_{\mathbf{m}} \Box A$ iff $b \Vdash_{\mathbf{m}} A$ for all b, such that aRb
4. $a \Vdash_{\mathbf{m}} \Diamond A$ iff $b \Vdash_{\mathbf{m}} A$ for some b, such that aRb

We can restrict our attention, without loss of generality, to modal implication models with the additional property that any two points verifying exactly the same formulae are identical.

DEFINITION 15. We say that a formula A is *verified* in a *modal implication model* M if it is verified at the identity point 1 of M. We also say that A is *verified* in a *frame* $\mathcal{F}$ if is verified in all modal implication models based on $\mathcal{F}$.

Let us denote by $S(a)$ *the sphere* of a, i.e. the set of all pieces of information accessible from a. We now define the two unary operators ! and ? as follows:

(68) $!a =_{\text{def}} \bigsqcap S(a)$

and

(69) $?a =_{\text{def}} \bigsqcup S(a)$

Now observe that (63) implies that the following property of R is satisfied in every frame:

(70) $(\forall a)(\exists b)aRb$.

In other words, for all a, $S(a) \neq \emptyset$. In fact, for every point a, both $!a$ and $?a$ are accessible from a, that is

(71) $(\forall a)aR!a$ and $aR?a$

In the theory of classical modal logics a property like (70) is known as *seriality*. So, in our approach *all* frames are serial. Using the ! and ? operators, this assumption is expressed by the following inequation which holds for all modal information frames:

(72) $!a \leq ?a$

It is not difficult to show that (66) and (67) above are respectively equivalent to the following conditions on ? and ! expressing the fact that the operators ? and ! are both *order-preserving*:

(73) $a \leq b \implies ?a \leq ?b$

(74) $a \leq b \implies !a \leq !b$

PROPOSITION 16. *In every modal information frame, (67) holds if and only if (74) holds, and (66) holds if and only if (73) holds.*

Given our definitions of the operators ! and ?, (which are reminiscent of the ones used in the algebraic semantics of classical modal logics, for which see [Bull and Segerberg, 1984]) the forcing clauses for $\Box$ and $\Diamond$ can be reformulated more concisely as below:

(75) $a \Vdash_{\mathbf{m}} \Box A$ iff $!a \Vdash_{\mathbf{m}} A$

and

(76) $a \Vdash_{\mathbf{m}} \Diamond A$ iff $?a \Vdash_{\mathbf{m}} A$

We can now exploit the new operators to express complex statements about the accessibility relation R concisely, in the form of simple inequalities of the form $\alpha \leq \beta$, where α and β are expressions built up from atomic terms by means of the operators $\circ$, ! and ?.

Let us consider some of the most familiar properties of the accessibility relation R (for a more detailed discussion see [D'Agostino *et al.*, 1997]).

Seriality A frame is *serial* if for every point a, $S(a) \neq \emptyset$, that is, for every point a there exists a b such that aRb. Observe that a frame is serial if and only if it verifies the formula $\Box A \to \Diamond A$. As mentioned above, in our approach all frames are serial, as a consequence of (63), and seriality corresponds to the assumption that $!a \leq ?a$ for all a.

Reflexivity A frame is said to be *reflexive* if aRa for all a. Observe that a frame is reflexive if and only if it verifies the formula $\Box A \to A$. In our notation reflexivity can be expressed by the following condition:

(Reflexivity) $(\forall a)(!a \leq a)$

Transitivity A frame is said to be *transitive* if the following condition holds:

$$(\forall a)(\forall b)(\forall c)(aRb \text{ and } bRc \Longrightarrow aRc).$$

A frame is transitive if and only if it verifies the following formula $\Box A \to \Box\Box A$. In our notation, transitivity is expressed by the following condition:

(Transitivity) $(\forall a)(!a \leq !!a)$

Symmetry A frame is said to be *symmetric* if the following condition holds:

$$(\forall a)(\forall b)(aRb \Longrightarrow bRa).$$

A frame is symmetric if and only if it verifies the formula $A \to \Box\Diamond A$. In our notation, symmetry is expressed by the following condition:

(Symmetry) $(\forall a)(a \leq ?!a)$

Euclideanism A frame is said to be *Euclidean* if the following condition holds true:

$$(\forall a)(\forall b)(\forall c)(aRb \text{ and } aRc \Longrightarrow bRc).$$

A frame is Euclidean if and only if it verifies the formula $\Diamond A \to \Box\Diamond A$. The condition corresponding to this property is the one stated below:

(Euclideanism) $(\forall a)(?a \leq ?!a)$.

Modal formulae	Conditions on R	Modal constraints
$\Box A \to A$	$(\forall a)aRa$	$!a \leq a$
$\Box A \to \Box\Box A$	$(\forall a)(\forall b)(\forall c)(aRb$ and $bRc \Longrightarrow aRc)$	$!a \leq !!a$
$A \to \Box\Diamond A$	$(\forall a)(\forall b)(aRb \Longrightarrow bRa)$	$a \leq ?!a$
$\Diamond A \to \Box\Diamond A$	$(\forall a)(\forall b)(\forall c)(aRb$ and $aRc \Longrightarrow bRc)$	$?a \leq ?!a$
$\Diamond\Box A \to \Box\Diamond A$	$(\forall a)(\forall b)(\forall c)(aRb$ and $aRc \Longrightarrow (\exists d)(bRd$ and $cRd))$	$!?a \leq ?!a$

Table 5. Correspondence between modal formulae, conditions on R and constraints on the modal information frames

Directedness A frame is *directed* if the following condition holds true:

$$(\forall a)(\forall b)(\forall c)(aRb \text{ and } aRc \Longrightarrow (\exists d)(bRd \text{ and } cRd))$$

which corresponds to:

(Directedness) $(\forall a)(!?a \leq ?!a)$.

Observe that a frame is directed if and only if it verifies the following formula $\Diamond\Box A \to \Box\Diamond A$. Our discussion so far amounts to a reformulation of the well-known correspondence theory in our new setting. This correspondence is summarized in Table 5.

We shall now state a crucial lemma. Let σ_1 and σ_2 be (possibly empty) strings of ! and ?. The *dual* of σ_i, denoted by σ_i' is the string obtained by interchanging ! and ? in σ_i. Moreover, let $\mathbf{M}^* \subseteq \mathbf{M}$, where $\mathbf{M}$ is the set of modal conditions listed above. We have the following *duality principle*:

THEOREM 17 (Duality Lemma). *Assume that $\sigma_1 a \leq \sigma_2 a$ holds for all a in all $\mathbf{M}^*$-frames, i.e. there is a chain of inequalities $\phi_0 \leq \cdots \leq \phi_n$ such that (i) $\phi_0 = \sigma_1 a$, (ii) $\phi_n = \sigma_2 a$, and (iii) each inequality $\phi_i \leq \phi_{i+1}$, with $i = 0, \ldots, n-1$, is one of of the primitive inequalities in $\mathbf{M}^*$. Then $\sigma_2' a \leq \sigma_1' a$ also holds for all a in all $\mathbf{M}^*$-frames.*

Proof. The proof is an easy induction on the length of the chain $\phi_0 \leq \cdots \leq \phi_n$. ■

Let now $\mathbb{L}'$ be a propositional language containing the two binary operators $\to_1$ and $\to_2$ plus the the two modal operators $\Box$ and $\Diamond$. (Recall that the two conditionals collapse in all the systems allowing the structural rule of Exchange, in which case we use the symbol $\to$ without subscripts.) By a *modal implication logic* — or **MIL** for short — we mean a (substructural) implication logic (see Section 1.2) over $\mathbb{L}'$ satisfying the following conditions on the operators $\Box$ and $\Diamond$:

$$(77) \quad \frac{A \vdash B}{\Box A \vdash \Box B}$$

and

$$(78)\quad \frac{A \vdash B}{\Diamond A \vdash \Diamond B}$$

In addition to (77) and (78) above, different **MIL**'s may also satisfy different subsets of the following *modal axioms*:

1. $\Box A \to \Diamond A$
2. $\Box A \to A$
3. $\Box A \to \Box\Box A$
4. $A \to \Box\Diamond A$
5. $\Diamond A \to \Box\Diamond A$
6. $\Diamond\Box A \to \Box\Diamond A$

In the sequel we restrict our attention to *serial* **MIL**'s, i.e. those satisfying the modal axiom $\Box A \to \Diamond A$.

Adding Modalities to the LKE System

In this section we show how the **LKE** method of Section 3.1 can be extended to deal with the modal operators in a very natural way.

Modal **LKE** *Rules*

We enrich the labelling algebra of Section 3.1 with the two unary operators ! and ? satisfying Conditions (72), (73), (74), and the duality principle (see the Duality Lemma above). Notice that this labelling algebra is sufficient for completeness, that is, as already stressed, we do not need all the complexity brought in by the potentially infinitary lattice operations which appear in the definition of information frames. The latter enter the picture only as intermediary for an explanation of the unary operator ! and ?. The completeness of the system (see [D'Agostino *et al.*, 1997]) guarantees that they do not play any additional role with respect to characterizing the set of valid formulas.

The valuation clauses (75) and (76) immediately imply simple **LKE**-style rules for the modal operators.

Elimination rules for $\Box$

$$\frac{T\,\Box A : a}{T\,A : !a}\ ET\Box \qquad\qquad \frac{F\,\Box A : a}{F\,A : !a}\ EF\Box$$

Elimination rules for $\diamond$

$$\frac{T \diamond A : a}{T\, A :?a}\ \mathrm{E}T\diamond \qquad\qquad \frac{F \diamond A : a}{F\, A :?a}\ \mathrm{E}F\diamond$$

The rules have a clear intuitive interpretation: if $\Box A$ is true at a point a, then it is true at all points accessible from it and, therefore, it is true at the special point $!a$ which verifies all and only the formulae which are true at all points in $S(a)$. Conversely, if $\Box A$ is false at a, then it must be false at $!a$.

Similarly, if $\diamond A$ is true at a point a, then it is true at some point accessible from it and, therefore, at the special point $?a$ which verifies all and only the formulae which are true at at least one point in $S(a)$. Conversely, if $\diamond A$ is false at a, then it must be false at $?a$.

By adding these rules to the implication rules we obtain an **LKE**-system for modal substructural implication. Different systems will be characterized by adding to the labelling algebra different sets of structural and modal constraints to be used in checking for branch-closure. So, this extended method preserves the 'separation-by-closure' property of the original method. Moreover, it also preserves the atomic closure property (see Section 3.1 above). Notice that, by virtue of the duality principle, whenever the constraint $\sigma_1 a \leq \sigma_2 b$ is derivable in the algebra of the labels, so is its dual $\sigma_2' b \leq \sigma_1' a$.

In the next subsection we shall see some examples of proofs in this modal extension of the **LKE** system.

Examples

EXAMPLE 18. The axiom T, $\Box P \rightarrow P$ is valid in all reflexive frames.

$$F\,\Box P \rightarrow P : 1$$
$$|$$
$$T\,\Box P : a$$
$$|$$
$$F\, P : 1 \circ a (= a)$$
$$|$$
$$T\, P :!a$$

This one-branch tree is closed in all reflexive frames since $!a \leq a$. Observe that, by the duality principle, the constraint $a \leq ?a$ also holds in all reflexive frames and, by means of this, the dual of T, $P \rightarrow \diamond P$ can be proved.

EXAMPLE 19. The axiom D, $\Box P \rightarrow \Diamond P$ is valid in all serial frames:

$$F\ \Box P \rightarrow \Diamond P : 1$$
$$|$$
$$T\ \Box P : a$$
$$|$$
$$F\ \Diamond P : a$$
$$|$$
$$T\ P : !a$$
$$|$$
$$F\ P : ?a$$

This tree is closed in all serial frames because $!a \leq ?a$.

EXAMPLE 20. The axiom 5, $\Diamond P \rightarrow \Box\Diamond P$ is valid in all Euclidean frames.

$$F\ \Diamond P \rightarrow \Box\Diamond P : 1$$
$$|$$
$$T\ \Diamond P : a$$
$$|$$
$$F\ \Box\Diamond P : 1 \circ a (= a)$$
$$|$$
$$T\ P : ?a$$
$$|$$
$$F\ \Diamond P : !a$$
$$|$$
$$F\ P : ?!a$$

A similar tree, making use of the dual constraint $!?a \leq !a$, shows the validity of the dual axiom $\Diamond\Box A \rightarrow \Box A$.

EXAMPLE 21. The formula

$$(P \rightarrow (P \rightarrow Q)) \rightarrow (\Box P \rightarrow Q)$$

is valid in all contractive and reflexive frames.

$$F\ (P \to (P \to Q)) \to (\Box P \to Q) : 1$$

$$T\ P \to (P \to Q) : a$$

$$F\ \Box P \to Q : a$$

$$T\ \Box P : b$$

$$F\ Q : a \circ b$$

$$T\ P :!b$$

$$T\ P \to Q : a\circ!b$$

$$T\ Q : a\circ!b\circ!b$$

In all contractive frames $!b\circ!b \leq !b$, so $a\circ!b\circ!b \leq a\circ!b$. Moreover, in all reflexive frames $!b \leq b$. So, in all contractive and reflexive frames $a\circ!b\circ!b \leq a \circ b$.

EXAMPLE 22. The formula

$$\Box P \to (Q \to \Box\Box P)$$

is valid in all frames which are both monotonic and transitive.

$$F\ \Box P \to (Q \to \Box\Box P) : 1$$

$$T\ \Box P : a$$

$$F\ Q \to \Box\Box P : a$$

$$T\ Q : b$$

$$F\ \Box\Box P : a \circ b$$

$$T\ P :!a$$

$$F\ \Box P :!(a \circ b)$$

$$F\ P :!!(a \circ b)$$

This one-branch tree is closed for all monotonic-transitive. frames. For, $a \leq a \circ b$ by monotonicity; therefore, $!a \leq !(a \circ b)$ by (74); moreover, by transitivity, $!(a \circ b) \leq !!(a \circ b)$; so, $!a \leq !!(a \circ b)$ and the branch is closed.

EXAMPLE 23. The formula

$$P \to (P \to \Box\Diamond P)$$

is valid in all frames which are expansive and symmetric.

$$F\, P \to (P \to \Box\Diamond P) : 1$$
$$|$$
$$T\, P : a$$
$$|$$
$$F\, P \to \Box\Diamond P : a$$
$$|$$
$$F\, \Box\Diamond P : a \circ a$$
$$|$$
$$F\, \Diamond P : !(a \circ a)$$
$$|$$
$$F\, P : ?!(a \circ a)$$

This one-branch tree is closed in all expansive-symmetric frames. For, $a \leq a \circ a$ by expansion; $(a \circ a) \leq ?!(a \circ a)$ by symmetry. These two inequalities imply that $a \leq ?!(a \circ a)$.

EXAMPLE 24. The formula $\Diamond(P \to P)$ is valid in all monotonic frames.

$$F\, \Diamond(P \to P) : 1$$
$$|$$
$$F\, P \to P : ?1$$
$$|$$
$$T\, P : a$$
$$|$$
$$F\, P : ?1 \circ a$$

In all monotonic frames, $a \leq ?1 \circ a$ and the branch is closed.

EXAMPLE 25. The proof rule

$$\frac{\vdash A \to B}{\vdash \Box A \to \Box B}$$

is valid in all modal information frames.

This is shown by the following closed **LKE**-tree:

$$
\begin{array}{c}
F\ \Box A \to \Box B : 1 \\
| \\
T\ \Box A : a \\
| \\
F\ \Box B : a \\
| \\
T\ A :!a \\
| \\
F\ B :!a \\
\diagup \qquad \diagdown \\
\begin{array}{cc}
T\ A \to B : 1 & F\ A \to B : 1 \\
| & \mathcal{T} \\
T\ B :!a &
\end{array}
\end{array}
$$

where $\mathcal{T}$ is a closed **LKE**-tree for $F\ A \to B : 1$ which exists by hypothesis.

EXAMPLE 26. The proof rule

$$\frac{\vdash A}{\vdash \Box A}$$

is valid in all monotonic frames.

This is shown by the following closed **LKE**-tree:

$$
\begin{array}{c}
F\ \Box A : 1 \\
| \\
F\ A :!1 \\
\diagup \qquad \diagdown \\
\begin{array}{cc}
T\ A : 1 & F\ A : 1 \\
 & \mathcal{T}
\end{array}
\end{array}
$$

where $\mathcal{T}$ is a closed **LKE**-tree for $FA : 1$ which exists by hypothesis. Here the left-hand brach is closed because $1 \leq 1 \circ !1$ by the monotonicity of $\circ$.

EXAMPLE 27 (Regular modal implication systems). We define the class of *regular* **MIL**'s, as the class of all **MIL**'s closed under the additional axiom

$$\Box(A \to B) \to (\Box A \to \Box B).$$

(Recall that all **MIL**'s considered in this chapter satisfy the other condition that — in the context of classical modal logic — is used to characterized regular modal logics, namely Condition 77.) The class of *regular frames*, corresponding to regular **MIL**'s, is characterized by the following constraint:

$$!a \circ !b \leq !(a \circ b)$$

The reader can easily check that the above axiom becomes provable if this constraint is added to the algebra of the labels. The tree in Figure 13 shows that the following formula

$$\vdash \Box(\Diamond P \to \Diamond Q) \to (\Box\Diamond P \to \Diamond\Box Q) \to R) \to R$$

is verified in all frames which are both *regular* and *directed* (we use x, y, z as label-variables and skip the first few nodes in the construction of the tree). Since this example involves an application of the branching rule of generalized bivalence, it can serve also the purpose of illustrating the use of *variables in the labels* hinted at the end of Section 3.1.

Since the left-hand branch of this tree contains the pair of LS-formulae $T\ R : x \circ b$ and $F\ R : a \circ b$, it is closed if $x \circ b \leq a \circ b$. On the other hand, the right-hand branch is closed if $?(!a \circ !c) \leq !?(x \circ c)$, because of the pair of LS-formulae $T\ Q : ?(!a \circ !c)$ and $F\ Q : !?(x \circ c)$. Hence, this tree is closed for a given class of frames if the system consisting of the two inequations:

$$x \circ b \leq a \circ b \qquad ?(!a \circ !c) \leq !?(x \circ c)$$

has a solution in the corresponding algebra of the labels. In this simple case it is easy to work out the solution $x = a$ for the algebra of the labels corresponding to directed regular frames. To verify this is a solution of the inequation generated by the right-hand branch, it is sufficient to go through the following steps:

$!a \circ !c \leq !(a \circ c)$	by regularity
$?(!a \circ !c) \leq ?!(a \circ c)$	by the monotonicity of ?
$?!(a \circ c) \leq !?(a \circ c)$	by directedness and duality
$?(!a \circ !c) \leq !?(a \circ c)$	by transitivity of $\leq$

A general technique for solving systems of inequations generated by **LKE**-trees (with variables) within a given algebra of the labels is crucial for developing decision algorithms based on our method. This problem will be discussed in a subsequent work. Here we only observe that our approach separates each proof in two components: (i) a logical component, which depends only on the "universal" meaning of the logical operators as defined by the inference rules which are a straightforward generalization of the classical ones, and (ii) an algebraic component, which consists in solving a system of inequations within a given labelling algebra that mirrors the specific properties of the information system in which the inferential process takes place. While algorithms for the first component can

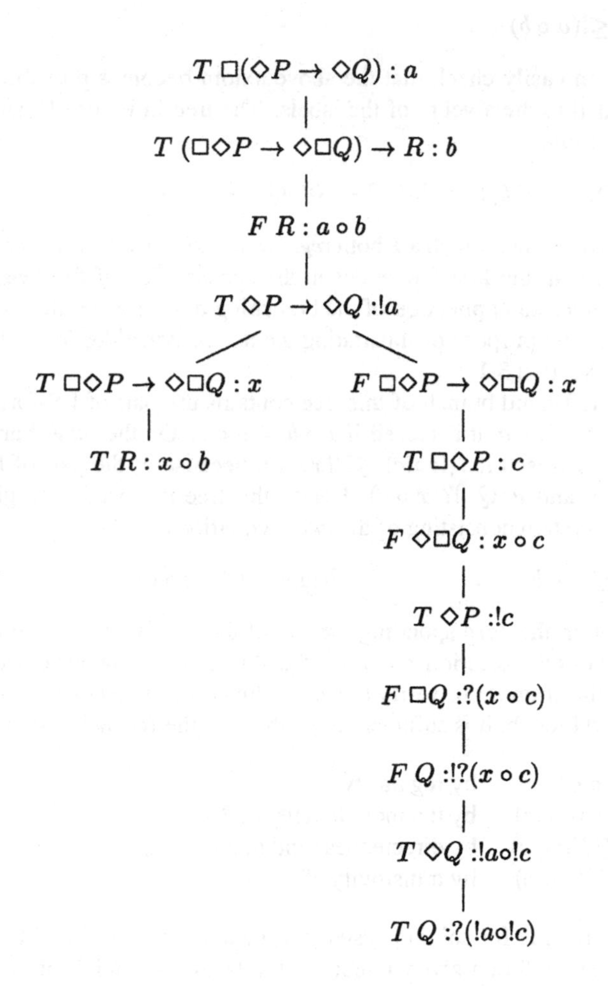

Figure 13. This tree is closed for all regular and directed frames, under the substitution $x = a$

be devised as straightforward extensions of known algorithms for classical logic (simply by shifting from formulae to labelled formulae), algorithms for the second component can be devised by generalizing existing algebraic methods for solving similar problems in related areas (e.g., unification, term-rewriting, constraint programming, etc. see [Broda *et al.*, 1997] for one approach).

Marcello D'Agostino
Università di Ferrara, Italy.

Dov Gabbay
King's College, London.

Krysia Broda
Imperial College, London.

REFERENCES

[Abrusci, 1990] V. M. Abrusci. Noncommutative intuitionistic linear propositional logic. *Zeitschr. f. Math. Logik u. Grundlagen d. Math.*, 36:297–318, 1990.

[Abrusci, 1991] V. M. Abrusci. Phase semantics and sequent calculus for pure noncommutative classical linear propositional logic. *Journal of Symbolic Logic*, 56:1403–1451, 1991.

[Allwein and Dunn, 1993] G. Allwein and J. M. Dunn. Kripke models for linear logic. *the Journal of Symbolic Logic*, 58:514–545, 1993.

[Amati and Pirri, 1993] G. Amati and F. Pirri. A uniform tableau method for intuitionistic modal logics I. Manuscript, 1993.

[Anderson and Belnap, 1975] A. R. Anderson and N. D. Belnap Jr. *Entailment: the Logic of Relevance and Necessity*, Princeton University Press, Princeton, 1975.

[Anderson *et al.*, 1992] A. R. Anderson, N. D. Belnap Jr and J. M. Dunn. *Entailment: The Logic of Relevance and Necessity*. Princeton University Press, Princeton, 1992.

[Avron, 1988a] A. Avron. The semantics and proof theory of linear logic. *Theoretical Computer Science*, 57:161–184, 1988.

[Avron, 1988b] A. Avron. The semantics and proof theory of linear logic. *Theoretical Computer Science*, 57:161–184, 1988.

[Avron, 1991] A. Avron. Simple consequence relations. *Journal of Information and Computation*, 92:105–139, 1991.

[Beckert *et al.*, 1993] B. Beckert, R. Hähnle, and P. Schmitt. The even more liberalized δ-rule in free-variable semantic tableaux. In G. Gottlob, A. Leitsch, and Daniele Mundici, editors, *Proceedings of the 3rd Kurt Gödel Colloquium, Brno, Czech Republic*, number 713 in Lecture Notes in Computer Science, pages 108–119. Springer-Verlag, 1993.

[Božić and Došen, 1984] M. Božić and K. Došen. Models for normal intuitionistic modal logics. *Studia Logica*, 43:217–245, 1984.

[Broda *et al.*, 1997] K. Broda, M. Finger and A. Russo. LDS-Natural Deduction for Substructural Logics. Technical Report DOC 97-11. Department of Computing, Imperial College, 1987. Short version presented at WOLLIC 96, *Journal of the IGPL*, 4:486–491.

[Broda *et al.*, 1998] K. Broda, M. D'Agostino, and A. Russo. Transformation methods in LDS. In Hans Jürgen Ohlbach and Uwe Reyle, editors, *Logic, Language and Reasoning. Essays in Honor of Dov Gabbay*. pp. 347–390. Kluwer Academic Publishers, 1997. To appear.

[Bull and Segerberg, 1984] R. Bull and K. Segerberg. Basic modal logic. In Dov Gabbay and Franz Guenthner, editors, *Handbook of Philosophical Logic*, volume II, chapter II.1, pages 1–88. Kluwer Academic Publishers, 1984.

[D'Agostino and Gabbay, 1994] M. D'Agostino and D. M. Gabbay. A generalization of analytic deduction via labelled deductive systems.Part I: Basic substructural logics. *Journal of Automated Reasoning*, 13:243–281, 1994.

[D'Agostino and Gabbay, 1999] M. D'Agostino and D. M. Gabbay. A generalization of analytic deduction via labelled deductive systems. Part II: full substructural logics. Forthcoming, 1999.

[D'Agostino *et al.*, 1997] M. D'Agostino, D. M. Gabbay, and A. Russo. Grafting modalities onto substructural implication systems. *Studia Logica*, 59:65–102, 1997.

[Dôsen, 1988] Kosta Dôsen. Sequent systems and groupoid models I. *Studia Logica*, 47:353–385, 1988.

[Dôsen, 1989] K. Dôsen. Sequent systems and groupoid models II. *Studia Logica*, 48:41–65, 1989.

[Dôsen, 1993] K. Dôsen. A historical introduction to substructural logics. In P. Schroeder Heister and Kosta Dôsen, editors, *Substructural Logics*, pages 1–31. Oxford University Press, 1993.

[Dunn, 1986] M. Dunn. Relevance logic and entailment. In D. M. Gabbay and F. Guenthner, editors, *Handbook of Philosophical Logic*, volume III, chapter 3, pages 117–224. Kluwer Academic Publishers, 1986.

[Fischer-Servi, 1977] G. Fischer-Servi. On modal logic with an intuitionistic base. *Studia Logica*, 36:141–149, 1977.

[Fitting, 1983] M. Fitting. *Proof Methods for Modal and Intuitionistic Logics.* Reidel, Dordrecht, 1983.

[Gabbay, 1992] D. M. Gabbay. How to construct a logic for your application. In Hans Juergen Ohlbach, editor, *GWAI-92: Advances in Artificial Intelligence (LNAI 671)*, pages 1–30. Springer, 1992.

[Gabbay, 1996] D. M. Gabbay. *Labelled Deductive Systems, Volume 1 - Foundations.* Oxford University Press, 1996.

[Garson, 1989] J. Garson. Modularity and relevant logic. *Notre Dame Journal of Formal Logic*, 30:207–223, 1989.

[Gentzen, 1935] G. Gentzen. Unstersuchungen über das logische Schliessen. *Math. Zeitschrift*, 39:176–210, 1935. English translation in [Szabo, 1969].

[Girard, 1987] J.-Y. Girard. Linear logic. *Theoretical Computer Science*, 50:1–102, 1987.

[Kowalski, 1979] R. Kowalski. *Logic For Problem Solving.* North-Holland, Amsterdam, 1979.

[Lambek, 1958] J. Lambek. The mathematics of sentence structure. *Amer. Math. Monthly*, 65:154–169, 1958.

[MacCaull, 1996] W. MacCaull. Relational semantics and a relational proof system for full Lambek calculus. Technical report, Dept. Mathematics and Computing Sciences, St. Francis Xavier University, 1996.

[McRobbie and Belnap, 1979] M. A. McRobbie and N. D. Belnap. Relevant analytic tableaux. *Studia Logica*, XXXVIII:187–200, 1979.

[McRobbie and Belnap, 1984] M. A. McRobbie and N. D. Belnap. Proof tableau formulations of some first-order relevant orthologics. *Bulletin of the Section of Logic, Polish Academy of the Sciences*, 13:233–240, 1984.

[Meyer *et al.*, 1995] R. K. Meyer, M. A. McRobbie, and N. D. Belnap. Linear analytic tableaux. In Peter Baumgartner, Reiner Hähnle, and Joachim Posegga, editors, *Proceedings of Tableaux '95, 4th conference on Theorem Proving with Analytic Tableaux and Related Methods, St. Goar*, volume 918 of *Lecture Notes in Artificial Intelligence*, pages 278–293. Springer, 1995.

[Ono and Komori, 1985] H. Ono and Y. Komori. Logics without the contraction rule. *The Journal of Symbolic Logic*, 50:169–201, 1985.

[Ono, 1993] H. Ono. Semantics for substructural logics. In Peter Schroeder-Heister, editor, *Substructural Logics*, pages 259–291. Oxford University Press, 1993.

[Plotkin and Stirling, 1986] G. D. Plotkin and C. P. Stirling. A framework for intuitionistic modal logic. In J. Y. Halpern, editor, *Theoretical Aspects of Reasoning About Knowledge*, pages 399–406, 1986.

[Routley and Routley, 1972] R. and V. Routley. The semantics of first degree entailment. *Noûs*, VI:335–359, 1972.

[Routley and Meyer, 1973] R. Routley and R. K. Meyer. The semantics of entailment, I. In H. Leblanc, editor, *Truth, Syntax and Semantics*. North Holland, 1973.

[Russo, 1996] A. Russo. Generalising propositional modal logics using labelled deductive systems. In *Proceedings of FroCoS'96*, 1996.

[Sambin, 1993] G. Sambin. The semantics of pretopologies. In Peter Schroeder-Heister, editor, *Substructural Logics*, pages 293–307. Oxford University Press, 1993.

[Sympson, 1993] A. K. Sympson. *The Proof Theory and Semantics of Intuitionistic Modal Logic.* PhD thesis, University of Edinburgh, 1993.

[Tarski, 1930a] A. Tarski. Fundamentalc bcgriffc dcr mcthodologie der deduktiven wissenschaften, I. *Monatshefte für Mathematik und Physik*, 37:361–404, 1930.

[Tarski, 1930b] A. Tarski. Über einige fundamentale begriffe der metmathematik. *Comptes Rendus des Séances de la Société des Sciences et des Lettres de Varsovie*, 23:22–29, 1930.

[Thistlewaite *et al.*, 1988] P. B. Thistlewaite, M. A. McRobbie, and B. K. Meyer. *Automated Theorem Proving in Non Classical Logics*. Pitman, 1988.

[Urquhart, 1972] A. Urquhart. Semantics for relevant logic. *The Journal of Symbolic Logic*, 37:159–170, 1972.

[Urquhart, 1984] A. Urquhart. The undecidability of entailment and relevant implication. *Journal of Symbolic Logic*, 49:1059–1073, 1984.

[van Benthem, 1991] J. van Benthem, *Language in Action: Categories, Lambdas and Dynamic Logic*, North-Holland, Amsterdam, 1991.

[Wagner, 1991] G. Wagner. Logic programming with strong negation and inexact predicates. *Journal of Logic and Computation*, 1:835–859, 1991.

[Wansing, 1993] H. Wansing. *The Logic of Information Structures*. Number 681 in Lecture Notes in Artificial Intelligence. Springer-Verlag, Berlin, 1993.

[William, 1980] A. H. William. The formuale-as-types notion of construction. In J.P. Seldin and J.R. Hindley, editors, *To H.B. Curry: Essays on Combinatory Logics, Lambda Calculus and Formalism*. Academic Press, London, 1980.

NICOLA OLIVETTI

TABLEAUX FOR NONMONOTONIC LOGICS

1 INTRODUCTION

In the last 15 years there has been a huge amount of research on logical formalization of commonsense reasoning. One of the major difficulties against the use of standard logics for this purpose is that commonsense reasoning requires the capability of dealing with information which is not complete and is subject to change. In many situations, one cannot simply be stuck by the incompleteness of the available information and has 'to jump' to plausible conclusions, despite they are not logically valid. New incoming data might defeat plausible conclusions of the reasoning agent. In such a case, the reasoning agent must be prepared to withdraw its conclusions and to block its inferences. Traditional logics, even non classical ones, are not suitable to express and formalize revisable inferences. This is not surprising, as in almost application of logics, from mathematical reasoning to program verification, one never has to jump to some conclusion from insufficient data. In mathematics, for instance, all the hypotheses of a theorem must be explicit in order to prove it; if the data are not sufficient to prove the theorem, one just gives up the proof, or proves another theorem.

As it is largely acknowledged, the property of traditional logics which makes them unsuitable to formalize the process of 'jumping to conclusions' is their *monotonicity*. Any deductive system (identified with a consequence relation $\vdash$), is called *monotonic* if adding more hypotheses, one cannot get less theorems, that is:

$$\Delta \vdash \phi \text{ and } \Delta \subseteq \Gamma \text{ implies } \Gamma \vdash \phi.$$

Some non classical logics, such as relevance logics do not satisfy this property, still we will not classify them as nonmonotonic logics. The characteristic feature of so-called nonmonotonic systems is that they allow one to draw conclusions which are stronger than logical consequences. Nonmonotonic consequences are not logically valid. One may doubt whether representing nonmonotonic reasoning requires specific logics, or it just matters how logic is used. At an intuitive level, all attempts of formalizing nonmonotonic reasoning start from the idea of enabling some inferences because of lacking of contrary evidence.

Many AI tasks and applications require nonmonotonic reasoning capabilities. We list a few well-known examples.

- *Frame problem*: when reasoning about successive states of an object subject to change, we must be able to conclude that the object preserves its properties, unless their change can be inferred from the available information.

M. D'Agostino et al. (eds.), Handbook of Tableau Methods, 469–528.

- *Qualification problem*: all the relevant preconditions for bringing about an action, or for solving a problem, are those which can be inferred from the data.
- *Closed world reasoning*: facts which are not explicitly asserted, nor deducible are false.
- *Reasoning about typicality and prototypes*: an individual has some typical property, unless its exceptional behaviour can be inferred from the available data.

Frame and qualification problems arise in the context of planning and reasoning about actions. Closed world reasoning is a convention usually adopted in database query-answering; a database can store only positive information, and yet there must be a way of inferring negative information. The need of reasoning about typicality arises when we represent properties of objects in a knowledge base; the information is usually organized in the form hierarchies of (classes of) objects and their properties; we expect properties be inherited by a subclass of objects, unless we know that it makes an exception.

In all cases discussed above, subsequent information may lead to withdraw conclusions. That is why a deductive system capable of performing such kinds of reasoning has to be nonmonotonic. All approaches to formalization of nonmonotonic reasoning address the above kinds of reasoning, although substantially differing in the type of formalization.

In this context we cannot even attempt to give an account of more than 15 years of research in the area. The seminal papers in the area, representative of the main approaches to formalization of nonmonotonic reasoning, were collected in a special issue of *Artificial Intelligence* (Vol. 13, 1980); for an introduction to the area we refer to Reiter's paper [1987], and to some books, such as [Brewka, 1985; Etherington, 1988], and [Besnard, 1989].

We may distinguish the approaches to nonmonotonic reasoning in two types which we respectively call the *fixpoint* approach and the *semantic preference* approach. We give below an intuitive description of the two approaches, further details are contained in the next section. We present both of them, by discussing a simple example.

Suppose that we have the following information: '(*) Greeks are usually dark-haired', 'Kostas is Greek', 'albinos and balds are not dark-haired'. Let first try to express the knowledge by first order axioms:

$$\forall x(greek(x) \rightarrow dark\_haired(x)),$$
$$greek(kostas),$$
$$\forall x(albino(x) \vee bald(x) \rightarrow \neg dark\_haired(x)).$$

Since we have no evidence that Kostas is albino or bald we want to infer that he is dark-haired. From the axioms, we can infer in first-order logic that Kostas is dark-haired, and there are neither albinos nor bald Greeks. Thus, if we later learn that Kostas is either albino or bald, the information becomes inconsistent and we waste everything.

However, once we have learned that Kostas is bald, we just want to withdraw the conclusion that he is dark-haired, without retreating the general knowledge. How can it be done?

One way of reading a statement as (*) is 'if x is greek and we do not have evidence that x is not dark-haired, we can conclude that x actually is'. That is, we conclude that x is dark-haired by the unability to conclude that x is not. Since unable to prove $\neg\phi$, means that ϕ is consistent, one possibility is to introduce a modality, M, whose meaning is consistency: we can represent the (*) by

$$\forall x(greek(x) \wedge M(dark\_haired(x)) \to dark\_haired(x)).$$

To use this rule, we must be able to make inferences such as

(NM-rule): from $\not\vdash \neg\phi$ infer $\vdash M\phi$.

It is easily seen that we cannot constructively define a deduction relation, (on the top of classical logic) incorporating such a deductive rule: any attempt to define provability by means of non provability leads to an inconsistent circularity[1]. To define what is derivable by means of a such NM-rule from a set of formulas Σ, the only thing we can do is to say what condition must satisfy a set S to be regarded as the set consequence of Σ under NM-rule. The condition is the following:

$$S = \{\phi : \ \Sigma \cup \{M\psi : S \not\vdash \neg\psi\} \vdash \phi\}.$$

A set S satisfying the condition above can be seen as a fixpoint of a transformation on sets of formulas. Such sets S may not exist, but if they do, nonmonotonic inferences from Σ can be defined as monotonic (ordinary) inferences from any such set S (we can chose of considering just one, or all of them). This was the first proposal of a nonmonotonic logic by McDermott and Doyle [1980]. It was later elaborated in a modal context [McDermott, 1982]. But certain anomalies of Doyle and McDermott's logic, led Moore to propose a better behaved variant of it. According to Moore, his logic is capable of capturing another form of nonmonotonic reasoning, he called *autoepistemic*. Autoepistemic reasoning is the kind of reasoning we use (quoting Moore) to conclude that 'I don't have an older brother since otherwise I would know it'. Moore's proposal consists, as explained in the next section, in introducing a belief operator L ('it is believed that'), dual of the consistency operator M of Doyle and McDermott, i.e. $L\phi = \neg M\neg\phi$. To express the inference about the brother, we would use the axiom:

$$\neg L\ brother \to \neg brother.$$

[1] Unless we put some bound on deduction resources.

But, to conclude that I don't have a brother, I still need a way of inferring $\neg L\psi$, from the unability of inferring ψ. As in Doyle's and McDermott's logic, autoepistemic consequences are defined by means of non-constructive extensions S of the initial data, satisfying a fixpoint condition. We postpone a more detailed exposition of Moore's proposal to the next section.

A similar, although technically different, proposal was put forward by Reiter in [1980]. Reiter's purpose is to deal directly with inference rules, whose conclusions can be inferred by 'lack of contrary evidence'. Instead of expanding the language by introducing belief or consistency operators, he has proposed to extend classical provability by means of inference rules which may contain consistency assumptions. These rules are called *default rules* (defaults), and have the form

$$\frac{\alpha : \beta}{\gamma}$$

whose meaning is: if α can be inferred and $\neg\beta$ cannot be inferred, then conclude γ. For instance the rule concerning Greeks would be expressed by the default:

$$\frac{greek(x) : dark\_haired(x)}{dark\_haired(x)} \quad .$$

Once more, the set of derivable formulas by means of default rules cannot be described constructively, but only by means of fixpoint extensions induced by the initial data together with the set of defaults.

To conclude, we have gathered under the *fixpoint approach* all proposals, in which nonmonotonic inferences are sanctioned by non-provability. A common feature of these approaches is that the only way of characterizing the set of nonmonotonic consequences is by means of non-constructive fixpoint extensions.

To explain the semantic preference approach, we go back to the initial example. One way to formalize the fact that 'normally, Greeks are dark haired' is

$$\forall x(greek(x) \wedge \neg exceptional(x) \rightarrow dark\_haired(x)).$$

That is, Greeks are dark-haired, unless they are unnormal under a certain respect. From the set of axioms given above we can infer:

$$greek(x) \wedge (albino(x) \vee bald(x)) \rightarrow exceptional(x).$$

Even if we knew that Kostas is neither bald nor albinos, we would still be unable to infer that he is dark-haired. We would need to know that the *only* exceptions to having dark hair are to be bald or albino. Semantically, this would correspond to consider only models in which there are no more exceptional individuals than it is required by the axioms. In other words, we would like restrict the models of our axioms, to those in which the extension of the predicate $exceptional(x)$ is as small as possible. In this case, they are the models which make true:

$$exceptional(x) \leftrightarrow greek(x) \wedge (bald(x) \vee albino(x)).$$

If we restrict our concern to such *minimal* models and if we have the additional information

$\neg bald(kostas), \neg albino(kostas),$

we can conclude $dark\_haired(kostas)$. That is, this conclusion holds in all minimal models of the axioms.

Another use of minimal entailment is closed-world reasoning in databases. Suppose we have a database as shown in Figure 1.

student	advisor
ann	paul
kate	linda
tom	paul

Figure 1.

It is clear that we cannot answer to a query such as 'does every student have an advisor?' represented by the formula:

$$\forall x(student(x) \rightarrow \exists y\, advisor(x, y)),$$

unless we assume that the are no other student than those listed in the database. Minimal entailment, in its variations, can be used to automatically generate the necessary closure conditions involved in such database reasoning.

What we have just sketched is what we call the minimal entailment approach to nonmonotonic reasoning and is based on the idea of restricting the notion of logical consequence to a subset of minimal or preferred models of the axioms. These models represent the intended meaning of the axioms. Minimal entailment is nonmonotonic since a minimal model of a set of formulas Σ might be no longer minimal with respect to subsets of Σ. This semantical approach to nonmonotonic reasoning find its origin in the notion of Circumscription introduced by McCarthy [1980].

The two approaches we have described do not exhaust all possible attempts to formalize various kind of nonmonotonic reasoning which have been proposed so far; for instance, conditional logics have received some interest to formalize reasoning by default [Delgrande, 1988]; but, the fixpoint approach and the semantic preference approach are the most developed and well-known, so our attention will concentrate on them. This chapter is intended to be a survey of tableau methodologies applied to nonmonotonic reasoning. We will mainly treat the propositional case, and postpone the discussion of the first-order case to the last section.

2 THE FIXPOINT APPROACH

2.1 The Modal Approach

In this section we expose Moore's *autoepistemic logics* as the most significant example of modal nonmonotonic logics. Moore proposes a logics which aims to model the beliefs of an ideal agent, capable of reflecting upon its own beliefs. Let us denote by L the belief operator, so that $L\phi$ means 'ϕ is believed'. The language of autoepistemic logic is a conventional propositional language augmented by the operator L. The primary object of interest in autoepistemic logics are sets of formulas which are intended to represent the belief of a self-reflecting agent.

In order to formalize the beliefs of an ideal agent, Moore makes two fundamental assumptions: (1) the agent is capable of both positive and negative introspection, (2) the agent is logically omniscient. The two assumptions determine the following conditions on the belief set S of an agent:

- *deductive closure* if $S \vdash B$, then $B \in S$ ($\vdash$ denotes classical provability);
- *positive introspection* if $\phi \in S$, then $L\phi \in S$;
- *negative introspection* if $\phi \notin S$, then $\neg L\phi \in S$.

An infinite set of formulae S satisfying the conditions above is called *stable.*[2] Notice that if S is a stable set, for all formulas ψ either $L\psi \in S$ or $\neg L\psi \in S$.

Usually we are interested in belief sets which extend or complete, so to say, a given set of premises Σ. We come to the main definition.

DEFINITION 1. Given a set of formulas Σ, we say that a set of formulas S is a *stable expansion* of Σ iff

$$S = Th(\Sigma \cup \{L\phi : \phi \in S\} \cup \{\neg L\phi : \phi \notin S\}).$$

It is immediate to see that a stable expansion S of Σ is stable. Moore gives an equivalent semantical definition of stable expansions, but for the present purpose we do not need it.

Let us call *ordinary sentences* formulas which do not involve the belief operator. It can be proved that any stable expansion is completely determined by the set of ordinary formulas it contains in the sense that if T and T' are stable sets which contain the same set of ordinary formulas, then $T = T'$. Let us denote by $Ord(S)$ the subset of ordinary formulas contained in a set of formulas S and by $Th(S)$ the deductive closure of S.

As the next examples show a set Σ can have zero, one, or more stable expansions.

[2]The notion of stable set was previously introduced by Stalnaker.

EXAMPLE 2.

1. Let $\Sigma = \{\neg Lq \rightarrow \neg q\}$. Then Σ has exactly one stable expansion S and
$$S = Th(\Sigma \cup \{\neg q\}).$$

2. Let $\Sigma = \{\neg Lp \rightarrow q, \neg Lq \rightarrow p\}$. Then Σ has two extensions, namely
$$S_1 = Th(\Sigma \cup \{p\}), \quad S_2 = Th(\Sigma \cup \{q\}).$$

3. Let $\Sigma = \neg Lp \rightarrow p$. Suppose that S is an extension of Σ. If $p \in S$, then $Lp \in S$, but $\Sigma, Lp \not\vdash p$. Thus, it must be $p \notin S$, and hence $\neg Lp \in S$. But $\Sigma, \neg Lp \vdash p$, so that it must be $p \in S$. We have a contradiction. We can conclude that Σ has no extensions.

Regarding to logical consequence, in autoepistemic logics there are two different notions which naturally arise, respectively called *skeptical* and *credulous*. Given a set of premises Σ we define:

- (*skeptical derivability*) $\Sigma \vdash_s \phi$ iff ϕ belongs to all autoepistemic expansions of Σ;
- (*credulous derivability*) $\Sigma \vdash_c \phi$ iff there exists an autoepistemic expansion T of Σ which contains ϕ.

The difference clearly arises when (1) Σ has no stable expansions[3], or (2) Σ has more than one stable expansion. The first notion corresponds to the point of view of an external observer who, not being able to determine which is the belief set of the agent, considers any possible belief set. The second notion corresponds the point of view of the agent who has its own belief set. The two definitions were explicitly introduced by Niemelä, who proposed a decision method for autoepistemic logic based on analytic tableaux. His method will be exposed in section 5.

More recent work has pointed out that the notion of stable expansion is somewhat too weak to adequately represent a belief set. For instance, let $\Sigma = \{Lp \rightarrow p\}$, then Σ has two stable expansions S_0 containing both p and Lp, and S_1 containing $\neg Lp$ and not containing p. But S_0 is hardly justified as a belief set: in S_0 p can be inferred only because the agent has chosen to believe p, that is to put Lp in S_0. To overcome this problem Konolige has proposed two tighter notions of extension respectively called *moderately* grounded and *strongly* grounded. Moderately grounded extensions are minimal extensions of a set of premises, i.e. extensions whose ordinary part is not included in the ordinary part of any other extension. It is easily seen that only S_1 is moderately grounded but not S_0. Strongly grounded extensions eliminate some further anomalies of moderately grounded extensions,

[3]In this case, for any $\phi, \Sigma \vdash_s \phi$

which cannot be ruled out by the simple minimality requirement. We will not give the definition and we will refer to [Moore, 1982] for a complete development. Strongly grounded extensions find their interest in the fact that they are the exact counterpart of default extensions [Konolige, 1988], introduced in the next section.

To conclude this review of the modal approach, we go back to the previous proposal by McDermott and Doyle. If we turn the consistency operator M into its dual L, ($M\phi = \neg L \neg \phi$), it is easy to see that extensions in McDermott and Doyle's logic can be defined as fixed points of the following equation:

$$(*)\ E = Th(\Sigma \cup \{\neg L\psi : E \not\vdash \psi\})$$

Thus, the difference with a stable expansion is that an extension is not required to be closed with respect to positive introspection:

if $E \vdash \psi$, then $L\psi \in E$

For this reason McDermott and Doyle's logic has a quite odd behaviour; for instance the set

$$\{p, \neg Lp\}$$

admits an extension, against all intuitions. This and other anomalies, lead McDermott to the attempt of re-formulating his and Doyle's logic in a modal setting. This amounts to change $\vdash$ in equation $(*)$, from provability in classical logic to provability in some modal system. The move to a modal setting has opened a new realm of investigation with somewhat unexpected results. For instance the closure condition of equation (*) turns out to be very strong: it makes collapse the notion of extension as based on every modal logic between K and S5. System S5 by itself is not a good choice: McDermott has shown that skeptical nonmonotonic provability based on S5 coincide with ordinary, monotonic S5-provability. Autoepistemic logic finds its place in this modal setting [Schwarz, 1990]: *consistent* stable expansions happen to be exactly extensions in the sense of equation (*), once that we consider modal logic K45 as the underlying system.

In Section 5, for the sake of completeness, we will give a tableau procedure for McDermott and Doyle nonmonotonic modal logic.

2.2 The Inferential Approach

As we have seen in the previous section, the primary object of interest in the modal approach are certain infinite sets of formulas representing the beliefs of an agent. These sets are defined by a sort of fixpoint equation. The conditions involved in the fixpoint equation completely constrain extensions, irrespectively to the exact interpretation of modality.[4] All the work is already done in the metalanguage by imposing conditions on extensions. Thus, one can formalize the notion of extension, or

[4]The inessentiality of modal operators can be seen also from the fact that two extensions, agreeing on ordinary formulas, coincide.

belief set without using modality. The first proposal in this sense was Reiter's Default logics. Variants of Reiter's logic have been explored in [Łukasiewicz, 1990; Brewka, 1992; Giordano and Martelli, 1994; Schaub, 1992].

Reiter's Default Logic

The idea of Reiter is to extend propositional provability by *inference* rules called *defaults* of the form

$$\frac{\alpha : \beta}{\gamma}$$

where α, β and γ are formulas,[5] whose meaning is as follows: if α is derivable and β can be consistently assumed (that is $\neg\beta$ is not derivable), then γ is derivable. Knowledge is represented by *default theories* $T = (\Sigma, D)$, where Σ is a set of formulas and D is a set of defaults of the above form.

As in autoepistemic logic, the main object of interest in default logic are sets of formulas closed both under propositional derivability and under defaults. These sets are called *extensions* and can be defined in a number of equivalent ways as fixpoints satisfying certain conditions. Given any set of formulas S, let $\mathbf{\Gamma}(S)$ the smallest set of formulas which satisfies the following conditions:

1. $\Sigma \subseteq \mathbf{\Gamma}(S)$;
2. if $\mathbf{\Gamma}(S) \vdash \phi$ then $\phi \in \mathbf{\Gamma}(S)$;
3. for all defaults $d = \frac{\alpha : \beta}{\gamma} \in D$ if $\alpha \in \mathbf{\Gamma}(S)$ and $\neg\beta \notin S$, then $\gamma \in \mathbf{\Gamma}(S)$.

Given S, such smallest set $\mathbf{\Gamma}(S)$ always exists. A set of formulas E is an extension of a default theory $\Delta = (\Sigma, D)$ if it is a fixpoint of Γ, i.e. $E = \mathbf{\Gamma}(E)$.

Let us introduce some notions and abbreviations. Given a default $d = \frac{\alpha : \beta}{\gamma}$, we call α the *prerequisite* of d, (denoted by $pre(d)$), β the *justification* of d, (denoted by $just(d)$) and γ the consequent of d (denoted by $cons(d)$). Given a set of defaults D, we also define:

$$PRE(D) = \{pre(d) : d \in D\}$$
$$JUST(D) = \{just(d) : d \in D\}$$
$$CONS(D) = \{cons(d) : d \in D\}.$$

[5] In this chapter, we only consider default rules with a single consistency assumption (β), rather than more general ones of the form

$$\frac{\alpha : \beta_1, \ldots, \beta_n}{\gamma}$$

However, the extension of the techniques and results we present to default rules of this form is straightforward.

Given a set D of defaults and a set of formulas S, we define the subset $GD(D,S)$ of defaults of D which can be used to infer their consequents by S:

$$GD(D,S) = \left\{ d = \frac{\alpha : \beta}{\gamma} \in D : \alpha \in S \text{ and } \neg\beta \notin S \right\}.$$

Using this notation, it can be shown that E is an extension of $\Delta = (\Sigma, D)$ if and only if

$$E = Th(\Sigma \cup CONS(GD(D,E))).$$

This characterization of extensions points out that every extension is the deductive closure of the base theory Σ together with the consequents of a subset of defaults in D. We can give an alternative characterization of extensions, which is closer to an inductive definition: E is an extension of a default theory $\Delta = (\Sigma, D)$ if $E = \bigcup_i^\infty E_i$, where

$$\begin{array}{rcl} E_0 & = & \Sigma, \\ E_{i+1} & = & Th(E_i) \cup \left\{ \gamma : \dfrac{\alpha : \beta}{\gamma} \in D, \alpha \in E_i \text{ and } \neg\beta \notin E \right\}. \end{array}$$

We will see in Section 6 that the tableau methods for default logic, developed so far, have been inspired by either one or the other characterization of extensions as given above.

Similarly to autoepistemic logic, a default theory can have zero, one, or more extensions. Here are some examples.

EXAMPLE 3. Let $\Delta = (\Sigma, D)$ be a default theory with $\Sigma = \emptyset$, and $D = \{d_1, d_2\}$, where

$$d_1 = \frac{: a}{a}, \quad d_2 = \frac{: \neg a}{\neg a}.$$

Then, Δ has two extensions,

$$E_1 = Th(\{a\}),\ E_2 = Th(\{\neg a\}).$$

EXAMPLE 4. Let $\Delta = (\Sigma, D)$, where, $\Sigma = \emptyset$, and $D = \{\frac{: \neg a}{a}\}$. Suppose that E is an extension, then we have

$$\begin{array}{l} a \in E \leftrightarrow \neg\neg a \notin E, \text{ whence} \\ a \in E \leftrightarrow a \notin E, \end{array}$$

since E is deductively closed. We have a contradiction, thus Δ has no extensions.

EXAMPLE 5. Let $\Delta = (\Sigma, D)$, with $\Sigma = \emptyset$, and $D = \{d_1, d_2\}$, where

$$d_1 = \frac{a\ : b}{b}, \quad d_2 = \frac{b\ : a}{a}.$$

It is easily seen that $Th(\emptyset)$ satisfies the conditions for being an extension, and hence Δ has a unique extension $E = Th(\emptyset)$.

Two properties, already proved by Reiter are worth mentioning: an inconsistent default theory (Σ, D), (i.e. where Σ is inconsistent) has exactly one extension, namely the set of all formulas. Moreover, extensions are minimal: if E_1 and E_2 are two extensions of (Σ, D) and $E_1 \subseteq E_2$, then $E_1 = E_2$.

In the literature, particular classes of defaults have received a special attention. Two important ones are: *normal defaults* which have the form:

$$\frac{\alpha : \beta}{\beta},$$

and *semi-normal defaults* which have the form:

$$\frac{\alpha : \beta}{\gamma},$$

where $\models \beta \rightarrow \gamma$. Normal default theories have at least one extension. This does not hold for seminormal default theories.

As in autoepistemic logic, two notions of inference naturally arise. Given a default theory $\Delta = (\Sigma, D)$ we can define:

- $\Delta \vdash_s \phi$ iff ϕ holds in every extension of Δ[6], and
- $\Delta \vdash_c \phi$ iff there is an extension E of Δ in which ϕ holds.

Variants of Default Logics

Although Reiter's notion of default inference rule is natural and reasonable, his default logic has some arguable features, which have lead to alternative formulations. We will briefly review some of them, starting from their motivating issues.

A first problem—as we have seen— is that a default theory may fail to have any extension; this is the case, for instance, of (Σ, D), where $\Sigma = \{p \rightarrow \neg q\}$ and $D = \frac{true\ :q}{p}$. What happens here is that the justification q is consistent with $Th(\Sigma)$, hence the default can and must be applied; we are forced to add its consequent p; but now we can prove $\neg q$, that is the justification becomes inconsistent and the default is no longer applicable. This is why the theory has no extensions.

Łukaszewicz [1990] proposes to strengthen and correct Reiter's criterium of applicability of a default to prevent the case we have just seen: in addition to the conditions on the justification and the prerequisite, a default rule is applicable only if its conclusion does not contradict the justification of an already applied default (including its own one).

[6] In particular, $\Delta \vdash_s \phi$, for any ϕ, if Δ has no extensions.

In order to capture this restriction, Łukaszewicz introduces an alternative notion of extension, (called *m-extensions*), by means of a two operator fixpoint definition. For our purposes, we will rather present m-extensions by means of their equivalent semi-inductive characterization (due to Łukaszewicz himself).

We define when a set of formulas E is an m-extension of a default theory $\Delta = (\Sigma, D)$, with respect to a set of formulas F. Intuitively, F is the set of the justifications of the defaults from D, used in the construction of E; we have seen that the modified applicability criterium requires we keep track of them in the construction of E. We can then define: E is a m-extension of Δ if there is a set F, such that E is a m-extension of Δ w.r.t. F.

DEFINITION 6 (cfr. Theorem 4.4, [Łukasiewicz, 1990]). Given $\Delta = (\Sigma, D)$, E is a m-extension with respect to a set of formulas F, iff $E = \bigcup_i^\infty E_i$, and $F = \bigcup_i^\infty F_i$, where

$$\begin{aligned}
E_0 &= \Sigma, \quad F_0 = \emptyset, \\
E_{i+1} &= Th(E_i) \cup \Big\{\gamma : \frac{\alpha : \beta}{\gamma} \in D, \alpha \in E_i \text{ and for each} \\
&\qquad \phi \in F \cup \{\beta\},\ E \cup \{\gamma\} \not\vdash \neg\phi\Big\} \\
F_{i+1} &= F_i \cup \Big\{\beta : \frac{\alpha : \beta}{\gamma} \in D, \alpha \in E_i \text{ and for each} \\
&\qquad \phi \in F \cup \{\beta\},\ E \cup \{\gamma\} \not\vdash \neg\phi\Big\}.
\end{aligned}$$

For instance the theory above $(\{p \rightarrow \neg q\}, \{\frac{true\ :\ q}{p}\})$, has exactly one m-extension, namely $E = Th(\emptyset)$, (with respect to $F = \emptyset$).

EXAMPLE 7. A more concrete example, (due to Łukaszewicz [1990]): is the following: 'On Sunday (su) I usually go fishing (fi), except if I am tired (ti). Usually, if I worked hard yesterday (wh) I am tired, except if I wake up late today (wkl). On holidays (ho) I usually wake up late, except if I go fishing'. We can represent this information by the set D of semi-normal defaults:

$$d1 : \frac{su\ : fi \wedge \neg ti}{fi} \quad d2 : \frac{wh : ti \wedge \neg wkl}{ti} \quad d3 : \frac{ho : wkl \wedge \neg fi}{wkl}.$$

Suppose now that it's Sunday, holidays and that I worked hard yesterday, that is $\Sigma = \{su, ho, wh\}$. This theory has no Reiter extensions. To see this suppose we apply first d_1, so that we add fi; this blocks d_3, whereas d_2 is not blocked, for its justification is consistent; then we add its consequent ti, which contradicts the justification of the already applied d_1. Thus, we don't get an extension. If we consider the defaults in different orders, we obtain a similar situation.

On the opposite, according to Łukaszewicz notion of applicability, once we have applied d_1, both d_2 as much as d_3 cannot be applied anymore. The application of one default blocks the other two. Consequently, the theory has three m-extensions corresponding to the conclusions of each default rule.

Remarkable properties of Łukaszewicz default logics are that (1) every default theory has at least one m-extension; (2) every Reiter extension is also an m-extension; (3) m-extensions monotonically depend on the set of defaults of a default theory (semimonotonicity).

Although Łukaszewicz reformulation ensures the existence of extensions, it leaves unsolved other problems of Reiter default logic. In Łukaszewicz's variant there is no concern about the consistency of justifications of applied defaults among themselves. This may lead to very unintuitive results, as the following example [Brewka, 1992] shows. Let $\Delta = (\Sigma, D)$, where

$$\Sigma = \{broken(rightarm) \vee broken(leftarm)\},$$

and $D = \{d_1, d_2\}$, with

$$d_1 = \frac{: usable(rightarm) \wedge \neg broken(rightarm)}{usable(rightarm)}$$
$$d_2 = \frac{: usable(leftarm) \wedge \neg broken(leftarm)}{usable(leftarm)}.$$

The above theory has exactly one Reiter extension (which is also its only m-extension) E that contains both $usable(leftarm)$ and $usable(rightarm)$. The reason is that we check separately the consistency of the justifications of the involved defaults. If we interpret justifications as assumptions, we should not make two contradictory assumptions at the same time, such as $\neg broken(leftarm)$ and $\neg broken(rightarm)$. This problem is known in the literature as that one of *commitment to assumptions*, for in Reiter logic (as well as in Łukaszewicz's variant) given two justification/assumptions a, $\neg a$, we may accept both, refusing to commit to one of the two.

Another problem of Reiter's logic (as well as Łukaszewicz's variant) is the *failure of cumulativity*. By *cumulativity*, we intend the following property: if we add a formula which holds in one or more extensions of a given theory to the theory, we do not get more extensions than from the original theory.[7] Let us consider for instance $\Delta = (\Sigma, D)$, where $\Sigma = \emptyset$, and $D = \{\frac{: a}{a}, \frac{a \vee b : \neg a}{\neg a}\}$). Then Δ has a unique extension (m-extension) E which contains a, and hence also $a \vee b$. Let Δ'

[7] This notion of cumulativity is somewhat different, but related to the usual definition of cumulativity [Makinson, 1989], which is a property of a consequence relation $\vdash$, and states that

$$\text{if } \Gamma \vdash \phi, \text{ then } \Gamma \vdash \psi \text{ if and only if } \Gamma \cup \{\phi\} \vdash \psi.$$

In case of default logic, it is not so clear where to add a derived formula ϕ. If we consider the addition of ϕ to Σ, the base knowledge, and the consequence relation $\vdash_s$, we have that no variant of default logic is cumulative. Brewka and Schaub's variant are cumulative at the price of considering a more complex notion of 'addition'. See below.

be the theory (Σ', D), where $\Sigma' = \{a \vee b\}$. Δ' has *two* extensions, one containing a and other containing $\neg a$. The trouble is that when we add $a \vee b$ to Σ (i.e. we step to Σ'), we forget that $a \vee b$ was derived from a, so that we can make a contradictory assumption $\neg a$ which enables the second default.

Brewka [1992] has proposed a cumulative variant of Reiter logic. His reformulation takes into account as basic data, *assertions*, that are pairs $\langle \phi, J \rangle$, where ϕ is a formula, and J is a set of formulas, called the *support* of ϕ, the idea is that J represents the set of formulas used to prove ϕ. By keeping track of the support of a formula, Brewka's reformulation of default logic enjoys cumulativity. With regard to the previous example, the theory Δ and $\Delta' = (\Sigma', D)$, where now $\Sigma' = \{\langle a \vee b, \{a\} \rangle\}$ have exactly the same extensions as Δ, namely only one extension which contains a.

Brewka's variant is not only cumulative, but it also has the properties of commitment to assumptions and that one of the existence of extensions, moreover it is semimonotonic.

For our purposes, we will not present it, for it requires to introduce all the technicalities involved in the assertional formalism. We will briefly describe an almost equivalent formulation by Schaub, called *constrained default logic*, which does not need the machinery of assertions.[8] A constrained extension of a default theory, (*c-extension* in Schaub's terminology), is a pair of sets of formulas (E, C) satisfying a fixpoint equation; intuitively E contains the consequent of each applicable default, whereas C contains both the justification and the consequent of each applicable default. As in the case of Łukaszewic's variant we give a semi-inductive equivalent characterization of extensions, rather than their fixpoint definition.

DEFINITION 8 (cfr. Theorem 2.1, [Schaub, 1992]). Given $\Delta = (\Sigma, D)$, (E, C) is a c-extension of Δ, iff $E = \bigcup_i^\infty E_i$, and $C = \bigcup_i^\infty C_i$, where

$$\begin{aligned} E_0 &= \Sigma, \quad F_0 = \Sigma, \\ E_{i+1} &= Th(E_i) \cup \left\{\gamma : \frac{\alpha : \beta}{\gamma} \in D, \alpha \in E_i \text{ and } C \not\vdash \neg(\beta \wedge \gamma)\right\} \\ C_{i+1} &= Th(C_i) \cup \left\{\beta \wedge \gamma : \frac{\alpha : \beta}{\gamma} \in D, \alpha \in E_i \text{ and } C \not\vdash \neg(\beta \wedge \gamma)\right\}. \end{aligned}$$

Looking at the *broken arm* example, it is easily seen that we get two symmetric c-extensions, (E_1, C_1) and (E_2, C_2), such that E_1 contains $usable(leftarm)$, and C_1 additionally contains $\neg broken(leftarm)$, $broken(rightarm)$; E_2 contains $usable(rightarm)$, and C_2 additionally contains $\neg broken(rightarm)$, $broken$ $(leftarm)$.

Schaub and Brewka's variant solves the problem of cumulativity, provided one

[8]Brewka and Schaub variants are equivalent in the sense that there is a bijective correspondence between Schaub extensions and Brewka extensions, modulo the identification of logically equivalent supports.

opportunely defines what it means to add a formula ϕ which holds in one extension E to a theory.[9]

For a broader discussion on the properties of default logic, we refer to [Delgrande *et al.*, 1994].

3 THE MINIMAL ENTAILMENT APPROACH

The preferential approach is based on the idea of restricting the notion of logical consequence to a subset of privileged or preferred models. Instead of considering what holds in every model of a given theory, we only consider what holds in the subset of preferred models. This approach to nonmonotonic reasoning finds its root in McCarthy's notion of *circumscription* [McCarthy, 1980]. The starting point is to define a preference ordering among interpretations. Conforming to the standard terminology, we will consider as preferred models of a theory the set of its *minimal* models according to the given preference relation. Then we define,

$$\Sigma \models_m \phi \text{ iff } \phi \text{ holds in all minimal models of } \Sigma.$$

The above entailment relation is called *minimal entailment*. This abstract setting has been put forward by Shoham [1987], in order to give a unifying account of nonmonotonic systems. In this respect, several proposals such as McCarthy's circumscription, Reiter's *closed world assumption* [10] and Bossu–Siegel subimplication [1985], can be seen as different attempts to formalize the notion of minimal entailment, arising from a specific preference relation among models.

The first notion introduced by McCarthy was that of *domain circumscription*, or domain minimization as a formalization of a form of Okham's razor: the only objects which are assumed to exist are those which are forced to exist by our knowledge. Semantically, this notion corresponds to regard as preferred, models which have a minimal (w.r.t. set-inclusion) domain. This notion was later developed by McCarthy's and others in the notion of circumscription. The basic idea of circumscription (and its semantical counterpart, minimal entailment)

---

[9] In the case of Brewka, it means to add ϕ to its support (forming an assertion). In case of Schaub, it means to add a default rule of the form:

$$d_\phi = \frac{: \bigwedge_i \beta_i \wedge \bigwedge \gamma_i}{\phi},$$

where the β_i and γ_i are the justifications and the consequents of those defaults involved in the construction of E, used to prove ϕ (remember that the E-part of a c-extension is always the deductive closure of Σ together with some default consequents). Schaub shows that given a c-extension (E, C) of a theory $\Delta = (\Sigma, D)$ and a formula $\phi \in E$, if d_ϕ is the default rule built as above, then Δ and $\Delta' = (\Sigma, D \cup \{d_\phi\})$ have the same c-extensions. In this sense 'adding ϕ', which is derived from E, to the theory, does not change the set of extensions of the original theory.

[10] The relationship between circumscription and closed world assumption are investigated in [Lifschitz, 1985; Olivetti, 1989].

is to minimize the extension of some predicates, that is to assume that the information we have on the instances of some predicate is complete. Circumscription by itself is a syntactic approximation to the semantic notion of minimal entailment, and can be formulated either as a first-order schema of formulas, or as a second-order axiom. It can be added to complete a set of formulas, in the same way as induction (schema or axiom) completes the other Peano's axioms. Indeed it can be seen as a sort of generalization of induction to arbitrary theories. This observation explains why, in general, circumscription is sound but not complete with respect to minimal entailment. For the present purpose, we will take as primitive the semantic notion of minimal entailment, and we will not give circumscription definition (the reader is referred to [McCarthy, 1980; McCarthy, 1986] for details).

We first define minimal entailment in a first-order setting and then in the propositional case as a special case. Let L be a first-order language, let $\mathbf{R}$ be the set of all predicate symbols in L, and let $\mathbf{P}$ and $\mathbf{Q}$ be two disjointed subsets of $\mathbf{R}$. Given two L-structures M and N, we define the following ordering relations $M \leq N$, $M \leq_{\mathbf{P}} N$ and $M \leq_{\mathbf{P},\mathbf{Q}} N$. Consider the following conditions:

(a) M and N have the same functional structure, that is the same domain and the same interpretation of constant and function symbols.

(b) the following holds:

- (i) for all $p \in \mathbf{P}$, $M(p) \subseteq N(p)$,
- (ii) for all $r \in \mathbf{R} - (\mathbf{P} \cup \mathbf{Q})$, $M(r) = N(r)$.

Then we define (1) $M \leq_{\mathbf{P},\mathbf{Q}} N$ iff (a) and (b) holds, (2) $M \leq_{\mathbf{P}} N$ iff (a) and (b) holds, and $\mathbf{P} \cup \mathbf{Q} = \mathbf{R}$, (3) $M \leq N$ iff (a) and (b) holds and $\mathbf{P} = \mathbf{R}$. Predicates in $\mathbf{P}$ are assumed to be minimized, predicates in $\mathbf{Q}$ are assumed to change arbitrarily to make the extension of predicates in $\mathbf{P}$ as smaller as possible. Finally, predicates in $\mathbf{R} - (\mathbf{P} \cup \mathbf{Q})$ are assumed to be fixed. To each semantic ordering corresponds a notion of minimal entailment, which we will denote respectively as, $\models_{\mathbf{P},\mathbf{Q}}$ corresponding to (1), $\models_{\mathbf{P}}$ corresponding to (2), and $\models_m$ corresponding to (3).

We will mainly deal with the more restricted case of $\models_m$, that is when all predicates are minimized. The reason is that the problems and the techniques are better illustrated in this basic case; furthermore minimal entailment under the latter more general preferential relations (that is $\models_{\mathbf{P},\mathbf{Q}}$ and $\models_{\mathbf{P}}$) can be reduced to the basic case of minimal entailment $\models_m$, where all predicates are minimized.

A consistent theory may have zero, one, or more minimal models. For certain classes of consistent theories, minimal models always exist. Theories enjoining this property are, for instance, those which have finite models. Another class is that one of theories for which the preference relation is well-founded on the class of their models [Lifschitz, 1986]. Such theories are called *well-founded*; a sufficient condition for well-foundness of a theory Σ is that Σ only contains universal axioms.

In the propositional case, the role played by sets of predicates is played by sets of atoms. For instance, we have that $V \leq V'$ iff for all atoms p, $V(p) = 1$, implies $V'(p) = 1$. Since the partial order on evaluations of propositional theories is well-founded, consistent propositional theories always have minimal models.

To conclude this section, we go back to *domain minimal entailment.* The semantic relation is the following: given two interpretations M and N we say that M is d-smaller than N if M is a *substructure* of N. We denote by $\models_d$ the corresponding notion of minimal entailment.

For instance, $\exists x\, p(x) \models_d \forall x\, p(x)$.

As for the case of predicate minimization, a consistent theory may have zero, one or more domain minimal models. The condition of well-foundness ensures the existence of minimal models (for consistent theories). Notice that a universal theory is obviously well-founded and its class of minimal models is, up to isomorphism, the class of its Herbrand models. We recall that existential formulas (and hence ground formulas) are preserved moving upward on the substructure relationship. It turns out that for well-founded theories, domain minimization is a very weak nonmonotonic entailment, for we easily obtain that if Σ is well-founded and ψ is an existential formula, then

$$\Sigma \models_d \psi \quad - \quad \Sigma \models \psi.$$

By this result, we cannot infer any new existential (and hence ground) information by domain-minimization from a well-founded theory. In particular, we cannot infer new equalities between ground terms.

This weakness has lead Lorenz [1994] to introduce a stronger notion of domain minimization which he calls *variable domain minimal entailment.* We can introduce this notion of minimal entailment by changing the preference relation. Given two structures N and M for a language $\mathcal{L}$, we define $N \leq_{v-d} M$ if

1. $\mid N \mid \;\subseteq\; \mid M \mid$,[11]
2. for each term $t \in \mathcal{L}$, $t^N = t^M$.

The difference with the substructure relation, which underlies domain minimization, is that we do not require that for any predicate p, $p^N = p^M \cap \mid N \mid$, that is to say, we allow the extension of any predicate to vary to make the domain as smaller as possible. For instance, given the formula:

$$\phi :\ \exists x p(x) \wedge \exists x q(x),$$

we have that $\exists x(p(x) \wedge q(x))$ is a *variable* domain minimal consequence of ϕ, but it is not a domain minimal consequence of ϕ.

[11] $\mid M \mid$ denotes the domain of M.

Other versions of minimal entailment, together with their syntactical counterpart of circumscription schema/axiom, have been introduced in the literature: we just recall *prioritized* and *pointwise* circumscription [Lifschitz, 1985a]. In [Lin, 1990] the idea of circumscription is extended to modal logics. We will not be concerned with these further developments.

4 TABLEAUX AS A GENERAL METHODOLOGY

Although the use of tableau techniques in nonmonotonic reasoning has not been fully explored yet, tableaux are probably the most flexible and promising proof method in this area. There are at least two good reasons in favour of tableau methods. First, tableaux can be used to check both *provability* and *consistency*. Checking the latter by other proof methods is not as easy as checking it by tableaux. As we have seen in the previous section, consistency (unprovability) has the same importance as provability for fixpoint based nonmonotonic logics.

Furthermore, tableaux can be profitably used also for preference based nonmonotonic logics. Open branches of a completed tableau (partially) describes the set of models of the input formulas. One can often define a preference criterion among open branches which mirrors the intended semantic preference. The preference criteria on open branches can then be used to eliminate unwanted models.

Tableaux, as any other standard proof-method is monotonic in the following sense: if T is a tableau which contains an initial set of formulas Σ and T results to be closed by the rules of tableau expansion, then whatever we add to T, (i.e. in any branch of T) T remains closed. How can a tableau methodology be used to represent nonmonotonic mechanisms? If we look to the work in the area we can see that some general principles have been employed to adapt tableau technology to nonmonotonic logics:

- *Operation on branches*: addition and removal of formulas from completed branches, according to some conditions.
- *Relative closure*: the closure of a tableau may depend on another related tableau being *open*.
- *Negative closure conditions*: some branches in a completed tableau are forced to be 'closed' because they do not contain some formulas.
- *Selection*: open branches of a tableau are ruled out because of an external selection criterion. This is what happens, for instance, in nonmonotonic logic based on preferential semantic.

We assume familiarity with the terminology and the basic construction of analytic (signed) tableaux.[12] We identify a tableau with a set of sets of signed formu-

[12] We refer the reader to the first chapter of this volume for the background knowledge on tableau systems.

las, each one of these sets is called a *branch*. Initially a tableau contains just one set with the input formulas. Signed formulas are divided in two types: α- type and β-type. The rules of tableau expansion allows one to replace a branch containing a formula ϕ by one or more branches obtained by dropping ϕ, and then adding some signed subformulas of ϕ, according to the type of ϕ.

Given a set of formulas Σ and a formula ϕ, we say that a tableau T is *for* (Σ, ϕ) if T contains as initial data $\mathbf{T}\psi$ for every $\psi \in \Sigma$ and $\mathbf{F}\phi$.[13] We also say that ϕ is the goal formula of T.

Given a set of formulas Σ, we also write $\mathbf{T}\Sigma$, $\mathbf{F}\Sigma$ to denote respectively the set of $\mathbf{T}$-signed and $\mathbf{F} - signed$ formulas from Σ. Moreover, given a set of signed formulas Q we denote by $Tab(Q)$ any *completed* tableau with input formulas Q.

To simplify notation, we write $Tab(\mathbf{T}\phi)$, (and $Tab(\mathbf{F}\phi)$), rather than $Tab(\{\mathbf{T}\phi\})$, (and $Tab(\{\mathbf{F}\phi\})$). Moreover, if $T = Tab(Q)$ and T just contains one branch B, we usually write $T = B$, instead of $T = \{B\}$.

We finally introduce (see [Schwind, 1990]) a useful composition operation on tableaux. Given two tableaux $T1$ and $T2$ we let

$$T_1 \otimes T2 = \{X \cup Y : X \in T1 \wedge Y \in T_2\}.$$

5 TABLEAUX FOR AUTOEPISTEMIC LOGIC

In this section we expose the method by Niemelä [1988]. The method is similar to the previous McDermott and Doyle's method for their *nonmonotonic modal logic*. To check if ψ is in some (all) stable expansions of a given set of premises Σ, we start building a tableau to show $\Sigma \rightarrow \psi$. In the tableau construction formulas of the kind $L\phi$ may occur. The basic idea is that $L\phi$ must be put in a stable expansion of the initial set of premises, if ϕ is provable from that expansion, and $\neg L\phi$ is to be put in the stable expansion if ϕ is not provable from that expansion. To check $L\phi$ we start a new tableau for $\Sigma \rightarrow \phi$. This tableau may involve other formulas of the form $L\chi$, and thus the process must be iterated. Since the provability of a formula depends on the provability of other formulas, this process requires an arbitrary decision about provability of all formulas ϕ such that $L\phi$ (or $\neg L\phi$) occur in a given tableau. What is important is that decisions are made in a consistent way. This amount to say that having declared ϕ provable ('non-provable'), the addition of $L\phi$ (respectively $\neg L\phi$) to the tableau does not change the provability status of the goal formula of any tableau involved in the construction. Each consistent decision about provability exactly corresponds to a stable expansion.

We describe below the tableau method in greater details. To know whether

[13]A tableau for (Σ, ϕ) is what is obtained by a two-step expansion of a tableau initialized by $\mathbf{F}(\bigwedge \Sigma \rightarrow \phi)$.

$\Sigma \vdash_s \phi$ and $\Sigma \vdash_c \phi$, we build up the following tableau-structure

$$(\Sigma, \phi, T_0, X),$$

where T_0 is a tableau for (Σ, ϕ), and X is the least set satisfying the following conditions:

- $T_0 \in X$;
- if $\mathbf{T}L\psi$ or $\mathbf{F}L\psi$ occurs in an open branch of some tableau in X, then X contains a tableau for (Σ, ψ).

A *labelling* l for (Σ, ϕ, T_0, X) is a function which assigns a label $OPEN$ or $CLOSED$ to each tableau in X. The procedure for deciding both whether $\Sigma \vdash_s \phi$ and $\Sigma \vdash_c \phi$, can be described as follows.

1. Build up the structure (Σ, ϕ, T_0, X), first completing T_0, and then completing every tableau in X.
2. For each labelling l and tableau $T \in X$, we let the tableau $upd(l, T, X)$ be obtained from T by adding (1) $\mathbf{T}L\psi$ to every branch of T if there is a tableau $T' \in X$ for (Σ, ψ) such that $l(T') = CLOSED$, and (2) $\mathbf{F}L\psi$ to every branch of T if there is a tableau $T' \in X$ for (Σ, ψ) such that $l(T') = OPEN$.
3. for each labelling l we check whether l is *admissible*. We say that l is admissible for (Σ, ϕ, T_0, X) whenever, for every $T \in X$, it holds that:
   - if $l(T) = OPEN$ then $upd(l, T, X)$ contains an open branch;
   - if $l(T) = CLOSED$ then all branches of $upd(l, T, X)$ are closed.
4. As a last step we set that
   - $\Sigma \vdash_s \phi$ iff T_0 is labelled $CLOSED$ in all admissible labellings for (Σ, ϕ, T_0, X)[14];
   - $\Sigma \vdash_c \phi$ if there is an admissible labelling for (Σ, ϕ, T_0, X) in which T_0 is labelled $CLOSED$. [15]

[14] In particular, $\Sigma \vdash_s \phi$, for any ϕ, if there are no admissible labellings for (Σ, ϕ, T_0, X).

[15] This definition is not correct if Δ has no extensions. The following counterexample is by C. Schwind. Let $\Delta = \{\mathbf{T}Lq\}$ and let ϕ be any propositional tautology. Suppose we build a tableau structure (Δ, ϕ, T_0, X), where T_0 is for (Δ, ϕ). Since T_0 is closed, we have $X = \{T_0\}$. Let l be a labelling with $l(T_0) = CLOSED$; obviously, l is admissible. But Δ has no extensions. To get the correct answer, we must check whether Δ has an extension by itself, independently from ϕ. To this purpose, we can build a tableau structure (Δ, T_0, X), in which T_0 only contains the formulas of Δ and no goal formula; then we check if there is an admissible labelling for this tableau structure.

EXAMPLE 9. Let $\Sigma = \{\neg Lp \rightarrow \neg p\}$. We show that $\Sigma \vdash_s \neg p$. We first build a tableau T_0 for $(\Sigma, \neg p)$. T_0 has two completed branches:

$$T_0 = \quad \{\{\mathbf{T}Lp, \mathbf{T}p\},\ \{\mathbf{F}p, \mathbf{T}p\}\}.$$

Since Lp occurs in some open branch of T_0, we build a tableau T_1 for (Σ, p). No other tableau is needed, that is to say $X = \{T_0, T_1\}$. T_1 has the following two branches:

$$T_1 = \quad \{\{\mathbf{T}Lp, \mathbf{F}p\},\ \{\mathbf{F}p\}\}.$$

Let $l_1(T_1) = OPEN$, we must add $\mathbf{F}Lp$, in every branch of every tableau. Then T_1 remains open, (more precisely, $upd(l_1, T_1, X)$ is open), whereas T_0 becomes closed ($upd(l_1, T_0, X)$ is closed). Hence, if $l_1(T_0) = CLOSED$, then l_1 is admissible, otherwise it is not. Let $l_2(T_1) = CLOSED$, we have to add $\mathbf{T}Lp$ everywhere. But doing so, T_1 is still open, thus l_2 is not admissible. We can conclude that there is only one admissible labelling l_1, with $l_1(T_1) = OPEN$ and $l_1(T_0) = CLOSED$. Since T_0 is closed in l_1 we can conclude $\Sigma \vdash_s \neg p$.

EXAMPLE 10. Let $\Sigma = \{\neg Lp \rightarrow q, \neg Lq \rightarrow p\}$. We show that $\Sigma \vdash_s p \vee q$. We start building a tableau T_0 for $(\Sigma, p \vee q)$. T_0 has the following four branches:

$$B_{0,1} = \{\mathbf{F}p, \mathbf{F}q, \mathbf{T}Lp, \mathbf{T}Lq\},\ B_{0,2} = \{\mathbf{F}p, \mathbf{F}q, \mathbf{T}Lp, \mathbf{T}p\},$$

$$B_{0,3} = \{\mathbf{F}p, \mathbf{F}q, \mathbf{T}q, \mathbf{T}Lq\},\ B_{0,4} = \{\mathbf{F}p, \mathbf{F}q, \mathbf{T}q, \mathbf{T}p\}.$$

Notice that $B_{0,2}$, $B_{0,3}$, and $B_{0,4}$ are closed. T_0 is open. To complete the construction, we have to build a tableau T_1 for (Σ, p), and a tableau T_2 for (Σ, q). T_1 has the following branches:

$$B_{1,1} = \{\mathbf{F}p, \mathbf{T}Lp, \mathbf{T}Lq\},\ B_{1,2} = \{\mathbf{F}p, \mathbf{T}Lp, \mathbf{T}p\},$$

$$B_{1,3} = \{\mathbf{F}p, \mathbf{T}q, \mathbf{T}Lq\},\ B_{1,4} = \{\mathbf{F}p, \mathbf{T}q, \mathbf{T}p\},$$

where $B_{1,2}$ and $B_{1,4}$ are closed. T_1 is open. T_2 has the following branches:

$$B_{2,1} = \{\mathbf{F}q, \mathbf{T}Lp, \mathbf{T}Lq\},\ B_{2,2} = \{\mathbf{F}q, \mathbf{T}Lp, \mathbf{T}p\},$$

$$B_{2,3} = \{\mathbf{F}q, \mathbf{T}q, \mathbf{T}Lq\},\ B_{2,4} = \{\mathbf{F}q, \mathbf{T}q, \mathbf{T}p\},$$

where $B_{2,3}$ and $B_{2,4}$ are closed. T_2 is open. If a label l assigns $CLOSED$ to both T_1 and T_2, then we have to add $\mathbf{T}Lp$ and $\mathbf{T}Lq$, but tableaux T_1 and T_2 remain open; hence l is not admissible. If l assigns $OPEN$ to both T_1 and T_2, then we add $\mathbf{F}Lp$ and $\mathbf{F}Lq$, so that both of them become closed; thus l is not admissible. It easily seen that the only admissible labellings are l_1 and l_2 with $l_i(T_0) = CLOSED$, for $i = 1, 2$, and $l_1(T_1) = CLOSED$, $l_1(T_2) = OPEN$, and $L_2(T_1) = OPEN, l_2(T_2) = CLOSED$. Since T_0 is $CLOSED$ under both of them, we can conclude that $\Sigma \vdash_s p \vee q$.

EXAMPLE 11.

Let $\Sigma = \{\neg Lp \rightarrow p\}$. We build T_0 for Σ:

$$T_0 = \{\{\mathbf{T}Lp\}, \{\mathbf{T}p\}\},$$

and T_1 for (Σ, p):

$$T_1 = \{\{\mathbf{T}Lp, \mathbf{F}p\}, \{\mathbf{T}p, \mathbf{F}p\}\}.$$

Suppose that $l(T_1) = CLOSED$, then we add $\mathbf{T}Lp$ and T_1 remains open; thus T_1 cannot be admissible. Suppose that $l(T_1) = OPEN$, then we add $\mathbf{F}Lp$ and T_1 becomes closed. Thus, there are not admissible labellings.

Niemelä proves the following facts:

PROPOSITION 12. *If S is a stable expansion of Σ, then there is an admissible labelling l of the tableau structure (Σ, ϕ, T_0, X), such that*

$$\phi \in S \quad \textit{iff} \quad l(T_0) = CLOSED.$$

PROPOSITION 13. *Let l be an admissible labelling of a tableau structure (Σ, ϕ, T_0, X), then there is a stable expansion S of Σ such that for every tableau $T \in X$ for (Σ, ψ),*

$$\psi \in S \quad \textit{iff} \quad l(T) = CLOSED.$$

Since the tableau structure construction always terminates (for a finite set of premises Σ), and there are a only finitely many labellings to check for admissibility, by Proposition 1 and 2 we obtain correctness and completeness of Niemelä method, and decidability of provability in autoepistemic logic.

The method is also important for theoretical reasons. Autoepistemic expansions are infinite theories. Thus, it is important to find a finite representation of these sets. Intuitively, the method described above shows that each expansion is uniquely determined by a finite set of subformulas of the initial set of premises. Such sets are called *full sets* (see Section 9) in [Niemelä, 1991], where a finite characterization of expansions is studied.

From the point of view of efficiency, we notice that the number k of tableaux that have to be constructed is linear in the size of (Σ, ϕ), and there are obviously 2^k labellings to be checked for admissibility. However, it is unlikely that a more efficient, yet general method exists. Niemelä [1991] and then Gottlob [Gottlob, 1992] have provided a complete analysis of complexity problems concerning autoepistemic logic, their results situate autoepistemic logic in the second level of the polynomial hierarchy. In particular $\vdash_s$ is $\Pi_2 - complete$, and $\vdash_c$ is $\Sigma_2 - complete$. The tableau method we have presented intuitively accounts for such a complexity, for we operate on two levels: first we build the various tableaux, and then we check the admissible labellings.

Tableaux for McDermott and Doyle's Logic

We can easily modify the procedure given above for McDermott and Doyle non-monotonic logic. We work with tableau structures of the form

$$(\Sigma, \phi, T_0, X),$$

where T_0 is a tableau for (Σ, ϕ), and X is the least set satisfying the following conditions:

- $T_0 \in X$;
- if $\mathbf{T}L\psi$ occurs in an open branch of some tableau in X, then X contains a tableau for (Σ, ψ).

The procedure for checking both skeptical and credulous derivability is similar to that one for autoepistemic logic, the only difference is in the definition of the tableau *upd*(l, T, X), given a labelling l:

upd(l, T, X) is obtained from T by adding $\mathbf{F}L\psi$ to every branch of T, if there is a tableau $T' \in X$ for (Σ, ψ) such that $l(T') = OPEN$.

As a difference with respect to the procedure for the case of autoepistemic modal logics, notice the asymmetry between provability and non-provability that we have pointed out in Section 2.1: (1) we *only* have to build another tableau to evaluate $\mathbf{T}L\psi$ if it occurs in an open branch of some tableau in X; This means that we *cannot* assume $\neg L\psi$ only if we derive ψ; (2) if ϕ is not derivable, in the update step, we add $\mathbf{F}L\psi$, but we do not care about the dual case, (i.e., even if we derive ψ we do *not* add $\mathbf{T}L\psi$).

For instance, let us see that there is an extension containing both p and $\neg Lp$. Let

$$T_0 = \{\mathbf{T}p, \mathbf{F}Lp\}$$

T_0 is open, we do not need to generate another tableau, that is $X = \{T_0\}$. Let $l(T_0) = OPEN$. Then, *upd*$(l, T_0, X) = T_0$, which is open and hence l is admissible.

The procedure we have sketched has been extended by McDermott to his modal nonmonotonic logics (in the case of S4). We refer to [McDermott, 1982] for details.

6 TABLEAUX FOR DEFAULT LOGIC

The first attempt of using a tableau based methodology to address default logic has been developed by Schwind and Risch. An alternative proposal has been put forward by Amati *et al.* in [1996]. Both the Schwind and Risch methodology, on the one hand, and Amati *et al.*'s methodology, on the other, are well suited to accommodate several variants of default logics.

The two approaches differs substantially. Any extension is the deductive closure of the initial theory Σ together with a set of consequents of defaults. The idea pursued by Schwind and Risch is then to add all together the consequents of defaults to the initial theory. This may result in an inconsistent theory, but we can split the set of consequents of all defaults in maximal consistent subsets. Each such subset is a candidate to represent an extension. Further processing is required to find the actual subsets of consequents which represent extensions.

On the contrary, the construction provided by Amati *et al.* is incremental: starting from the initial theory we add one consequent of a default at a time (when possible), until we reach a set which contains all consequents of applicable defaults. This set, provided it satisfies some additional condition, represents an extension. Different extensions may be generated by considering defaults in a different order, and all extensions can be obtained in this way.

6.1 Tableaux for Reiter's Default Logics

We start the exposition from Schwind's and Risch's method. In [Schwind, 1990] it is proposed a method for normal defaults, which is extended in [Schwind and Risch, 1994] to arbitrary defaults theories. The authors have given for the first time a necessary and sufficient condition for the existence of extensions. We recall that E is an extension of $\Delta = (\Sigma, D)$ if and only if

$$E = Th(\Sigma \cup CONS(GD(D, E))).$$

This means that every extension E corresponds to a subset of defaults D. However, not all subsets correspond to any extension. As a first approximation, we consider only those sets of defaults such that their prerequisites can be established using Σ, or using the consequents of other defaults, but avoiding circular dependencies. More formally we introduce the following definition.

DEFINITION 14. Given a theory Σ, a set of defaults D is *grounded* in Σ, if for each $d \in D$ there is a sequence $d_1, \ldots, d_n = d$ of elements of D of the form

$$\frac{\alpha_i : \beta_i}{\gamma_i},$$

such that

$\Sigma \vdash \alpha_1$,
for $i = 1, \ldots, n-1$, $\Sigma \cup \{\gamma_1, \ldots, \gamma_i\} \vdash \alpha_{i+1}$.

It easy to see that the set of generating defaults $GD(D, E)$ of any extension is grounded. The opposite partially holds: to each grounded subset of defaults, satisfying an additional property, it corresponds an extension.

THEOREM 15. *Let (Σ, D) be a default theory, then E is an extension of (Σ, D) if and only if there is a subset $D' \subseteq D$, such that*

- *D' is grounded in Σ;*
- *$E = Th(\Sigma \cup CONS(D'))$;*
- *for each $d \in D$ of the form $\frac{\alpha : \beta}{\gamma}$, we have*
  1. *if $d \in D'$ then $\alpha \in E$ and $\neg\beta \notin E$;*
  2. *if $d \notin D'$ then either $\alpha \notin E$ or $\neg\beta \in E$.*

We can see condition (1) as a correctness condition on D': D' cannot contain defaults which are unapplicable to E; we can see condition (2) as a maximality condition on D': each default belongs to D', unless it is unapplicable to E. Notice that conditions (1) and (2) are equivalent to

- (1′) if $d \in D'$ then $\Sigma \cup CONS(D') \vdash \alpha$ and $\Sigma \cup CONS(D') \not\vdash \neg\beta$;
- (2′) if $d \notin D'$ then either $\Sigma \cup CONS(D') \not\vdash \alpha$ or $\Sigma \cup CONS(D') \vdash \neg\beta$.

In this way, the task of checking whether a default theory $\Delta = (\Sigma, D)$ admits extensions is reduced to checking conditions (1′) and (2′) on all possible subsets of D. In particular, if for every subset D' of D there is a default d such that d satisfies neither (1′) nor (2′), then Δ does not have any extension. Obviously conditions (1′) and (2′) can be checked by any proof method, including tableaux. But tableaux can be used to generate possible candidates, that is to say subsets of D, to be checked against conditions (1′) and (2′).

In principle, the algorithm to check and generate extensions of a default theory (Σ, D) may be logically structured in three (high level) steps:

(a) Find all maximal subsets D_j of D such that $\Sigma \cup CONS(D_j)$ is consistent.

(b) For each D_j, find all subsets D_j^l of D_j which satisfy conditions (1) and (2).

(c) For each D_j^l check whether it is grounded.

By Theorem 15, there is one-to-one mapping between grounded subsets D_j^l and extensions.

We now discuss how to implement steps (a) and (b) by using tableaux.

For (a), we build a tableau T for $\Sigma \cup CONS(D)$. If the tableau is closed, then it is possible to open it by removing from one closed branch any signed atom from $CONS(D)$ which is responsible of the closure.

More precisely, let $T = Tab(\mathbf{T}\Sigma) \otimes Tab(\mathbf{T}CONS(D))$. Delete closed branches whose closure does not depend on $CONS(D)$. Set $Def(T) = D$. For any

$d \in Def(T)$, if some signed atom coming from $cons(d)$[16] is responsible of the closure of some branch of T, then (i) remove any formula, coming from $cons(d)$, from T, and (ii) remove d from $Def(T)$. Repeat this step until the resulting tableau T' is open (that is *one* branch of T' is open). Since there may be more than one way to open T by removing some signed atom, there will be many 're-opened' sub-tableaux T_j of T. We generate all of them. We let $D_j = Def(T_j)$= defaults whose consequents are in T_j. This concludes step (a).

With regard to step (b), we start checking whether D_j satisfies conditions (1′) and (2′). We have to check that: for each $d \in D_j$ with $d = \frac{\alpha : \beta}{\gamma}$

- the tableau $T_j \otimes Tab(\{\mathbf{F}\alpha\})$ is closed and
- the tableau $T_j \otimes Tab(\{\mathbf{T}\beta\})$ is open.

If it is not the case, we check whether the second condition holds. That is, let T_j' be obtained by removing $cons(d)$ from T_j, we check whether

- the tableau $T_j' \otimes Tab(\{\mathbf{F}\alpha\})$ is open, or
- the tableau $T_j' \otimes Tab(\{\mathbf{T}\beta\})$ is closed.

If the latter test succeeds, then we drop d from D_j (and $cons(d)$ from T_j), if it fails, we rule out D_j. Notice that if D_j is the only candidate, in the case of failure, we have determined that there are no extensions. We repeat the process on D_j until either D_j is ruled out, or a maximal subset of D_j for which the conditions (1′) and (2′) hold is found. Notice that there may be more than one maximal subset D_j^l satisfying conditions (1′) and (2′), according to the order in which defaults are checked. For instance, if $\Sigma = \emptyset$ and $D = \{d_1, d_2\}$, where

$$d_1 = \frac{: \neg a}{b}, \quad d_2 = \frac{: \neg b}{a},$$

then $T = Tab(\mathbf{T}\Sigma) \otimes Tab(\mathbf{T}CONS(D)) = \{\{a, b\}\}$, which is consistent, thus the only candidate D_j is D. D does not satisfy condition (1′). However, there are *two* maximal subsets of D which satisfy condition (1′) and (2′), namely $\{d_1\}$ and $\{d_2\}$.

In [Schwind and Risch, 1994] it is described an algorithm which implements the method we have described. The algorithm reorganizes the above steps in order to avoid the repetition of justification and prerequisite tests. Such optimization is based on the following observations.

- Let D_j be a candidate for being an extension, if $d \in D_j$ and $just(d)$ is consistent with T_j, then it remains consistent with every subset D_j' of D_j,

[16] To this purpose, as suggested in [Schwind, 1990] we can label each atom occurring in $CONS(D)$ with the name of the default it comes from.

(more exactly, with $Tab(\mathbf{T}\Sigma) \otimes Tab(\mathbf{T}CONS(D'_j))$. This means that we do not need to check more than once the justification condition involved in (1').

- The groundness test subsumes the prerequisite test involved in (1'). Thus, there is no point in checking first the prerequisite condition, and then groundness.

By exploiting these observations, we can re-organize the procedure as follows:

Algorithm for computing Reiter's extensions

(i) Find all maximal subsets D_j of D such that $\Sigma \cup CONS(D_j)$ is consistent.

(ii) Given D_j, find all its maximal subsets D_j^m such that

$$Th(\Sigma \cup CONS(D_j^m)) \cap \{\neg just(d) : d \in D_j^m\} = \emptyset,$$

that is all maximal D_j^m such that, letting

$$T_j^m = Tab(\mathbf{T}\Sigma) \otimes Tab(\mathbf{T}CONS(D_j^m)),$$

for all $d \in D_j^m$,

$$T_j^m \otimes Tab(\mathbf{T}just(d))$$

is open.

(iii) For all D_j^m, find its maximal grounded subset D_j^l (See below).

(iv) For all D_j^l, check whether it satisfies condition (2), that is, letting $T_j^l = Tab(\mathbf{T}\Sigma) \otimes Tab(\mathbf{T}CONS(D_j^l))$, check whether, for all $d \notin D_j^l$, either

$T_j^l \otimes Tab(\mathbf{F}pre(d))$ is open, or
$T_j^l \otimes Tab(\mathbf{T}just(d))$ is closed.

If there exists one $d \notin D_j^l$ for which the above condition does not hold, delete D_j^l.

(v) Output all D_j^l, if any, otherwise, print 'NO- EXTENSIONS'.

Extensions, if any, correspond exactly to subsets D_j^l of the original D which are left after step (iv). To find maximal subsets in steps (i) and (ii), we use the re-opening technique as discussed above. To find a grounded subset D with respect to Σ, (step (iii)), we pick up a default $d \in D$, and we check (i) whether $\Sigma \cup CONS(D - \{d\}) \vdash prereq(d)$, and then (ii) whether $D - \{d\}$ is grounded; we must find a sequence of all elements of D, such that the prerequisite of the last default is proved by Σ alone. In the search of a maximal grounded subset of D, elements not belonging to any such sequence are deleted from D. In [Schwind and Risch, 1994] it is described a recursive algorithm to find maximal grounded subsets, we refer to that paper for details. The authors observe that checking groundness for a set of defaults D is not more difficult than checking the prerequisite condition.

Once that we have determined each grounded subset D^l_j , then we can use them for inference, this amounts to check whether a formula holds in

$$Tab(\mathbf{T}\Sigma) \otimes Tab(\mathbf{T}CONS(D^l_j)),$$

for each D^l_j if any, (respectively, for some D^l_j, in case of credulous inference).
EXAMPLE 16.

Let $\Delta = (\Sigma, D)$ be a default theory with $\Sigma = \emptyset$, and $D = \{d_1, d_2\}$, where

$$d_1 = \frac{:a}{a}, \quad d_2 = \frac{:\neg a}{\neg a}.$$

We start building a tableau T for $\Sigma \cup \{a, \neg a\}$:

$$T = \{\mathbf{T}a, \mathbf{F}a\}$$

we can re-open T, by deleting $\mathbf{T}a$, that corresponds to delete the first default or by deleting $\mathbf{F}a$, that corresponds to delete the second default: Thus, we obtain

$$T_1 = \{\mathbf{T}a\}, \quad T_2 = \{\mathbf{F}a\}.$$

We must check conditions (1′) and (2′) on both tableaux, that is to say

$T_1 \otimes Tab(\mathbf{T}a)$ is open, for condition (1′) w.r.t. T_1;
$T_1 \otimes Tab(\mathbf{T}\neg a)$ is closed, for condition (2′) w.r.t. T_1;
$T_2 \otimes Tab(\mathbf{T}\neg a)$ is open, for condition (1′) w.r.t. T_2;
$T_2 \otimes Tab(\mathbf{T}a)$ is closed, for condition (2′) w.r.t. T_2.

All of them are true, thus Δ has two extensions corresponding to the two subsets $D_1 = \{d_1\}$ and $D_2 = \{d_2\}$, namely

$$E_1 = Th(\{a\}),\ E_2 = Th(\{\neg a\}).$$

EXAMPLE 17. Let $\Delta = (\Sigma, D)$, where, $\Sigma = \emptyset$, and $D = \{d\}$

$$d = \frac{:\neg a}{a}.$$

The tableau for $\Sigma \cup CONS(D)$ is simply

$$T = \{\mathbf{T}a\},$$

which is open. Since $T \otimes Tab(\mathbf{T}\neg a)$ is closed, d satisfies condition (2′), and then we must eliminate it, and consider $T' = T - \{\mathbf{T}a\} = \emptyset$. Now we have that $T' \otimes \{\mathbf{T}\neg a\}$ is open, so that d does no longer satisfy condition (2′). Thus the theory has no extensions.

EXAMPLE 18. Let $\Delta = (\Sigma, D)$, with $\Sigma = \emptyset$, and $D = \{d_1, d_2\}$, where

$$d_1 = \frac{a\ :b}{b}, \quad d_1 = \frac{b\ :a}{a}.$$

Let us build a tableau T for $\Sigma \cup CONS(D)$, we have:

$$T = \{\mathbf{T}a, \mathbf{T}b\},$$

which is consistent. Since, both $T \otimes Tab(\mathbf{T}a)$ and $T \otimes Tab(\mathbf{T}b)$ are open, both satisfy condition (1′). But D is not grounded. The only grounded subset is the empty subset. Letting $T' = T - \{\mathbf{T}a, \mathbf{T}b\} = \emptyset$, then both d_1 and d_2 satisfy condition (2′) with respect to T'. Thus the only extension is $Th(\emptyset)$.

A last remark concerning efficiency, if the initial tableau for $\Sigma \cup CONS(D)$ is closed, we have to open it by removing some atoms (corresponding to some defaults). There may be many ways to re-open a tableau, and we must consider all of them. In general, as the next example shows if Δ contains n defaults, there may be 2^n different ways to re-open the initial tableau. Let $\Delta = (\Sigma, D)$, where $\Sigma = \{b\}$ and D contains

$$\frac{:a_1}{a_1}, \frac{:\neg a_1}{\neg a_1}, \ldots, \frac{:a_n}{a_n}, \frac{:\neg a_n}{\neg a_n}, \frac{b\ :c}{c}.$$

Then the tableau T for $\Sigma \cup CONS(D)$ contains just one set B:

$$B = \{\mathbf{T}a_1, \mathbf{F}a_1, \ldots, \mathbf{T}a_n, \mathbf{F}a_n, \mathbf{T}c, \mathbf{T}b\}.$$

There are 2^n ways of opening T corresponding to each choice of deleting either $\mathbf{T}a_i$ or $\mathbf{F}a_i$ for $i = 1, \ldots, n$ from B. On the other hand we can easily see that all opened tableaux still contain $\mathbf{T}c$, and $\mathbf{T}b$, and this suffices to know that c will be in all extensions of Δ. Thus, if the purpose is inference, one may wonder if the explicit construction and inspection of all candidates T' is really needed. In principle, the answer seems affirmative: by Gottlob's results all problems concerning default reasoning lay on the second level of the polynomial hierarchy and are complete in their respective classes. Still in practical applications, some heuristics are to be developed to keep the construction feasible.

As we already remarked at the beginning of this section, the approach pursued by Amati et al. is totally different, and is based on an incremental construction of each extension at a time.

A default rule can be seen as a tableau rule which prescribes to add its consequent to a tableau whenever its preconditions are satisfied by the tableau. The authors call the general pattern of these tableau rules *default tableau restriction rule* to emphasize that the addition of a default consequent restrict possible countermodels of the set of input formulas. Let Γ be a set of formulas, for any default:

$$d = \frac{\alpha\ : \beta}{\gamma}$$

the corresponding tableau rule can be stated as follows:

$$\frac{\Gamma \vdash \alpha \quad \Gamma, \beta \not\vdash \bot}{\Gamma \cup \{\gamma\}}$$

That is, if $\Gamma \vdash \alpha$ and $\Gamma \cup \{\beta\}$ is consistent, then add γ to Γ. We say that the rule is applicable if the two conditions are satisfied. Furthermore let us say that Γ is Δ-saturated if for all applicable default tableau rules

$$\frac{\Gamma \vdash \alpha \quad \Gamma, \beta \not\vdash \bot}{\Gamma \cup \{\gamma\}}$$

we have $\Gamma \vdash \gamma$.

Given a default theory (Σ, D), the default tableau rule is applied to build up a sequence of tableaux, starting from a tableau for Σ up to a saturated theory. The construction proceeds as follows:

- (step 0) Let $T_0 = Tab(\mathbf{T}\Sigma)$.
- (step i+1). Suppose we have built T_i; let $d = \frac{\alpha\ : \beta}{\gamma} \in D$, check whether

  $T_i \otimes Tab(\mathbf{F}\alpha)$ is closed and
  $T_i \otimes Tab(\mathbf{T}\beta)$ is open.

  This is done by opening a special subtableau. If the test succeed, that is d is applicable to T_i, we let

  $$T_{i+1} = T_i \otimes Tab(\mathbf{T}\gamma).$$

  That is $\mathbf{T}\gamma$ is added to every open branch of T_i.

If the test fails from T_i we can try with another default $d \in D$. In this way we generate a (finite[17]) sequence of tableaux $T_0, \ldots, T_n$. The sequence terminates

[17]Provided D is finite.

with T_n, if T_n is Δ-saturated in the sense that for every default $d = \frac{\alpha : \beta}{\gamma} \in D$, if d is applicable to T_n then $T_n \otimes Tab(\mathbf{F}\gamma)$ is closed. The authors prove that there is a complete mapping between such Δ-saturated tableaux T_n satisfying an additional stability condition and extensions.

The *stability condition* says that for every default $d = \frac{\alpha : \beta}{\gamma} \in D$, which has been used in some step T_i of the construction, $T_n \otimes Tab(\mathbf{T}\beta)$ is open. This condition ensures that a default applied at a certain stage of the construction cannot become unapplicable at a later stage.

THEOREM 19. *Given a default theory* (Σ, D), *E is a consistent extension of* (Σ, D) *if and only if there is a finite sequence of tableaux* $T_0 = Tab(\mathbf{T}\Sigma), \ldots, T_n$, *such that* T_n *is* Δ*-saturated, satisfies the stability condition, and*

$$E = Th(\{\phi : \mathbf{T}\phi \in T_n\} \cup \{\neg\psi : \mathbf{F}\psi \in T_n\}).$$

EXAMPLE 20. Let $\Delta = (\Sigma, D)$ be a default theory with $\Sigma = \emptyset$, and $D = \{d_1, d_2\}$, where

$$d_1 = \frac{: a}{a}, d_2 = \frac{: \neg a}{\neg a}.$$

We start building a tableau T_0 for Σ:

$$T_0 = \emptyset,$$

then, we consider the first default d_1 and we check whether

$T_0 \otimes Tab(\mathbf{T}a)$ is open.

(the other condition is trivially satisfied as there are no prerequisites). Since it is open, we put

$$T_1 = T_0 \otimes Tab(\mathbf{T}a) = \{\mathbf{T}a\}.$$

We now consider d_2, and we check whether

$T_1 \otimes Tab(\mathbf{T}\neg a)$ is open,

but it is closed. Thus, we cannot add the consequent. It is easy to see that T_1 is both saturated and stable. Hence it represents an extension. It is clear that if we consider d_2 first, and then d_1 we obtain the other extension containing $\neg a$.

Let $\Delta = (\Sigma, D)$, where, $\Sigma = \emptyset$, and $D = \{d\}$

$$d = \frac{: \neg a}{a}.$$

The tableau for T_0 for Σ is empty. We check whether $T_0 \otimes Tab(\mathbf{T}\neg a)$ is open. Since it is, we put

$$T_1 = T_0 \otimes Tab(\mathbf{T}a) = \{\mathbf{T}a\}.$$

It is easy to see that T_1 is saturated, but since $T_1 \otimes Tab(\mathbf{T}\neg a)$ is closed, T_1 is not stable. Thus, the theory has no extensions.

Finally, let $\Delta = (\Sigma, D)$, with $\Sigma = \emptyset$ and $D = \{d_1, d_2\}$, where

$$d_1 = \frac{a\ :b}{b}, \quad d_1 = \frac{b\ :a}{a}.$$

The tableau T_0 for Σ is empty, we consider d_1 and we check whether:

$T_0 \otimes Tab(\mathbf{T}a)$ is closed, and
$T_0 \otimes Tab(\mathbf{T}b)$ is open.

The first condition fails. If we consider d_2, we have the same result. Thus T_0 is saturated, and obviously stable. Hence $Th(\emptyset)$ is the only extension of Δ.

As the first example shows, the final Δ- saturated tableau depend on the order in which defaults are selected and applied along the construction. To each possible ordering of defaults, it corresponds to a possibly different final tableau. Thus in order to generate all extensions and check a formula in them we must carry on the construction for all possible ordering of defaults.

6.2 *Tableaux for Variants of Default Logic*

The two methods we have examined for Reiter's logic can be both adapted to variant of default logic, such as Łukaszewicz' and Schaub's ones. It only requires a few minor modifications.

The extension of the Risch–Schwind approach to these latter variants, is based on the following theorem on extensions [Risch, 1996].

THEOREM 21. *Let $\Delta = (\Sigma, D)$ be a default theory, then*

- *E is a m-extension of Δ with respect to F, if and only if there is a* maximal *grounded subset $D' \subseteq D$, such that $E = Th(\Sigma \cup CONS(D'))$, $F = JUST(D')$ and it satisfies:*

  (i_L) *for all $d \in D'$, $d = \frac{\alpha : \beta}{\gamma}$, $\alpha \in E$, and $\neg\beta \notin E$.*

- *(E, C) is a c-extension for Δ if and only if there is a* maximal *grounded subset $D' \subseteq D$, such that $E = Th(\Sigma \cup CONS(D'))$, $C = Th(\Sigma \cup CONS(D') \cup JUST(D'))$ and it satisfies:*

  (i_S) *for all $d \in D'$, $d = \frac{\alpha : \beta}{\gamma}$, $\alpha \in E$, and $E \cup JUST(D')$ is consistent.*

¿From the theorem, we also have that $Th(\Sigma \cup CONS(D'')$ is a c-extension iff there is an m-extension $Th(\Sigma \cup CONS(D'))$, such that D'' is a maximal grounded subset of D' such that $JUST(D'')$ is consistent.

The tableau method for computing m-extensions is similar to that one for Reiter's default logic; namely, it is simpler for we do not require that defaults which do not participate to an extension must be unapplicable with respect to it; (this is the role of condition 2 in Theorem 15). Thus, the algorithm for computing m-extensions is obtained from that one for Reiter's extensions, by simply omitting Step (iv).

EXAMPLE 22. (a) We consider the default theory (Σ, D) of Example 7. By applying the algorithm, we first compute

$$Tab(\mathbf{T}\Sigma) \otimes Tab(\mathbf{T}CONS(D)) = \{\mathbf{T}su, \mathbf{T}wh, \mathbf{T}ho, \mathbf{T}fi, \mathbf{T}ti, \mathbf{T}wkl\},$$

which is open. Then, we have to find all maximal subsets D_i of D, which are consistent with Σ and with their own justifications (each justification is considered separately). We get

$$D_1 = \{d_1\},\ D_2 = \{d_2\}, D_3 = \{d_3\}.$$

Each subset is obviously grounded, thus we have three m-extensions as expected.

(b) Let us consider the theory $(\emptyset, \{d\})$, where $d = \frac{:\neg a}{a}$, then $Tab(\mathbf{T}\Sigma) \otimes Tab(\mathbf{T}cons(d)) = \{\mathbf{T}a\}$, which is open, but if we add $\mathbf{T}just(d)$, it becomes closed. Thus, we must remove d, and the only m-extension is given by $Th(\emptyset)$.

Let us turn to Schaub's variant. In the light of the previous remark, we just have to consider the algorithm for computing m-extensions and modify the justification test performed in step (ii):

> Given D_j, find all its maximal subsets D_j^m such that
>
> $$Tab(\mathbf{T}\Sigma) \otimes Tab(\mathbf{T}CONS(D_j^m)) \otimes Tab(\mathbf{T}JUST(D_j^m))$$
>
> is open.

EXAMPLE 23 (Brewka's *broken arm* example). Let $\Delta = (\{br(ri) \vee br(le)\}, \{d_1, d_2\})$, where

$$d_1 = \frac{: us(ri) \wedge \neg br(ri)}{us(ri)} \quad d_2 = \frac{: us(le) \wedge \neg br(le)}{us(le)}\}.$$

We first compute $Tab(\mathbf{T}\Sigma) \otimes Tab(\mathbf{T}CONS(\{d_1, d_2\}))$ (we label both the consequent and the justification) so that we obtain:

$$T = \{\{\mathbf{T}br(ri), \mathbf{T}us(ri)_1, \mathbf{T}us(le)_2\}, \{\mathbf{T}br(le), \mathbf{T}us(ri)_1, \mathbf{T}us(le)_2\}\}$$

which is open. Then we have the justification test. To this purpose, we compute $T \otimes Tab(\mathbf{T}JUST(D))$ and we get:

$$\{\{\mathbf{T}br(ri), \mathbf{T}us(ri)_1, \mathbf{T}us(le)_2, \mathbf{F}br(ri)_1, \mathbf{F}br(le)_2\},$$
$$\{\mathbf{T}br(le), \mathbf{T}us(ri)_1, \mathbf{T}us(le)_2, \mathbf{F}br(ri)_1, \mathbf{F}br(le)_2\}\},$$

both branches are closed. We can re-open it in two ways: either removing the first default (we re-open the first branch), or the second one (we re-open the second branch). The two resulting sets are grounded. Consequently, we get two c-extensions corresponding to the conclusions of the two. Notice the difference with the justification test in Łukaszewicz's variant. In the latter case we would check that

$$T \otimes Tab(\mathbf{T}just(d_1)) \text{ and } T \otimes Tab(\mathbf{T}just(d_1)) \text{ are open.}$$

Since both of them are open, we would get just one m-extension corresponding to $Th(\Sigma \cup CONS(d_1, d_2))$.

The tableau methodology of Aiello *et al.* can be extended to these variants of default logic as well. For Łukaszewic's variant, a default $\frac{\alpha : \beta}{\gamma}$, is interpreted as the following inference rule:

$$\frac{\Gamma \vdash \alpha \quad \Gamma \cup \{\beta, \gamma\} \not\vdash \bot}{\Gamma \cup \{\gamma\}} \quad .$$

We introduce a new criterium of applicability, given a default $d = \frac{\alpha : \beta}{\gamma}$, we say that d is *applicable* to Γ iff

1. $Tab(\mathbf{T}\Gamma) \otimes Tab(\mathbf{F}\alpha)$ is closed
2. $Tab(\mathbf{T}\Gamma) \otimes Tab(\mathbf{T}(\beta \wedge \gamma))$ is open.

Given a default theory (Σ, D), default rules are applied to build up a sequence of tableaux T_i. We also need to record what defaults have been applied at each step (by means of sets D_i).

- (step 0) We let $T_0 = Tab(\mathbf{T}\Sigma)$ and $D_0 = \emptyset$.
- (step k) Let d be a default, if d is applicable to T_{k-1} and all $d' \in D_{k-1}$ are applicable to $T_{k-1} \otimes Tab(\mathbf{T}\gamma)$, then we let

  $$T_k = T_{k-1} \otimes Tab(\mathbf{T}\gamma),$$
  $$D_k = D_{k-1} \cup \{d\}.$$

The incremental construction is carried on until a saturated tableau T_n is reached. The authors prove that the deductive closure of T_n is an m-extension. The F-part of the extension is given by $JUST(D_n)$. Moreover, all m-extensions can be constructed in this way.

EXAMPLE 24. We consider the theory of 7. Let $D_0 = \emptyset$ and

$$T_0 = Tab(\mathbf{T}\Sigma) = \{\mathbf{T}su, \mathbf{T}wh, \mathbf{T}ho\}.$$

Let us consider d_1; to see whether it is applicable, we must check that

1) $T_0 \otimes Tab(\mathbf{F}pre(d_1)) = T_0 \cup \{\mathbf{F}su\}$ is closed,
2) $T_0 \otimes Tab(\mathbf{T}cons(d_1), \mathbf{T}just(d_1)) = T_0 \cup \{\mathbf{T}fi, \mathbf{F}ti\}$ is open,
3) for all $d \in D_0$, $T_0 \otimes Tab(\mathbf{T}cons(d_1)) \otimes Tab(\mathbf{T}just(d))$ is open.

The three tests succeed, thus we can add the consequent of d_1 to T_0 and we let

$$T_1 = T_0 \cup \{\mathbf{T}fi\}, \quad D_1 = \{d_1\}.$$

We consider d_2, we check whether

1) $T_0 \otimes Tab(\mathbf{F}pre(d_2)) = T_1 \cup \{\mathbf{F}wh\}$ is closed,
2) $T_1 \otimes Tab(\mathbf{T}cons(d_2), \mathbf{T}just(d_2)) = T_1 \cup \{\mathbf{T}ti, \mathbf{F}wkl\}$ is open,
3) $T_1 \otimes Tab(\mathbf{T}cons(d_2)) \otimes Tab(\mathbf{T}just(d_1)) = T_1 \cup \{\mathbf{T}ti, \mathbf{T}fi, \mathbf{F}ti\}$ is open.

The first two succeed, but the third fails, thus we cannot apply d_2 to T_1. Let us consider d_3, we have that

$T_1 \otimes Tab(\mathbf{F}pre(d_3)) = T_1 \cup \{\mathbf{F}ho\}$ is closed, but also
$T_1 \otimes Tab(\mathbf{T}cons(d_3), \mathbf{T}just(d_3)) = T_1 \cup \{\mathbf{T}wkl, \mathbf{F}fi\}$ is closed,

whereas it should be open; thus d_3 is not applicable. T_1 is obviously saturated, and represents one m-extension. The other two can be generated by considering the defaults in a different order.

For Schaub's variant, we have to expand the basic machinery by taking into account *pairs* of sets of formulas (Γ^+, Γ^-), and hence pairs of tableaux. The reason is that c-extensions are *pairs* of sets of formulas. A default rule $\frac{\alpha : \beta}{\gamma}$ is interpreted as the following inference rule:

$$\frac{\Gamma^+ \vdash \alpha \quad \Gamma^- \cup \{\beta, \gamma\} \not\vdash \bot}{\Gamma^+ \cup \{\gamma\}, \Gamma^- \cup \{\beta \wedge \gamma\}} .$$

The meaning is that: if $\Gamma^+ \vdash \alpha$ and $\beta \wedge \gamma$ is consistent with Γ^-, then add γ to Γ^+ and $\beta \wedge \gamma$ to Γ^-.

In tableaux terms, we say that a default d is applicable to a pair (Γ^+, Γ^-) iff

1. $Tab(\mathbf{T}\Gamma^+) \otimes Tab(\mathbf{F}\alpha)$ is closed
2. $Tab(\mathbf{T}\Gamma^-) \otimes Tab(\mathbf{T}\beta \wedge \gamma)$ is open.

We redefine accordingly the notion of saturation. We say that (Γ^+, Γ^-) is saturated with respect to a set of defaults D iff for all $d = \frac{\alpha : \beta}{\gamma} \in D$, if d is applicable to (Γ^+, Γ^-), then $Tab(\mathbf{T}\Gamma^+) \otimes Tab(\mathbf{F}\gamma)$ and $Tab(\mathbf{T}\Gamma^-) \otimes Tab(\mathbf{F}\gamma \wedge \beta)$ are closed.[18] The same definition of applicability and saturation make sense if (Γ^+, Γ^-) is already a pair of expanded tableaux.

Given a default theory (Σ, D), default rules are applied to build up a sequence of *pairs* of tableaux $\langle T_i^+, T_i^- \rangle$.

- (step 0) We let $T_0^+ = T_0^- = Tab(\mathbf{T}\Sigma)$.
- (step k) Let $d = \frac{\alpha : \beta}{\gamma}$ be a default, if d is applicable to (T_{k-1}^+, T_{k-1}^-), then we let

$$T_k^+ = T_{k-1}^+ \otimes Tab(\mathbf{T}\gamma),$$
$$T_k^- = T_{k-1}^- \otimes Tab(\mathbf{T}\beta \wedge \gamma).$$

The incremental construction is carried on until a saturated pair of tableaux (T_n^+, T_n^-) is reached. The authors prove that the deductive closure of the two components gives a c-extension (E, C), namely $E = Th(T_n^+)$ and $C = Th(T_n^-)$. Moreover, all c-extensions may be generated this way.

EXAMPLE 25 (Brewka's *broken arm* example). Let $\Delta = (\{br(ri) \vee br(le)\}, \{d_1, d_2\})$, where

$$d_1 = \frac{: us(ri) \wedge \neg br(ri)}{us(ri)} \quad d_2 = \frac{: us(le) \wedge \neg br(le)}{us(le)}\}.$$

We have that

$$T_0^+ = T_0^- = \{\{\mathbf{T}br(ri)\}, \{\mathbf{T}br(le)\}\}.$$

Consider d_1, we check whether

1) $T_0^+ \otimes Tab(\mathbf{F}pre(d_1))$ is closed, and
2) $T_0^- \otimes Tab(\mathbf{T}just(d_1), \mathbf{T}cons(d_1)) =$

[18] The authors make use of pairs of theories and of pairs of formulas in order to give a uniform presentation of all variants of default logic. In a theory (Γ^+, Γ^-), the first component Γ^+ represents the actual knowledge, and Γ^- its support, that is additional information, such as consistent assumptions. Formulas are also considered in this paired fashion: given $\phi = \langle \phi^+, \phi^- \rangle$, ϕ^+ is the asserted formula and ϕ^- its justification (a consistency assumption). This machinery is somewhat reminiscent of Brewka's assertional language. A default rule is presented in the general form:

$$\frac{\alpha^+ : \alpha^-}{\gamma^+, \gamma^-},$$

and its meaning, is the following: given a theory (Γ^+, Γ^-), if Γ^+ proves α^+, and α^- is consistent with Γ^-, then add γ^+ to Γ^+ and γ^- to Γ^-. The conclusion is split in two parts, one is added to Γ^+ and the other to Γ^-, the support of Γ^+. However, for Reiter's as well as for Łukaszewicz's logic , we have that $\gamma^+ = \gamma^-$, and the two components Γ^+, Γ^- can be identified, having the same role. This is why we have chosen a simplified presentation in these two cases.

$= \{\{\mathbf{T}br(ri), \mathbf{T}us(ri), \mathbf{F}br(ri)\}, \{\mathbf{T}br(le), \mathbf{T}us(ri), \mathbf{F}br(ri)\}\}$
is open.

Both tests succeed, hence we let

$$\begin{aligned} T_1^+ &= T_0^+ \otimes Tab(\mathbf{T}cons(d_1)) \\ &= \{\{\mathbf{T}br(ri), \mathbf{T}us(ri)\}, \{\mathbf{T}br(le), \mathbf{T}us(ri)\}\} \\ T_1^- &= T_0^- \otimes Tab(\mathbf{T}cons(d_1), \mathbf{T}just(d_1)) \\ &= \{\mathbf{T}br(le), \mathbf{T}us(ri), \mathbf{F}br(ri)\}. \end{aligned}$$

We now consider d_2,

1) $T_1^+ \otimes Tab(\mathbf{F}pre(d_2))$ is closed, but
2) $T_1^- \otimes Tab(\mathbf{T}just(d_2), \mathbf{T}cons(d_2)) =$
$= \{\mathbf{T}br(le), \mathbf{T}us(ri), \mathbf{F}br(ri), \mathbf{T}us(le), \mathbf{F}us(le)\}$
is closed,

whereas it should be open. Thus d_2 is not applicable. Since (T_1^+, T_1^-) is saturated, it gives a c-extension which contain $us(ri)$ and $\neg us(le)$. If we consider d_2 first and then d_1, we obtain the symmetric one.

7 TABLEAUX FOR MINIMAL ENTAILMENT

In this section, we will treat tableaux for the propositional case, and we will defer the first order case as well as domain minimization (only meaningful in a first-order context) to the next section. We first consider the case when all atoms are minimized. Our intent is to modify the standard tableau construction so that any tableau for (Δ, ϕ) is closed iff $\Delta \models_m \phi$. We divert from the classical case in two respects.

First, we reformulate the rules of tableau expansion in order to eliminate the sign-switching rules for negation. That is, instead of eliminating negation by switching the sign of a formula $\neg\phi$, and then processing ϕ, we process directly $\neg\phi$ without changing its sign. In Figure 2 we reformulate the table rules for splitting formulas.

The reason for this reformulation is that $\mathbf{T}p$ and $\mathbf{F}\neg p$ will turn out to have a different meaning. Then, we reformulate accordingly the usual notion of completed branch, by distinguishing the case of $\mathbf{T}$-signed formulas and $\mathbf{F}$-signed formulas.

DEFINITION 26. We say that a branch B is $\mathbf{T}$-completed ($\mathbf{F}$- completed) if all its $\mathbf{T}$-signed ($\mathbf{F}$-signed formulas) are literals. A branch B is completed if it is both $\mathbf{T}$- completed and $\mathbf{F}$-completed.

We introduce a new definition of closed branch.

DEFINITION 27. We say that a branch B is $\mathbf{T}$-closed ($\mathbf{F}$-closed) if it contains a $\mathbf{T}\phi$ and $\mathbf{T}\neg\phi$, for some formula ϕ, (respectively, $\mathbf{F}\phi$ and $\mathbf{F}\neg\phi$). We say that

α	α_1	α_2
$\mathbf{T}(\phi \wedge \psi)$	$\mathbf{T}\phi$	$\mathbf{T}\psi$
$\mathbf{T}\neg(\phi \vee \psi)$	$\mathbf{T}\neg\phi$	$\mathbf{T}\neg\psi$
$\mathbf{T}\neg(\phi \rightarrow \psi)$	$\mathbf{T}\phi$	$\mathbf{T}\neg\psi$
$\mathbf{T}\neg\neg\phi$	$\mathbf{T}\phi$	
$\mathbf{F}(\phi \vee \psi)$	$\mathbf{F}\phi$	$\mathbf{F}\psi$
$\mathbf{F}\neg(\phi \wedge \psi)$	$\mathbf{F}\neg\phi$	$\mathbf{F}\neg\psi$
$\mathbf{F}(\phi \rightarrow \psi)$	$\mathbf{F}\neg\phi$	$\mathbf{F}\psi$
$\mathbf{F}\neg\neg\phi$	$\mathbf{F}\phi$	

β	β_1	β_2
$\mathbf{T}(\phi \vee \psi)$	$\mathbf{T}\phi$	$\mathbf{T}\psi$
$\mathbf{T}\neg(\phi \wedge \psi)$	$\mathbf{T}\neg\phi$	$\mathbf{T}\neg\psi$
$\mathbf{T}(\phi \rightarrow \psi)$	$\mathbf{T}\neg\phi$	$\mathbf{T}\psi$
$\mathbf{F}(\phi \wedge \psi)$	$\mathbf{F}\phi$	$\mathbf{F}\psi$
$\mathbf{F}\neg(\phi \vee \psi)$	$\mathbf{F}\neg\phi$	$\mathbf{F}\neg\psi$
$\mathbf{F}\neg(\phi \rightarrow \psi)$	$\mathbf{F}\phi$	$\mathbf{F}\neg\psi$

Figure 2.

a branch B is *ordinary closed* if is either **T**-closed, or **F**-closed, or it contains a formula $\mathbf{T}\phi$ and $\mathbf{F}\phi$.

We say that a branch B is *m-closed* if

- it is **T**-completed;
- for some literal $\neg p$, we have $\mathbf{F}\neg p \in B$, but $\mathbf{T}p \notin B$.

Finally, we say that a branch B is *closed* if it is either ordinary-closed or m-closed.

The intuitive meaning of m-closure is as follows: suppose that B is m-closed because of $\mathbf{F}\neg q$. Then, for every model M (if any) which makes true all **T**-formulas of B, there is a smaller model N which makes true all **T**-formulas of B and makes q false. Thus we can consider B as closed, for it cannot provide a minimal countermodel.

Notice that the condition of m-closure is a negative condition, and hence it is nonmonotonic. If we add some formulas to an m-closed branch B, the resulting branch might no longer be m-closed. We introduce another condition which allows one to get rid of non-minimal branches. Given a branch B we let:

$$At(B) = \{p : \mathbf{T}p \in B \text{ and } p \text{ is an atom}\}.$$

DEFINITION 28. Given a branch B in a tableau T, we say that B is *ignorable* if there is another branch $B' \in T$, such that (a) B' is **T**-completed, (b) it is not **T**-closed, and (c)

$$At(B') \subset At(B).$$

Clauses (a) and (b) ensures that B' is a (partial) model of the initial T-formulas, clause (c) expresses the preference for the smaller one.

A branch which is neither closed, nor ignorable is called a counterexample branch. Finally, we say that a tableau T is closed iff all branches of T are either closed or ignorable.

EXAMPLE 29.

We check if $p \vee q \models_m \neg p \vee \neg q$. The initial tableau contains:

$$\{\mathbf{T}p \vee q, \mathbf{F}\neg p \vee \neg q\},$$

which generates the final two branches:

$$\{\mathbf{T}p, \mathbf{F}\neg p, \mathbf{F}\neg q\}, \quad \{\mathbf{T}q, \mathbf{F}\neg p, \mathbf{F}\neg q\}.$$

Both of them are m-closed, thus the tableau is closed.

Let us check if $p \to q \models_m \neg q$. The initial tableau contains:

$$\{\mathbf{T}p \to q, \mathbf{F}\neg q\},$$

which generates the following branches:

$$\{\mathbf{T}\neg p, \mathbf{F}\neg q\}, \quad \{\mathbf{T}q, \mathbf{F}\neg q\},$$

the first branch is m-closed the second one is ignorable, thus the tableau is closed.

We finally check whether

$$p \vee q, p \to r \models_m \neg r.$$

The initial tableau contains:

$$\{\mathbf{T}p \vee q, \mathbf{T}p \to r, \mathbf{F}\neg r\},$$

from which we obtain the following three branches (we omit the **T**-closed one).

$$B_1 = \{\mathbf{T}p, \mathbf{T}r, \mathbf{F}\neg r\}, \quad B_2 = \{\mathbf{T}q, \mathbf{T}\neg p, \mathbf{F}\neg r\}, \quad B_3 = \{\mathbf{T}q, \mathbf{T}r, \mathbf{F}\neg r\}.$$

Branch B_2 is m-closed, branch B_3 is ignorable because of B_2, but B_1 is a counterexample branch. It is easily seen that the evaluation V, with $V(p) = V(r) = 1$, is a minimal model of $(p \vee q) \wedge (p \to r)$, which falsifies $\neg r$.

The method is sound and complete with respect to minimal entailment.

THEOREM 30. *For any finite set of formulas Γ and formulas ϕ, it holds that $\Gamma \models_m \phi$ if and only if any tableau for (Γ, ϕ) is closed.*

The method can be easily adapted to the case of minimal entailment with variable atoms (that one we denoted by $\models_{\mathbf{P}}$, where $\mathbf{P}$ are the atoms to be minimized). All we need to do is to change the notion of m-closure, and the ignorability condition as follows. For the former, we say that B is $\mathbf{P}$-m-closed if it is $\mathbf{T}$-completed, and for some literal $\neg p$, such that $p \in \mathbf{P}$, we have $\mathbf{F}\neg p \in B$, but $\mathbf{T}p \notin B$.

For the latter, we let $\mathbf{P}(B)$ be the set of atoms $p \in \mathbf{P}$ such that $\mathbf{T}p \in B$.

Given a branch B in a tableau T, we say that B is $\mathbf{P}$-*ignorable* if there is another branch $B' \in T$, such that (a) B' is $\mathbf{T}$-completed, (b) it is not $\mathbf{T}$-closed, and (c) $\mathbf{P}(B') \subset \mathbf{P}(B)$.

Let us consider the following example (a similar one is discussed in [Brewka, 1985]).

Let Σ be the conjunction of the following formulas:

$$german(tom) \land \neg abnormal(tom) \to drink\_beer(tom),$$
$$german(tom),$$
$$eats\_cake(tom) \to \neg drink\_beer(tom).$$

We minimize all atoms, but $drinks\_beer(tom)$, since we want to determine whether Tom drinks beer or not. That is we take:

$$\mathbf{P} = \{german(tom), abnormal(tom), eats\_cake(tom)\}.$$

The initial tableau T contains $\mathbf{T}\Sigma$ and $\mathbf{F}drink\_beer(tom)$. The following branches are generated (we only list those ones which are not $\mathbf{T}$-closed):

$$B_1 = \{\mathbf{T}abn(tom), \mathbf{T}ger(tom), \mathbf{T}\neg eats\_cake(tom), \mathbf{F}drink\_b(tom)\},$$

$$B_2 = \{\mathbf{T}abn(tom), \mathbf{T}ger(tom), \mathbf{T}\neg drink\_b(tom), \mathbf{F}drink\_b(tom)\},$$

$$B_3 = \{\mathbf{T}drink\_b(tom), \mathbf{T}ger(tom), \mathbf{T}\neg eats\_cake(tom), \mathbf{F}drink\_b(tom)\}.$$

We have that B_3 is closed; B_1 and B_2 are $\mathbf{P}$-ignorable because of B_3. Hence the tableau proves that $\Sigma \models_{\mathbf{P}} drink\_beer(tom)$.

The general case, with fixed atoms is slightly more complex, although we recall that fixed-atoms can be always eliminated. This case has been studied in [Kuhna, 1993], where a tableau method to generate $\leq_{\mathbf{P},\mathbf{Q}}$ minimal models is proposed. The main result of that paper is a characterization of minimal models in terms of tableau branches for the general case. Given two disjointed sets of atoms $\mathbf{P}$, $\mathbf{Q}$, let $Fixed$ be the set of atoms not in $\mathbf{P} \cup \mathbf{Q}$. For a tableau branch B, we use the notation $Fixed(B)$ to denote the set of $\mathbf{T}$-signed literals in B, whose atoms belongs to $Fixed$. Let Σ be a set of formulas, given an open branch $B \in Tab(\mathbf{T}\Sigma)$ (namely, a set of $\mathbf{T}$-signed literals), we consider the following property:

> Let $B_1, \ldots, B_k \in Tab(\mathbf{T}\Sigma)$ be all open branches such that $\mathbf{P}(B_i) \subset \mathbf{P}(B)$. If $k > 0$, then for $i = 1, \ldots, k$, there are literals $l_i \in$

$Fixed(B_i)$, such that $B \cup \{\mathbf{T}l_1^c, \ldots, \mathbf{T}l_k^c\}$ [19] is open.

There is a complete correspondence between open branches B which satisfy the above property and $\leq_{\mathbf{P},\mathbf{Q}}$ minimal models of Σ: if B is open and satisfies the above property for some literals $l_1, \ldots, l_k$, then every maximal consistent extension of

$$B \cup \{\neg p : p \notin \mathbf{P}(B)\} \cup \{\mathbf{T}l_1^c, \ldots, \mathbf{T}l_k^c\}$$

is a $\leq_{\mathbf{P},\mathbf{Q}}$ minimal model of Σ. On the other hand, to every $\leq_{\mathbf{P},\mathbf{Q}}$ minimal model of Σ it corresponds to an open branch B which satisfies the above property. This result suggests a discipline to check whether an open branch B represents a minimal model or not. We have to complete B by adding the complement of a literal in $Fixed(B')$ for each potential $\mathbf{P}$-smaller branch B'. If we can make a consistent choice for every such B', then B represent a minimal model, otherwise it does not. In the paper, the author discusses how to implement the method with some efficiency.

We conclude this section by a brief discussion of efficiency issues. From a theoretical point of view, Gottlob's results classify all types of propositional minimal entailment to the second level of the polynomial hierarchy; thus the complexity of this form of nonmonotonic inference is the same as default reasoning and autoepistemic logic. The critical point of the method is the fact that in order to decide whether to keep or not a branch we must compare it, in the worst case, with any other else.

Although, in practice reasonable heuristics can be developed (often a branch can be detected as ignorable, without a complete expansion of the tableau), we can still wonder whether there is or not an alternative to branch comparison. In the next subsection, we will show that this problem is connected to that one of finding a *cut free* sequent calculus for minimal entailment.

A Gentzen Calculus for Minimal Entailment

In this section we transform the tableau method for the basic form of minimal entailment into a Gentzen calculus. The calculus we present below, called MLK, is equivalent to that one in [Olivetti, 1992], although it does not contain an explicit rule for negation, but is closer to the tableau method of the previous section. This calculus is a useful tool for proof-theoretical investigation on minimal entailment. In [Olivetti, 1992] it is shown how to extend the calculus to minimal entailment with variable and fixed atoms; here we will present the calculus only for the basic case of minimization.

Sequents have the form

$$\Gamma \vdash \Delta,$$

[19]By l^c we denote the complementary literal of l.

where Γ and Δ are sets of formulas. We write $\Gamma, \phi_1, \ldots \phi_n$ instead of $\Gamma \cup \{\phi_1, \ldots, \phi_n\}$, and Δ_1, Δ_2 instead of $\Delta_1 \cup \Delta_2$.

As usual, there are two type of rules: logical and structural; logical rules are the standard ones and can be read as an upside version of tableau rules. In order to state them concisely we refer to the splitting table in Figure 2, with the following understanding:

> if a formula ϕ occurs in the left (right) part, then it is treated as it were $\mathbf{T}\phi$ ($\mathbf{F}\phi$).

Thus, the rules are as follows:

- *Logical rules*

$$\frac{\Gamma, \phi_1, \phi_2 \vdash \Delta}{\Gamma, \phi \vdash \Delta} \qquad \frac{\Gamma \vdash \Delta, \phi_1, \phi_2}{\Gamma \vdash \Delta, \phi}$$

  if ϕ is of type α and ϕ_1, ϕ_2 are the α-subformulas of ϕ.

$$\frac{\Gamma, \psi_1, \vdash \Delta \quad \Gamma, \psi_2, \vdash \Delta}{\Gamma, \psi \vdash \Delta} \qquad \frac{\Gamma \vdash \Delta, \psi_1 \quad \Gamma \vdash \Delta, \psi_2}{\Gamma \vdash \Delta, \psi}$$

  if ψ is of type β and ψ_1, ψ_2 are the β-subformulas of ψ.

- *Structural rules*

$$(CM) \quad \frac{\Gamma \vdash \phi \quad \Gamma \vdash \Delta}{\Gamma, \phi \vdash \Delta}$$

$$(CUT) \quad \frac{\Gamma \vdash \phi, \Delta_1 \quad \Gamma, \phi \vdash \Delta_2}{\Gamma \vdash \Delta_1, \Delta_2}$$

- *Initial sequents*

$$\Gamma \vdash \Delta,$$

  provided either

  (a) $\Gamma \cap \Delta \neq \emptyset$, or for some ϕ, both $\phi, \neg\phi \in \Gamma$, or $\phi, \neg\phi \in \Delta$, or

  (b) Γ and Δ are sets of literals and for some literal $\neg p \in \Delta, p \notin \Gamma$.

Let MLK be the calculus containing the above rules. We can make a few observations. First, notice that the logical rules are just the upside version of the tableau rules, for instance, we have:

$$\frac{\Gamma \vdash \Delta, \phi \quad \Gamma \vdash \Delta, \neg\psi}{\Gamma \vdash \Delta, \neg(\phi \rightarrow \psi)}$$

Rule CM is usually called *cautious monotony*. Notice that the first premise of cautious monotony is restricted to contain just a single consequent. Cut rule requires that the left part of the second premise contains exactly all formulas of the left part of the first premise (Γ), together with the cut formula (ϕ). Regarding to initial sequents, we notice that condition (a) of the definition of initial sequents corresponds to the ordinary tableau closure, whereas condition (b) corresponds to m-closure in tableaux. It can be shown that the rule of *right weakening* is admissible in this system; the corresponding rule on the left does not hold, being equivalent to monotonicity. The calculus we have presented satisfies the axioms for *preferential* entailment, as defined in [Kraus *et al.*, 1990], in the context of an axiomatic treatment of nonmonotonic logics.

EXAMPLE 31. We show that $p \vee q \models_m \neg p \vee \neg q$, by a derivation in MLK:

$$\cfrac{\cfrac{p \vdash \neg p, \neg q}{p \vdash \neg p \vee \neg q} \quad \cfrac{q \vdash \neg p, \neg q}{q \vdash \neg p \vee \neg q}}{p \vee q \vdash \neg p \vee \neg q}$$

We show that $p \rightarrow q \models_m \neg q$, by the following derivation:

$$\cfrac{\cfrac{\vdash \neg p, q}{\vdash p \rightarrow q} \quad \vdash \neg q}{p \rightarrow q \vdash \neg q}$$

The last step of the derivation above is by CM rule.

The calculus is a sound and complete formalization of minimal entailment.

THEOREM 32. *Let Δ be a set of formulas, and ϕ a formula we have that*

$$\Delta \models_m \phi \;-\; \Delta \vdash \phi \textit{ is derivable in MLK.}$$

In MLK, cut is not eliminatable. This can be seen by the following example.

EXAMPLE 33. It is easy to see that:

$$(p \wedge q) \vee (p \wedge \neg q) \models_m \neg q.$$

By the previous theorem we have that the sequent

$$(*)\; (p \wedge q) \vee (p \wedge \neg q) \vdash \neg q$$

is derivable in MLK, but any cut-free derivation of the above sequent can have as a last step only (a) the application of the logical rule for $\vee$ in the left part (a case of

β-rule), or (b) the application of CM. In case (a), we should be able to derive both

$$p \wedge q \vdash \neg q \text{ and } p \wedge \neg q \vdash \neg q.$$

But, it is obvious (by the correctness of MLK) that the first sequent cannot be derived. In case (b), we should be able to derive both

$$\vdash (p \wedge q) \vee (p \wedge \neg q) \text{ and } \vdash \neg q.$$

But, again, the first sequent cannot be derived by the correctness of MLK. On the other hand, there is a derivation of sequent (*), whose last steps are:

$$\dfrac{\dfrac{\vdots}{(p \wedge q) \vee (p \wedge \neg q) \vdash p} \qquad \dfrac{\dfrac{\vdots}{p \vdash (p \wedge q) \vee (p \wedge \neg q)} \quad p \vdash \neg q}{p, (p \wedge q) \vee (p \wedge \neg q) \vdash \neg q}}{(p \wedge q) \vee (p \wedge \neg q) \vdash \neg q}$$

Given the correspondence between the tableau rules and MLK rules, we can give a partial answer to the issue of branch comparison in tableaux we were concerned in the previous section. The next proposition highlights the connection between cut-elimination in MLK and branch comparison in tableaux.

PROPOSITION 34. *The following are equivalent:*

1. *No tableau proof of* $\Gamma \models_m \phi$ *requires branch comparison;*[20]

2. *There is a MLK derivation of* $\Gamma \vdash \phi$, *which makes use neither of CM rule, nor of CUT rule.*

Although cut-elimination fails in MLK, one can still hope that some form of *analytic* cut suffice, as in the example above.[21]

8 THE FIRST ORDER CASE

In this section we will shortly review the problems concerning the extension of the methods we have presented to first-order languages. We will concentrate on the preferential approach.

First, we notice that all nonmonotonic formalisms we have presented become seriously uneffective at the first-order level. This is clear for fixpoint-based formalisms which presuppose the ability of deciding unprovability, and hence at the

[20] That is to say, no ignorable branch is generated in the tableau construction.

[21] An analytic calculus for minimal entailment has been recently proposed in [Bonatti and Olivetti, 1997].

first-order level are not even recursive-enumerable. The same is true for preferential-based methods such as minimal entailment. Every relevant question about minimal models of an arbitrary theory may be not even arithmetic (existence of countable minimal models is Σ^1_2-complete, and hence truth in all countable minimal models is Π^1_2-complete in the analytical hierarchy [Schlipf, 1987]).

But intractability is not the only problem. For methods based on the fixpoint approach, there is no complete agreement on what is supposed to be their first-order formulation. For instance, it is not clear how to interpret in autoepistemic logic a formula such as $\exists x L p(x)$. Similarly, in default logic, given the base theory $\exists x p(x)$, it is not evident what may be the meaning of an open default rule such as

$$\frac{p(x) : q(x)}{r(x)}.$$

For all these reasons, in most of the cases severe restrictions are put on the use of first-order logic. Usually, the treatment is confined to universal function-free theories. This avoids the difficulty about the existential quantifier and skolemization. On the other hand, these restrictions give a decidable fragment of first-order logic.

There are two additional assumptions which are taken into consideration when dealing with first-order nonmonotonic logics. The first one is the so-called *domain closure axiom*. Given a function-free set of formulas Δ, whose constants are $a_1, \ldots, a_n$, $DCA(\Delta)$ is

$$\forall x(x = a_1 \vee \ldots \vee x = a_n).$$

The axiom states that the only existing objects are those named by the constants. By this axiom we can replace quantified formulas by propositional disjunctions and conjunctions.

Another critical issue is the identity of constants: for instance, given the only information $p(a)$ we can infer by minimal entailment $\neg p(b)$, only if $a \neq b$. For a similar example in default logic, let us consider the theory containing only the following default rule:

$$\frac{: \neg p(a)}{p(b)}$$

The theory has an extension iff $a = b$ cannot be proved. To deal with this kind of problems, it is usually made the simplifying assumption that different constants denote different objects. In the function-free case, this assumption can be represented by the axiom

$$\bigwedge_{i \neq j} a_i \neq a_j,$$

where a_i, a_j range over all constants of the language. This assumption is called *unique name assumption* (UNA). The two restrictions DCA and UNA have been first introduced for the logical analysis of databases and allow us to reduce a first-order theory to a propositional theory. When dealing with first-order nonmonotonic reasoning these restrictions are often implicitly assumed. For instance in [Schwind and Risch, 1994] the tableau method is said to be applicable to first-order default theories with open defaults. The authors consider function-free universal theories and suggest to interpret an open-default rule as the set of its propositional instances. This is tantamount to assume the DCA restriction.

To conclude, we are faced with the following situation: on the one hand first-order nonmonotonic methods are uncomputable, and for fixpoint methods there is not even agreement on their formulation, on the other hand we can make rather strong assumptions which trivialize the first-order case. Because of this situation, we have chosen to confine our treatment to minimal entailment. The first-order formulation of minimal entailment (in all variants) has a clear semantics which is not a later development, but it is its original definition.

8.1 *Minimal Entailment*

In contrast with other formalisms, minimal entailment has originally been introduced in a first-order setting and it has a well-defined meaning. In the previous section we have developed a tableau procedure for the propositional case. We wonder whether and how it can be extended to the first-order case. There are several difficulties. In view of the previous considerations, we cannot expect to have an effective and complete method. This is already seen from the fact that the method we have presented requires that some branch be completed in order to check both the m-closure and the ignorability condition. But a first-order tableau may fail to terminate, by leading to generate infinite branches. We must therefore restrict our concern to a subset of first-order logic for which a standard tableau procedure is ensured to terminate. Namely, we must restrict to sets of formulas Δ having a finite model property of the sort: if $M \models \Delta$, then there is a *finite* substructure M' of M such that $M' \models \Delta$. For instance, this can be ensured by considering function-free Σ_2-formulas. Thus, a tableau may only contain function-free formulas of type Σ_2 with label $\mathbf{T}$, and of type Π_2 with label $\mathbf{F}$. In this way, we ensure that the tableau construction terminates.

But termination, and hence completeness, is not the only difficulty. It is not obvious how to extend the propositional procedure in a *sound* manner. For instance, the formula $\phi = \forall x(p(x) \rightarrow r(x))$ is *not* a minimal consequence of $\Delta = \{\exists x(p(x) \wedge q(x)), \exists x(q(x) \wedge r(x))\}$. However, let us build a tableau for (Δ, ϕ) as shown in Figure 3.

The tableau has only one branch. Since the branch contains $\mathbf{F}\neg p(c)$, but not $\mathbf{T}p(c)$, it is m-closed, and hence the tableau is closed. The problem is that, in the tableau construction, we have implicitly assumed that every *skolem* constant, i.e. introduced by the quantifier rules ($\mathbf{T}\exists$, $\mathbf{F}\forall$), denotes a different individual. This is not correct. The $\mathbf{T}$-labelled formulas contained in the last node minimally entail $\neg p(c)$ only if $a \neq c$. Indeed, a counterexample is given by a minimal model M

$$\{\mathbf{T}\exists x(p(x) \wedge q(x)), \mathbf{T}\exists(q(x) \wedge r(x)), \mathbf{F}\forall x(p(x) \rightarrow r(x))\}$$

$$\{\mathbf{T}(p(a) \wedge q(a)), \mathbf{T}\exists x(q(x) \wedge r(x)), \mathbf{F}\forall x(p(x) \rightarrow r(x))\}$$

$$\{\mathbf{T}p(a), \mathbf{T}q(a), \mathbf{T}\exists x(q(x) \wedge r(x)), \mathbf{F}\forall x(p(x) \rightarrow r(x))\}$$

$$\{\mathbf{T}p(a), \mathbf{T}q(a), \mathbf{T}(q(b) \wedge r(b)), \mathbf{F}\forall x(p(x) \rightarrow r(x))\}$$

$$\{\mathbf{T}p(a), \mathbf{T}q(a), \mathbf{T}q(b), \mathbf{T}r(b), \mathbf{F}\forall x(p(x) \rightarrow r(x))\}$$

$$\{\mathbf{T}p(a), \mathbf{T}q(a), \mathbf{T}q(b), \mathbf{T}r(b), \mathbf{F}p(c) \rightarrow r(c)\}$$

$$\{\mathbf{T}p(a), \mathbf{T}q(a), \mathbf{T}q(b), \mathbf{T}r(b), \mathbf{F}\neg p(c), \mathbf{F}r(c)\}$$

Figure 3.

of the initial theory Δ, where $a^M = c^M = d_1$, $b^M = d_2$, $p^M = \{d_1\}$, $q^M = \{d_1, d_2\}$, $r^M = \{d_2\}$. A possible solution to this problem would be to branch according to $a = c \vee a \neq c$, for every pair of constants a and c.[22] Alternatively, we can modify the rules for ($\mathbf{T}\exists$, $\mathbf{F}\forall$) to incorporate such a branching. We give the rule for $\mathbf{T}\exists$:

Modified existential rule

> If $\mathbf{T}\exists x\phi[x]$ occurs in a branch, we create $n+1$ branches, where $a_1, \ldots, a_n$ are the constants already occurring in the branch; in each one of the first n branches $\mathbf{T}\exists x\phi[x]$ is replaced by $\mathbf{T}\phi[a_i]$, and in the last $n+1$-one $\mathbf{T}\exists x\phi[x]$ is replaced by $\mathbf{T}\phi[c]$, for a new constant c.

This modification of existential quantifier rule is due to Hintikka who proposed a tableau method for a form of domain minimization, (see next section).

But, even in the case we do not have to deal with existentially quantified formulas the method does not work correctly. Let Δ_1 be the following set of formulas:

$$\Delta_1;\ \{\forall x(p(x) \vee q(x)), \forall x(q(x) \rightarrow r(a)), \neg q(a)\}.$$

It is easy to see that Δ_1 does not minimally entail $\neg r(a)$. A counterexample may

[22] If we assume the *Unique Name Assumption* (UNA) we can restrict this branching to the case in which a is a *Skolem* constant and c is an arbitrary constant.

be easily found by considering minimal models whose domains have $n \geq 2$ elements. However, let us generate a tableau:

$$\{\mathbf{T}\forall x(p(x) \vee q(x)), \mathbf{T}\forall(q(x) \rightarrow r(a)), \mathbf{T}\neg q(a), \mathbf{F}\neg r(a)\}$$

$$\{\mathbf{T}(p(a) \vee q(a)), \mathbf{T}\forall x(q(x) \rightarrow r(a)), \mathbf{T}\neg q(a), \mathbf{F}\neg r(a)\}$$

$$\{\mathbf{T}(p(a) \vee q(a)), \mathbf{T}(q(a) \rightarrow r(a)), \mathbf{T}\neg q(a), \mathbf{F}\neg r(a)\}$$

$$\{\mathbf{T}p(a), \mathbf{T}(q(a) \rightarrow r(a)), \mathbf{T}\neg q(a), \mathbf{F}\neg r(a)\}$$

$$\{\mathbf{T}p(a), \mathbf{T}\neg q(a), \mathbf{F}\neg r(a)\} \qquad \{\mathbf{T}p(a), \mathbf{T}r(a), \mathbf{T}\neg q(a), \mathbf{F}\neg r(a)\}$$

Figure 4.

Figure 4 only shows branches which are not $\mathbf{T}$-closed. Branch

$$B_1 = \{\mathbf{T}p(a), \mathbf{T}\neg q(a), \mathbf{T}\neg q(a), \mathbf{F}\neg r(a)\}$$

is m-closed, whereas

$$B_2 = \{\mathbf{T}p(a), \mathbf{T}r(a), \mathbf{T}\neg q(a), \mathbf{F}\neg r(a)\}$$

is ignorable, hence the tableau is closed. The formula $\neg r(a)$ would be a minimal consequence of Δ_1 just in case a were the only object of the universe.

In other words, the tableau procedure deals correctly with minimization, but only on *Herbrand models*. But minimal entailment has the pathology that, even in the case of a *universal* theory, Herbrand models are not sufficient for determining minimal consequences. For ordinary logical consequence we know that given a universal formula ϕ and an existential formula ψ, if ψ holds in all Herbrand models of ϕ, then $\phi \models \psi$. A corresponding property does not hold for minimal entailment. The reason is that the class of *minimal* models of a universal theory is not necessarily closed under *substructures*. This problem seems fatal for the development of a general and clean method even for *universal* theories.

On the other hand, there are syntactic restrictions which prevent such a pathology, for instance the *allowedness* condition, well-known from database and logic programming literature [Lloyd, 1984]. It turns out that for universal, *allowed*, function-free theories ϕ and existential formulas ψ, the tableau method is sound and complete, under the *unique name assumption* [Olivetti, 1992].

8.2 Domain Minimization

Hintikka [1988] has proposed a modification of standard first-order tableaux as a proof-theoretical approximation of domain-minimization. His tableau method implements domain minimization in the following way. Suppose that we want to check whether ϕ 'minimally entails' ψ.

1. We generate the tableau for $\mathbf{T}\phi$.
2. The rule for existential quantifier is modified as described in the previous section, to the purpose of *reusing constants* as much as possible, instead of introducing new ones.
3. Branches are compared as follows: given two completed open branches B_1 and B_2, $B_1 << B_2$ iff B_1 has been completed exactly the same as B_2, except that it has re-used some old constants, instead of introducing one or more new ones (in $\exists$-elimination steps). This relation determines a partial order on branches, so that we can consider minimal elements wih respect to it.
4. To each minimal branch B, we add the axiom $DCA(B)$.
5. We check ψ in every minimal branch, that is, we add $\mathbf{F}\psi$, and we expand the tableau.

Lorenz [1994] has shown that Hintikka's entailment notion is not a good approximation of domain minimization, for it does not preserve logical equivalence. Let us consider the following example: $p(a)$ is logically equivalent to $p(a) \vee \exists x(x \neq a \wedge p(a))$. In Figure 5 it is shown a tableau for the latter formula.

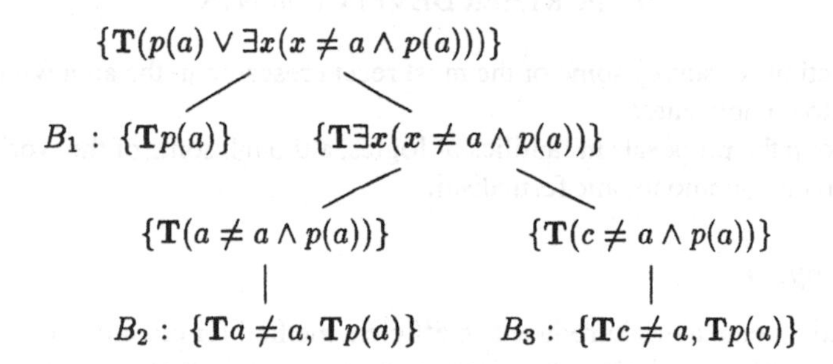

Figure 5.

In the tableau, branch B_2 is closed and hence disregarded. Both B_1 and B_3 are minimal according to Hintikka's definition, since they are uncomparable. We can conclude that, whereas the former formula $P(a)$ entails in Hintikka sense the formula $\forall x\, x = a$ (as it is immediate to see), the equivalent formula $P(a) \vee \exists x(x \neq a \wedge P(a))$ does not.

Lorenz has shown how to modify Hintikka's method to obtain a tableau procedure for his stronger notion of *variable* domain minimal entailment. The point is to allow the comparison of branches which arise also from different logical alternatives provided by the tableau expansion rules. We call two completed open branches B_1 and B_2 *comparable* if it is not the case that $a = b \in B_1$, but $a \neq b \in B_2$. Let us denote by $SKO(B)$ the set of skolem constants introduced in a branch B. For the rest, the method is the same as Hintikka's one. Lorenz calls the entailment determined by his tableau procedure sk-consequence.

The comparison relation becomes: $B_1 << B_2$ iff

1. B_1 and B_2 are comparable;
2. there exists an *injective* mapping $G : SKO(B_1) \to SKO(B_2)$, such that $range(G) \subset SKO(B_2)$.

The previous example is now handled correctly, since branches B_1 and B_3 can now be compared and it is easy to see that $B_1 << B_3$.

In [Lorenz, 1994] there is a soundness and completeness result for finitely well-founded theories, that is well-founded theories whose variable domain minimal models are finite.

THEOREM 35. *If Σ is finitely well-founded, then ϕ is a variable domain minimal consequence of Σ if and only if ϕ is an sk-consequence of Σ.*

It remains open the problem of finding an exact characterization by tableaux of the original notion of domain minimization.

9 FURTHER DEVELOPMENTS

In this section we survey some of the most recent research in the area without any pretence to completeness.

We group the proposals in families of logics, although some of the works cover more than one nonmonotonic formalism.

Circumscription

In [1996a], Niemelä has introduced an efficient method for circumscription based on a *hypertableaux* calculus (see [Baumgartner *et al.*, 1996]). We describe the method for set of clauses, in the basic case of propositional minimization, Niemelä has then extended it to other form of minimal entailment/circumscription [Niemelä, 1996]. We write a clause K as follows

$$a_1 \wedge \ldots \wedge a_m \to b_1 \vee \ldots \vee b_n,$$

we allow $m, n = 0$ and we define $Head(K) = \{b_1, \ldots, b_n\}$, $Body(K) = \{a_1, \ldots, a_m\}$. The tableau procedure is based on the following characterization of minimal models: let Σ be a set of clauses, M be an interpretation and

$$N_\Sigma(M) = \{\neg a \mid \exists K \in \Sigma, a \in Head(K), M \not\models a\}.$$

Then we have

> M is a minimal model of Σ iff $M \models \Sigma$ and for every atom a, $M \models a$ implies $\Sigma \cup N_\Sigma(M) \models a$.

The above characterization of minimal models can be re-stated in terms of tableau branches, allowing to check their minimality without explicitly comparing them.

We first describe the basic hypertableaux construction. A tableau for $\Gamma \models_m \psi$ is constructed as follows. The root of the tableau is labelled with $\Gamma \cup \bar{\Delta}$, where $\bar{\Delta} = \{\, \bar{C}_1, \ldots, \bar{C}_k \,\}$ is a set of clauses representing $\neg\psi$ in CNF. Tableau branches are extended according to the following two rules:

$$\frac{\begin{array}{c} a_1 \wedge \ldots \wedge a_m \rightarrow b_1 \vee \ldots \vee b_n \\ a_1, \ldots, a_m \\ \neg b_1, \ldots, \neg b_{j-1}, \neg b_{j+1}, \ldots, \neg b_n \end{array}}{b_j}$$

$$\frac{\begin{array}{c} a_1 \wedge \ldots \wedge a_m \rightarrow b_1 \vee \ldots \vee b_n \\ a_1, \ldots, a_m \end{array}}{\neg b_j \mid b_j}$$

A branch B is *closed* if it contains a clause $b_1 \wedge \ldots \wedge b_n \rightarrow$ $(n \geq 1)$ together with the atoms $b_1, \ldots, b_n$. A branch B is *finished* if, whenever $a_1 \wedge \ldots \wedge a_m \rightarrow b_1 \vee \ldots \vee b_n \in B$, and $a_1, \ldots a_m \in B$, then some $b_j \in B$. A branch B is *ungrounded in* Γ if for some atom a on B, $\Gamma \cup N_\Gamma(B) \not\models a$, where

$$N_\Gamma(B) = \{\, \neg a \mid \exists K \in \Sigma, a \in Head(K) - B \,\}.$$

A tableau T for (Σ, ϕ) is *MM-closed* iff every branch is closed or ungrounded in Γ. It can be shown that

THEOREM 36. $\Sigma \models_m \phi$ *iff any tableau for* (Σ, ϕ) *is MM-closed.*

We can observe that ungrounded branches correspond to branches which are either m-closed or ignorable according to the tableau procedure of section 7. We can replace the ignorability test by the ungroundedness test. The ungroundedness test, by itself, is a provability test which can be implemented, for instance, by another tableau construction. The main advantage of this approach is that ungroundness is a local condition, whence in order to check it we do not need to keep other branches for comparison. This leads to a polynomial space complexity which is not ensured by the other method of section 7.

The use of hypertableaux for minimal model generation has been independently explored by [Bry and Yahya, 1996] along somewhat different lines.

Autoepistemic and NonMonotonic Modal Logics

Donini, Massacci *et al.* in [1996] have developed a uniform tableau method for the whole class of nonmonotonic modal logics based on a modal preferential semantics developed by Schwarz [1992]. Their tableau procedure is very general and is parametric with respect to the underlying modal logic and the preference criteria on Kripke structures. Within their framework they can capture Autoepistemic logic, the Logic of Minimal Knowledge [Halpern and Moses, 1985], and default logic via its modal translation. Their tableau systems are refutational as usual; open branches of a tableau, which are possible counterexamples, are checked to see whether they correspond to most-preferred models. This check involves the construction of another tableau, whose open branches (if any) may specify more preferred models (called *opponent models*), in this case the candidate counterexample branch of the original tableau can be ruled out or ignored. The required semantic notions are rather complex to give a short exposition of their work, we refer the reader to [Donini *et al.*, 1996].

In [1994], Niemelä has proposed several algorithms for autoepistemic logic, its variants, and default logic; these procedures are rather different from the tableau procedure presented in section 5. His proof-procedures are not strictly related to tableaux, although they are analytic, and can be implemented using a variety of theorem prover mechanisms. However, tableaux may still play a significant role. We limit the exposition to autoepistemic logic. The starting notion is the one of *full set* (see end of section 5), which provides a finitary characterization of expansions.

Let Σ be a finite set of formulas, by $Sub(\Sigma)$ we denote the set of subformulas of Σ, we also define:

$$Sf^L(\Sigma) = \{L\phi \in Sub(\Sigma)\} \text{ and } \neg Sf^L(\Sigma) = \{\neg L\phi \mid L\phi \in Sf^L(\Sigma)\}.$$

We say that a subset $B \subseteq Sf^L(\Sigma) \cup \neg Sf^L(\Sigma)$ is a *full set* for Σ, if for every $L\phi \in Sf^L(\Sigma)$

- $L\phi \in B \;-\; \Sigma \cup B \vdash_{AE} \phi$;
- $\neg L\phi \in B \;-\; \Sigma \cup B \not\vdash_{AE} \phi$;

where $\vdash_{AE}$ is the trivial extension of propositional calculus in which L-formulas are just treated as propositional atoms.

Full sets exactly correspond to stable expansions. To explain the precise connection, we first extend the relation $\vdash_{AE}$ according to the provability meaning of the L-operator. We define (recursively)

$$\Sigma \vdash_{AEL} \phi \;-\; \Sigma \cup SB_\Sigma(\phi) \vdash_{AE} \phi \text{ where}$$

$$SB_{\Sigma}(\phi) = \{L\chi \in S^{IL}(\phi) \mid \Sigma \vdash_{AEL} \chi\} \cup \cup\{\neg L\chi \mid \chi \in S^{IL}(\phi) \wedge \Sigma \nvdash_{AEL} \chi\}$$

$S^{IL}(\phi)$ denotes the set of immediate L- subformulas of ϕ, that is not in the scope of another L-operator.

THEOREM 37 ([Niemelä, 1994]). *The mapping E*

$$E_{\Sigma}(B) = \{\psi \mid \Sigma \cup B \vdash_{AEL} \psi\}$$

is a bijection between full sets *and* stable expansions.

The relation $\Sigma \vdash_{AEL} \psi$ is decidable. It is relatively easy to devise a tableau-based algorithm for it. In order to check whether $\Sigma \vdash_{AEL} \psi$, one builds a tableau system in a similar way to the one described in section 5. A hierarchy of auxiliary tableaux is created in order to evaluate the L-subformulas of ψ (on the opposite, the L-subformulas of Σ are treated just as propositional atoms), until tableaux for subformulas of ψ with no occurrences of L are reached. These terminal tableaux can be completed in the ordinary way, and according to their status (open or closed) we recursively proceede backwards to update the other tableaux, as described in the method of 5. At the end we have that $\Sigma \vdash_{AEL} \psi$ iff the (updated) tableau for (Σ, ψ) is closed.

In order to perform autoepistemic reasoning we need a way of generating the full sets. This can be done by the method described in section 5; as we have remarked, each admissible labelling corresponds to a full set. But there are more efficient ways to generate full sets than guessing the admissible labellings. In figure 6, we present the procedure $FULL\_FIND$, which builds a full set of a given set of formulas Σ, if there exists one.

We say that B *covers* Σ iff for every $\psi \in Sf^{L}(\Sigma)$, either $L\psi \in B$ or $\neg L\psi \in B$. The procedure $FULL\_FIND(\Sigma, B, F)$ makes use of an auxiliary procedure $EXTEND$,[23] B represent the temptative full set being created, F represent the set of $L\psi$ yet uncovered by B which should be added to B and are, so to say, 'frozen'. The initial call is $FULL\_FIND(\Sigma, \emptyset, \emptyset)$ and it returns either a full set B of Σ or **false** if it does not exist.

Intuitively, the procedure EXTEND expands B by 'positive introspection' as much as possible. When B can no longer be extended, we check whether B contains some $\neg\chi$ which should not be in B (this happens if $\Sigma \cup B \cup F \vdash_{AE} \chi$). In this case B cannot be extended to a full set and we report failure. If this is not the case we check whether all L-subformulas of Σ are covered by $B \cup F$. If this is the case and $F \subseteq B$, then B covers Σ and we report success, otherwise (if $F \nsubseteq B$) some subformula of Σ cannot be covered by B and we report failure. If some formula $L\chi$ is not covered by $B \cup F$, we try to add $\neg L\chi$ to B, and see whether B can be extended to a full set. If it cannot we retry putting $L\chi$ in F.

[23] In Niemelä's paper [1994] the procedure EXTEND is called POS_M and the procedure FULL_FIND is called DER_M.

$EXTEND(\Sigma, B, F)$

repeat $B' := B$

  for all $L\phi \in Sf^L(\Sigma)$

  $B := B \cup \{L\phi\}$ if $\Sigma \cup B \cup F \vdash_{AE} \phi$

until $B = B'$

return B

$FULL\_FIND(\Sigma, B, F)$

$B := EXTEND(\Sigma, B, F)$

if for some $\neg L\chi \in B, \Sigma \cup B \cup F \vdash_{AE} \chi$ **then return false**

else if $Sf^L(\Sigma) \subseteq Sf^L(B) \cup F$ **then**

  if $F \subseteq B$ **then return true**

  else return false;

else let $L\chi$ such that $L\chi \notin B$ and $\neg L\chi \notin B$

  if $FULL\_FIND(\Sigma, B' \cup \{\neg L\chi\}, F) =$ **false**

  then

   return $FULL\_FIND(\Sigma, B, F \cup \{L\chi\})$

  else return true

Figure 6.

The algorithm can be adapted to obtain a uniform decision method for both skeptical and credulous consequence. By definition we have

- $\Sigma \vdash_c \phi$ – $\exists$B full set : $\Sigma \cup B \vdash_{AEL} \phi$;
- $\Sigma \vdash_s \phi$ – $\neg\exists$B full set : $\Sigma \cup B \not\vdash_{AEL} \phi$;

In the credulous case we must find a full set B satisfying $\Sigma \cup B \vdash_{AEL} \phi$, and we succeed as soon as we find it; in the skeptical case we try to build a "counterexample" full set B such that $\Sigma \cup B \not\vdash_{AEL} \phi$ and we succeeds if the search of such full set B fails. The case are dual and being able to decide whether $\Sigma \cup B \vdash_{AEL} \phi$, we can modify the procedure FULL_FIND in a way that instead of returning **true**, it returns the outcome of $\vdash_{AEL}$-derivability test $\mathbf{test}(\Sigma, B, \phi)$ with respect to the full set B being constructed.

A similar method have been devised by Niemelä [1995] for default logic, in which the role of the full sets is played by sets of justifications.

Default logic

A different line of research, started by Bonatti, aims to a pure proof-theoretical reconstruction of the major families of nonmonotonic formalisms based on sequent calculi, which are independent from any search strategy. The methodology has been applied so far to default logic, (both in the credulous and skeptical case) [Bonatti and Olivetti, 1997a], to autoepistemic logic (in the credulous case for normal theories) [Bonatti, 1996], and to circumscription [Bonatti and Olivetti, 1997]. The main features of the calculi are that they are *analytic*, they comprise an *axiomatic rejection method* (in the form of a sequent calculi by itself), and they make use of *provability assumptions*, that can be either used as facts, or be confuted by means of the rejection method. These calculi can also be used to investigate the power and the efficiency of nonmonotonic reasoning (for instance they have been used to prove non-elementary speed-up results with respect to classical logic [Egly and Tompits, 1997]). In the following, we briefly sketch the calculus for skeptical default inference contained in [Bonatti and Olivetti, 1997a].

The calculus is organized in three levels and is shown in Figure 7. The base level is the one of the classical sequent calculus and the anti-sequent calculus. The latter is a calculus for non-theoremhood. Each derivation in the anti-sequent calculus is nothing else than a completed open branch in a tableau proof turned upside down. An *anti-sequent* is a pair of sets of sentences, denoted by $\Gamma \not\vdash \Sigma$, and its intended meaning is: there exists a model of Γ where all the sentences of Σ are false. An anti-sequent $\Gamma \not\vdash \Sigma$ is an axiom of the anti-sequent calculus if, and only if, Γ and Σ are disjoint sets of propositional variables. In fig 7(a) for brevity we give the rules only for negation and conjunction, the rules for classical provability are omitted. The anti-sequent calculus preserves many properties of the standard sequent calculus (permutation, symmetry, subformula properties).

The second level and the third level are specific to default logic. The distinction between these two levels is not necessary in principle, but it enhances and somewhat simplify the presentation. The intermediate level is the one of monotonic inference rule, which are called *residues*. Such rules are default rules without justification, that is inference rules of the form α/β, whose meaning is: if α is derivable, then also β is derivable. Let $\mathcal{L}$ denote the propositional language, we define $\mathcal{L}^{res}$, by stipulating:

$$\mathcal{L}^{res} = \mathcal{L} \cup \{\alpha/\beta \mid \alpha, \beta \in \mathcal{L}\}.$$

Given a subset S of $\mathcal{L}^{res}$, we can define the deductive closure of S under classical provability *and* residues, denoted by $\mathrm{Th}^{res}(S)$, as the least set $S' \subseteq \mathcal{L}$ which satisfies the following conditions:

a) $S \cap \mathcal{L} \subseteq S'$;

b) if $S' \vdash \alpha$, then $\alpha \in S'$;

c) if $\alpha \in S'$, and $\alpha/\gamma \in S$, then $\gamma \in S'$.

In can be easily shown that for any $S \subseteq \mathcal{L}^{res}$, $\mathrm{Th}^{res}(S)$ exists. A sound and complete (anti)-sequent calculus with respect to Th^{res}-(un)derivability is obtained by adding rules *(Re1)-(Re4)* of Figure 7 (b) to the classical sequent and antisequent calculus restricted to $\mathcal{L}$.

The third level is the one of the skeptical sequent calculus for default logic. The calculus makes use of *constraints* of the form $\mathbf{M}\alpha$ or $\mathbf{L}\alpha$, where $\alpha \in \mathcal{L}$. We say that a set of sentences E *satisfies* a constraint $\mathbf{M}\alpha$ if $E \not\vdash \neg\alpha$; we say that E satisfies $\mathbf{L}\alpha$ if $E \vdash \alpha$.

A *skeptical default sequent* is a 3-tuple $\langle \Sigma, \Gamma, \Delta \rangle$, denoted by $\Sigma; \Gamma \vdash \Delta$, where Σ is a set of constraints, Γ is a propositional default theory (i.e. a set of propositional formulas and defaults), and Δ is a set of propositional sentences. The intended meaning of the above sequent is: the disjunction of the formulas in Δ, $\bigvee \Delta$, belongs to all the extensions of Γ that satisfy the constraints Σ. When this is the case, we say that the sequent is *true*. We give a quick explanation of the rules 7 (c). Rule (Sk1) can be read backwards as a default-elimination rule; it performs a case analysis on the applicability of the default being eliminated, the first premise considers the case when the justification β is consistent, so that the default can be replaced by its residue; the second premise consider the case when the justification is inconsistent, so that the default itself can be eliminated. Rules (Sk2) and (Sk3) assert that the lower sequent is vacously true if some constraint (respectively of the form $\mathbf{M}\alpha$, $\mathbf{L}\alpha$) cannot be satisfied. Finally (Sk4) asserts that skeptical default provability extends classical provability.

THEOREM 38 (Soundness-Completeness). *A sequent is derivable if and only if it is true.*

We can notice that the rules for residues cannot be applied until all *proper* defaults—that is, defaults with nonempty justification—have been eliminated. This causes proof trees to be exponentially large in the size of the default theory. However, in general, it is not necessary to consider every default, in order to derive a skeptical conclusion. It is possible to give a sound (and obviously complete) generalization of rules (Sk2)-(Sk4) which may be used to reduce dramatically the proof size. In the generalized rules Γ may contain proper defaults. For instance the generalization of (Sk4) is the following rule:

$$\frac{\Sigma', \Gamma' \vdash \Delta}{\Sigma; \Gamma \vdash \Delta}$$

where $\Sigma' \subseteq \{\alpha \mid \mathbf{L}\alpha \in \Sigma\}$, $\Gamma' \subseteq \Gamma^{res} \stackrel{\text{def}}{=} \Gamma \cap \mathcal{L}^{res}$. The idea under the generalization is that each extension of Γ that satisfies Σ, contains both the propositional sentences of Γ and all the sentences α such that $\mathbf{L}\alpha \in \Sigma$; moreover, these sentences are closed under classical entailment and the residues occurring in Γ. Therefore, the sentences in this closure can be used to prove the conclusion of a skeptical sequent. The other rules are generalized similarly, see [Bonatti and Olivetti, 1997a].

$$\frac{\Gamma \nvdash \Sigma, \alpha}{\Gamma, \neg\alpha \nvdash \Sigma}\ (\neg \nvdash) \qquad \frac{\Gamma, \alpha \nvdash \Sigma}{\Gamma \nvdash \Sigma, \neg\alpha}\ (\nvdash \neg)$$

$$\frac{\Gamma, \alpha, \beta \nvdash \Sigma}{\Gamma, \alpha \wedge \beta \nvdash \Sigma}\ (\wedge \nvdash) \qquad \frac{\Gamma \nvdash \Sigma, \alpha}{\Gamma \nvdash \Sigma, \alpha \wedge \beta}\ (\nvdash \bullet\wedge) \qquad \frac{\Gamma \nvdash \Sigma, \beta}{\Gamma \nvdash \Sigma, \alpha \wedge \beta}\ (\nvdash \wedge\bullet)$$

(a) Rules of the anti-sequent calculus restricted to $\mathcal{L}$.

$$\frac{\Gamma \vdash \Delta}{\Gamma, \alpha/\gamma \vdash \Delta}\ (\mathbf{Re1}) \qquad \frac{\Gamma \vdash \alpha \quad \Gamma, \gamma \vdash \Delta}{\Gamma, \alpha/\gamma \vdash \Delta}\ (\mathbf{Re2})$$

$$\frac{\Gamma \nvdash \Delta \quad \Gamma \nvdash \alpha}{\Gamma, \alpha/\gamma \nvdash \Delta}\ (\mathbf{Re3}) \qquad \frac{\Gamma, \gamma \nvdash \Delta}{\Gamma, \alpha/\gamma \nvdash \Delta}\ (\mathbf{Re4})$$

(b) Rules for residues restricted to $\mathcal{L}^{res}$.

$$\frac{\mathbf{M}\beta, \Sigma; \Gamma, \alpha/\gamma \vdash \Delta \quad \mathbf{L}\neg\beta, \Sigma; \Gamma \vdash \Delta}{\Sigma; \Gamma, \frac{\alpha : \beta}{\gamma} \vdash \Delta}\ (\mathbf{Sk1})$$

$$\frac{\Gamma \vdash \neg\alpha}{\mathbf{M}\alpha, \Sigma; \Gamma \vdash \Delta}\ (\mathbf{Sk2})\ (*) \qquad \frac{\Gamma \nvdash \alpha}{\mathbf{L}\alpha, \Sigma; \Gamma \vdash \Delta}\ (\mathbf{Sk3})\ (*)$$

$$\frac{\Gamma \vdash \Delta}{\Sigma; \Gamma \vdash \Delta}\ (\mathbf{Sk4})\ (*)$$

(c) Rules for defaults and constraints. (*) It must be $\Gamma \subseteq \mathcal{L}^{res}$.

Figure 7. Skeptical sequent calculus

To conclude, we point out what we regard as the main novelties of the works we have described in this section.

First, we can notice the use of efficient and specific proof-procedures such as the hypertableaux calculus to optimize the search of minimal models [Niemelä, 1996a]. One idea arising in different contexts of nonmonotonic logics which admit a preferential semantic, is that, in order to see whether a tableau branch represents a preferred model, the branch is, so to say, 'challenged' by another tableau construction which may determine a more preferred model. This is what happens in Niemelä's method for circumscription with the ungroundedness test, and also in the tableau procedures for nonmonotonic modal logics by Donini *et al.* [1996]. By such an auxiliary tableau construction, one can decide whether a branch is most-preferred without comparing it to any other.

Another aspect which is worth mentioning is that there has been a conceptual progress in dealing with *skeptical* reasoning. The first approaches to skeptical reasoning were simply enumerative: one had to build all extensions/expansions and check the goal formula in each one of them. In contrast, the new method by Niemelä for autoepistemic reasoning and the skeptical sequent calculus presented in this section, perform skeptical reasoning in a rather different way. The former, performs skeptical reasoning by searching for a counterexample full-set which witness the non-derivability of the goal formula. In principle one does not need to carry on the construction of each full set to the end, for one can often discover in advance that it will not lead to a counterexample. The latter performs skeptical reasoning by a case analysis on default applicability which may save the actual examination of each default rule and the construction of each default extension.

As a final remark, we can compare the state of the art of tableau systems for nonmonotonic reasoning to the state of the art of tableau-based deduction in other areas. The impression is that there is still much work to be done, not only in the design of efficient algorithms and search strategies, but also in the achievement of a uniform tableau methodology. A substantial progress on this side could come from the development of a well-accepted and unifying semantics for all the main nonmonotonic formalisms, which at present is still under investigation.

ACKNOWLEDGEMENTS

I am indebted to Camilla Schwind for her careful checking of the previous versions of the manuscript and for her fruitful suggestions. I also thank Piero Bonatti, Laura Giordano and Vincent Risch for their advices and for the useful discussions we had.

Università di Torino, Italy.

REFERENCES

[Amati *et al.*, 1996] G. Amati, L. Carlucci Aiello, D. Gabbay, and F. Pirri. A proof-theoretical approach to default reasoning I: tableaux for default logic. *Journal of Logic and Computation*, 6:205–232, 1996.

[Baumgartner *et al.*, 1996] P. Baumgartner, U.furbach and I. Niemelä. Hyper tableaux. In *Logic in AI. Proc. Europ.workshop on Logics in AI*, LNAI n. 1126, pp. 1–17. Springer-Verlag, 1996.

[Besnard, 1989] P. Besnard. *An Introduction to default-logic*. Springer-Verlag, 1989.

[Bonatti, 1996] P. A. Bonatti. Sequent calculus for default and autoepistemic logic. In *TABLEAUX'96, LNAI n. 1071*, pages 127–142. Springer-Verlag, 1996.

[Bonatti and Olivetti, 1997] P. A. Bonatti and N. Olivetti. A sequent calculus for circumscription. In *Preliminary Proceedings of CSL'97*, pages 95–107, Aarhus, 1997. To appear in *Proceedings of CSL '97*, LNCS, Springer, 1998.

[Bonatti and Olivetti, 1997a] P. A. Bonatti and N. Olivetti. A sequent calculus for skeptical default logic. In *TABLEAUX'97, LNAI n. 1227*, pages 107–121, 1997.

[Bossu and Siegel, 1985] G. Bossu and P. Siegel. Saturation, non-monotonic reasoning and the closed-world-assumption. *Artificial Intelligence*, 25:13–63, 1985.

[Brewka, 1985] G. Brewka. *Non-monotonic reasoning: logical foundations of commonsense*. Cambridge Tracts in Theoretical Computer Science 12, Cambridge University Press, 1985.

[Brewka, 1992] G. Brewka. Cumulative default logic: in defence of nonmonotonic inference rules. *Artificial Intelligence*, 50:183–206, 1992.

[Bry and Yahya, 1996] F. Bry and A. Yahya. Minimal model generation with positive hyper- resolution tableaux. In *TABLEAUX'96, LNAI n. 1071*, pages 143–159. Springer-Verlag, 1996.

[Delgrande, 1988] J. P. Delgrande. A first-order conditional logics for prototypical properties. *Artificial Intelligence*, 36:63–90, 1988.

[Delgrande *et al.*, 1994] J. P. Delgrande, T. Schaub, and W. K. Jackson. Alternative approaches to default logic. *Artificial Intelligence*, 70:167–237, 1994.

[Donini *et al.*, 1996] M. Donini, F. Massacci, D. Nardi, and R. Rosati. A uniform tablcaux method for nonmonotonic modal logics. In *Logic in AI, (Proc. Europ. Work. on Logics in AI), LNAI 1126*, pages 87–103. Springer-Verlag, 1996.

[Egly and Tompits, 1997] U. Egly and H. Tompits. Non-elementary speed-ups in default reasoning. In *FAPR'97, LNAI n. 1244*, pages 237–251, 1997.

[Etherington, 1988] D. Etherington. *Reasoning with incomplete information*. Pitman, 1988.

[Giordano and Martelli, 1994] L. Giordano and A. Martelli. On cumulative default reasoning. *Artificial Intelligence*, 50:183–206, 1994.

[Gottlob, 1992] G. Gottlob. Complexity results for non-monotonic logics. *J. of Logic and Computation*, 2:397–425, 1992.

[Halpern and Moses, 1985] J. Y. Halpern and Y. Moses. Toward a theory of knowledge and ignorance: Preliminary report. Technical report, CD-TR 92/34, IBM, 1985.

[Hintikka, 1988] J. Hintikka. Model minimization - an alternative to circumscription. *J. of Automated Reasoning*, 4:1–13, 1988.

[Konolige, 1988] K. Konolige. On the relation between default and autoepistemic logic. *Artificial Intelligence*, 35:343–382, 1988.

[Kraus *et al.*, 1990] S. Kraus, D. Lehmann, and M. Magidor. Nonmonotonic reasoning, preferential models and cumulative logics. *Artificial Intelligence*, 44:167–207, 1990.

[Kuhna, 1993] P. Kuhna. Circumscription and minimal models for propositional logics. In *Proc. of the First Workshop on Theorem Proving with Analytic Tableaux and Related Methods*, Marseille, 1993.

[Lifschitz, 1985] V. Lifschitz. Closed world databases and circumscription. *Artificial Intelligence*, 27:229–235, 1985.

[Lifschitz, 1985a] V. Lifschitz. Computing circumscription. In *Proc. of IJCAI'85*, pages 121–129. Morgan Kaufmann, 1985.

[Lifschitz, 1986] V. Lifschitz. On the satisfability of circumscription. *Artificial Intelligence*, 28:17–27, 1986.

[Lin, 1990] F. Lin. Circumscription in a modal logic. In *Proc. of TARK88*, pages 193–219. Morgan Kaufmann, 1990.

[Lloyd, 1984] J. W. Lloyd. *Foundations of Logic Programming*. Springer-Verlag, 1984.

[Lorenz, 1994] S. Lorenz. A tableau prover for domain minimization. *J. of Automated Reasoning*, 13:375–390, 1994.

[Łukasiewicz, 1990] W. Łukasiewicz. Considerations on default logic. In *Proc. of TARK90*, pages 219–193. Morgan Kaufmann, 1990.

[Makinson, 1989] D. Makinson. General theory of cumulative inference. In *Proceedings of the Second International Workshop on Non-Monotonic Reasoning, LNAI 346*, pages 1–18. Springer-Verlag, 1989.

[McCarthy, 1980] J. McCarthy. Circumscription, a form of non- monotonic reasoning. *Artificial Intelligence*, 13:27–39, 1980.

[McCarthy, 1986] J. McCarthy. Applications of circumscription, to formalizing commonsense knowledge. *Artificial Intelligence*, 28:89–118, 1986.

[McDermott, 1982] D. McDermott. Non-monotonic logic ii: non-monotonic modal theories. *J. of the ACM*, 29:33–57, 1982.

[McDermott and Doyle, 1980] D. McDermott and J. Doyle. Non-monotonic logic i. *Artificial Intelligence*, 13:41–72, 1980.

[Moore, 1982] R. Moore. Semantical considerations on non- monotonic logics. *Atificial Intelligence*, 29:33–57, 1982.

[Niemelä, 1988] I. Niemelä. Decision procedures for autoepistemic logic. In *Proc. of the Nineth Conference on Automated Deduction (CADE-88), LNCS n. 310*, pages 541–546. Springer-Verlag, 1988.

[Niemelä, 1991] I. Niemelä. Towards automatic autoepistemic reasoning. In *Logic in AI, (Proc. Europ. Work. on Logics in AI), LNAI 478*. Springer-Verlag, 1991.

[Niemelä, 1994] I. Niemelä. A decision method for nonmonotonic reasoning based on autoepistemic reasoning. *Journal of Automated Reasoning*, 14:3–42, 1994.

[Niemelä, 1995] I. Niemelä. Toward efficient default reasoning. In *Proc. of IJCAI'95*, pages 312–318. Morgan Kaufmann, 1995.

[Niemelä, 1996] I. Niemelä. Implementing circumscription using a tableau method. In *Proc. of ECAI'96*, pages 80–84. Pitman Publishing, 1996.

[Niemelä, 1996a] I. Niemelä. A tableaux calculus for mininal model reasoning. In *TABLEAUX'96, LNAI n. 1071*, pages 278–294. Springer-Verlag, 1996.

[Olivetti, 1989] N. Olivetti. Circumscription and closed world assumption. *Atti della Accademia delle Scienze di Torino*, 123 Fasc. 5-6, 1989.

[Olivetti, 1992] N. Olivetti. Tableaux and sequent calculus for minimal entailment. *J. of Automated Reasoning*, 9:99–139, 1992.

[Reiter, 1980] R. Reiter. A logic for default reasoning. *Artificial Intelligence*, 13:81–132, 1980.

[Reiter, 1987] R. Reiter. Nonmonotonic reasoning. *Ann. Rev. Computer Science*, 2:147–186, 1987.

[Risch, 1996] V. Risch. Analytic tableaux for default logics. *J. of Applied Non-Classical Logics*, 6:71–88, 1996.

[Schaub, 1992] T. Schaub. On constrained default theories. In *Proc. of ECAI-92*, pages 304–308. Wiley, 1992.

[Schlipf, 1987] J. S. Schlipf. Decidability and definability with circumscription. *Annals of Pure and Applied Logic*, 35:173–191, 1987.

[Schwarz, 1990] G. Schwarz. Autoepistemic modal logics. In *Proc. of TARK90*, pages 97–109. Morgan Kaufmann, 1990.

[Schwarz, 1992] G. Schwarz. Minimal model semantics for nonmonotonic modal logics. In *Proc. of KR'92*, pages 581–590. Morgan Kaufmann, 1992.

[Schwind, 1990] C. Schwind. A tableau-based theorem prover for a decidable subset of default logic. In *Proc. of the Tenth Conference on Automated Deduction (CADE-90)*, pages 541–546. Springer-Verlag, 1990.

[Schwind and Risch, 1994] C. Schwind and V. Risch. Tableau-based charachterization and theorem-proving for default logic. *J. of Automated Reasoning*, 13:223–242, 1994.

[Shoham, 1987] Y. Shoham. Non-monotonic logics: meaning and utility,. In *Proc. of IJCAI-87*. Morgan Kaufmann, 1987.

REINER HÄHNLE

TABLEAUX FOR MANY-VALUED LOGICS

1 INTRODUCTION

This article reports on research done in the intersection between many-valued logics and logical calculi related to tableaux. A lot of important issues in many-valued logic, such as algebras arising from many-valued logic, many-valued function minimization, philosophical topics, or applications are not discussed here; for these, we refer the reader to general monographs and overviews such as [Rosser and Turquette, 1952; Rescher, 1969; Urquhart, 1986; Bolc and Borowik, 1992; Malinowski, 1993; Hähnle, 1994; Panti, to appear]. More questionable, perhaps, than the omissions is the need for a handbook chapter on *tableaux* for many-valued logics in the first place.

Two objections can readily be raised: (i) complete, generic sequent and tableau systems for arbitrary finitely-valued first-order logics were developed (for example) in [Rousseau, 1967; Takahashi, 1967] and [Carnielli, 1987; Carnielli, 1991], one could leave it at that; (ii) why not concentrate on other inference procedures such as resolution?

In my opinion, of course, both objections can be defeated. First, applications in computer science demand, as usually, *efficiently automatizable* calculi. The papers cited under (i) above only provide the merest theoretical foundations. Much work needs to be done before these can be turned into actual code. We hope to demonstrate that tableau systems are a particularly well-suited starting point for the development of computational insights into many-valued logics. Second, there has been no efficient theorem proving framework for infinitely-valued logics until recently—the solution turns out to be an enriched tableau calculus. Another pertinent reason for choosing tableau calculi is, of course, the lack of normal forms in many-valued logics in general.[1] In fact, we will see that tableau procedures even can be viewed as algorithms for normal form translation and in this sense they enable the use of resolution based approaches for many-valued logics in the first place.[2] Also certain deduction tasks (for example, in program verification) inhibit the use of normal forms, in which case one is forced to resort to genuine non-clausal proof procedures.

There is another, more indirect, reason for investigating tableau calculi for many-valued logics: we claim that a close interplay between model theoretic and

[1] This does not contradict the fact that for certain logics normal forms exist as is witnessed by the remarkable result of Mundici [1994].

[2] There are non-clausal resolution methods for many-valued logics [Stachniak and O'Hearn, 1990; O'Hearn and Stachniak, 1992], but, as already their classical counterparts, they suffer from efficiency problems.

M. D'Agostino et al. (eds.), Handbook of Tableau Methods, 529–580.

proof theoretic tools is necessary and fruitful during the development of proof procedures for non-classical logics. Since many-valued logics have a simpler syntax and semantics than, say, modal or temporal logics, the development is perhaps more advanced here than in other non-classical logics. Therefore, we hope that the present chapter can serve as a paradigmatic example showing that both model theoretic and proof theoretic analyses have an impact on each other and that building proof procedures for non-classical logics cannot dispense with either of them.

Finally, as will be pointed out, there has been a considerable amount of redundancy in research related to sequent and tableau calculi for many-valued logics. The same techniques were invented repeatedly without much advance being made. Hopefully, the present chapter, and the highly visible context in which it stands, will help to avoid further waste of time.

A somewhat fuller, but not quite up-to-date, account on automated reasoning in many-valued logics in general is [Hähnle, 1994]. A very recent, although condensed, overview is [Hähnle and Escalada-Imaz, 1997].

2 MANY-VALUED LOGIC

In this section we collect basic definitions and concepts from many-valued logics as well as some examples.

2.1 *Syntax*

DEFINITION 1. Let Σ_0 be a *propositional signature*, that is, a denumerable set of propositional variables $\{p_0, p_1, \ldots\}$.

A *first-order signature* Σ is a triple $\langle \mathbf{P}_\Sigma, \mathbf{F}_\Sigma, \alpha_\Sigma \rangle$, where $\mathbf{P}_\Sigma$ is a non-empty family of *predicate symbols*, $\mathbf{F}_\Sigma$ is a possibly empty family of *function symbols* disjoint from $\mathbf{P}_\Sigma$, and α_Σ assigns a non-negative arity to each member of $\mathbf{P}_\Sigma \cup \mathbf{F}_\Sigma$.

Let Term_Σ be the set of Σ-*terms* over *object variables* $\mathrm{Var} = \{x_0, x_1, \ldots\}$ and let Term^0_Σ be the set of ground terms.

DEFINITION 2. In the propositional case we define the set of *atomic formulas* (or *atoms* for short) to be Σ_0, in the first-order case let

$$\mathrm{At}_\Sigma = \{p(t_1, \ldots, t_n) | p \in \mathbf{P}_\Sigma, \alpha_\Sigma(p) = n, t_i \in \mathrm{Term}_\Sigma\} \ .$$

DEFINITION 3. A *propositional language* is a pair $\mathbf{L}^0 = \langle \Theta, \alpha \rangle$, where Θ is a finite or denumerable set of *logical connectives* and α defines the arity of each connective. Connectives with arity 0 are called *logical constants*.

The set L^0_Σ of $\mathbf{L}^0$-formulas over Σ is inductively defined as the smallest set with the following properties:

1. $\Sigma \subseteq L^0_\Sigma$.

2. If $\theta \in \Theta$ and $\alpha(\theta) = 0$ then $\theta \in L^0_\Sigma$.

3. If $\phi_1, \ldots, \phi_m \in L^0_\Sigma$, $\theta \in \Theta$ and $\alpha(\theta) = m$ then $\theta(\phi_1, \ldots, \phi_m) \in L^0_\Sigma$.

NOTATION 4. We denote propositional languages $\langle \Theta, \alpha \rangle$ with finite sets of connectives as $\langle \theta_1/\alpha(\theta_1), \ldots, \theta_r/\alpha(\theta_r) \rangle$, where $\Theta = \{\theta_1, \ldots, \theta_r\}$. Moreover, we make the usual conventions on bracketing and infix notation for well-known connective symbols.

EXAMPLE 5. The language of *propositional Łukasiewicz* logic is given by $\mathbf{L}_{\text{Łuk}} = \langle \neg/1, \supset/2 \rangle$. Examples of well-formed formulas are: $p \supset (q \supset p), \neg(p \supset p)$.

DEFINITION 6. A *first-order language* is a triple $\mathbf{L} = \langle \Theta, \Lambda, \alpha \rangle$ such that $\langle \Theta, \alpha \rangle$ is a propositional language and Λ is a finite or denumerable set of *quantifiers*. The set L_Σ of $\mathbf{L}$-formulas over Σ is inductively defined as the smallest set with the following properties:

1. $\mathrm{At}_\Sigma \subseteq L_\Sigma$.

2. If $\phi_1, \ldots, \phi_m \in L_\Sigma$, $\theta \in \Theta$ and $\alpha(\theta) = m$ then $\theta(\phi_1, \ldots, \phi_m) \in L_\Sigma$.

3. If $\lambda \in \Lambda$, $\phi \in L_\Sigma$, and $x \in \mathrm{Var}$ occurs not in ϕ then $(\lambda x)\phi \in L_\Sigma$.

We extend our notation for propositional languages in the obvious way.

EXAMPLE 7. The language of first-order *Kleene logic* coincides with the language of classical first-order logic and is defined by $\mathbf{L}_{\text{Kle}} = \langle \neg/1, \vee/2, \wedge/2, \forall, \exists \rangle$.

2.2 Semantics

DEFINITION 8. The *set of truth values* N is either the unit interval on the rational numbers, denoted with $[0, 1]$, or it is a finite set of rational numbers of the form $\{0, \frac{1}{n-1}, \ldots, \frac{n-2}{n-1}, 1\}$, where $n \in \mathbb{N}$. In either case $|N|$ denotes the cardinality of N.

DEFINITION 9. If $\mathbf{L}^0 = \langle \Theta, \alpha \rangle$ is a propositional language then we call a pair $\mathbf{A} = \langle N, A \rangle$, where N is a set of truth values and A assigns to each $\theta \in \Theta$ a function $A(\theta) : N^{\alpha(\theta)} \to N$ a *propositional matrix* for $\mathbf{L}^0$.

When no confusion can arise we use the same symbol for θ and $A(\theta)$.

DEFINITION 10. If $\mathbf{L} = \langle \Theta, \Lambda, \alpha \rangle$ is a first-order language then we call a triple $\mathbf{A} = \langle N, A, Q \rangle$, where N is finite, $\langle N, A \rangle$ is a propositional matrix and Q assigns to each $\lambda \in \Lambda$ a function $Q(\lambda) : \mathcal{P}^+(N) \to N$ a *first-order matrix* for $\mathbf{L}$ (we abbreviate $2^N \backslash \{\emptyset\}$ by $\mathcal{P}^+(N)$). $Q(\lambda)$ is called the *distribution function* of the quantifier λ.

This generalized notion of a quantifier is due to Rosser and Turquette [1952, Chapter IV]. The above, simplified, definition is due to Mostowski [1957]. The phrase *distribution quantifier* for referring to quantifiers of this kind was coined by Carnielli [1987].

DEFINITION 11. A pair $\mathcal{L} = \langle \mathbf{L}, \mathbf{A} \rangle$ consisting of a propositional (first-order) language and a propositional (first-order) matrix for it is called *$|N|$-valued propositional (first-order) logic.*

Sometimes a logic is equipped with a non-empty subset D of the set of truth values called the *designated truth values.*

EXAMPLE 12. Let N be arbitrary and $n = |N|$. Then we define the family of n-valued Łukasiewicz logics to be the propositional logics with language $\mathbf{L}_{\text{Łuk}}$, designated truth values $D = \{1\}$, and the matrix defined by:

$$\neg i = 1 - i \tag{1}$$

$$i \supset j = \min\{1, 1 - i + j\} \tag{2}$$

The propositional part of the family of n-valued Kleene logics relative to the language $\mathbf{L}_{Kle}$ is defined by

$$\neg i = 1 - i \tag{3}$$

$$i \vee j = \max\{i, j\} \tag{4}$$

$$i \wedge j = \min\{i, j\} \tag{5}$$

In each case min, max, $+, -$ are interpreted naturally on N. First-order n-valued Kleene logics can be defined by stipulating $Q(\forall) = \min$, $Q(\exists) = \max$.

DEFINITION 13. Let $\mathcal{L}$ be a propositional logic. A *propositional interpretation* over Σ is a function $\mathbf{I} : \Sigma \to N$. $\mathbf{I}$ is extended to arbitrary $\phi \in L^0_\Sigma$ in the usual way:

1. If ϕ is a logical constant then $\mathbf{I}(\phi) = A(\phi)$.
2. If $\phi = \theta(\phi_1, \ldots, \phi_r)$, then $\mathbf{I}(\theta(\phi_1, \ldots, \phi_r)) = A(\theta)(\mathbf{I}(\phi_1), \ldots, \mathbf{I}(\phi_r))$.

DEFINITION 14. Let $\mathcal{L}$ be an n-valued propositional logic. $\mathcal{L}$ is *functionally complete* iff for every $k \geq 0$ and for every function $f : N^k \to N$ there is a formula $\phi_f \in L_\Sigma$ such that for all $\{p_1, \ldots, p_k\} \subseteq \Sigma$ and for all Σ-interpretations $\mathbf{I}$: $\mathbf{I}(\phi_f(p_1, \ldots, p_k)) = f(\mathbf{I}(p_1), \ldots, \mathbf{I}(p_k))$.

DEFINITION 15. Let $\mathcal{L}$ be a first-order logic. A *first-order structure* $\mathbf{M}$ over Σ is a pair $\langle \mathbf{D}, \mathbf{I} \rangle$, where $\mathbf{D}$ is a non-empty set, called the *domain* and an *interpretation* $\mathbf{I}$ that maps $p \in \mathbf{P}_\Sigma$ to functions $\mathbf{I}(p) : D^{\alpha(p)} \to N$ and $f \in \mathbf{F}_\Sigma$ to functions $\mathbf{I}(f) : D^{\alpha(f)} \to D$.

A *variable assignment* is a function $\beta : \text{Var} \to D$. For $d \in D$ the *d-variant at x* of β is defined as

$$\beta^d_x(y) = \begin{cases} d & x = y \\ \beta(y) & x \neq y \end{cases}$$

Given a structure $\mathbf{M}$ and a variable assignment β we define a *valuation function* $v_{\mathbf{M},\beta} : Term \to D$ on arbitrary terms t as usual:

1. If $t \in \text{Var}$ then $v_{\mathbf{M},\beta}(t) = \beta(t)$.
2. If $t = f(t_1, \ldots, t_n)$ then $v_{\mathbf{M},\beta}(t) = \mathbf{I}(f)(v_{\mathbf{M},\beta}(t_1), \ldots, v_{\mathbf{M},\beta}(t_n))$.

$v_{\mathbf{M},\beta}$ is extended on arbitrary formulas ϕ to $v_{\mathbf{M},\beta} : L_\Sigma \to N$ via:

1. If $\phi = p(t_1, \ldots, t_n)$ then $v_{\mathbf{M},\beta}(\phi) = \mathbf{I}(p)(v_{\mathbf{M},\beta}(t_1), \ldots, v_{\mathbf{M},\beta}(t_n))$.
2. If $\phi = \theta(\phi_1, \ldots, \phi_m)$ then $v_{\mathbf{M},\beta}(\phi) = A(\theta)(v_{\mathbf{M},\beta}(\phi_1), \ldots, v_{\mathbf{M},\beta}(\phi_m))$.
3. The *distribution* of ϕ (at x) is $d_{\mathbf{M},\beta,x}(\phi) = \{v_{\mathbf{M},\beta^d_x}(\phi) | d \in D\}$.[3]
   If $\phi = (\lambda x)\psi$ then $v_{\mathbf{M},\beta}(\phi) = Q(\lambda)(d_{\mathbf{M},\beta,x}(\psi))$.

DEFINITION 16. Let $S \subseteq N$. A propositional formula ϕ is said to be *propositionally S-satisfiable* iff there is an interpretation $\mathbf{I}$ such that $\mathbf{I}(\phi) \in S$. We say then that $\mathbf{I}$ is an *S-model* of ϕ. ϕ is an *S-tautology*, in symbols $\models \phi$, iff every interpretation S-satisfies ϕ.

DEFINITION 17. Let $S \subseteq N$. A first-order formula ϕ is said to be *(first-order) S-satisfiable*[4] iff there is a structure $\mathbf{M}$ and a variable assignment β such that $v_{\mathbf{M},\beta}(\phi) \in S$. We say that a structure $\mathbf{M}$ is an *S-model* of ϕ iff $v_{\mathbf{M},\beta}(\phi) \in S$ for all variable assignments β. ϕ is *S-valid*, in symbols $\models \phi$, iff every structure is an S-model of ϕ.

DEFINITION 18. Let $S \subseteq N$. A (propositional or first-order) formula ψ is a *logical S-consequence* of ϕ, in symbols $\phi \models \psi$ iff every S-model of ϕ is an S-model of ψ. ϕ and ψ are *logically S-equivalent* iff each is a logical S-consequence of the other.

If a logic is equipped with designated truth values D, and if D is obvious from the context, then we say *satisfiable* instead of D-satisfiable, *model* instead of D-model etc.

2.3 Representation of Finitely-valued Connectives

In the present subsection we state and prove a fundamental result regarding the representation of finitely-valued connectives or, equivalently, of functions over finite domains. The result is straightforward, but nevertheless it constitutes the theoretical basis of tableaux and Gentzen calculi for finitely-valued propositional logics.

NOTATION 19. We use $\langle i, \phi \rangle$ for the fact that $\mathbf{I}(\phi) = i$ under a fixed, but arbitrary interpretation $\mathbf{I}$.

[3] Note that the value of $d_{\mathbf{M},\beta,x}$ is always a non-empty set.

[4] The concepts of S-satisfiability and S-validity seem to appear in [Kirin, 1966] for the first time.

THEOREM 20. *Let $\phi = \theta(\phi_1, \ldots, \phi_m)$ $(m \geq 1)$ be a formula from an n-valued logic $\mathcal{L}$, and let $i \in N$ be such that i is in the range of $A(\theta)$. Then there are numbers $M_1, M_2 \leq n^m$, index sets $I_1, \ldots, I_{M_1}, J_1, \ldots, J_{M_2} \subseteq \{1, \ldots, m\}$, and truth values i_{rs}, j_{kl} with $1 \leq r \leq M_1$, $1 \leq k \leq M_2$ and $s \in I_r$, $l \in J_k$ such that*

$$\langle i, \phi\rangle \textit{ holds iff } \bigvee_{r=1}^{M_1} \bigwedge_{s \in I_r} \langle i_{rs}, \phi_s\rangle \textit{ holds iff } \bigwedge_{k=1}^{M_2} \bigvee_{l \in J_k} \langle j_{kl}, \phi_l\rangle \textit{ holds.}$$

We call the first a DNF representation *of $\langle i, \phi\rangle$, the second a* CNF representation *of $\langle i, \phi\rangle$.*[5] *The numbers M_1, M_2 are the* size *of the representation.*

Proof sketch. To obtain a DNF representation let

$$A(\theta)^{-1}(i) = \{\langle i_1, \ldots, i_m\rangle \in N^m \mid A(\theta)(i_1, \ldots, i_m) = i\} ,$$

$M_1 = |A(\theta)^{-1}(i)|$, $I_1 = \cdots = I_{M_1} = \{1, \ldots, m\}$, and let $\langle i_{r1}, \ldots, i_{rm}\rangle$ be the r-th tuple in $A(\theta)^{-1}(i)$ according to an arbitrary enumeration. It is easy to check that this indeed is a DNF representation of $\langle i, \phi\rangle$.

The proof of the CNF part is similar. ∎

A full proof of the theorem can be found (or follows from results), for instance, in [Rousseau, 1967; Takahashi, 1967; Carnielli, 1987; Zach, 1993; Baaz and Fermüller, 1995a].

EXAMPLE 21. A CNF representation of three-valued Łukasiewicz implication (2) and the truth value assignment $\frac{1}{2}$ is :

$$\left(\langle\frac{1}{2}, p\rangle \vee \langle\frac{1}{2}, q\rangle\right) \wedge \left(\langle 1, p\rangle \vee \langle 0, q\rangle\right)$$

A DNF representation of the same expression is:

$$\left(\langle\frac{1}{2}, p\rangle \wedge \langle 0, q\rangle\right) \vee \left(\langle 1, p\rangle \wedge \langle\frac{1}{2}, q\rangle\right)$$

3 EARLY WORK AND PRECURSORS

The oldest proof theoretic characterizations of many-valued logic are, of course, the Hilbert style systems given by Post and Łukasiewicz. They can be found,

[5] Some authors [Rosser and Turquette, 1952; Zach, 1993; Baaz and Fermüller, 1995a] prefer the expression *i-th partial normal form* which we do not adopt, because it does not give an indication of whether it is based on a CNF or DNF. Indeed, Rosser and Turquette worked with CNF representations only.

for example, in [Rescher, 1969; Bolc and Borowik, 1992]. But after Gentzen had introduced his sequent calculi, many-valued versions of them appeared in due course. We would like to mention the work of Schröter [1955], Kirin [1963; 1966], Rousseau [1967; 1970] and Takahashi [1967; 1970]. The papers [Rousseau, 1967; Takahashi, 1967] already consider finitely-valued logics with generic connectives; Takahashi [1970] even considers infinitely-valued logics with continous distribution quantifiers, where the truth value set may be an arbitrary compact Hausdorff space. We describe briefly the most important variants of sequent systems for many-valued logics.

3.1 Sequent Systems with Meta Connectives

The first question that arises in connection with sequents for many-valued logics is how to associate subformulas with the truth values they are supposed to take on. Recall that the meaning of a classical sequent of the form $\Gamma \Rightarrow \Delta$ is: one of the formulas in Γ is false or one of the formulas in Δ is true (or equivalently: if all the formulas in Γ are true, then at least one of the formulas in Δ is true). If we introduce *signs* or *meta connectives* that assert the truth value of a formula we can rewrite this sequent into

$$\langle 0, \gamma_1 \rangle \vee \cdots \vee \langle 0, \gamma_n \rangle \vee \langle 1, \delta_1 \rangle \vee \cdots \vee \langle 1, \delta_m \rangle \ ,$$

where $\Gamma = \{\gamma_1, \ldots, \gamma_n\}$, $\Delta = \{\delta_1, \ldots, \delta_m\}$ and the meaning of a pair $\langle i, \phi \rangle$ is as defined in Section 2.3. A more compact representation of the above sequent would be [Takahashi, 1967]

$$(\{0\} \times \Gamma) \cup (\{1\} \times \Delta) \ ,$$

where Cartesian products are used to abbreviate sets of truth value assertions, and a sequent is a set of such assertions which must be thought as a disjunction of its elements.

In order to arrive at inference rules one uses Theorem 20 in Section 2.3 to characterize a truth value assertion to a complex many-valued formula as a CNF representation over truth value assertions to its direct subformulas. From the CNF representation of a truth value assignment to a formula with leading connective θ one can immediately derive an introduction rule by noting that the set $\overline{D_k}$ of disjuncts occurring in each disjunction $D_k = \bigvee_{l \in J_k} \langle j_{kl}, \phi_l \rangle$ is (part of) a many-valued sequent

$$\frac{\Gamma \cup \overline{D_1} \quad \cdots \quad \Gamma \cup \overline{D_M}}{\Gamma} \ ,$$

whenever $\bigwedge_{k=1}^{M} D_k$ is a CNF representation of some $\langle i, \theta(\phi_1, \ldots, \phi_m) \rangle \in \Gamma$. From Theorem 20 in Section 2.3 it is obvious that this rule is invertible, that is, the conjunction of its premisses holds if and only if its conclusion holds.

EXAMPLE 22. The introduction rule for three-valued Łukasiewicz implication and the truth value assignment $\frac{1}{2}$ is (cf. Example 21 in Section 2.3):

$$\frac{\Gamma \cup \{\langle \frac{1}{2}, p\rangle, \langle \frac{1}{2}, q\rangle\} \quad \Gamma \cup \{\langle 1, p\rangle, \langle 0, q\rangle\}}{\Gamma \cup \{\langle \frac{1}{2}, p \supset q\rangle\}}$$

Like in two-valued logic one can design *dual sequent systems* in which sequents correspond to conjunctions, axioms to contradictory formulas, and introduction rules are obtained from DNF representations of truth value assignments. The details are completely straightforward and are left as an exercise to the reader.

There is another usage of introduction rules based on DNF representations *not* within dual, but within standard sequent systems: if $\bigvee_{r=1}^{M} C_r$ is a DNF representation of $\langle i, \theta(\phi_1, \ldots, \phi_m)\rangle \in \Gamma$, then *for each* $C_r = \bigwedge_{s \in I_r} \langle i_{rs}, \phi_s\rangle$ there is a sound introduction rule

$$\frac{\Gamma \cup \{\langle i_{rs_1}, \phi_{s_1}\rangle\} \quad \cdots \quad \Gamma \cup \{\langle i_{rs_{|I_r|}}, \phi_{s_{|I_r|}}\rangle\}}{\Gamma}$$

The consequence is that one obtains many introduction rules for each connective and the rules are no longer invertible. But there are advantages: certain infinitely-valued logics still have finite proof trees [Takahashi, 1970], although the calculus contains infinitely many introduction rules and sequents in general may contain infinitely many truth assertions.

EXAMPLE 23. The DNF based introduction rules for three-valued Łukasiewicz implication and the truth value assignment $\frac{1}{2}$ are (cf. Example 21 in Section 2.3):

$$\frac{\Gamma \cup \{\langle \frac{1}{2}, p\rangle\} \quad \Gamma \cup \{\langle 0, q\rangle\}}{\Gamma \cup \{\langle \frac{1}{2}, p \supset q\rangle\}} \qquad \frac{\Gamma \cup \{\langle 1, p\rangle\} \quad \Gamma \cup \{\langle \frac{1}{2}, q\rangle\}}{\Gamma \cup \{\langle \frac{1}{2}, p \supset q\rangle\}}$$

Finally we mention that there is a version of sequent calculi for many-valued logics that employs n-ary sequents rather than explicit truth value assignments [Rousseau, 1967]. First observe that, clearly, each sequent can be written in the form

$$(\{0\} \times \Gamma_0) \cup (\{\tfrac{1}{n-1}\} \times \Gamma_{\frac{1}{n-1}}) \cup \cdots \cup (\{1\} \times \Gamma_1)$$

for suitable sets of formulas $\Gamma_0, \ldots, \Gamma_1$ (where some of the Γ_i may be empty). Instead we can use Rousseau's [1967] notation

$$\Gamma_0 \mid \Gamma_{\frac{1}{n-1}} \mid \cdots \mid \Gamma_1$$

in which the i-th slot contains the formulas being asserted truth value i. We stress that all those notations for many-valued sequents are slight variants of each other.

Note that if logical constants are a part of the logical language, then further axioms of the form

$$\mid \mid \cdots \mid \{c\} \mid \cdots \mid \mid$$

are needed, where c constantly evaluates to i and it occurs in the i-th slot.

In any notation axioms and weakening rules are completely straightforward. Axiomatic sequents in Rousseau's notation, for example, are of the form

$$\{\phi\} \mid \{\phi\} \mid \cdots \mid \{\phi\}$$

while cut rules[6] (in Takahashi's notation) would be as follows:

$$\text{(6)} \qquad \frac{\Gamma \cup \{\langle i, \phi \rangle\} \quad \Gamma \cup \{\langle j, \phi \rangle\}}{\Gamma} \qquad \text{for } i \neq j$$

The translation of axioms into a different notation, the DNF version of axioms for 0-ary connectives, and the construction of a suitable weakening rule is left as an exercise to the reader.

An immediate consequence of Theorem 20 in Section 2.3 is that (just as in classical logic) in any finitely-valued logic one can give a cut-free and complete sequent system *without any elimination rules* that characterizes its tautologies, because *any* combination of connective and truth value is expressible.

Quantifier rules will not be treated in this section. Later, we will give quantifier rules for tableaux from which sequent rules can easily be derived.

Finally, we remark that, exactly as in classical sequent calculi, one has the choice between separate weakening rules versus combining weakening and introduction rules, between representing sequents as sets versus representing them as lists (contraction and exchange rules are needed then), and so on.

All approaches mentioned so far employ meta connectives in one way or another to characterize more than two truth values. On other words, information that is present in the semantics, namely the number of truth values and, through DNF/CNF representation of connectives, also the truth tables of connectives are explicitly introduced into the proof theory.

[6] Cut elimination is possible just as in classical logic [Takahashi, 1970].

3.2 Other Calculi

From the point of view of traditional proof theory it is more natural, however, to characterize logics by the means that are already present in classical sequent calculi. Obviously, this cannot be achieved for arbitrary many-valued logics, but there exist partial results in this direction that include traditional sequent systems for several well-known many-valued logics. Those systems typically work by weakening or modifying the structural rules [Avron, 1991; Hösli, 1993] (contraction, for instance, is not valid in Łukasiewicz logic) or by using elimination rules which cannot be replaced by introduction rules [Hoogewijs and Elnadi, 1994]. Such sequent systems cannot be transformed into tableau systems which is why we do not discuss them any further here, but refer to the cited papers.

There is an intermediate way between a purely proof theoretical characterization and the explicit use of meta connectives: in functionally complete logics for each truth value $i \in N$ there is a formula ϕ_i such that $\mathbf{I}(\phi_i) = i$ for all interpretations $\mathbf{I}$. Such formulas can substitute meta connectives. If, in addition, a suitable connective that expresses equality on N is defined, then one can, in principle, design complete calculi without the need to introduce meta connectives. Kirin [1963; 1966], Saloni [1972], and Orłowska [1985] give treatments of variants of finitely-valued Post logic in this spirit.[7]

Even for logics that are *not* functionally complete (such as the family of Kleene logics) refutation based proof systems such as resolution can be obtained by working with so-called *verifiers* instead of working with formulas that correspond to truth values; verifiers essentially are minimal sets of formulas that suffice to express inconsistency. In [Stachniak, 1988; Stachniak and O'Hearn, 1990] a version of non-clausal resolution [Murray, 1982] is given that does not use meta connectives. There seems, however, nothing to be gained from not using meta connectives in a tableau framework and we do not pursue this topic further.

4 SIGNED TABLEAUX

4.1 Simple Tableau Systems

In D'Agostino's Chapter we saw that classical (signed) tableau systems correspond in a one-to-one manner to cut-free dual (DNF based) sequent systems without elimination rules. This correspondence extends in a straightforward manner to many-valued dual sequent systems and many-valued signed tableaux, respectively, if we associate each truth value assignment $\langle i, \phi \rangle$ that occurs in a dual sequent with a *signed formula* $i\ \phi$.

[7] The syntax of Post logics can be defined with $\langle \text{shift}/1, \vee/2 \rangle$ designated truth values $\{m, \ldots, 1\}$ for some $m > 0$ and matrix $i \vee j = \max\{i, j\}$, $\text{shift}(i) = \min\{1, i + \frac{1}{n-1}\}$. Post logics are functionally complete and, therefore, are often defined with different sets of connectives.

EXAMPLE 24. The sequent rules given in Example 23 in Section 3.1 translate into the tableau rule shown below on the left. On the right we give the rule for sign 0.

(7)
$$\begin{array}{c|c} \multicolumn{2}{c}{\frac{1}{2}\,(\phi \supset \psi)} \\ \hline \frac{1}{2}\,\phi & 1\,\phi \\ 0\,\psi & \frac{1}{2}\,\psi \end{array} \qquad\qquad \begin{array}{c} 0\,(\phi \supset \psi) \\ \hline 1\,\phi \\ 0\,\psi \end{array}$$

Each DNF based introduction rule corresponds to one rule extension. The side formulas Γ of the sequent rule correspond to the current tableau branch on which the rule is applied.

DEFINITION 25. If $\bigvee_{r=1}^{M} C_r$ is a DNF representation of $\langle i, \theta(\phi_1, \ldots, \phi_m)\rangle$ $(m \geq 1)$ with $C_r = \bigwedge_{s \in I_r} \langle i_{rs}, \phi_s\rangle$ $(1 \leq r \leq M)$, then a *many-valued tableau rule* for $i\,\theta(\phi_1, \ldots, \phi_m)$ is defined as

$$\begin{array}{c|c|c} \multicolumn{3}{c}{i\,\theta(\phi_1, \ldots, \phi_m)} \\ \hline \overline{C_1} & \cdots & \overline{C_M} \end{array},$$

where $\overline{C_r} = \{i_{rs}\,\phi_s \mid s \in I_r\}$.

The notion of a tableau for a signed formula $i\,\phi$ is defined exactly as in the classical case, but with respect to many-valued rules. We must, however, change the notion of closure:

DEFINITION 26. Let $\mathbf{T}$ be a many-valued tableau and B one of its branches. B is *closed* iff one of the following conditions holds:

1. There are signed formulas $i\,\phi$, $j\,\phi$ on B such that $i \neq j$.
2. There is a signed formula $i\,\theta(\phi_1, \ldots, \phi_m)$ $(m \geq 0)$ on B such that i does not occur in the range of $A(\theta)$.

$\mathbf{T}$ is closed iff each of its branches is closed.

THEOREM 27 ([Surma, 1974]). *Let ϕ be a formula in a finitely-valued logic and let $S \subseteq N$. Then ϕ is S-valid iff there exist finite, closed many-valued tableaux for each $i\,\phi$ with $i \in N \backslash S$.*

EXAMPLE 28. In Łukasiewicz logic a formula is valid iff it is $\{1\}$-valid. We prove that $p \supset (q \supset p)$ is a valid formula in three-valued Łukasiewicz logic. We must construct a finite, many-valued tableau for each of $0\,p \supset (q \supset p)$ and $\frac{1}{2}\,p \supset (q \supset p)$ using the rules from Example 24. Such tableaux are depicted in Figure 1.

Many-valued tableau systems as given in the present subsection are of considerable generality (they work for all finitely-valued first-order logics with distribution quantifiers), but they have a number of drawbacks:

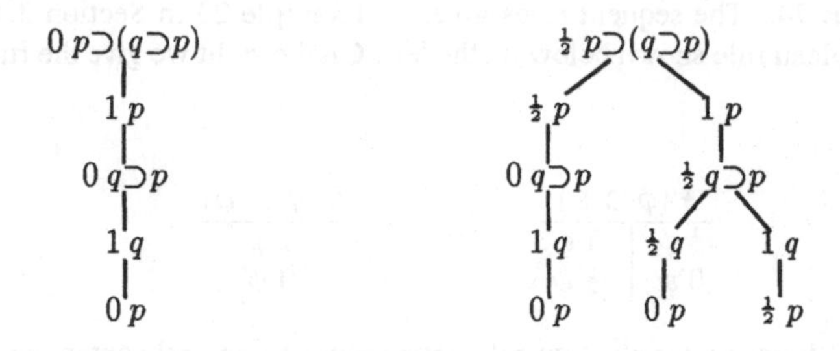

Figure 1. Tableau proof trees of Example 28

1. The tableaux contain a large amount of redundancy. One observes that the tree on the left is isomorphic to the tree that results when $\supset$ is interpreted as classical implication. Moreover, modulo the signs it occurs as a subtree of the tree on the right.

2. In general, several tableaux must be built for just one proof (the number of tableaux is the cardinality of the set of non-designated truth values).

3. Infinitely-valued logics cannot be handled effectively with this approach as infinitary branching trees would result.

Historical Note In two-valued logic, signed tableau systems have a long tradition, see the introductory article by Fitting in this handbook.

Signed tableaux for many-valued logics were first considered in a paper by Suchón [1974], where a signed tableau system for the family of finitely-valued Łukasiewicz logics is given. Generic signed tableau systems that correspond to dual sequent systems as sketched above first were described by Surma [1974]; other sources are [Carnielli, 1985; Carnielli, 1987; Carnielli, 1991; Baaz and Zach, 1992; Zabel, 1993; Bloesch, 1994]. While the work of Carnielli generalizes the approach of Surma [1974] to finitely-valued first-order logic with distribution quantifiers and Zabel considers also infinitely-valued logics, the other papers apparently were written independently.[8]

[8]It must also be noted that, with the exception of Zabel [1993] none of the papers is fully aware of the earlier work done for sequent systems mentioned in the previous section and they give their own soundness and completeness proofs. This is a pity, because the key ingredients of many-valued tableau systems are already inherently present in sequent systems, for example, signs, n-ary trees, and the DNF representation of many-valued connectives. From the correspondence between dual sequent systems and tableaux it is obvious that the latter are sound and complete. None of the cited papers developed computationally sophisticated representations or refinements; instead the basics of many-valued tableaux were re-invented time and again.

4.2 *Sets as Signs*

It turns out that a simple device, which looks like a mere representational optimization at first sight, has far reaching consequences for the efficiency and generality of tableau methods for many-valued logic.

Consider, for example, Łukasiewicz Logic: if we want to prove with the tableau method that a formula ϕ is valid in this logic, then we must refute the following statement:

$$\left.\begin{array}{c}\phi \text{ can evaluate to } 0\\ \text{or}\\ \dots\\ \text{or}\\ \phi \text{ can evaluate to } \frac{n-2}{n-1}\end{array}\right\} \text{ for a suitable interpretation } \mathbf{I}$$

corresponding to the construction of $n-1$ closed tableaux for $0\,\phi, \dots, \frac{n-2}{n-1}\,\phi$, respectively.

As an abbreviation for the disjunctive statement above one can introduce the following notation:

$$\{0, \dots, \tfrac{n-2}{n-1}\}\,\phi \tag{8}$$

Rousseau [1967] and Takahashi [1967] in fact used a very similar notation to write down their sequent rules. The crucial observation, however, is that we can use truth value sets as signs systematically to optimize many-valued tableau systems.

This immediately eliminates the need to build more than one tableau in order to prove a many-valued formula as valid by simply starting with a signed formula like (8).

But tableau rules become considerably more concise as well. In three-valued Łukasiewicz logic, , for example, we start a tableau with $\{0, \frac{1}{2}\}\,\phi$ which characterizes non-designatedness of ϕ. Now assume that ϕ has $\supset$ as its top-level connective. That is we ask: under which conditions on ϕ_1, ϕ_2 does $\mathbf{I}(\phi_1 \supset \phi_2) \neq 1$ hold?

Graphically, this can be described as characterizing the shaded entries in the truth table of $\supset$:

(9)

$\supset$	0	$\frac{1}{2}$	1
0	1	1	1
$\frac{1}{2}$	$\frac{1}{2}$	1	1
1	0	$\frac{1}{2}$	1

Naïvely, this can be achieved by disjunctively combining the two rules that result from a DNF representation of $0\,\phi_1 \supset \phi_2$ and $\frac{1}{2}\,\phi_1 \supset \phi_2$, respectively, as they are given in Example 24 in Section 4.1.

If, however, one allows formulas that are signed with truth values sets to appear in the conclusion as well, then a single and smaller rule is sufficient:

(10)
$$\frac{\{0,\frac{1}{2}\}\,\phi_1 \supset \phi_2}{\begin{array}{c|c} \{\frac{1}{2},1\}\,\phi_1 & \{1\}\,\phi_1 \\ \{0\}\,\phi_2 & \{0,\frac{1}{2}\}\,\phi_2 \end{array}}$$

Each rule extension covers two of the shaded entries in the truth table.

EXAMPLE 29. We redo Example 28 using rule (10):

$$\{0,\tfrac{1}{2}\}\; p\supset(q\supset p)$$

Left branch: $\{\frac{1}{2},1\}\;p$ — $\{0\}\;q\supset p$ — $\{1\}\;q$ — $\{0\}\;p$

Right branch: $\{1\}\;p$ — $\{0,\frac{1}{2}\}\;q\supset p$, which splits into:
- $\{\frac{1}{2},1\}\;q$ — $\{0\}\;p$
- $\{1\}\;q$ — $\{0,\frac{1}{2}\}\;p$

One of the two tableaux needed in Example 28 of Section 4.1 is represented implicitly in the single tableau shown above.

The previous example suggests that sets of truth values as signs can be seen as a kind of *semantic structure sharing*, schematically depicted in Figure 2. Another explanation of their representational power is that a DNF sets-as-signs rule has the metalogical structure $\bigvee\bigwedge\bigvee i\;\phi$, where i is a truth value that is it contains an additional level of nesting of connectives as compared to a simple rule. See also Section 7.2, where DNF sets-as-signs rules are interpreted as OR-AND-OR circuits.

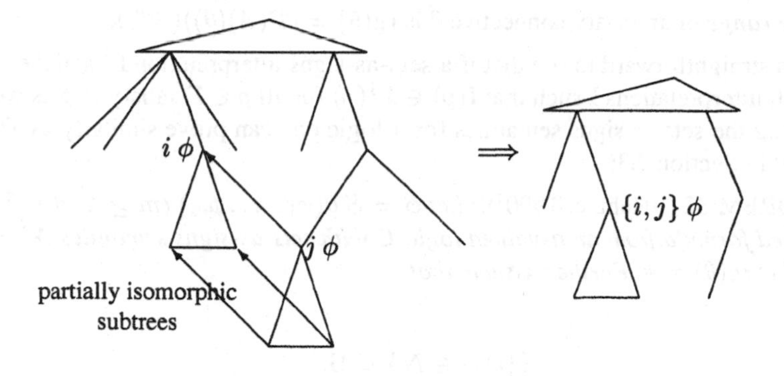

Figure 2. Schema of semantic structure sharing using sets-as-signs

Now we restate the above considerations more formally:

DEFINITION 30. Let $\mathbf{S}\subseteq\mathcal{P}^+(N)$ be a family of truth value sets. Let ϕ be a formula and $S\in\mathbf{S}$. Then we call the expression $S\ \phi$ *signed formula.* If p is an atomic formula, then $S\ p$ is a *signed literal.*

Propositional formulas signed with truth value sets can be directly given a semantics as follows: instead of the matrix

$$\mathbf{A} = \langle N, A\rangle$$

in order to define a many-valued logic $\mathcal{L}$ we take the *power algebra* induced by some $\mathbf{S}$ (cf. Definition 11 in Section 2.2):

$$\mathbf{A}^{\mathbf{S}} = \langle \mathbf{S}, \mathcal{P}(A)\rangle,$$

where

$$\mathcal{P}(A)(\theta)\colon \begin{cases} \mathcal{P}^+(N)^{\alpha(\theta)} \to \mathcal{P}^+(N) \\ \langle S_1,\ldots,S_{\alpha(\theta)}\rangle \mapsto \bigcup\{A(\theta)(i_1,\ldots,i_{\alpha(\theta)}) \mid i_j\in S_j,\ 1\le j\le \alpha(\theta)\} \end{cases}$$

Let us call $\mathbf{A}^{\mathbf{S}}$ a *sets-as-signs semantics* of $\mathcal{L}$ (wrt $\mathbf{S}$).

DEFINITION 31. Let $\mathcal{L}$ be a propositional logic defined wrt a sets-as-signs semantics $\mathbf{A}^{\mathbf{S}}$. A *propositional sets-as-signs interpretation*[9] over Σ is a function $\mathbf{I}^{\mathbf{S}}:\Sigma\to\mathbf{S}$. $\mathbf{I}^{\mathbf{S}}$ is extended on arbitrary $\phi\in L^0_\Sigma$ just as $\mathbf{I}$, but wrt $\mathbf{A}^{\mathbf{S}}$.

A signed formula $S\ \phi$ with $S\in\mathbf{S}$ is **satisfiable** iff $\mathbf{I}^{\mathbf{S}}(\phi)\subseteq S$ for some $\mathbf{I}^{\mathbf{S}}$ iff $\mathbf{I}(\phi)\in S$ for some $\mathbf{I}$.

[9] This is called *extended interpretation* by Lu [1996].

The *range* of an m-ary connective θ is $\text{rg}(\theta) = (\mathcal{P}(A)(\theta))(N^m)$.

It is straightforward to see that if a sets-as-signs interpretation $\mathbf{I}^{\mathbf{S}}$ satisfies $S\ \phi$ then all interpretations $\mathbf{I}$ such that $\mathbf{I}(p) \in \mathbf{I}^{\mathbf{S}}(p)$ for all $p \in \Sigma$ satisfy $S\ \phi$ as well.

Using the sets-as-signs semantics for a logic one can prove similarly as Theorem 20 in Section 2.3:

THEOREM 32 ([Hähnle, 1990]). *Let $\phi = S\ \theta(\phi_1, \ldots, \phi_m)$ $(m \geq 1, S \in \mathbf{S})$ be a signed formula from an n-valued logic $\mathcal{L}$ with sets-as-signs semantics $\mathbf{A}^{\mathbf{S}}$ such that $S \cap rg(\theta) \neq \emptyset$. Further assume that*

$$\{\{i\} \mid i \in N\} \subseteq \mathbf{S} \tag{11}$$

holds. Then there are numbers $M_1, M_2 \leq n^m$, index sets $I_1, \ldots, I_{M_1}, J_1, \ldots, J_{M_2} \subseteq \{1, \ldots, m\}$, and signs $S_{rs}, S_{kl} \in \mathbf{S}$ with $1 \leq r \leq M_1$, $1 \leq k \leq M_2$ and $s \in I_r$, $l \in J_k$ such that

ϕ is satisfiable iff $\bigvee_{r=1}^{M_1} \bigwedge_{s \in I_r} (S_{rs}\ \phi_s$ is satisfiable$)$

iff $\bigwedge_{k=1}^{M_2} \bigvee_{l \in J_k} (S_{kl}\ \phi_l$ is satisfiable$)$,

where $\bigvee$ and $\bigwedge$ denote classical meta disjunction and conjunction, respectively, and have their usual meaning. We call the first expression a sets-as-signs DNF representation *of ϕ, the second a* sets-as-signs CNF representation *of ϕ.*

In the previous theorem the condition '$S \in \mathbf{S}$' can be relaxed to '$S = S_1 \cup \cdots \cup S_t$ for some $\{S_1, \ldots, S_t\} \subseteq \mathbf{S}$' by applying the theorem separately to each S_i. If quantifiers with arbitrary signs are to be processed this in fact becomes necessary. With the relaxed condition it is trivially possible to give representations of arbitrary $S \subseteq N$ by (11).

Sets-as-signs representations of many-valued functions can either be computed manually as exemplified in (10) or the problem can be reformulated as a many-valued function minimization problem and can be computed with methods from that field: for instance, an optimal sets-as-signs DNF (CNF) representation of a many-valued function can be obtained by computing a minimal covering set of prime implicates (implicants) over signed clauses of that function [Hähnle, 1994, Section 4.5]. Further connections between tableau rules and logic design are discussed in Sections 7.1 and 7.2.

Of course, not every set $\mathbf{S}$ is suitable to obtain optimal representations. Making $\mathbf{S}$ large usually allows for short representations. Neither condition (11) nor that $\mathbf{A}^{\mathbf{S}}$ is an algebra is necessary to obtain representations for all $S \in \mathbf{S}$ (the latter is only needed to have a well-defined semantics). On the other hand, taking all possible $2^{|N|} - 1$ signs is usually not feasible. Hähnle [1994] gives sufficient conditions on $\mathbf{S}$ guaranteeing that representations for all $S \in \mathbf{S}$ can be obtained. One example for a weaker condition by which (11) can be replaced is the following:

(12) For all $i \in N$ there are $S_1, \ldots, S_k \in \mathbf{S}$ such that $\bigcap_{j=1}^{k} S_j = \{i\}$

In the following we will only consider such sets of signs that satisfy this condition.

DEFINITION 33. If $\bigvee_{r=1}^{M} \bigwedge_{s \in I_r} S_{rs}\, \phi_s$ is a sets-as-signs DNF representation of $\phi = S\, \theta(\phi_1, \ldots, \phi_m)$ $(m \geq 1)$, then a *many-valued sets-as-signs tableau rule* for ϕ is defined as

$$\frac{S\, \theta(\phi_1, \ldots, \phi_m)}{\overline{C_1} \quad \Big| \quad \cdots \quad \Big| \quad \overline{C_M}},$$

where $\overline{C_r} = \{S_{rs}\, \phi_s \mid s \in I_r\}$.

Again, the notion of a tableau for a signed formula $S\, \phi$ is defined exactly as in the classical case, but with respect to many-valued sets-as-signs rules. Some care must be spent on the definition of closure:

DEFINITION 34. Let $\mathbf{T}$ be a many-valued sets-as-signs tableau and B one of its branches. B is *closed* iff one of the following conditions holds:

1. There are signed formulas $S_1\, \phi, \ldots, S_r\, \phi$ on B such that $\bigcap_{i=1}^{r} S_i = \emptyset$. In this case we say that (the set of signed formulas on) B is *inconsistent*.[10]

2. There is a signed formula $S\, \theta(\phi_1, \ldots, \phi_m)$ $(m \geq 0)$ on B such that $S \cap \text{rg}(\theta) = \emptyset$.

$\mathbf{T}$ is closed iff each of its branches is closed. A branch that is not closed is called *open*.

THEOREM 35 ([Hähnle, 1990]). *Let ϕ be a formula in a finitely-valued logic and let $\emptyset \neq S \subseteq N$. Then ϕ is S-valid iff there exists a finite, closed many-valued sets-as-signs tableau for $(N \backslash S)\, \phi$.*

Proof sketch. Soundness is completely straightforward by repeated application of Theorem 32 in Section 4.2. For completeness, assume ϕ is S-valid and consider an open branch B in a many-valued sets-as-signs tableau for $(N \backslash S)\, \phi$ in which all possible rules have been applied.

Now we need the many-valued sets-as-signs version of a well-known concept from classical logic:

DEFINITION 36. A *many-valued sets-as-signs Hintikka set* is a set of signed formulas $\mathbf{H}$ such that:

[10]Just as in classical logic it is sufficient to restrict the definition to the case when ϕ is atomic. Accordingly one says that B is *atomically inconsistent*.

1. $\mathbf{H}$ is open in the sense of Definition 34 (where branches are identified with sets of signed formulas).

2. If $\phi = S\,\theta(\phi_1, \ldots, \phi_m) \in \mathbf{H}$ and $\bigvee_{r=1}^{M} C_r$ is a sets-as-signs DNF representation of ϕ then $\overline{C_r} \subseteq \mathbf{H}$ for at least one $1 \leq r \leq M$.

THEOREM 37 (Hintikka Lemma). *Every many-valued sets-as-signs Hintikka set* $\mathbf{H}$ *has a model.*

Proof. From $\mathbf{H}$ we construct directly a satisfying sets-as-signs interpretation over $\mathbf{A}^{\mathcal{P}^+(N)}$. For each $p \in \Sigma$ let $\mathbf{H}(p)$ be the set of signed literals with atom p in $\mathbf{H}$.

$$\mathbf{I^S}(p) = \begin{cases} \bigcap_{Sp \in \mathbf{H}(p)} S & \mathbf{H}(p) \neq \emptyset \\ N & \mathbf{H}(p) = \emptyset \end{cases}$$

$\mathbf{I^S}$ is well-defined as $\mathbf{H}$ is atomically consistent. By structural induction on the depth of formulas one proves that $\mathbf{I^S}$ indeed satisfies $\mathbf{H}$. The literal case follows directly from the definition of $\mathbf{I^S}$. Assume $\mathbf{I^S}$ satisfies smaller formulas than the complex formula $S\,\psi \in \mathbf{H}$. By Definitions 36(1), 34(2) and Theorem 32 of Section 4.2 $S\,\psi$ has a sets-as-signs DNF representation $\bigvee_{r=1}^{M} C_r$. Hence, $\overline{C_r} \subseteq \mathbf{H}$ for some $1 \leq r \leq M$ by Definition 36(2). By the induction hypothesis, $\mathbf{I^S}$ satisfies $\overline{C_r}$ and again by Theorem 32 of Section 4.2, $\mathbf{I^S}$ satisfies $S\,\psi$. ■

We proceed with the proof of Theorem 35. Obviously, B is a Hintikka set (we may assume $S \neq N$ for otherwise the theorem holds trivially), hence it has a model. This means that $\mathbf{I}(\phi) \in (N \backslash S)$ for some $\mathbf{I}$ contradicting S-validity of ϕ. ■

The use of sets-as-signs rules has not only the structure sharing effect among refutations of different truth value assertions as shown in Example 29, a sets-as-signs rule can even possess a lower branching factor than each of the single-value rules corresponding to its sign. To see this, consider the connective $\circ$, whose truth table is shown in Figure 3. The many-valued tableau rules for 0 and $\frac{1}{2}$, and the sets-as-signs rule for $\{0, \frac{1}{2}\}$ are depicted as well. The latter rule has only one extension, whereas each of the other rules has two extensions. Connectives can occur nested in a formula, thus for each $n \geq 3$ there exist classes of formulas Φ_m and n-valued logics for which the smallest sets-as-signs tableau is exponentially shorter than the smallest many-valued tableau (with respect to the number of nodes).

Even if the number of needed signs may be large (up to $2^n - 1$) it still pays to use sets-as-signs, because the information stored in the signs has to be stored in the tableau anyway. To see this, consider a signed formula $S\,\phi$ with $S \subseteq N$. Now, if S is not available then one has to find signs $S_1, \ldots, S_k$ such that $S = S_1 \cup \cdots \cup S_k$ and instead of $S\,\phi$ one has k branches or k separate tableaux containing $S_1\,\phi, \ldots, S_k\,\phi$, causing $k-1$ additional copies of ϕ. Hence, if $2^n - 1$ different signs are needed in order to obtain minimal representations in a given logic this simply reflects the

$\circ$	0	$\frac{1}{2}$	1
0	1	1	1
$\frac{1}{2}$	$\frac{1}{2}$	0	1
1	0	$\frac{1}{2}$	1

$$\frac{\frac{1}{2}\,(\phi \circ \psi)}{\begin{array}{c|c} \frac{1}{2}\,\phi & 1\,\phi \\ 0\,\psi & \frac{1}{2}\,\psi \end{array}} \qquad \frac{0\,(\phi \circ \psi)}{\begin{array}{c|c} 1\,\phi & \frac{1}{2}\,\phi \\ 0\,\psi & \frac{1}{2}\,\psi \end{array}} \qquad \frac{\{0, \frac{1}{2}\}\,(\phi \circ \psi)}{\begin{array}{c} \{\frac{1}{2}, 1\}\,\phi \\ \{0, \frac{1}{2}\}\,\psi \end{array}}$$

Figure 3. Truth table and some tableau rules for $\circ$

complexity of the given logic and with using less signs the size of proofs will be even larger.

One could also generate signs and tableau rules 'on demand', because — depending on the query — only a small portion of all signs might be needed.

Tautological formulas Assume $S = \mathrm{rg}(\theta)$. Obviously, each signed formula of the form $S_0\,\theta(\phi_1, \ldots, \phi_r)$ such that $S \subset S_0$ is a tautology. As it cannot contribute to the closure of any branch, such formulas may be deleted whenever they occur in a tableau.

Complexity of Tableau Rules The size of a tableau is largely determined by the maximal size of CNF/DNF representations (the branching factor of tableau rules). In [Hähnle, 1994] (for arity 2), [Rousseau, 1967; Rousseau, 1970; Zach, 1993; Baaz and Fermüller, 1995a] (for general arity r) several results about worst-case complexity of rules are stated and proved which are collected in the following table:

$n = \|N\|$, arity r	DNF	CNF
singleton signs	n^r	n^{r-1}
sets-as-signs	n^{r-1}	n^{r-1}

All bounds are sharp. The same connective $\oplus$ can serve for all cases:

$$\oplus(i_1, \ldots, i_r) = \frac{((i_1 + \cdots + i_r) \cdot (n-1)) \bmod n}{n-1}$$

Historical note Sets-as-signs were first introduced by Hähnle [1990], a fuller account can be found in [Hähnle, 1994]. Similar, though slightly less general ideas were independently expressed by Doherty [1990; 1991] and Murray and Rosenthal [1991]. Signs that implicitly correspond to sets of truth values also appear in [Suchoń, 1974; Fitting, 1991; Zabel, 1993; Nait Abdallah, 1995]. Rousseau [1970] proved the CNF-case of Theorem 32 in Section 4.2 when only signs of the form $\{0, \ldots, i\}$ or $\{j, \ldots, 1\}$ occur (see also Section 6.1). In the same paper also an intuitionistic version of this result is proved.

4.3 First-order Logic

DEFINITION 38. A signed first-order formula $S\ \phi$ is satisfied by a structure $\mathbf{M}$ and variable assignment β iff ϕ is S-satisfied by $\mathbf{M}$ and β. $\mathbf{M}$ is a model of $S\ \phi$ iff it is an S-model of ϕ.

NOTATION 39. The notation $(\lambda x)\phi(x)$ emphasizes that the body of the quantified formula possibly contains free occurrences of the quantified variable. If $\phi(x)$ is the body of a quantified variable and t is any term, then the expression $\phi(t)$ denotes the result of replacing all occurences of x in $\phi(x)$ with t.

Recall that the semantics of a first-order quantifier λ in many-valued logic is defined via a distribution function $Q(\lambda) : \mathcal{P}^+(N) \rightarrow N$ which maps each distribution of truth values to a truth value. Just as in the propositional case, where a DNF representation of a signed formula $i\ \theta(\phi_1, \ldots, \phi_m)$ or $S\ \theta(\phi_1, \ldots, \phi_m)$ led to a tableau rule for the given sign and connective, we may use a suitable DNF representation of the distribution function of a quantifier to obtain a tableau rule. Informally, we must characterize the distributions that are mapped to one of the truth values that occur in the sign of the premiss.

PROPOSITION 40. *Let* $(Q(\lambda))^{-1}(S) = \{\emptyset \neq I \subseteq N \mid Q(\lambda)(I) \in S\}$. *Then a signed quantified formula* $S(\lambda x)\phi(x)$ *is satisfiable iff there is an* $I \in (Q(\lambda))^{-1}(S)$ *such that:*

1. *for each* $i \in I$*, there is a ground term* c_i *not occurring in* $\phi(x)$ *such that all* $\{i\}\ \phi(c_i)$ *and*
2. $I\ \phi(t)$ *for all ground terms* t

are simultaneously satisfiable.

Proof. Let $\mathbf{M}$, β be such that $Q(\lambda)(d_{\mathbf{M},\beta,x}(\phi(x))) = v_{\mathbf{M},\beta}((\lambda x)\phi(x)) \in S$. Let $I = d_{\mathbf{M},\beta,x}(\phi(x))$. By definition of $d_{\mathbf{M},\beta,x}$, for any $i \in I$ there is a $d_i \in D$ with $v_{\mathbf{M},\beta_x^{d_i}}(\phi(x)) = i$. Let $\mathbf{M}'$ be exactly as $\mathbf{M}$, but $v_{\mathbf{M}',\beta}(c_i) = d_i$. Then for all $i \in I$: $v_{\mathbf{M}',\beta}(\phi(c_i)) = v_{\mathbf{M}',\beta_x^{d_i}}(\phi(x)) = v_{\mathbf{M},\beta_x^{d_i}}(\phi(x)) = i$ which gives (1). On the other hand, let t be any ground term and assume $v_{\mathbf{M},\beta}(t) = d$. (2) is implied by $v_{\mathbf{M}',\beta}(\phi(t)) = v_{\mathbf{M}',\beta_x^d}(\phi(x)) = v_{\mathbf{M},\beta_x^d}(\phi(x)) \in d_{\mathbf{M},\beta,x}(\phi(x)) = I$. The other direction is similar. ■

Conditions (1) and (2) in the previous proposition informally tell that for each distribution $I \in (Q(\lambda))^{-1}(S)$: (1) there is a witness c_i for each $i \in I$ assuring that the truth value i is reached by ϕ and (2) no other truth values than I are reached by ϕ. They can be conveniently expressed in rule format using Skolem constants and arbitrary ground terms exactly as in the quantifier rules for classical logic (see Letz's Chapter). The following rule is sound and complete by the previous proposition:

$$(13)\qquad \frac{S\ (\lambda x)\phi(x)}{\begin{array}{c|c|c} \{i_{11}\}\ \phi(c_1) & \cdots & \{i_{m1}\}\ \phi(c_1) \\ \vdots & & \vdots \\ \{i_{1k_1}\}\ \phi(c_{k_1}) & \cdots & \{i_{mk_m}\}\ \phi(c_{k_m}) \\ I_1\ \phi(t_1) & \cdots & I_m\ \phi(t_m) \end{array}}$$

Here $(Q(\lambda))^{-1}(S) = \{I_1, \ldots, I_m\}$, $I_j = \{i_{j1}, \ldots, i_{jk_j}\}$, the $c_1, c_2, \ldots$ are new Skolem constants and the $t_1, \ldots, t_m$ are arbitrary ground terms.

As an immediate simplification we note that if $I_j = \{i_{j1}\}$ for some j, then in the corresponding extension it is sufficient to list merely the signed formula $I_j\,\phi(t)$. Moreover, one can always delete tautological signed formulas of the form $N\ \phi(t)$.

In the Chapter by Letz, free variable first-order rules are given, where instead of an arbitrary ground term t a free variable X is introduced which is to be instantiated later. Skolem constants then have to be parameterized with the free variables that are present in the premise. The free variable version of (13) is:

$$(14)\qquad \frac{S\ (\lambda x)\phi(x)}{\begin{array}{c|c|c} \{i_{11}\}\ \phi(f_{11}(X_1, \ldots, X_r)) & \cdots & \{i_{m1}\}\ \phi(f_{m1}(X_1, \ldots, X_r)) \\ \vdots & & \vdots \\ \{i_{1k_1}\}\ \phi(f_{1k_1}(X_1, \ldots, X_r)) & \cdots & \{i_{mk_m}\}\ \phi(f_{mk_m}(X_1, \ldots, X_r)) \\ I_1\ \phi(Y_1) & \cdots & I_m\ \phi(Y_m) \end{array}}$$

Here $\{X_1, \ldots, X_r\}$ are the free variables of $(\lambda x)\phi(x)$, $\{Y_1, \ldots, Y_m\}$ are new free variables, and the f_{jk} are new r-ary Skolem function symbols; the rest is as above. Completeness of this rule is established by a straightforward combination of the techniques in [Fitting, 1996; Hähnle, 1994], soundness by a suitable adaption of [Hähnle and Schmitt, 1994] to the many-valued case. Further liberalizations along the lines of [Beckert *et al.*, 1993; Baaz and Fermüller, 1995] are possible.

In the presence of quantifiers the usefulness of sets-as-signs becomes even more striking. The reason is that condition (2) in Proposition 40 is awkward and complicated to represent, when only singleton signs are available, cf. [Carnielli, 1991; Zabel, 1993]. Again, the use of sets-as-signs can lead to exponential speed-ups.

If we compute the tableau rule with the premise $\{0, \frac{1}{2}\}(\forall x)\phi(x)$ in three-valued first-order Kleene logic according to the method suggested by Proposition 40, then we obtain the rule shown in Figure 4.

This rule is obviously not the simplest possible one, for example, the rule dis-

$$
\begin{array}{c|c|c|c|c|c}
\multicolumn{6}{c}{\{0, \frac{1}{2}\}\ (\forall x)\phi(x)} \\
\hline
\{0\}\ \phi(c) & \{0\}\ \phi(c) & \{0\}\ \phi(c) & \{0\}\ \phi(c) & & \{\frac{1}{2}\}\ \phi(c) \\
 & \{\frac{1}{2}\}\ \phi(d) & \{\frac{1}{2}\}\ \phi(d) & & \{\frac{1}{2}\}\ \phi(d) & \\
 & & \{1\}\ \phi(e) & \{1\}\ \phi(e) & & \{1\}\ \phi(e) \\
 & \{0, \frac{1}{2}\}\ \phi(t_2) & & \{0,1\}\ \phi(t_4) & & \{\frac{1}{2}, 1\}\ \phi(t_6)
\end{array}
$$

Figure 4. Tableau rule for $\{0, \frac{1}{2}\}\ (\forall x)\phi(x)$ in three-valued logic

played in Figure 5(c) is also sound and complete.[11] In Proposition 40 we encoded each truth value set in $(Q(\forall))^{-1}(\{0, \frac{1}{2}\})$ with Skolem conditions. If we turn this process around, and ask ourselves which distributions of truth values can be encoded using conjunctions of Skolem conditions of the form $I\,\phi(c)$ or $J\,\phi(t)$ (where $I, J \subseteq N$, c is 'new' and t is arbitrary) then we see that this 'Skolem language' is quite powerful. In order to appreciate the following results some background in lattice theory is necessary, see, for example [Davey and Priestley, 1990].

THEOREM 41 ([Hähnle, 1998]).[12] *Let N be a finite set of truth values, $S \subseteq N$. If $(Q(\lambda))^{-1}(S)$ is a filter*[13] *of the Boolean set lattice $\mathbf{2^N}$ generated by $F = \{i_1, \ldots, i_r\} \subseteq N$, then the rule displayed in Figure 5(a) is sound and complete. If $(Q(\lambda))^{-1}(S) \cup \{\emptyset\}$ is an ideal of the Boolean set lattice $\mathbf{2^N}$ generated by $J \subseteq N$, then the rule displayed in Figure 5(b) is sound and complete.*

$$
\begin{array}{c}
S\ (\lambda x)\phi(x) \\
\hline
\{i_1\}\ \phi(c_1) \\
\vdots \\
\{i_r\}\ \phi(c_r)
\end{array}
\qquad\qquad
\begin{array}{c}
S\ (\lambda x)\phi(x) \\
\hline
J\ \phi(t)
\end{array}
\qquad\qquad
\begin{array}{c}
\{0, \frac{1}{2}\}\ (\forall x)\phi(x) \\
\hline
\{0, \frac{1}{2}\}\ \phi(t)
\end{array}
$$

(a) (b) (c)

Figure 5. Tableau rules when $(Q(\lambda))^{-1}(S)$ is a filter or an ideal

The significance of the preceding theorem stems from the fact that lattice-based quantifiers have exactly the required form of distributions.

[11] The rule in Figure 5(c) appears already in [Saloni, 1972] in the context of first-order Post logics.

[12] A similar theorem, but for singleton signs only, is proved by Zabel [1993, Section 1.3.3].

[13] Note that in a finite lattice each filter (ideal) is principal, hence generated by a single element as its upset (downset).

THEOREM 42 ([Hähnle, 1998]). [14] *Let the truth value set N be partially ordered such that it constitutes a finite distributive lattice $L = \langle N, \sqcap, \sqcup \rangle$ and define quantifiers via $Q(\Pi) = \sqcap$, $Q(\Sigma) = \sqcup$. Then the following tableau rules are sound and complete ($\Uparrow i$ and $\Downarrow i$ denote upsets and downsets in L, respectively):*

If i is meet-irreducible: *If i is join-irreducible:*

$$\frac{\{i\}\,(\Pi x)\phi(x)}{\begin{array}{c}\{i\}\,\phi(c)\\ \Uparrow i\;\phi(t)\end{array}} \qquad \frac{\Downarrow i\,(\Pi x)\phi(x)}{\Downarrow i\;\phi(c)} \qquad \frac{\{i\}\,(\Sigma x)\phi(x)}{\begin{array}{c}\{i\}\,\phi(c)\\ \Downarrow i\;\phi(t)\end{array}} \qquad \frac{\Uparrow i\,(\Sigma x)\phi(x)}{\Uparrow i\;\phi(c)}$$

For all i:

$$\frac{\Uparrow i\,(\Pi x)\phi(x)}{\Uparrow i\;\phi(t)} \qquad \frac{\Downarrow i\,(\Sigma x)\phi(x)}{\Downarrow i\;\phi(t)}$$

This theorem gives a direct justification for the rule in Figure 5(c), because $\forall$ is defined via the meet operator on the totally ordered chain N in which all elements but 1 are meet-irreducible.

More general, but less compact, rules can be found that work for non-irreducible elements [Hähnle, 1998] and even for non-distributive semi-lattices and interval-shaped signs [Salzer, 1996a]. There is also an algorithm for computing minimial CNF/DNF representations of distribution quantifiers developed by Salzer [1996a].

Fitting [1991] gives signed tableau rules for the case when the set of truth values forms a certain type of *bilattice* [Ginsberg, 1988], however, as Fitting requires his rules to follow Smullyan's uniform notation (cf. D'Agostino's Chapter and [Smullyan, 1995]), the class of logics for which they work is somewhat restricted.

Like γ-rules in first-order classical tableaux (cf. Letz's Chapter) a many-valued quantifier rule is not invertible whenever a universal expression (that is a free variable or an arbitrary ground term) occurs in its conclusion. It is, therefore, necessary to apply such rules arbitrarily often in a fair manner on the branches containing their premisses.

An interesting complication as compared to the classical case is that Skolemization of formulas seems not easily to be possible as a preprocessing step that is merely by applying suitable term substituions and without application of any extension rules. In other words, in contrast to classical logic quantifier elimination in many-valued logic can blow up a formula exponentially.

[14] A less general result for singleton signs was shown by Zach [1993, Section 1.7] and Baaz and Fermüller [1995a, Example 4.20].

5 TABLEAUX AS TRANSLATION PROCEDURES

5.1 Translation to Signed Clauses

In the previous sections we saw that DNF (CNF) representations of complex signed formulas correspond to tableau (sequent) rules. A tableau (sequent) proof is nothing else than the recursive application of rules until a contradiction (an axiom) is reached in each branch (end sequent). If we start with a satisfiable (a non-tautological) formula, then we end up with some branches (end sequents) that cannot be closed (are not axioms). In the propositional case each formula in each branch (sequent) needs only once to have a rule applied to it (cf. D'Agostino's Chapter), hence a propositional tableau (sequent) proof is always finite. It is well known that in classical logic the disjunction (conjunction) over the conjunctively (disjunctively) connected literals in each open branch (non-axiomatic end sequent) constitutes a DNF (CNF) of the original formula. This relation extends to many-valued tableau (sequent) proofs in a straightforward manner.

In the following we describe the computation of CNF representations of arbitrary formulas from any finitely-valued logic. The DNF case is essentially dual for the propositional level. Anyway, only a CNF is interesting to compute in the first-order case as universal quantifiers distribute over conjunctions, but not over disjunctions.

In this subsection we concentrate on the translation into CNF for the sets-as-signs case. A thorough and systematic treatment of the CNF singleton signs case which includes full proofs as well as many complexity results on the branching factor of rules can be found in [Baaz and Fermüller, 1995a].

Instead of sequents for sake of readability and to stress the similarity of sequent rules and tableau rules we use a tableau-like notation that we call *dual tableaux* (cf. [Kapetanović and Krapež, 1989] for classical dual tableaux). We use double vertical bars || in the conclusion to indicate dual tableau rules. In this notation, the rule from Example 22 in Section 3.1 is written as below. On the right is an example for a dual sets-as-signs rule (also for three-valued Łukasiewicz implication).

$$
\begin{array}{c|c}
\multicolumn{2}{c}{\frac{1}{2}\,(\phi \supset \psi)} \\ \hline
\frac{1}{2}\,\phi & 1\,\phi \\
\frac{1}{2}\,\psi & 0\,\psi
\end{array}
\qquad\qquad
\begin{array}{c|c|c}
\multicolumn{3}{c}{\{0, \frac{1}{2}\}\,(\phi \supset \psi)} \\ \hline
\{\frac{1}{2}, 1\}\,\phi & \{1\}\,\phi & \\
 & \{0\}\,\psi & \{0, \frac{1}{2}\}\,\psi
\end{array}
$$

As to the quantifier rules, first we provide a CNF version of Proposition 40 in Section 4.3. The main idea for it is to express 'characterize one of the sets I in $(Q(\lambda))^{-1}(S)$' equivalently with 'exclude all of the sets $\overline{I}$ not in $(Q(\lambda))^{-1}(S)$'. For a distribution $d_{\mathbf{M},\beta,x}(\phi)$ not to be equal to $\overline{I}$ there are two possibilities: (1) there is a witness c such that the truth value of $\phi(c)$ lies outside of $\overline{I}$ or (2) $d_{\mathbf{M},\beta,x}(\phi)$ is a proper subset of $\overline{I}$.

PROPOSITION 43. *Let* $\overline{(Q(\lambda))^{-1}(S)} = \mathcal{P}^{+}(N)\backslash(Q(\lambda))^{-1}(S)$. *Then a signed quantified formula* $S\ (\lambda x)\phi(x)$ *is satisfiable iff for all the sets* $\overline{I} = \{i_1, \ldots, i_r\} \in$

$\overline{(Q(\lambda))^{-1}(S)}$:

1. *there is an* $i \in (N\backslash\overline{I})$ *and a constant term* c *not occurring in* $\phi(x)$ *such that* $\{i\}\ \phi(c)$ *is satisfiable (iff* $(N\backslash\overline{I})\ \phi(c)$ *is satisfiable) or*
2. *there is a proper (non-empty) subset* J *of* $\overline{I}$ *such that* $J\ \phi(t)$ *is satisfiable for all ground terms* t.

As before, from this proposition one can derive tableau rules for distribution quantifiers. Hereby, condition (2) is slightly reformulated as:

2′. *there are* $\mathbf{M}$, β *such that for at least one* $i \in \overline{I}$, $d_{\mathbf{M},\beta,x}(\phi(x))$ *is a non-empty subset of* $\overline{I}\backslash\{i\}$.

Note that in the case when $|\overline{I}| = 1$ (2′) is not satisfiable and can be dropped. A generic dual free variable quantifier rule then is as follows:

$$(15)\qquad \frac{S\ (\lambda x)\phi(x)}{\begin{array}{c|c|c} (N\backslash\overline{I_1})\ \phi(f_1(X_1,\ldots,X_r)) & \cdots & (N\backslash\overline{I_m})\ \phi(f_m(X_1,\ldots,X_r)) \\ (\overline{I_1}\backslash\{i_{11}\})\ \phi(Y_1) & \cdots & (\overline{I_m}\backslash\{i_{m1}\})\ \phi(Y_1) \\ \vdots & & \vdots \\ (\overline{I_1}\backslash\{i_{1k_1}\})\ \phi(Y_{k_1}) & \cdots & (\overline{I_m}\backslash\{i_{mk_m}\})\ \phi(Y_{k_m}) \end{array}}$$

Here $\overline{(Q(\lambda))^{-1}(S)} = \{\overline{I_1},\ldots,\overline{I_m}\}$, $\overline{I_j} = \{i_{j1},\ldots,i_{jk_j}\}$, $\{X_1,\ldots,X_r\}$ are the free variables of $(\lambda x)\phi(x)$, $\{Y_1,\ldots,Y_{\max\{k_1,\ldots,k_m\}}\}$ are new free variables, and the f_i are new r-ary Skolem function symbols.

As an example, let us compute the CNF version of the rule displayed in Figure 4. We have $\overline{(Q(\forall))^{-1}(\{0,\frac{1}{2}\})} = \{\{1\}\}$ producing the rule:

$$\frac{\{0,\frac{1}{2}\}\ (\forall x)\phi(x)}{\{0,\frac{1}{2}\}\ \phi(f(X_1,\ldots,X_r))}$$

A singleton signs version of (15) can be easily obtained by splitting a formula of the form $\{j_1,\ldots,j_r\}\ \phi$ that occurs in an extension into r disjuncts $\{j_k\}\ \phi$. This is possible, because in the CNF case conditions on formulas in extensions are disjunctively combined. There is no such simple representation for quantifiers with singleton signs in the DNF case.

Dual tableaux can be used to obtain a simple algorithm for converting formulas from finitely-valued first-order logic into a CNF over signed first-order atoms.

DEFINITION 44. Let $S\ \phi$ be a signed formula from a finitely-valued first-order logic. A *dual sets-as-signs tableau* for it is defined exactly as a sets-as-signs tableau with the exception that each formula is removed from those branches in which it had a rule applied to it.

Even quantified formulas that create universal expressions are deleted in a dual tableau. This is legitimate, however, as we are working with a CNF and free variables remain implicitly universally quantified (cf. Definition 46 below). It is obvious that after a finite number of steps from each signed formula a dual tableau is reached to which no more rules can be applied. We call such a tableau *completed*. All formulas occurring in a completed tableau either are signed literals or they satisfy condition (2) in Definition 34, Section 4.2. In the latter case they are unsatisfiable and can be dropped. Tautological branches need not be considered at all:

DEFINITION 45. A branch B in a dual tableau is called *tautological* if either

1. there are signed formulas $S_1\ \phi, \ldots, S_r\ \phi$ on B such that $\bigcup_{i=1}^{r} S_i = N$ or
2. there is a signed formula $S\ \theta(\phi_1, \ldots, \phi_r)$ on B such that $\mathrm{rg}(\theta) \subseteq S$ or
3. there is a signed formula $S\ (\lambda x)\phi(x)$ on B such that $\mathrm{rg}(Q(\lambda)) \subseteq S$.

DEFINITION 46. An expression of the form $\bigvee_{i=1}^{r} L_i$, where the L_i are signed literals is a *signed clause*. It is satisfiable iff there is a structure $\mathbf{M}$ such that for all variable assignments β at least one of the L_i is satisfied by $\mathbf{M}$ and β.[15] An expression of the form $\bigwedge_{j=1}^{m} C_j$, where the C_j are signed clauses, is a *signed CNF formula*. It is satisfiable iff all C_j are satisfiable. The empty clause ($r = 0$) is always unsatisfiable.

Let $S\ \phi$ be a signed formula from a finitely-valued first-order logic and let T be a completed dual sets-as-signs tableau for it. Let $\overline{B_1}, \ldots, \overline{B_m}$ be the sets of literals occurring in the non-tautological branches $B_1, \ldots, B_m$ of T. Then the expression $\bigwedge_{i=1}^{m} \bigvee_{L \in \overline{B_i}} L_i$ is a *signed CNF* of $S\ \phi$.

Using Proposition 43 one proves

THEOREM 47. *For any finitely-valued first-order logic there is an effective procedure f that maps any signed formula $S\phi$ into a signed CNF formula $\psi = f(S\phi)$ such that for all ϕ, S: $S\ \phi$ is satisfiable iff ψ is satisfiable.*

The signed CNF of a many-valued formula has a simple and uniform structure which can serve as the basis for resolution style proof procedures. As this is a handbook of *tableau* methods we only mention the relevant literature: [Murray and Rosenthal, 1991a; Hähnle, 1996; Baaz and Fermüller, 1995a].

The signed CNF of a many-valued formula is finite, but it can be exponential in size wrt the input. In classical logic the situation can be improved by using a *structure preserving CNF transformation* [Plaisted and Greenbaum, 1986] (as

[15] So just as in classical logic, clauses are universally quantified implicitly.

opposed to a *language preserving* transformation). It works by extending the signature of the given formula with suitable predicates which serve as abbreviations for subformulas. In rule form this can be conveniently expressed as:

$$(16) \qquad \frac{\mathbf{T}(\phi)}{\mathbf{T}(p_\phi) \;\Big\|\; \begin{array}{c} p_\phi \\ \neg\phi \end{array} \;\Big\|\; \begin{array}{c} \neg p_\phi \\ \phi \end{array}}$$

This rule is applied to a dual tableau $\mathbf{T}$ in which a complex formula ϕ does occur as a proper subformula in at least one node. Each occurrence of ϕ is replaced with a new propositional variable p_ϕ (in the first-order case with an atom $p_\phi(x_1, \ldots, x_r)$, where $x_1, \ldots, x_r$ are the free variables of ϕ) and the tableau is extended with two new branches as shown. This process terminates when all formulas in the tableau are at most of depth 2.

As ϕ does not occur anymore in $\mathbf{T}$ as a proper subformula the 'elimination' of each subformula creates at most two new branches. The number of subformulas of a given formula is linear in its size. Therefore, a signed CNF which is linear in the size of the input can be obtained by first applying rule (16) as long as possible and then the standard dual tableau rules.

The two right branches in the conclusion of (16) ensure that $\models p_\phi \leftrightarrow \phi$ holds. This observation leads immediately to a many-valued version of the algorithm:

$$(17) \qquad \frac{\mathbf{T}(\phi)}{\mathbf{T}(p_\phi) \;\Big\|\; \begin{array}{c} \{0\}\, p_\phi \\ N\backslash\{0\}\, \phi \end{array} \;\Big\|\; \begin{array}{c} \cdots \\ \cdots \end{array} \;\Big\|\; \begin{array}{c} \{1\}\, p_\phi \\ N\backslash\{1\}\, \phi \end{array}}$$

Here, ϕ may occur in several signed formulas in $\mathbf{T}$ with possibly differing sign. Not merely two, but $|N| = n$ additional extensions are generated. The n right extensions express: for all $i \in N$: $\{i\}$-satisfiability of p_ϕ implies $\{i\}$-satisfiability of ϕ iff (for all $S \subseteq N$: p_ϕ is S-satisfiable iff ϕ is S-satisfiable) iff for all $\mathbf{I}$: $\mathbf{I}(p_\phi) = \mathbf{I}(\phi)$.

Complexity of Tableau Rules The branching factor of sets-as-signs tableau rules for distribution quantifiers is bounded by $2^n - 2$. It is easy to see that this bound is sharp for the representation used in Propositions 40 in Section 4.3 and 43. It is unknown whether better representations exist in general, however, for the special case of a CNF representation and of singleton signs an improved bound of 2^{n-1} was obtained in [Zach, 1993; Baaz and Fermüller, 1995a] which is also proved to be sharp.

5.2 Polarity

In classical logic a well-known fact is used to simplify structure-preserving CNF's:

If ϕ occurs only with positive (negative) polarity in $\mathbf{T}(\phi)$ then $\mathbf{T}(\phi)$ is satisfiable iff $\mathbf{T}(p_\phi) \wedge (p_\phi \rightarrow \phi)$ $(\mathbf{T}(p_\phi) \wedge (p_\phi \leftarrow \phi))$ is satisfiable.

In the many-valued case one can take $\{S \subseteq N \mid S\ \phi \text{ occurs in } \mathbf{T}(\phi)\}$ as the polarity of an unsigned subformula. Based on this notion of polarity it is possible to prove:

THEOREM 48 ([Hähnle, 1994b]). *Assume ϕ occurs in $\mathbf{T}(\phi)$ with polarity $R = \{S_1, \ldots, S_m\}$ then $\mathbf{T}(\phi)$ is satisfiable iff $\{\mathbf{T}(p_\phi),\ \{1\}\ (p_\phi \Rightarrow_R \phi)\}$ is satisfiable, where*

$$i \Rightarrow_R j = \begin{cases} 0 & \text{if } i \notin S_k,\ j \in S_k, \text{ for any } S_k \in R \\ 1 & \text{otherwise} \end{cases}$$

A signed CNF with polarity optimization is obtained by applying the following rule and a minimal rule for $\{1\}\ (p_\phi \Rightarrow_R \phi)$ as long as possible and then applying minimal dual sets-as-signs rules to the resulting dual tableau (the rules must be applied to maximal subformulas so that R is always known):

$$\text{(18)} \qquad \frac{\mathbf{T}(\phi)}{\mathbf{T}(p_\phi)\ \|\ \{1\}\ (p_\phi \Rightarrow_R \phi)}$$

Here R is the polarity of ϕ in $\mathbf{T}(\phi)$.

In Section 6.1 a coarser and much simpler notion of polarity for a subclass of signed CNF formulas will be introduced.

5.3 Translation to Mixed Integer Programs

In this section we strictly focus on propositional logic. In addition, some background knowledge of linear optimization is useful as provided, for example, in [Schrijver, 1986]. It is a well known fact (see, for example, [Hooker, 1988; Jeroslow, 1988]) that propositional classical CNF formulas correspond to certain 0-1 integer programs. More precisely, given a set Φ of classical propositional clauses over a propositional signature Σ one transforms each clause

$$\text{(19)} \qquad p_1 \vee \cdots \vee p_k \vee \neg p_{k+1} \vee \cdots \vee \neg p_{k+m}$$

into a linear inequation

$$\text{(20)} \qquad \Sigma_{i=1}^{k} p_i - \Sigma_{j=k+1}^{m} p_j \geq 1 - m$$

The variables from Σ are then interpreted as *arithmetical variables* over $\{0,1\}$. It is easy to see that the resulting set of inequations is solvable iff Φ is satisfiable.

In the remainder of this section we show that with a combination of the arithmetical reduction just sketched and a suitable structure preserving CNF translation a polynomial time reduction from satisfiability in infinitely-valued propositional logics to 0-1 mixed integer programming is obtained.

With the expression *linear inequation* we mean in the following always a term of the form $a_1p_1 + \cdots + a_mp_m \geq c$, where $a_1, \ldots, a_m, c \in \mathbb{Z}$ and the p_i are variables over N or over $\{0,1\}$ (w.l.o.g. we may prefer an equivalent representation of the inequation without explicit mention), where N either is finite or infinite as in Definition 8, Section 2.2. An expression of the form $a_1p_1 + \cdots + a_mp_m$ is called a *linear term*.

DEFINITION 49. Let $\mathbf{J}$ be a finite set of linear inequations and K a linear term. Let Σ be the set of variables occurring in $\mathbf{J}$ and K. Assume N is finite. Then $\langle \mathbf{J}, K\rangle$ is a *(bounded) integer program* (IP).[16] If N is infinite, then we have a *(bounded) 0-1 mixed integer program* (MIP). When all variables in Σ run over infinite N we have a *(bounded) linear program* (LP).

A variable assignment to Σ such that all inequations in $\mathbf{J}$ are satisfied is called a *feasible solution* of $\langle \mathbf{J}, K\rangle$. A variable assignment to Σ such that the value of K is minimal among all feasible solutions is called an *optimal solution*. $\langle \mathbf{J}, K\rangle$ is *feasible* iff there are feasible solutions.

PROPOSITION 50 (see [Schrijver, 1986]). *Each of the problems to check whether a 0-1 MIP (resp., an IP) has feasible solutions and to find an optimal/feasible solution is NP-complete. The problem to check whether an LP has feasible solutions and to find an optimal/feasible solution is in P.*

DEFINITION 51. Let $\boxed{\geq i}$ denote the set $\{j \in N \mid j \geq i\}$ and let $\boxed{\leq i}$ denote the set $\{j \in N \mid j \leq i\}$. If a sign S is equal to either $\boxed{\geq i}$ or $\boxed{\leq i}$ for some $i \in N$, then it is called *regular sign*.

PROPOSITION 52. $\boxed{\geq i}\ \phi$ *is satisfiable iff* $\phi \geq i$ *is solvable when* ϕ *and* i *are considered as arithmetical variables. Similary,* $\boxed{\leq i}\ \phi$ *is satisfiable iff* $\phi \leq i$ *is solvable.*

We extend dual sets-as-signs tableaux to constraint tableaux:

DEFINITION 53. A *constraint tableau* is a dual sets-as-signs tableau over regular signs of the form $\boxed{\geq I}$, $\boxed{\leq I}$, where I is a linear term. In addition, nodes may be linear inequations instead of signed formulas.

Consider the following signed rule obtained from a structure preserving CNF rule, where β is any classical disjunctive formula (cf. (16)):

[16] The term *integer* is justified, because the elements of N can w.l.o.g. assumed to be of the form $\{0,1,\ldots,n-1\}$.

(21)
$$\frac{\boxed{\geq i}\ \beta}{\boxed{\geq i}\ p_\beta \;\Big\|\; \boxed{\geq 1}\ p_\beta \vee \neg\beta \;\Big\|\; \boxed{\geq 1}\ \neg p_\beta \vee \beta}$$

Assume the rule is applied only to formulas β occuring on the top-level, that is with positive polarity. Then we can delete the middle extension according to the polarity optimization introduced in the previous section. The remaining extension $\boxed{\geq 1}\ \neg p_\beta \vee \beta$ claims that $\neg p_\beta \vee \beta$ is satisfiable. Using (20) and the fact that β is equivalent to $\beta_1 \vee \beta_2$ one obtains:

$$\frac{\boxed{\geq i}\ \beta}{\boxed{\geq i}\ p_\beta \;\Big\|\; \beta_1 + \beta_2 - p_\beta \geq 0}$$

Renaming $p_\beta - \beta_1$ into j and applying Proposition 52 yields:

$$\frac{\boxed{\geq i}\ \beta}{\boxed{\geq i-j}\ \beta_1 \;\Big\|\; \boxed{\geq j}\ \beta_2}$$

Essentially the same technique can be used to derive a rule for, say, Łukasiewicz implication (2) and the sign $\boxed{\leq i}$. The first steps are very similar as above and yield the rule below on the left. With Proposition 52 the rule below on the right is obtained.

$$\frac{\boxed{\leq i}\ \phi \supset \psi}{\boxed{\leq i}\ p_\supset \;\Big\|\; \boxed{\geq 1}\ p_\supset \vee (\phi \supset \psi)} \qquad\qquad \frac{\boxed{\leq i}\ \phi \supset \psi}{\boxed{\leq i}\ p_\supset \;\Big\|\; \boxed{\leq p_\supset}\ \phi \supset \psi}$$

Recall that $\phi \supset \psi = \min\{1, 1 - \phi + \psi\}$. In order to express this definition with linear inequations we introduce a $\{0,1\}$-valued variable y which is supposed to be smaller or equal than i. Under this condition $1 - \phi + \psi - y \leq p_\supset + y$ holds iff $\boxed{\leq p_\supset}\ \phi \supset \psi$ is satisfiable. Thus we can rewrite the rule as follows:

$$\frac{\boxed{\leq i}\ \phi \supset \psi}{\boxed{\leq i}\ p_\supset \;\Big\|\; y \leq i \;\Big\|\; 1 - \phi + \psi - y \leq p_\supset + y}$$

Renaming $p_\supset + y + \phi - 1$ into j and applying Proposition 52 finally yields:

(22)

$$\frac{\boxed{\leq i}\ \phi \supset \psi}{\boxed{\geq 1-i+j-y}\ \phi \ \Big\| \ y \leq i \ \Big\| \ \boxed{\leq j+y}\ \psi}$$

Note that i and j run over N while y is two-valued. As no step of the translation relies on the size of N, this rule is sound and complete for arbitrary N including infinite N. A rule for $\boxed{\geq i}$ is derived similarly:

(23)

$$\frac{\boxed{\geq i}\ \phi \supset \psi}{\boxed{\leq 1-i+j}\ \phi \ \Big\| \ \boxed{\geq j}\ \psi}$$

Now assume we want to test tautologyhood of a formula ϕ in Łukasiewicz logic. First one recursively applies (22) and (23) to $\boxed{\leq c}\ \phi$ until only linear inequations and signed literals are left. The latter are turned into inequations with Proposition 52. Let us call the set of resulting inequations $\mathbf{J}$. The form of the rules guarantees that (i) each branch contains exactly one inequation (ii) the number of branches is linear in the size of ϕ. Soundness and completeness of the rules implies that $\boxed{\leq c}\ \phi$ is satisfiable iff $\mathbf{J}$ considered as part of an (M)IP is feasible.

When N is finite we set $c = \frac{n-2}{n-1}$. Then $\mathbf{J}$ is infeasible iff $\mathbf{I}(\phi) = 1$ for all $\mathbf{I}$ iff ϕ is a tautology. When N is infinite, consider the 0-1 MIP $\langle \mathbf{J}, c \rangle$. Then the optimal solution of $\langle \mathbf{J}, c \rangle$ is 1 iff $\mathbf{I}(\phi) = 1$ for all $\mathbf{I}$ iff ϕ is a tautology. Satisfiability problems can be expressed similarly, see [Hähnle, 1994a].

The same reduction technique to 0-1 MIP works for other infinitely-valued logics as well. The main requirement is that connectives θ with $\alpha(\theta) = k$ can be represented with a 0-1 MIP over $[0,1]^{k+m} \times \{0,1\}^r$ for some $m, r \geq 0$ [Jeroslow, 1988].

EXAMPLE 54. To show that $p \supset (q \supset p)$ is a tautology of infinitely-valued Łukasiewicz logic apply (22) and (23), then Proposition 52.

$$\frac{\boxed{\leq c}\ p \supset (q \supset p)}{\dfrac{\boxed{\geq 1-c+j-y}\ p}{p-j+c+y \geq 1} \ \Big\| \ y \leq c \ \Big\| \ \dfrac{\boxed{\leq j+y}\ q \supset p}{\dfrac{\boxed{\geq 1-j+k-z}\ q}{q+j-k+z \geq 1} \ \Big\| \ z \leq j+y \ \Big\| \ \dfrac{\boxed{\leq k+z}\ p}{-p+k+z \geq 0}}}$$

We show that for every optimal solution of the resulting 0-1 MIP problem $c = 1$: assume there was a feasible solution such that $c < 1$, hence as y is two-valued, by $y \leq c$, $y = 0$. If $j-c$ were greater than 0 from $y = 0$ and $p-j+c+y \geq 1$ one could infer $p \geq 1+j-c > 1$, a contradiction to $p \in [0,1]$; thus, $j \leq c$ which gives $z = 0$

observing $z \leq j + y$ and $c < 1, y = 0$, z two-valued. From $N = [0, 1]$ we know that $-q \geq -1$. From this and $q+j-k \geq 1$ infer $j-k \geq 0$. Adding this to $-p+k \geq 0$ gives $-p+j \geq 0$ which together with $p-j+c \geq 1$ gives $c \geq 1$—contradiction. On the other hand, $\boxed{\leq 1}\, p \supset (q \supset p)$ is trivially satisfiable, thus $c = 1$ yields indeed a solution.

From Proposition 50 and the observations above one has the following theorem:

THEOREM 55 ([Mundici, 1987]). *The problem of deciding tautologyhood (satisfiability) in infinitely-valued Łukasiewicz logic is in Co-NP (is in NP).*

We close with two final remarks: first, Co-NP-, respectively, NP-hardness is easy to show: the idea is to construct a polynomial-size (in p) formula $c(p)$ which forces p to take on classical truth values merely, see [Mundici, 1987] for details (this technique is applicable to other infinitely-valued logics as well); second, similar results follow routinely, once sound and complete constraint tableau rules like (22) and (23) are available for a given logic.

Historical Note A constraint tableau calculus was developed independently by Hähnle [1992; 1994a] and Zabel [1993, Chapter 8]. We followed the notation of [Hähnle, 1994a], but chose a more compact derivation. The main differences to [Zabel, 1993] are: the latter does not employ a structure preserving translation, hence he does neither obtain a reduction to 0-1 MIP nor Theorem 55. On the other hand, the resulting constraints contain no integer variables, thus the required constraint language is simple enough to be polynomially decidable. Moreover, constraints are simplified as soon as they are generated in Zabel's calculus.

The complexity result for infinitely-valued Łukasiewicz logic is due to Mundici [1987]. Whereas the (NP-containment part of the) latter relies on particular properties of Łukasiewicz logic the tableau-based technique applies to any so-called *bMIP-representable* logic, see [Hähnle, 1994a] for details.

6 EFFICIENT DEDUCTION IN MVL

Up to now the main concern were non-clausal tableaux. The translation procedures from the previous section offer to consider many-valued clausal tableau as well.[17] Many computational refinements are a lot simpler to define in the clausal case. Moreover, the intricacies of many-valued quantifiers have been dealt with in the translation rules. Once a signed CNF is obtained the usual lifting property holds. Therefore, we consider mainly ground formulas in the present section.

---

[17] Once one has obtained a (signed) CNF, it is also possible to develop resolution rules. See [Lu *et al.*, 1993; Murray and Rosenthal, 1994; Baaz and Fermüller, 1995a; Hähnle, 1996] for some results along these lines.

6.1 Regular Formulas

DEFINITION 56. Let Φ be a signed CNF formula. If all signs that occur in Φ are regular, then Φ is called a **regular formula**. Literal occurrences of the form $\boxed{\geq i}\, p$ ($\boxed{\leq j}\, q$) in a regular formula are said to have *positive (negative) polarity.*

The significance of regular formulas is justified by the following

PROPOSITION 57 ([Murray and Rosenthal, 1994; Hähnle, 1996]). *If the number of truth values is finite then for every signed CNF formula Φ there is a regular formula Ψ such that Φ and Ψ are logically equivalent.*

Note, however, that the 'regularization' of a signed CNF formula can grow exponentially long.

Combining Proposition 57 with Theorem 47 from Section 5.1 one sees that regular formulas are sufficient to express any satisfiability problem of finitely-valued first-order logic.

The main advantage of regular formulas is that they have a CNF structure and all literals have either positive or negative polarity. In addition, any inconsistent set of regular literals has an inconsistent subset of cardinality two. This makes it possible to define similar refinements of many-valued tableaux as there are of classical clause tableaux. In the following sections we describe some important cases, but first we would like to mention that the concept of regular formulas can be lifted to the non-clausal case:

Consider an arbitrary signed formula $S\ \phi$. Even if S is regular, in general the signs occurring in a CNF/DNF representation need not. On the other hand, a DNF representation in which only regular signs occur can always be easily obtained: simply compute a singleton signs representation for $S\ \phi$ and replace each formula of the form $\{i\}\ \psi$ with $\boxed{\geq i}\, \psi$ and $\boxed{\leq j}\, \psi$.

In general a *minimal* representation of a signed formula involves non-regular signs. For some logics, however, at least one of the minimal representations is also regular. One such class of logics was defined in [Hähnle, 1991] called, not surprisingly, *regular logics*. Regular logics have an even stronger property: the tableau rule derived from a regular minimal DNF representation follows one of the uniform notation rule schemata α, β, γ, or δ from *classical* logic (cf. the Chapters by D'Agostino and Letz in this volume). The consequence is that formulas from regular logics have no larger tableau proofs than classical problems. Moreover, classical theorem proving tools can be used with only slight modifications for regular logics as well. Still, regular logics are not a trivial subclass of finitely-valued logics as is demonstrated in [Hähnle, 1994].

Historical Note Post logics (see footnote on p. 538) are often equipped with unary connectives D_i for $i \in N$ having the matrix

$$D_i(j) = \begin{cases} 1 & j \geq i \\ 0 & j < 1 \end{cases} ,$$

in particular, when investigated in connection with Post algebras, see [Epstein, 1960]. Obviously, $D_i(\phi)$ is $\{1\}$-satisfiable iff $\boxed{\geq i}\ \phi$ is satisfiable. Thus they can replace regular signs. Used in this spirit, a sequent system for first-order Post logics (where $\forall$ and $\exists$ are defined as in Kleene logic) which is equivalent to a dual sets-as-signs tableau system with regular signs was given by Saloni [1972]. The strong properties of Post algebras which can be seen as a natural generalization of Boolean algebras explain to some extent the usefulness of regular signs.

6.2 Tableau Variants Based on Regular Formulas

Many-valued Horn Formulas, KE and DPL Procedures

DEFINITION 58. A *many-valued Horn formula* is a regular formula in which each clause contains at most one positive literal.

THEOREM 59 ([Hähnle, 1996]). *Satisfiability of propositional many-valued Horn formulas Φ can be decided in $\mathcal{O}(|\Phi|)$ steps when N is fixed and finite.*

Many deduction procedures implicitly or explicitly rely on the existence of a simple procedure for the Horn case. The above notion of a many-valued Horn formula allows to extend such procedures naturally to the case of multiple truth values.

Consider the following version of the KE system (cf. D'Agostino's Chapter in this volume) intended for finitely-valued propositional regular formulas:[18]

$$(24)\quad \frac{\begin{array}{l} S\,p \\ C_1 \vee S'\,p \vee C_2 \end{array}}{C_1 \vee C_2} \quad \text{if } S \cap S' = \emptyset \qquad\qquad \frac{}{\boxed{\leq i}\ p \ \Big|\ \boxed{\geq i + \frac{1}{n-1}}\ p} \quad \text{if } i < 1$$

By a straightforward adaption of the classical proofs the following results can be seen to hold:

- The rules are sound and complete for regular formulas.
- The non-branching rule alone is complete for many-valued Horn formulas.

Recall that in the classical case, KE for propositional formulas in CNF is virtually indistinguishable from the Davis-Putnam-Loveland procedure if unary rules are preferred. The non-branching rule can be seen as a variant of *many-valued unit resolution*. The branching rule is, of course, an instance of many-valued analytic cut, but in the context of KE systems the term *principle of bipartition (PBP)* (of the set of truth values) is more appropriate. Heuristics for choosing the branching literal that are successful in the classical case can be extended to the many-valued

[18] A branch is closed when it contains the empty clause.

case, see [Hähnle, 1996]. Note that in the many-valued case the value of i must be supplied by such a heuristic as well.

The rules above are not complete for signed CNF formulas as can be easily seen with the inconsistent three-valued example $\Phi = \{\{0,\frac{1}{2}\}\,p, \{0,1\}\,p, \{\frac{1}{2},1\}\,p\}$. Unit resolution is not applicable and PBP always yields a literal that is already present in one extension. On the other hand, consider the regularization of Φ: $\{\boxed{\leq\frac{1}{2}}\,p, \boxed{\leq 0}\,p \vee \boxed{\geq 1}\,p, \boxed{\geq\frac{1}{2}}\,p\}$. One unit resolution step yields $\boxed{\leq 0}\,p$ which renders the KE tableau closed. An alternative to regularization would be to introduce an additional rule that allows *residuation* such as the following:[19]

$$\frac{\begin{array}{l} S\,p \\ C_1 \vee S'\,p \vee C_2 \end{array}}{C_1 \vee (S \cap S')\,p \vee C_2} \quad \text{if } S \cap S' \neq \emptyset$$

Obviously, we have a trade-off between the cost of computing a normal form and the simplicity of the required deduction rules. On the one hand, the regularization of a formula can blow it up exponentially, on the other hand, a simple format allows for efficient implementation.

Connection Tableaux

DEFINITION 60. A *many-valued clause tableau* for a signed CNF ground formula Φ is a finitely branching tree whose nodes are signed literals constructed by the following rules:

1. The single node $\top$ is a many-valued clause tableau.
2. If $\mathbf{T}$ is already a many-valued clause tableau, $\bigvee_{i=1}^{r} L_i$ is a signed clause in Φ, then a many-valued clause tableau can be obtained by appending r new nodes containing exactly the L_i to $\mathbf{T}$.

One is tempted to define many-valued connection tableaux analogously to the classical case (cf. Letz's Chapter) as

> a many-valued clause tableau in which every inner node forms an inconsistent set with one or more of its immediate successors.

The problem is that this does not constitute a complete family of tableaux for signed CNF as the following example shows:

$$\begin{aligned} \Psi = \{&\{0,\tfrac{1}{2}\}\,p \vee \{0,\tfrac{1}{2}\}\,q,\ \{0,\tfrac{1}{2}\}\,p \vee \{0,1\}\,q,\ \{0,\tfrac{1}{2}\}\,p \vee \{\tfrac{1}{2},1\}\,q, \\ &\{\tfrac{1}{2},1\}\,p \vee \{0,\tfrac{1}{2}\}\,q,\ \{\tfrac{1}{2},1\}\,p \vee \{0,1\}\,q,\ \{\tfrac{1}{2},1\}\,p \vee \{\tfrac{1}{2},1\}\,q, \\ &\{0,1\}\,p \vee \{0,\tfrac{1}{2}\}\,q,\ \{0,1\}\,p \vee \{0,1\}\,q,\ \{0,1\}\,p \vee \{\tfrac{1}{2},1\}\,q\} \end{aligned}$$

[19] Compare the present discussion to the one in [Lu *et al.*, 1993] which is restricted to resolution, but nevertheless calls upon the same problems. Residuation is called *reduction* in [Kifer and Subrahmanian, 1992], see also Definition 63 below.

Ψ is inconsistent, but simple inspection shows that there is no closed connection tableau for it. After the first expansion of any clause no further extension is possible. The problem is that the connection condition cannot easily be defined when more than two-element sets are needed to establish inconsistency.

In the case of regular formulas, however, we obtain a complete refinement[20] and, moreover, their definition can be slightly simplified:

DEFINITION 61. A *many-valued connection tableau* for a regular formula is a many-valued clause tableau in which every inner node forms an inconsistent set with *one* of its immediate successors.

6.3 *Regular Tableaux and Factorized Tableaux*

For the rest of this section we work with arbitrary signed CNF formulas, not just with regular formulas. Note that the notion of a regular tableau, to be defined below, has nothing in common with the previously introduced regular formulas and they should not be confused. Recall from D'Agostino's Chapter that in a regular clause tableau no branch contains more than one occurrence of the same literal. For the many-valued case we must extend this definition to accomodate signs. Obviously, one can do better than to simply exclude identical signed literals, because a literal of the form $\boxed{\geq i}\, p$ is implied by $\boxed{\geq j}\, p$ if $j \geq i$ and should not be repeated.

In fact, an even more restrictive version of regularity is complete:[21] assume that $S_1\, p, \ldots, S_r\, p$ are on a branch B and we can construct a model for B. Then one has a model as well for all $S\, p$ such that $(\bigcap_{i=1}^{r} S_i) \subseteq S \subseteq N$. This consideration leads to the following definitions:

DEFINITION 62. A branch B in a many-valued clause tableau is regular iff for all sets of signed literals $S_1\, p, \ldots, S_r\, p$ $(r \geq 1)$ on B, if $(\bigcap_{i=1}^{r} S_i) \subseteq S \subseteq N$ then $S\, p$ does not occur on B below $S_1\, p, \ldots, S_r\, p$.

A *regular many-valued clause tableau* is a many-valued clause tableau in which all branches are regular.

DEFINITION 63. We define the *reduction* of a set of signed literals C as the set that is obtained when each subset of the form $S_1\, p, \ldots, S_r\, p$ is replaced by $(\bigcap_{i=1}^{r} S_i)$.

DEFINITION 64. Let C, D be sets of signed literals. We say that D *subsumes*[22] C iff for each signed literal $S\, p$ in D there is a literal $S'\, p$ in the reduction of C such that $S' \subseteq S$.

---

[20]The completeness proof of the classical result (see Letz's chapter) can be adapted in a straightforward manner.

[21]For regular formulas, however, the following does not add anything to the condition stated in the previous paragraph.

[22]Here we define subsumption between conjunctively connected sets of literals. Subsumption between signed clauses is defined dually.

We work with the reduction of C rather than with C itself, because we want, for instance, $\{\{0,\frac{1}{2}\}\,p, \{\frac{1}{2},1\}\,p\}$ to be subsumed by $\{\{\frac{1}{2}\}\,p\}$.

Note that without sets-as-signs the definition would become substantially more complicated.

It is straightforward to prove that if C subsumes D then $\{D\} \models C$.

We can use subsumption, for instance, to make many-valued clause tableaux more powerful similar as in the classical case:

DEFINITION 65. A many-valued clause tableau is *closed with factorization* iff each branch B either is closed or there is a closed branch B' such that B is subsumed by an initial segment B'' of B' which is not an initial segment of B as well. Branches closed by factorization are marked. Closures by factorization must not be cyclic.[23]

6.4 Many-valued Analytic Cuts

The singleton signs cut rule (see, e.g. [Takahashi, 1967]) is very simple. Let us consider the DNF version of (6):

$$\text{(25)} \qquad \overline{\;0\,\phi \;\big|\; \cdots \;\big|\; 1\,\phi\;}$$

In the sets-as-signs case many other variants of a cut rule are possible, see, for example, (24). Moreover, considerable care must be taken to control their application. An analysis of many-valued cut rules in the sets-as-signs case was done in [Hähnle, 1994, Section 6.1]. The following is a much simplified and improved presentation.

Recall from D'Agostino's Chapter that the deductive power of (i) tableau rules with local lemmas, (ii) tableaux with analytic cut and (iii) KE systems was exactly the same that is they are equivalent w.r.t. relative proof length complexity. In Figure 6 it is demonstrated that exactly the same branches can be generated with either one of the mentioned formalisms.

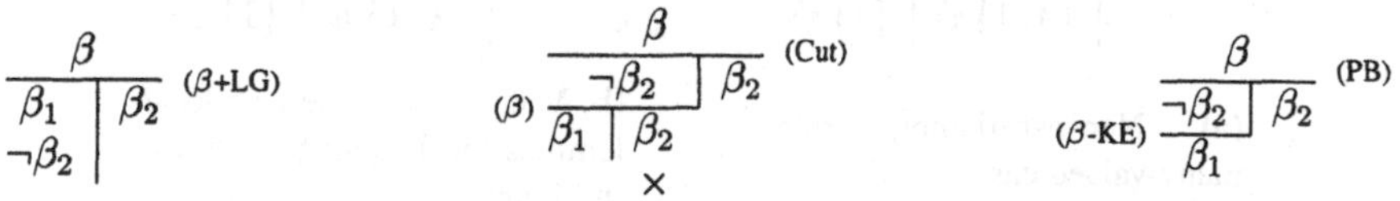

(i) β-rule with local lemmas:

$$\frac{\beta}{\beta_1 \;\big|\; \beta_2 \atop \neg\beta_2 \quad\;} \;(\beta\text{+LG})$$

(ii) β-rule and analytic cut:

$$\frac{\beta}{\dfrac{\neg\beta_2}{\beta_1 \;\big|\; \beta_2 \atop \times}\,(\beta) \;\Big|\; \beta_2} \;(\text{Cut})$$

(iii) Principle of Bivalence and KE β-rule:

$$\frac{\beta}{\dfrac{\neg\beta_2}{\beta_1}\,(\beta\text{-KE}) \;\Big|\; \beta_2} \;(\text{PB})$$

Figure 6. Equivalent extensions of classical tableau rules

[23] See D'Agostino's Chapter for a formal definition.

In truth table (9) we saw how the union of the extensions of a sets-as-signs rule correspond to a complete covering of the truth table entries that occur in the sign of the premise. This covering is not necessarily a partition that is some entries are possibly covered in more than one extension as, for example, it is the case with the field containing 0 in (9).

With analytic cut, respectively, with local lemmas one can enforce that the extensions of a rule form a partition of the entries to be covered. For the classical case this can be seen in Figure 7(a) in which the lighter shaded area corresponds to the left extensions in Figure 6 and the darker shaded area to the right extensions.

$\vee$	0	1
0	0	1
1	1	1

(a) A partitioning of classical disjunction

$\supset$	0	$\frac{1}{2}$	1
0	1	1	1
$\frac{1}{2}$	$\frac{1}{2}$	1	1
1	0	$\frac{1}{2}$	1

(b) A partitioning of Łukasiewicz implication

Figure 7. Partitions of truth table entries that generate analytic cuts

In Figure 7(b) a partition of the entries that contain the truth value 1 in the truth table of Łukasiewicz implication is displayed. In Figure 8(a) a non-partitioning rule for truth value 1 and Łukasiewicz implication is shown. The rule corresponding to the partition in Figure 7(b) is displayed in Figure 8(b). Formally, we define partitioning rules as follows:

$$\frac{\{1\}\,\phi \supset \psi}{\{0\}\,\phi \;\Big|\; \begin{array}{c}\{0,\frac{1}{2}\}\,\phi\\ \{\frac{1}{2},1\}\,\psi\end{array} \;\Big|\; \{1\}\,\psi}$$

(a) Non-partitioning rule, many-valued case

$$\frac{\{1\}\,\phi \supset \psi}{\{0\}\,\phi \;\Big|\; \begin{array}{c}\{\frac{1}{2}\}\,\phi\\ \{\frac{1}{2},1\}\,\psi\end{array} \;\Big|\; \begin{array}{c}\{1\}\,\phi\\ \{1\}\,\psi\end{array}}$$

(b) Many-valued rule with local lemmas for the partition shown in Figure 7(b).

Figure 8. Equivalent extensions of many-valued tableau rules

DEFINITION 66. Let $\Phi = \bigvee_{r=1}^{M} C_r$ be a sets-as-signs DNF representation of $\phi = S\,\theta(\phi_1,\ldots,\phi_m)$ $(m \geq 1)$. We call Φ a *partitioning DNF representation* or a *DNF representation with local lemmas* of ϕ iff for any two conjuncts C_i, C_j with

$i \neq j$ the set of literals $C_i \cup C_j$ is inconsistent. *Partitioning sets-as-signs rules,* respectively, *sets-as-signs rules with local lemmas* are sets-as-signs rules based on a partitioning DNF representation.

Just as in classical logic (cf. Figure 6) it is possible to *derive* many-valued local lemma rules from arbitrary ones with the help of many-valued analytic cut:

$$(26) \quad \frac{}{S_1\,\phi \mid \cdots \mid S_m\,\phi} \quad m \geq 2, \{S_1, \ldots, S_m\} \text{ set partition of } N$$

Note that in the light of our previous analysis it suffices to define the cut rule based on *partitions* of N rather than on the broader class of *coverings* in order to serve as the proper generalization of Takahashi's and Rousseau's cut rules for singleton signs. This sharpens the analysis given in [Hähnle, 1994, Section 6.1.4].

The partition displayed in Figure 7(b) obviously corresponds to the following instance of the many-valued cut rule: $\{0\}\,\phi \mid \{\frac{1}{2}\}\,\phi \mid \{1\}\,\phi$. With it we can derive the rule in Figure 8(b) from the rule in Figure 8(a). One first applies the cut rule and then in each extension the rule in Figure 8(a):

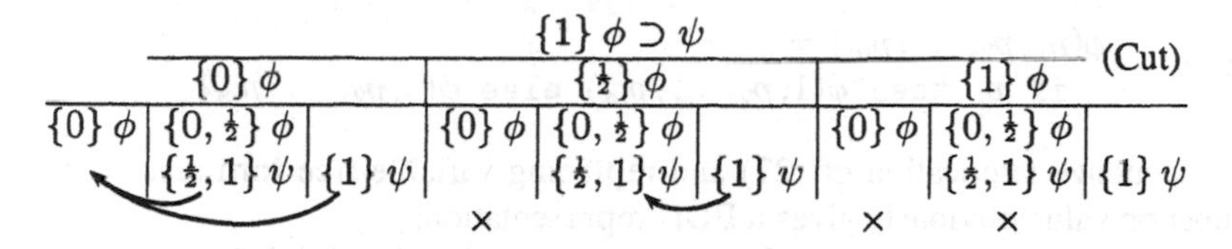

It is an immediate consequence of Definition 64 that if the literals on a branch B subsume the literals on a branch B' then whenever B' can be closed B can be closed as well. Therefore, it is only necessary to keep extensions which are not subsumed any other extension (subsumed extensions are denoted with arrows in the rule above). If we delete as well inconsistent extensions then the reductions of the remaining extensions are exactly those of the rule in Figure 8(b). With a different cut rule we would have obtained a different partitioning rule.

7 CONNECTIONS AND APPLICATIONS

7.1 Many-valued Decision Diagrams

Binary decision diagrams (BDDs) are a family of efficient data structures for representation of Boolean formulas; a standard reference is [Bryant, 1986]. Their main strength is that they can represent the models of very large satisfiable formulas in an efficient manner. Moreover, insertion of new formulas and combination of BDDs can be done quickly as well. There exist very efficient packages for BDD manipulation [Brace *et al.*, 1990] whose use is very popular in hardware verification [Burch *et al.*, 1990].

Basically, BDDs are a representation of Boolean functions based on the three-place `if-then-else` connective:

$$\texttt{if } i \texttt{ then } j \texttt{ else } k = \begin{cases} j & \text{if } i = 1 \\ k & \text{if } i = 0 \end{cases}$$

Every Boolean function can be expressed with a formula that contains no connective but `if-then-else` and logical constants 0 and 1, where atomic formulas occur as the first argument of `if-then-else` and nowhere else. For instance, $p \wedge q$ is equivalent to

$$\texttt{if } p \texttt{ then (if } q \texttt{ then 1 else 0) else } 0 \ .$$

Such a representation of a formula is called a BDD. A systematic way to obtain a BDD representation of a formula or logical function ϕ is provided by the *Boole-Shannon expansion*.[24] Assume that the atoms occurring in ϕ are $\{p_1, \ldots, p_m\}$ and denote this with $\phi(p_1, \ldots, p_m)$. Then

$$\phi(p_1, p_2, \ldots, p_m) = \texttt{if } p_1 \texttt{ then } \phi(1, p_2, \ldots, p_m) \texttt{ else } \phi(0, p_2, \ldots, p_m) \tag{27}$$

Recursive application of (27) and replacing variable-free formulas with their function value obviously gives a BDD representation.

Usually, BDDs are assumed to be *reduced* and *ordered* (ROBDDs). Reduced means that the syntactic tree of a BDD is turned into a graph by identifying isomorphic subtrees and applying a simplification step based on the equation

$$(\texttt{if } i \texttt{ then } j \texttt{ else } j) = j \ .$$

Ordered means that relative to a given total ordering $\prec$ of the atoms, whenever q occurs in the body of `if` p ... then $p \prec q$ must hold. An important property of ROBDDs is that two ROBDDs of the same Boolean function are identical up to isomorphism that is they are a *strong normal form* for Boolean functions.

The relevance of BDDs to the present chapter comes from the facts that first, there is a close relationship between BDDs and tableaux with local lemmas [Posegga, 1993] and second, they can be extended to *many-valued decision diagrams* and finitely-valued logics in a natural way by simply replacing the `if-then-else` with an $(n+1)$-ary `case-of` connective in n-valued logic:

[24] In the BDD and function minimization literature this equation is usually attributed to Shannon [1938], however, it appears already in [Boole, 1854]. Expansions are sometimes called *decompositions*.

$$\begin{array}{l}\texttt{case}\ i\ \texttt{of}\\ \quad 0:\ j_0;\\ \quad \frac{1}{n-1}:\ j_{\frac{1}{n-1}};\\ \quad \cdots\ \cdots\\ \quad 1:\ j_1\\ \texttt{esac}\end{array} = \begin{cases} j_0 & \text{if } i = 0 \\ j_{\frac{1}{n-1}} & \text{if } i = \frac{1}{n-1} \\ \cdots & \cdots \\ j_1 & \text{if } i = 1 \end{cases}$$

Orłowska [1967] gave a proof procedure for propositional Post logic based on an MDD-like structure, but she used a different notation. MDDs were rediscovered in [Thayse *et al.*, 1979; Srinivasan *et al.*, 1990] in connection with the growing interest in BDD methods. There it is also shown that like their binary counterparts n-valued MDDs are functionally complete and they can be computed with a generalized Boole-Shannon-expansion.

$$\phi(p_1, p_2, \ldots, p_m) = \begin{array}{l}\texttt{case}\ i\ \texttt{of}\\ \quad 0:\ \phi(0, p_2, \ldots, p_m);\\ \quad \frac{1}{n-1}:\ \phi(\frac{1}{n-1}, p_2, \ldots, p_m);\\ \quad \cdots\ \cdots\\ \quad 1:\ \phi(1, p_2, \ldots, p_m)\\ \texttt{esac}\end{array} \tag{28}$$

and that ROMDDs (called *canonical function graphs* in [Srinivasan *et al.*, 1990]) are a strong normal form representation of n-valued functions.

As already mentioned, BDDs and tableaux with local lemmas (or, more generally, tableaux with partitioning rules as introduced in Definition 66 in the previous section) bear a close relationship. Consider the tableau rule and BDD depicted in Figure 9(a) and (b).

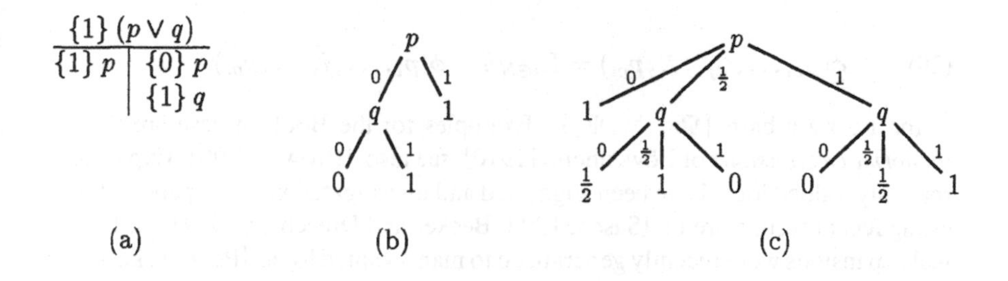

Figure 9. Signed tableaux with local lemmas versus BDDs and MDDs

In the BDD `then` and `else` branches are labelled with 1 and 0, respectively. An edge labelled with i that comes out of a node p can be seen as an assertion of the truth value i to p, in other words a signed formula $\{i\}\ p$. Now the following relationship between tableaux with local lemmas and BDDs holds (cf. [Posegga, 1993]): for each set of signed literals corresponding to the edges on a path in a BDD for ϕ that ends with 1 there is an open branch in any tableau with local lemmas for $\{1\}\ \phi$ containing exactly the same literals and vice versa.

This relationship extends to singleton signs tableaux with partitioning rules and MDDs in the following way: for each set of signed literals corresponding to the edges on a path in an MDD for ϕ that ends with j there is an open branch in any singleton sets tableau with partitioning rules for $\{j\}\ \phi$ containing exactly the same literals and vice versa. For instance, the reduced MDD for $\{\frac{1}{2}\}\ (p \supset q)$, the three-valued Łukasiewicz implication defined in (2), is displayed in Figure 9(c); the paths ending with $\frac{1}{2}$ correspond to the extensions of the tableau rule from Example 24 in Section 4.1.

A BDD or MDD representation (or the literals on the open branches of a partitioning tableau) can be seen as a Boolean polynomial over signed literals with truth values as coefficients. Let us write a signed literal of the form $\{i\}\ p$ as p^i, 'and' as '$\cdot$', 'or' as '$+$'. Then, by (27), for example,

$$\begin{aligned} p \vee q &= p^1 \cdot (1 \vee q) + p^0 \cdot (0 \vee q) \\ &= p^1 \cdot 1 + p^0 \cdot (q^1 \cdot (0 \vee 1) + q^0 \cdot (0 \vee 0)) \\ &= p^1 \cdot 1 + p^0 \cdot q^1 \cdot 1 + p^0 \cdot q^0 \cdot 0 \end{aligned}$$

Polynomial representations can be generalized. In the many-valued case one may, of course, consider arbitrary signed literals of the form $S\ p$. Written as p^S they are well-known in logic circuit synthesis and optimization [Sasao, 1981], sometimes under the name *set literal* or *universal literal*.[25] On the other hand, there is no reason to restrict oneself to unary functions for the base of a polynomial representation. Of course one needs to make restrictions lest the resulting representations are useless. Equation (28), for example, might be generalized to

$$\text{(29)} \qquad \phi(p_1, \ldots, p_j, \ldots, p_m) = \Sigma_{i \in N} \Psi_i \cdot \phi(p_1, \ldots, i, \ldots, p_m)$$

for a certain base $\{\Psi_0, \ldots, \Psi_1\}$. Examples for the Boolean case are the *orthonormal expansions* of Löwenheim [1910], see also [Brown, 1990]. Expansions for many-valued logic have been suggested and investigated in many papers. Interesting recent results are in [Sasao, 1992; Becker and Dreschler, 1994]. Orthonormal expansions were recently generalized to many-valued logic [Perkowski, 1992] in an attempt to systematize the plethora of existing subclasses. Certain orthonor-

[25] In logic optimization often restrictions on the form of S are imposed, for example, a popular class of literals are *window literals*, where S is of the form $[i, j]$ with $i \leq j$. Recently, however, the use of completely general literals has been advocated [Dueck and Butler, 1994].

mal expansions based on unary functions result in partitioning sets-as-signs rules as introduced in Definition 66 in the previous section. More general expansions have no representation in rule form; on the other hand, arbitrary sets-as-signs tableau rules do not necessarily correspond to any expansion of the form (29). The reason for this is that tableau rules work by analysing the leading connective of a complex formula, whereas expansion schemas as used in MDDs and logic synthesis work by analysing a certain variable of the formula. The exact relationship between both approaches remains to be investigated.

7.2 *Logic Synthesis*

In this section we strengthen the links between tableau-based deduction and logic design by demonstrating that OR-AND-OR implementations can be synthesized via sets-as-signs tableaux. As a background reading on logical circuit synthesis we suggest [Brayton *et al.*, 1984].

EXAMPLE 67. (cf. [Sasao, 1993]) Let the Boolean 4-ary function $f(z, w, x, y)$ be defined by the Karnaugh map in Figure 10(a). It is possible to interpret f as a *four-valued binary function* $f(Y, X)$ (displayed in Figure 10(c)) via the correspondence given in Figure 10(b). A minimal sets-as-signs DNF tableau rule for $\{1\}\ f(Y, X)$ is easily obtained:

$\{1\}\ f(Y,X)$

$\{0,\frac{1}{3},\frac{2}{3}\}\,X$	$\{\frac{2}{3},1\}\,X$	$\{\frac{1}{3}\}\,X$	$\{0,1\}\,X$
$\{0\}\,Y$	$\{\frac{1}{3}\}\,Y$	$\{\frac{2}{3}\}\,Y$	$\{1\}\,Y$

Decoding the variables in the extensions into (z, w, x, y) one obtains:

$1\ f(z,w,x,y)$

$0\,x \vee 1\,y$	$1\,x$	$0\,x$	
		$1\,y$	$0\,y$
$0\,z$	$0\,z$	$1\,z$	$1\,z$
$0\,w$	$1\,w$	$1\,w$	$0\,w$

This rule, however, can directly serve as a specification for the OR-AND-OR circuit displayed in Figure 11.

In general, the specification of a function f can be nested, thus several rule applications may be needed. The structure of the open branches of the resulting tableau gives the AND-OR or PLA (see below) part of the circuit, while the signed literals in the extensions represent another level of ORs (only one OR gate is needed in the example). Instead of a whole level of ORs one can also supply a fixed structure: an n-bit decoder (in the example n is 2) mapping a group of n two-valued variables to an 2^n-valued variable. This class of circuits is also known as *programmable logic arrays* (PLA) with n-bit decoder.

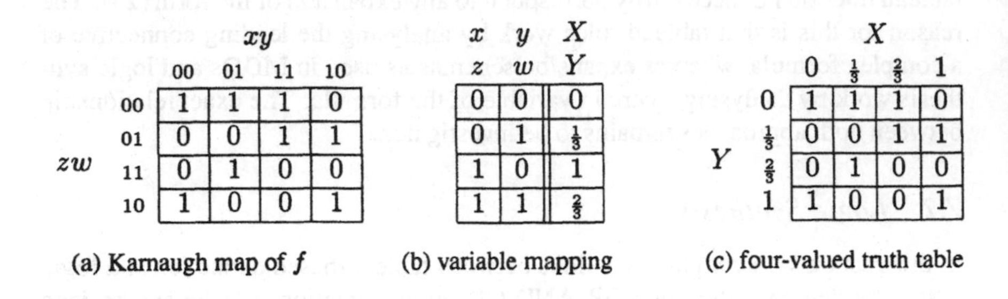

(a) Karnaugh map of f (b) variable mapping (c) four-valued truth table

Figure 10. The function from Example 67 as Karnaugh map and four-valued truth table

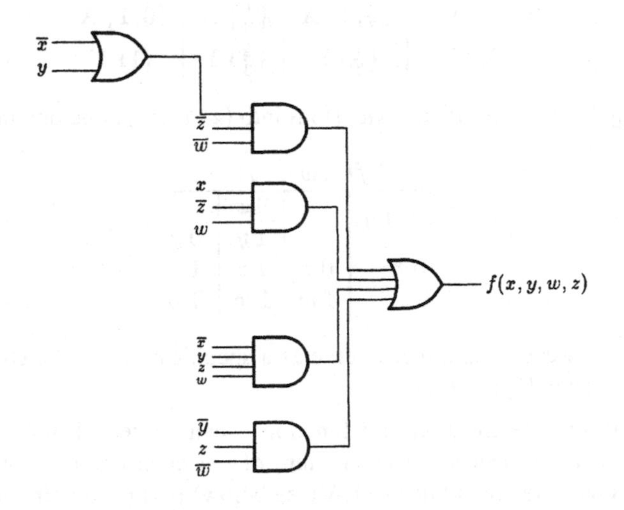

Figure 11. OR-AND-OR circuit

It is shown in [Sasao, 1981] that AND-OR realizations with decoders and, even more so, OR-AND-OR realizations can be consideraby smaller than mere AND-OR realizations. As is to be expected, among other factors the choice of the mapping between binary and many-valued variables influences the quality of the result [Sasao, 1984].

Although we exhibited many parallels and correspondences between many-valued calculi and methods from logic design, in both areas rather different goals are pursued: The methods developed for obtaining minimal representations of many-valued functions in logic design are often of a heuristical[26] nature which means they yield in general only a near minimal solution; on the other hand, they can deal with rather large inputs. The functions to be modelled as a circuit are, of course, typically not constant, hence neither tautologies nor unsatisfiable. And finally, the specification of logic design problems is almost always purely propositional. Still we think it would be fruitful if both fields became better aware of each other.

7.3 Implementations

The main purpose of this handbook is to provide a reference for the theoretical foundations of the tableau method. Implementational issues are discussed in a generic way in a separate chapter. Any kind of information on actual implementations or on applications is doomed to become quickly outdated. Despite this we feel that our presentation would be incomplete without even a short glance on existing implementations and tableau-specific applications as provided in the present and in the following subsection.

Theorem Provers $_3\mathcal{T}^A\mathcal{P}$ [Beckert *et al.*, 1996a; Beckert *et al.*, 1996] is a Prolog-based implementation[27] of a generic tableau-based theorem prover for full order-sorted first-order finitely-valued logic. Its two-valued instance has special routines for treating equality. It requires Sicstus Prolog and a Unix machine to run.

The prover *Deep Thought*[28] [Gerberding, 1996] offers a subset of the functionality of $_3\mathcal{T}^A\mathcal{P}$ and essentially is a reimplementation in C. It runs on Amigas and various Unix platforms.

The system KARNAK [Hoogewijs and Elnadi, 1994] is a theorem prover for a variant of Kleene's three-valued logic, cf. Example 7 in Section 2.1. As the underlying calculus does not employ explicit meta connectives it is an example for the line of development sketched in Section 3.2. Another

[26]There are exact methods, though: see, for example, [Brayton *et al.*, 1993].

[27]$_3\mathcal{T}^A\mathcal{P}$ is available without charge from `http://i12www.ira.uka.de/~threetap/`.

[28]*Deep Thought* is available without charge from
`http://kirmes.inferenzsysteme.informatik.th-darmstadt.de/~stefan/dt.html`.

implementation of a reasoning system for an extended three-valued Kleene logic is reported in [Doherty, 1991a].

An implementation of the method outlined in Section 5.3 with Constraint Logic Programming languages such as Eclipse or CLP($\mathcal{R}$) is provided in [Hähnle, 1997].

MDD and Function Minimization Packages In the light of the links between MDDs, logic minimization and tableaux established in Sections 7.1 and 7.2 we provide two pointers to implementations in that area: [Srinivasan *et al.*, 1990] reports the implementation of an MDD package in the spirit of the well-known BDD implementation by Brace et al. [1990] while [Rudell and Sangiovanni-Vincentelli, 1987] describes a more specialized package for many-valued function minimization.

Miscellaneous Tools The system MULTLOG [Salzer, 1996; Vienna Group, 1996] automatically derives singleton signs tableau, sequent (dual tableau), and natural deduction rules for given finitely-valued connectives and quantifiers and can be considered as an automation of the corresponding constructive proofs in [Baaz and Fermüller, 1995a].

7.4 Applications Specific to Many-valued Tableaux

In the following we restrict ourselves to material with direct relevance for tableau-like systems. A more complete list of (potential) applications of many-valued deduction can be found in [Hähnle, 1994, Chapter 7].

- As is evident from D'Agostino *et al.*'s Chapter in this volume, tableaux are particularly suitable to characterize various non-monotonic and default logics. Doherty [1990; 1991; 1991a] gives a tableau formulation of a non-monotonic three-valued logic using sets-as-signs rules.
- The use of many-valued logic to deal with partiality in mathematics and program verification has been suggested several times, for example, in the specification languages VDM [Blikle, 1991] and Spectrum [Broy *et al.*, 1993]. As many program verification systems are based on sequent calculi [Heisel *et al.*, 1991], tableau-based many-valued deduction can improve the degree of automation in such systems. A tableau system for a sorted many-valued logic modelling partially defined mathematical expressions was suggested by [Kerber and Kohlhase, 1996].
- MDD methods (cf. Section 7.1, [Srinivasan *et al.*, 1990]) can be used in formal hardware verification [Bryant and Seger, 1990; Hähnle and Kernig, 1993; Sasao, 1996]. If their expressivity is to be increased by introducing quantification it might be advantageous to view them as variants of tableaux.

- A many-valued tableau system with applications in knowledge representation was reported in [Straccia, 1997].
- Moscato *et al.* [1994; 1995] use many-valued signed tableaux to improve deduction in intuitionistic and modal logic.

7.5 *Many-valued Logics and Substructural Logics*

In this section we briefly sketch how many-valued logics and substructural logics as discussed in D'Agostino *et al.*'s Chapter can be related.

Consider singleton-signed formulas $\{i\}\ \phi$ in some many-valued logic $\mathcal{L}$ with truth values N. It is straightforward to define a *frame semantics* $\mathcal{F}$ for $\mathcal{L}$ by letting N be the points of $\mathcal{F}$ and by forcing ϕ at i, whenever $\{i\}\ \phi$ is valid in $\mathcal{L}$. In a similar way, the semantic conditions of the connectives are determined by their meaning in $\mathcal{L}$ (note that this construction does in general not work in the opposite direction: connectives determined by a frame semantics might simply be not truth functional).

The resulting frame will typically be very unnatural unless the many-valued connectives fulfill some of the algebraic properties commonly present in frame semantics. This is the case for such logics as Łukasiewicz logic or Post logic for which meaningful algebraic semantics exist.

A closer investigation of the correspondances between many-valued and substructural logics could turn out to be fruitful indeed. Among the many possible perspectives are the following: if a substructural logic *can* be (partially) represented as a many-valued logic, then the well-understood proof theory of signed many-valued logic opens a door to its automation; or transfer the idea of using sets of truth values as signs to the substructural case thus arriving at sets of points. Like in the many-valued case a lot theoretical and practical advantages could be gained from doing so.

8 INSTEAD OF A CONCLUSION

The investigation of the tableau case answered the question for the basic computational components of many-valued logic such as: what are atomically inconsistent sets, what are saturated sets, which meta connectives are needed for an adequate representation of many-valued semantics, what is polarity, what is a link etc.

Once these questions are answered, one can design many-valued versions of other deduction paradigms than tableaux as well.

In our experience, redundancies in non-classical proof procedures tend to be more obvious in tableau frameworks than in other procedures (for instance, the excessive duplication of formulas when only singleton signs are present is a lot harder to see in resolution based systems). Thus we hope that the present chapter can serve both as an argument for using tableaux in non-classical deduction as well as a paradigmatic example for problems and solutions in non-classical deduction.

ACKNOWLEDGEMENTS

I would like to thank Verena Rose and Daniele Mundici (second reader of this chapter) for many useful comments on a draft of this text. I also profited from several stimulating discussions with the Vienna Group on Many-Valued Logic (Matthias Baaz, Christian Fermüller, Gernot Salzer, and Richard Zach).

University of Karlsruhe, Germany.

REFERENCES

[Avron, 1991] A. Avron. Natural 3-valued logics—characterization and proof theory. *Journal of Symbolic Logic*, 56(1):276–294, 1991.

[Baaz and Fermüller, 1995] M. Baaz and C. G. Fermüller. Nonelementary speedups between different versions of tableaux. In Peter Baumgartner, Reiner Hähnle, and Joachim Posegga, editors, *Proc. 4th Workshop on Deduction with Tableaux and Related Methods, St. Goar, Germany*, volume 918 of *LNCS*, pages 217–230. Springer-Verlag, 1995.

[Baaz and Fermüller, 1995a] M. Baaz and C. G. Fermüller. Resolution-based theorem proving for many-valued logics. *Journal of Symbolic Computation*, 19(4):353–391, April 1995.

[Baaz and Zach, 1992] M. Baaz and R. Zach. Note on calculi for a three-valued logic for logic programming. *Bulletin of the EATCS*, 48:157–164, 1992.

[Becker and Dreschler, 1994] B. Becker and R. Drechsler. Efficient graph-based representation of multi-valued functions with an application to genetic algorithms. In *Proc. 24th International Symposium on Multiple-Valued Logic, Boston/MA*, pages 65–72. IEEE Press, Los Alamitos, May 1994.

[Beckert *et al.*, 1996] B. Beckert, R. Hähnle, K. Geiß, P. Oel, C. Pape, and M. Sulzmann. The Many-Valued Tableau-Based Theorem Prover $_3T^AP$, Version 4.0. Interner Bericht 3/96, Universität Karlsruhe, Fakultät für Informatik, 1996.

[Beckert *et al.*, 1996a] B. Beckert, R. Hähnle, P. Oel, and M. Sulzmann. The tableau-based theorem prover $_3T^AP$, version 4.0. In Michael McRobbie and John Slaney, editors, *Proc. 13th Conference on Automated Deduction, New Brunswick/NJ, USA*, volume 1104 of *LNCS*, pages 303–307. Springer-Verlag, 1996.

[Beckert *et al.*, 1993] B. Beckert, R. Hähnle, and P. H. Schmitt. The *even more* liberalized δ-rule in free variable semantic tableaux. In G. Gottlob, A. Leitsch, and D. Mundici, editors, *Proceedings of the third Kurt Gödel Colloquium KGC'93, Brno, Czech Republic*, volume 713 of *LNCS*, pages 108–119. Springer-Verlag, August 1993.

[Blikle, 1991] A. Blikle. Three-valued predicates for software specification and validation. *Fundamenta Informaticae*, XIV:387–410, 1991.

[Bloesch, 1994] A. Bloesch. Tableau style proof systems for various many-valued logics. Technical Report 94-18, Software Verification Research Center, Dept. of Computer Science, University of Queensland, April 1994.

[Bolc and Borowik, 1992] L. Bolc and P. Borowik. *Many-Valued Logics. 1: Theoretical Foundations.* Springer-Verlag, 1992.

[Boole, 1854] G. Boole. *An Investigation of the Laws of Thought.* Walton, London, 1854. Reprinted by Dover Books, New York, 1954.

[Brace *et al.*, 1990] K. S. Brace, R. L. Rudell, and R. E. Bryant. Efficient implementation of a BDD package. In *Proc. 27th ACM/IEEE Design Automation Conference*, pages 40–45. IEEE Press, Los Alamitos, 1990.

[Brayton *et al.*, 1984] R. K. Brayton, G. D. Hachtel, C. T. McMullen, and A. L. Sangiovanni-Vincentelli. *Logic Minimization Algorithms for VLSI Synthesis.* Kluwer, Boston, 1984.

[Brayton *et al.*, 1993] R. K. Brayton, P. C. McGeer, J. V. Sanghavi, and A. L. Sangiovanni-Vincentelli. A new exact minimizer for two-level logic synthesis. In Tsutomu Sasao, editor, *Logic Synthesis and Optimization*, chapter 1, pages 1–32. Kluwer, Norwell/MA, USA, 1993.

[Brown, 1990] F. M. Brown. *Boolean Reasoning.* Kluwer, Norwell/MA, USA, 1990.

[Broy *et al.*, 1993] M. Broy, C. Facchi, R. Grosu, R. Hettler, H. Hussmann, Dieter Nazareth, Franz Regensburger, and Ketil Stølen. The requirement and design specification language Spectrum, an informal introduction, version 1.0. Technical report, Institut für Informatik, Technische Universität München, March 1993.

[Bryant and Seger, 1990] R. E. Bryant and C.-J. H. Seger. Formal verification of digital circuits using symbolic ternary system models. In E. M. Clarke and R. P. Kurshan, editors, *Computer-Aided Verification: Proc. of the 2nd International Conference CAV'90*, LNCS 531, pages 33–43. Springer-Verlag, 1991.

[Bryant, 1986] R. E. Bryant. Graph-based algorithms for Boolean function manipulation. *IEEE Transactions on Computers*, C-35:677–691, 1986.

[Burch *et al.*, 1990] J. R. Burch, E. M. Clarke, K. L. McMillan, and D. L. Dill. Sequential circuit verification using symbolic model checking. In *Proc. 27th Design Automation Conference (DAC 90)*, pages 46–51, 1990.

[Carnielli, 1985] W. Carnielli. An algorithm for axiomatizing and theorem proving in finite many-valued propositional logics. *Logique et Analyse*, 28(112):363–368, 1985.

[Carnielli, 1987] W. A. Carnielli. Systematization of finite many-valued logics through the method of tableaux. *Journal of Symbolic Logic*, 52(2):473–493, June 1987.

[Carnielli, 1991] W. A. Carnielli. On sequents and tableaux for many-valued logics. *Journal of Non-Classical Logic*, 8(1):59–76, May 1991.

[Davey and Priestley, 1990] B. A. Davey and H. A. Priestley. *Introduction to Lattices and Order*. Cambridge Mathematical Textbooks. Cambridge University Press, Cambridge, 1990.

[Doherty, 1990] P. Doherty. Preliminary report: NM3 — a three-valued non-monotonic formalism. In Z. Raś, M. Zemankova, and M. Emrich, editors, *Proc. of 5th Int. Symposium on Methodologies for Intelligent Systems, Knoxville, TN*, pages 498–505. North-Holland, 1990.

[Doherty, 1991] P. Doherty. A constraint-based approach to proof procedures for multi-valued logics. In *First World Conference on the Fundamentals of Artificial Intelligence WOCFAI-91, Paris*, 1991.

[Doherty, 1991a] P. Doherty. *NML3 — A Non-Monotonic Formalism with Explicit Defaults.* PhD thesis, University of Linköping, Sweden, 1991.

[Dueck and Butler, 1994] G. W. Dueck and J. T. Butler. Multiple-valued logic operations with universal literals. In *Proc. 24th International Symposium on Multiple-Valued Logic, Boston/MA*, pages 73–79. IEEE Press, Los Alamitos, May 1994.

[Epstein, 1960] G. Epstein. The lattice theory of Post algebras. *Transactions of the American Mathematical Society*, 95(2):300–317, May 1960.

[Fitting, 1991] M. C. Fitting. Bilattices and the semantics of logic programming. *Journal of Logic Programming*, 11(2):91–116, August 1991.

[Fitting, 1996] M. C. Fitting. *First-Order Logic and Automated Theorem Proving*. Springer-Verlag, New York, 1996. First edition, 1990.

[Gerberding, 1996] S. Gerberding. DT—an automated theorem prover for multiple-valued first-order predicate logics. In *Proc. 26th International Symposium on Multiple-Valued Logics, Santiago de Compostela, Spain*, pages 284–289. IEEE Press, Los Alamitos, May 1996.

[Ginsberg, 1988] M. L. Ginsberg. Multi-valued logics. *Computational Intelligence*, 4(3), 1988.

[Hähnle, 1990] R. Hähnle. Towards an efficient tableau proof procedure for multiple-valued logics. In Egon Börger, Hans Kleine Büning, Michael M. Richter, and Wolfgang Schönfeld, editors, *Selected Papers from Computer Science Logic, CSL'90, Heidelberg, Germany*, volume 533 of *LNCS*, pages 248–260. Springer-Verlag, 1991.

[Hähnle, 1991] R. Hähnle. Uniform notation of tableaux rules for multiple-valued logics. In *Proc. International Symposium on Multiple-Valued Logic, Victoria*, pages 238–245. IEEE Press, Los Alamitos, 1991.

[Hähnle, 1992] R. Hähnle. A new translation from deduction into integer programming. In Jacques Calmet and John A. Campbell, editors, *Proc. Int. Conf. on Artificial Intelligence and Symbolic Mathematical Computing AISMC-1, Karlsruhe, Germany*, volume 737 of *LNCS*, pages 262–275. Springer-Verlag, 1992.

[Hähnle, 1994] R. Hähnle. *Automated Deduction in Multiple-Valued Logics*, volume 10 of *International Series of Monographs on Computer Science*. Oxford University Press, 1994.

[Hähnle, 1994a] R. Hähnle. Many-valued logic and mixed integer programming. *Annals of Mathematics and Artificial Intelligence*, 12(3,4):231–264, December 1994.

[Hähnle, 1994b] R. Hähnle. Short conjunctive normal forms in finitely-valued logics. *Journal of Logic and Computation*, 4(6):905–927, 1994.

[Hähnle, 1996] R. Hähnle. Exploiting data dependencies in many-valued logics. *Journal of Applied Non-Classical Logics*, 6(1):49–69, 1996.

[Hähnle, 1997] R. Hähnle. Proof theory of many-valued logic—linear optimization—logic design: Connections and interactions. *Soft Computing—A Fusion of Foundations, Methodologies and Applications*, 1(3):107–119, September 1997.

[Hähnle, 1998] R. Hähnle. Commodious axiomatization of quantifiers in multiple-valued logic. *Studia Logica*, 61(1):101–121, 1998. Special Issue on Many-Valued Logics, their Proof Theory and Algebras.

[Hähnle and Escalada-Imaz, 1997] R. Hähnle and G. Escalada-Imaz. Deduction in many-valued logics: a survey. *Mathware & Soft Computing*, IV(2):69–97, 1997.

[Hähnle and Kernig, 1993] R. Hähnle and W. Kernig. Verification of switch level designs with many-valued logic. In Andrei Voronkov, editor, *Proc. LPAR'93, St. Petersburg, Russia*, volume 698 of *LNCS*, pages 158–169. Springer-Verlag, 1993.

[Hähnle and Schmitt, 1994] R. Hähnle and P. H. Schmitt. The liberalized δ-rule in free variable semantic tableaux. *Journal of Automated Reasoning*, 13(2):211–222, October 1994.

[Heisel *et al.*, 1991] M. Heisel, W. Reif, and W. Stephan. Formal software development in the KIV system. In L. McCartney, editor, *Automating Software Design*. AAAI Press, 1991.

[Hoogewijs and Elnadi, 1994] A. Hoogewijs and T. M. Elnadi. KARNAK: an automated theorem prover for PPC. The CAGe Reports 10, University of Gent, Computer Algebra Group, 1994.

[Hooker, 1988] J. N. Hooker. A quantitative approach to logical inference. *Decision Support Systems*, 4:45–69, 1988.

[Hösli, 1993] B. Hösli. *Robuste Logik*. PhD thesis, Eidgenössische Technische Hochschule Zürich, 1993.

[Jeroslow, 1988] R. G. Jeroslow. *Logic-Based Decision Support. Mixed Integer Model Formulation*. Elsevier, Amsterdam, 1988.

[Kapetanović and Krapež, 1989] M. Kapetanović and A. Krapež. A proof procedure for the first-order logic. *Publications de l'Institut Mathematiqu, nouvelle série*, 59(45):3–5, 1989.

[Kerber and Kohlhase, 1996] M. Kerber and M. Kohlhase. A tableau calculus for partial functions. In *Collegium Logicum. Annals of the Kurt-Gödel-Society*, volume 2, pages 21–49. Springer-Verlag, Wien New York, 1996.

[Kifer and Subrahmanian, 1992] M. Kifer and V. S. Subrahmanian. Theory of generalized annotated logic programming and its applications. *Jornal of Logic Programming*, 12:335–367, 1992.

[Kirin, 1963] V. G. Kirin. On the polynomial representation of operators in the n-valued propositional calculus (in Serbocroatian). *Glasnik Mat.-Fiz. Astronom. Društvo Mat. Fiz. Hravtske Ser. II*, 18:3–12, 1963. Reviewed in MR 29 (1965) p. 420.

[Kirin, 1966] V. G. Kirin. Gentzen's method of the many-valued propositional calculi. *Zeitschrift für mathematische Logik und Grundlagen der Mathematik*, 12:317–332, 1966.

[Löwenheim, 1910] L. Löwenheim. Über die Auflösung von Gleichungen im logischen Gebietekalkül. *Mathematische Annalen*, 68:169–207, 1910.

[Lu, 1996] J. J. Lu. Logic programming with signs and annotations. *Journal of Logic and Computation*, 6(6):755–778, 1996.

[Lu *et al.*, 1993] J. J. Lu, N. V. Murray, and E. Rosenthal. Signed formulas and annotated logics. In *Proc. 23rd International Symposium on Multiple-Valued Logics*, pages 48–53. IEEE Press, Los Alamitos, 1993.

[Malinowski, 1993] G. Malinowski. *Many-Valued Logics*, volume 25 of *Oxford Logic Guides*. Oxford University Press, 1993.

[Miglioli *et al.*, 1994] P. Miglioli, U. Moscato, and M. Ornaghi. An improved refutation system for intuitionistic predicate logic. *Journal of Automated Reasoning*, 13(3):361–374, 1994.

[Miglioli *et al.*, 1995] P. Miglioli, U. Moscato, and M. Ornaghi. Refutation systems for propositional modal logics. In P. Baumgartner, R. Hähnle, and J. Posegga, editors, *Proc. 4th Workshop on Deduction with Tableaux and Related Methods, St. Goar, Germany*, volume 918 of *LNCS*, pages 95–105. Springer-Verlag, 1995.

[Mostowksi, 1957] A. Mostowski. On a generalization of quantifiers. *Fundamenta Mathematicæ*, XLIV:12–36, 1957.

[Mundici, 1987] D. Mundici. Satisfiability in many-valued sentential logic is NP-complete. *Theoretical Computer Science*, 52:145–153, 1987.

[Mundici, 1994] D. Mundici. A constructive proof of McNaughton's Theorem in infinite-valued logic. *Journal of Symbolic Logic*, 59(2):596–602, June 1994.

[Murray, 1982] N. V. Murray. Completely non-clausal theorem proving. *Artificial Intelligence*, 18:67–85, 1982.

[Murray and Rosenthal, 1991] N. V. Murray and E. Rosenthal. Improving tableau deductions in multiple-valued logics. In *Proceedings 21st International Symposium on Multiple-Valued Logic, Victoria*, pages 230–237. IEEE Press, Los Alamitos, May 1991.

[Murray and Rosenthal, 1991a] N. V. Murray and E. Rosenthal. Resolution and path-dissolution in multiple-valued logics. In *Proceedings International Symposium on Methodologies for Intelligent Systems, Charlotte*, LNAI. Springer-Verlag, 1991.

[Murray and Rosenthal, 1994] N. V. Murray and E. Rosenthal. Adapting classical inference techniques to multiple-valued logics using signed formulas. *Fundamenta Informaticae*, 21(3):237–253, 1994.

[Nait Abdallah, 1995] M. Areski Nait Abdallah. *The Logic of Partial Information*. Monographs in Theoretical Computer Science — An EATCS Series. Springer-Verlag, 1995.

[O'Hearn and Stachniak, 1992] P. O'Hearn and Z. Stachniak. A resolution framework for finitely-valued first-order logics. *Journal of Symbolic Computing*, 13:235–254, 1992.

[Orłowska, 1967] E. Orłowska. Mechanical proof procedure for the n-valued propositional calculus. *Bull. de L'Acad. Pol. des Sci., Série des sci. math., astr. et phys.*, XV(8):537–541, 1967.

[Orłowska, 1985] E. Orłowska. Mechanical proof methods for Post Logics. *Logique et Analyse*, 28(110):173–192, 1985.

[Panti, to appear] G. Panti. Multi-valued logics. In D. Gillies and P. Smets, editors, *Handbook of Defensible Reasoning and Uncertainty Management Systems*, volume 1, chapter 2. Kluwer, to appear.

[Perkowski, 1992] M. A. Perkowski. The generalized orthonormal expansion of functions with multiple-valued inputs and some of its application. In *Proc. 22nd International Symposium on Multiple-Valued Logic*, pages 442–450. IEEE Press, Los Alamitos, May 1992.

[Plaisted and Greenbaum, 1986] D. A. Plaisted and S. Greenbaum. A structure-preserving clause form translation. *Journal of Symbolic Computation*, 2:293–304, 1986.

[Posegga, 1993] J. Posegga. *Deduktion mit Shannongraphen für Prädikatenlogik erster Stufe*. PhD thesis, University of Karlsruhe, 1993. diski 51, infix Verlag.

[Rescher, 1969] N. Rescher. *Many-Valued Logic*. McGraw-Hill, New York, 1969.

[Rosser and Turquette, 1952] J. B. Rosser and A. R. Turquette. *Many-Valued Logics*. North-Holland, Amsterdam, 1952.

[Rousseau, 1967] G. Rousseau. Sequents in many valued logic I. *Fundamenta Mathematicæ*, LX:23–33, 1967.

[Rousseau, 1970] G. Rousseau. Sequents in many valued logic II. *Fundamenta Mathematicæ*, LXVII:125–131, 1970.

[Rudell and Sangiovanni-Vincentelli, 1987] R. Rudell and A. Sangiovanni-Vincentelli. Multiple-valued minimization for PLA optimization. *IEEE Transactions on Computer-Aided Design*, 6(5):727–750, September 1987.

[Saloni, 1972] Z. Saloni. Gentzen rules for m-valued logic. *Bull. de L'Acad. Pol. des Sci., Série des sci. math., astr. et phys.*, XX:819–825, 1972.

[Salzer, 1996] G. Salzer. MUltlog: an expert system for multiple-valued logics. In *Collegium Logicum. Annals of the Kurt-Gödel-Society*, volume 2. Springer-Verlag, Wien New York, 1996.

[Salzer, 1996a] G. Salzer. Optimal axiomatizations for multiple-valued operators and quantifiers based on semilattices. In Michael McRobbie and John Slaney, editors, *Proc. 13th Conference on Automated Deduction, New Brunswick/NJ, USA*, volume 1104 of *LNCS*, pages 688–702. Springer-Verlag, 1996.

[Sasao, 1981] T. Sasao. Multiple-valued decomposition of generalized Boolean functions and the complexity of programmable logic arrays. *IEEE Transactions on Computers*, C-30:635–643, September 1981.

[Sasao, 1984] T. Sasao. Input variable assignment and output phase optimization of PLA's. *IEEE Transactions on Computers*, C-33(10):879–894, October 1984.

[Sasao, 1992] T. Sasao. Optimization of multi-valued AND-XOR expressions using multiple-place decision diagrams. In *Proc. 22nd International Symposium on Multiple-Valued Logic*, pages 451–458. IEEE Press, Los Alamitos, May 1992.

[Sasao, 1993] T. Sasao. Logic synthesis with EXOR gates. In Tsutomu Sasao, editor, *Logic Synthesis and Optimization*, chapter 12, pages 259–286. Kluwer, Norwell/MA, USA, 1993.

[Sasao, 1996] T. Sasao. Ternary decision diagrams and their applications. In Tsutomu Sasao and Masahiro Fujita, editors, *Representations of Discrete Functions*, chapter 12, pages 269–292. Kluwer, Norwell/MA, USA, 1996.

[Schrijver, 1986] A. Schrijver. *Theory of Linear and Integer Programming*. Wiley-Interscience Series in Discrete Mathematics. John Wiley & Sons, 1986.

[Schröter, 1955] K. Schröter. Methoden zur Axiomatisierung beliebiger Aussagen- und Prädikatenkalküle. *Zeitschrift für math. Logik und Grundlagen der Mathematik*, 1:241–251, 1955.

[Shannon, 1938] C. E. Shannon. A symbolic analysis of relay and switching circuits. *AIEE Transactions*, 67:713–723, 1938.

[Smullyan, 1995] R. M. Smullyan. *First-Order Logic*. Dover Publications, New York, second corrected edition, 1995. First published 1968 by Springer-Verlag.

[Srinivasan *et al.*, 1990] A. Srinivasan, T. Kam, S. Malik, and R. E. Brayton. Algorithms for discrete function manipulation. In *Proc. IEEE International Conference on CAD, Santa Clara/CA, USA*, pages 92–95. IEEE Press, Los Alamitos, November 1990.

[Stachniak, 1988] Z. Stachniak. The resolution rule: An algebraic perspective. In *Proc. of Algebraic Logic and Universal Algebra in Computer Science Conf.*, pages 227–242. Springer LNCS 425, Heidelberg, 1988.

[Stachniak and O'Hearn, 1990] Z. Stachniak and P. O'Hearn. Resolution in the domain of strongly finite logics. *Fundamenta Informaticae*, XIII:333–351, 1990.

[Straccia, 1997] U. Straccia. A sequent calculus for reasoning in four-valued description logics. In Didier Galmiche, editor, *Proc. International Conference on Automated Reasoning with Analytic Tableaux and Related Methods, Pont-à-Mousson, France*, volume 1227 of *LNCS*, pages 343–357. Springer-Verlag, 1997.

[Suchoń, 1974] W. Suchoń. La méthode de Smullyan de construire le calcul n-valent de Łukasiewicz avec implication et négation. *Reports on Mathematical Logic, Universities of Cracow and Katowice*, 2:37–42, 1974.

[Surma, 1974] S. J. Surma. An algorithm for axiomatizing every finite logic. In David C. Rine, editor, *Computer Science and Multiple-Valued Logics*, pages 143–149. North-Holland, Amsterdam, second edition, 1984. Selected Papers from the International Symposium on Multiple-Valued Logics 1974.

[Takahashi, 1967] M. Takahashi. Many-valued logics of extended Gentzen style I. *Science Reports of the Tokyo Kyoiku Daigaku, Section A*, 9(231):95–116, 1967. MR 37.39, Zbl 172.8.

[Takahashi, 1970] M. Takahashi. Many-valued logics of extended Gentzen style II. *Journal of Symbolic Logic*, 35(4):493–528, December 1970.

[Thayse *et al.*, 1979] A. Thayse, M. Davio, and J.-P. Deschamps. Optimization of multiple-valued decision diagrams. In *Proc. International Symposium on Multiple-Valued Logics, ISMVL'79, Rosemont/IL, USA*, pages 171–177. IEEE Press, Los Alamitos, 1979.

[Urquhart, 1986] A. Urquhart. Many-valued logic. In D. Gabbay and F. Guenthner, editors, *Handbook of Philosophical Logic, Vol. III: Alternatives in Classical Logic*, chapter 2, pages 71–116. Reidel, Dordrecht, 1986.

[Vienna Group, 1996] Vienna Group for Multiple Valued Logics. MUltlog 1.0: Towards an expert system for many-valued logics. In Michael McRobbie and John Slaney, editors, *Proc. 13th Conference on Automated Deduction, New Brunswick/NJ, USA*, volume 1104 of *LNCS*, pages 226–230. Springer-Verlag, 1996.

[Zabel, 1993] N. Zabel. *Nouvelles Techniques de Déduction Automatique en Logiques Polyvalentes Finies et Infinies du Premier Ordre*. PhD thesis, Institut National Polytechnique de Grenoble, April 1993.

[Zach, 1993] R. Zach. Proof theory of finite-valued logics. Master's thesis, Institut für Algebra und Diskrete Mathematik, TU Wien, September 1993. Available as Technical Report TUW-E185.2-Z.1-93.

JOACHIM POSEGGA AND PETER H. SCHMITT

IMPLEMENTING SEMANTIC TABLEAUX

What is it people want to hear about an implementation? Most likely they will be content to hear that it works, or better that it works extremely well. Who cares on the sun deck of a cruise ship what happens in the engine-room? There is some excuse for this attitude: implementation is often associated with nitty-gritty details, with cumbersome work-arounds caused by insuffficencies of the programming language, or with genial short cuts through several levels of abstraction in the specification. This is definitely not what we want to present in this chapter. It is tempting to join the group in their deck chairs and talk elegantly about how one would theoretically realize an implementation, step-by-step refining the top level specification and carefully weighing-up all design decisions. This is a possible approach. For this time we decided on a presentation half-way between the two portrayed alternatives. We will present runnable code, but in an easily accessible language that also has the advantage that some of the important procedures used in theorem proving algorithms are already available as built-ins or library functions. We are speaking of Prolog. The reader is invited to type the theorem proving programs he will find in this chapter into his favorite Prolog system and enjoy playing around with them.

The programs are based on, or inspired by, the the leanT^AP theorem prover [Beckert and Posegga, 1994]. The idea behind leanT^AP is to implement logical calculi by minimal means. This has two advantages: Firstly, the resulting programs are small, which makes them easier to understand them. Second, they provide an ideal starting point as they can be easily modified or adapted to specific needs. Furthermore, they are more than mere toy systems and surprisingly fast.

We will provide extensive comments on these programs and in one case also a complete soundness and correctness proof. Different alternatives for representing the tableaux and for organizing the proof search will be considered and exemplified by small Prolog programs.

A draw back of this approach is that the reader will be required to understand Prolog. But let us hasten to assure that acquaintance with the basic ideas of Prolog will suffice, all of which may be found e.g. on the first 22 pages of [Clocksin and Mellish, 1981]. In addition we will need in the soundness and correctness proof a formal semantics of the underlying programming language. To this end we will review below the basic computation model of Prolog, the computation tree. This offers another possibility for the reader to acquire an understanding for this language or to consolidate it.

The plan for this chapter is as follows: Section 1 fixes a couple of assumptions we will make for discussing our approaches to implementing tableaux. This involves some Prolog-oriented issues, as well as certain points about the tableaux calculi underlying our implementations. In Section 2 an algorithm for deriving Skolemized negation normal form is presented. The input formulæ for the programs given subsequently will be in Skolemized negation normal form. Section 3

M. D'Agostino et al. (eds.), Handbook of Tableau Methods, 581–629.

presents the first and simplest version of our theorem prover. The program has been proposed in [Beckert and Posegga, 1994]; here we recall it and prove its soundness and completeness. Section 4 extends the program by a powerful heuristic called *universal formulæ*. Building upon a theorem prover that represents tableaux as graphs in Section 5, we present a compilation-based approach to tableau-based deduction in Section 6. Section 7 discusses lemmata in tableau calculi which leads us to Binary Decision Diagrams (BDDs). Section 8 functions as a conclusion by giving some ideas on how one can build upon the presented programs when working towards his or her own implementation.

1 PRELIMINARIES

There are a couple of issues one needs to consider when carrying out implementations of deduction systems. Clearly, the concrete calculus that is to be implemented and the language chosen for an implementation are of most importance.

We have chosen to use Prolog as the implementation language used in this chapter. The reason for this is pragmatic: Prolog is a very convenient language for implementing first-order reasoning, since the primitives of Prolog are already quite close to first-order logic. This allows one to program in a very elegant and short style, as we will see in the sequel. Nevertheless there are some subtle points to be considered if we want to obtain efficient code:

As we want to achieve efficient code, we will want, to take advantage of the strengths of Prolog systems. One is that Prolog's depth-first search with backtracking is usually implemented very efficiently. Fortunately, this is also a well-suited search strategy for implementing deduction. Unfortunately, Prolog's search strategy is incomplete, since it chooses whatever comes first in the database instead of having a *fair* selection scheme. Since we want to implement logically *complete* deduction systems, we will have to overcome this drawback; one way to tackle it is switching to a bounded depth-first search. The desired completeness can be obtained by successively increasing a depth bound.[1]

It is also important to observe that Prolog's efficiency is strongly enhanced by indexing on the first argument position of the clause head. Thus, putting the right information in the first argument pays off.

The Prolog code we will give in the sequel is standard Prolog (in Edinburgh syntax) and should run on most Prolog systems.[2] We assume that the following Prolog predicates, user defined or otherwise, are available:

[1]The other choice would have been to implement fairness. But given the facts that no convincing fairness criteria are known and the difficulty in changing Prolog's search strategy without losing efficiency, this is not a viable alternative.

[2]The code was developed and tested with Sicstus Prolog , but runs without changes with Quintus Prolog and Eclispe ; other Prolog dialects might require little changes to our programs.

`append/3`. `append(L1,L2,L3)` succeeds if `L3` is the result of appending the lists `L1` and `L2`.

`unify/2`. `unify(T1,T2)` unifies the Prolog terms `T1` and `T2` by *sound* unification.

Logical formulæ will be represented by Prolog terms as follows:

Prolog atom	atomic formula
`-`	negation
`;`	disjunction
`,`	conjunction
`all(X,F)`	universal quantification with `X` a Prolog variable and `F` the scope of quantification.

Thus `(p(a),all(X,(-p(X);p(f(X)))))` stands for

$$p(a) \wedge \forall x(\neg p(x) \vee p(f(x))).$$

Furthermore, we assume that no variable is used twice for quantification within a set of input formulæ, e.g. a formula of the form

$$\forall x(q(x) \rightarrow q(x)) \wedge \forall x(p(x))$$

should be avoided. This assumption is in fact not neccessary for using the programs we will present, but it makes them more consise and easier to understand.

As we now have presented a rough idea of the means by which we will implement deductions, let us say a word on the calculi we will implement.

We will consider theorem provers for classical first-order logic without equality and will use a tableau calculus in its free-variable version, see Section 4 in 'First-order Tableau Methods'. Usually tableau calculi are set up for general formulæ with many logical connectives. We decided from a presentational perspective to use only formulæ in negation normal form using only universal quantifiers.[3] Arbitrary formulæ will be reduced to this format in a preprocessing step.

The issue whether preprocessing or normalization by tableau rules during tableau expansion is the better choice is not resolved at the moment, and maybe there will never be a definite answer. But separating preprocessing from the actual proof search certainly leads to a much clearer presentation. It is of course possible to extend the implementations we will give below to non-negation normal form formulæ without preprocessing, if one wishes to do so.

[3]Note, that we do not require formulæ to be in prenex normal form.

2 PREPROCESSING

2.1 Computing a Negation Normal Form

Recall that the conversion into negation normal form is linear w.r.t. the length of a formula not containing equivalences.[4] Most operations for deriving negation normal form are straightforward. What is not straightforward is coming up with a good Skolemization; this is one reason we give a complete Prolog implementation of the conversion. The second is that we show how to optimize the negation normal form without extra cost by changing the order of disjunctively connected formulæ.

The predicate used for computing a negation normal form is

```
nnf(+Fml,+FreeV,-NNF,-Paths)
```

`Fml` is the formula to be transformed, `FreeV` is the list of free variables occurring in `Fml`, `NNF` is bound to the Prolog term representing the computed negation normal form of `Fml`, and `Paths` is bound to the number of disjunctive paths in `NNF` (resp. `Fml`). We will see soon what this latter information is good for.

We implement a convenient syntax for first-order formulæ, using as logical connectives '`v`' (disjunction), '`&`' (conjunction), '`=>`' (implication), and '`<=>`' (equivalence).

The Prolog query we are going to use for computing the negation normal form of a closed formula bound to `Fml` is `nnf(Fml,[],NNF,_)`.[5] The corresponding program is given in Figure 1. The first clause of the predicate `nnf` (lines 1–11) corresponds to the standard rules in semantic tableaux; nothing exciting is done—we just use tautologies for rewriting formulæ. For universally quantified formulæ, we add the quantified variable to `FreeV` to compute the negation normal form of the scope (12–13).

Skolemization has to be carried out very carefully, since straightforwardly Skolemizing can easily hinder finding a proof: In the first edition of [Fitting, 1996] Skolem-termsw containing all variables that appear free on a branch are inserted; this is correct, but too restrictive: it often prevents inconsistent branches from closing. The current state of the art [Beckert *et al.*, 1993] is less restrictive: It suffices to use a Skolem-term that is unique (up to variable renaming) to the existentially quantified formula; this term only needs to hold the free variables occurring in the formula. An ideal candidate for such a term is the formula itself. This way of Skolemization has actually been known for more than fifty years: it resembles the ϵ-formulæ described in [Hilbert and Bernays, 1939, §1]. Lines 14–17 show how this can be elegantly implemented in Prolog.

We generate a copy `Fml1` of `Fml`; only the variable `X` gets renamed, since all others are contained in `FreeV` and `X` is replaced by `Fml`. Note, that we could **not** have

[4] If the formula contains equivalences, its negation normal form becomes exponential when computed in a naive way; more clever algorithms result in an at most quadratic NNF [Eder, 1992].

[5] The symbol _, called anonymous variable, is a convenient way to name Prolog variables, you don't care about

% Rewriting logical connectives:

```
1  nnf(Fml,FreeV,NNF,Paths) :-
2      (Fml = -(-A)        -> Fml1 = A;
3       Fml = -all(X,F)    -> Fml1 = ex(X,-F);
4       Fml = -ex(X,F)     -> Fml1 = all(X,-F);
5       Fml = -(A v B)     -> Fml1 = -A & -B;
6       Fml = -(A & B)     -> Fml1 = -A v -B;
7       Fml = (A => B)     -> Fml1 = -A v B;
8       Fml = -(A => B)    -> Fml1 = A & -B;
9       Fml = (A <=> B)    -> Fml1 = (A & B) v (-A & -B);
10      Fml = -(A <=> B) -> Fml1 = (A & -B) v (-A & B)),!,
11      nnf(Fml1,FreeV,NNF,Paths).
```

% Universal Quantification:

```
12 nnf(all(X,F),FreeV,all(X,NNF),Paths) :- !,
13     nnf(F,[X|FreeV],NNF,Paths).
```

% Skolemization:

```
14 nnf(ex(X,Fml),FreeV,NNF,Paths)  :- !,
15     copy_term((X,Fml,FreeV),(Fml,Fml1,FreeV)),
16     copy_term((X,Fml1,FreeV),(ex,Fml2,FreeV)),
17     nnf(Fml2,FreeV,NNF,Paths).
```

% Conjunctions:

```
18 nnf(A & B,FreeV,(NNF1,NNF2),Paths)  :- !,
19     nnf(A,FreeV,NNF1,Paths1),
20     nnf(B,FreeV,NNF2,Paths2),
21     Paths is Paths1 * Paths2.
```

% Disjunctions:

```
22 nnf(A v B,FreeV,NNF,Paths)  :- !,
23     nnf(A,FreeV,NNF1,Paths1),
24     nnf(B,FreeV,NNF2,Paths2),
25     Paths is Paths1 + Paths2,
26     (Paths1 > Paths2 -> NNF = (NNF2;NNF1);
27                         NNF = (NNF1;NNF2)).
```

% Literals:

```
28 nnf(Lit,_,Lit,1).
```

Figure 1. Computing negation normal form

used unification `X = Fml` since this would have created a cyclic term. The formula `Fml1` still contains `X`. The second `copy_term` goal substitutes the (arbitrary) constant `ex` for `X`. Here is an example: the goal `nnf(ex(X,p(X,Y)),[Y],NNF,_)` will succeed and bind `Fml` to `p(X,Y)`, `Fml1` to `p(p(X,Y),Y)`, `Fml2` to `p(p(ex,Y),Y)`. We have to pay a price for not introducing a new function symbol by reusing the formula: the set of function symbols has to be disjoint from the set of predicate symbols. Let us mention at this point another restriction on the input formula: it should not contain the same variable for different quantifiers, e.g. $\forall X(q(X) \wedge \exists X(r(X))$ is not allowed. But these are pathological formulas anyhow and we may remedy the situation by renaming bound variables.

In Sicstus Prolog `copy_term` behaves as if it were defined by

```
copy_term(X,Y) :-
        assert('copy of'(X)),
        retract('copy of'(Y)).
```

If `t` is a prolog term with the Prolog variables `X,A,B,C` then the query `copy_term((X,t,[A,B,C]),(X1,Y,[A,B,C]))` succeeds binding `Y` to a copy of `t` where `X` is replaced by `X1`. The variable `X1` is new, in the sense that any subsequent binding of `X` does not affect `X1`. For example, the query

```
copy_term(f(X),f(Y)),X=a,Y=b.
```

succeeds.

The free variables `FreeV` must appear as an argument in both parameters in line 15 and 16, since we do *not* want to rename them.

From a logical point of view, this might look a bit odd, as we turn predicate symbols into function symbols when Skolemizing in this way. However, it works under the assumption that disjoint sets of predicate and function symbols are used. This is usually the case; if not, we can simply 'wrap' the inserted scope in a new function symbol.

The next clause (17–20) is routine, besides counting disjunctive paths. The number of disjunctive paths in a formula (i.e. the number of branches a fully expanded tableau for it will have) is used when handling disjunctions (12–26): we put the less branching formula to the left. That way the number of choice points during the proof search is reduced, since leanT^AP expands the left formula first.

The last clause will match literals and is again straightforward.

3 A SIMPLE AND EFFICIENT TABLEAUX-BASED THEOREM PROVER

In this section we present a simple Prolog implementation of the tableau method for formulæ in Skolemized negation normal form. This is not only a pedagogical device, it really works: type the code from Figure 2 — we will refer to this program by the symbol leanT^AP in the following — into your favorite Prolog system and it will run extremely well at least on small examples. A second objective of this section is the proof that leanT^AP is a correct implementation. More precisely, we will eventually show

THEOREM 1.

1. *The* leanT^AP *program terminates on all inputs.*
2. *If the query* `prove(fml,h,[],[],d)` *to the program* leanT^AP *returns success as an answer, where* fml *is a formula,* h *is a list of formulæ and* d *is a natural number, than* ¬fml *is a logical consequence of* h.
3. *If* ¬fml *is a logical consequence of* h *then there is a natural number* d *such that the query* `prove(fml,h,[],[],d)` *to* leanT^AP *terminates with success.*

In the proof of this theorem we will not deal with the logical consequence relation directly but make use of the completeness theorem established in a previous chapter:

$$\neg fml \text{ is a logical consequence of } h$$
$$\text{iff}$$
$$\text{there is a closed tableau for } [fml \mid h].$$

A tableau T for a list L of formulæ starts with a non ramifying branch B, the nodes in B being labelled by the formulæ in L in some order.

Before going into details we will briefly outline the working principle of the leanT^AP program:

`prove(Fml,[],[],[],VLim)` succeeds if `Fml` can be proven inconsistent without using more than `VLim` free variables on each branch.

The proof proceeds by considering individual branches (from left to right) of a tableau; the parameters `Fml`, `UnExp`, and `Lits` represent the current branch: `Fml` is the formula being expanded, `UnExp` holds a list of formulæ not yet expanded, and `Lits` is a list of the literals present on the current branch. `FreeV` is a list of the free variables on the branch. A positive integer `VarLim` is used to initiate backtracking; it is an upper bound for the length of `FreeV`.

We will number clauses of the leanT^AP-program by the line in which they start in the listing of Figure 2. Clause 1 handles conjunctions: the first conjunct is selected, the other is put in the list of not yet expanded formulæ. Handling disjunctions, clause 3 splits the current branch and two new goals have to be proven.

Universally quantified formulæ require a little more effort. Clause 6 uses the built-in predicate `length(X,Y)`, which succeeds if `X` is a list of length `Y`. The symbol `\+` denotes negation in Sicstus Prolog. The built-in predicate `copy_term` has already been explained above.

Application of clause 6 initiates backtracking if the depth bound `VarLim` is reached. Otherwise, we generate a 'fresh' instance of the formula with `copy_term`,

```
% Conjunction:

1  prove((A,B),UnExp,Lits,FreeV,VarLim) :- !,
2      prove(A,[B|UnExp],Lits,FreeV,VarLim).

% Disjunction:

3  prove((A;B),UnExp,Lits,FreeV,VarLim) :- !,
4      prove(A,UnExp,Lits,FreeV,VarLim),
5      prove(B,UnExp,Lits,FreeV,VarLim).

% Universal Quantification:

6  prove(all(X,Fml),UnExp,Lits,FreeV,VarLim) :- !,
7      \+ length(FreeV,VarLim),
8      copy_term((X,Fml,FreeV),(X1,Fml1,FreeV)),
9      append(UnExp,[all(X,Fml)],UnExp1),
10     prove(Fml1,UnExp1,Lits,[X1|FreeV],VarLim).

% Closing Branches:

11 prove(Lit,_,[L|Lits],_,_) :-
12     (Lit = -Neg; -Lit = Neg) ->
13     (unify(Neg,L); prove(Lit,[],Lits,_,_)).

% Extending Branches:

14 prove(Lit,[Next|UnExp],Lits,FreeV,VarLim) :-
15     prove(Next,UnExp,[Lit|Lits],FreeV,VarLim).
```

Figure 2. A complete and sound tableau prover

without renaming the free variables in FreeV. The original γ-formula is stored for subsequent use, and the renamed scope becomes the current formula.

Clause 11 closes branches; it is the only one which is not determinate. Note that it will only be entered with a literal as its first argument. Neg is bound to the negated literal and sound unification is tried against the literals on the current branch. The clause calls itself recursively and traverses the list in its second argument; no other clause will match since UnExp is set to the empty list. Clause 11 incorporates the design decision to look for complementary formulas only at the level of literals. This suffices for completeness, but closure with arbitrary complementary formulas can be faster.[6]

[6]As an example, take $\Phi \wedge \neg\Phi$, where Φ is an arbitrarily complex formula. In practice, however, such phenomena occur very rarely.

The last clause is reached if the current branch cannot be closed. We add the current formula (always a literal) to the list of literals on the branch and pick a formula waiting for expansion.

3.1 Proving Completeness & Correctness of leanTAP

To reason about the execution of the program leanTAP we need an operational semantics of Prolog as a programming language. We will use for this purpose the computation tree, T_P, associated with a Prolog program P. This concept will be explained in detail below. Computation trees are a very simple model for an operational semantics of Prolog and other descriptions are available, see e.g. [Börger and Rosenzweig, 1994]. But leanTAP uses the cut '!' only in obvious ways, negation only in line 7 and none of the meta programming features at all. Only the '->' construct in line 12 may not be completely standard. Thus, the overhead for a deeper operational model does not pay off.

To begin our explanation of the concept of the computation tree, T_P, we remark that the nodes of this tree are labelled by the states of computation that arise during the execution of a Prolog program P. At this level of abstraction a state of computation consists of a list, *goallist*, of atomic formulæ, called goals, and a substitution σ of the Prolog variables occurring in this list. If a goal list $[F_1, \ldots, F_k]$ and a substitution σ are attached to a node $\mathbf{n}$, then it is really the list of formulæ $[\sigma(F_1), \ldots, \sigma(F_k)]$ that we consider. But it will prove convenient to separate this information into a formula part and a substitution part.

Now we turn to the question, what are the successor nodes of a node $\mathbf{n}$ labelled with a pair $(goallist, \sigma)$? Among the list of goals still to be solved Prolog always chooses the first one. Putting aside for the moment the case of the empty list of goals we write $goallist = [goal \mid restgoals]$. An attempt is made to unify $\sigma(goal)$ with the head of a clause in the program P; to be precise the variables in the clause are first renamed to guarantee that it has no common variables with $\sigma(goal)$. Assume that a most general unifier μ exists for $\sigma(goal)$ and the head *head* of a clause *head*:- *body* $\in P$, then there will be a successor node $\mathbf{n}_1$ of $\mathbf{n}$ labelled with $(body + restgoals, \mu \circ \sigma)$. Here, '+' denotes the concatenation of two lists and $\circ$ composition of substitutions. For each successful unification a successor node of $\mathbf{n}$ will be created in this way from left to right in the order of appearance in P. Branches in T_P will always be called *computations* to avoid confusion with branches in other tree structures, e.g. branches in a tableau. A computation terminates successfully if it is a finite branch in T_P and its last node $\mathbf{n}_f$ is labelled with the empty list of goals. The substitution in the label of $\mathbf{n}_f$ is called the *answer substitution* of the computation. A computation fails if its last node is labelled with $([goal \mid restgoals], \sigma)$, such that $\sigma(goal)$ is not unifiable with the head of any clause in P. The root of T_P is labelled with the initial list of goals, $goallist_0$, i.e. the query entered by the user, and the empty substitution. Since the shape of the computation tree also depends on $goallist_0$, we should strictly speak of the computation tree $T_{P,query}$ for a program P and a query *query*. We will use T_P whenever *query* is clear from the context.

The computation tree provides only a static picture of the evaluation of a Prolog program. The dynamic behaviour is easily explained: evaluation starts with the root node of the tree. Whenever there is branching, Prolog chooses the leftmost continuation. If a computation fails, Prolog backs up to the next branching point and then continues along the leftmost continuation that has not yet been explored. This is called backtracking in Prolog terminology. If all backtracking alternatives at all branching points have been exhausted without reaching the empty list of goals the evaluation fails. Of course there is, as with all programming languages, the possibility that your program was not written carefully enough and evaluation runs into an infinite loop.

After this general description of computation trees we will take a closer look at the computation trees associated with the leanT^{A}P program. The root node will be labelled with the one-element goal list

```
[prove(fml,h,[],[],d)]
```

with `fml` being a formula, `h` a list of formulæ and `d` a natural number. Any node `n` with goal list

```
[prove((fml1,fml2),h,lits,freev,d) | Restgoals ]
```

will have a successor node arising from clause 1. The same goal also unifies with the head of the clauses 11 and 14 and there should be corresponding successor nodes in the computation tree. This is the right time to explain the meaning of the symbol '`!`', called cut. When '`!`' is reached during the evaluation of the body of a clause with head *head* all alternative clauses that satisfy *head* will be cut off. In the case at hand here there is consequently no branching at node `n` in the computation tree. The same remark applies to nodes labelled with

```
[prove((fml1;fml2),h,lits,freev,d) | Restgoals ].
```

The cut in clause 3 prevents branching. Note also that this time the length of the list of goals is increased. The universal quantifier case requires more explanations. Here we look at a node `n` labelled with

```
[prove(all(X,fml),h,lits,freev,d) | Restgoals]
```

and clause 6 is called. `\+length(freev,d)` succeeds if the length of the list `freev` is strictly less than the number `d`.[7] `length/2` is a built-in predicate in most Prolog systems. It may not be available in your system and you will have to program it yourself. The same may also be true for the predicate `copy_term/2`. The append predicate has already been mentioned above. Clause 6 is the only clause

[7] Recall that '\+' denotes Prologs negation as failure.

that changes the value of the fourth argument of the `prove` predicate. Because of the cut '`!`' in the body of clause 6 node n has only one successor node. Since the input formulæ do not contain existential quantifiers or negation signs in front of composite subformulæ it only remains to consider nodes labelled with

```
[prove(fml,h,lits,freev,d) | Restgoals ]
```

for `fml` a literal. These nodes may have two successors, the first and leftmost arising from an application of clause 11 and the second arising from clause 14. We only comment on clause 11. A goal of the form `G1 -> G2` is resolved by Prolog's evaluation mechanism by first calling goal `G1`. If this fails, `G1 -> G2` also fails. In the special case of clause 11 this will never happen, see below. If `G1` succeeds then the goal `G2` is called. On backtracking only alternative solutions to `G2` are considered, `G1` is not considered again. This is crucial in the case of clause 11. Here `G1 = (fml = -Neg ; -fml = Neg)` is a disjunctive goal: first `(fml = -Neg)` is tried and only if this fails is `(-fml = Neg)` considered. If `fml` is a positive literal the first subgoal fails and `Neg` gets bound to `-fml`, which is the dual of `fml`. If `fml` is a negative literal, `-fml0`, then the first subgoal succeeds binding the Prolog variable `Neg` to `fml0`, the dual of `fml`. Without the Prolog-implication '`->`', backtracking would call the second subgoal and this would yield the unwanted solution `Neg` = `-fml`. Since the disjunctive goal in front of the '`->`' sign in the body of clause 11 succeeds exactly once we will not mention it in the computation tree. The leftmost successor node will instead be labelled with the goal sequence whose head is the disjunctive goal

```
(unify(Neg,L) ; prove(Lit,[],Lits,_,_))
```

where the substitution of this node binds `Neg` to the dual of the literal `fml`. If `unify(Neg,L)` succeeds then clause 11 has been successful and Prolog's evaluation mechanism will start to work on the next goal in the sequence *restgoals*. The answer substitution is the most general unifier of the two literals. On backtracking all possible most general unifiers between `lit` and some literal in the list of literals `lits` will be produced. If `unify(Neg,L)` fails then the clause on line (11) will recursively be called with the third argument `Lits` instead of `[Lit | Lits]`. Note also that the second argument is now set to `[]`.

Let us consider as a specific example the leanT^4P proof of $(p \vee (q \wedge r)) \rightarrow ((p \vee q) \wedge (p \vee r))$, which is just one half of the distributive law of propositional logic. The query submitted to leanT^4P is $\mathtt{prove}(F, [], [], X, d)$ where F is the negation of the formula to be proved, explicitely given in Figure 3. In this and the following figures of computation trees we only show goals of the form $\mathtt{prove}(fml, h, lits, varlist, d)$, and sometimes also of the form $\mathtt{unify}(l_1, l_2)$. A successful computation of the corresponding computation tree is shown in Figure 3. To prevent the picture from becoming too confusing we did not show the failed computations to the left of the successful path nor the branchings to the right that were not explored. This is (almost) made up for in Figure 4 on page 593. Figure 5

input formula	F	=	(F_1, F_2)
with	F_1	=	$(p; (q, r))$
and	F_2	=	$(F_{21}; F_{22})$
with	F_{21}	=	$(-p, -q)$
and	F_{22}	=	$(-p, -r)$

1 $\langle prove(F, [], []) \rangle$
2 $\langle prove(F_1, [F_2], []) \rangle$
3 $\langle prove(p, [F_2], []), prove((q,r), [F_2], []) \rangle$
4 $\langle prove(F_2, [], [p]), prove((q,r), [F_2], []) \rangle$
5 $\langle prove(F_{21}, [], [p]), prove(F_{22}, [], [p]), prove((q,r), [F_2], []) \rangle$
6 $\langle prove(-p, [-q], [p]), prove(F_{22}, [], [p]), prove((q,r), [F_2], []) \rangle$
7 $\langle prove(F_{22}, [], [p]), prove((q,r), [F_2], []) \rangle$
8 $\langle prove(-p, [-r], [p]), prove((q,r), [F_2], []) \rangle$
9 $\langle prove((q,r), [F_2], []) \rangle$
10 $\langle prove(q, [r, F_2], []) \rangle$
11 $\langle prove(r, [F_2], [q]) \rangle$
12 $\langle prove(F_2, [], [r, q]) \rangle$
13 $\langle prove(F_{21}, [], [r, q]), prove(F_{22}, [], [r, q]) \rangle$
14 $\langle prove(-p, [-q], [r, q]), prove(F_{22}, [], [r, q]) \rangle$
15 $\langle prove(-q, [], [-p, r, q]), prove(F_{22}, [], [r, q]) \rangle$
16 $\langle prove(F_{22}, [], [r, q]) \rangle$
17 $\langle prove(-p, [-r], [r, q]) \rangle$
18 $\langle prove(-r, [], [-p, r, q]) \rangle$
19 $\langle \rangle$

We have suppressed the last two arguments of the *prove* predicate since they are not relevant for propositional formulas.

Figure 3. A successful computation for the leanTAP program with a propositional input formula

shows the computatation tree for leanTAP with the first-order input formula

$$p(a) \land \neg p(f(f(a))) \land \forall x(p(x) \to p(f(x)))$$

Proof. Theorem 1, part 1.
By definition any computation tree is finitely branching. Thus a computation tree is finite if all its computations are finite. Let T be the computation tree for leanTAP and the $query$ = `prove(fml,h,[],[],d)`. By $gs(\mathbf{n}) = \langle g_1, \ldots, g_k \rangle$ we denote the goal sequence attached to node $\mathbf{n}$ in T. For each individual goal g = `prove`$(fml, h, lits, varlist,$
$d)$ we define a complexity measure $c(g)$ which is the quadruple $\langle c_1(g), c_2(g),$

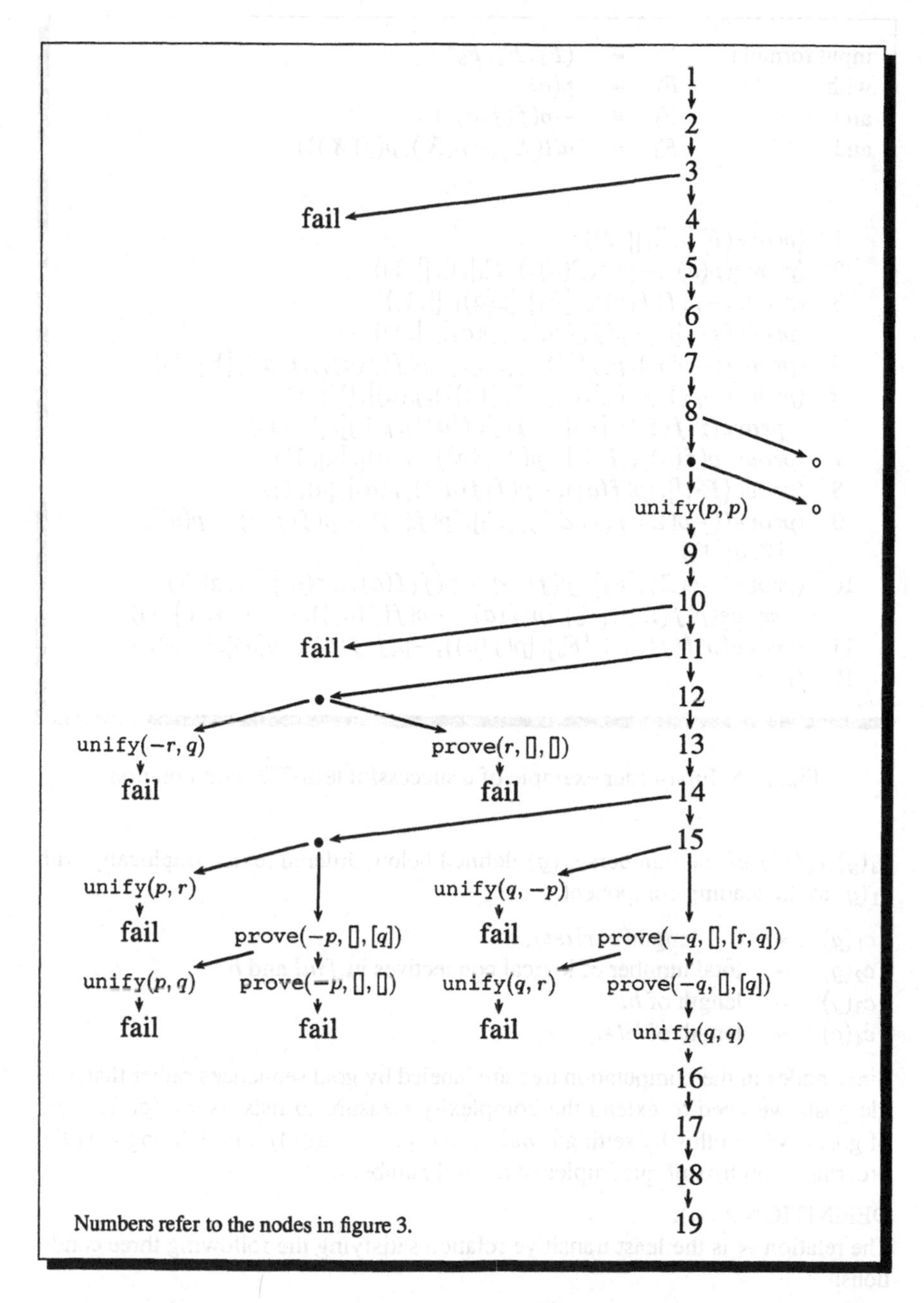

Figure 4. The computation tree for the leanT^AP program with a propositional input formula

input formula	F	=	(F_1, F_2, F_3)
with	F_1	=	$p(a)$
and	F_2	=	$-p(f(f(a)))$
and	F_3	=	$all(X, (-p(X); p(f(X)))$

1 $\langle prove(F, [], [], [], 2)\rangle$
2 $\langle prove(p(a), [-p(f(f(a))), F_3], [], [], 1)\rangle$
3 $\langle prove(-p(f(f(a))), [F_3], [p(a)], [], 1)\rangle$
4 $\langle prove(F_3, [], [-p(f(f(a))), p(a)], [], 1)\rangle$
5 $\langle prove((-p(Y); p(f(Y))), [F_3], [-p(f(f(a))), p(a)], [Y], 1)\rangle$
6 $\langle prove(-p(Y), [F_3], [-p(f(f(a))), p(a)], [Y], 1),$
$prove(p(f(Y)), [F_3], [-p(f(f(a))), p(a)], [Y], 1)\rangle$
7 $\langle prove(p(f(a)), [F_3], [-p(f(f(a))), p(a)], [a], 1)\rangle$
8 $\langle prove(F_3, [], [p(f(a)), -p(f(f(a))), p(a)], [a], 1)\rangle$
9 $\langle prove((-p(Z); p(f(Z))), [F_3], [p(f(a)), -p(f(f(a))), p(a)],$
$[Z, a], 1)\rangle$
10 $\langle prove(-p(Z), [F_3], [p(f(a)), -p(f(f(a))), p(a)], [Z, a], 1),$
$prove(p(f(Z)), [F_3], [p(f(a)), -p(f(f(a))), p(a)], [Z, a], 1)\rangle$
11 $\langle prove(p(f(f(a))), [F_3], [p(f(a)), -p(f(f(a))), p(a)], [f(a), a], 1)\rangle$
12 $\langle\rangle$

Figure 5. First-order example of a successful leanTAP computation

$c_3(g), c_4(g)\rangle$ of the numbers $c_i(g)$ defined below ordered lexicographically with $c_1(g)$ as the leading component.

$$
\begin{aligned}
c_1(g) &= d - \text{length}(varlist),\\
c_2(g) &= \text{total number of logical connectives in } fml \text{ and } h,\\
c_3(g) &= \text{length of } h,\\
c_4(g) &= \text{length of } lits.
\end{aligned}
$$

Since nodes in the computation tree are labeled by goal sequences rather than single goals we need to extend the complexity measure to lists $gs = \langle g_1, \ldots, g_k\rangle$ of goals. We do this by setting $c(gs) = \langle c(g_1), \ldots, c(g_k)\rangle$ and defining a partial ordering $\prec$ on lists of quadruples of natural numbers.

DEFINITION 2 ($\prec$).
The relation $\prec$ is the least transitive relation satisfying the following three conditions:

$$
\begin{array}{lcll}
\langle c^2, \ldots, c^n\rangle & \prec & \langle c^1, c^2, \ldots, c^n\rangle & \\
\langle e^1, c^2, \ldots, c^n\rangle & \prec & \langle c^1, c^2, \ldots, c^n\rangle & \text{if } e^1 < c^1\\
\langle e^1, e^2, c^2 \ldots, c^n\rangle & \prec & \langle c^1, c^2 \ldots, c^n\rangle & \text{if } e^1 < c^1 \text{ and } e^2 < c^1
\end{array}
$$

A goal sequence gs^1 is of smaller complexity than goal sequence gs^2 iff

$$c(gs^1) \prec c(gs^2)$$

The crucial observation, which is easily checked by looking at the leanTAP program, is that for any application of a program clause leading from the goal sequence gs^1 to the immediate successor gs^2 we have $c(gs^2) \prec c(gs^1)$. Thus finiteness of computations will follow when we can show that $\prec$ does not allow infinite descending chains. ■

LEMMA 3 (Wellfoundedness of $\prec$).
The ordering $\prec$ on lists of quadruples of natural numbers does not allow infinite descending chains.

Proof.
Assume to the contrary that an infinite descending chain $s^1 \succ \dots s^n \succ s^{n+1} \dots$ exists. Each element s^i in this chain is a list. Let m be the least number occuring as the length of some s^i. We must have $m > 0$ since the empty list $\langle\rangle$, which is the least element with respect to $\succ$, cannot occur among the s^i. Choose i_0 such that $s^{i_0} = \langle s_1^0, \dots s_m^0 \rangle$ has length m. We consider $s^i = \langle s_1^i, \dots s_r^i, s_2^0, \dots s_m^0 \rangle$ and $s^j = \langle s_1^j, \dots s_k^j, s_2^0, \dots s_m^0 \rangle$ for $j > i \geq i_0$. By choice of m and i_0 we must have $r, k > 0$. It is easily checked that all elements $s_1^j, \dots s_k^j$ are strictly smaller with respect to $<$ than all elements $s_1^i, \dots s_r^i$. From this it follows that the first elements in the lists s^i with $i \geq i_0$ form a descending chain contradicting the wellfoundedness of the lexicographical ordering on quadruples of natural numbers. ■

We now turn to the proof of correctness and completeness. for both parts we will use the fact that the tableaux that can be reached from an initial set $\{fml\} \cup h$[8] of formulæ can be retrieved from the goal sequences in the computation tree starting with $\texttt{prove}(fml, h, [], [], d)$. It turns out that this is not directly possible for all nodes. If at a node $\mathbf{n}$ execution of clause 11 is possible two successor nodes will be created, one corresponding to the $\texttt{unify}(l_1, l_2)$ predicate in the body of the clause and the other to the $\texttt{prove}(fml, [], lits, varlist, d)$ predicate. Nodes of the second type will be called *exception nodes*. At exception nodes the correspondence with the tableau structure breaks down. We first associate with every non-exception node $\mathbf{n}$ of the computation tree for a leanTAP-program the structure $tab_0(\mathbf{n})$ and secondly define a tableau structure $(tab(\mathbf{n}), \sigma(\mathbf{n}))$ for all nodes $\mathbf{n}$.

For our purposes a tableau will simply be a set of branches and a branch is simply a set of formulæ.

DEFINITION 4 (Tableau at nodes of the computation tree).
Let $\mathbf{n}$ be a node in the computation tree with attached goal sequence $gs(\mathbf{n}) =$

[8]Strictly speaking h is a list, we assume without mentioning that h is converted into the set of its elements whenever h is used as an argument for a set theoretic operation like $\cup$.

$\langle g_1, \ldots, g_r \rangle$ with $g_i = \texttt{prove}(fml_i, h_i, lits_i, varlist, d)$ Then

$$\begin{array}{lcl} branch(g_i) & = & \{fml_i\} \cup h_i \cup lits_i \\ tab_0(\mathbf{n}) & = & \{branch(g_i) \mid 1 \leq i \leq r\} \end{array}$$

The definition of $tab(\mathbf{n})$ and $\sigma(\mathbf{n})$ proceed by induction on the nodes in T. For the root node $\mathbf{n_0}$ we have

$$\begin{array}{lcl} tab(\mathbf{n_0}) & = & tab_0(\mathbf{n_0}) \\ \sigma(\mathbf{n_0}) & = & \emptyset \end{array}$$

If node $\mathbf{n_1}$ is reached from node $\mathbf{n}$ with $gs(\mathbf{n}) = \langle g_1, \ldots, g_r \rangle$ by an application of the `unify` goal in the body of clause 11 then

$$\begin{array}{lcl} tab(\mathbf{n_1}) & = & tab_0(\mathbf{n_1}) \\ \sigma(\mathbf{n_1}) & = & \sigma_u \circ \sigma(\mathbf{n}) \end{array}$$

Here σ_u is the most general unifier computed in the successful execution of `unify`. If node $\mathbf{n_1}$ is an exception node reached from $\mathbf{n}$ then

$$\begin{array}{lcl} tab(\mathbf{n_1}) & = & tab(\mathbf{n}) \\ \sigma(\mathbf{n_1}) & = & \sigma(\mathbf{n}) \end{array}$$

In all other cases where node $\mathbf{n_1}$ is reached from node $\mathbf{n}$

$$\begin{array}{lcl} tab(\mathbf{n_1}) & = & tab_0(\mathbf{n_1}) \\ \sigma(\mathbf{n_1}) & = & \sigma(\mathbf{n}) \end{array}$$

LEMMA 5. *For every node* $\mathbf{n}$ *in the computation tree of the query* $\texttt{prove}(fml, h, [], [], d)$ *there is a tableau* T *for* $\{fml\} \cup h$ *such that the tableau* $\sigma(\mathbf{n})(tab(\mathbf{n}))$ *contains all open branches of* T.

Proof. (of the lemma)
The proof proceeds by induction on the nodes in the computation tree and is obviously true for the root node. Let $\mathbf{n}$ be a node and T a tableau such that $\sigma(\mathbf{n})(tab(\mathbf{n}))$ contains all open branches of T. When $\mathbf{n_1}$ is reached from $\mathbf{n}$ by a program clause 1, 3 or 6 then $\sigma(\mathbf{n_1})(tab(\mathbf{n_1}))$ contains all open branches of the tableau T_1, T_3 or T_6 respectively, where T_i is reached from T by an application of an α, β or γ rule respectively. If $\mathbf{n_1}$ is reached from $\mathbf{n}$ by program clause 14 tab and σ remain unchanged and there is nothing to show. The same is true when $\mathbf{n_1}$ is an exception node. It remains to consider the case when $\mathbf{n_1}$ is reached from $\mathbf{n}$ by the `unify` predicate in the body of program clause 11. If $g = \texttt{prove}(fml, h, lits, varlist, d)$ is the leftmost goal in $gs(\mathbf{n})$ then $tab(\mathbf{n_1}) = tab(\mathbf{n}) \setminus \{branch(g)\}$. But since $\sigma_u(branch(g))$ is closed the statement of the lemma remains true for node $\mathbf{n_1}$. ■

Proof. (Theorem 1 Part 2)
When leanT^AP terminates successfully at node $\mathbf{n}$ then $tab(\mathbf{n})$ is the empty set. This implies by the previous lemma that there is a tableau for the initial set of formulæ with no open branches. Thus correctness of leanT^AP is proved. ■

Proof. Theorem 1 (Part 3).
This part of the proof starts from the assumption that the formula F is a logical consequence of the list of formulæ H. Thus there exists a closed tableau T for $[F \mid H]$. This T may, on the face of it, not match very well with the tableau that the leanTAP program tries to construct. From the completeness proof of the tableau calculus (Letz's Chapter) we know already more: not only does there exist a closed tableau, but **every** fair expansion strategy will eventually produce a closed tableau. It thus remains to show the tableaux $(tab(\mathbf{n}), \sigma(\mathbf{n})$ that are associated to leanTAP's computation tree constitutes a fair tableau search strategy.
Proof sketch:
There are only two kinds of branching points in the computation tree of leanTAP. The first reflects the alternatives to solve a goal of the form `prove`$(lit, h, lits, varlist, d)$ where lit is a literal. Either clasue 11 or clause 14 are applicable. Clause 11 is tried first and if this does not lead to successful termination finiteness of the computation tree will force backtracking and clause 14 will be taken. This shows that for any node $\mathbf{n}$ and any branch $b \in tab(\mathbf{n})$ any formula $fml \in b$, that is not a literal, will eventually be expanded, unless the computation has in the meantime already ended successfully.
The second kind of branching points occur in the execution of the body of clause 11; there the alternative to unify lit with the first element in the list $lits$ or to shorten the list $lits$ arises. First unification is pursued. If this does not lead to successful termination finiteness of the computation tree again forces the second alternative to be considered. In this way all possibilities to close a branch will be explored.
It remains to observe that by increasing the bound d in the initial query to the leanTAP-program the number of occurences of a particular universal formula on each branch may be arbitrarily increased.
Altogether this shows that leanTAP pursues a fair search strategy. ■

4 INCLUDING UNIVERSAL FORMULAE

There are many known heuristics which can be included into a tableau-based theorem prover. Including heuristics usually means either strengthening the underlying calculus for gaining shorter proofs, or directing the proof search in order to avoid useless search. Both are not universally good ideas: the fact that proof lengths decrease does not say anything on the difficulty of actually *finding* these shorter proofs, and guiding the proof search usually involves some effort for computing the particular guidelines. Heuristics are not a panacea: one must carefully analyze whether it really pays off to include a concrete heuristic. The more focussed an application of a theorem prover, the better the chances that appropriate heuristics increase the efficiency of an implementation.

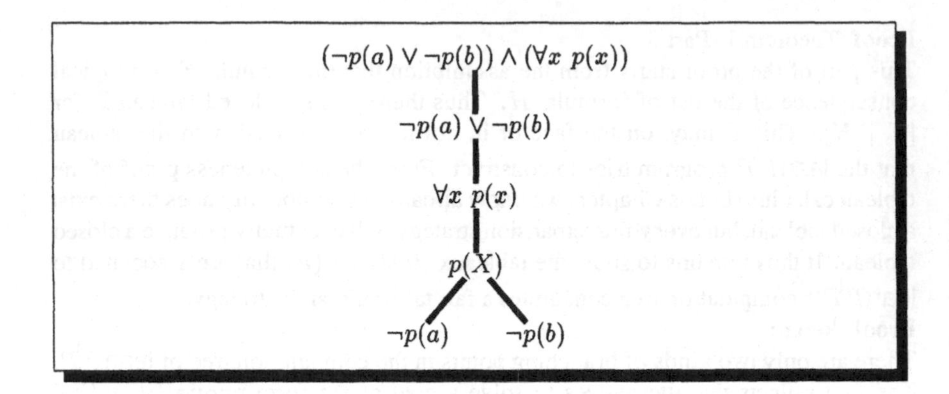

Figure 6. An example for the use of universal formulae

From some heuristics, however, most application areas benefit and it is generally a good idea to give them at least a try. One of these domain-independent heuristics is called *universal formulae*. The idea behind it is the following:

Universally quantified formulae are often used more than once for closing a tableau, and each time a different substitution for the free variables is needed. The standard procedure in semantic tableaux for this is to apply the γ-rule to the corresponding formulae more than once and generate several instances of them. Each instance contains different free variables, which can be bound differently for closing branches. The problem is that the more instances of γ-formulae are created, the bigger the tableau (i.e. the search space) grows. Here it helps to recognize 'universal' formulae; these can be used arbitrarily often in a proof with different substitutions for some of their free variables.

DEFINITION 6. Suppose ϕ is a formula on some tableau branch B. ϕ is *universal* with respect to the variable x if the following holds for every model $\mathbf{M}$ and every ground substitution σ:

$$\text{If } \mathbf{M} \models B\sigma, \text{ then } \mathbf{M} \models ((\forall x)\phi)\sigma.$$

Notational agreement: If we want to refer to a variable x which is universal w.r.t. a certain formula on some branch, and it is clear which branch and formula are meant, we will simply write 'the universal variable x' in the sequel.

A more detailed discussion of universal formulae can be found in [Beckert and Posegga, 1995] and also in [Beckert and Hähnle, 1992]; we give a slightly simplified account here. Figure 6 gives an example: The variable X is universal to both branches and thus they can be closed without applying the γ-rule again.

The problem of recognizing universal formulae is undecidable in general. However, a wide and important class of universal formulæ can be recognized quite easily: assume there is a sequence of tableau rule applications that does not contain

```
% Conjunction:

1  prove((A,B),UnExp,Lits,DisV,FreeV,UnivV,VarLim) :- !,
2    prove(A,[(UnivV:B)|UnExp],Lits,DisV,FreeV,UnivV,VarLim).

% Disjunction:

3  prove((A;B),UnExp,Lits,DisV,FreeV,UnivV,VarLim) :- !,
4    copy_term((Lits,DisV),(Lits1,DisV)),
5    prove(A,UnExp,Lits,(DisV+UnivV),FreeV,[],VarLim),
6    prove(B,UnExp,Lits1,(DisV+UnivV),FreeV,[],VarLim).

% Universal Quantification:

7  prove(all(X,Fml),UnExp,Lits,DisV,FreeV,UnivV,VarLim) :- !,
8   \+ length(FreeV,VarLim),
9   copy_term((X,Fml,FreeV),(X1,Fml1,FreeV)),
10  append(UnExp,[(UnivV:all(X,Fml))],UnExp1),
11  prove(Fml1,UnExp1,Lits,DisV,[X1|FreeV],[X1|UnivV],VarLim).

% Closing Branches:

12 prove(Lit,_,[L|Lits],_,_,_,_) :-
13   (Lit = -Neg; -Lit = Neg ) ->
14   (unify(Neg,L); prove(Lit,[],Lits,_,_,_,_)).

% Extending Branches:

15 prove(Lit,[(UnivV:Next)|UnExp],Lits,DisV,FreeV,_,VarLim) :-
16   prove(Next,UnExp,[Lit|Lits],DisV,FreeV,UnivV,VarLim).
```

Figure 7. leanT^AP with universal variables

a disjunctive rule (i.e. the tableau does not branch). All formulae that are generated by this sequence are universal w.r.t. the free variables introduced within the sequence. Substitutions for these variables can be ignored, since the sequence could be repeated arbitrarily often for generating new copies of these variables — without generating new branches.

Including this optimization in the previously discussed implementation is not a major undertaking: we simply collect a list of 'universal' variables for each formula. For this, the arity of prove is extended from 5 to 7:

```
prove(Fml,UnExp,Lits,DisV,FreeV,UnivV,VarLim)
```

UnivV and DisV are new parameters, the use of all others remains unchanged. UnivV is a list of the universal variables in the current formula Fml. DisV represents something like their counterpart: it is a Prolog term containing all variables on the current branch which are *not* universal in one of the formulae (we will call these the 'disjunctive variables').[9] Each unexpanded formula in UnExp will have the list of its universal variables attached. The Prolog functor ':' is used for this purpose.

The prover is now started with the goal

```
prove(Fml,[],[],[],[],[],VarLim)
```

for showing the inconsistency of Fml. We will discuss the extended program by explaining the differences to our previous version.

```
| ?- prove((all(X,p(X)),(-(p(a));-(p(b)))),[],[],[],[],[],1) .
Call: prove((all(X1,p(X1)),(-(p(a));-(p(b)))),[],[],[],[],[],1)
Call: prove(all(X1,p(X1)),[[]:(-(p(a));-(p(b)))],[],[],[],[],1)
Call: prove(p(X2),[[]:(-(p(a));-(p(b))),[]:all(X1,p(X1))],[],
                                             [],[X2],[X2],1)
Call: prove((-(p(a));-(p(b))),[[]:all(X1,p(X1))],[p(X2)],
                                             [],[X2],[],1)
Call: prove(-(p(a)),[[]:all(X1,p(X1))],[p(X2)],[]+[],[X2],[],1)
Exit: prove(-(p(a)),[[]:all(X1,p(X1))],[p(a)],[]+[],[a],[],1)
 Call: prove(-(p(b)),[[]:all(X1,p(X1))],[p(X3)],[]+[],[a],[],1)
 Exit: prove(-(p(b)),[[]:all(X1,p(X1))],[p(b)],[]+[],[a],[],1)
Exit: prove((-(p(a));-(p(b))),[[]:all(X1,p(X1))],[p(a)],
                                             [],[a],[],1)
Exit: prove(p(a),[[]:(-(p(a));-(p(b))),[]:all(X1,p(X1))],[],
                                             [],[a],[a],1)
Exit: prove(all(X1,p(X1)),[[]:(-(p(a));-(p(b)))],[],[],[],[],1)
Exit: prove((all(X1,p(X1)),(-(p(a));-(p(b)))),[],[],[],[],[],1)
```

Figure 8. Trace of problem shown in Fig. 6

All universal variables of a conjunction are universal for each component (lines 1 and 2 in Figure 6).[10]

When dealing with disjunctions (3–6), we exploit universality of variables and rename these variables on both branches. Experiments have shown that for most examples it is best to only rename the variables in the literals. We could also rename the universal variables in UnExp, but this requires an extra effort which does not pay off in many cases.

[9]To be precise: DisV holds the variables which are not universal w.r.t. a formula on the current branch, whereas UnivV holds the variables universal w.r.t. the current formula.

[10]The implementation given here results in both conjunctions sharing universal variables. This is correct but not necessary: the variables could be renamed in one conjunct.

Besides this, disjunctions also destroys universality: the universal variables of a disjunction are therefore not universal to its components. The tableau is split and these variables become non-universal on both resulting branches. We therefore add them to `DisV` by creating a new Prolog term.[11] Universal variables of other formulae on the right-hand branch are renamed by `copy_term`. This allows universal variables to be instantiated differently on the two resulting branches.

When introducing a new variable by the quantifier rule (7–11), this variable becomes universal for the scope (it may loose that status if a disjunction in the scope is expanded, see above).

The next clause (lines 12–14) remains unchanged, besides having two more parameters.

Recall that the sixth parameter of `prove` holds the universal variables of the current formula (not of the whole branch). Thus, when extending branches in the last clause we must change this argument.

Figure 8 shows a trace of the program when run on the example in Figure 6.

4.1 Performance

It is interesting to compare the performance of the two programs we have presented in Figures 2 and 7. Table 1 gives the respective figures for some of Pelletier's problems [Pelletier, 1986]. The negated theorem has been placed in front of the axioms and the program of Figure 1 for computing negation normal form was applied as a preprocessing step.

Some of the theorems, like Problem 38, are quite hard: the $_3T^AP$ prover [Beckert *et al.*, 1992], for instance, needs more than ten times as long. Schubert's Steamroller (Pelletier No. 47) cannot be solved; this is no surprise, since the problem is designed for forward chaining systems based on conjunctive normal form. It can only be proven in tableau-based systems if good heuristics for selecting γ-formulae are used. Using a queue, as in our case, is not sufficient. We console ourselves with Problems No. 34 and 38, which are barely solvable in a comparable time by CNF-based provers unless sophisticated algorithms for deriving conjunctive normal forms are applied. Pelletier No. 34 (also called 'Andrew's Challenge') is not solvable without universal formulae, either. This example demonstrates the usefulness of the heuristic for complex problems. The use of universal formulae, however, also has disadvantages: the runtime for other problems (like 38) increased, as there is some overhead involved with maintaining universal variables.[12]

---

[11] We could use a list, but creating a new term by '+' (an arbitrary functor) is faster.

[12] The overhead, however, is not dramatic: in practice, an implementation is slowed down by a constant factor of about 2. On the other hand, exploiting universal formulae can result in an exponential speedup.

Table 1. Performance of the programs given in Figure 6/Figure 7 for Pelletier's problem set (the runtime has been measured on a SUN SPARC 10 workstation with SICStus Prolog 2.1; '0 *msec*' means 'not measurable')

No.	Limit `VarLim`	Branches closed	Closings tried	Time *msec*
32	3/3	10/10	10/10	10/10
33	1/1	11/11	11/11	0/10
34	??/5	–/79	–/79	∞/109
35	4/2	1/1	1/1	0/0
36	6/6	3/3	3/3	0/0
37	7/7	8/8	8/8	9/30
38	4/4	90/90	101/101	210/489
39	1/1	2/2	2/2	0/0
40	3/3	4/4	5/5	0/0
41	3/3	4/4	5/5	0/9
42	3/3	5/5	5/5	9/19
43	5/5	18/18	18/18	109/179
44	3/3	5/5	5/5	10/19
45	5/5	17/17	17/17	39/79
46	5/5	53/53	63/63	59/189

5 TABLEAUX AS GRAPHS

Taken literally, the theoretical treatment of semantic tableaux seems to suggest that tableaux are trees. However, the programs presented so far as well as the completeness proof consider tableaux as sets of branches. One standard approach to improve an implementation is to look for efficient data structures. Using trees would be an impovement over sets of branches, but acyclic graphs are even better. Since graphs use structure sharing, i.e. multiple occurences of the same subtree are replaced by pointers to only one occurence of the subtree, they allow for a very compact representation of the branches of a tableau which may in extreme cases be exponentially smaller than a tree representation.

This section investigates such a graphical representation of tableaux. The underlying idea is to reduce the amount of computation required during deduction by moving some of the effort for expanding formulae into preprocessing. The preprocessing computes a graph representation of a partially extended tableau, where α- and β-formulae are already fully expanded and need not be considered during the proof search any more.

We begin with an explanation of the syntax used to describe graphs. The simplest tableau graph consists of one node labelled with the atom '1'. This atom is used as a marker for the end of branches, i.e. it is the last entry in all branches of a tableau graph. If F and G are graphs, then $F \vee G$ will be the graph with a top node labelled with the connective $\vee$ from which two edges lead to the top nodes of graph F and graph G respectively. In addition we will use the graph constructor $\wedge$ which is particular to the class of graphs considered here. If F and G are graphs then $F \wedge G$ denotes the graph obtained from G by adding a new top node n^0 above the original top node of G. Node n^0 is labelled by F. This offers the possibility to represent graphs inside of graphs and we use it for treating universal quantifiers.

The following function maps a formula in negation normal form into a graphical representation of its partially expanded tableau:

DEFINITION 7. (*Mapping Formulæ to Tableau Graphs*)
Let F be a first-order formula in Skolemized negation normal form, and let '1' denote the atomic truth constant[13]

$$tgraph(F) \stackrel{\text{def}}{=} \begin{cases} F \wedge 1 & \textit{if } F \textit{ is a literal} \\ tgraph(A)\left[\frac{1}{tgraph(B)}\right] & \textit{if } F = (A \wedge B) \\ tgraph(A) \vee tgraph(B) & \textit{if } F = (A \vee B) \\ (\forall x\; tgraph(F')) \wedge 1 & \text{if } F = \forall x\, F' \end{cases}$$

where

$$F\left[\frac{G}{G'}\right] \stackrel{\text{def}}{=} \begin{cases} A \wedge (B\left[\frac{G}{G'}\right]) & \text{if } F = A \wedge B \\ (A\left[\frac{G}{G'}\right]) \vee (B\left[\frac{G}{G'}\right]) & \text{if } F = A \vee B \\ G' & \text{if } F = G \end{cases}$$

The replacement function used to deal with a conjunctions $A \wedge B$ has the effect of appending the graph for B at the end of all branches of the $tgraph(A)$.

Universally quantified formulae are represented by deriving a representation for a fully expanded tableau of the scope of the formula and putting it into one node together with a reference to the quantified variable. In this way we use nesting of graphs for treating quantified formulae. Note that an application of the replacement mapping does never touch these nested graphs.

Figure 9 contains two examples. The left tableau graph shows the principle of structure sharing: the graph is smaller than a fully expanded tableau for the formula below, but the paths in the graph correspond to the branches of such a tableau. The graph on the right-hand side is a tableau graph with a universally quantified formula: the quantified formula is represented as a subgraph.

It may at first seem difficult to realize structure sharing in Prolog since it does not support the use of pointers. The key to the solution is to use difference lists for the implementation of the replacement operation: literally, an expression '$F\left[\frac{1}{G}\right]$' means that each occurrence of '1' is replaced by an instance of G. If instead of 1 we use a Prolog variable X then assigning G to X will have the same effect. Figure 10 shows a literal translation of Definition 7 into Prolog. A goal `tgraph(Fml,G1)`

[13] W.l.g. we assume that '1' does not occur in F.

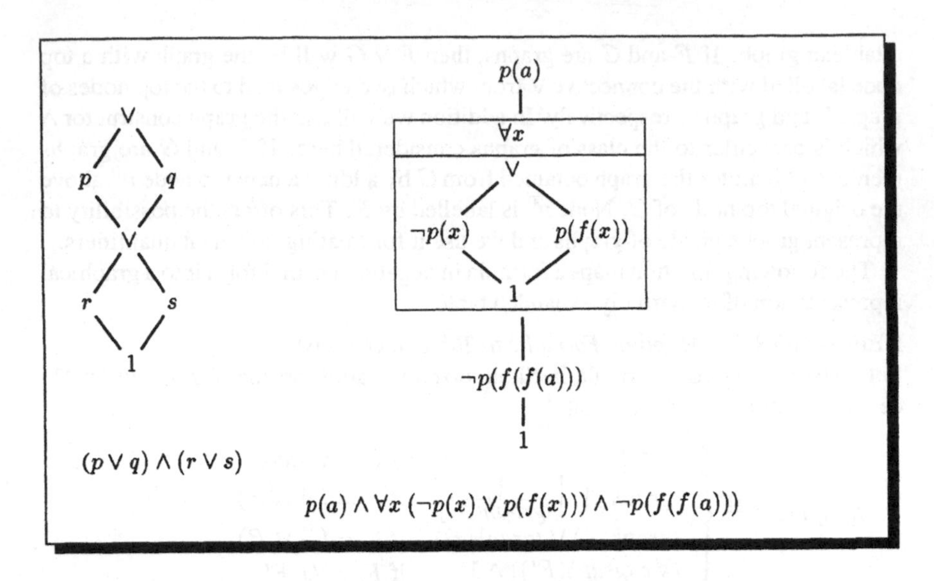

Figure 9. Sample tableau graphs

will succeed with binding `G1` to a tableau graph for `Fml`. Note, that the computation of `tgraph(Fml,G1)` requires only linear effort w.r.t. to the length of the input formula `Fml`. For the propositional formula in Figure 9 the `tgraph` procedure outputs the term

$$G = (p, ((r, \_); (s, \_)); (q, ((r, \_); (s, \_))$$

which seems to duplicate the subterm $((r, \_); (s, \_))$, but this happens only when the term is printed. Internally it is represented something like

$$\begin{aligned} G &= X_1; X_2 & \qquad Z &= (Z_1; Z_2) \\ X_1 &= (p, Z) & Z_1 &= (r, \_) \\ X_2 &= (q, Z) & Z_2 &= (s, \_) \end{aligned}$$

Figure 11 lists a prover that takes such a tableau graph as input. More precisely, the branch end markers will first be set to `true` and the difference list construct will be removed. The combination of the two programs thus looks like

```
tgraph(Fml,Graph/true) , gprove(Graph,[],[],[],d)
```

For showing that the original formula is inconsistent, we must check that each path in the represented tableau is closed. This is done by recursively descending the graph and constructing paths. A path is closed if there exists a substitution that generates a contradiction within the literals of the path. The proof search succeeds when all paths are closed.

```
% Conjunction:

1 tgraph((A,B),GraphA/GraphEnd):-!,
2   tgraph(A,GraphA/GraphB),
3   tgraph(B,GraphB/GraphEnd).

% Disjunction:

4 tgraph((A;B),(GraphA;GraphB)/GraphEnd) :-!,
5   tgraph(A,GraphA/GraphEnd),
6   tgraph(B,GraphB/GraphEnd).

% Universal Quantification:

7 tgraph(all(X,F),(all(X,GraphF),TEnd)/TEnd):- !,
8   tgraph(F,GraphF/true).

% Literals:

9 tgraph(Literal,(Literal,End)/End).}
```

Figure 10. Implementing Definition 7

If a path cannot be closed, we must perform the equivalent of applying a γ-rule in a standard tableau; recall that universally quantified formulae are represented as nested subgraphs, which contain a tableau for the scope of the quantified formula. We can simulate the application of a γ-rule by appending a copy of the subgraph to the branch we are currently considering.[14] We implement this by collecting the entry-points to such subgraphs until we end up at the end of a branch. If it is not closed, we select one of the subgraphs for expansion.

The implementation shown in Figure 11 uses the predicate

`gprove(TGraph,Gammas,Lits,FreeV,VarLim)`

where `TGraph` is a tableau graph, computed by the predicate `tgraph/2` as explained above. `VarLim` limits the number of free variables on every branch during the proof search (analogously to the previous programs). The other arguments, which are initially set to the empty list, represent the currently considered branch: `Gamma` will hold all universally quantified sub-tableau graphs that are valid on the current branch, and `Lits` will hold all literals on it. Free variables that have been introduced are collected in `FreeV`.

The procedure for proving that the tableau graph given as input is inconsistent mainly consists of expanding the individual branches or paths in the graph. The

[14]In more formal notion this means: if $\forall x\Phi$ is on a branch, we conjunctively add Φ' to it, where Φ' is a copy of Φ with x being renamed. This is correct, as $(\forall x\Phi) \rightarrow \Phi'$ is valid.

```
1  memberunify(X,[H|T]) :- unify(X,H);memberunify(X,T).

% Disjunction:

2  gprove((A;B),Gammas,Lits,FreeV,VarLim) :- !,
3        gprove(A,Gammas,Lits,FreeV,VarLim),
4        gprove(B,Gammas,Lits,FreeV,VarLim).

% Noticing Universal Quantification:

5  gprove((all(X,Gr),Rest),Gammas,Lits,FreeV,VarLim) :- !,
6        gprove(Rest,[all(X,Gr)|Gammas],Lits,FreeV,VarLim).

% Applying Universal Quantification:

7  gprove(true,[all(X,Gr)|Gammas],Lits,FreeV,VarLim) :- !,
8        \+ length(FreeV,VarLim),
9        copy_term((X,Gr,FreeV),(X1,Gr1,FreeV)),
10       append(Gammas,[all(X,Gr)],Gammas1),
11       gprove(Gr1,Gammas1,Lits,[X1|FreeV],VarLim).

% Closing Branches:

12 gprove((Lit,Rest),Gammas,Lits,FreeV,VarLim) :-
13       (Lit = -Neg; -Lit = Neg) -> memberunify(Neg,Lits)
14             ; gprove(Rest,Gammas,[Lit|Lits],FreeV,VarLim).
```

Figure 11. Deduction with tableau graphs

first and the last clauses shown in Figure 11 work in a way similar to that explained in previous versions of the leanTAP program: the first clause corresponds to the implementation of a β-rule, whilst the last one closes a branch. Just for a change, we have encoded the latter clause slightly differently using a predicate `memberunify/2`: it does what its name suggests: namely implements the standard member predicate, but using sound unification.

The only part in the program that is a bit tricky are clauses 5 and 7: these implement the treatment of universally quantified formulæ. As it is a good heuristic to postpone the expansion of γ-formulae as long as possible, we first just collect them in Gamma when decending a branch in the tableau graph. This is what the second clause does in a quite obvious way. Whenever we reach a leaf (denoted by `true`) of a branch, the subgraphs representing γ-formulae come in again: program clause 7 selects the first one in the list unless the limit for free variables is reached (line 8) and replaces all free variables in the subgraph (line 9). For fairness reasons, the formula just expanded is moved to the end of the list Gammas (line 10), and the proof search continues with one more free variable on the branch.

The program is textually not much smaller than the previous version working with sets of branches as the data structure for tableaux, but there is less work to be carried out during the proof search. On the other hand one cannot expect too much of a speed up, since during proof search all branches have to be considered, no matter how succinctly they have been represented. We have observed a typical increase of the performance of the graphical version of about 25%. The main reason for this is that multiple expansion of formulae is avoided and less applications of tableau rules are needed. As an example consider a formula of the type $(A \vee B) \wedge (C \vee D)$. lean$T^4P$ will expand one of the conjuncts twice. This is avoided here.

6 COMPILING THE PROOF SEARCH

The tableau graphs introduced in the last section provide a very compact representation of fully expanded tableaux; they can also be used for further preprocessing, like computing information about contradictory literals in advance, applying propositional simplifications, etc. One optimization we will further investigate is the *compilation* of semantic tableaux.

Compilation-based provers have been introduced by Stickel [1988]; the idea of this approach is to translate formulæ into executable programs that carry out the proof search during run time; it is well known that this can increase the efficiency of the proof search considerably. The reason is basically the following: rather than using a meta-interpreter that handles tableaux (or representations thereof), we compile this interpreter down into the language we used for implementing the meta-interpreter. The result is a program that carries out the proof search for a particular set of axioms, in contrast to the meta-interpreter, which must be able to handle *any* set of axioms. This is directly comparable to interpreter and compilers in standard programming languages: Whilst an interpreter must be able to handle any program in the language, the result of compiling it is machine code for one particular program, and therefore potentially much more efficient.

Compilation-based approaches to theorem proving are usually carried out for model elimination-based calculi, only. They work by mapping formulæ in clausal form into Prolog programs, thus taking advantage of the fact that Prolog programs are Horn clauses. The resulting programs can be understood as logical variants of the clauses they have been generated for, where all contrapositives of the clauses have been created.[15] The approach presented here works differently; its principle was described in [Posegga, 1993a] and builds upon the following idea:

Instead of using a program as `gprove` for descending the graphically represented tableau, we *generate* a program that performs the search procedure. Thus,

[15] Variants exist that avoid the use of contrapositives, but require a more complex deduction algorithm [Baumgartner and Furbach, 1994].

```
 1 % Auxiliary predicate that instantiates a list of variables
 2 % to a list of integers.
 3 %
 4 instantiate(_,[]).
 5 instantiate(N,[N|Tail]) :- N1 is N + 1,instantiate(N1,Tail).
 6
 7 tgraph(Formula,Graph) :-
 8         tgraph(Formula,IDs/[],Graph/(0:true)),
 9         instantiate(1,IDs).
10
11 tgraph((A,B),IDs/IDsTail,GrA/GrEnd):-!,
12         tgraph(A,IDs/IDsB,GrA/GrB),
13         tgraph(B,IDsB/IDsTail,GrB/GrEnd).
14
15 tgraph((A;B),[N|IDs]/IDsTail,(N:(GrA;GrB))/GrEnd)  :-!,
16         tgraph(A,IDs/IDsB,GrA/GrEnd),
17         tgraph(B,IDsB/IDsTail,GrB/GrEnd).
18
19 tgraph(all(X,F),[N|IDs]/IDsT,
20        (N:(all(X,GrF),GrEnd))/GrEnd)  :-!,
21         tgraph(F,IDs/IDsT,GrF/(0:true)).
22
23 tgraph(Literal,[N|IDs]/IDs,(N:(Literal,End))/End).
```

Figure 12. Deriving tableau graphs with labelled nodes

we move from interpreting the graphical representation of a tableau to compiling it into a program and executing the generated code. This is the main difference from compilation-based model-elimination mentioned above: the latter transforms clauses into declaratively equivalent Prolog clauses, whereas our approach generates a procedurally equivalent Prolog program. This can, in principle, be carried out in any high-level programming language. We will, however, follow the line of this chapter and describe how to program a compiler in Prolog that generates Prolog programs.

Before we explain the idea of the compilation procedure we need a preparatory step: we have to extend the tableau-graph generation of Figure 10 by adding labels to the nodes in the graph. These will eventually serve as unique names for the generated Prolog clauses, and are necessary to control the search process and avoid duplication of Prolog clauses.

Figure 12 shows how the program for deriving tableau graphs from Figure 10 can be extended to generating graphs with labels; it works in the same way as the previous program, but additionally collects a list of Prolog variables. When the construction of the graph is completed procedure `instantiate/2` instantiates

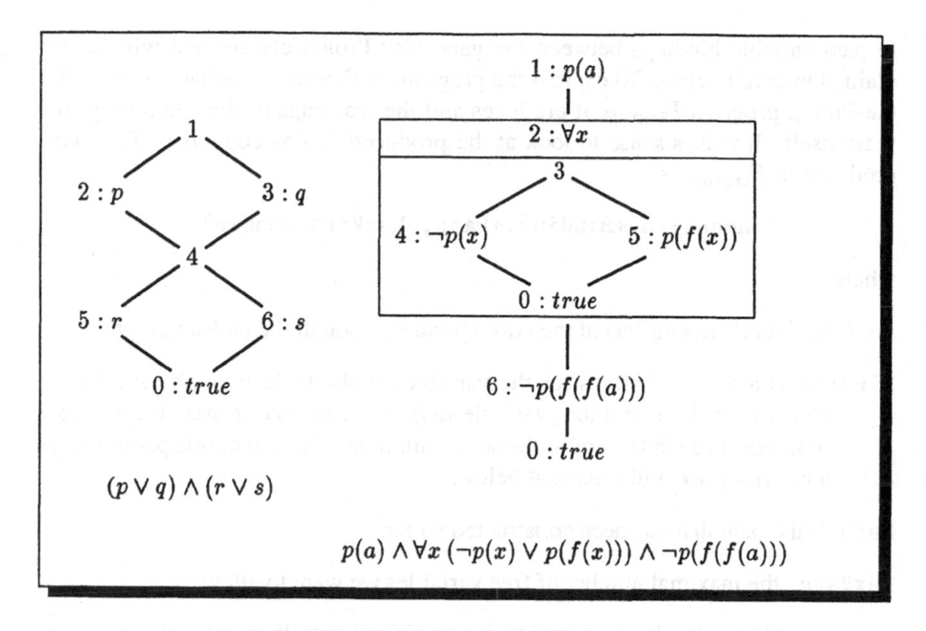

Figure 13. Sample labelled tableau graphs

these variables to integers starting with 1. These lists of integers will serve as labels. In principle, it does not matter what sort of labels are used; integers are just a convenient choice. Note, that the label 0 is used as the label of the leaf *true*. The formulas from Figure 9 yield the result

```
G =
  1:(2:(p,4:(5:(r,0:true);6:(s,0:true)));
  3:(q,4:(5:(r,0:true);6:(s,0:true))
```

respectively

```
G =
  1:(p(a),2:(all(X,3:(4:(-(p(X)),0:true);5:(p(f(X)),0:true))),
  6:(p(f(f(a))),0:true)))
```

for the tgraph-procedure. A graphical representation of these outputs is shown in Figure 13.

Figure 18 lists a program that compiles labelled tableau graphs into Prolog programs. Its main predicate is `comp/3`. For a labelled tableau graph `ltgraph` the call `comp(ltgraph,X,Y)` produces a Prolog program $P_{\mathtt{ltgraph}}$. Technically this is achieved by using the built-in `assert` predicate that adds the clauses of $P_{\mathtt{ltgraph}}$ to the global database. The second and third arguments of `comp` are occupied by uninstantiated variables at the first call. They implement a sophisticated encoding to pass variable bindings between the generated Prolog clauses and will be explained in detail below. To explain the program in Figure 18 we have to describe the Prolog program $P_{\texttt{ltgraph}}$ it produces and the workings of the compiling program itself. It makes sense to look at the produced Prolog code first. The main predicate in $P_{\texttt{ltgraph}}$ is

```
node(+Id,+Binding,+Path,+MaxVars,+Gamma).
```

where

`Id` is the label[16] (identifier) of the corresponding node in the tableau graph.

`Binding` is a list of bindings for the variables in the tableau graph. As Prolog clauses are, by definition, variable disjoint, it is used to pass the variable bindings through the `node`-clauses at run time. The use of this parameter is a bit tricky, we will discuss it below.

`Path` is the path that has been constructed so far.

`MaxVars`, the maximal number of free variables we want to allow.

`Gamma` is a list of labels of nodes which contain universally quantified subgraphs.

The `node/5` clauses will succeed if the tableau graph below its label is inconsistent. This test is performed by considering the individual paths in the graph and by showing that all of them can be closed with a common substitution for the free variables appearing in the paths.

The `Binding`-parameter will be instantiated to a list: for each universally quantified variable in the initial tableau graph, there is a fixed position in the list that holds the current binding for the corresponding variable. The variable that was first encountered will always correspond to the first position in the binding list, the second encountered variable to the second position and so on. The third, forth and fifth arguments are used as in `gprove`: the argument `Path` holds the current path in the considered tableau (as a list of literals), `MaxVars` limits the number of free variables on a branch, and `Gamma` collects applicable γ-formulae.

Figure 14 gives the generated Prolog code for our running example: Lines 1–11 are not generated by the `comp`-procedure. They are part of the `node`-definition independently of the input graph. Lines 1–3 define `close`, a simple predicate which closes branches. Line 5 defines the top level predicate for starting the proof search, where the only parameter serves as a gamma limit. The clause starting in line 7 defines the action of the search procedure when a leaf of the tableau graph is reached. If the current path can be closed the calling goal succeeds. If `MaxVars`

[16]These labels appear as the first argument, since most Prolog systems perform indexing on the first argument of a clause. Thus, the identifier is at the right position to allow fast access to clauses by their labels.

```
1  close(Lit,[L|Lits]) :-
2          (Lit = -Neg; -Lit = Neg) ->
3             (unify(Neg,L); close(Lit,Lits)).
4
5  start(N) :- node(1,_,[],N,[]),!.
6
7  node(0,B,P,MaxVars,[Id|Gamma]):-
8          MaxVars > 0, MaxVars1 is MaxVars - 1,
9          append(Gamma,[Id],NewGamma),
10         node(Id,B,P,MaxVars1,NewGamma).
11
12 node(1, A, B, C, D) :-
13         (   close(p(a), B)
14         ;   node(2, A, [p(a)|B], C, D)  ).
15 node(2, A, B, C, D) :-
16         node(6, A, B, C, [3|D]).
17 node(3, [_|E], A, B, C) :-
18         node(4, [D|E], A, B, C),
19         node(5, [D|E], A, B, C).
20 node(4, [D|E], A, B, C) :-
21         ( close(-(p(D)), A)
22           ; node(0, [D|E], [-(p(D))|A], B, C) ).
23 node(5, [D|E], A, B, C) :-
24         ( close(p(f(D)), A)
25           ; node(0, [D|E], [p(f(D))|A], B, C) ).
26 node(6, [D|E], A, B, C) :-
27         ( close(-(p(f(f(a)))), A)
28           ; node(0, [D|E], [-(p(f(f(a))))|A], B, C) ).
```

Figure 14. Prolog Code for $p(a) \wedge \forall X(p(x) \rightarrow p(f(X))) \wedge \neg p(f(f(a)))$

is reached the goal fails. Otherwise the next universally quantified subgraph from the list Gamma is enter at its top node.

Lines 12–28 is the compiled code for the tableau graph in Figure 13: the clauses `node(1,...)` and `node(6,...)` correspond to $p(a)$ and $p(f(f(a)))$ in the graph. The universally quantified subgraph is implemented by the clause `node(2,...)`. `node(3,...)` implements the disjunction, and `node(4,...)` and `node(5,...)` correspond to the literals in the disjunct.

The trace of the program from Fig. 14 for `Maxvars = 2` is shown in Fig. 15

In our running example there is only one variable, this is not enough to see how the `bindings` parameter works in general. Let us consider as a second example the labelled tableau graph in Figure 16 and the prolog programm that is compiled

```
1    C: start(2)
2    C: node(1,X,[],2,[])
3    C: close(p(a),[])
3    F: close(p(a),[])
3    C: node(2,X,[p(a)],2,[])
4    C: node(6,X,[p(a)],2,[3])
5    C: close(-(p(f(f(a)))),[p(a)])
5    F: close(-(p(f(f(a)))),[p(a)])
5    C: node(0,[Y|Z],[-(p(f(f(a)))),p(a)],2,[3])
9    C: node(3,[Y|Z],[-(p(f(f(a)))),p(a)],1,[3])
10   C: node(4,[U|Z],[-(p(f(f(a)))),p(a)],1,[3])
11   C: close(-(p(U)),[-(p(f(f(a)))),p(a)])
11   E: close(-(p(a)),[-(p(f(f(a)))),p(a)])
10   E: node(4,[a|Z],[-(p(f(f(a)))),p(a)],1,[3])
26   C: node(5,[a|Z],[-(p(f(f(a)))),p(a)],1,[3])
27   C: close(p(f(a)),[-(p(f(f(a)))),p(a)])
27   F: close(p(f(a)),[-(p(f(f(a)))),p(a)])
27   C: node(0,[a|Z],[p(f(a)),-(p(f(f(a)))),p(a)],1,[3])
31   C: node(3,[a|Z],[p(f(a)),-(p(f(f(a)))),p(a)],0,[3])
32   C: node(4,[W|Z],[p(f(a)),-(p(f(f(a)))),p(a)],0,[3])
33   C: close(-(p(W)),[p(f(a)),-(p(f(f(a)))),p(a)])
33   E: close(-(p(f(a))),[p(f(a)),-(p(f(f(a)))),p(a)])
32   E: node(4,[f(a)|Z],[p(f(a)),-(p(f(f(a)))),p(a)],0,[3])
50   C: node(5,[f(a)|Z],[p(f(a)),-(p(f(f(a)))),p(a)],0,[3])
51   C: close(p(f(f(a))),[p(f(a)),-(p(f(f(a)))),p(a)])
51   E: close(p(f(f(a))),[p(f(a)),-(p(f(f(a)))),p(a)])
50   E: node(5,[f(a)|Z],[p(f(a)),-(p(f(f(a)))),p(a)],0,[3])
31   E: node(3,[a|Z],[p(f(a)),-(p(f(f(a)))),p(a)],0,[3])
. . . .
1    E: start(2)

C = Call, E = Exit, F = Fail
```

Figure 15. Trace of the program from Figure 14

from it in Figure 17.[17]

The `binding` parameter will in this case be a three-element list which holds the bindings of the universal variables X,Y,Z of the input formula. In lines 12 and 15 of Figure 17 the first element of the binding list is accessed and used to instantitate $p(X)$ respectively $q(X)$. In lines 23 and 26 the first two positions are accessed, and the second position is used to instantiate $q(Y)$ respectively $r(Y)$. Finally in lines 34 and 37 all three positions are accessed, but only the third position is used to instantiate $r(Z)$ respectively $p(f(Z))$. The first two positions are passed on to the subsequent calls of the predicate `node`.

The open tail R avoids annecessary overhead, if only the first variable is needed the remaining positions will be lumped together in the remaining list R and passed on without change.

In line 9 of Figure 17 the Prolog variable _ that occupies the first position in the binding list does not occur in the body of the clause. The value of _ will not be passed on to the calls of `node(5,[A|R],X3,X4,X5)` and `node(6,[A|R],X3,X4,X5)`. Instead, a new variable A is introduced. This is exactly the effect of a γ-rule application for variable X. This also explains why the length of the `binding`-list equals the number of quantified variables in hte original tableau graph no matter how often the γ-rule is invoked. In lines 20 and 31 the same happens for the second and third variable, in our case Y and Z. Note that the values of the other positions of the binding list are passed on unchanged.

Now we have a look at the compilation program, see Figure 18 on page 616. Its main predicate is

```
comp(+TableauGraph,+BindIn,+BindOut),
```

where `TableauGraph` is a tableau graph with the leaf `0:true` and `BindIn` and `BindOut` are initially the empty list.

We will refer to the binding list in the head of a `node`-clause as the *inbound binding*, and the to binding list in the body of a `node`-clause as the *outbound binding*. The compiling predicate `comp` constructs both lists while descending the tableau graph; the lists for inbound and outbound binding are passed through the calls of `comp` *without* the trailing Prolog variable for a possible tail. The position of a variable's binding in the list is determined by the order in which the variables occur in the input graph. The only place where new variables are introduced is when compiling a universally quantified node. Then, the inbound and outbound lists are extended by one slot at their ends (lines 17,18 of Figure 18).

Line 1 terminates the recursion if a leaf is reached. The clause in line 3 succeeds if a `node`-clause with the same `id`-number has already been added. No further action will follow in this case.

Lines 6–13 compile a disjunction: in 6 and 7, we append a tail to the inbound and outbound binding and assert a `node`-clause which implements a disjunction:

[17] For brevity we have omitted that part of the code that does not depend on the input graph, cf. Figure 14

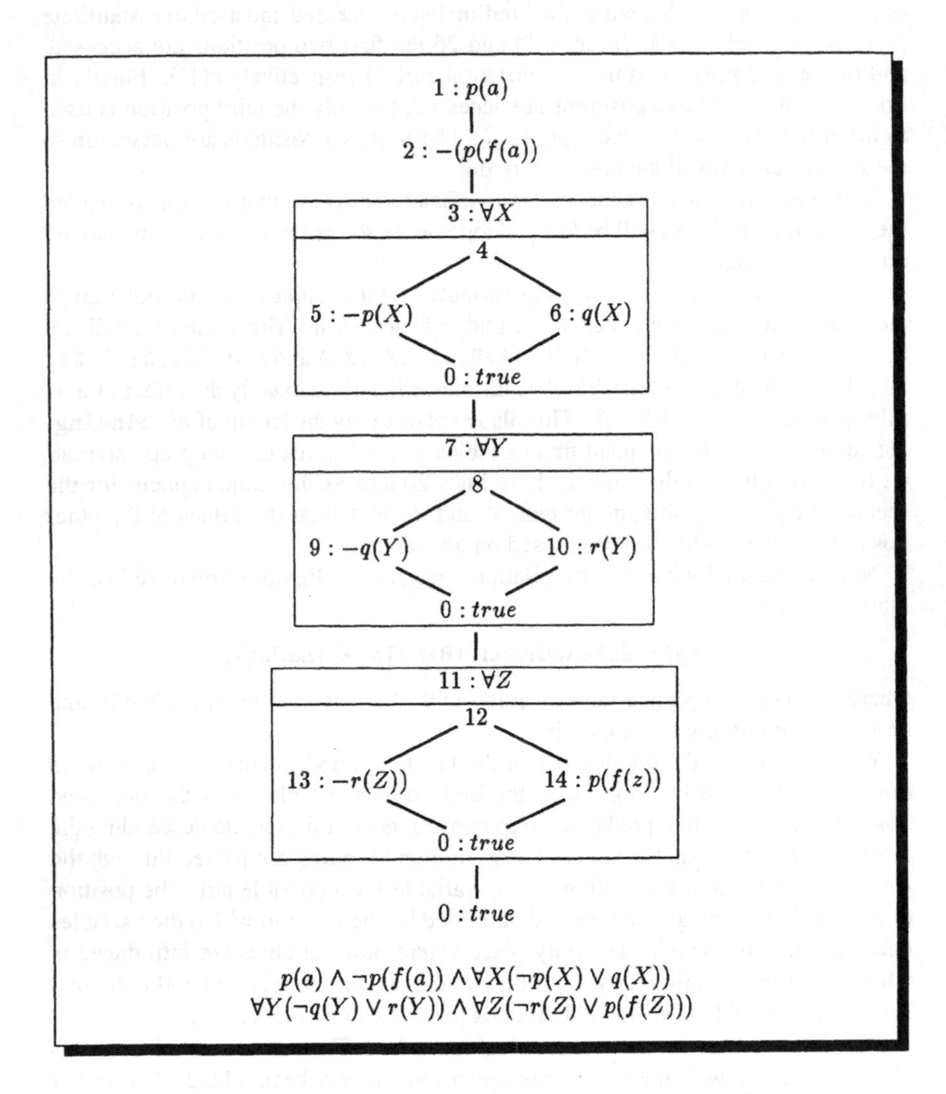

Figure 16. Third example of a labelled tableau graph

```
1  node(1,X2,X3,X4,X5) :- (close(p(a),X3);
2                          node(2,X2,[p(a)|X3],X4,X5)).
3
4  node(2,X2,X3,X4,X5) :- (close(-(p(f(a))),X3);
5                          node(3,X2,[-(p(f(a)))|X3],X4,X5)).
6
7  node(3,X2,X3,X4,X5) :- node(7,X2,X3,X4,[4|X5]).
8
9  node(4,[_|R],X3,X4,X5) :- node(5,[A|R],X3,X4,X5),
10                           node(6,[A|R],X3,X4,X5).
11
12 node(5,[A|R],X3,X4,X5) :- (close(-(p(A)),X3);
13              node(0,[A|R],[-(p(A))|X3],X4,X5)).
14
15 node(6,[A|R],X3,X4,X5) :- (close(q(A),X3);
16              node(0,[A|R],[q(A)|X3],X4,X5)).
17
18 node(7,X2,X3,X4,X5) :-  node(11,X2,X3,X4,[8|X5]).
19
20 node(8,[A,_|R],X3,X4,X5) :- node(9,[A,B|R],X3,X4,X5),
21                             node(10,[A,B|R],X3,X4,X5)).
22
23 node(9,[A,B|R],X3,X4,X5) :- (close(-(q(B)),X3);
24              node(0,[A,B|R],[-(q(B))|X3],X4,X5)).
25
26 node(10,[A,B|R],X3,X4,X5) :- (close(r(B),X3);
27              node(0,[A,B|R],[r(B)|X3],X4,X5)).
28
29 node(11,X2,X3,X4,X5) :- node(0,X2,X3,X4,[12|X5]).
30
31 node(12,[A,B,_|R],X3,X4,X5) :- node(13,[A,B,C|R],X3,X4,X5),
32              node(14,[A,B,C|R],X3,X4,X5).
33
34 node(13,[A,B,C|R],X3,X4,X5) :- (close(-(r(C)),X3);
35              node(0,[A,B,C|R],[-(r(C))|X3],X4,X5)),
36
37 node(14,[A,B,C|R],X3,X4,X5) :- (close(p(f(C)),X3);
38              node(0,[A,B,C|R],[p(f(C))|X3],X4,X5)),
```

Figure 17. Code for third example of a labelled tableau graph

```
1  comp(0:true,_,_):- !.
2
3  comp((Id:_),_,_) :- clause(node(Id,_,_,_,_),_),!.
4
5  % Disjunctions:
6  comp(Id:((LeftId:Left);(RightId:Right)),BindIn,BindOut):-!,
7      append(BindIn,BTail,BI),
8      append(BindOut,BTail,BO),
9      assert((node(Id,BI,P,MaxVars,Gamma) :-
10             node(LeftId,BO,P,MaxVars,Gamma),
11             node(RightId,BO,P,MaxVars,Gamma))),
12     comp(LeftId:Left,BindOut,BindOut),
13     comp(RightId:Right,BindOut,BindOut).
14
15 % Univ. quantification:
16 comp(Id:(all(X,(ScId:Scope)),SuccId:Succ),BindIn,BindOut):-!,
17     append(BindIn,[_],ScBindIn),
18     append(BindOut,[X],ScBindOut),
19     assert((node(Id,Bind,P,MaxVars,Gamma) :-
20             node(SuccId,Bind,P,MaxVars,[ScId|Gamma]))),
21     comp(ScId:Scope,ScBindIn,ScBindOut),
22     comp(SuccId:Succ,ScBindOut,ScBindOut).
23
24 % Literals:
25 comp(Id:(Lit,SuccId:Succ),BindIn,BindOut) :-!,
26     append(BindIn,BTail,BI),
27     append(BindOut,BTail,BO),
28     assert((node(Id,BI,Path,MaxVars,Gamma):-
29             close(Lit,Path)
30               ;node(SuccId,BO,[Lit|Path],MaxVars,Gamma))),
31     comp(SuccId:Succ,BindOut,BindOut).
```

Figure 18. Compiling tableau graphs

two goals that correspond to both generated branches must be solved. After asserting the clause, we continue to compile both disjuncts.

Lines 16–22 compile universally quantified subgraphs; we generate inbound and outbound bindings for compiling the scope of the universal quantification, as discussed above. The compiled code for the current tableau graph node does not use these: the only action we perform is to add the address of this node to the list `Gamma`, and continue by calling the node *after* the universally quantified subgraph. Thus, at runtime, we do *not* enter the universally quantified subgraph on the first transversal of the graph, but just 'jump' over it ignoring its contents. The actual renaming of the quantified variable takes place if we enter the code for the subgraph during runtime: line 21 calls the compiler for this subgraph, which will generate the corresponding code. Line 22 calls the compiler for the next node after the subgraph. Note, that the changed inbound and outbound binding is only relevant for compiling these clauses. The current clause we compile does not refer to any variables, so it is sufficient to pass the bindings with a simple Prolog variable. Here, it is called `Bind`.

The purpose of the last clause is again quite obvious: when compiling a literal, we either close the current branch or (if this fails) call the clause for the next node.

Compiling formulae into Prolog code yields another speedup compared to the version dealing with tableau graphs: depending on the quality of your Prolog compiler, speed can easily double. An interesting point is that the compilation principle can be integrated into the 'interpreting' versions of our tableau provers: One could, for instance, compile theories that are often used in advance and load the corresponding Prolog code when required.

Compilation is a powerful technique for theorem provers, but it also has its limits. It is important to understand that it does not increase the efficiency of a calculus, but 'just' the efficiency of its implementation. For achieving the former we must modify the underlying calculus; one possible way is by using lemmata.

7 INCLUDING LEMMATA

Lemmata have already been mentioned in Section 4 of the Chapter 'Tableaux Methods for Classical Propositional Logic'. Here, we will treat the technical aspects of including lemmata into the proof search, after having recalled the basic idea behind lemmata.

7.1 *What are Lemmata?*

Lemmata can be seen as one way to strengthen a tableau calculus. Although the use of lemmata can shorten proofs, this does not mean that the shorter proofs are also easier to find. In a sense, this reduces the depth of the search space for the price of broadening it. However, if we consider certain classes of theorems that have

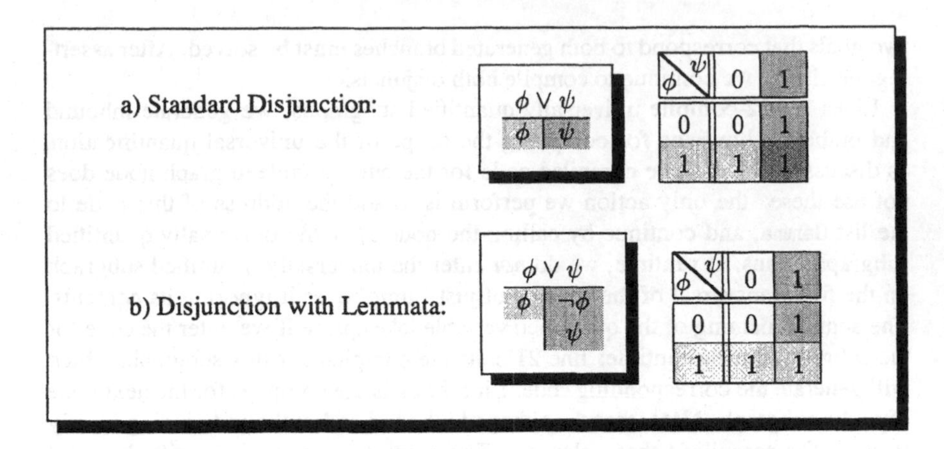

Figure 19. Lemmata from the truth-table point of view

exponential proof length (like, for instance, the pigeon-hole formulæ), including lemmata is often the only way to cope with such problems.

There are two perspectives from which one can look at lemmata; one is the truth table point of view, the other is a more 'operational' point of view. The former is illustrated in Figure 19: Part a) gives the truth table for the usual treatment of disjunction in a tableau calculus, Part b) for a disjunction with lemmata. Each of the two branches in a tableau resulting from expanding a disjunction corresponds to one of the shaded areas in the truth table. In a), the entry where both subformulæ are true (the bottom-right entry) is covered twice: this means, that this entry in the truth table is 'covered' by two branches.

This entry is covered only once with a disjunction rule with lemmata. In terms of tableaux, this means that we add the information that the other disjunct is false to one branch when decomposing a disjunction. By this, we cover each entry only on one branch. Thus, the information on branches is more specific than in the case above: no interpretation for the two disjuncts will satisfy both branches. The difference also becomes clear if we consider a fully expanded tableau for a given formula: Whilst with the standard rule for disjunctions, all paths form a DNF for the formula, we will get an XOR-normal form if lemmata are included.

Figure 20 presents this semantic consideration from another, more operational point of view: here, a lemma can be seen as a shorthand for a subproof, which would have to be carried out multiply during the proof if we did not use lemmata. As such subproofs can be arbitrarily complex, lemmata can considerably decrease the length of proofs.

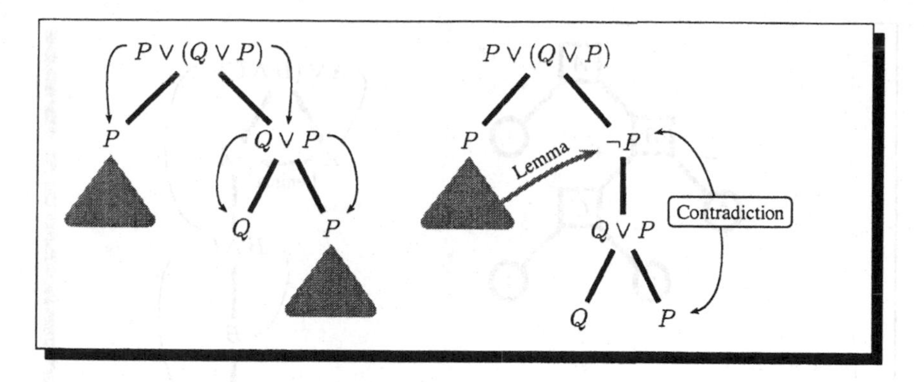

Figure 20. Lemmata from an operational point of view

7.2 *Integrating Lemmata into our Framework*

Lemmata are easily included into a tableau calculus by modifying the β-rules according to Figure 19: the right branch will hold the negation of the disjunct that went to the left, i.e., we recall on the right branch that the left branch is closed.[18] Unfortunately, this has two nasty side effects: Firstly, it is not a very good representation, as one disjunct appears twice in the expansion: We would need to represent a formula and its corresponding lemma separately. Secondly, we cannot restrict a prover to negation normal form anymore, since the required negation for building the lemma would destroy the normal form during run time.[19]

Fortunately, there are well known solutions to this problem: we can take out the screw driver and manipulate the basic representation underlying our prover: instead of building upon the disjunctive normal form-representation for tableaux, we change to an *if-then-else* representation:

DEFINITION 8. The set of *if-then-else* expressions $\mathcal{SH}$ is the smallest set such that:

1. $\{0,1\} \subset \mathcal{SH}$
2. If A is atomic and $\mathcal{B}_-, \mathcal{B}_+ \in \mathcal{SH}$ then $sh(A, \mathcal{B}_-, \mathcal{B}_+) \in \mathcal{SH}$.
3. If $\mathcal{B}, \mathcal{B}_-, \mathcal{B}_+ \in \mathcal{SH}$, then $sh((\forall x\ \mathcal{B}), \mathcal{B}_-, \mathcal{B}_+) \in \mathcal{SH}$

The semantics of $sh(A, C, B)$ is defined as: *if A then B else C*, i.e.: $(A \wedge B) \vee (\neg A \wedge C)$.

[18]Note, that this does not require that the left branch is actually closed before proceeding with the right branch.

[19]The latter point can be resolved if lemma generation is moved into preprocessing, e.g. if we extended the derivation of negation normal form appropriately. However, the problem of considerably increasing the size of tableaux remains.

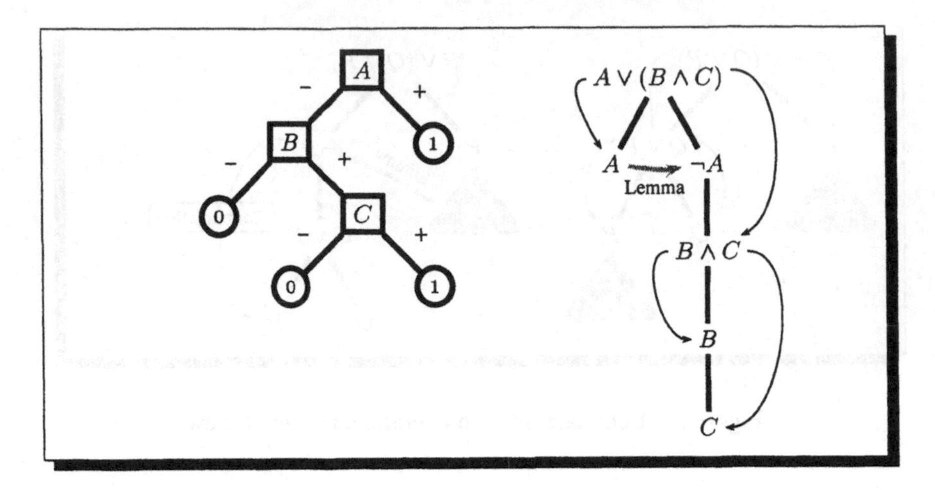

Figure 21. BDD and tableau for $A \vee (B \wedge C)$

Definition 8 defines a class of formulae which, when represented graphically, are called *BDDs* or *Binary Decision Diagrams*.[20] These formulae are built solely by atomic formulae, an *if-then-else*–connective and the atomic truth constants 1 and 0.

BDDs are usually defined for propositional logic, only. For handling quantifiers, we use nested *if-then-else*-expressions, analogously to tableau graphs in Definition 7. As with tableau graphs, we can easily map first-order formulae into BDDs:

DEFINITION 9. Let F be a first-order formula in Skolemized negation normal form; then

$$\mathit{f2Sh}(F) = \begin{cases} \mathit{f2Sh}(A)\left[\frac{1}{\mathit{f2Sh}(B)}\right] & \mathit{if}\ F = A \wedge B \\ \mathit{f2Sh}(A)\left[\frac{0}{\mathit{f2Sh}(B)}\right] & \mathit{if}\ F = A \vee B \\ \mathit{sh}((\forall x\ \mathit{f2Sh}(A)), 0, 1) & \mathit{if}\ F = \forall x\ A \\ \mathit{sh}(F, 0, 1) \text{ or } \mathit{sh}(F, 1, 0) & \mathit{if}\ F\ \mathit{is\ a\ Literal} \end{cases}$$

Figure 21 shows the BDD $\mathit{f2Sh}(A \vee (B \wedge C))$ and a corresponding tableaux with lemmata. If we apply the same trick as used for tableau graphs to BDDs, we can derive a graphical representation instead of a tree: the replacement operation is carried out analogously to that which was explained for Definition 7 by replacing

[20]Note, that the notion 'BDD' is often used in the literature to refer to ordered, reduced BDDs (ROBDDs); this is not meant here: we use non-ordered, and non-reduced BDDs. ROBDDs are considered later in Section (7.3).

edges, rather than nodes.

The motivation for BDDs is to handle lemmata in a better way than by modifying the β-rule; the paths to 1-leaves in BDDs are indeed nothing but a representation of branches in a tableau with disjunctive rules that incorporate lemmata. The reader is invited to verify this by examining Figure 21.

It is beyond the scope of this chapter to formalize this in detail, so we will just give the basic idea (see [Posegga, 1993b] for details), restricted to propositional logic: The key to understanding it is to compare branches in fully expanded tableaux to paths in BDDs:

Assume we have a fully expanded tableau $\mathcal{F}$ for a propositional formula F; we can then interpret the branches in $\mathcal{F}$ as conjunctions of literals. Then, the disjunction of all branches in $\mathcal{F}$ is a DNF for F.

Paths in BDDs can be regarded analogously: In propositional logic, each node in a BDD is labelled with an atomic formula. Thus, a path can be seen as a sequence of signed atoms. The signs denote which 'exit' was chosen at each node: if the *then*-part was used, the sign is positive, otherwise it is negative. Analogously to branches in tableaux, we can regard these paths as a conjunction of literals, where the sign attached to the atoms denotes whether the literal is negated or not. The difference to tableaux is that there are two kinds of paths, namely paths to 1-leafs and paths to 0-leafs. The paths to 1-leaves play the same role as the branches in tableaux, i.e., they are a DNF for the underlying formula. The 0-paths, however, build a DNF for the *negated* formula, i.e., the conjunction of all 0-paths in *f2Sh*(F) is a DNF for $\neg F$.

To summarize, tableaux represent models of formulæ, whilst BDDs represent models *and* counter models. Note, that both our graphical representation for fully expanded tableaux (cf. Definition 7) and BDDs (Definition 9) can be computed linearly w.r.t. to the length of the negation normal form of a formula. With BDDs, we thus get the additional information of counter models more or less 'for free'.

It remains to show that BDDs actually fulfill their intended purpose, i.e. that they represent tableau with lemmata. We shall argue informally: Figure 20 shows how lemmata can be integrated when decomposing disjunctions in a tableau calculus; we add the negated left conjunct to the right branch. This means, the right branch will contain information about all counter models of the left disjunct when the tableau is fully expanded. The disjunction rule for computing a BDD for a formula acts similarly: for $A \vee B$, *f2Sh*(B) is inserted for the 0-leaf of *f2Sh*(A). As the 0-paths of *f2Sh*(A) represent counter models of A, the 1-paths of the resulting graph will contain these. It is not very hard to show formally that the 1-paths of a BDD for a formula F are identical to the branches in corresponding tableau for F: by induction over the structure of F, we relate 1-paths to branches of a tableau for F, and 0-paths to branches of a tableau for $\neg F$. The proof is left as an exercise to the reader.

When implementing deduction based on BDDs, the first step required is to translate formulae into the graphs. Based on the mapping given in Definition 9, we can implement this very elegantly in Prolog; Figure 22 shows a simple program

```
1  f2bdd((A,B),True_B,False,BDD_A) :-!,
2      f2bdd(A,BDD_B,False,BDD_A),
3      f2bdd(B,True_B,False,BDD_B).

4  f2bdd((A;B),True,False_B,BDD_A):-!,
5      f2bdd(A,True,BDD_B,BDD_A),
6      f2bdd(B,True,False_B,BDD_B).

7  f2bdd(all(X,Fml),True,False,
8          (all(X,BDD_Fml) -> True; False)) :-!,
9      f2bdd(Fml,1,0,BDD_Fml).

10 f2bdd(Literal,True,False,BDD):-
11     (Literal= -Lit) -> BDD = (Lit -> False; True)
12                ; BDD= (Literal -> True; False).
```

Figure 22. Implementing Definition 9

which is nearly a literal translation of Definition 9:

We use the prolog *if-then-else* construct '`... -> ... ; ...`' to denote *if-then-else*. The clause

`f2bdd(Formula,True,False,BDD)`

succeeds if `BDD` is a BDD for `Formula` with the *true*-leaf `True` and the *false*-leaf `False`.

The first clause handles conjunctions. We recursively compute graphs for `A` and `B`, and insert the latter for the *true*-leaf of the graph for `A`. This corresponds to the first case in Definition 9.

Disjunctions work analogously, but the graph for `B` goes to the false-leaf of the graph for `A`.

Universal quantification is handled as with tableau graphs. Note, that the leaves inside universally quantified subgraphs are instantiated to the constants `true` and `false`.

Figure 23 gives an example: it shows a graphical representation of the binding of `BDD` after successful termination of the Prolog query:

```
f2bdd((p(a),-p(f(f(a))),all(X,(-p(X);p(f(X))))),true,false,BDD)
```

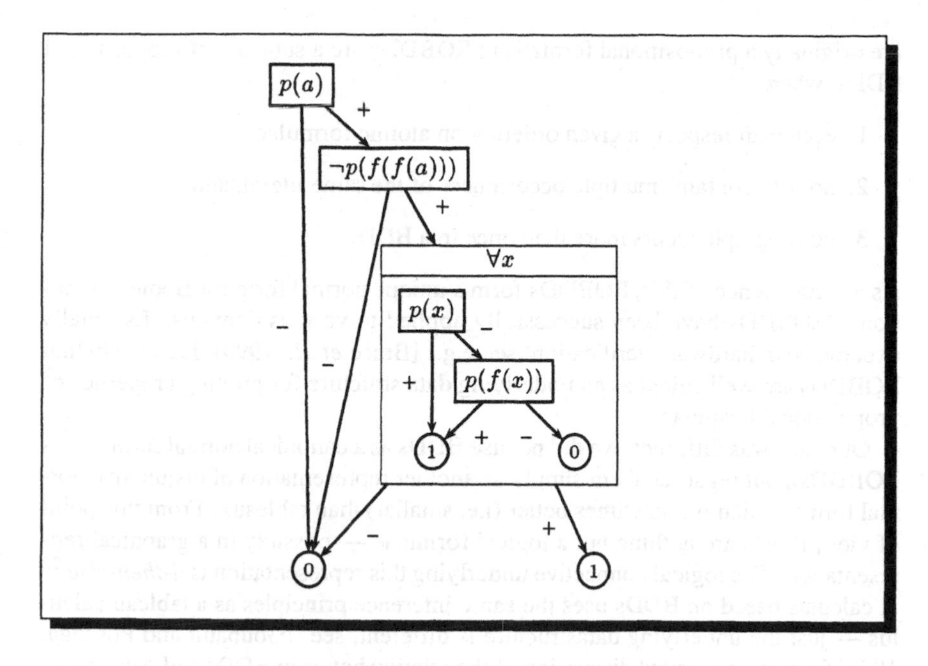

Figure 23. An example BDD

Deduction with BDDs

From what we have seen about BDDs, it should be clear how the presented algorithms for deduction with tableau graphs can be adapted to BDDs; the underlying principle is the same: BDDs represent a disjunctive normal form and the paths in BDDs are the analog to branches in tableaux. Thus, when trying to show that a given BDD represents an inconsistent formula, we inspect its paths and try to find contradictory literals on each of them. Extension steps for applying universal quantification work in the same way as for tableau graphs. It is not difficult to modify the Prolog programs given for tableau graphs such that they work on BDDs instead.

7.3 *Reduced, Ordered Binary Decision Diagrams*

The reader being familiar with Binary Decision Diagrams will have noticed that we use BDDs in their non-reduced, non-ordered form.[21] In the literature, however, BDDs appear mostly as reduced, ordered BDDs (ROBDDs) [Bryant, 1986; Bryant, 1992; Goubault and Posegga, 1994]. One reason for this is that BDDs

[21] These are also called *free BDDs* by some authors, *Shannon graphs* by others.

are originally a propositional formalism; ROBDDs are a subclass of propositional BDDs, where

1. each path respects a given ordering on atomic formulae,
2. no path contains multiple occurrences of the same literal, and
3. no subgraph occurs more than once in a BDD.

As a consequence of this, ROBDDs form a unique normal form for Boolean functions. ROBDDs have been successfully applied to various domains. Especially experience in hardware verification (see e.g. [Brace *et al.*, 1990]) has shown that ROBDDs are well suited as an underlying data structure for proving properties of propositional formulae.

Our view was different: we did not use BDDs as a canonical normal form (ie: as ROBDDs), but regarded them simply as another representation of disjunctive normal forms, which is sometimes better (i.e. smaller) than tableaux. From this point of view, BDDs are nothing but a logical formulae — possibly in a graphical representation. The logical connective underlying this representation is *if-then-else*.[22] A calculus based on BDDs uses the same inference principles as a tableau calculus — just the underlying datastructure is different, see [Goubault and Posegga, 1994] for a more detailed discussion of the relation between BDDs and Automated Reasoning.

Furthermore, we used BDDs for representing first-order formulae, and showing that they are inconsistent. This is also not a standard use of ROBDDs: these have been designed for *representing* Boolean functions, rather than for showing that the function never evaluates to 1. The main purpose of ordering atoms and maintaining a reduced graphical representation in ROBDDs is, however, to ease the representation of Boolean functions: it results in a *unique* representation (w.r.t. the ordering).

It is clear that a unique normal form for first-order logic is not computable, since the language of first-order logic is undecidable. This might appear as an argument against the use of ordered in BDDs for first-order logic, but it is not: it just says that we will not achieve a unique normal form, but does not tell anything about the efficiency of an ordered, reduced format w.r.t. the unordered BDDs we have used.[23] It might well be the case that a first-order calculus based on ROBDDs works more efficient for a certain class of formulae than one based on BDDs. The opposite, however, can also be the case. The answer to 'what should I choose?' is not context-free.

[22]Such formulæ have already been considered in 1854 by George Boole [Boole, 1958]; Alonzo Church showed about one century later that *if-then-else* is a primitive basis for propositional logic [Church, 1956, §24, pp. 129ff].

[23]The use of ROBDDs in a tableau-like setting can, from a purely logical point of view, be seen as using regular tableaux. Consult Letz's Chapter for details.

8 A GLIMPSE INTO THE FUTURE

Automated Deduction is at present neither an engineering discipline, nor pure mathematics. The key to successfully applying Automated Deduction is careful analysis and experimenting. Both are equally important and depend on each other. The above considerations on BDDs vs. ROBDDs stress an important point in working on Automated Deduction: There is no panacea. For nearly each heuristic, or modification to a calculus, there is a counterexample where things become worse than before. The language of first-order logic is not decidable, and it is highly unlikely that this will ever change. This makes the field hard, but it also makes it interesting: we will never run out of problems.

Our motivation for writing this chapter as it is was to support experimenting: we presented a couple of implementation techniques which preserve the openess of tableau calculi. It is unlikely (although not impossible) that one of programs we presented will exactly fit for a concrete application one has. But the reader is likely to find a starting point in this chapter.

It is hard to give any reasonable predictions of the future course of automated theorem proving in general and tableau-based automated theorem proving in particular. But it is pretty clear that the distinctions between fully automated theorem provers and interactive ones will fade. Interactive components will be added to upto now fully automated systems and automated systems will be integrated in interactive proof development systems. As one example for an interactive prover based on a sequent calculus one may name IMPS (Interactive Mathematical Proof System) described in [Farmer *et al.*, 1992] and [Farmer *et al.*, 1993]. Instead of implementing one prover for one logical calculus it has also been tried to develop shells that help realize a custom-made logical system. Within the family of sequent calculi such an approach has been undertaken in [Richards *et al.*, 1994].

9 A BRIEF HISTORICAL SURVEY ON TABLEAU-BASED PROVERS

The following list of implementations of theorem provers can certainly not claim to be exhaustive. Apart from the difficulty of locating the relevant information it is also not clear where to draw the line between tableau-based theorem provers and those that are not. We tended to include programs based on sequent calculi because of their close relationship to tableaux, but left out systems based on natural deduction. We also did not consider provers for propositional logic only. The following account is for the greatest part gleaned from [Beckert and Posegga, 1995].

The first tableau-based theorem prover that we know of was developed in the late fifties by Dag Prawitz, Håkan Prawitz, and Neri Voghera [Prawitz *et al.*, 1960]. It ran on a computer named Facit EDB (manufactured by AB Åtvidabergs Industrier). The tableau calculus implemented was already quite similar to today´s versions; it did not, however, use free variables. This prover was perhaps the earliest for first-order logic at all.[24]

At about the same time, Hao Wang implemented a prover for first-order logic, that was based on a sequent calculus similar to semantic tableaux [Wang, 1960]. The program ran on IBM 704-computers.

Ewa Orłowska implemented a calculus that can be seen as tableau-based in 1967 on a GIER digital computer[25]. The calculus was based on deriving *if-then-else* normal forms rather than disjunctive normal forms. Only the propositional part of the calculus was implemented.

We are not aware of any implementation-oriented research around tableaux in the seventies; there have been a number of theoretic contributions to tableau calculi but nothing seems to have been implemented.

In the eighties, the research lab of IBM in Heidelberg, Germany was a major driving force of tableau-based deduction: Wolfgang Schönfeld developed a prover within a project on legal reasoning [Schönfeld, 1985]. It was based on free-variable semantic tableaux and used unification for closing branches. A few years later Peter Schmitt developed the THOT theorem prover at IBM [Schmitt, 1987]; this was also an implementation of free-variable tableaux and part of a project aiming at natural language understanding. Both implementations have been carried out in Prolog. Based on experiences with the THOT theorem prover, the development of the $_3T^AP$ system started around 1990 at Karlsruhe University [Beckert *et al.*, 1992b]; the project was funded by IBM Germany and carried out by Peter Schmitt and Reiner Hähnle. The $_3T^AP$ prover was again written in Prolog and implemented a calculus for free-variable tableaux, both for classical first-order logic with equality as well as for multi-valued logics. This program can bee seen as the direct ancestor of leanT^AP.

Besides the line of research outlined above there was also other work on tableau-based deduction in the eighties: Oppacher and Suen published their well-known paper on the HARP theorem prover in 1988 [Oppacher and Suen, 1988]. This prover was implemented in LISP and is probably the best-known instance of a tableau-based deduction system. Another implementation, the Helsinki Logic Machine (HLM), is a Prolog program that actually implements about 60 different calculi, among them semantic tableaux for classical first-order logic, non-monotonic logic, dynamic logic, and autoepistemic logic. Approximately at the same time a tableau-based prover was implemented at Karlsruhe University by Thomas Käufl [Käufl and Zabel, 1990]; the system, called 'Tatzelwurm', implemented classical first-order logic with equality, but did not use a calculus based on free variables. Its main purpose was to be used as an inference engine in a program verification system.

---

[24]Actually, Prawitz *et al.* implemented a calculus for first-order logic without function symbols; that, however, has the same expressiveness as full first-order logic.

[25]The GIER (Geodaetisk Instituts Elektroniske Regnemaskine) was produced by Regnecentralen in Copenhagen (Denmark) in the early sixties.

Since 1990, the interest in tableau-based deduction continuously increased, and we will not try continue our survey beyond this date. From 1992 onwards, the activities of the international tableau community are quite well documented, as annual workshops were started; we refer the interested reader to the workshop proceedings of these workshops [Fronhöfer *et al.*, 1992; Basin *et al.*, 1993; Broda *et al.*, 1994].[26] Another interesting source of information on implementations are the system abstracts in the proceedings of the CADE conferences since 1986. Among the newer developments let us mention the sequent calculus based prover called GAZER [Barker-Plummer and Rothenberg, 1992]. GAZER is implemented in Prolog.

Joachim Posegga
Deutsche Telekom AG, Research Center, Darmstadt, Germany.
Peter Schmittt
Universität Karlsruhe, Germany.

REFERENCES

[Broda *et al.*, 1994] K. Broda, M. D'Agostino, R. Goré, R. Johnson, and S. Reeves. 3rd workshop on theorem proving with analytic tableaux and related methods. Technical Report TR-94/5, Imperial College London, Department of Computing, London, England, April 1994. (Workshop held in Abingdon, England).

[Baumgartner and Furbach, 1994] P. Baumgartner and U. Furbach. Model Elimination without Contrapositives. In A. Bundy, editor, *12th Conference on Automated Deduction*, volume 814 of *Lecture Notes in Artificial Intelligence*, pages 87–101. Springer, 1994.

[Basin *et al.*, 1993] D. Basin, B. Fronhöfer, R. Hähnle, J. Posegga, and C. Schwind. 2nd workshop on theorem proving with analytic tableaux and related methods. Technical Report 213, Max-Planck-Institut für Informatik, Saarbrücken, Germany, May1993. (Workshop held in Marseilles, France).

[Beckert *et al.*, 1992] B. Beckert, S. Gerberding, R. Hähnle, and W. Kernig. The tableau-based theorem prover $_3T^AP$ for multiple-valued logics. In *11th Conference on Automated Deduction*, Lecture Notes in Computer Science, pp. 758–760, Springer-Verlag. 1992.

[Beckert *et al.*, 1992b] B. Beckert, S. Gerberding, R. Hähnle, and W. Kernig. The tableau-based theorem prover $_3T^AP$ for multiple-valued logics. In *11th International Conference on Automated Deduction (CADE)*, Lecture Notes in Computer Science, pages 758–760, Albany, NY, Springer-Verlag, 1992.

[Beckert and Hähnle, 1992] B. Beckert and R. Hähnle. An improved method for adding equality to free variable semantic tableaux. In Depak Kapur, editor, *11th Conference on Automated Deduction*, Lecture Notes in Computer Science, pages 507–521, Albany,NY, 1992. Springer-Verlag.

[Beckert *et al.*, 1993] B. Beckert, R. Hähnle, and P. H. Schmitt. The even more liberalized δ–rule in free variable semantic tableaux. In G. Gottlob, A. Leitsch, and D. Mundici, editors, *Proceedings of the 3rd Kurt Gödel Colloquium (KGC)*, Lecture Notes in Computer Science, pages 108–119, Brno, Czech Republic, 1993. Springer-Verlag.

[Boole, 1958] G. Boole. *An investigation of the laws of thought, on which are founded the mathematical theories of logic and probabilities*. Dover, New York, January1958. (First Edition 1854).

[Beckert and Posegga, 1994] B. Beckert and J. Posegga. leanT^AP: lean, tableau-based theorem proving. In A. Bundy, editor, *12th Conference on Automated Deduction*, volume 814 of *Lecture Notes in Artificial Intelligence*, Nancy, France, June/July 1994. Springer-Verlag.

[26]Proceedings of subsequent workshops will be published within Springer's LNCS series.

[Beckert and Posegga, 1995] B. Beckert and J. Posegga. lean$T^A P$: Lean tableau-based deduction. *Journal of Automated Reasoning*, 15(3):339–358, 1995.

[Barker-Plummer and Rothenberg, 1992] D. Barker-Plummer and A. Rothenberg. The GAZER theorem prover. In *11th International Conference on Automated Deduction (CADE)*, Lecture Notes in Computer Science, pages 726–730, Albany, NY, 1992. Springer-Verlag.

[Börger and Rosenzweig, 1993] E. Börger and D. Rosenzweig. Full prolog in a nutshell. In D. S. Warren, editor, *Proceedings of the 10th International Conference on Logic Programming*, page 832. MIT Press, 1993.

[Börger and Rosenzweig, 1994] E. Börger and D. Rosenzweig. A mathematical definition of full Prolog. *Science of Computer Programming*, 1994. See also [Börger and Rosenzweig, 1993].

[Brace *et al.*, 1990] K. S. Brace, R. L. Rudell, and R. E. Bryant. Efficient implementation of a BDD package. In *Proc. 27th ACM/IEEE Design Automation Conference*, pages 40 – 45. IEEE Press, 1990.

[Bryant, 1986] R. Y. Bryant. Graph–based algorithms for Boolean function manipulation. *IEEE Trans. Computers*, C–35, 1986.

[Bryant, 1992] R. Y. Bryant. Symbolic boolean manipulation with ordered binary decision diagrams. Technical report, CMU, 1992.

[Church, 1956] A. Church. *Introduction to Mathematical Logic*, volume 1. Princeton University Press, Princeton, New Jersey, 1956. Sixth printing 1970 .

[Clocksin and Mellish, 1981] W. F. Clocksin and C. S. Mellish. *Programming in Prolog*. Springer-Verlag, 1981.

[Eder, 1992] E. Eder. *Relative Complexities of First-Order Calculi*. Artificial Intelligence. Vieweg Verlag, 1992.

[Farmer *et al.*, 1992] W. M. Farmer, J. D. Guttman, and F. J. Thayer. IMPS: System description. In *11th International Conference on Automated Deduction (CADE)*, Lecture Notes in Computer Science, pages 701–705, Albany, NY, 1992. Springer-Verlag.

[Farmer *et al.*, 1993] W. M. Farmer, J. D. Guttman, and F. J. Thayer. IMPS: An interactive mathematical proof system. *Journal of Automated Reasoning*, 11(2):213–248, October 1993.

[Fronhöfer *et al.*, 1992] B. Fronhöfer, R. Hähnle, and T. Käufl. Workshop on theorem proving with analytic tableaux and related methods. Technical Report 8/92, Universität Karlsruhe, Fakultät für Informatik, Karlsruhe, Germany, Mar1992. (Workshop held in Lautenbach, Germany).

[Fitting, 1996] M. C. Fitting. *First-Order Logic and Automated Theorem Proving*. Springer-Verlag, 1996. First edition, 1990.

[Goubault and Posegga, 1994] J. Goubault and J. Posegga. BDDs and automated deduction. In *Proc. 8th International Symposium on Methodologies for Intelligent Systems (ISMIS)*, Lecture Notes in Artificial Intelligence, Charlotte, NC, October1994. Springer-Verlag.

[Hilbert and Bernays, 1939] D. Hilbert and P. Bernays. *Grundlagen der Mathematik II*, volume 50 of *Die Grundlehren der mathematischen Wissenschaften in Einzeldarstellungen mit besonderer Berücksichtigung der Anwendungsgebiete*. Springer-Verlag, 1939.

[Käufl and Zabel, 1990] T. Käufl and N. Zabel. Cooperation of decision procedures in a tableau-based theorem prover. *Revue d'Intelligence Artificielle*, 4(3), 1990.

[Oppacher and Suen, 1988] F. Oppacher and E. Suen. HARP: A tableau-based theorem prover. *Journal of Automated Reasoning*, 4:69–100, 1988.

[Pelletier, 1986] F. J. Pelletier. Seventy-five problems for testing automatic theorem provers. *Journal of Automated Reasoning*, 2:191–216, 1986.

[Posegga, 1993a] J. Posegga. Compiling proof search in semantic tableaux. In *Proc. 7th International Symposium on Methodologies for Intelligent Systems (ISMIS)*, volume 671 of *Lecture Notes in Computer Science*, pages 67–77, Trondheim, Norway, June1993. Springer-Verlag.

[Posegga, 1993b] J. Posegga. *Deduktion mit Shannongraphen für Prädikatenlogik erster Stufe*. Infix Verlag, Sankt Augustin, Germany, 1993.

[Prawitz *et al.*, 1960] D. Prawitz, H. Prawitz, and N. Voghera. A mechanical proof procedure and its realization in an electronic computer. *Journal of the ACM*, 7(1–2):102–128, 1960.

[Richards *et al.*, 1994] B. L. Richards, I. Kraan, A. Smaill, and G. A. Wiggins. Mollusc: A general proof-development shell for sequent-based logics. In ALAN Bundy, editor, *12th International Conference on Automated Deduction (CADE)*, volume 814 of *LNAI*, pages 826–830. Springer, 1994.

[Schönfeld, 1985] W. Schönfeld. Prolog extensions based on tableau calculus. In *9th International Joint Conference on Artificial Intelligence, Los Angeles*, volume 2, pages 730–733, 1985.

[Schmitt, 1987] P. H. Schmitt. The THOT theorem prover. Technical Report 87.9.7, IBM Germany, Scientific Center, Heidelberg, Germany, 1987.

[Stickel, 1988] M. E. Stickel. A Prolog Technology Theorem Prover. *Journal of Automated Reasoning*, 4(4):353–380, 1988.

[Wang, 1960] H. Wang. Toward mechanical mathematics. *IBM Journal of Research and Development*, 4(1), January1960.

GRAHAM WRIGHTSON

A BIBLIOGRAPHY ON ANALYTIC TABLEAUX THEOREM PROVING

We have attempted to make this bibliography as comprehensive as possible with respect to published works relating to automated reasoning with analytic tableaux. The emphasis has been on the automated aspect rather than the tableaux. The literature published by logicians and philosophers relating to tableaux as a means for demonstrating valid arguments in logic is quite large and such literature, with few exceptions, has not been included in this collection. This leaves us with about 200 entries almost all of which are concerned with the attempts of computer scientists to automate formal reasoning using analytic tableaux and to apply the techniques to various problem areas. The entries have been divided into several categories: Early Work, Books and Proceedings, Classical Logic, Non-Classical Logic, Implementations, and Applications.

Naturally any entry might fall into one or more of these categories, such as some paper dealing with tableaux in temporal logic applied to real-time system verification. In these cases the author used his judgement and entered it into one or the other categories rather than have two entries of the some work.

EARLY WORK

This section includes some of the early papers on tableaux theorem proving by computer. It also has a few of the early works on tableaux as a validity testing technique in logic.

[1] I. Anellis. From semantic tableaux to Smullyan trees: the history of the falsifiability tree method. *Modern Logic*, 1(1):36–69, June 1990.

This paper gives a good overview of the early work on tableaux automation.

[2] I. Anellis. Erratum, From semantic tableaux to Smullyan trees: the history of the falsifiability tree method. *Modern Logic*, 2(2):219, Dec. 1991.

[3] E. W. Beth. Semantic entailment and formal derivability. *Mededelingen der Kon. Ned. Akad. v. Wet.*, 18(13), 1955. new series.

Everet Beth was one of the first to come up with the idea of tableaux. This paper is now difficult to obtain but it has been reprinted in various collections including K. Berka and L. Kreiser, editors, Logik-Texte. Kommentierte Auswahl zur Geschichte der modernen Logik, pages 262–266. Akademie-Verlag, Berlin, 1986

[4] E. W. Beth. On Machines Which Prove Theorems. *Simon Stevin Wisen Naturkundig Tijdschrift*, 32:49–60, 1958.

M. D'Agostino et al. (eds.), Handbook of Tableau Methods, 631–655.

Beth quickly saw the possibility of using computers to prove theorems and actually started the (presumably) first research group trying to do it with tableaux. This paper also appeared in J.Siekmann and G.Wrightson, editors, Automation of Reasoning, volume 1, pp.79-90, Springer Verlag, Berlin, 1983.

[5] R. W. Binkley and R. L. Clark. A Cancellation Algorithm for Elementary Logic. *Theoria*, 33:79–97, 1967.

Binkley and Clark developed an algorithm based on a tableaux-like approach. It was later used by others in an attempt to apply it to modal logic - presumably the first such attempt. Corrections to this paper appeared in the same journal, p.85, 1968. This paper and its correction also appeared in J.Siekmann and G.Wrightson, editors, Automation of Reasoning, volume 2, pp.27-47, Springer Verlag, Berlin, 1983.

[6] J. K. J. Hintikka. Form and content in quantification theory. *Acta Philosophica Fennica*, VIII, 1955.

[7] S. Kanger. Provability in logic. *Acta Universitatis Stockolmiensis, Stockolm studies in Philosophy*, 1, 1957.

[8] V. G. Kirin. Gentzen's method of the many-valued propositional calculi. *Zeitschrift für mathematische Logik und Grundlagen der Mathematik*, 12: 317–332, 1966.

[9] S. Kripke. Semantical analysis of modal logic I: Normal modal propositional calculi. *Zeitschrift für Mathematische Logik und Grundlagen der Mathematik*, 9:67–96, 1963.

Kripke was the first to develop a semantics for modal logics, and hence for many other non-classical logics, through the introduction of the 'possible worlds', a notion which saw the development of tableaux consisting of many sub-tableaux - one for each possible world. We bring these three papers of Kripke's because of their significance even though they are not concerned with the automation aspect. Some modal logic systems are still difficult to automate.

[10] S. Kripke. Semantical considerations on modal logic. *Acta Philosophica Fennica, Proceedings of a colloquium on modal and many-valued logics 1962*, 16:83–94, 1963.

[11] S. Kripke. Semantical analysis of modal logic II: Non-normal modal propositional calculi. In *Symposium on the Theory of Models*, pages 206–220. North-Holland, Amsterdam, 1965.

[12] G. Mints. Proof theory in the USSR 1925 – 1969. *Journal of Symbolic Logic*, 56(2):385–424, 1991.

[13] R. J. Popplestone. Beth-tree methods in automatic theorem-proving. In N. L. Collins and D. Michie, editors, *Machine Intelligence*, pages 31–46. American Elsevier, 1967.

[14] D. Prawitz. Advances and Problems in Mechanical Proof Procedures. In Meltzer and Michie, editors, *Machine Intelligence 4*, volume 2, pages 73–89. Edinburgh University Press, Edinburgh, 1969.

Prawitz was also one of the first to begin automation of logic and did some significant work which later lead to important breakthroughs by others. This paper also appeared in J.Siekmann and G.Wrightson, editors, Automation of Reasoning, volume 2, pp.283-297, Springer Verlag, Berlin, 1983.

[15] D. Prawitz, H. Prawitz, and N. Voghera. A Mechanical Proof Procedure and its Realization in an Electronic Computer. *J. ACM*, 7:102–128, 1960.

This paper also appeared in J.Siekmann and G.Wrightson, editors, Automation of Reasoning, volume 1, pp.202-228, Springer Verlag, Berlin, 1983.

[16] K. Schröter. Methoden zur Axiomatisierung beliebiger Aussagen- und Prädikatenkalküle. *Zeitschrift für math. Logik und Grundlagen der Mathematik*, 1:241–251, 1955.

[17] R. M. Smullyan. A unifying principle in quantification theory. *Proceedings of the National Academy of Sciences*, 49(6):828–832, June 1963.

Smullyan was the first to introduce a unified notation for the format of reduction rules in tableaux.

[18] R. M. Smullyan. Analytic natural deduction. *Journal of Symbolic Logic*, 30:123–139, 1965.

[19] R. M. Smullyan. Trees and nest structures. *Journal of Symbolic Logic*, 31:303–321, 1966.

BOOKS AND PROCEEDINGS

In this section we list some of the books covering the field as well as proceedings of the four international Workshops on Analytic Tableaux and Related Methods.

[1] P. B. Andrews. *An Introduction to Mathematical Logic and Type Theory: To Truth Through Proof.* Academic Press, Orlando, Florida, 1986.

[2] D. Basin, R. Hähnle, B. Fronhöfer, J. Posegga, and C. Schwind, editors. *Workshop on Theorem Proving with Analytic Tableaux and Related Methods*, MPI-I-93-213, Saarbrücken, 1993. Max-Planck-Institut für Informatik.

[3] P. Baumgartner, R. Hähnle, and J. Posegga, editors. *4th Workshop on Theorem Proving with Analytic Tableaux and Related Methods*, LNAI 918. Springer Verlag, 1995.

Proceedings of the 4th Workshop on Theorem Proving with Analytic Tableaux and Related Methods held in St.Goar, Germany.

[4] J. L. Bell and M. Machover. *A Course in Mathematical Logic*. North-Holland, Amsterdam, 1977.

[5] E. W. Beth. *The Foundations of Mathematics*. North-Holland, Amsterdam, 1959. Revised Edition 1964.

[6] W. Bibel. *Automated Theorem Proving*. Vieweg, Braunschweig, second revised edition, 1987.

[7] W. Bibel and E. Eder. Methods and calculi for deduction. In D. M. Gabbay, C. J. Hogger, and J. A. Robinson, editors, *Handbook of Logic in Artificial Intelligence and Logic Programming*, volume 1: Logical Foundations, pages 67–182. Oxford University Press, Oxford, 1992.

In this overview article Bibel and Eder present several of the main approaches to automated theorem proving.

[8] J. C. Bradfield. *Verifying Temporal Proporties of Systems*. Birkhaeuser, Boston, 1992.

One of the main areas of application of automated reasoning is concerned with automatically demonstrating that a given complex system has certain desirable properties.

[9] K. Broda, M. D'Agostino, R. Goré, R. Johnson, and S. Reeves, editors. *3rd Workshop on Theorem Proving with Analytic Tableaux and Related Methods*, TR-94/5, London, 1994. Imp. College of Science, Technology and Medicine.

Proceedings of the 3rd Workshop on Theorem Proving with Analytic Tableaux and Related Methods held in Abingdon, United Kingdom. Selected papers from this conference have now appeared in a special issue of the Journal of the Interest Group in Pure and Applied Logics (IGPL), Volume 3, Number 6, October 1995, with guest editors: K. Broda, M. D'Agostino, R. Gore, R. Johnson, S. Reeves.

[10] C. C. Chang and H. J. Keisler. *Model Theory*. North-Holland Publishing Company, third edition, 1990.

[11] R. J. G. B. de Queiroz and D. Gabbay. *An introduction to labelled natural deduction*. Oxford University Press, Oxford, 1992.

In this book a recently developed technique based on referencing of formulas is presented. This is cutting-edge research.

[12] E. Eder. *Relative Complexities of First-Order Calculi*. Artificial Intelligence. Vieweg Verlag, 1992.

Eder shows the relative computational complexities of several calculi using simulation.

[13] M. C. Fitting. *Intuitionistic Logic Model Theory and Forcing*. North-Holland Publishing Co., Amsterdam, 1969.

[14] M. C. Fitting. *Proof Methods for Modal and Intutionistic Logics*. Reidel, Dordrecht, 1983.

This book is not concerned with automation but is listed because it gives a good overview of many non-classical logics and tableaux techniques as used in these logics. This has been reviewed by R. A. Bull in:
Review of 'Melvin C. Fitting, Proof Methods for Modal and Intutionistic Logics, Synthese Library, Vol. 169, Reidel, 1983'. Journal of Symbolic Logic, 50:855–856, 1985.

[15] M. C. Fitting. *First-Order Logic and Automated Theorem Proving*. Springer, New York, 1996. First edition, 1990.

This is probably the best book in recent years to provide a good introduction to the tableaux method and to resolution. It is excellent for beginners in the field. More importantly, from a research point of view, Fitting's tableaux permit the use of free variables, thus providing a way in which unification can be used in tableaux. This is surely the most significant breakthrough for the tableaux method in recent years.

[16] B. Fronhöfer, R. Hähnle, and T. Käufl, editors. *Workshop on Theorem Proving with Analytic Tableaux and Related Methods, Lautenbach/Germany*. University of Karlsruhe, Dept. of Computer Science, Internal Report 8/92, Mar. 1992.

Proceedings of the 1st Workshop on Theorem Proving with Analytic Tableaux and Related Methods held in Lautenbach, Germany.

[17] J. H. Gallier. *Logic for Computer Science: Foundations of Automated Theorem Proving*. Harper and Row, New York, 1986.

[18] R. Hähnle. *Automated Deduction in Multiple-Valued Logics*, volume 10 of *International Series of Monographs on Computer Science*. Oxford University Press, 1994.

Hähnle has specialised in applying tableaux to multiple-valued logics. This book is his doctoral dissertation and is the most comprehensive work on the topic so far.

[19] R. Hähnle. Short conjunctive normal forms in finitely-valued logics. *Journal of Logic and Computation*, 4(6):905–927, 1994.

[20] A. Heyting. *Intuitionism, an Introduction*. North-Holland, Amsterdam, 1956. Revised Edition 1966.

[21] J. Hintikka. *Knowledge and Belief.* Cornell University Press, 1962.

[22] G. E. Hughes and M. J. Cresswell. *An Introduction to Modal Logic*. Methuen and Co., London, 1968.

[23] R. C. Jeffrey. *Formal Logic: Its Scope and Limits*. McGraw-Hill, New York, 1967.

Although rather dated now, this book gives a good and easy introduction to tableaux. It is not concerned with computerisation.

[24] S. C. Kleene. *Introduction to Metamathematics*. D. Van Nostrand, North-Holland, P. Noordhoff, 1950.

[25] D. W. Loveland. *Automated Theorem Proving. A Logical Basis*, volume 6 of *Fundamental Studies in Computer Science*. North-Holland, Amsterdam, 1978.

Loveland's book covers resolution and his invention, model elimination. Model elimination has many close connections to the tableaux technique and it is often implemented in automated reasoning software.

[26] Z. Manna and R. Waldinger. *The Logical Basis for Computer Programming*. Addison-Wesley, 1990. 2 vols.

[27] N. Rescher and A. Urquhart. *Temporal Logic*. Springer-Verlag, 1971.

[28] J. Siekmann and G. Wrightson, editors. *Automation of Reasoning*, volume 1 and 2. Springer-Verlag, New York, 1983.

This is a collection of many of the papers from the early work in automated reasoning up to 1970.

[29] R. Smullyan. *First-Order Logic*. Springer, New York, 1968.

Although this work has nothing to do with automation it is one of the classical works on analytical tableaux by one of the early researchers in the field and it is often quoted. It is highly recommended. A reprinting has appeared in Dover Press, 1995.

[30] M. E. Szabo, editor. *The Collected Papers of Gerhard Gentzen*. North-Holland, Amsterdam, 1969.

[31] A. Tarski. *Logic, Semantics, Metamathematics*. Oxford, 1956. J. H. Woodger translator.

[32] P. B. Thistlewaite, M. A. McRobbie, and B. K. Meyer. *Automated Theorem Proving in Non Classical Logics*. Pitman, 1988.

Another book on non-classical logic theorem proving by three people who have been very active in this field, particularly regarding relevant logic.

[33] S. Toledo. *Tableau Systems for First Order Number Theory and Certain Higher Order Theories*, volume 447 of *Lecture Notes in Mathematics*. Springer-Verlag, Berlin, 1975.

[34] L. Wallen. *Automated Deduction in Non-Classical Logics*. The MIT Press, Cambridge, Mass., 1990.

[35] G. Wrightson. Special issue on analytic tableaux. *Journal of Automated Reasoning*, 13(2), 1994.

This is the first of two special issues on analytic tableaux.

[36] G. Wrightson. Special issue on analytic tableaux. *Journal of Automated Reasoning*, 13(3), 1994.

This is the second of two special issues on analytic tableaux.

CLASSICAL LOGIC

In this section we present papers which are concerned primarily with the use of tableaux in propositional or first-order predicate logic.

[1] P. B. Andrews. Theorem proving through general matings. *JACM*, 28:193–214, 1981.

Andrews' paper introduces the idea of a mating, a set of pairs of complementary and unifiable literals in set of formulae which would demonstrate the unsatisfiability of the set of formulae. Although not directly concerned with tableaux it has since turned out that this idea can be applied in several of the automated reasoning approaches.

[2] K. Aspetsberger and S. Bayerl. Two Parallel Versions of the Connection Method for Propositional Logic on the L-Machine. In H. Stoyan, editor, *GWAI'85*, volume 118 of *Informatik Fachberichte*, Dassel/Solling, FRG, September 23–28, 1985. Springer Verlag. Also: Technical Report 85-8, RISC-Linz, Johannes Kepler University, Linz, Austria, 1985.

[3] U. Assmann. Combining Model Elimination and Resolution Techniques. In K. Broda, M. D'Agostino, R. Goré, R. Johnson, and S. Reeves, editors, *3rd Workshop on Theorem Proving with Analytic Tableaux and Related Methods*, TR-94/5, pages 3–11, London, 1994. Imperial College of Science Technology and Medicine.

[4] O. Astrachan and M. Stickel. Caching and lemmaizing in model elimination theorem provers. In D. Kapur, editor, *Proceedings, 11th Conference on Automated Deduction (CADE), Albany/NY, USA*, pages 224–238. Springer LNAI 607, 1992.

[5] A. Avron. Gentzen-type systems, resolution and tableaux. *Journal of Automated Reasoning*, 10(2):265–281, 1993.

[6] M. Baaz and C. G. Fermüller. Non-elementary Speedups between Different Versions of Tableaux. In P. Baumgartner, R. Hähnle, and J. Posegga, editors, *4th Workshop on Theorem Proving with Analytic Tableaux and Related Methods*, LNAI 918, pages 217–230, Berlin, 1995. Springer Verlag.

This is a continuation of the work in [18] and [55].

[7] M. Baaz, C. G. Fermüller, and R. Zach. Dual systems of sequents and tableaux for many-valued logics. *Bulletin of the European Association for Theoretical Computer Science*, 51, 1993.

[8] P. Baumgartner. A Model Elimination Calculus with Built-in Theories. In H.-J. Ohlbach, editor, *Proceedings of the 16-th German AI-Conference (GWAI-92)*, pages 30–42. Springer Verlag, 1992. LNAI 671.

Just as for the equality predicate, special theories have special features which need special consideration when being built into the theorem proving algorithm.

[9] P. Baumgartner. Refinements of Theory Model Elimination and a Variant without Contrapositives. In A. Cohn, editor, *11th European Conference on Artificial Intelligence, ECAI 94*. Wiley, 1994. (Long version in: Research Report 8/93, University of Koblenz, Institute for Computer Science, Koblenz, Germany).

[10] P. Baumgartner and U. Furbach. Consolution as a Framework for Comparing Calculi. *Journal of Symbolic Computation*, 16(5), 1993. Academic Press.

Consolution was introduced by Eder and is used here to compare various calculi.

[11] P. Baumgartner and U. Furbach. Model Elimination Without Contrapositives and Its Application to PTTP. *Journal of Automated Reasoning*, 13(3): 339–360, 1994.

[12] P. Baumgartner, U. Furbach, and F. Stolzenburg. Model Elimination, Logic Programming and Computing Answers. In *14th International Joint Conference on Artificial Intelligence (IJCAI 95)*, volume 1, pages 335–340, 1995. (Long version in: Research Report 1/95, University of Koblenz, Germany).

[13] P. Baumgartner and F. Stolzenburg. Constraint Model Elimination and a PTTP-Implementation. In P. Baumgartner, R. Hähnle, and J. Posegga, editors, *4th Workshop on Theorem Proving with Analytic Tableaux and Related Methods*, LNAI 918, pages 201–216, Berlin, 1995. Springer Verlag.

[14] B. Beckert. Adding Equality to Semantic Tableaux. In K. Broda, M. D'Agostino, R. Goré, R. Johnson, and S. Reeves, editors, *3rd Workshop on Theorem Proving with Analytic Tableaux and Related Methods*, TR-94/5, pages 29–41, London, 1994. Imperial College of Science Technology and Medicine.

[15] B. Beckert. Semantic tableaux with equality. *Journal of Logic and Computation*, 1996. To appear.

[16] B. Beckert, S. Gerberding, and R. Hähnle. The tableau-based theorem prover 3TAP for multiple-valued logics. In D. Kapur, editor, *CADE*, pages 507–521. Springer-Verlag, LNCS 607, 1992.

[17] B. Beckert and R. Hähnle. Deduction by combining semantic tableaux and integer programming. Presented at Computer Science Logic, CSL'95, Paderborn, Germany, Sept. 1995.

[18] B. Beckert, R. Hähnle, and P. Schmitt. The even more liberalized δ-rule in free-variable semantic tableaux. In G. Gottlob, A. Leitsch, and D. Mundici, editors, *Proceedings of the 3rd Kurt Gödel Colloquium, Brno, Czech Republic*, LNCS 713, pages 108–119. Springer-Verlag, 1993.

[19] B. Beckert and J. Posegga. lean$T^A P$: Lean Tableau-based Deduction. *Journal of Automated Reasoning*, 15(3):339–358, 1995.

[20] C. Belleannée and R. Vorc'h. A Tableau Proof of Linear Size for the Pigeon Formulas using Symmetry. In K. Broda, M. D'Agostino, R. Goré, R. Johnson, and S. Reeves, editors, *3rd Workshop on Theorem Proving with Analytic Tableaux and Related Methods*, TR-94/5, pages 43–49, London, 1994. Imperial College of Science Technology and Medicine.

[21] E. W. Beth. On Padoa's method in the theory of definition. *Indag. Math.*, 15:330–339, 1953.

[22] E. W. Beth. Some consequences of the theorem of Löwenheim-Skolem-Gödel-Malcev. *Indag. Math.*, 15, 1953.

[23] W. Bibel. Tautology testing with a generalized matrix method. *Theoretical Computer Science*, 8:31–44, 1979.

Bibel introduces his matrix method in this paper.

[24] W. Bibel. Matings in matrices. *Communications of the ACM*, 26:844–852, 1983.

Here Bibel combines his matrix approach with Andrews' matings.

[25] W. Bibel. *Automated Theorem Proving*. Vieweg, Braunschweig, second revised edition, 1987.

[26] W. Bibel. Short proofs of the pigeonhole formulas based on the connection method. *Journal of Automated Reasoning*, 6(3):287–298, Sept. 1990.

When links are introduced to matrices, Bibel then calls this the connection method.

[27] W. Bibel, S. Brüning, U. Egly, D. Korn, and T. Rath. Issues in Theorem Proving Based on the Connection Method. In P. Baumgartner, R. Hähnle, and J. Posegga, editors, *4th Workshop on Theorem Proving with Analytic Tableaux and Related Methods*, LNAI 918, pages 1–16, Springer Verlag, Berlin, 1995.

[28] W. Bibel and E. Eder. Methods and calculi for deduction. In D. M. Gabbay, C. J. Hogger, and J. A. Robinson, editors, *Handbook of Logic in Artificial Intelligence and Logic Programming*, volume 1: Logical Foundations, pages 67–182. Oxford University Press, Oxford, 1992.

In this overview article Bibel and Eder present several of the main approaches to automated theorem proving.

[29] K. Bowen. Programming with full first-order logic. In Hayes, Michie, and Pao, editors, *Machine Intelligence*, volume 10, pages 421–440. 1982.

[30] K. Broda. *The application of semantic tableaux with unification to automated deduction.* PhD thesis, Department of Computing, Imperial College, University of London, 1991.

[31] S. Brüning. Exploiting Equivalences in Connection Calculi. *Journal of the Interest Group in Pure and Applied Logics (IGPL)*, 3(6):857–886,1995.

[32] R. Caferra and N. Zabel. An application of many-valued logic to decide propositional S5 formulae: a strategy designed for a parameterised tableaux-based theorem prover. In P. Jorrand and V. Sgurev, editors, *Artificial Intelligence iV: Methodology, Systems, Applications (AIMSA)*, pages 23–32. Elsevier Science Publishers B.V. (North-Holland), 1990.

[33] R. Caferra and N. Zabel. A tableaux method for systematic simultaneous search for refutations and models using equational problems. *Journal of Logic and Computation*, 3(1):3–26, 1993.

[34] M. Clausen. Multivariate polynomials, standard tableaux, and representations of symmetric groups. *Journal of Symbolic Computation*, 11(5 and 6):483–522, May/June 1991.

[35] R. Cleaveland. Tableau-based model checking in the propositional mu-calculus. *Acta Informatica*, 27, 1990.

[36] J. Coldwell. *Paradigms for improving the control structure of analytic tableaux.* PhD thesis, University of Newcastle, Australia, 1995.

[37] J. Coldwell and G. Wrightson. Some deletion rules for analytic tableaux. In *Annual Australasian Logic Conference.* University of Sydney, 1990. Abstract in Journal of Symbolic Logic, vol. 56(3), Sept 1991, pp. 1108-1114.

[38] J. Coldwell and G. Wrightson. A truncation technique for clausal analytic tableaux. *Information Processing Letters*, 42:273–281, 1992.

[39] J. Coldwell and G. Wrightson. Lemmas and links in analytic tableau. In *Proceedings of the 7th Australian Joint Conference on Artificial Intelligence*, pages 275–282, Armidale, 1994. University of New England, World Scientific.

Links can be introduced into tableaux just as they were used in Kowalski's connection graph resolution and Bibel's connection method.

[40] J. Coldwell and G. Wrightson. The modified a-rule for link inheritance. In *Proceedings of the 8th Australian Joint Conference on Artificial Intelligence*, Canberra, 1995. Australian Defence Force Academy.

[41] R. H. Cowen. A characterization of logical consequence in quantification theory. *Notre Dame Journal of Formal Logic*, 16:375–377, 1973.

[42] R. H. Cowen. Solving algebraic problems in propositional logic by tableau. *Archiv für Mathematische Logik und Grundlagenforschung*, 22(3–4):187–190, 1982.

[43] R. H. Cowen. Combinatorial analytic tableaux. *Reports on Mathematical Logic*, 27:29–39, 1993.

[44] M. D'Agostino. Are tableaux an improvement on truth tables? Cut-free proofs and bivalence. *Journal of Logic, Language and Information*, 1:235–252, 1992.

[45] M. D'Agostino and M. Mondadori. The taming of the cut. *Journal of Logic and Computation*, 4:285–319, 1994.

[46] G. V. Davydov. Synthesis of the resolution method with the inverse method. *Journal of Soviet Mathematics*, 1:12–18, 1973. Translated from Zapiski Nauchnykh Seminarov Leningradskogo Otdeleniya Mathematicheskogo Instituta im. V. A. Steklova Akademii Nauk SSSR, vol. 20, pp. 24–35, 1971.

[47] E. Eder. Consolution and its relation with resolution. In *Proceedings, International Joint Conference on Artificial Intelligence (IJCAI)*, pages 132–136, 1991.

Consolution provides a certain abstraction of resolution and some other calculi to give further insight into the techniques.

[48] M. C. Fitting. Partial models and logic programming. *Theoretical Computer Science*, 48:229–255, 1987.

[49] M. C. Fitting. Tableaux for Logic Programming. *Journal of Automated Reasoning*, 13(2):175–188, Oct. 1994.

This is an excellent paper by Fitting illustrating the use of tableaux in logic programming through the recursive call of tableaux.

[50] M. C. Fitting. A program to compute Gödel-Löb fixpoints. To appear, 1995.

[51] G. Gentzen. Untersuchungen über das logische Schliessen. *Mathematische Zeitschrift*, 39:176–210, 405–431, 1935. English translation, "Investigations into logical deduction," in M. E. Szabo, The Collected Papers of Gerhard Gentzen, North-Holland, Amsterdam, 1969.

[52] J. Goubault. A BDD-based Skolemization Procedure. *Journal of the Interest Group in Pure and Applied Logics (IGPL)*, 3(6):827–855, 1995.

[53] M. Grundy. *Theorem Prover Generation Using Refutation Procedures*. PhD thesis, Department of Computer Science, University of Sydney, June 1990.

[54] M. Grundy. A Stepwise Mapping from Resolution to Tableau Calculi. In K. Broda, M. D'Agostino, R. Goré, R. Johnson, and S. Reeves, editors, *3rd Workshop on Theorem Proving with Analytic Tableaux and Related Methods*, TR-94/5, pages 89–103, London, 1994. Imperial College of Science Technology and Medicine.

[55] R. Hähnle and P. H. Schmitt. The Liberalized δ-Rule in Free Variable Semantic Tableaux. *Journal of Automated Reasoning*, 13(2):211–222, Oct. 1994.

This paper shows how to slacken the constraints for the use of the δ-rule.

[56] P. Hertz. Über Axiomensysteme für beliebige Satzsysteme. *Mathematische Annalen*, 101:457–514, 1929.

[57] J. Hintikka. A new approach to sentential logics. *Soc. Scient. Fennica, Comm. Phys.-Math.*, 17(2), 1953.

[58] C. A. Johnson. Factorization and circuit in the connection method. *Journal of the ACM*, 40(3):536–557, July 1993.

[59] R. Johnson. Tableau Structure Prediction. In K. Broda, M. D'Agostino, R. Goré, R. Johnson, and S. Reeves, editors, *3rd Workshop on Theorem Proving with Analytic Tableaux and Related Methods*, TR-94/5, pages 113–127, London, 1994. Imperial College of Science Technology and Medicine.

[60] S. Klingenbeck. Generating Finite Counter Examples with Semantic Tableaux. In P. Baumgartner, R. Hähnle, and J. Posegga, editors, *4th Workshop on Theorem Proving with Analytic Tableaux and Related Methods*, LNAI 918, pages 31–46, Berlin, 1995. Springer Verlag.

[61] S. Klingenbeck and R. Hähnle. Semantic tableaux with ordering restrictions. In A. Bundy, editor, *Proceedings, 12th International Conference on Automated Deduction (CADE), Nancy/France*, LNAI 814, pages 708–722. Springer Verlag, 1994. To appear in Journal of Logic and Computation, 1996.

[62] J. Komara and P. J. Voda. Syntactic Reduction of Predicate Tableaux to Propositional Tableaux. In P. Baumgartner, R. Hähnle, and J. Posegga, editors, *4th Workshop on Theorem Proving with Analytic Tableaux and Related Methods*, LNAI 918, pages 231–246, Berlin, 1995. Springer Verlag.

[63] R. Letz. *First-Order Calculi and Proof Procedures for Automated Deduction*. PhD thesis, TH Darmstadt, June 1993.

[64] R. Letz, K. Mayr, and C. Goller. Controlled Integration of Cut Rule into Connection Tableau Calculi. *Journal of Automated Reasoning*, 13(3):297–338, Dec. 1994.

[65] R. Letz, J. Schumann, S. Bayerl, and W. Bibel. SETHEO: A high-perfomance theorem prover. *Journal of Automated Reasoning*, 8(2):183–212, 1992.

[66] R. Li and A. Sernadas. Reasoning about objects using a tableau method. *Journal of Logic Computing*, 1(5):575–611, Oct. 1991.

[67] Z. Lis. Wynikanie Semantyczne a Wynikanie Formalne (Logical consequence, semantic and formal). *Studia Logica*, 10:39–60, 1960. Polish, with Russian and English summaries.

[68] K. Mayr. Link Deletion in Model Elimination. In P. Baumgartner, R. Hähnle, and J. Posegga, editors, *4th Workshop on Theorem Proving with Analytic Tableaux and Related Methods*, LNAI 918, pages 169–184, Berlin, 1995. Springer Verlag.

[69] G. L. Mogilevskii and D. A. Ostroukhov. A mechanical propositional calculus using Smullyan's analytic tables. *Cybernetics*, 14:526–529, 1978. Translation from *Kibernetika*, 4, 43–46 (1978).

[70] M. Mondadori. Efficient Inverse Tableaux. *Journal of the Interest Group in Pure and Applied Logics (IGPL)*, 3(6):939–953, 1995.

[71] N. V. Murray and E. Rosenthal. Dissolution: Making paths vanish. *Journal of the ACM*, 3(40):504–535, 1993.

[72] N. V. Murray and E. Rosenthal. On the relative merits of path dissolution and the method of analytic tableaux. *Theoretical Computer Science*, 131, 1994.

Dissolution is another proof technique which has resolution as a special case.

[73] D. Nardi. Formal synthesis of a unification algorithm by the deductive-tableau method. *Journal of Logic Programming*, 7(1):1–43, July 1989.

[74] G. Neugebauer. Reachability analysis for the extension procedure — a topological result. In Y. Deville, editor, *Logic Program Synthesis and Transformation. Proceedings of LOPSTR'93*, Workshops In Computing. Springer Verlag, 1994.

[75] G. Neugebauer and U. Petermann. Specifications of Inference Rules and Their Automatic Translation. In P. Baumgartner, R. Hähnle, and J. Posegga, editors, *4th Workshop on Theorem Proving with Analytic Tableaux and Related Methods*, LNAI 918, pages 185–200, Berlin, 1995. Springer Verlag.

[76] G. Neugebauer and T. Schaub. A pool-based connection calculus. In C. Bozşahin, U. Halıcı, K. Oflazar, and N. Yalabık, editors, *Proceedings of Third Turkish Symposium on Artificial Intelligence and Neural Networks*, pages 297–306. Middle East Technical University Press, 1994.

[77] M. Ohnishi and K. Matsumoto. A system for strict implication. *Annals of the Japan Assoc. for Philosophy of Science*, 2:183–188, 1964.

[78] U. Petermann. How to build in an open theory into connection calculi. *Journal on Computer and Artificial Intelligence*, 11(2):105–142, 1992.

[79] U. Petermann. Completeness of the pool calculus with an open built in theory. In G. Gottlob, A. Leitsch, and D. Mundici, editors, *3rd Kurt Gödel Colloquium '93*, volume LNCS 713. Springer Verlag, 1993.

[80] U. Petermann. A framework for integrating equality reasoning into the extension procedure. In *Proceedings Workshop on Theorem Proving with Analytic Tableaux and Related Methods, Marseille, 1993*, pages 195–207, 1993.

[81] U. Petermann. A complete connection calculus with rigid E-unification. In *JELIA94*, volume LNCS 838, pages 152–167. Springer Verlag, 1994.

[82] D. Prawitz. An improved proof procedure. *Theoria*, 26, 1960. Reprinted in *Automation of Reasoning*, Jörg Siekmann and Graham Wrightson, Springer-Verlag, (1983), vol 1, pp 244 – 264., vol. 1, pp 162 – 199.

[83] D. Pym and L. Wallen. Investigations into proof-search in a system of first-order dependent function types. In *10th International Conference on Automated Deduction, Kaiserslautern, FRG, July 24–27, 1990*, LNAI 449, pages 236–250, Berlin, 1990. Springer Verlag.

[84] H. Rasiowa. Algebraic treatment of the functional calculi of Heyting and Lewis. *Fundamenta Mathematica*, 38, 1951.

[85] H. Rasiowa. Algebraic models of axiomatic theories. *Fundamenta Mathematica*, 41, 1954.

[86] S. V. Reeves. *Theorem-proving by Semantic Tableaux.* PhD thesis, University of Birmingham, 1985.

[87] S. V. Reeves. Adding equality to semantic tableaux. *Journal of Automated Reasoning*, 3:225–246, 1987.

[88] S. V. Reeves. Semantic tableaux as a framework for automated theorem-proving. In C. S. Mellish and J. Hallam, editors, *Advances in Artificial Intelligence (Proceedings of AISB-87)*, pages 125–139. Wiley, 1987.

[89] K. Schneider, T. Kropf, and R. Kumar. Accelerating tableaux proofs using compact representations. *Journal of Formal Methods in System Design*, pages 145–176, 1994.

[90] K. Schneider, R. Kumar, and T. Kropf. Efficient representation and computation of tableau proofs. In L.J.M. Claesen and M.J.C. Gordon, editors, *International Workshop on Higher Order Logic Theorem Proving and its Applications*, pages 39–58, Leuven, Belgium, Sept. 1992. North-Holland.

[91] J. Schröder. Körner's criterion of relevance and analytic tableaux. *Journal of Philosophical Logic*, 21(2):183–192, 1992.

[92] G. Sidebottom. Implementing CLP (B) using the Connection Theorem Proving Method and a Clause Management System. *Journal of Symbolic Computation*, 15(1):27–48, Jan. 1993.

[93] R. M. Smullyan. Abstract quantification theory. In A. Kino, J. Myhill, and R. E. Vesley, editors, *Intuitionism and Proof Theory, Proceedings of the Summer Conference at Buffalo N. Y. 1968*, pages 79–91. North-Holland, Amsterdam, 1970.

[94] M. E. Stickel. Upside-Down Meta-Interpretation of the Model Elimination Theorem-Proving Procedure for Deduction and Abduction. *Journal of Automated Reasoning*, 13(2):189–2120, Oct. 1994.

[95] A. Tarski. Der Aussagenkalkül und die Topologie. *Fundamenta Mathematica*, 31:103–34, 1938. Reprinted as 'Sentential calculus and topology' in A. Tarski, Logic, Semantics, Metamathematics, Oxford 1956.

[96] V. Vialard. Handling Models in Propositional Logic. In K. Broda, M. D'Agostino, R. Goré, R. Johnson, and S. Reeves, editors, *3rd Workshop on Theorem Proving with Analytic Tableaux and Related Methods*, TR-94/5, pages 213–219, London, 1994. Imperial College of Science Technology and Medicine.

[97] R. Vorc'h. A Connection-based Point of View of Propositional Cut Elimination. In K. Broda, M. D'Agostino, R. Goré, R. Johnson, and S. Reeves, editors, *3rd Workshop on Theorem Proving with Analytic Tableaux and Related Methods*, TR-94/5, pages 221–231, London, 1994. Imperial College of Science Technology and Medicine.

[98] K. Wallace and G. Wrightson. Regressive Merging in Model Elimination Tableau-based Theorem Provers. *Journal of the Interest Group in Pure and Applied Logics (IGPL)*, 3(4):921–937, 1995.

[99] K. J. Wallace. *Proof Truncation Techniques in Model Elimination Tableaux*. PhD thesis, University of Newcastle, Australia, 1994.

[100] L. A. Wallen. Generating connection calculi from tableau and sequent based proof systems. In A. Cohn and J. Thomas, editors, *Artificial Intelligence and its Applications*, pages 35–50, Warwick, England, 1986. AISB85, John Wiley & Sons. Edinburgh University, Edinburgh EH1 2QL, U.K.

[101] H. Wang. Toward mechanical mathematics. *IBM Journal for Research and Development*, 4:2–22, 1960. Reprinted in *A Survey of Mathematical Logic*, Hao Wang, North-Holland, (1963), pp 224 – 268, and in *Automation of Reasoning*, Jörg Siekmann and Graham Wrightson, Springer-Verlag, (1983), vol 1, pp 244 – 264.

[102] C. Weidenbach. First Order Tableaux with Sorts. *Journal of the Interest Group in Pure and Applied Logics (IGPL)*, 3(4):887–906, 1995.

[103] G. Wrightson. Semantic tableaux with links. In *AI'87 Conference*, 1987.

[104] G. Wrightson and J. Coldwell. Truncating analytic tableaux. In *13th Australian Computer Science Conference*, pages 412–420, Melbourne, 1990. Monash University.

NON-CLASSICAL LOGIC

Here we list papers on tableaux theorem proving in various non-classical logics modal, intuistionistic, relevant, linear, default, multi-valued.

[1] G. Amati, L. C. Aiello, D. Gabbay, and F. Pirri. A proof-theoretical approach to default reasoning I: Tableaux for default logic. *Journal of Logic and Computation*, to appear.

[2] M. Baaz and C. G. Fermüller. Resolution-based theorem proving for many-valued logics. *Journal of Symbolic Computation*, 19(4):353–391, Apr. 1995.

[3] B. Beckert, S. Gerberding, R. Hähnle, and W. Kernig. The many-valued tableau-based theorem prover 3TAP. In D. Kapur, editor, *Proceedings, 11th International Conference on Automated Deduction (CADE), Albany/NY*, pp. 758–760. Springer LNAI 607, 1992.

[4] N. D. Belnap Jr. A useful four-valued logic. In J. M. Dunn and G. Epstein, editors, *Modern Uses of Multiple-Valued Logic*, pages 8–37. D. Reidel, Dordrecht and Boston, 1977.

[5] E. W. Beth. Semantic construction of intuitionistic logic. *Mededelingen der Kon. Ned. Akad. v. Wet.*, 19(11), 1956. new series.

[6] A. Bloesch. *Signed Tableaux – a Basis for Automated Theorem Proving in Nonclassical Logics*. PhD thesis, University of Queensland, Brisbane, Australia, 1993.

[7] A. Buchsbaum and T. Pequeno. Automated Deduction with Non-classical Negations. In K. Broda, M. D'Agostino, R. Goré, R. Johnson, and S. Reeves,

editors, *3rd Workshop on Theorem Proving with Analytic Tableaux and Related Methods*, TR-94/5, pages 51–63, London, 1994. Imperial College of Science, Technology and Medicine.

[8] R. A. Bull. Review of 'Melvin Fitting, Proof Methods for Modal and Intuitionistic Logics, Synthese Library, Vol. 169, Reidel, 1983'. *Journal of Symbolic Logic*, 50:855–856, 1985.

[9] P. Bystrov. Tableau variants of some modal and relevant systems. *Polish Acad. Sci. Inst. Philos. Sociol. Bull. Sectl Logic*, 17:92–103, 1988.

[10] R. Caferra and N. Zabel. An application of many-valued logic to decide propositional S_5 formulae: a strategy designed for a parameterized tableaux-based theorem prover. In *Proceedings, Artificial Intelligence—Methodology Systems Application (AIMSA'90)*, pages 23–32, 1990.

[11] W. A. Carnielli. Systematization of finite many-valued logics through the method of tableaux. *Journal of Symbolic Logic*, 52(2):473–493, June 1987.

[12] W. A. Carnielli. On sequents and tableaux for many-valued logics. *Journal of Non-Classical Logic*, 8(1):59–76, May 1991.

[13] H. F. Chau. A proof search system for a modal substructural logic based on labelled deductive systems. In A. Voronkov, editor, *Proceedings, 4th International Conference on Logic Programming and Automated Reasoning (LPAR'93), St. Petersburg*, pages 64–75. Springer Verlag, LNAI 698, 1993.

[14] M. D'Agostino and D. M. Gabbay. A Generalization of Analytic Deduction via Labelled Deductive Systems. Part I: Basic Substructural Logics. *Journal of Automated Reasoning*, 13:243–281, 1994.

[15] B. Davidson, F. C. Jackson, and R. Pargetter. Modal trees for t and $s5$. *Notre Dame Journal of Formal Logic*, 18(4):602–606, 1977.

'Trees' is often used synonymously with 'tableaux'.

[16] P. de Groote. Linear Logic with Isabelle: Pruning the Proof Search Tree. In P. Baumgartner, R. Hähnle, and J. Posegga, editors, *4th Workshop on Theorem Proving with Analytic Tableaux and Related Methods*, LNAI 918, pages 263–277, Berlin, 1995. Springer Verlag.

[17] S. Demri. Using Connection Method in Modal Logics: Some Advantages. In P. Baumgartner, R. Hähnle, and J. Posegga, editors, *4th Workshop on Theorem Proving with Analytic Tableaux and Related Methods*, LNAI 918, pages 63–78, Berlin, 1995. Springer Verlag.

[18] P. Doherty. A constraint-based approach to proof procedures for multi-valued logics. In *First World Conference on the Fundamentals of Artificial Intelligence WOCFAI–91, Paris*, 1991.

[19] J. M. Dunn. Intuitive semantics for first-degree entailments and 'coupled trees'. *Philosophical Studies*, 29:149–168, 1976.

[20] J. M. Dunn. Relevance logic and entailment. In D. Gabbay and F. Guenthner, editors, *Handbook of Philosophical Logic*, volume 3, chapter III.3, pages 117–224. Kluwer, Dordrecht, 1986.

[21] E. A. Emerson. Automata, tableaux and temporal logics (extended abstract). In *Proceedings, Conference on Logics of Programs, Brooklyn*, pages 79–87. Springer, LNCS 193, 1985.

[22] E. A. Emerson and J. Y. Halpern. Decision procedures and expressiveness in the temporal logic of branching time. *Journal of Computer and System Sciences*, 30:1–24, 1985.

[23] R. Feys. *Modal Logics*. Number IV in Collection de Logique Mathématique, Sèrie B. E. Nauwelaerts (Louvain), Gauthier-Villars (Paris), 1965. Joseph Dopp, editor.

[24] M. C. Fitting. A tableau proof method admitting the empty domain. *Notre Dame Journal of Formal Logic*, 12:219–224, 1971.

[25] M. C. Fitting. Tableau methods of proof for modal logics. *Notre Dame Journal of Formal Logic*, 13:237–247, 1972.

[26] M. C. Fitting. A modal logic analog of Smullyan's fundamental theorem. *Zeitschrift für mathematische Logik und Gründlagen der Mathematik*, 19:1–16, 1973.

[27] M. C. Fitting. A tableau system for propositional S5. *Notre Dame Journal of Formal Logic*, 18:292–294, 1977.

[28] M. C. Fitting. First-order modal tableaux. *Journal of Automated Reasoning*, 4:191–213, 1988.

[29] M. C. Fitting. Bilattices in logic programming. In *20th International Symposium on Multiple-Valued Logic, Charlotte*, pages 238–247. IEEE Press, Los Alamitos, 1990.

[30] M. C. Fitting. Modal logic should say more than it does. In J.-L. Lassez and G. Plotkin, editors, *Computational Logic, Essays in Honor of Alan Robinson*, pages 113–135. MIT Press, Cambridge, MA, 1991.

[31] M. C. Fitting. A modal Herbrand theorem. To appear, 1995.

[32] M. C. Fitting. Tableaus for many-valued modal logic. *Studia Logica*, 55:63–87, 1995.

[33] M. C. Fitting, W. Marek, and M. Truszczynski. The pure logic of necessitation. *Journal of Logic and Computation*, 2:349–373, 1992.

[34] R. Goré. Semi-analytic tableaux for modal logics with applications to non-monotonicity. *Logique et Analyse*, 133-134:73–104, 1991.

[35] R. Goré. The cut-elimination theorem for Diodorean modal logics. In *Proc. of the Logic Colloquium*, Veszprem, Hungary, 1992. Abstract in the *Journal of Symbolic Logic, vol. 58, number 3, pp. 1141–1142, 1993.*

[36] R. Goré. *Cut-free sequent and tableau systems for propositional normal modal logics*. PhD thesis, Computer Laboratory, University of Cambridge, England, 1992.

[37] R. Goré. Cut-free sequent and tableau systems for propositional Diodorean modal logics. *Studia Logica*, 53:433–457, 1994.

[38] G. Governatori. Labelled Tableaux for Multi-Modal Logics. In P. Baumgartner, R. Hähnle, and J. Posegga, editors, *4th Workshop on Theorem Proving with Analytic Tableaux and Related Methods*, LNAI 918, pages 79–94, Berlin, 1995. Springer Verlag.

[39] C. Groeneboer and J. P. Delgrande. Tableau-based theorem proving in normal conditional logics. In *Proceedings, AAAI'88, St. Paul, MN*, pages 171–176, 1988.

[40] R. Hähnle. Towards an efficient tableau proof procedure for multiple-valued logics. In *Proceedings, Workshop on Computer Science Logic (CSL), Heidelberg*, pages 248–260. Springer, LNCS 533, 1990.

[41] R. Hähnle. Uniform notation of tableaux rules for multiple-valued logics. In *Proceedings, International Symposium on Multiple-Valued Logic (ISMVL), Victoria*, pages 238–245. IEEE Press, 1991.

[42] R. Hähnle. Efficient deduction in many-valued logics. In *Proc. International Symposium on Multiple-Valued Logics, ISMVL'94, Boston/MA, USA*, pages 240–249. IEEE Press, Los Alamitos, 1994.

[43] R. Hähnle. Many-valued logic and mixed integer programming. *Annals of Mathematics and Artificial Intelligence*, 12(3,4):231–264, Dec. 1994.

[44] R. Hähnle and O. Ibens. Improving temporal logic tableaux using integer constraints. In D. Gabbay and H.-J. Ohlbach, editors, *Proceedings, International Conference on Temporal Logic (ICTL), Bonn*, LNCS 827, pages 535–539. Springer-Verlag, 1994.

[45] J. Hintikka. Modality and quantification. *Theoria*, 27:110–128, 1961.

[46] J. Hintikka. Model minimization - an alternative to circumscription. *J. of Automated Reasoning*, 4:1–13, 1988.

[47] J. Hudelmaier. On a Contraction-free Sequent Calculus for the Modal Logic S4. In K. Broda, M. D'Agostino, R. Goré, R. Johnson, and S. Reeves, editors, *3rd Workshop on Theorem Proving with Analytic Tableaux and Related Methods*, TR-94/5, pages 105–111, London, 1994. Imperial College of Science, Technology and Medicine.

[48] J. Hudelmaier and P. Schroeder-Heister. Classical Lambek Logic. In P. Baumgartner, R. Hähnle, and J. Posegga, editors, *4th Workshop on Theorem Proving with Analytic Tableaux and Related Methods*, LNAI 918, pages 247–262, Berlin, 1995. Springer Verlag.

[49] R. Johnson. A blackboard approach to parallel temporal tableaux. In Jorrand, P. and V. Sgurev, editors, *Artificial Intelligence, Methodologies, Systems, and Applications (AIMSA)*, pages 183–194, Singapore, 1994. World Scientific.

[50] J. R. Kenevan and R. E. Neapolitan. A model theoretic approach to propositional fuzzy logic using Beth tableaux. In L. A. Zadeh and J. Kacprzyk, editors, *Fuzzy Logic for the Management of Uncertainty*, pages 141–158. John Wiley & Sons, 1992.

[51] M. Kohlhase. Higher-Order Tableaux. In P. Baumgartner, R. Hähnle, and J. Posegga, editors, *4th Workshop on Theorem Proving with Analytic Tableaux and Related Methods*, LNAI 918, pages 294–309, Berlin, 1995. Springer Verlag.

This is one of the few papers in recent years which considers tableaux for higher-order logic. Kohlhase is a member of a group researching higher order logic theorem proving.

[52] S. Kripke. A completeness theorem in modal logic. *Journal of Symbolic Logic*, 24:1–14, 1959.

[53] P. Kuhna. Circumscription and minimal models for propositional logics. In *Proc. of the 2nd Workshop on Theorem Proving with Analytic Tableaux and Related Methods*, pages 143–155, Marseille, 1993. Tech. Report MPI-I-93-213.

[54] S. Lorenz. A Tableau Prover for Domain Minimization. *Journal of Automated Reasoning*, 13(3):375–390, Dec. 1994.

[55] J. J. Lu, N. V. Murray, and E. Rosenthal. Signed formulas and annotated logics. In *Proceedings, International Symposium on Multiple-Valued Logics (ISMVL)*, pages 48–53, 1993.

Another paper looking at labelled deduction systems.

[56] F. Massacci. Strongly analytic tableaux for normal modal logics. In A. Bundy, editor, *12th International Conference on Automated Deduction*, LNAI 814, pages 723–737, Nancy, France, 1994. Springer-Verlag.

[57] K. Matsumoto. Decision procedure for modal sentential calculus S3. *Osaka Mathematical Journal*, 12:167–175, 1960.

[58] M. C. Mayer and F. Pirri. First-order abduction via tableau and sequent calculi. *Bulletin of the IPGL*, 1(1):99–117, 1993.

[59] M. C. Mayer and F. Pirri. Propositional Abduction in Modal Logic. *Journal of the Interest Group in Pure and Applied Logics (IGPL)*, 3(6):907–919, 1995.

[60] T. McCarty. Clausal intuitionistic logic: I. Fixed-point semantics. *Journal of Logic Programming*, 5:1–31, 1988.

[61] T. McCarty. Clausal intuitionistic logic: II. Tableau proof procedures. *Journal of Logic Programming*, 5:93–132, 1988.

[62] M. A. McRobbie and N. D. Belnap. Relevant analytic tableaux. *Studia Logica*, XXXVIII:187–200, 1979.

This is one of the first papers on automated reasoning using tableaux applied to relevant logic.

[63] R. K. Meyer, M. A. McRobbie, and N. D. Belnap Jr. Linear Analytic Tableaux. In P. Baumgartner, R. Hähnle, and J. Posegga, editors, *4th Workshop on Theorem Proving with Analytic Tableaux and Related Methods*, LNAI 918, pages 278–293, Berlin, 1995. Springer Verlag.

It turns out that there is a close connection between linear logic and relevant logic.

[64] P. Miglioli, U. Moscato, and M. Ornaghi. An Improved Refutation System for Intuitionistic Predicate Logic. *Journal of Automated Reasoning*, 13(3):361–374, Dec. 1994.

[65] P. Miglioli, U. Moscato, and M. Ornaghi. How to avoid Duplications in Refutation Systems for Intuitionistic and Kuroda Logic. In K. Broda, M. D'Agostino, R. Goré, R. Johnson, and S. Reeves, editors, *3rd Workshop on Theorem Proving with Analytic Tableaux and Related Methods*, TR-94/5, pages 169–187, London, 1994. Imperial College of Science, Technology and Medicine.

[66] P. Miglioli, U. Moscato, and M. Ornaghi. Refutation systems for propositional modal logics. In P. Baumgartner, R. Hähnle, and J. Posegga, editors, *4th Workshop on Theorem Proving with Analytic Tableaux and Related Methods*, LNAI 918, pages 95–105, Berlin, 1995. Springer Verlag.

[67] G. Mints. Proof theory in the USSR 1925 – 1969. *Journal of Symbolic Logic*, 56(2):385–424, 1991.

[68] G. L. Mogilevskii and D. A. Ostroukhov. A mechanical propositional calculus using Smullyan's analytic tables. *Cybernetics*, 14:526–529, 1978. Translation from *Kibernetika*, 4, 43–46 (1978).

[69] C. G. Morgan and E. Orłowska. Kripke and relational style semantics and associated tableau proof systems for arbitrary finite valued logics. In *Proceedings, 2nd Workshop on Theorem Proving with Tableau-Based and Related Methods, Marseille*. Tech. Report, MPII Saarbrücken, 1993.

[70] N. V. Murray and E. Rosenthal. Improving tableau deductions in multiple-valued logics. In *Proceedings, 21st International Symposium on Multiple-Valued Logic, Victoria*, pages 230–237. IEEE Computer Society Press, Los Alamitos, May 1991.

[71] N. V. Murray and E. Rosenthal. Resolution and path-dissolution in multiple-valued logics. In *Proceedings, International Symposium on Methodologies for Intelligent Systems (ISMIS), Charlotte*, LNCS 542, pages 570–579, Berlin, 1991. Springer Verlag.

[72] A. Nakamura and H. Ono. On the size of refutation Kripke models for some linear modal and tense logics. *Studia Logica*, 34:325–333, 1980.

[73] I. Niemelä. Decision procedure for autoepistemic logic. In *Proceedings of the 9th International Conference on Automated Deduction*, pages 675–684, Argonne, USA, May 1988. Springer-Verlag.

[74] Z. Ognjanovic. A tableau-like proof procedure for normal modal logics. *Theoretical Computer Science*, 129, 1994.

[75] M. Ohnishi. Gentzen decision procedures for Lewis's systems S2 and S3. *Osaka Mathematical Journal*, 13:125–137, 1961.

[76] M. Ohnishi and K. Matsumoto. Gentzen method in modal calculi I. *Osaka Mathematical Journal*, 9:113–130, 1957.

[77] M. Ohnishi and K. Matsumoto. Gentzen method in modal calculi II. *Osaka Mathematical Journal*, 11:115–120, 1959.

[78] N. Olivetti. Tableaux and sequent calculus for minimal entailment. *Journal of Automated Reasoning*, 9(1):99–139, Aug. 1992.

[79] E. Orłowska. Mechanical proof methods for Post Logics. *Logique et Analyse*, 28(110):173–192, 1985.

[80] J. Otten and C. Kreitz. A Connection Based Proof Method for Intuitionistic Logic. In P. Baumgartner, R. Hähnle, and J. Posegga, editors, *4th Workshop on Theorem Proving with Analytic Tableaux and Related Methods*, LNAI 918, pages 122–137, Berlin, 1995. Springer Verlag.

[81] R. Pliuškevičius. The Saturated Tableaux for Linear Miniscoped Horn-like Temporal Logic. *Journal of Automated Reasoning*, 13(3):391–408, Dec. 1994.

[82] V. Risch. Analytic tableaux for default logics. In *AAAI Fall Symposium on Automated deduction in Nonstandard Logics*, pages 149–155, 1993.

[83] V. Risch. Une caracterisation en termes de tableaux semantiques pour la logique des defauts au sens de Lukaszewicz. *Revue d'intelligence artificielle*, 25(5):347–418, 1993.

[84] V. Risch and C. B. Schwind. Tableau-Based Characterization and Theorem Proving for Default Logic. *Journal of Automated Reasoning*, 13(2):243–281, Oct. 1994.

This is one of the few publications using tableaux in non-monotonic logics.

[85] G. Rousseau. Sequents in many valued logic I. *Fundamenta Mathematica*, 60:23–33, 1967.

[86] P. H. Schmitt. Perspectives in multi-valued logic. In R. Studer, editor, *Proceedings International Scientific Symposium on Natural Language and Logic, Hamburg*, LNCS 459, pages 206–220, Berlin, 1989. Springer Verlag.

[87] P. H. Schmitt and W. Wernecke. Tableau calculus for order-sorted logic. In K. H. Bläsius, U. Hedstück, and C.-R. Rollinger, editors, *Sorts and Types in Artificial Intelligence, Proc. of the workshop, Ehringerfeld, 1989*, volume 418 of *LNCS*, pages 49–60, Berlin, 1990. Springer Verlag.

[88] S. Schmitt and C. Kreitz. On Transforming Intuitionistic Matrix Proofs into Standard-Sequent Proofs. In P. Baumgartner, R. Hähnle, and J. Posegga, editors, *4th Workshop on Theorem Proving with Analytic Tableaux and Related Methods*, LNAI 918, pages 106–121, Berlin, 1995. Springer Verlag.

[89] C. B. Schwind. A tableaux-based theorem prover for a decidable subset of default logic. In M. E. Stickel, editor, *10th International Conference on Automated Deduction*, LNAI 449, pages 528–542, Kaiserslautern, FRG, July 24–27, 1990. Springer-Verlag.

[90] N. Shankar. Proof search in the intuitionistic sequent calculus. In D. Kapur, editor, *Automated Deduction — CADE-11*, number 607 in Lecture Notes in Artificial Intelligence, pages 522–536. Springer-Verlag, Berlin, 1992.

[91] R. L. Slaght. Modal tree constructions. *Notre Dame Journal of Formal Logic*, 18(4):517–526, 1977.

[92] R. M. Smullyan. A generalization of intuitionistic and modal logics. In H. Leblanc, editor, *Truth, Syntax and Modality, Proceedings of the Temple University Conference on Alternative Semantics*, pages 274–293. North-Holland, Amsterdam, 1973.

[93] W. Suchoń. La méthode de Smullyan de construire le calcul n-valent des propositions de Łukasiewicz avec implication et négation. *Reports on Mathematical Logic, Universities of Cracow and Katowice*, 2:37–42, 1974.

[94] S. J. Surma. An algorithm for axiomatizing every finite logic. In D. C. Rine, editor, *Computer Science and Multiple-Valued Logics*, pages 143–149. North–Holland, Amsterdam, 1984.

Second Edition

[95] M. Takahashi. Many-valued logics of extended Gentzen style I. *Sci. Rep. Tokyo Kyoiku Daigaku Sect. A*, 9:271, 1967.

[96] J. Underwood. Tableau for Intuitionistic Predicate Logic as Metatheory. In P. Baumgartner, R. Hähnle, and J. Posegga, editors, *4th Workshop on Theorem Proving with Analytic Tableaux and Related Methods*, LNAI 918, pages 138–153, Berlin, 1995. Springer Verlag.

[97] L. A. Wallen. *Automated Proof Search in Non-Classical Logics: Efficient Matrix Proof Methods for Modal and Intuitionistic Logics*. PhD thesis, University of Edinburgh, 1987.

[98] L. A. Wallen. Matrix Proof Methods for Modal Logics. In *Proc. of IJCAI 87*, pages 917–923. Morgan Kaufman, 1987.

[99] G. Wrightson. Non-classical logic theorem proving. *Journal of Automated Reasoning*, 1:5–48, 1985.

[100] Zabel. Analytic tableaux for finite and infinite post logics. In A. M. Borzyszkowski and S. Sokotowski, editors, *Symposium on Mathematical Foundations of Computer Science*, LNCS 711, pages 767–776, Berlin, 1993. Springer Verlag.

[101] N. Zabel. *Nouvelles Techniques de Déduction Automatique en Logiques Polyvalentes Finies et Infinies du Premier Ordre*. PhD thesis, Institut National Polytechnique de Grenoble, Apr. 1993.

IMPLEMENTATIONS

The section lists publications describing various implementations of the tableaux method or similar approaches such as the connection method or model elimination.

[1] O. Astrachan. METEOR: Exploring Model Elimination Theorem Proving. *Journal of Automated Reasoning*, 13(3):283–296, Dec. 1994.

[2] P. Baumgartner and U. Furbach. PROTEIN: A *PRO*ver with a *T*heory *E*xtension *I*nterface. In A. Bundy, editor, *Automated Deduction – CADE-12*, volume 814 of *LNAI*, pages 769–773. Springer Verlag, 1994.

[3] L. Catach. TABLEAUX: A general theorem prover for modal logics. *Journal of Automated Reasoning*, 7(4):489–510, Dec. 1991.

[4] H. C. M. de Swart and W. M. J. Ophelders. Tableaux versus Resolution a Comparison. *Fundamentae Informaticae*, 18:109–127, 1993.

[5] E. Eder. An implementation of a theorem prover based on the connection method. In W. Bibel and B. Petkoff, editors, *Proceedings, Artificial Intelligence—Methodology Systems Application (AIMSA'84)*, pages 121–128. North-Holland, Sept. 1984.

[6] C. Goller, R. Letz, K. Mayr, and J. M. P. Schumann. SETHEO V3.2: Recent developments. In A. Bundy, editor, *12th International Conference on Automated Deduction, Nancy, France, June/July 1994*, LNAI 814, pages 778–782, Berlin, 1994. Springer Verlag.

[7] M. Koshimura and R. Hasegawa. Modal Propositional Tableaux in a Model Generation Theorem-prover. In K. Broda, M. D'Agostino, R. Goré, R. Johnson, and S. Reeves, editors, *3rd Workshop on Theorem Proving with Analytic Tableaux and Related Methods*, TR-94/5, pages 145–151, London, 1994. Imperial College of Science, Technology and Medicine.

[8] Y. Malach and Z. Manna. TABLOG: The deductive-tableau programming language. In *Conference Record of the 1984 ACM Symposium on LISP and Functional Programming, Austin, TX*, pages 323–330, New York, NY, 1984. ACM.

[9] F. Oppacher and E. Suen. HARP: A tableau-based theorem prover. *Journal of Automated Reasoning*, 4(1):69–100, Mar. 1988.

[10] J. Schumann. Tableau-based Theorem Provers: Systems and Implementations. *Journal of Automated Reasoning*, 13(3):409–421, 1994.

This is a list of tableaux based theorem provers indicating the contacts for the implementers and a brief description of the systems.

INDEX